Positioning and Navigation in Complex Environments

Kegen Yu
Wuhan University, China

A volume in the Advances in Wireless Technologies and Telecommunication (AWTT) Book Series

IGI Global
DISSEMINATOR OF KNOWLEDGE

Published in the United States of America by
IGI Global
Information Science Reference (an imprint of IGI Global)
701 E. Chocolate Avenue
Hershey PA, USA 17033
Tel: 717-533-8845
Fax: 717-533-8661
E-mail: cust@igi-global.com
Web site: http://www.igi-global.com

Library of Congress Cataloging-in-Publication Data

Names: Yu, Kegen, editor.
Title: Positioning and navigation in complex environments / Kegen Yu, editor.
Description: Hershey, PA : Information Science Reference, 2018.
Identifiers: LCCN 2017019263| ISBN 9781522535287 (hardcover) | ISBN 9781522535294 (ebook)
Subjects: LCSH: Geographic information systems. | Global Positioning System. | Radio frequency identification systems. | Indoor positioning systems (Wireless localization)
Classification: LCC G70.212 .P685 2018 | DDC 910.285--dc23 LC record available at https://lccn.loc.gov/2017019263

This book is published in the IGI Global book series Advances in Wireless Technologies and Telecommunication (AWTT) (ISSN: 2327-3305; eISSN: 2327-3313)

British Cataloguing in Publication Data
A Cataloguing in Publication record for this book is available from the British Library.

For electronic access to this publication, please contact: eresources@igi-global.com.

Advances in Wireless Technologies and Telecommunication (AWTT) Book Series

Xiaoge Xu
Xiamen University Malaysia, Malaysia

ISSN:2327-3305
EISSN:2327-3313

Mission

The wireless computing industry is constantly evolving, redesigning the ways in which individuals share information. Wireless technology and telecommunication remain one of the most important technologies in business organizations. The utilization of these technologies has enhanced business efficiency by enabling dynamic resources in all aspects of society.

The **Advances in Wireless Technologies and Telecommunication Book Series** aims to provide researchers and academic communities with quality research on the concepts and developments in the wireless technology fields. Developers, engineers, students, research strategists, and IT managers will find this series useful to gain insight into next generation wireless technologies and telecommunication.

Coverage

- Global Telecommunications
- Telecommunications
- Mobile Technology
- Digital Communication
- Network Management
- Mobile Communications
- Virtual Network Operations
- Broadcasting
- Mobile Web Services
- Radio Communication

IGI Global is currently accepting manuscripts for publication within this series. To submit a proposal for a volume in this series, please contact our Acquisition Editors at Acquisitions@igi-global.com or visit: http://www.igi-global.com/publish/.

The Advances in Wireless Technologies and Telecommunication (AWTT) Book Series (ISSN 2327-3305) is published by IGI Global, 701 E. Chocolate Avenue, Hershey, PA 17033-1240, USA, www.igi-global.com. This series is composed of titles available for purchase individually; each title is edited to be contextually exclusive from any other title within the series. For pricing and ordering information please visit http://www.igi-global.com/book-series/advances-wireless-technologies-telecommunication/73684. Postmaster: Send all address changes to above address.

Titles in this Series

For a list of additional titles in this series, please visit: www.igi-global.com/book-series

Advanced Mobile Technologies for Secure Transaction Processing Emerging Research and Opportunities
Raghvendra Kumar (LNCT Group of Colleges, India) Preeta Sharan (The Oxford College of Engineering, India) and Aruna Devi (Surabhi Software, India)
Information Science Reference • ©2018 • 177pp • H/C (ISBN: 9781522527596) • US $130.00 (our price)

Examining Developments and Applications of Wearable Devices in Modern Society
Saul Emanuel Delabrida Silva (Federal University of Ouro Preto, Brazil) Ricardo Augusto Rabelo Oliveira (Federal University of Ouro Preto, Brazil) and Antonio Alfredo Ferreira Loureiro (Federal University of Minas Gerais (UFMG), Brazil)
Information Science Reference • ©2018 • 330pp • H/C (ISBN: 9781522532903) • US $195.00 (our price)

Graph Theoretic Approaches for Analyzing Large-Scale Social Networks
Natarajan Meghanathan (Jackson State University, USA)
Information Science Reference • ©2018 • 355pp • H/C (ISBN: 9781522528142) • US $225.00 (our price)

Powering the Internet of Things With 5G Networks
Vasuky Mohanan (Universiti Sains Malaysia, Malaysia) Rahmat Budiarto (Albaha University, Saudi Arabia) and Ismat Aldmour (Albaha University, Saudi Arabia)
Information Science Reference • ©2018 • 304pp • H/C (ISBN: 9781522527992) • US $215.00 (our price)

Routing Protocols and Architectural Solutions for Optimal Wireless Networks and Security
Dharm Singh (Namibia University of Science and Technology, Namibia)
Information Science Reference • ©2017 • 277pp • H/C (ISBN: 9781522523420) • US $205.00 (our price)

Sliding Mode in Intellectual Control and Communication Emerging Research and Opportunities
Vardan Mkrttchian (HHH University, Australia) and Ekaterina Aleshina (Penza State University, Russia)
Information Science Reference • ©2017 • 128pp • H/C (ISBN: 9781522522928) • US $140.00 (our price)

Emerging Trends and Applications of the Internet of Things
Petar Kocovic (Union – Nikola Tesla University, Serbia) Reinhold Behringer (Leeds Beckett University, UK) Muthu Ramachandran (Leeds Beckett University, UK) and Radomir Mihajlovic (New York Institute of Technology, USA)
Information Science Reference • ©2017 • 330pp • H/C (ISBN: 9781522524373) • US $195.00 (our price)

IGI Global
DISSEMINATOR OF KNOWLEDGE

701 East Chocolate Avenue, Hershey, PA 17033, USA
Tel: 717-533-8845 x100 • Fax: 717-533-8661
E-Mail: cust@igi-global.com • www.igi-global.com

Editorial Advisory Board

Table of Contents

Detailed Table of Contents

This chapter focuses on positioning based on received signal strength measurements and radio path-loss modeling. Typically, receiver signal strength from a device such as a smart phone is used to calculate the path-loss and thus estimate distances using a loss model calibrated in an offline process. With known positions and ranges to WiFi access points or simple devices using Bluetooth or Zigbee for data communications, the mobile device position can be estimated. However, due to the complex multipath propagation indoor environment, distance estimation and position determination using current methods are not very accurate. Based on knowledge of the nature of indoor signal propagation and algorithms especially designed for mobile applications, new methods show that positional accuracy of a few meters is possible, even with non-line-of-sight propagation through many intervening walls. Given the current widespread deployment of WiFi indoors, simple software-only solutions are feasible for applications such as general personal navigation and tracking within buildings.

The advancements in wireless communication technologies have enabled new sensing possibilities where the channel measurements of the radio are used for inferring physical changes in the surrounding environment. Relating the channel measurements to the location and actions of people has been of particular interest due to the wide range of application opportunities enabled by such a sensing capability. As an example, the low-amplitude received signal measurements of low-cost wireless communication systems have been used to detect the presence of a person, to locate and track them, identify gestures and activities of the person, and even monitor their vital signs. This chapter aims to give a deep insight on how people influence radio signals, how these effects are observed at the receiver antenna, and how the measurement system impacts the recorded measurements. These topics are presented to shed light on the relation between the location of people and signal strength measurements of narrowband radios.

Indoor positioning systems have received increasing attention for supporting location-based services in indoor environments. Received signal strength (RSS), mostly utilized in Wi-Fi fingerprinting systems, is known to be unreliable due to two reasons: orientation mismatch and variations in hardware. This chapter introduces an approach based on histogram equalization to compensate for orientation mismatch in robust Wi-Fi localization. The proposed method involves converting the temporal-spatial radio signal strength into a reference function (i.e., equalizing the histogram). This chapter also introduces an enhanced positioning feature, which is called delta-fused principal strength, to enhance the robustness of Wi-Fi localization against the problem of heterogeneous hardware. This algorithm computes the pairwise delta RSS and then integrates with RSS using principal component analysis. The proposed methods effectively and efficiently improve the robustness of location estimation in the presence of mismatch orientation and hardware variations, respectively.

Radio frequency identification (RFID) is a technique using two-way radio transmission pattern to transmit information through the device of interrogator (also called reader) and tag. It is considered to be one of the most popular techniques for internet of things (IOT). In this chapter, the authors study indoor localization techniques based on passive UHF RFID, which works around the frequency of 900MHz. Passive RFID has the advantage of reasonable reading distance, non-contact, easy deployment, and low cost. The tags do not need battery and it can harvest power through wireless charging. Due to those advantages, passive UHF RFID positioning has always been an active research area in the past few decades. This chapter discusses the key techniques in passive UHF RFID positioning, which include range-based, range-free, tag-based (device-based), tag-free (device-free), and improved positioning methods. All the techniques studied are suited to be implemented in RFID systems, each of which can be accommodated to a specific application scenario.

This chapter focuses on estimating the azimuth and elevation angles of a sound emitter based on time-difference-of-arrival (TDOA) measurements using an array of acoustic sensors. The TDOA-based direction-finding problem is appropriate because in a range of scenarios the source only emits a transient signal and TDOA measurements provide a simple method of finding the direction of the received signal. Given the measurement of TDOA, three methods for calculating the actual bearing of an acoustic source are considered—algebraic calculations based on trigonometric functions, linear least squares, and nonlinear least squares—and these results are also compared with the Cramer-Rao lower bound (CRLB). In this chapter, a comprehensive analysis of TDOA-based direction-finding methods is presented with regard to different application conditions, while their estimation performances are analysed with both simulation and field experimental results produced by 3-D microphone array.

Chapter 6

Yunlong Wang, Tsinghua University, China
Yang Cai, Tsinghua University, China
Yuan Shen, Tsinghua University, China

This chapter describes cooperative localization in wireless networks, where mobile nodes with unknown positions jointly infer their positions through measuring and exchanging information with each other. The technique of cooperation localization, efficiently even in harsh propagation environment, enables amounts of location-based services that rely on high-accuracy position information of mobile nodes. After a brief introduction of cooperative localization, the Cramer-Rao lower bound is given as a standard metric for performance. Then the information in the temporal and spatial domain is illustrated with geometrical interpretations. Two classes of cooperative localization algorithms, namely, centralized and distributed algorithms, are presented to show the implementation of the cooperative localization in a wireless network. Then the performance of cooperative localization under non-line-of-sight condition is analyzed. Lastly, numerical results are given to illustrate the performance of cooperative localization algorithms.

Chapter 7

Eva Lagunas, University of Luxembourg, Luxembourg
Monica Navarro, Centre Tecnologic de Telecomunicacions de Catalunya (CTTC), Spain
Pau Closas, Northeastern University, USA
Montse Najar, Universitat Politecnica de Catalunya, Spain
Ricardo Garcia-Gutierrez, Universitat Politecnica de Catalunya, Spain
Genís Floriach, Universitat Politecnica de Catalunya, Spain

IR-UWB has emerged as a promising candidate for positioning passive nodes in wireless networks due to its extremely short time domain transmitted pulses. The two-step approaches in which first different TOAs are estimated and then fed into a triangulation procedure are suboptimal in general. This is because in the first stage of these methods, the measurements at distinct anchors are independent and ignore the constraint that all measurements must be consistent with a single emitter location. In this chapter, the authors investigate two techniques to overcome this issue. First, a two-step procedure based on multi-TOA estimation is proposed. Second, a positioning approach omitting the intermediate known as DPE is presented. Complementarily, the authors explore the CS-based modeling of both approaches so that the temporal sparsity of the UWB received signal and the consequent sparseness of the discrete spatial domain are exploited to select the most significant TOAs and to reduce the amount of information to be sent to a central fusion unit in the DPE approach.

The chapter studies the positioning techniques based on ultra wideband (UWB) and low cost inertial measurement unit (IMU) with a focus on the fundamental theories of integrated positioning based on UWB and IMU. Details are provided for multilateral positioning theory of UWB, UWB calibration method, IMU error analysis, inertial navigation algorithm, and Kalman filter (KF) theory. Particularly, to mitigate the influence of non-line-of-sight (NLOS) propagation on positioning accuracy, a NLOS mitigation method based on fuzzy theory is presented. Meanwhile, the detailed data fusion processes of loosely coupled and tightly coupled systems are introduced and performance evaluation is conducted using experimental data.

With the increasing ubiquity of smartphones and tablets, users are now routinely carrying a variety of sensors with them wherever they go. These devices are enabling technologies for ubiquitous computing, facilitating continuous updates of a user's context. They have built-in MEMS-based accelerometers for ubiquitous activity monitoring and there is a growing interest in how to use these together with gyroscopes and magnetometers to build dead reckoning (DR) systems for location tracking. Navigation in complex environments is needed mainly by consumer users, private vehicles, and pedestrians. Therefore, the navigation system has to be small, easy to use, and have reasonably low levels of power consumption and price. The technologies and techniques discussed here include the fusion of inertial navigation (IN) and other sensors, positioning based on signals from wireless networks (such as Wi-Fi), image-based methods, cooperative positioning systems, and map matching (MM). The state-of-the-art of MEMS-based location sensors and their integration into modern navigation systems are also presented.

The art of multi-sensor processing, or "sensor-fusion," is the ability to optimally infer state information from multiple noisy streams of data. One major application area where sensor fusion is commonly used is navigation technology. While global navigation satellite systems (GNSS) can provide centimeter-level location accuracy worldwide, they suffer from signal availability problems in dense urban environment and they hardly work indoors. While several alternative backups have been proposed, so far, no single sensor or technology can provide the desirable precise localization in such environments under reasonable costs and affordable infrastructures. Therefore, to navigate through these complex areas, combining sensors is beneficial. Common sensors used to augment/replace GNSS in complex environments include inertial measurement unit (IMU), range sensors, and vision sensors. This chapter discusses the design and implementation of tightly coupled sensor fusion of GNSS, IMU, and light detection and ranging (LiDAR) measurements to navigate in complex urban and indoor environments.

This chapter presents an alternative framework to the traditional centralised Kalman filtering (CKF) approach for implementing the optimal state estimation algorithm in support of multisensor integration. The data fusion algorithm is implemented via a series of transformations of vectors in the so-called information space (iSpace). This chapter describes how the conventional decentralised Kalman filtering (DKF) algorithm can be derived using a unified approach. A new global optimal fusion (GOF) algorithm is derived using the iSpace approach. Two case studies are presented to illustrate applications of the multisensor algorithms for GNSS/Locata/INS and GNSS/WiFi/INS integration.

High sensitivity and fast acquisition are two important goals that must be considered in the development of signal processing techniques for a GNSS acquisition function to meet the demands for LBS in GNSS-challenged environments, such as indoor and urban canyon. This chapter introduces the fundamentals of GNSS acquisition functions, GNSS acquisition techniques for new GNSS signals, and GNSS acquisition techniques achieving high sensitivity and fast acquisition. Therefore, this chapter contains useful information for engineers who study the fundamentals and principles of GNSS acquisition and the state-of-the-art GNSS signal acquisition techniques for weak signals.

This chapter examines sources of global navigation satellite system (GNSS) vulnerability, identifying the broad range of topics that comes under this heading, and cites some key references in each category area. GNSS vulnerability has been a very productive area for GNSS researchers in recent years and this chapter sets out to be a comprehensive review of the different ways that the operation of GNSS can be degraded by outside influences, from the high (system) to the low (receiver component) level.

Preface

Positioning and navigation is about making use of a device, network or system to determine the positions of one or multiple objects such as assets, people, animals and vehicles, as well as exploiting the position information for navigation purpose. It basically involves one or more of the three major aspects: algorithm, software, and hardware, depending on application scenario. Positioning and navigation in complex environments such as indoors, underground mines and tunnels, and city canyons is vital to a wide range of applications and services such as automated mining, security management, personal monitoring, entertainment, shopping guidance, and intelligent home.

In the design or selection of a positioning system or device, one needs to be aware of whether the system is intended for use indoors, outdoors, or both. For outdoor applications, global navigation satellite system (GNSS) is typically used. For instance, smart phones are usually embedded with a GNSS receiver chip that can periodically calculates user's position which is displayed on an electronic map so that the user knows where he or she is and how to get to the destination such as in a city by walking along a street at the correct direction. Similarly, a driver usually relies on a GNSS navigation device to travel from one place to another even in a big city where he or she has lived for many years; the device frequently tells direction messages such as the distance to turn left or turn right or keeping going straight. On the other hand, in an indoor environment, GNSS usually is not able to provide useful position information since GNSS signals are blocked by building roofs, walls, and nearby high-rise buildings and other structures. As a consequence, a non-GNSS local positioning system is required. Different from outdoor scenarios where GNSS is dominantly used, a range of positioning systems is developed for indoor scenarios, which make use of different types of signals such as radio, ultrasonic, magnetic, visual light, and inertial sensor signal.

This book focuses on positioning and navigation in complex environments. Both GNSS and non-GNSS systems and approaches are studied. There are thirteen chapters, dealing with a good range of topics in the field. Each of the topics is written by one or more experts who are experienced researchers in the field. Basically, each chapter presents both fundamental concepts and in-depth studies for a specific topic. Although there is some overlapping between a few chapters due to similar topics, the topics are dealt with in different ways and different results are presented, so these chapters are complementary to each other.

CHALLENGING ISSUES

Over the past three decades, positioning and navigation has been extensively investigated in terms of theory, practice, or application, mainly due to its tremendous importance to our modern society and daily life. Although great advances have been made in both indoor and outdoor positioning and navigation and huge literature (books, journal and conference articles) is available, positioning and navigation are currently still a very active research area. This is mainly because a range of challenging issues still exist especially related to application in complex and harsh environments. For instance,

1. Multipath and non-line-of-sight propagation usually prevail indoors, which is a troubling issue for such as radio and light positioning systems.
2. Interference can be a serious issue and different interference sources may exist, which can be radio interference, light interference, or magnetic interference. Many of current ground-based positioning systems do not consider the effect of such interference or do not treat it seriously.
3. In some cases jamming and spoofing is a serious problem, which may cause positioning system ill-functioning. Suitable anti-jamming and anti-spoofing approaches are needed to enable secure and robust positioning and navigation.
4. Seamless positioning is an important issue when a mobile moves from an outdoor to an indoor environment or visas versa. It is desirable to provide positioning service without any interruption when positioning environment changes abruptly.
5. To achieve desirable performance such as accuracy, coverage, and reliability, it is best to use two or more types of systems simultaneously. But the question is which systems should be used and what method should be employed to combine them to achieve the required performance, when taking cost and complexity into account.

This book is intended to present theoretical frameworks and recent advances in positioning and navigation in complex environments, to discuss possible solutions to the various challenging issues in the field, including those mentioned above.

TARGET AUDIENCE

The target audience of this book will be composed of professionals and researchers working in a range of fields, including positioning and navigation engineering, electronic and electrical engineering, telecommunications, computer engineering, and geospatial engineering. The book can also be used as a reference for senior undergraduate students and postgraduate students who are studying in the above-mentioned fields and are interested in positioning and navigation. It can be used as a quick reference as well as for in-depth study on the-state-of-the-art approaches and systems in the field.

ORGANIZATION OF THE BOOK

Chapter 1 focuses on RSS-based positioning by dealing with the challenging issue of inaccurate modeling of path-loss mainly caused by dynamic multipath propagation and shadowing effect leading to both fast and slow fading. Different from the conventional path-loss modeling with a constant path-loss exponent, this chapter utilizes the excess loss due to penetrating through walls. The excess loss is modeled as a linear function of number of walls on the propagation path. Based on the excess loss per wall and the number of walls, an estimate of the excess loss is obtained, hence the path-loss is corrected to generate a more accurate distance estimate. Two new position determination schemes are proposed, which are range-based and propagation loss-based iterative least-squares methods. The model parameters are obtained by processing data collected with a ZigBee network in an office building about 2000 m^2.

Chapter 2 investigates RSS modeling for scenarios where there is no communication between the mobile target and base stations or access points probably because the target does not hold a compatible communication device or deliberately switches off the device. In fact the model-based and fingerprint-based approaches including those presented in most of chapters of the book are associated with scenarios where the mobile can communicate with a group of base stations either in one-way or two-way mode. The parameters of the signal transmitted from one side can be estimated through reception and signal processing on the other side. In this chapter deterministic and statistical RSS models are developed from data collected with a RSS measurement system. A fixed base station transmits a radio signal, while another fixed base station receives the signal. The RSS varies as the mobile moves within the scene. Such models have the potential for applications such as regional border crossing detection, link-line monitoring, and low complexity radio tomographic imaging, as demonstrated in the chapter.

Chapter 3 investigates the robustness issues in RSS based positioning. When using RSS fingerprints for position determination, one major factor affecting position accuracy is the difference between the propagation environment during offline calibration or database establishment and that during online localization. The difference can be caused by adding or removing walls, relocating furniture, and randomness in the number and movement of pedestrians within a building. The difference can also be produced by mismatch in the orientation and attitude of both user's body and device. For instance, the offline fingerprint at a position was collected when the user was facing North, but it is unlikely that the user will face exactly the same direction at the same position or nearby during localization. With and without body shadowing, the difference in RSS can be many decibels. Position accuracy may also be affected by mismatch in the type of device, such as mobile phones manufactured by different companies. In this chapter, two types of mismatch issues are analyzed in details, which are orientation mismatch and heterogeneous hardware. Different solutions for solving the mismatch problems are discussed.

Chapter 4 studies the use of radio frequency identification (RFID) for indoor localization with a focus on passive RFID localization techniques. Five different algorithms are described in detail, which are the range-based, range-free, tag-based, tag-free, and improved positioning methods. A RFID system consists of one or more readers and a number of tags. In a passive RFID system the tags are passive and do not have their own power supply or battery. A passive tag harvests the energy through electromagnetic induction when the tag is within the radio range of the reader, which is typically rather small such as several meters. Since passive tags have the advantages of small size and low cost, passive RFID localization technique has been investigated for applications in many different scenarios. In particular, it is a favorite candidate among the various localization techniques for internet of things (IOT). This is intuitively reasonable because a huge number of items and objects will be attached with tags especially passive ones designed for identification and tracking in the future.

Chapter 5 presents techniques for direction finding of a sound source such as a gunshot and bomb explosion by use of an acoustic array of acoustic sensors. Locating such an acoustic source is particularly important during military operations. For instance, during Iraq war American soldiers used the Boomerang, a device consisting of seven acoustic sensors, to find the direction of a sniper, greatly reducing the casualties. The sound generated by a bomb explosion is a transient signal which resembling a pulse signal. Beamforming, a popular signal processing technique, is widely used for direction finding. However, it is difficult to apply beamforming to a pure pulse signal. The basic but also effective way to tackle the problem is to use the time-of-arrival (TOA) or time-difference-of-arrival (TDOA) measurements. In this chapter two evolutionary algorithms (genetic algorithm and particle swamp optimization) and expectation-maximization (EM) algorithm are investigated. Typically, the ratio of the range between the source and the array center over the diameter of the array would be dozens and even greater than one hundred. It is thus usually infeasible to obtain valid position estimate of the source with a single array. By using multiple arrays or adding a distance measurement unit, valid source position information can be obtained.

Chapter 6 focuses on cooperative localization techniques in wireless networks. This chapter provides an extensive review on localization theory and algorithms as well as spatial and temporal cooperation schemes. A number of centralized localization algorithms and several distributed localization algorithms are described in detail. Cooperative localization under NLOS propagation is also investigated. Typically, in a wireless network, mobile devices communicate independently with base stations; uplink and/or downlink parameter measurements are used for position determination. However, if peer-to-peer or device-to-device communication can be enabled so that additional measurements such as range and angle can be made, positional accuracy can be enhanced considerably. In addition to this spatial cooperation, temporal cooperation can be exploited to achieve a performance gain, which makes use of information of previous states (i.e. the historical data). Cooperative localization may involve the issue of privacy, but if both sides have mutual agreement, this will not be a problem and data sharing can be reached such as through Bluetooth sharing.

Chapter 7 studies TOA-based impulse-radio Ultra-wideband (IR-UWB) positioning techniques. A frequency domain receiver structure is proposed to realize direct position estimation without involving the issue of threshold selection. Power position profile and compressive sensing technique are exploited. UWB usually refers to a wireless technology with a signal frequency bandwidth greater than 500 MHz or a fractional bandwidth larger than 20%. UWB is widely considered for short-range wireless communications due to a number of nice properties such as low power consumption, good interference suppression, and difficult to detect or intercept. It can be used for precise ranging since the signal has rather high temporal resolution, which is attractive for positioning in rich multipath indoor environments. Although the communication range of an early UWB system was rather small such as around 10 meters, current UWB systems may be able to transmit signals over a much longer distance (e.g. a few hundred meters in LOS condition) due to the use of advanced signal design and signal processing techniques. Such relatively long-distance communication capability would greatly expand UWB applications.

Chapter 8 also focuses on UWB positioning but integrated with MEMS inertial sensors (gyroscope and accelerator). A UWB positioning system can produce position estimates with accuracy of better than half a meter in an indoor environment where the propagation is multipath and LOS as well. However, in a complex indoor environment, the received signal may be lack of a direct path component and only consists of multipath components. Alternatively, the received signal does have a direct path component but it penetrated through one or more walls. In these cases, position accuracy would be significantly degraded if NLOS mitigation is not considered. Another case is that a mobile can only communicate with less than three (or four) base stations for 2D (or 3D) positioning, so that it is infeasible to obtain a unique position estimate. This chapter is intended to seek innovative solutions to tackle the above issues. A fuzzy theory based NLOS mitigation UWB positioning method is presented. Both loosely-coupled and tightly-coupled integration methods are developed and evaluated based on data through conducting experiments in two different geographic locations.

Chapter 9 studies the various sensors embedded in smartphones, including accelerometers, gyroscopes, magnetic field sensors, and barometric pressure sensors. All the sensors can be employed to realize different positioning and navigation techniques such as dead reckoning, inertial navigation, and map-matching. They can be used to determine positon, velocity, altitude, and to detect activity and step. A comprehensive comparison between a range of location sensors in smartphones and other location sensors is provided. Combination of MEM sensors and other sensor systems to form an integrated navigation system is investigated. Those sensor systems include Wi-Fi, Bluetooth, cellular network, and camera. Since almost everyone carries a smartphone indoors or outdoors either during working hours or spare-time activities, it is very useful to exploit the smartphone sensors for positioning and navigation purpose. The sensors in smartphones have the advantages of small size and low cost, but those advantages create the drawbacks of large noise and biases in the measurements. It is thus desirable to enhance the location performance of the sensors in smartphone such as by sensor integration and advanced signal processing.

Chapter 10 studies the integration of GNSS, inertial measurement unit (IMU) and light detection and ranging (LiDAR) for navigation in complex urban and indoor environments. Kinematic modeling of IMU is studied with details. A range of major error sources in the measurements provided by inertial sensors are analyzed and a great deal of experimental results is demonstrated. Error sources in GNSS are also studied and the basic principle of standard GNSS positioning service is also described. To mitigate the effect of the errors, integration of GNSS and INS sensors in typically considered, which in fact has been extensively studied and is a rather mature technology. On the other hand, integration of GNSS/INS with other sensors requires further investigations. This chapter investigates the fusion of LiDAR with GNSS/INS sensors to achieve a significant performance gain in complex propagation environments. Both theoretical analysis and experimental results are presented. Chapter 11 will focus on two other different integration schemes.

Chapter 11 also investigates integration of GNSS/INS and additional sensors which are WiFi and Locata. WiFi is a popular technology with a huge number of access points deployed indoors. Such established infrastructure can be exploited for positioning and navigation as extensively investigated by researchers. Locata is a wireless positioning system which uses techniques similar to GNSS in signal design, transmission and reception, but is ground-based. This chapter first studies information space theory and data fusion algorithms, followed by optimal multisensor data fusion, which can be realized through centralized Kalman filter, Federated Kalman filter, and global optimal fusion. Both GNSS/Locata/INS integration and GNSS/WiFi/INS integration are described with details. System testing were conducted for both integration schemes and the system performance were verified by experimental results. The GNSS/Locata/INS testing was conducted on Sydney Harbor. The INS, GNSS receiver and Locata user unit were all fixed on a vessel, while the Locata base stations (i.e. LocataLites) were installed on the shore and on the Bridge.

Chapter 12 focuses on acquisition of weak GNSS signal in indoor environments. A range of issues and techniques associated with signal acquisition are studied. Weak GNSS signal acquisition is a challenging problem, which may occur in urban canyons, under forests, and indoors. The usage of GNSS can be considerably improved, especially when integrated with other location sensors as discussed in the previous chapters. This chapter first studies the fundamentals including signal integration techniques, search strategies, detection schemes, error sources, and challenging issues. Then, fast acquisition techniques with low computational complexity are presented, followed by channel combining algorithms. Assisted acquisition technique can be used to enhance acquisition sensitivity and reduce the time to first fix, which is also discussed. Finally, techniques to achieve high sensitivity GNSS receiver are discussed, by introducing a multisensor-based indoor positioning platform which uses extended Kalman filter to integrate data from GNSS (GPS and Galileo), MEMS IMU, magnetometer, barometer, and WiFi. Doppler frequency hypothesis testing technique is also studied, which was proposed by the author and can be used to achieve fast and low-complexity acquisition.

Chapter 13 studies GNSS vulnerability with a focus on the categorization and detailed description of the various vulnerabilities. In addition to signal interference in the user segment, GNSS vulnerability also occurs in the space and control segments. The former is mainly related to actual satellite lifetime being shorter than the designed one, launched into incorrect orbits, launch failure, clock grifting or jumping, as well as solar radiation, missile attack, nuclear detonation or electromagnetic pulse. The latter may result in problems such as uploads of low quality data and GNSS outage for a short period of time. This chapter provide detailed information on user segment signal interference, including system and natural interference and man-made interference. The former includes intra-satellite, intra-system, inter-system and extra-system interferences, while the latter includes intentional and unintentional interferences. To achieve the desired functions and performance for positioning, navigation and timing as well as for other applications, all the major issues associated with GNSS vulnerability should be taken into account in the design and maintenance of GNSS and in the manufacture of GNSS receivers.

Kegen Yu
Wuhan University, China

Acknowledgment

The editors would like to thank all the authors for their valuable time and effort to contribute the thirteen nice chapters of the book, who are Mohamed Atia, Yang Cai, Pau Closas, Xunxue Cui, Andrew G Dempster, Shih-Hau Fang, Genís Floriach, Zheng Gao, Ricardo García, Riku Jantti, Wei Jiang, Ossi Kaltiokallio, Allison Kealy, Seung-Hyun Kong, Eva Lagunas, Yong Li, Songsheng Lu, Yongtao Ma, Montse Nájar, Monica Navarro, Guenther Retscher, Chris Rizos, Ian Sharp, Yuan Shen, Yunlong Wang, Kai Wen, Ling Yang, Hüseyin Yiğitler, and Yang Zhao. Most of the authors and the book editorial advisory board (EAB) members participated in the review process. The editors also wish to acknowledge the valuable contributions of the reviewers regarding the improvement of quality, coherence, and content presentation of chapters. Without the support from the authors, reviewers, and EAB members, this book would not have become a reality.

Kegen Yu
Wuhan University, China

Chapter 1
New Methods for Improved Indoor Signal Strength Positioning

Ian Sharp
Independent Researcher, Australia

Kegen Yu
Wuhan University, China

ABSTRACT

This chapter focuses on positioning based on received signal strength measurements and radio path-loss modeling. Typically, receiver signal strength from a device such as a smart phone is used to calculate the path-loss and thus estimate distances using a loss model calibrated in an offline process. With known positions and ranges to WiFi access points or simple devices using Bluetooth or Zigbee for data communications, the mobile device position can be estimated. However, due to the complex multipath propagation indoor environment, distance estimation and position determination using current methods are not very accurate. Based on knowledge of the nature of indoor signal propagation and algorithms especially designed for mobile applications, new methods show that positional accuracy of a few meters is possible, even with non-line-of-sight propagation through many intervening walls. Given the current widespread deployment of WiFi indoors, simple software-only solutions are feasible for applications such as general personal navigation and tracking within buildings.

INTRODUCTION

A range of technologies have been utilized for positioning, but radio positioning is probably the most important one, especially due to the various military and civilian applications and services associated with global navigation satellite systems (GNSS). In radio positioning, five basic signal parameters have been widely considered for position determination, namely signal strength, time-of-arrival, frequency difference-of-arrival, carrier phase, and angle-of-arrival. The methods based on these different measured parameters have different strengths and weaknesses, so in general there must be a tradeoff between

DOI: 10.4018/978-1-5225-3528-7.ch001

complexity, the associated cost and positional accuracy. Receiver signal strength positioning is simple to implement and can use existing radio infrastructure such as WiFi access points, but using current techniques the positional accuracy and reliability is too poor to be of practical use in non-line-of-sight (NLOS) severe multipath conditions at typical ranges required of 30 meters or more. However, new techniques described in this chapter show that positional accuracy of 2-3 meters in indoor situations should be possible, thus opening up new applications of indoor positioning and navigation.

Signal time-of-arrival can be used to generate pseudo-range measurements such as for GNSS-based coarse positioning, and precise ultra-wide band positioning can exploit round-trip time measurements. Time-difference-of-arrival can be simply derived from time-of-arrival data to account for the unknown time offset between the two communicating nodes (typically a base station and a mobile device) (Chan & Ho, 1994)). Frequency difference-of-arrival (also called differential Doppler) can be applied for scenarios where there exists relative motion between two nodes so that significant Doppler shifts can be observed, such as by an airborne receiver. The first navigation satellite system (also known as the TRANSIT system) provided navigation for naval ships, commercial shipping and private light aircraft, and was also used by surveyors for determining coordinates of remote ground marks between the late 1960s and early 1990s.

Angle-of-arrival measurement requires a directional antenna and/or antenna array. A two-dimensional position of a mobile device can be estimated using two angle-of-arrival measurements made at two different base stations. A single base station can be used to estimate a mobile position by using both angle-of-arrival and distance measurements. The more accurate VHF omnidirectional range (VOR) used in aircraft for route navigation is based on determining the phase of a specially encoded signal transmitted from a ground station. Another more recent application of precise carrier phase measurement enables precise position determination such as in GNSS (Juan *et al.*, 2012). For instance, the GPS L1 signal has a wavelength of about 19cm. If the phase measurement accuracy is of the order of one-fifth of the wavelength, the ranging accuracy can be as good as 4cm. However, in many cases including GNSS, determining the integer cycle ambiguity is an issue that must be resolved. Furthermore, multipath interference can corrupt the phase information, hence applications are limited to outdoor applications in clear open areas, such as in the case of surveying.

Nearly all modern communication devices or receivers include Receiver Signal Strength Indicator (RSSI) functionality. Typical implementation is a diode detector in the IF of the receiver, and thus is cheap to implement. However, note that RSSI only gives an *indication* of the signal strength, and is not intended to be an accurate measurement of the signal strength at the antenna port of the device. As the measurement is not directly of the RF signal, some form of calibration is required. Specifications for standard common consumer radio systems such as WiFi, Bluetooth and Zigbee (as used for measurements described in this chapter) only require the measurement is within a specified range; for example, the Zigbee IEEE 802.15.4 specification has an accuracy range of ± 5 dB. Such imprecision in raw RSSI measurements must be taken into account if RSSI positioning is to achieve a reasonable accuracy.

In many indoor environments such as offices, hospitals, hotels, shopping centers and other similar large buildings, there are often many WiFi access points (AP) and Bluetooth devices which can provide short-range wireless communications. Note that these technologies are designed for data communications, with indoor signal range being limited to less than 10 m for Bluetooth, and perhaps 30-40 m for WiFi. Devices such as mobile phones include a RSSI measurement as a standard part of the API access to the device operating system software. This software access can be exploited by radiolocation applications on the device. There are two main types of such applications. One is based on using multiple

RSSI measurements to determine position using methods similar to those used for GNSS, and is the main topic of this chapter. The second method, referred to as "fingerprinting", is based on a fine grid of survey points with the RSSI measurements from (usually) WiFi access points stored in a database. The position is determined by matching the mobile device RSSI data with the database. Because of problems with both of these methods, indoor radiolocation using RSSI is currently little used.

This chapter studies the received signal strength based position determination with a focus on the model-based approach. Database or fingerprinting-based positioning is not considered in this chapter, but is the subject of Chapter 3 of the book. After introducing a number of basic concepts, signal propagation loss measurements in a typical office building are presented. These measurements are based on determining the packet data reception performance of Zigbee radios in NLOS conditions, and were not specifically related to positioning. However, the experimental data included the RSSI received from 24 locations at a centrally located Zigbee receiver. The analysis in this chapter includes the standard method universally used in the literature, and is shown to perform poorly for even moderately accurate indoor positioning with NLOS propagation paths. Using the well-known characteristics of indoor radio propagation, methods are described that improve the quality of the raw RSSI data so that they can be used for more accurate position determination. While standard methods use a propagation loss model to estimate range, the new method described uses the "corrected" RSSI data directly to determine the position. Indeed, the method is remarkable in what information is not required. In typical cases these include the base station transmitter power, antenna gains, RSSI calibration, and ranges to base stations. The only additional information required (apart from the RSSI measurements) is a map of the building with the location of the base stations indicated, and architectural details (internal walls) of the building. As positions without a map are meaningless, the map is not actually an additional requirement. RSSI data from a building covering about 2000 m^2 in area show that using NLOS measurements from 24 Zigbee radios the positional accuracy is about 1 m, and using just five measurements (typical of the number WiFi AP in range) the accuracy is about 3 m. Because of the desirable characteristics of the method, with minimal or no additional information required except for the mobile device's RSSI measurements, and the relatively good accuracy, the method is well suited to the development of mobile phone applications to provide general purpose positioning in any building that has sufficient WiFi AP. This compares with other current methods which require extensive additional data, surveying and individual calibration of each mobile device to work effectively.

BACKGROUND

The analysis and the development of a new RSSI positioning method was based on the desire to take advantage of existing radio communications infrastructure for position determination inside buildings. While other methods such as time-of-arrival systems (Sharp, Yu & Hedley, 2012) can work well with accuracies better than 1 m, these systems require specially designed and installed base stations and mobile devices. The most universal signal characteristic available in a wide range of devices is the signal strength as reported by the device RSSI. However, all current methods using RSSI perform poorly in NLOS indoor situations, and typically have only been demonstrated in experimental situations where all the equipment has been specially calibrated. The challenge is to develop a method that is both more accurate and easy to use without any special requirement to calibrate both base stations and mobile devices.

As part of a project to develop a new TOA indoor positioning system (Sathyan, Humphrey & Hedley, 2011), a series of experiments were performed using standard Zigbee radios in an indoor environment where all the paths are NLOS. These experiments were aimed at the data communications performance as a function of range, and were not aimed at RSSI positioning. However, as the measurements included the Zigbee radio RSSI, this data set was suitable for investigating the feasibility of a practical signal strength positioning system using a very simple (cheap) radio. While these experiments were performed using Zigbee radios (operating in the 2.4 GHz ISM band), the results should be directly applicable to other system (such as IEEE 802.11 – WiFi and IEEE 802.15.1 –Bluetooth)

The measurements used a MC13192 which is a ZigBee compliant radio developed by Freescale Semiconductor (formerly Motorola) and fits the profile of radio-chips potentially used in many common devices. The evaluation board has separate transmit and receiver ports which gives an opportunity to connect a Low Noise Amplifier (LNA) in the receive path and a Power Amplifier (PA) in the transmit path. This arrangement therefore can easily change parameters in the link budget.

The evaluation of the MC13192 radio in a typical indoor environment is necessary because performance specifications of “Freescale” are very vague. “Freescale” claims 400 m range (without LNA) and 600 m (with LNA) outdoors in line-of-sight (LOS) conditions. Typical indoor environment would introduce about 30 dB of additional loss (on average) on top of LOS loss.

The major specified parameters of this radio are:

- 16 frequency channels in 2400-2483.5 MHz band
- 0 dBm nominal, programmable up to 4 dBm typical maximum output power
- supports 250 kbps O-QPSK data in a 5.0 MHz channels and full spread-spectrum encode and decode (compatible with IEEE standard 802.15.4)
- Receiver sensitivity of -91 dBm (typical) at 1% packet error rate

Figure 1 shows the hardware used in the testing. Note the separate transmit and receiver antennas allows transmit and receiver signal paths to be individually changed to test various combinations of link budgets, such as including a transmitter power amplifier or a receiver low noise amplifier. The measurement points are marked on the map in Figure 2, with most having a range less than 30 m to the common receiver point. The receiver point is marked with a cross, and is located in an open space associated with an atrium. Note that all straightline paths pass through at least two walls. A computer program can analyze a map to easily determine the number of walls, as the walls on the map are black and free space is white. The thickness of the walls can also be determined, but was not used in the data processing.

The testing procedure was based on measuring the transmission of packets of data between two Zigbee radios, and logging the data on a PC. The data recorded was the receiver signal strength as measured by the chip radio RSSI, and the percentage of packets received without any transmission errors. The measurements were made in a building shown in Figure 2, which is typical of an office environment. The 24 propagation paths are all NLOS, so that straightline paths pass through at least two walls.

Most of the internal partitioning walls are plaster boards about 10 cm thick (~4 dB loss). The middle structure of the building (around toilets and the kitchen area) is made of thick concrete walls with propagation losses in the range 10-20 dB. This forces electromagnetic waves to diffract around the middle of the building, possibly finding a path along corridors. One particular point in the building at the end of a corridor ($range > 45m$) had a working radio link, even though the nominal straight line

Figure 1. M68EVM08GB60 development board ("Axiom") shown with MC13192 daughter board with transmitter path connected to PA and receiver path connected to LNA.

path passed through 10 walls. Note that actual transmission losses through walls also depends on the incidence angle and the wall thickness, but these details are not included in the simple wall loss modeling considered in this chapter; but rather this variability is considered as a random variable in the model.

The measurements were obtained from four independent experiments that recorded the received data from the 24 transmitter locations. The four experiments were conducted using the Zigbee radio without an LNA or PA, with a PA, and two different configurations with a PA. The data set from the experiments with the Zigbee radios form the basis for the analysis in the following sections.

INDOOR RADIO PROPAGATION

The characteristics of the radio signal are important in designing an indoor positioning system. To estimate the range from a transmitter to a receiver using the RSSI, the nature of large-scale signal variation with distance is required. While in free space a simple inverse-square law can be used to determine the signal strength as a function of range, in a NLOS environment the complexity of the environment makes calculations of the large-scale signal attenuation with distance difficult. Furthermore, while the average variation with distance is important, the medium-scale variation of the signal strength is also an important parameter in the design of indoor positioning systems. These medium-scale effects are typically on a scale of 5-50 wavelengths (0.5-5 meters at 2.4 GHz), and thus are associated with major structural elements of a building such as rooms. Small-scale signal variations can be considerable,

Figure 2. Ground floor layout of CSIRO ICT Centre showing measurement points. The range circles are 10 m, 20 m, and 30 m.

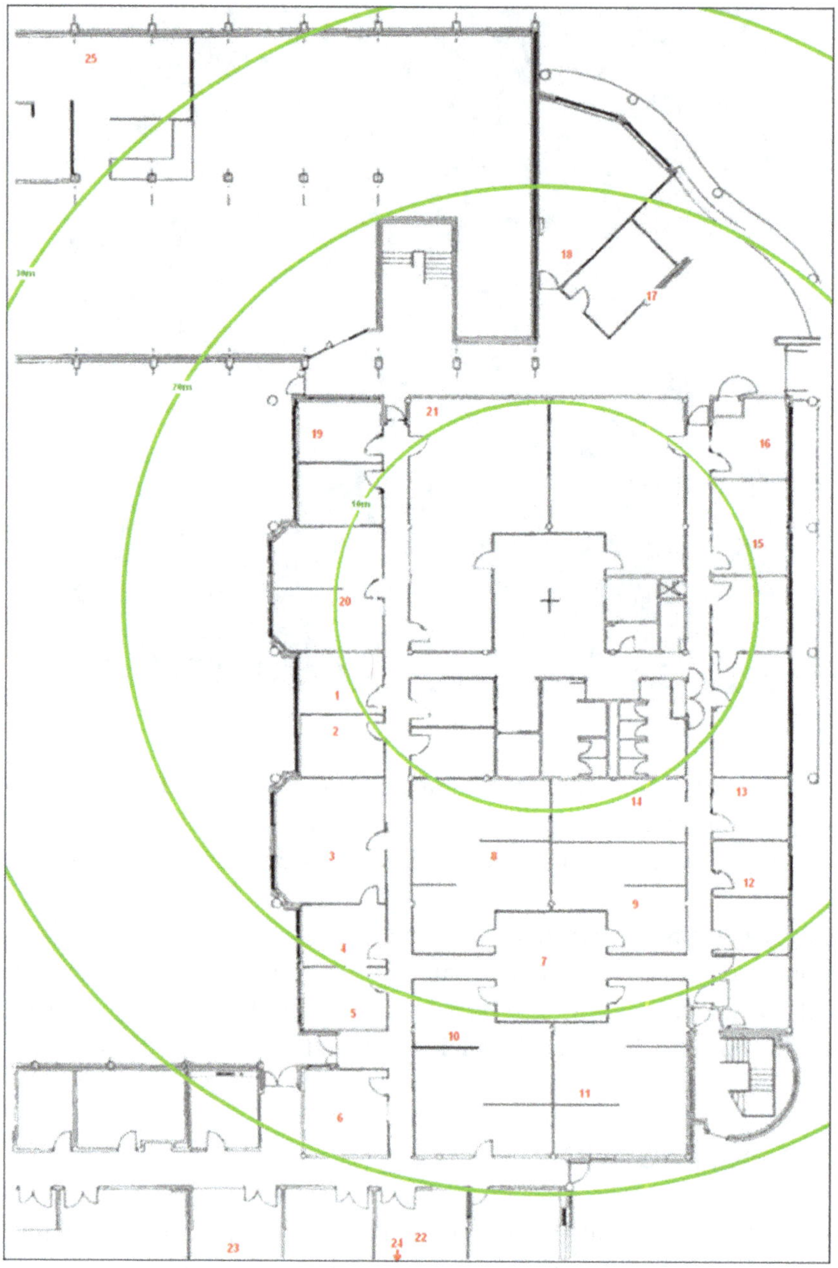

further complicating matters. In particular, constructive and more importantly destructive interference (Rayleigh and Rician statistical fading) occurs on the scale of the order of a wavelength (12.5 cm at 2.4 GHz). These dividing lines are only rough guidelines which are useful for engineering analysis of the complex propagation environment of indoor radio propagation at microwave frequencies.

In summary, small-scale variations in signal strength are associated with complex (amplitude and phase) constructive and destructive interference between the signal components, medium-scale effects are associated with signal scattering, diffraction and the absorption of RF energy by obstacles along the propagation path, and large-scale variation is associated with the free-space attenuation in addition to signal losses mainly associate with walls along the propagation path.

Link Budget

The basis of signal strength positioning is the determination of the propagation loss between a number of transmitters and a common receiver, and then use this information to determine the position of the receiver. The transmission process can be summarised in terms of a link budget, whereby all the major hardware components and associated parameters are defined, so that the propagation loss can be calculated. For calculations it is convenient to define the link budget in units of decibels, so that various elements in the budget can be added rather than multiplied. Thus the link budget can be expressed in the form

$$P_{rx} = P_{tx} + G_{tx} - L + G_{rx} \tag{1}$$

where

P_{rx} : received power (in dBm, or dB relative to a milliwatt)

P_{tx} : transmit power (3 dBm for Zigbee radios)

G_{tx} : transmitter antenna gain relative to an hypothetical omnidirectional antenna (-3 dBi)

L : propagation loss (in dB)

G_{rx} : receiver antenna gain (-3 dBi)

The numerical data above are associated with the Zigbee radios used in the experiments described, and may not be typical of other systems used for indoor RSS positioning. For example, WiFi will have higher transmitter powers and higher gain antennas, hence the Zigbee performance can be considered as a minimum that could be typically expected.

The basic starting point for loss determination is the free-space loss (also referred to as the LOS loss), as this represents the minimum possible loss. From Maxwell's equations it is known that the RF energy reduces as the square of the distance from a point radiating source, so it can be shown that the free-space path loss for a range R and a wavelength λ is given by[1]

$$\begin{aligned} l_{free\_space}(R) &= \left(4\pi \frac{R}{\lambda}\right)^2 \\ L_{free\_space}(R) &= \left(\frac{20}{\ln(10)}\right) \ln\left(4\pi \frac{R}{\lambda}\right) = \alpha_0 + \beta_0 \ln\left(\frac{R}{\lambda}\right) \end{aligned} \tag{2}$$

Applying (1) and (2) for the Zigbee radio at a range of 30 m gives a received power of -72 dB. As the ZigBee radio has a sensitivity of -92 dBm for a packet error rate of 1 percent, this radio will easily be able to operate in LOS conditions at 30 m range.

However, indoors the propagation environment is more complex, and usually there is no LOS, hence propagation occurs through multiple walls along the path from the transmitter to the receiver. At 30 m range, the propagation path loss excess is typically of the order of 30 dB. Under these circumstances P_{rx} is -102 dBm. With a receiver sensitivity of -92 dBm it is clear that this radio is not going to achieve desired range indoors without some modifications. Two changes are possible – increasing the transmitter power with the addition of an extra power amplifier, or using a low-noise amplifier to improve the noise figure of the Zigbee receiver. The RSSI from the basic Zigbee radio can be used to determine estimates of the actual range between the 24 transmitting points and common receiving point, with various combinations of these suggested enhancements. These measurements were made by moving the transmitting Zigbee module to all 24 points with the one particular operating mode first, and then repeated for the other modes at nominally the same transmitting points. In practice return to the exact same location cannot be achieved, but were within about 10 cm. This technique of four independent measurements per point is exploited in the data analysis.

Increasing the transmitter power is the simplest option if the transmitters are fixed base stations such as WiFi access points. In some potential applications the "base stations" are in fact mobile devices as well, so in this case increasing the transmitter power is not an option, as it will reduce the battery life. The second option is to increase the receiver sensitivity. The IEEE 802 15.4 specification (Zigbee) states that for a noise figure of 25 dB the receiver sensitivity shall be at least -85 dBm. The stated receiver sensitivity for the Zigbee devices used for the experimentation is -92 dBm, implying a noise figure of 18 dB. If (for example) a low noise amplifier with a noise figure of 3 dB and a gain of 20 dB were added, the effective noise figure would be 4 dB (14 dB less), and the receiver sensitivity improved to -106 dBm. Such a receiver sensitivity should be sufficient for 30 m range without increasing the transmitter power.

Large-Scale Propagation Characteristics

Large-scale propagation effects are the basis for determining the distance of transmitting nodes from a receiver, and hence the position. Given a radio system parameters such as the transmitter power, the receiver sensitivity and antenna gains, a link budget can be devised to estimate the propagation loss between the transmitter and receiver units. From this estimate and the large-scale attenuation characteristics, the range can be estimated using an appropriate range-loss model. As the propagation characteristics indoors are complex, the typical approach taken is for empirical generic equations to represent the propagation loss, with the parameters determined by appropriate fitting these equations to measured data.

The propagation loss as a function of distance for the Zigbee data is shown plotted in Figure 3. One common indoor propagation model based on an extension of (2) is of the form

$$l = l_0 \left(\frac{R}{\lambda}\right)^{\gamma} \tag{3}$$

Figure 3. Propagation loss as a function of range based on the raw RSSI signal strength data. The dashed line is a least-squares fit to a model of the form $L(R) = \alpha + \beta \ln(R / \lambda)$*, where* $\alpha = -35.7$ *dB and* $\beta = 25.3$ *so* $\gamma = 5.8$*. Also shown is the LOS propagation path loss. The model RMS fit error is 8.5 dB, and Pearson's correlation coefficient between range and loss is 0.71.*

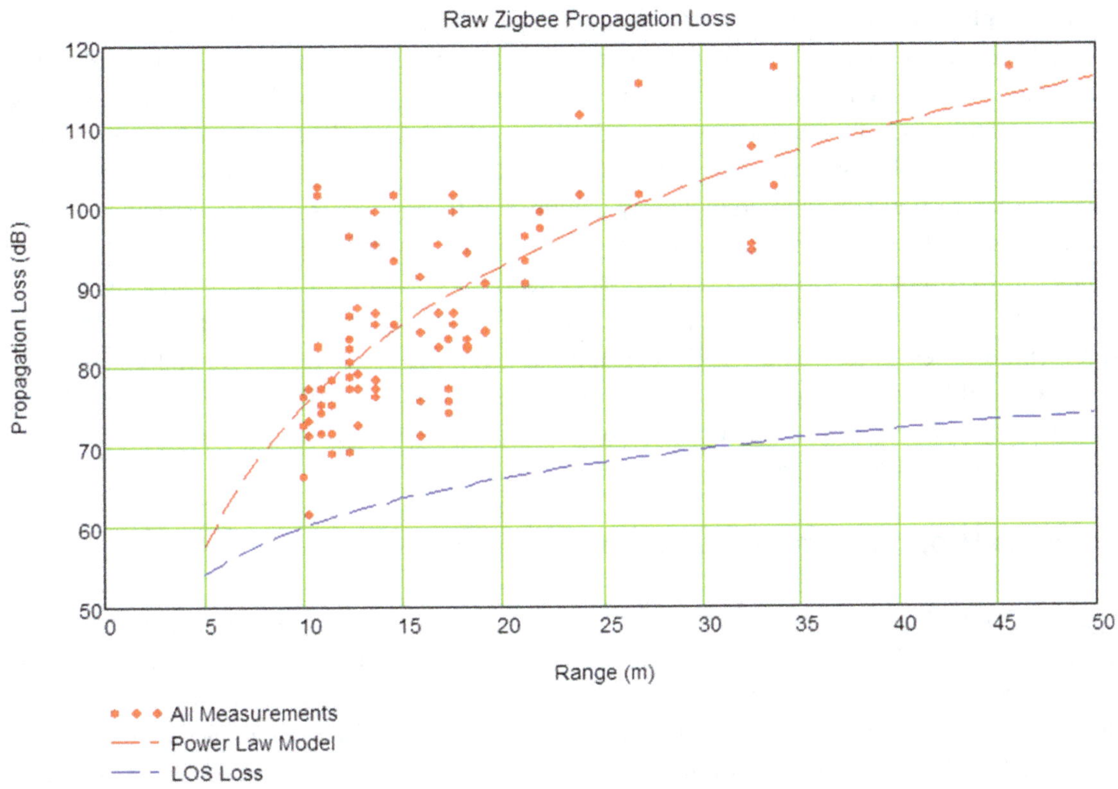

In this power-law model, the attenuation as a function of distance is controlled by the parameter "γ" with $\gamma > 2$ as the loss is greater than the free-space case, and l_0 the loss at one wavelength. The loss is assumed to be a function of the propagation distance measured in wavelengths, as with the free-space case. However, as in practice loss is measured in decibels, a more appropriate form for indoor NLOS large-scale propagation loss is

$$\begin{aligned} &L(\text{dB}) = \alpha + \beta \ln(R / \lambda) \qquad \left(L \geq 0 \quad \text{or} \quad R \geq R_{\min}\right) \\ &\alpha = \frac{10}{\ln(10)} \ln\left(l_0\right) \qquad \beta = 10\gamma / \ln\left(10\right) \qquad R_{\min} = e^{-\alpha/\beta} \end{aligned} \tag{4}$$

Notice that natural logarithms are used in (4), as it is more convenient in later mathematical developments. Figure 3 also shows the free-space loss for comparison, as well as the least-squares fit based on model (4). As can be observed, the measured loss is considerably greater than the free-space loss, and also exhibits random (medium-scale) variation. From the least-squares fit to these data the power law parameter γ is 5.6 (compared with 2 for free-space). As the power law exponent $\gamma = 5.6$ is much greater than the free-space exponent, the losses increase much more rapidly with propagation distance. While this model is perhaps a natural extension to the free-space law (based on the rigorous solution to Maxwell's equations), the model defined by equation (4) is not totally satisfactory. For example, when the propagation range is short LOS conditions will exist, implying that $\gamma = 2$ at short range, but a larger value applies at long ranges, with presumably a transitionary region at intermediate ranges. Further, at medium ranges the NLOS propagation indoors will typically be through a number of walls with a consequential loss at each wall, so this implies a linear increase (in dB) in the loss excess as a function of distance. As the range further increases and the losses through walls increases, there may be an alternative lower loss path along corridors in the building. Thus at long range the propagation loss (in dB) will increase at a slower rate than predicted by a simple wall loss excess model. Thus in summary, all these effects cannot be captured with a constant power law exponent.

Another criticism of the simple power law (4) is the difficulty in relating the power law parameters to the physical nature of the building. In practice the power law exponent γ can only be determined by measurements in specific buildings, with reported values in the range of 3-6 for indoor propagation. Because of the difficulty of applying (4) in practical situation in various buildings, an alternative model is proposed in the following section *Basic Loss Excess Modeling*, and can be related to each building based on architectural maps of the building.

Small-Scale Propagation Characteristics

While signal strength positioning depends on the large-scale variation of signal propagation loss with distance, the small-scale effects have a large contribution to the measurement signal strength *noise*. If the measurement noise is large, then the associated range noise is so large as to make signal strength positioning techniques non-viable for practical applications. Thus the first step in the positioning process must be to reduce RSSI measurement noise. Note that all the measurements in this experiment are associated with NLOS paths; LOS paths indoors are usually restricted to very short ranges, and thus are not of importance for any practical indoor positioning systems.

Based on the link budget equation (1), the raw estimate of the propagation loss is shown in Figure 3. There are several points to note from this figure. First as stated in *Link Budget* section, the propagation loss can be up to 40 dB greater than the LOS loss. For ranges greater than about 20 m the loss *excess* is at least 30 dB, while for shorter ranges there is a widespread variation, from near zero to 40 dB. Secondly, notice that range is a poor indicator of propagation loss. For example, at 10 m range the measured loss varies from 60 dB to about 100 dB. Conversely, a 100 dB loss can occur from about 10 m to about 35 m. Thus raw propagation loss data are not well correlated with propagation range, although of course there is a general trend of increasing loss with range. More formally, Pearson's correlation coefficient for the data in Figure 3 is 0.71, which shows there is a moderate correlation, but not sufficient to assist position determination.

One of the main reasons for measurement noise is associated with small-scale signal fading effects. In a NLOS indoor environment there are many possible paths between the transmitter and the receiver. At the receiver, the measured signal strength is the (complex) sum of all these signals, that is both the amplitude and the phase of the signal are important. The phase of the signal is closely related to the wavelength (360 degree phase variation), so phase effects can be expected on distance scales of the order of a wavelength. With no direct signal to dominate, the signal magnitude will be the complex sum of many signals scattered by the building walls and objects in rooms. These signals can combine either constructively or destructively, so that the measured signal can vary significantly from the local large-scale mean used for position determination. A detailed statistical analysis (Chapter 2, Yu, Sharp & Guo, 2009) shows that in severe NLOS conditions the small-scale signal amplitude (s) exhibits approximately Rayleigh distribution statistics, which has the following mathematical form for the probability density function (PDF) and the cumulative distribution function (CDF)

$$\begin{aligned} p_R(s) &= \frac{s}{\sigma^2}\exp\left[-s^2/2\sigma^2\right] \\ cdf_R(s) &= 1-\exp\left[-s^2/2\sigma^2\right] \end{aligned} \quad (5)$$

However, the RSSI measurements are in decibels (dBm), hence small changes in small signals can result in large changes in the measured RSSI signal. This effect explains some of the large variations in the computed losses shown in Figure 3.

The normalised Rayleigh CDF is shown in Figure 4a, where the signal amplitude is normalized by the signal mean and expressed in decibels, and the CDF is plotted on a log scale so that small probabilities can be observed. For example, the probability that a measured signal strength has a 10 dB fade relative to the mean signal is about 0.075, so that about 1 in 13 measurements will have a 10 dB fade. Note that the large scale mean signal strength is what is used to estimate range and hence positions, so that a 10 dB fade will result in a very considerable error in range estimation unless mitigation procedures are taken.

The data in Figure 3 can be tested to determine approximately if the signal strength data exhibits Rayleigh statistics. Rayleigh statistics are associated with signal fading relative to the local mean signal, which is also affected by unknown medium- and large-scale propagation attenuation effects. However, while the local mean signal at each measurement point is not known, an estimate can be made based on the multiple measurements near each position. This variation of the signal can be used to estimate the small-scale statistical variation, that is the normalized signal is the RSSI (in non-dB form) divided by the mean signal. In the case of the Zigbee measurements analyzed in this chapter there are four such measurements per location. For more details on the statistics of these measurements, refer to the Appendix. Note however that the normalized signal is the ratio of two random variables, as both the individual measurements and the derived mean are random variables. The results are shown in Figure 4b. The comparison between the measured and theoretical distributions is moderately accurate, but the measured data has a higher probability of occurrence at low signal amplitudes than the theoretical estimate.

As signal fades are the cause of measurement errors, significant signal fades should be avoided in any calculation of the mean signal level. A significant error is defined as fades greater than 6 dB. Because of the non-linearity associated with measurements in decibels, all calculations described below are in standard form, hence a significant fade is one in which the amplitude is less than a half of the mean signal. From Figure 4a the probability of such a 6 dB fade is 0.2 or 1 measurement in 5. In the case of Zigbee measurements with four independent measurements of the RSSI for each path, there is a good likelihood that one measurement will have a significant fade, but two faded measurements (with a probability of 0.04) are unlikely. Thus the proposed strategy is to test the weakest (of four) measurements, and if it is less than a half of the median of the four measurements, it is eliminated from further processing. The

Figure 4. a. Normalised Rayleigh cumulative distribution function plotted with the normalized signal level in dB. b. Cumulative distribution function derived from the measured raw data in Figure 3, and the theoretical expected distribution based on normalization with the mean signal. Also shown is the Rayleigh distribution.

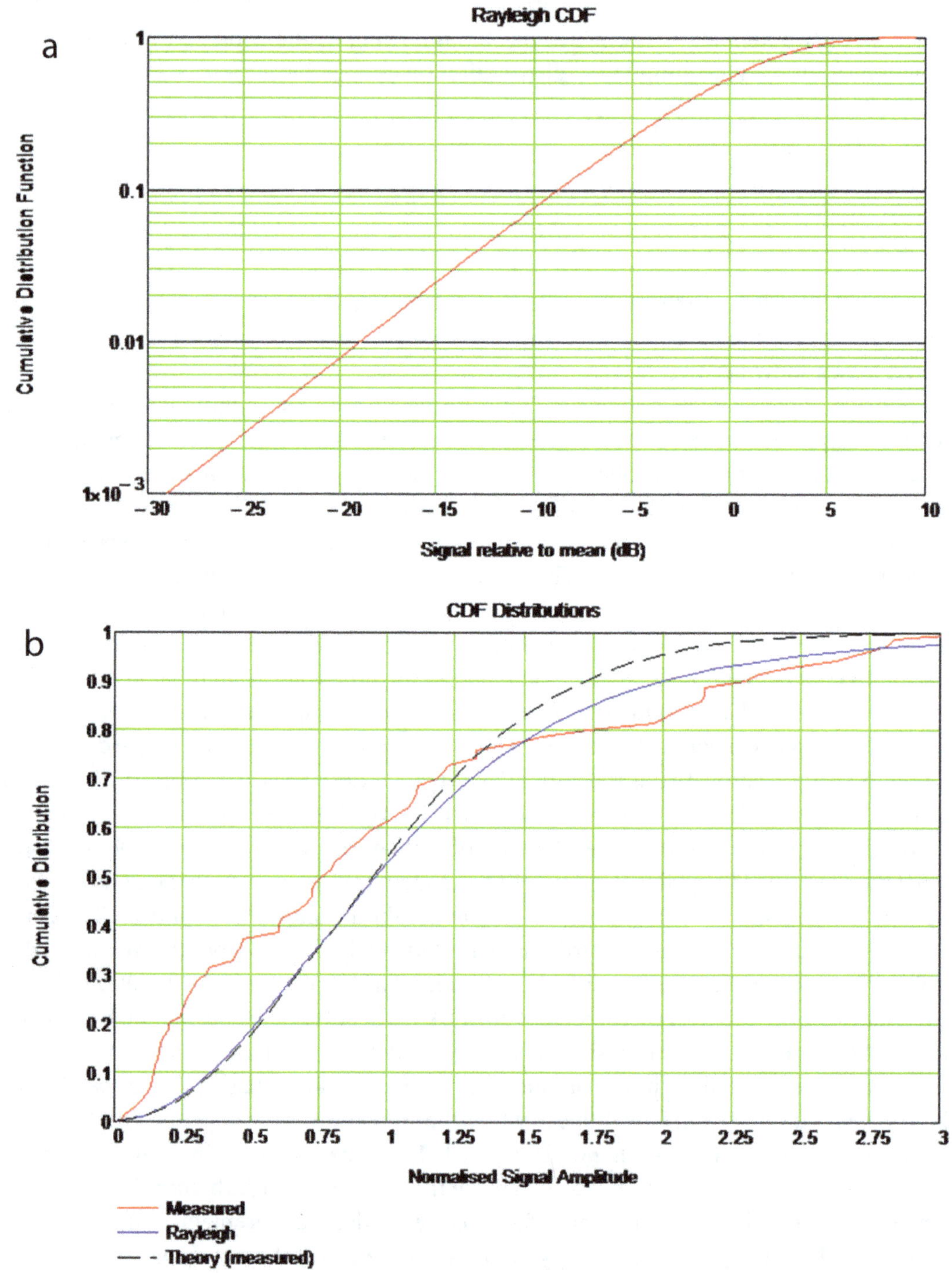

median rather than the mean is used, as it is less affected by "bad" data. Then the three of four samples are averaged. Finally the mean signal is converted back to the decibel version for subsequent processing.

The results of this processing are shown in Figure 5. The number of samples is reduced to 24, one for each path. The Pearson's correlation coefficient is increased from 0.71 to 0.85, which is reflected in the better correlation between path loss and range. However, there will be some residual measurement noise associated with the Rayleigh statistical variations in the measured data. Additional variation is associated with medium-scale effects, which are considered in the section *Medium Scale Propagation Characteristics.*

Basic Loss Excess Modeling

The basic indoor propagation model is the power law model described in *Large Scale Propagation Characteristics* section. This model is not based on any theoretical modeling, but is derived by numerically fitting the measured data to determine the power law parameters. As there is no theoretical basis, the

Figure 5. Propagation loss as a function of range based on signal strength corrected for Rayleigh fading losses. The dashed line is a least-squares fit to a model of the form $L(R) = \alpha + \beta \ln(R / \lambda)$, where $\alpha = -37.7$ dB and $\beta = 24.4$ so $\gamma = 5.6$. Also shown is the LOS propagation path loss. The model RMS fit error is 6.2 dB. Pearson's correlation coefficient between range and loss is 0.85.

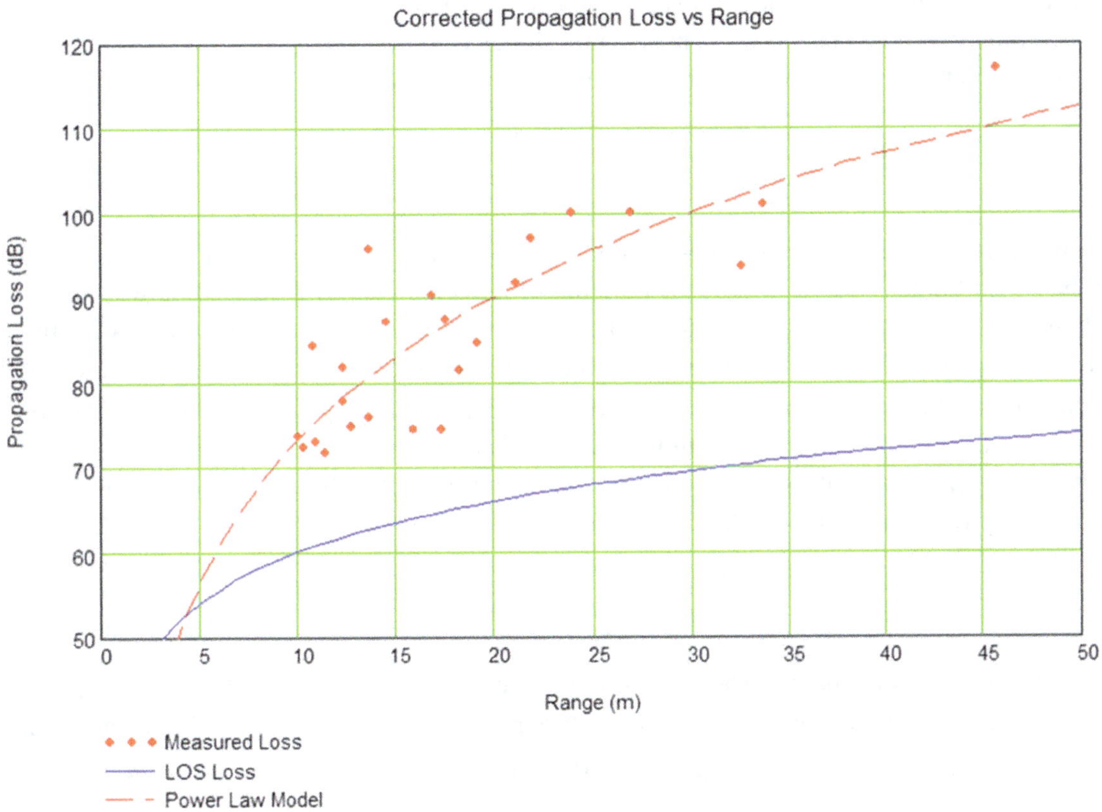

values of the model parameters are not easily determined in any particular building without extensive measurements, and thus it is difficult to apply as a general purpose model for signal strength positioning.

To obtain a model which can be more related to a particular building an alternative model is proposed, namely a *loss excess* model. In this model it is assumed that signal attenuation in NLOS conditions is a combination of the free-space losses between walls, and extra losses through walls along the propagation path. When measured in decibels these two components are added to form the total loss. Thus the model concentrates on modeling loss excesses, as the free-space losses have a simple well known mathematical model defined by (2). Therefore, the mean of these losses can be modeled systematically, while additional effects from signal scattering will remain as random noise.

For a simple loss excess model, the aim is to only approximate the basic characteristics of indoor radio propagation. At short ranges it is assumed that the path has LOS conditions, so that the loss excess will be zero up to some fixed range R_0. The value of this parameter will vary according to building type, but for the building associated with the data used in this chapter it is assumed to be about the size of typical rooms, or $R_0 = 5\,\text{m}$. At longer ranges it is assumed that the propagation is NLOS due to walls. Further, it is assumed up to medium ranges (say 15-20 m), the loss excess is approximately proportional to propagation range, as the number of walls will be approximately proportional to range. However, as ranges become longer, the straight line path from the transmitter to the receiver may not represent the lowest loss path, as lower loss but longer paths (such as along corridors) may exist. The mix of these two types of paths will be complex and thus beyond the scope of any simple modeling. However, the general characteristics are known, namely at shorter ranges the loss excess (in dB) is a linear function of range, but at longer distances the loss excess will become increasingly non-linear with distance. While many such functions could be chosen, a reasonable choice is to base the *loss excess* on the logarithm of range. In this model, the loss excess is thus represented by the model

$$L_{excess}(R) = L_0 \ln(R / R_0) \qquad \left(R \geq R_0\right) \tag{6a}$$

The measured loss data and the least-squares fit to model (6) is shown in Figure 6. Also shown is a model based on the loss excess as a linear function of range. Notice as described above, the loss data initially has an approximately linear relationship with range, hence for ranges up to about 25 m the two models are roughly equivalent. Thus to estimate the parameter L_0 the rate of increase of loss excess with range can be equated to the loss associated with walls, and the average distance between walls. This topic is considered in more detail in section *Modeling Based on Wall Losses* following.

If the average loss per wall is ω dB, and the average number of wall per unit distance is η, then the loss excess is given by the simple relationship

$$\Delta L = \eta \omega R \tag{6b}$$

As the two expressions (6a) and (6b) are modeling the same phenomena, performing an approximate match over a range interval allows the power law loss parameter L_0 to be expressed in terms of the wall loss parameters. In particular the integral over the range interval $(R_0, R_{\max})$ can be equated to zero, namely

Figure 6. Plot of the measured and two models of the propagation loss. The RMS fit error in both cases has a standard deviation of about 6 dB. The linear relationship with range is evident up to ranges up to about 25 m.

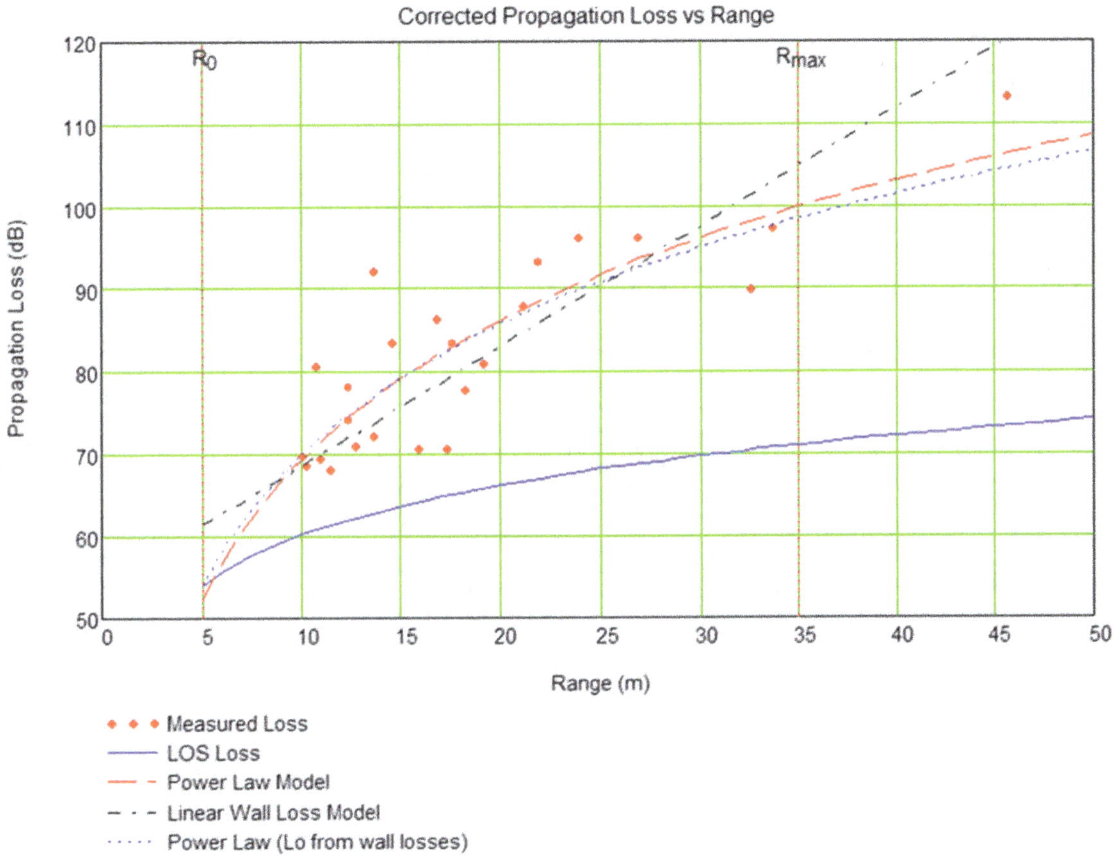

$$\int_{R_0}^{R_{max}} L_0 \ln\left(R / R_0\right) - \eta\omega R \, dR = 0 \tag{7a}$$

to solve for L_0. Evaluating the integral and solving (7a) gives

$$L_0 = \frac{\eta\omega}{2} \frac{R_{max}^{\ 2} - R_0^{\ 2}}{R_{max}\left(\ln\left(R_{max} / R_0\right) - 1\right) + R_0} \tag{7b}$$

Thus the power law loss parameter L_0 is estimated in terms of the wall loss parameters over a defined range interval. With $R_{max} = 35$ m and $\eta\omega$ estimated from Figure 6, (7b) gives $L_0 = 23$ dB. Figure 6 also shows the power law loss function using L_0 estimated using (7b); the close agreement with the corresponding curve based on a least-squares fit to the data shows that the power law parameters can be estimated from wall loss parameters, rather than relying on the model parameters derived from fitting to measured loss data.

Based on the above model of loss excess and (2), the total path loss is then given by

$$L_{path}(R) = \left(\frac{20}{\ln(10)}\right)\ln\left(4\pi\frac{R}{\lambda}\right) + L_0 \ln(R / R_0) \qquad \left(R \geq R_0\right) \tag{8a}$$

which can be manipulated into the form of (4)

$$L_{path}(R) = A + B\ln(R) \qquad \left(R \geq R_0\right) \tag{8b}$$

where the constant parameters are

$$A = \ln\left[\frac{1}{R_0^{L_0}}\left(\frac{4\pi}{\lambda}\right)^{20/\ln(10)}\right] \qquad B = L_0 + 20 / \ln(10) \tag{8c}$$

Equation (8b) can be further manipulated into the form of (4)

$$l(R) = l_0\left(R / \lambda\right)^{\gamma} \tag{8d}$$

where

$$l_0 = 10^{A/10}\lambda^{\gamma} \qquad \gamma = \ln(10)B / 10 \tag{8e}$$

Notice that (8e) is in the same format as (3), so this development has resulted in the power law model commonly used for indoor radio propagation modeling. However, the parameters $\left(A, B\right)$ and hence $\left(l_0, \gamma\right)$ are completely defined from known or measured subsidiary parameters by applying (7b).

Modeling Based of Wall Losses

While the path losses for indoor propagation can be complex, the dominant component of the loss excesses can be associated with the additional loss from transmission through walls. These losses are associated with the reflections at the boundaries of the walls due to their refractive index being greater than unity. Given knowledge of the orientation of the wall relative to the path, and the type of material of the wall, in theory the transmission loss can be calculated. The total loss (in dB) can be determined by summing all the individual wall losses. Note however that additional losses are possible due to scattering and diffraction effects from other objects along the propagation path. However, in practical situations such theoretical modeling is not feasible, so that simple empirical modeling methods must be used, with the model parameters matched to the situation in each building.

Figure 7 shows the loss excess as a function of the number of walls for the data in Figure 5, and the number of walls derived from the map in Figure 2. As can be observed, the measured losses are related in an approximately linear manner with the number of walls along the path, with some additional smaller random components. This result is consistent with the walls all being of similar material and construction. In this case, the loss per wall due to walls along a path can be expressed by the simple linear model

$$L_{wall}(W) = L_{w} + \omega W \tag{9}$$

For the data in Figure 7 $\omega = 4$ dB (average loss per wall). The offset loss parameter $L_{w} = 3.7$ dB is the loss excess with no intervening walls, and thus should be zero. Thus this offset is interpreted as a calibration offset in the RSSI measurements, and consequentially this procedure of estimating wall losses also allows the RSSI measurements to be calibrated. From the Zigbee specification the RSSI calibration window is ± 5 dB, so this measured offset is within the allowable RSSI calibration error. This RSSI calibration value is important in the standard method of range estimation, as calibration errors

Figure 7. Plot of loss excess versus the number of walls along the straight line propagation path, and the least-squares straight line fit to the data. The RMS fit error is 4.2 dB. The linear model is $3.7 + 4.1W$ *dB. Pearson's correlation coefficient between number of walls and loss excess is 0.88.*

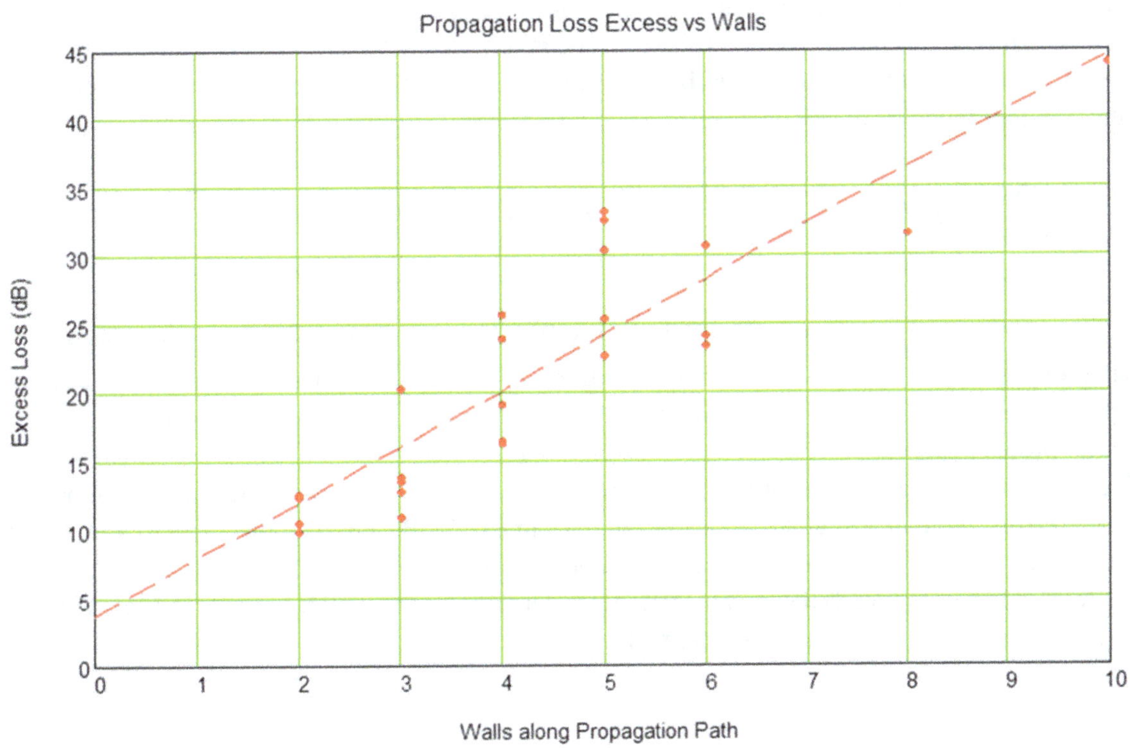

translate into range errors. However, it will be later shown that the new position determination algorithm is insensitive to this offset, so even though it can be estimated, it is not required for accurate position determination.

The above analysis and data are based on one particular building, but the general principle can be universally applied, namely that the loss excess is approximately linearly related to the number of walls along the propagation path. In more complex buildings with more variability in wall types the standard deviation may be more, but the above data would be representative of typical office environments.

Given a map of the building and the knowledge of the path, (9) can be used to estimate the loss excess from walls along the path, which then can be included in the range estimation process. However, knowledge of the path requires knowledge of the location of the mobile device whose position is being sort. Thus an iterative procedure is required. While the number of walls is not initially known, from the map of the building the average number of walls per unit distance can be computed, so the number of walls along a path can be estimated as

$$\hat{W} = \eta R \tag{10}$$

where η is the average walls per meter in the building. For the building in Figure 2 $\eta = 0.22$ walls per meter (walls are separated by about 5 m). Thus using (10), the wall loss excess is given by

$$L_{wall}(R) = L_w + \omega\left(\eta R\right) \tag{11}$$

which is solely a function of range, as is the LOS loss. As a consequence, this simplified model can be used to estimate the range from measured propagation losses. Based on this wall loss model, the estimated propagation loss is then given by the sum of (2) and either (9) or (11). In the latter case, the propagation loss model is then

$$\hat{L}\left(R\right) = \frac{20}{\ln(10)}\ln\left(4\pi R / \lambda\right) + L_{wall}(R) = \beta_0 \ln(R / \lambda) + \eta\omega R + \left(L_w + \alpha_0\right) \tag{12}$$

where $\beta_0 = 20 / \ln\left(10\right)$ and $\alpha_0 = 20\ln\left(4\pi\right) / \ln(10)$. The equation (12) gives the estimate total propagation loss solely as a function of range, with other parameters either known constants or can be determined from the building architecture. Thus by inverting (12) the range can be estimated given the measured propagation loss.

Figure 8 shows a comparison of the basic power law model (4) and the basic wall loss model (12). The RMS errors to the measured losses are about the same (6.2 dB), but the power law model has better matches at medium ranges, while the wall loss model better matches the data at long ranges. At short ranges (not shown in Figure 8) the wall loss model is superior, as the wall loss model reverts to the free-space model for ranges less than some threshold.

Figure 8. Comparison of basic power law range model and the basic wall loss model. Both models are a function of range only.

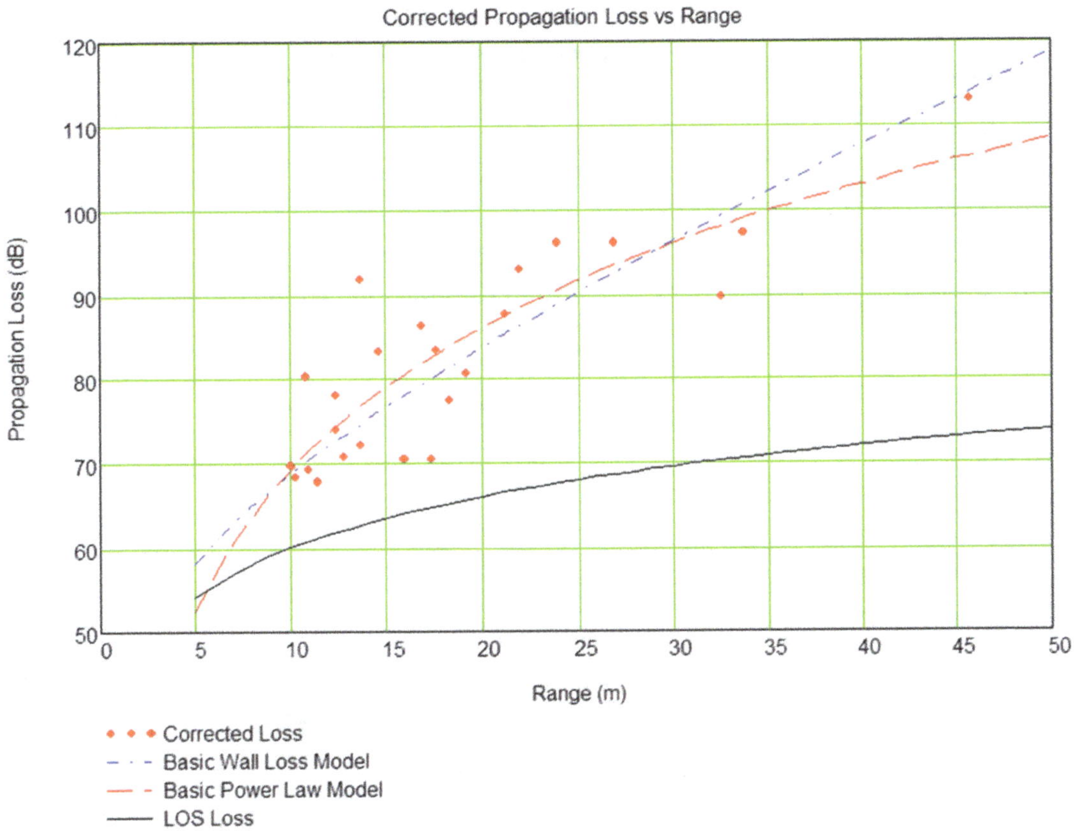

Medium-Scale Propagation Characteristics

From the introductory comments on indoor radio propagation, medium-scale variation is defined as being associated with energy scattering and absorption from large objects along the propagation path from the transmitter to the receiver. To measure the medium-scale energy, the small-scale signal variation must first be averaged over an area of the order of 10 wavelengths. The deviation between the measured loss (corrected for small-scale fading) and the large-scale model loss can be explained by the medium-scale random variations. As these variations are considered random due to the complex nature of the indoor environment, these variations place a lower limit in any deterministic modeling of the propagation loss, and by extension the range derived from the measured propagation losses. Thus for a positioning system based on signal strength measurements, there will be a lower limit on the accuracy that can be obtained. This lower limit is determined by the residual loss noise, such as shown in Figure 9.

Because of the complex nature of the propagation environment, direct deterministic calculation of the signal path attenuation variation on a medium-scale is not feasible, so that statistical methods must be adopted. Consider the simplified propagation diagram shown in Figure 10, which shows that the scattering/reflection coefficient (ρ) for each leg of a path from a transmitter to a receiver. Although only

Figure 9. Plot of the residual propagation loss noise after the removal of large-scale and small-scale effects. Pearson's correlation coefficient between range and residual loss is -0.004, so the residual noise appears to be independent of range.

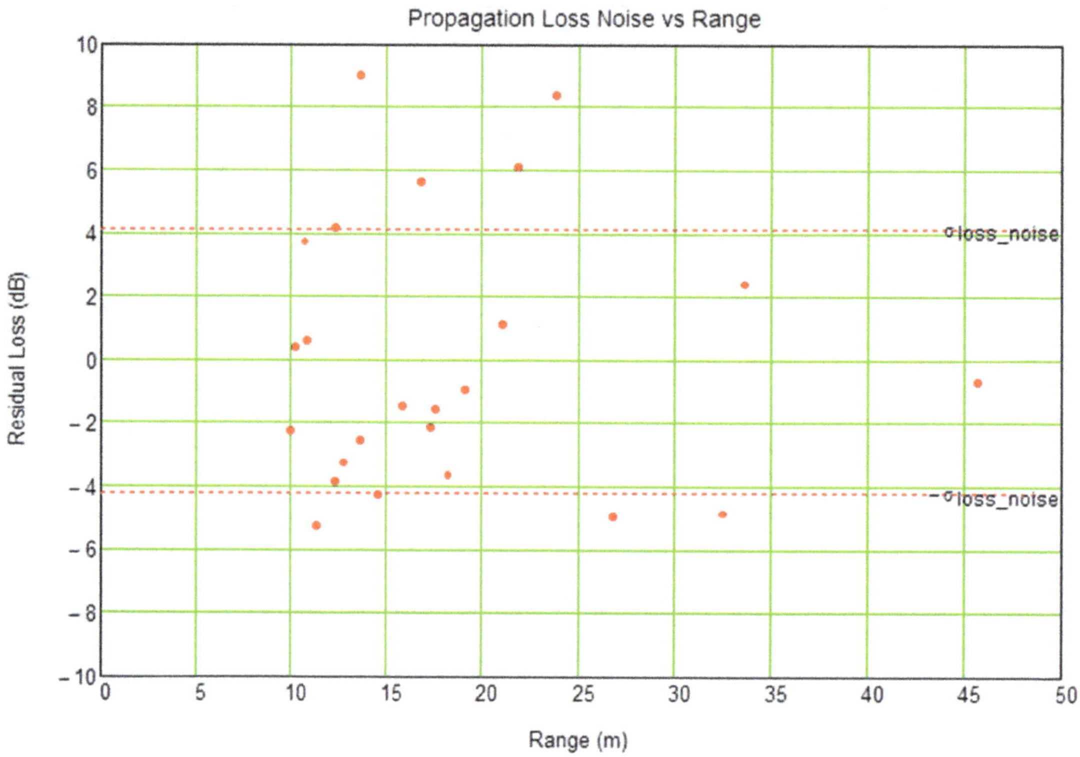

Figure 10. Simplified diagram showing the scattering of RF energy along a path from the transmitter to the receiver.

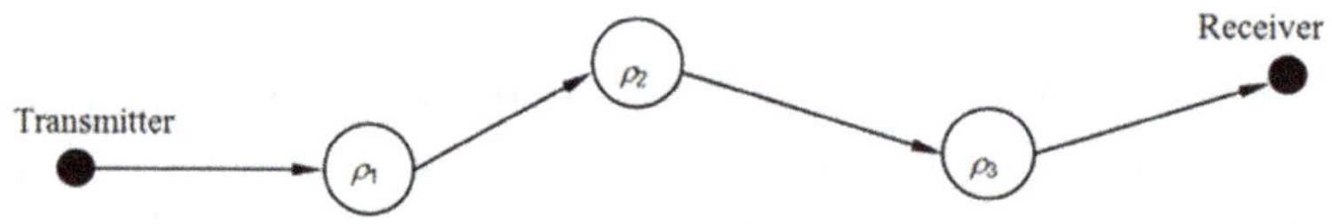

one such path is shown, in practice the received signal will be composed of many such paths (p), but initially only the dominant (smallest loss) path will be considered. From the transmitter power and the receiver power, the propagation loss[2] is given by

$$\frac{P_{tx}}{P_{rx}} = l_p = \prod_{n=1}^{N} \rho_n \tag{13}$$

where N is the number of significant obstacles along the path. For indoor propagation the significant obstacles along a propagation path are usually walls or losses associated with diffraction around corners (usually 90 degrees). The nature of the path attenuation and scattering losses is undefined, but are assumed to be random with undefined statistics. As the loss measurements will be defined in decibels, it is more useful to consider the log of l_p, so that

$$L_p\left(\text{dB}\right) = 10\log\left(l_p\right) = 10\sum_{n=1}^{N}\log\left(\rho_n\right) \tag{14}$$

Thus when expressed in decibels the loss is the sum of the scattering coefficients expressed in decibels. As the scattering coefficients are presumably independent random variables, their logarithms also will be random, so the path loss in decibels also will be a random variable. However, as the path loss in decibels is expressed as the sum of a moderately large number independent random variables, invoking the Central Limit Theorem implies as the number of scattering paths becomes large, the statistical distribution will approach the Normal distribution. Thus when the loss is measured in decibels, the expected distribution will be log-normal, at least for a simple dominant path. Note that the measurement of the medium-scale propagation loss requires the averaging out of the small-scale statistical fluctuations described in the Appendix where the area for averaging is over a few wavelengths. The smoothing of the medium-scale fluctuations over a larger area, say 50x50 wavelengths, results in the large-scale variation in the propagation loss with distance.

The above theoretical analysis indicates that the medium-scale variations in losses is expected to exhibit log-normal statistical characteristics. In Figure 9 the residual derived from the corrected loss measurement data in Figure 5 is shown plotted as a function of range. Based on the above simplified theory it is expected the residual noise will exhibit log-normal statistical characteristics, which is confirmed in Figure 11. The residual loss standard deviation is 4.2 dB. As these residual errors are random, this effect places a lower limit of the deterministic modeling of the indoors propagation loss, which in turn places a lower limit on the accuracy of any positioning system based on measured path propagation losses.

The data in Figure 10 show that the residual loss noise after accounting for losses associated with walls along the propagation path appear to be independent of range. This observation has some interesting consequences. As the residual noise is independent of range, the source of this residual noise is possibly associated with the effects of scattering near the transmitters and the common receiving point. As in this case the receiver is located in open space in an atrium, the conclusion is that the effect is local to the transmitters. The suggested mechanism is associated with the effects of scattering in the local vicinity, resulting in Rayleigh/Rician signal fading. In the processing of the Zigbee data this effect is partially compensated for, as four independent measures are combined to obtain the best estimate of the mean large-scale propagation loss, by eliminating measurements with significantly greater losses (signal fading) than the local median signal. This process is not perfect, but such processing is necessary in any practical implementation. This can be achieved in a couple of ways. The first is to implement spatial diversity by having at least two separate antennas, so that lowest amplitude signal can be bypassed by using the second antenna signal. This procedure is effective with stationary receiving units, but requires a more complex mobile unit. However, another approach is to measure the signal from a moving receiver unit, and sample the signal at a rate so that movement between samples is of the order of a wavelength. For example, for a 2.4 GHz WiFi signal and moving at 1 m/s (walking speed) a sample rate of about 10 per second would be required. If a one second update of data are used, any significant fades can be eliminated in the averaging process used to estimate the propagation loss. When a person is not

Figure 11. Comparison of the measured residual noise CDF with the Normal distribution CDF with the same standard deviation of 4.2 dB. The good match shows the residual medium-scale noise exhibits log-normal statistical characteristics.

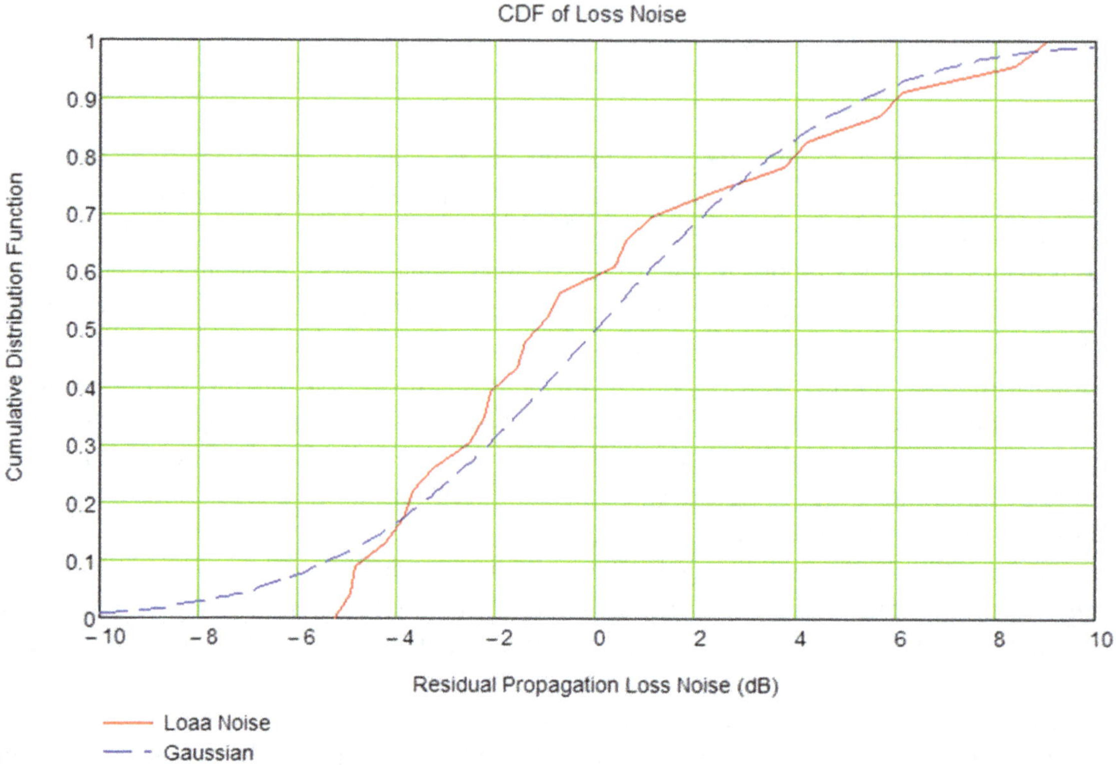

walking, simply moving in a waving pattern the handheld receiver unit (say a mobile phone) of the order of ± 15 cm would be effective in improving the accuracy of the position estimate.

The data in Figure 10 can be tested for log-normal statistics. The cumulative distribution function of the residual loss relative to the local (estimated) mean signal is shown in Figure 11. As can be observed there is a comparatively good correlation between the measured CDF and a least-squares fit to the data. The standard deviation of the log-normal distribution is 4 dB. This estimate is thus the approximate noise floor limit for these measurements of propagation loss, and can be used in estimates of the position accuracy, as described in the next section on range determination.

Range Determination From Propagation Loss

Given the calculated propagation loss derived from the measured receiver signal strength and the link budget, the range can be estimated using an appropriate model of the loss as a function of range; for more accurate estimates other path parameters, such as the number of walls along the path may also need to be included in the calculation. The following subsections provide various range estimates, from the simplest to the more complex models described previously.

Range From Basic Power Law Loss Model

The propagation model commonly described for indoor situations is the simple power law model (4), where the loss (in dB) is linear function of the logarithm of the range. This model can readily be inverted to determine the range, namely

$$R(L) = R_{\min} e^{L/\beta} \qquad \left(R > R_{\min},\ R_{\min} = \lambda e^{-\alpha/\beta}\right) \tag{15}$$

While (15) is simple to apply, the model parameters $\left(\alpha, \beta\right)$ can only be determined from a site survey and detailed knowledge of all the parameters in the link budget. Furthermore the model is limited in accuracy, particularly at short LOS ranges and long ranges near the limit of radio reception. From the data in Figure 5, $\alpha = -33.7$ dB and $\beta = 24.4$, so in this case $R_{\min} = 0.5$ m, and the standard deviation in the difference between the measured and model propagation loss is σ_L=6.2 dB.; while this figure is specific to this particular case, a standard deviation of around 6 dB is typical of indoor NLOS propagation using the power law model.

The accuracy of the range estimate (15) can be determined from the loss standard deviation and the derivative of the loss model (4), namely

$$\sigma_R = \sigma_L \,/\, \frac{dL(R)}{dR} = \sigma_L \frac{R}{\beta} \tag{16}$$

For example, using the above data at $R = 10$ m the range standard σ_R is 2.5 m, but at $R = 35$ m (suggested maximum distance for range determination) is 8.9 m. As the error standard deviation is proportional to the range, the basic power law model of NLOS propagation losses is rather poor for range estimation beyond rather short ranges. For these reasons the commonly used indoor propagation loss model is not suitable for signal strength positioning calculations.

In equation (15) the model parameters $\left(\alpha, \beta\right)$ are usually determined by a fitting process to the propagation loss, namely $L(R) = \alpha + \beta \ln(R \,/\, \lambda)$. This procedure seems to be invariably described in the literature on signal strength positioning, but as will be shown this results in sub-optimal performance. While the fitting of a linear equation to the loss data may be satisfactory for modeling losses, the process of using $\ln(R \,/\, \lambda)$ as the x-coordinate distorts the fitting process of determining the model parameters when range is required. An alternative and better approach is to use inverted function (15) directly to determine the model parameters $\left(\alpha, \beta\right)$ by fitting the model range to the true range data.

There is no simple procedure for this non-linear optimization process, however the least-squares fit procedure to determine the model parameters $\left(\alpha, \beta\right)$ can utilize an iterative procedure similar to that described in the P*osition Determination* section. Note that the underlying model is the same (power law model (4)) and the same loss versus range data are used in the fitting, but fitting using the direct range model (15) avoids the non-linearity of using the logarithm of range. For example, using the logarithm of range implies the data from ranges 1-10 m has equal weight as those from 10-100 m, even though the latter is far more important for position determination.

Range From Basic Wall Loss Model

To improve the accuracy of range estimation, the basic wall loss model (12) can be used. The analytical solution for the range involves the Lambert W function $W_0(x)$

$$R_{wall}\left(\hat{L}\right) = \lambda \exp\left[\frac{\hat{L} - \left(L_w + \alpha_0\right)}{\beta_0} - \mathrm{W}_0\left(\frac{\eta\omega\lambda}{\beta_0} e^{\frac{\hat{L} - \left(L_w + \alpha_0\right)}{\beta_0}}\right)\right] \tag{17a}$$

where $W_0(x)$ is the zeroth Lambert W function, which can be calculated numerically by an iterative procedure

$$\begin{aligned} W_{j+1} &= W_j - \Delta W_j \\ \Delta W_j &= \frac{W_j e^{W_j} - x}{\left(W_j + 1\right) e^{W_j}} \end{aligned} \tag{17b}$$

and $W_0 = \max\left(1, \ln\left(x\right)\right)$, and $\Delta W \rightarrow 0$. However, for practical calculations a simple numerical solution to (17a) may be more appropriate. The accuracy of this range estimate can be computed in a similar way to (16), namely

$$\sigma_R = \sigma_L \,/\, \frac{d\hat{L}(R)}{dR} = \frac{\sigma_L}{\beta_0 \,/\, R + \eta\omega} \tag{18}$$

Comparing (18) with (16) shows that for long ranges the standard deviation of the range using the basic wall loss model is capped at $\sigma_R = \sigma_L \,/\, \eta\omega$ for long ranges, so in this case range errors are lower than using the simple power law model.

Range From Wall Loss Model

The basic wall loss model assumes the number of walls along the path can be estimated from the average wall separation and the range, however by using a map of the building the *actual* number of walls along the path can be determined. However, for the number of walls to be determined the location of the mobile unit must be known. As this is not initially known, the procedure is to first use the basic wall loss model to estimate the position $\left(\hat{x}, \hat{y}\right)$, and then using this estimated position the number of walls along this initially estimated path can be determined. The correction for the number of walls along a path allows an estimate of a correction delta loss, namely

$$\Delta L = \omega\left(W\left(\hat{x}, \hat{y}\right) - \eta\omega R\left(\hat{x}, \hat{y}\right)\right) \tag{19a}$$

and the corrected wall model range is then given by applying (17a) using the adjusted loss, namely

$$R_{corrected}(L) = R_{wall}\left(L - \Delta L\right) \tag{19b}$$

POSITION DETERMINATION

This section describes methods for the determination of position using the receiver signal strength data from several base stations. Two different methods are described. The first method is the common approach described in the literature based on the range estimates derived from inverting a loss model of the form $L(R)$. Once ranges are determined, a standard least-squares fitting procedure is applied to the range data to determine the mobile position. The second method is to use the loss data directly without the need to first determine ranges. This method has several advantages over the range-based method, such as simplicity (no range calculations), improved positional accuracy and less data associated with the link budget is required. As this is a new method, full details are given, while only a brief summary of the standard range-based method is provided.

Iterative Least-Squares Method Using Range

Given measured receiver signal strength data (RSSI), a model of the propagation losses, and a link budget with parameters such as the transmitter power and antenna gains, the position can be determined. By adjusting the estimated mobile device position, the RMS error between the measured range data and the model range data is minimized, thus estimating the unknown position of the mobile unit. There are many procedures for performing such analysis, hence details are not included in this chapter. The results in this chapter are based on linearizing the error equation, and then performing a standard least-squares analysis to obtain the estimated position (Chapter 7, Yu, Sharp & Guo, 2009). The method requires an initial "guess" of the position of the mobile unit; this is assumed to be at the mean of the locations of the base stations in radio communications with the mobile unit. This method is known to be close to optimal (Teunissen, 2002), if the range errors are unbiased (best linear unbiased error, or BLUE). While Figures 10 and 11 show that the propagation loss errors meet these conditions, Figure 12 shows the ranges derived by inverting the loss model does have range bias errors, particularly at long ranges. This is true for both methods of range determination shown in Figure 12, but particularly when using the standard inversion procedure universally described in the literature of signal strength positioning. Thus this standard method does not meet the BLUE criteria, and consequentially this method will not result in best accuracy given the corrected RSSI data.

Iterative Least-Squares Method Using Propagation Loss Directly

The standard method of calculating a position is based on using a loss model which is a function of range (and possibly other variables such as the number of walls along the path), and then invert the function to estimate the range for each path. Using these ranges from "base stations" to the mobile unit, a range-based position determination algorithm can be applied to calculate the position. However, this method is problematic for signal strength positioning for several reasons.

Figure 12. Comparison of the range estimation based on the inverted loss model and directly from a range model.

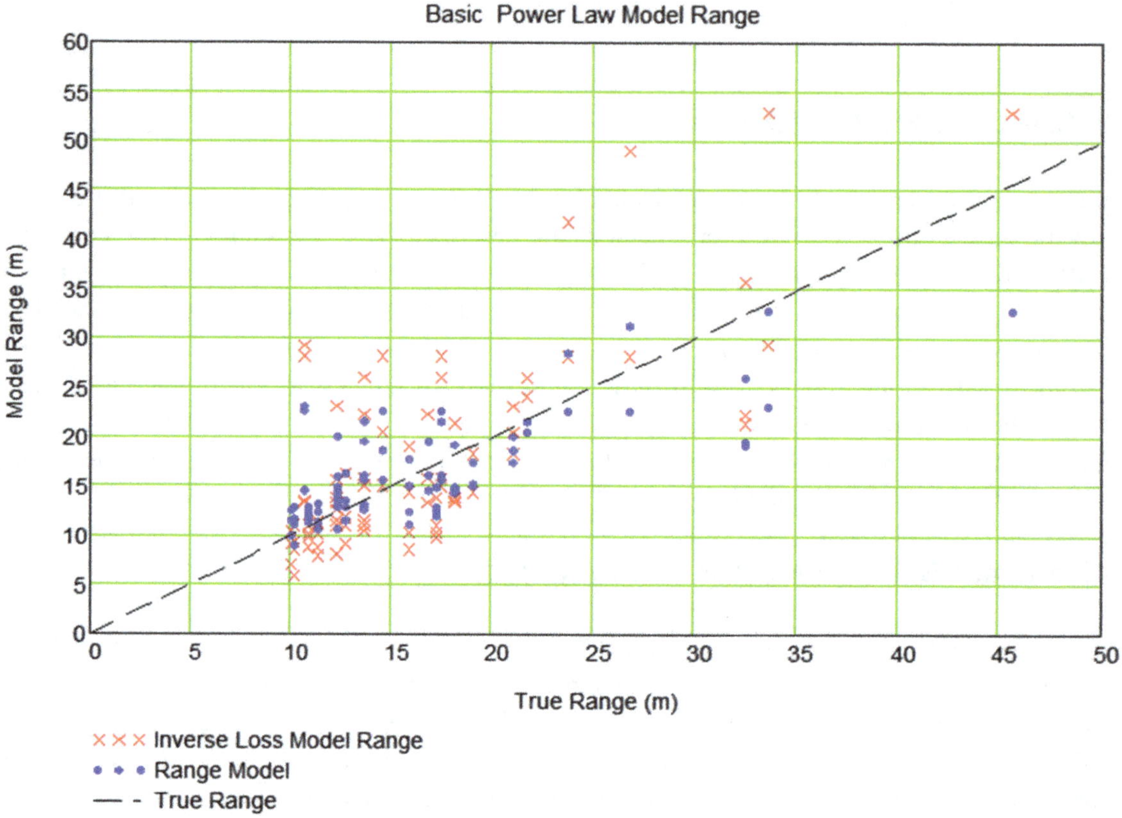

1. The process of inverting the loss function can be complex, involving iterations to correct for the loss excess which is not a simple function of range. In practice the method can only be approximated, thus introducing errors into the computed position.
2. The link budget parameters and other measurement calibrations factors must all be known to calculate an accurate range. In many practical situations these parameters may not be known or have poor accuracy, so that accurate ranges cannot be determined, and hence accurate positions are not possible.
3. The determination of range is not what is required, but rather the position of the mobile. Once the mobile position is calculated the ranges can be determined, but such information is not normally required in typical applications.

Thus a more appropriate approach is to calculate the position directly from the propagation loss data. This process can follow similar principles as for position determination using range, such as the iterative least-squares fitting method using linearized equations. Further, the fitting process can include "nuisance" parameters whose values are not known, but can be estimated as part of the fitting process.

This concept is similar to position determination using "pseudo-ranges" (range plus unknown offset common to all measurements) in technologies such as GNSS.

Based on the analysis of the propagation loss models in the *Loss Excess Modeling Based on Wall Loss* section, the loss model from the nth base station to the mobile position (x, y) can be expressed as

$$\hat{L}\left(R_n\left(x,y\right)\right) = \beta_0 \ln(R_n\left(x,y\right) / \lambda) + \eta\omega R_n\left(x,y\right) - \Delta Lwalls_n\left(x,y\right) + \left(L_w + \alpha_0\right) \tag{20}$$

Initially the position is not known, but for an iterative solution process an initial position (x_0, y_0) can be assumed, say at the mean of the base station positions. This initial estimate is appropriate, as typically the mobile unit will be "surrounded" by base stations which are within radio range. Also, as initially the exact number of walls along each path is not known, the wall loss correction $\Delta Lwalls$ is set to zero. For this initial (zeroth) estimate, the ranges to the nth base station can be expressed as

$$R^0{}_n = R_n^0\left(x_0, y_0\right) + \Delta R_n \tag{21}$$

where it is further assumed that $R_n^0\left(x_0, y_0\right) >> \Delta R_n$. Applying (21) to (20) with the above defined assumptions and using the approximation $\ln\left(1+x\right) \approx x$ gives

$$\hat{L}\left(R_n\left(x,y\right)\right) \approx \left[\beta_0 \ln(R_n^0 / \lambda) + \eta\omega R_n^0\right] + K_n^0 \Delta R_n + \left(L_w + \alpha_0\right) \tag{22a}$$

where the range from the nth base station to the mobile unit is

$$R_n = \sqrt{\left(x_n - x\right)^2 + \left(y_n - y\right)^2} \tag{22b}$$

and the geometric and wall loss parameter is

$$K_n = \beta_0 / R_n^0 + \eta\omega \tag{22c}$$

By differentiating (22b) the range increment is

$$\Delta R_n = \frac{\partial R_n}{\partial x}\Delta x + \frac{\partial R_n}{\partial y}\Delta y = \frac{x_n - x}{R_n}\Delta x + \frac{y_n - y}{R_n}\Delta y \tag{23}$$

Applying (23) to (22a) and equating to the loss equation for the nth base station gives

$$\Delta L_n = Loss_n - \left[\beta_0 \ln(R_n^0 / \lambda) + \eta\omega R_n^0 + \alpha_0\right] \approx \left(\gamma_n^0 \Delta x + \nu_n^0 \Delta y\right) + L_w \tag{24a}$$

where

$$\gamma_n^0 = K_n^0 \frac{x_n - x}{R_n^0} = K_n^0 \cos\left(\theta_n^0\right) \qquad \nu_n^0 = K_n^0 \frac{y_n - y}{R_n} = K_n^0 \sin\left(\theta_n^0\right) \tag{24b}$$

Observe from (24b) that the parameters (γ, ν) are related to the cosine and sine of the angle to the node multiplied by another factor K unique to the wall loss model which involves penetration losses through walls. Equations (24) represents a set of overly defined linear equations which can be solved in a least-squares sense to obtain $(\Delta x, \Delta y, L_w)$, which can be used to obtain a first order estimate of the position of the mobile unit, namely $(x_1, y_1) = (x_0 - \Delta x, y_0 - \Delta y)$, and an estimate of the walls loss excess constant parameter L_w. While L_w is determined as part of the fitting process shown in Figure 6, this loss with zero walls should be zero, so this parameter is interpreted as a RSSI calibration parameter, rather than being associated with walls. This process can be further iterated to converge to the estimated position of the mobile unit.

A further extension to the method is possible by applying the link budget equation (1) to typical real world applications. In a typical application the mobile unit could be a mobile (smart) phone, and the base stations are WiFi AP located throughout a building, mainly for data communications and access to the Internet. Applying (1) to the mobile device m and access point n gives

$$Loss_{n,m} = P_{tx} - RSSI_{n,m} + G^n{}_{tx} + G^m{}_{rx} = \left(P^0{}_{tx} + G^0{}_{tx} + G^0{}_{rx}\right) + \left(\Delta P^n{}_{tx} + \Delta G^n{}_{tx} + \Delta G^m_{rx}\right) - RSSI_{n,m} \tag{25}$$

where the superscript zeros indicate default values for the parameters, and the Δ terms a correction increment required for true values. While in experimental situations the parameters in the link budget are known or are carefully determined by a calibration process, in practical situations that is not the case. For example, mobile devices can have a wide range of parameters, including the antenna gain and the calibration (offset) for the RSSI measurement (Lui, Gallagher, Li, Dempster & Rizos, 2011). Further, the manner in which the mobile is held can affect the effective antenna gain. Similarly, the access points can have a variety of parameters, including the transmitter power and the antenna gain. However, a further simplification can be made by assuming all the access points in a building are of a similar type and hence have very similar characteristics. Applying these assumptions to (25) and default values to the parameters, the link budget loss equation can be written as

$$Loss_{n,m} = \left(P^0{}_{tx} + G^0{}_{tx} + G^0{}_{rx}\right) - RSSI_{n,m} + \left(\Delta P_{tx} + \Delta G_{tx} + \Delta G^m_{rx}\right) \tag{26}$$

where last bracketed term is a constant for all the measurements, and the first bracketed term is the default values which in general will be different from the actual vales, and may not be known by the application program. As the latest bracketed term accounts for calibration offsets which are not known, it is convenient to set the default values in the first bracketed term to zero. With these further assumptions and applying (26) to (24a), the loss equation becomes

$$
\begin{aligned}
&0 - RSSI_{n,m} - \left[\beta_0 \ln(R_n^0 / \lambda) + \eta\omega R_n^0\right] = \Delta L_n^0 = \left(\gamma_n^0 \Delta x + \nu_n^0 \Delta y\right) + c \\
&c = L_w - \left(\Delta P_{tx} + \Delta G_{tx} + \Delta G_{rx}^m\right)
\end{aligned} \tag{27}
$$

where c is a constant for all measurements and iterations. Equation (27) can be expressed in matrix form

$$
\mathbf{A\Delta} \approx \boldsymbol{\delta}\mathbf{L} \tag{28a}
$$

where

$$
\mathbf{A} = \begin{bmatrix} \gamma_1 & \nu_1 & 1 \\ \vdots & \vdots & \vdots \\ \gamma_N & \nu_N & 1 \end{bmatrix}, \quad \mathbf{\Delta} = \begin{bmatrix} \Delta x \\ \Delta y \\ c \end{bmatrix}, \quad \boldsymbol{\delta}\mathbf{L} = \begin{bmatrix} \Delta L_1 \\ \vdots \\ \Delta L_N \end{bmatrix}. \tag{28b}
$$

Notice that the delta loss measurements only include the RSSI measurements and other known or estimated parameters, and thus the link budget parameters associated with transmitter power and antenna gains are not required. The fitting process will estimate $\left(\Delta x, \Delta y, c\right)$ in the least-squares process, so this effectively calibrates all the unknown link budget parameters, so the four parameters (transmitter power, transmitter and receiver gains and RSSI calibration offset) do not need to be known for the position determination. This relaxing of the need to accurately calibrate the link budget components in a signal strength positioning system greatly extends its practical application.

The least-squares solution to (28a) is given by

$$
\mathbf{\Delta} = \tilde{\mathbf{A}}^{-1}\left(\mathbf{A}^T \boldsymbol{\delta}\mathbf{L}\right) \qquad \left(\tilde{\mathbf{A}} = \mathbf{A}^T\mathbf{A}\right) \tag{29}
$$

Note that (29) does not use a weighted fit (or more accurately all the weights are the same), as from Figure 10 the residual random loss noise appears to be independent of range, and hence all the adjusted loss measurements have (approximately) the same variance. As the optimum fitting of a linear system weights the data by the inverse of the measurement variance, the optimum weighting in this case is equal weighting.

The initial position uses an estimate of the number of walls along each propagation path based on the range and the average number of walls per unit length. For each subsequent iteration the corrections $\left(\Delta x, \Delta y, c\right)$ can be applied to converge to an estimate of the mobile position $\left(x, y\right)$. Using this estimated position and a map of the building, a better estimate of the number of walls can be determined, and the above process is repeated to obtain a better estimate of the mobile position. Note that the number of walls along the paths is only weakly related to the mobile position, hence position errors of a few meters will have minimal impact on the computed number of walls using the map and the straight line propagation path. Thus this process should converge rapidly. If the errors in the measurements of the loss are too large the iterations will diverge, and no useful position estimate is obtained.

The loss equation (27) includes parameters (β_0, η, ω). The first is a known constant, and the second can be calculated from a map of the building. The average loss per wall parameter ω can be determined if there is a monitoring point that can measure the RSSI from the base stations in the coverage area. Based on the above assumptions, this ω parameter can be determined by a least-squares analysis from the "quasi-loss" equation, namely

$$\left[\left(P^0_{tx} + G^0_{tx} + G^0_{rx}\right) - \beta_0 \ln(R_n / \lambda)\right] - RSSI_{n,m} = \eta\omega R_n \tag{30}$$

where R_n is the known range from the nth base station to the monitoring point. This means the method is readily useable in many practical situations with no expert setup required to calibrate the network of base stations.

GDOP for the Direct Loss Method

The concept of geometric dilution of precision (GDOP) is well known in positioning systems based on using ranges, but this concept can be readily extended to the case where propagation losses are used for position determination. This concept is useful, as it decouples errors in position from the errors in loss measurements and the geometry of the location of base stations.

The section *Iterative Least-Squares Method using Propagation Loss Directly* provides a deterministic estimate of the position as a function of geometry and propagation loss measurements. If this iterative procedure were to commence at the true position and one iteration used to determine a first order position estimate, the position error caused by the errors in the loss measurement can be determined. However, in practice the loss errors $\{\Delta L_n\}$ are random variables, so that the positional errors are also random variables. The following statistical properties are assumed for the path loss errors

$$\begin{aligned} &E\left[\Delta L_n\right] = 0, \quad n = 1,\ 2,\ \ldots,\ N \\ &E\left[\Delta L_n \Delta L_m\right] = 0 \\ &E\left[\Delta L_n^{\ 2}\right] = \sigma_L^2. \end{aligned} \tag{31}$$

The first equation states that the mean of the propagation loss errors is zero. While the above analysis showed that there can be a number of unknown offset losses, the fitting process incorporates these into an offset constant.

The second equation in (31) states that the loss errors are statistically independent. As the errors are associated with separate paths from the base stations to the mobile device, and they are physically well separated in the coverage area, it is reasonable to expect these errors are statistically independent. Finally, the third equation states that the variances of the loss errors for each path are all equal. The basis of this assumption is that the propagation environment is quasi-homogeneous, so that the statistical variance in the measurements is essentially independent of position. This assumption is only partially true, hence the actual error distributions will be somewhat different from this simplified model. However, this assumption is required for the simple geometric interpretation of GDOP developed below.

Now consider the statistical properties of the positional error Δx. From (27) and (29), it can be seen that

$$\begin{aligned}\Delta x &= [\tilde{\mathbf{A}}^{-1}]_{1,1}\sum_{n=1}^{N}\gamma_n\Delta L_n + [\tilde{\mathbf{A}}^{-1}]_{1,2}\sum_{n=1}^{N}\nu_n\Delta L_n + [\tilde{\mathbf{A}}^{-1}]_{1,3}\sum_{n=1}^{N}\Delta L_n \\ &= \sum_{n=1}^{N}\rho_n\Delta L_n\end{aligned} \tag{32a}$$

where

$$\rho_n = [\tilde{\mathbf{A}}^{-1}]_{1,1}\gamma_n + [\tilde{\mathbf{A}}^{-1}]_{1,2}\nu_n + [\tilde{\mathbf{A}}^{-1}]_{1,3}. \tag{32b}$$

The mean and variance of Δx are thus given by

$$\begin{aligned}E\left[\Delta x\right] &= E\left[\sum_{n=1}^{N}\rho_n\Delta L_{ni}\right] = \sum_{n=1}^{N}E[\rho_n\Delta L_n] = \sum_{n=1}^{N}\rho_n E[\Delta L_n] = 0 \\ \sigma_x^2 = E\left[\Delta x^2\right] &= E\left[\left(\sum_{n=1}^{N}\rho_n\Delta L_n\right)^2\right] = E\left[\sum_{n=1}^{N}(\rho_n\Delta L_n)^2\right] = \sigma_L^2\sum_{n=1}^{N}\rho_n^2\end{aligned} \tag{33}$$

In deriving the expression for the variance, use has been made of the statistical independence of the path losses, and that the mean of the loss errors is zero. Similar expressions apply for the statistics of the y-coordinate positional error, namely

$$\begin{aligned}&E\left[\Delta y\right] = 0 \\ &\sigma_y^2 = E\left[\Delta y^2\right] = \sigma_L^2\sum_{n=1}^{N}\kappa_n^2\end{aligned} \tag{34a}$$

where

$$\kappa_n = [\tilde{\mathbf{A}}^{-1}]_{2,1}\gamma_n + [\tilde{\mathbf{A}}^{-1}]_{2,2}\nu_n + [\tilde{\mathbf{A}}^{-1}]_{2,3}. \tag{34b}$$

The expression for the statistical covariance can also be determined as

$$\sigma_{xy} = E\left[\Delta x\Delta y\right] = \sigma_L^2\sum_{n=1}^{N}\rho_n\kappa_n \,. \tag{35}$$

Thus, in summary, the relationships can be expressed as

$$\frac{\sigma_x^2}{\sigma_L^2} = \sum_{n=1}^{N} \rho_n^2$$
$$\frac{\sigma_y^2}{\sigma_L^2} = \sum_{n=1}^{N} \kappa_n^2 \qquad (36)$$
$$\frac{\sigma_{xy}}{\sigma_L^2} = \sum_{n=1}^{N} \rho_n \kappa_n.$$

The propagation loss measurements will have both random and systematic components. Further, as can be observed from (29) the positional error calculation involves the sum of random measurement errors, thus the distribution will be approximately Gaussian due to the Central Limit Theorem. Also note that in general, as vectors $[\rho_1 \quad \rho_2 \quad \cdots \quad \rho_N]^T$ and $[\kappa_1 \quad \kappa_2 \quad \cdots \quad \kappa_N]^T$ are linearly dependent, their product is usually non-zero, which means that Δx and Δy will not be statistically independent. Finally, the Loss-GDOP is defined as

$$GDOP_L = \sqrt{\sigma_x^2 + \sigma_y^2} \Big/ \sigma_L = \sqrt{\sum_{n=1}^{N} \rho_n^2 + \sum_{n=1}^{N} \kappa_n^2}. \qquad (37)$$

Note that unlike the classical GDOP, the loss-based $GDOP_L$ has the dimension of length (m). In classical theory GDOP is related solely to the geometry of the base station locations relative to the mobile unit position. However, for this Loss-GDOP version it can be observed from (24b) that the (γ, ν) parameters are related to the term $K_n = \beta_0 / R_n + \eta\omega$ which has dimensions of dB/m. K is related to the geometry (range) as well as the LOS propagation parameter $\beta_0 = 20 / \ln(10)$ and parameters associated with propagation losses related to walls. However, these non-geometric parameters are system constants, and thus the variation of Loss-GDOP with position is solely related to geometry. Two specific geometries of practical interest are now considered.

As the mobile will always be "surrounded" by base stations as it moves through a building, it can be considered as a rough approximation for theoretical analysis that the mobile is at the center of a circle, and the base stations are equally-spaced around that circle. In this special case all the ranges are equal, and applying (37) in this case it can be shown that the Loss-GDOP for this circular geometry case is

$$GDOP_{circle} = GDOP_0 / K \text{ (m/dB)} \qquad (38)$$

where $GDOP_0 = 2 / \sqrt{N}$ is the classical GDOP with N base stations; and in fact the geometry corresponds to the minimum possible GDOP with N base stations. Note that Loss-GDOP is a function of the radius of the circle (range to base stations), and that the GDOP is smallest at short ranges. Conversely, the GDOP at long ranges approaches a constant GDOP of $GDOP_0 / \eta\omega$, which is related to the number of wall per unit distance and the wall loss (in dB). For the case studied in this chapter $\eta\omega = 0.97$ dB/m, so at long range

$$\sqrt{\sigma_x^2 + \sigma_y^2} = \frac{GDOP_0}{0.97}\sigma_L . \tag{39}$$

Thus if the target for the residual propagation loss noise standard deviation is $\sigma_L = 3\,\text{dB}$, then for the recommended minimum number of base stations of five, $\sqrt{\sigma_x^2 + \sigma_y^2} = 2.77$ m for long (limiting) ranges to the base stations. With a more appropriate base station range of (say) 25 m, $K = 1.32$ dB/m, then $\sqrt{\sigma_x^2 + \sigma_y^2} = 2.0$ m. Note that unlike the standard method of received signal strength positioning the theoretical estimate of position errors only increases slowly with range.

The second case considered is a building with the base stations (such as WiFi AP) on a square grid of side length S. It is assumed that the grid size is based on data communication, and what is important is the range to the closest access point. For reliable high speed communications this cannot be too large in a NLOS environment, say 20 m. For a square grid the maximum range to the closest access point is $S / \sqrt{2}$, so that $S = 28$ m. It is assumed that the signal strength positioning uses the same access points, but in this case what is important is the range to the most distant access point for the recommended minimum of five measurements for the position determination. For a square grid the minimum distance to the fifth access point is S, and the maximum distance is $\sqrt{2.5}S$. However, for signal strength positioning the maximum range is suggested to be 35 m, so that in this case $S = 22$ m, which is somewhat less than the grid size for data communications only. Thus receiving signals from five AP may be problematic. If the grid size is $S = 28$ m (data communications grid size), the maximum range to the fifth access point would be 44 m, which typically would exceed the maximum indoor range for reliable signal strength measurements.

The second case considered is a building with the AP on a square grid, and all base stations within range (up to 35 m) are used for position determination. With a mobile at the center of any square in the grid the closest access point has a range $R_1 = S / \sqrt{2}$, which is the distance from the center of the square to any of the four corners of the square. Thus this is the same geometry as the circle case considered above. For more base stations to be used in the position determination the next surrounding square must be considered, namely of size $3S$ with 12 nodes (base stations). The range to the eight closest nodes on this square is $\sqrt{2.5}S$, and again they all lie on a circle. Applying the above analysis to the combination of the inner and outer squares results in the $GDOP_L$ expression

$$GDOP_{2squares} = \frac{2\sqrt{N_1 + N_2}}{N_1 K_1 + N_2 K_2} \tag{40}$$

where $N_1 = 4$ is the number of nodes on the inner square and $N_2 = 8$ is the number of nodes on the outer square, $K_1 = \beta_0 / R_1 + \eta\omega$ and with a similar expression for K_2. For a grid spacing of $S = 22$ m, the maximum range is $R_2 \approx 35$ m, and the corresponding loss $GDOP_L$ is 0.46 m/dB. Again with $\sigma_L = 3\,\text{dB}$, the accuracy parameter is $\sqrt{\sigma_x^2 + \sigma_y^2} = 1.4$ m, with a total of 12 nodes within the maximum range of 35 m used for position determination. Real world situations will not have a uniform grid pattern, but this estimated accuracy would be approximately correct for a node density of one access point per 500 m^2, and building characteristics similar to that shown in Figure 2.

RESULTS

In previous sections the performance of receiver signal strength technology was analyzed, for both the conventional approach and the proposed enhanced methods. This section provides details of the performance based on a set of measurements obtained from testing a Zigbee radio system for data communications at 24 locations in a large office building. The measurements involve various combinations of transmitter powers and receiver noise figures, but for testing the positioning performance only the propagation loss is of relevance. The general approach taken in presenting the results is to show the effect of various data processing methods aimed at improving accuracy. There are three main areas considered, namely

- Modeling the propagation losses, and comparing the model with the measured losses.
- Using propagation losses to determine ranges.
- Position determination, both using ranges to base stations, and a new algorithm that bypasses range calculations by using loss data directly. Various combinations of base stations are considered.

In each case the results are presented in a manner that includes two indications of the performance. The first indicator is the standard deviation between the measured or computed result, or between the true and model result. The second indicator is Pearson's correlation coefficient, which gives a measure of how good the two data sets match, with the desired aim to achieve a correlation coefficient as close to unity as possible. The data are presented in charts and summary tables, which are organized according to increasingly better performance. Note that only the last data item in the tables is the actual performance suggested, while for comparison the first item is typically the performance as described in the literature on receiver signal strength positioning.

Propagation Loss Performance

The basis for all receiver signal strength positioning is the measured propagation loss as a function of range, and an associated model. An accurate agreement between model and measurements implies an accurate range estimate. The results on propagation loses as a consequence of various processing algorithms are summarised in Table 1. The measurements and results are organized according to the individual results of propagation loss for each of four experiments, for the combined loss data set, and for loss data processed to minimize signal fades. The initial results use raw loss data only and the basic power law loss model, but later items in the table use processed signal strengths to remove "bad" measurements. Also enhanced modeling is used, which provides better estimates of the NLOS losses, particularly the loss excesses through walls.

The first four items in Table 1 relate to the raw RSSI data processed according to the various link budget parameters to obtain the raw propagation loss. The raw data can be subjected to Rayleigh fading, which would increase the variability (standard deviation or STD) between the model and the measured losses. The model is the standard power law (4), with parameters defined by a least-squares linear fitting

Table 1. Summary of the loss data used in determination of the receiver signal strength positioning.

Signal Method	Maximum Range (m)	STD (dB)	Correlation Coefficient	Comment
Experiment 1 signals	18	5.7	0.57	No LNA or PA. One sample per location. Basic loss model.
Experiment 2 signals	34	4.7	0.82	LNA gain 22 dB, no PA. One sample per location. Basic loss model.
Experiment 3 signals	33	8.8	0.64	No LNA, PA gain 15 dB. One sample per location. Basic loss model.
Experiment 4 signals	46	8.7	0.73	No LNA, PA gain 25 dB. One sample per location. Basic loss model.
All valid loss data (80)	46	8.5	0.71	Up to four loss measurements per location. Basic loss model.
Mean signal for each point (24 points)	46	6.2	0.85	Up to 1 "bad" measurement rejected. Basic loss model.
Basic loss excess walls model (24 points)	46	6.3	0.73	Walls model based on mean distance between walls and range.
Loss excess walls model (24 points)	46	4.2	0.88	Walls model based number of walls along path as determined from a building map.

of the loss as a function of the logarithm of the range in wavelengths. The variability in the individual results (STD and correlation coefficient ρ) is typical of raw RSSI data. The results have a STD of 6-9 dB, and ρ in the range 0.6 to 0.7; however notice that Experiment 2 had considerably better results. Combining all valid (in radio range) data results in a total of 80 independent measurements, which should give a better overall estimate of the performance for ranges up to 46 m. This combined data set has a propagation loss STD of 8.5 dB and a correlation coefficient of 0.7. This can be used as a baseline measure of performance.

The first performance improvement is based on reducing the signal strength variability, by eliminating signal faded data in the (up to) four measurements per location, resulting in a more reliable estimate of the signal strength at each measurement point. After this data processing, the STD is reduced to 6.2 dB, and the correlation coefficient increased to 0.85. Note that using the same power law model for loss excess gives the same results, as shown in the *Basic Loss Excess Model* section.

The final two results in Table 1 are based on modeling the loss excesses associated with walls along the propagation path. The basic walls loss excess model calculates the number of walls based on the average wall separation in the building. This results in no improvement in performance over the basic power law model, but unlike the power law model, the loss excess model parameters do not need prior calibration, as the model parameters can be based on data from a map of the building and knowledge of the material in the walls. As a consequence this model is easier to apply in real world situations. The best result in the last item in Table 1 is based on using a map of the building to determine the correct number of walls along the propagation path. This results in the STD reduced to 4.2 dB, and the correlation coefficient increased to 0.88.

Thus, in summary, the best data processing and propagation model has a loss STD of 4.2 dB and the correlation coefficient of 0.88, compared with the basic model using raw RSSI data with a STD of 8.5 dB, and the correlation coefficient of 0.7. Figure 13 shows a comparison of the best model and measured losses. Notice that the model performs well over the entire range of losses, from 65 dB to 115 dB, covering ranges from 10 m to 46m.

Figure 13. Comparison of measured and modeled losses using the best processing algorithms.

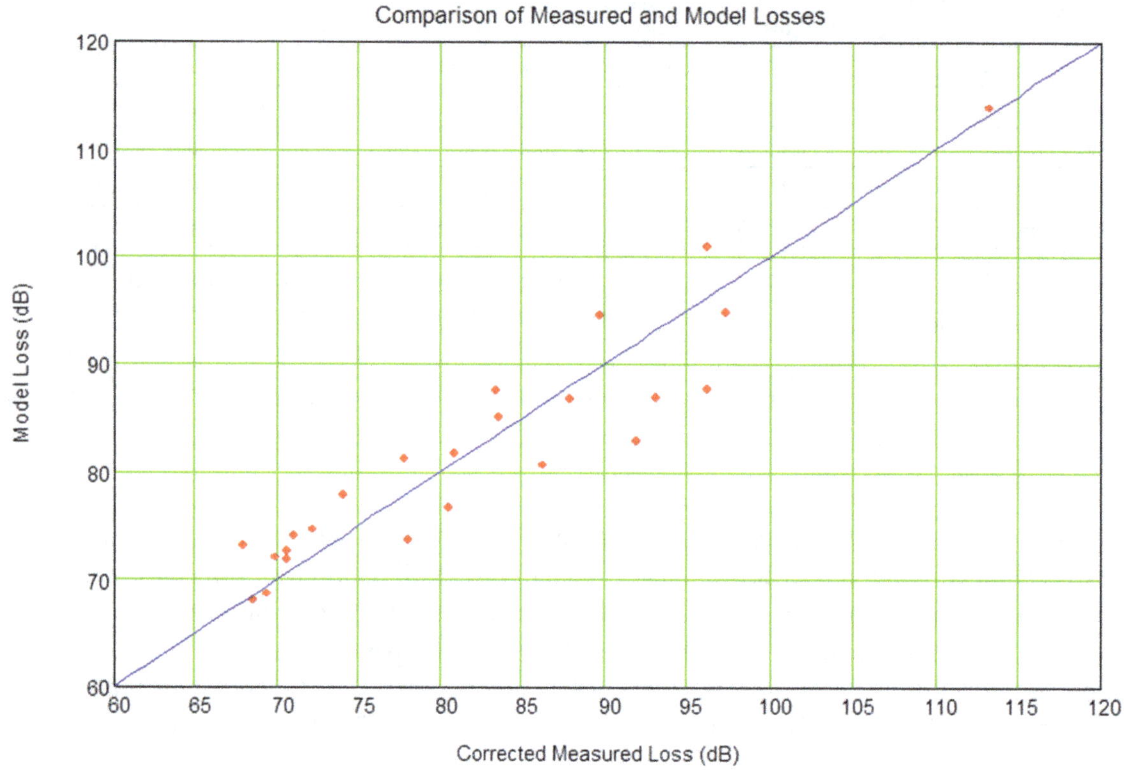

Ranging Performance

This section describes the results of various processing algorithms for determining the range, as summarised in Table 2. Again, the results proceed from the basic methods described in the literature, to the improved algorithms described in previous sections. Given a propagation loss model and the loss derived from RSSI measurements, the propagation range can be estimated.

Table 2. Summary of the range data derived from of the measured receiver signal strength.

Signal Method	STD (m)	Correlation Coefficient	Comment
All valid data (80) loss	11.4	0.75	Raw loss data and inverted power law function (15) to calculate range. Standard method in literature.
All valid data (80) loss with least-squares fit to range	4.6	0.74	Raw loss data and non-linear least-squares solution for range based on power law loss model.
Corrected loss data (24)	5.3	0.89	Corrected loss data and inverted power law function (15) to calculate range.
Corrected loss data with least-squares fit to range	4.0	0.88	Corrected loss data and non-linear least-squares solution for range based on power law loss model.
Basic loss excess walls model	4.6	0.86	Walls model based on mean distance between walls and range.
Loss excess walls model	2.9	0.94	Walls model based number of walls along path as determined from a building map.

The first item in Table 2 uses the standard method to determine the parameters of the power law model (4) using a least-squares fit to the measured loss data, and then inverts the model loss equation to obtain the range using (15). The results using all the valid raw measured RSSI data give a STD of 11.4 m, and a correlation coefficient of 0.75. These results represent a reference for better methods summarized in subsequent items in Table 2.

The second item in Table 2 tests the effectiveness of an improved algorithm to determine the range from the loss measurements. Basically the method uses (15) to determine the range as a function of loss, and performs a least-squares fit with the true range to determine the parameters $\left(\alpha, \beta\right)$. This optimizes the model parameters to get the best estimate of range, rather than optimizing for the loss and then deriving the range. This latter method is a sub-optimal for range estimation. The resulting range error STD reduces very substantially to 4.6 m, but the correlation coefficient is essentially the same at 0.74. The much smaller range error STD shows the superior nature of performing the least-squares directly on range to determine the power law parameters $\left(\alpha, \beta\right)$, rather than determining these parameters for the loss, and then inverting the power law expression to obtain the range.

The third item in Table 2 is based on the corrected RSSI data which minimizes the signal fades, but uses the same basic range determination equation (15) as used in the first item. This results in a significant reduction in the range STD to 5.3 m, and an increase in the correlation coefficient to 0.89. This result shows the importance of methods to mitigate signal fades. The fourth item in Table 2 uses the same RSSI data as item 3, but uses method of item two to calculate the range. This results in a significant reduction in the range error STD to 4.0 m, but the correlation coefficient of 0.88 is essentially the same as in item 3.

The comparison between the two methods (items three and four) of range determination are further explored in Figure 12, which shows the estimated ranges from the propagation losses shown in Figure 5. The "cross" data use the standard inversion technique, while the "dots" are based on the direct range model. It is clear that the later method results in significantly better range estimation, particularly for ranges up to 25 m. The standard method tends to significantly over-estimate the range for all ranges, but particularly for long ranges. In contrast the direct range model method tends to under-estimate ranges, particularly for ranges greater than 25 m.

The last two items in Table 2 show the effect of using the number of walls along the propagation path to estimate the loss excess. The simplest method estimates the number of walls from the range and the average distance between walls in the building. The range estimate is based on a least-squares fit of the model range (17a) to the actual range, thus determining estimates of the model parameters $\left(\eta, \omega, L_w\right)$. The range STD is 4.4 m, and the correlation coefficient 0.86. However, this result is merely a step towards the final range estimates based on the actual number of walls along the propagation path, using a map of the building. This method has a range error of 2.9 m and the correlation coefficient 0.94. A comparison of the estimated and actual ranges is shown in Figure 14. As can be observed the range estimates are good at all ranges up to 46 m, although there appears to be a somewhat increase in range errors with ranges beyond 20 m.

Thus, in summary, the basic power law algorithms resulted in a range STD of 11.4 m, whereas the best algorithm based on a loss excess model associated with wall penetration losses has a STD of 2.9 m. This shows the great benefit of using the enhanced processing algorithms over the standard method described in the literature of receiver strength positioning.

Figure 14. Plot of walls-corrected range estimate versus the true range. The STD in the range error is 2.9 m, and the correlation coefficient is 0.94.

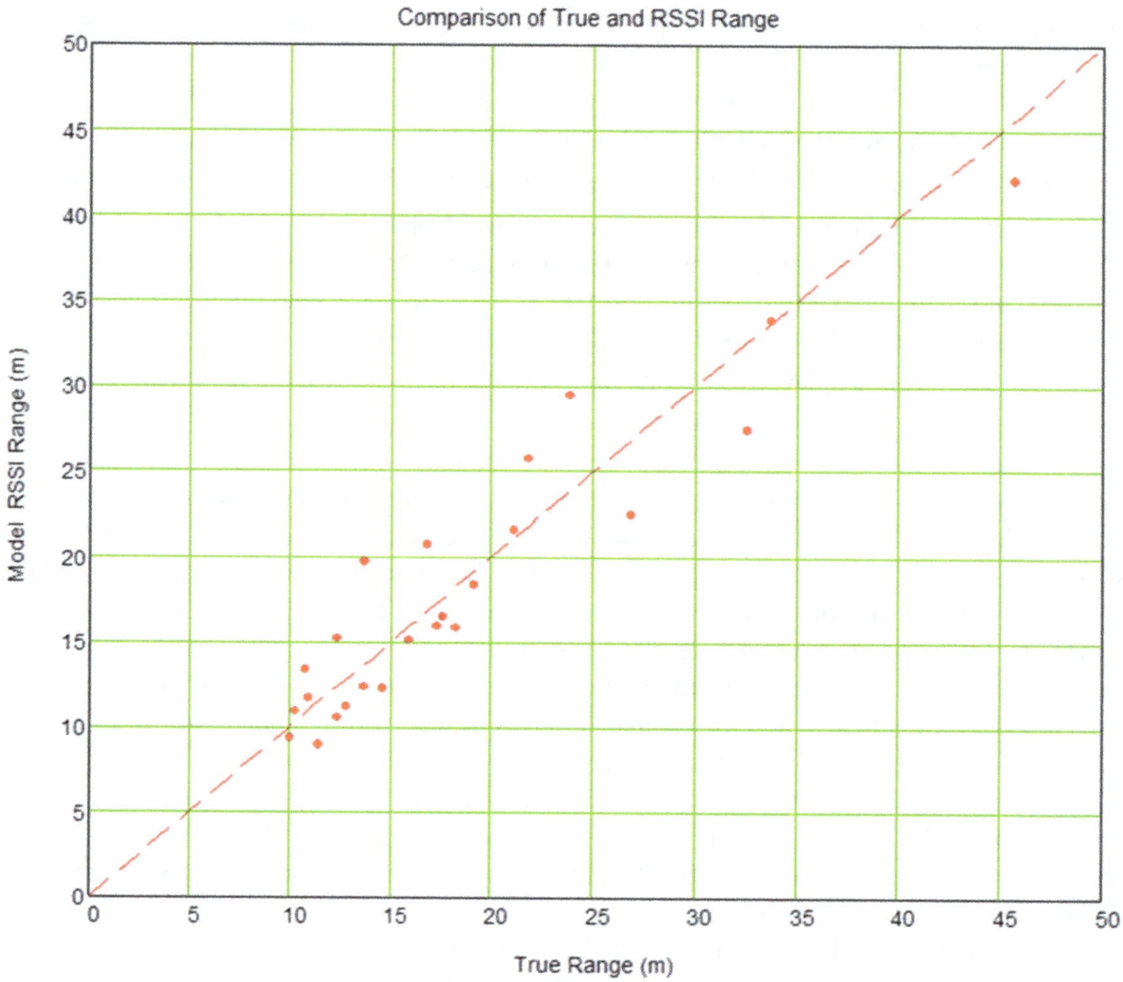

Positioning Performance

The propagation loss and range estimation performance considered above are not the main interest in assessing the performance, but are merely preliminary data that allow positions to be determined. The positioning performance can be considered in many ways, based on the one measured RSSI data set. Although the raw measurements were made from 24 transmitter locations to one receiver point, different combinations of transmitters (base stations) can be used to simulate the performance of a mobile device moving through the building. Thus the performance under more varied situations can be assessed from the one measurement data set.

There are two main methods that can be compared. The standard method is based on determining positions from range information, and a new method which determines positions directly from RSSI data with minimal extra information required apart from the locations of base stations.

Positioning Using All Measurement Points as Base Stations

The first test is based on using the RSSI data (or the associated estimated ranges) from all 24 base stations to determine the position of the receiver. Although this is not a realistic representation of the real world situation, it does show the relative merits of the different methods. The results are summarized in Table 3. The position is calculated either from range estimates or directly from the propagation loss estimates. For estimating the ranges, either the power law model or the wall loss excess model for loss is used. In the former case the range is either calculated by inverting the power law function (15), or the inverted function (15) is first used to calibrate the power law parameters by a least-squares fit to the true range, then the calibrated model is used to calculate the range. In all cases (except for the first which uses raw RSSI data) the RSSI data used is first corrected for signal fades.

However, as the data used for position determination have a random component, the computed position will also be a random variable, so that a single position estimate will have an unknown random component. Thus the single receiver point results should be treated with caution, as a small error may be the result of the random component cancelling the mean position error. Thus to get another estimate of the expected error, the GDOP and the STD of the range measurement errors are used, namely $GDOP * STD$, which is an estimate of the expected RMS position error. These data are also listed in the table.

The first item in Table 3 is the very basic model using raw RSSI measurement data, not corrected for signal fades. This method is commonly described in the literature, and thus these results data are a reference for other methods summarized in the table. The performance is very poor, and illustrates the need for more sophisticated methods if useful position information is to be obtained from the RSSI measurements.

The second item in Table 3 (and all the following items) uses RSSI data corrected for signal fading, and the power law loss model whose parameters were determined from a linear least-squares fit to the logarithm of the range in wavelengths. The positions are then determined by the standard iterative procedure described in (Chapter 7, Yu, Sharp & Guo, 2009). The measured error is 3.6 m, and $GDOP * STD = 2.5$ m.

The third item in Table 3 uses the inverted power law model to first perform a least-squares fit with the true range to determine the model parameters (α, β), and then uses these parameters to estimate the

Table 3. Summary of position error and other parameters associated with various methods.

Positioning Method	Error	GDOP	STD	GDOP*STD	Comment
Very basic RSSI data, range estimation and position determination	17.0 m	0.463	11.4 m	5.3 m	Standard method using raw RSSI data
Basic range determination from power law	3.6 m	0.463	5.3 m	2.5 m	Using (15) with loss LSF parameters
Direct range determination from power law	2.9 m	0.463	4.0 m	1.9 m	Using (15) with range LSF loss parameters
Basic walls model	3.5 m	0.463	4.4 m	2.0 m	Range uses (17a)
Walls model	1.0 m	0.463	2.9 m	1.3 m	Range uses (19b)
Direct loss position determination	1.2 m	0.303 m/dB	4.2 dB	1.3 m	Uses (28) to calculate position; Loss-GDOP

range using (15) and the fade-corrected loss data. The measured error is 2.9 m, and $GDOP * STD = 1.9$ m, which is significantly better than the standard method.

The fourth and fifth items in Table 3 are based on the walls loss excess model. The first method estimates the number of walls along the propagation path from the average wall separation, and the second method is based on the actual number of walls determined from a map of the building. The first approximate method determines the range using (17a), and then the position is determined using the same method as the previous two cases. The measured error is 3.5 m and 2.0 m, but is accurate enough to determine the number of walls along the propagation path from the map. After applying a correction for the true number of walls along each propagation path, the final measured error is 1.0 m, and $GDOP * STD = 1.3$ m. However note the previous comment on interpreting these statistical results.

The final item in Table 3 summarizes the performance of the position determination method based on direct use of the RSSI data without any need to first estimate ranges. Thus this method is simpler to implement, and as will be shown results in more reliable position estimates with less statistical scatter and position outliers. The details of this method are described previously in section *Iterative Least-Squares Method using Propagation Loss Directly*. The measured error is 1.2 m, and $GDOP * STD = 1.3$ m.

Positioning With 5 Base Stations

The positioning performance analyzed in the previous section is based on a large number (24) of "base station" measurements, which is not practical. To better demonstrate the statistical performance in real-world situations from the measured data set, positions derived using just five transmitting points (base stations) are investigated in this section. Five base stations were chosen as testing showed that this is the minimum number required for reasonably reliable position determination. This configuration would approximately simulate a mobile device moving through the building, with only five base stations in radio range at one time. As the mobile moves, these base stations will generally "surround" the mobile position, but the ranges and angles will change from point to point.

From Figure 2 it can be observed that the receiver point used in the measurements is surrounded by the transmitter positions. To obtain the five transmitter locations, these points are divided into four groups, summarised by the compass directions NE, SE, SW and NW. For the statistical simulation, one base station from each of these groups is selected. However, the pattern of points is not quasi-uniform, with less to the north than the south, so one extra position (number 18 in Figure 2) was also chosen in all the patterns of five base stations. The simulation is based on all combinations of such base stations, which results in 864 simulated positions, with ranges and geometry varying considerably.

The first simulation determines ranges based on the basic method described in the second item in Table 3. The scatter diagram of positions is shown in Figure 15a. The mean position error is (-0.9 m, 4.5 m) and the mean radial error is 4.5 m. Not all position attempts were successful, with 6 percent having such large range errors that no position could be determined. The mean GDOP is 1.06, so that from Table 3 $GDOP * STD = 5.6$ m. The maximum position error is about 12 m. The corresponding cumulative distribution function of position errors is shown in Figure 16. While the majority of positions are clustered about the mean position, there is a considerable scatter of positions more remote from the mean position. The mean position bias is associated with biases in the range estimates used for the position estimation. Without careful calibration of all the components in the link budget, such bias errors are inevitable, and this is a severe limitation of the standard method.

Figure 15. a. Scatter diagram of the position determined using five simulated base stations (selected from the 24 Zigbee measurement points) surrounding the receiver location. The positions are determined using basic range estimates derived from the measured RSSI data. The mean position (large dot) error is (-0.9 m, 4.5 m). b. Scatter diagram of the position determined using five simulated base stations using the new position determination method. The mean position error is (-0.80 m, 0.72 m). c. Scatter diagram of the position determined using five simulated base stations using the new position determination method. The mean position error is (-0.79 m, 0.94 m). Because of the large number of simulated position fixes only 25 percent are displayed in the figure.

a

b

c

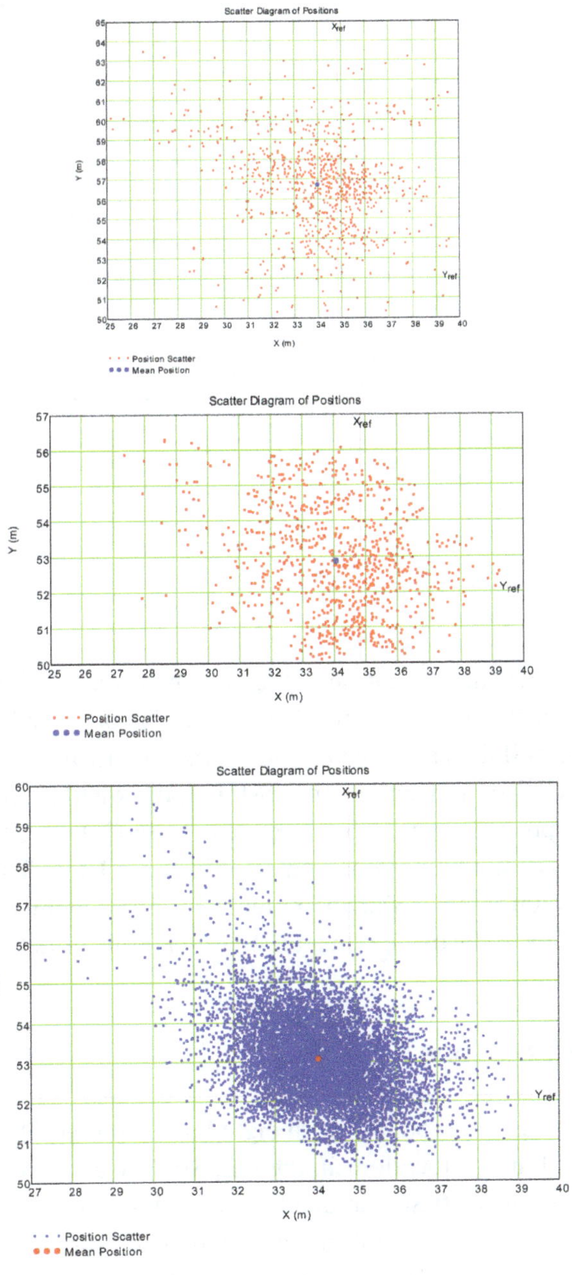

Figure 16. Cumulative distribution function of the position error data for the basic method (data in Figure 15a) and the new method (data in Figures 15b and 15c).

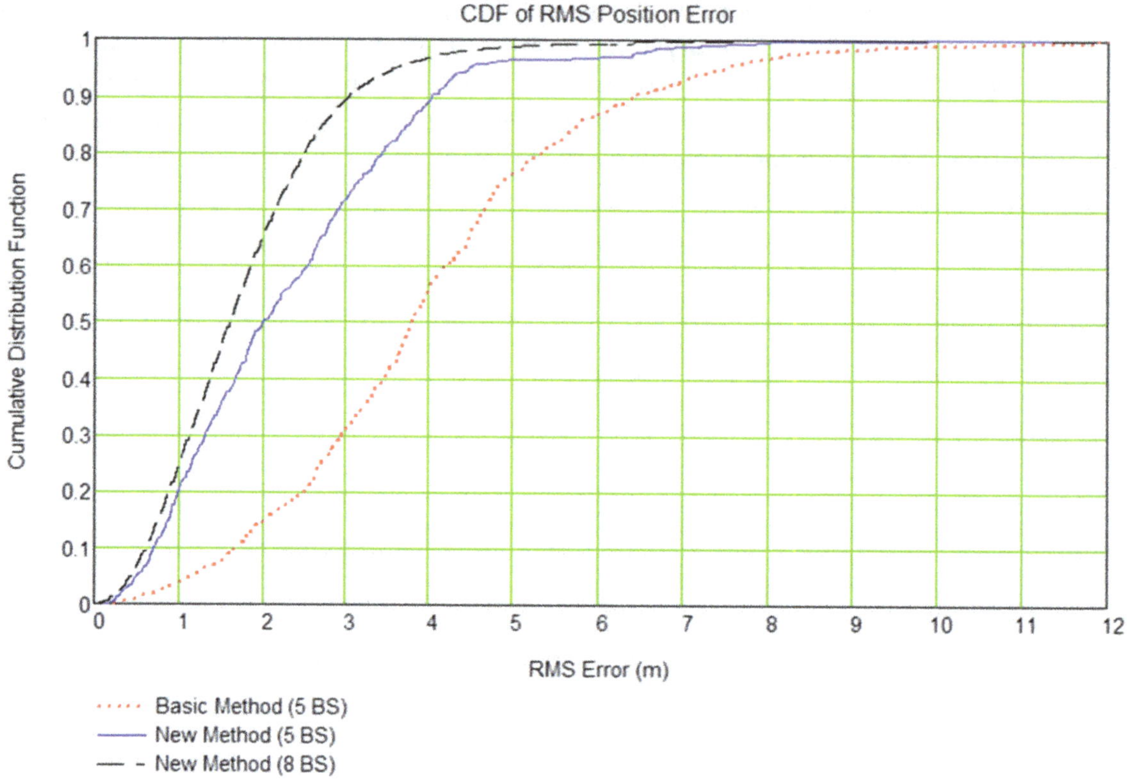

The second simulation determines positions based on the direct loss method described in the last item in Table 3. The results are shown in Figure 15b. As can be observed, the position scatter is much less, even though the same RSSI data are used in both cases. The mean position error is (-0.8 m, 0.7 m) and the RMS radial error is 2.3 m. Further, in contrast with the standard method shown in Figure 15a, 99 percent of position attempts were successful. The mean Loss-GDOP is 0.7 m/dB, so that from Table 3 $GDOP * STD = 0.7\,\text{m}$. The maximum position error is about 10 m. The corresponding cumulative distribution function of position errors is shown in Figure 16. The conclusion is that this direct positioning algorithm has much better performance than all other methods of determining position from RSSI data.

Positioning With 8 Base Stations

The final test uses the direct loss method, but with eight base stations. The selection process is similar to the five base station case, but two locations in the NE, SE, SW and NW groupings are selected. The simulation uses all combinations of base stations based on this selection criteria, resulting in 37,800 individual position determinations; this results in good statistical estimates of performance. Using eight base stations would likely be the upper limit of the number of nodes that could be expected from practical situations, such as WiFi AP, but is feasible for a grid of AP as discussed in the GDOP section. The resulting point scatter is shown in Figure 15c. The mean position of (-0.79 m, 0.94 m) is not significantly different from the five base stations case, but the RMS radial error of 1.76 m is significantly smaller, as is the scatter of positions which is more tightly grouped about the mean position. The associated cumulative distribution function is shown in Figure 16, which shows that median error is reduced from 2 m for 5 base stations to 1.5 m for 8 base stations, and the 90 percent probability is reduced from about 4 m for 5 base stations to about 3 m for 8 base stations.

The cumulative distribution function of position errors for the three different scenarios considered above are shown in Figure 16. The data show that the basic position method has a rather poor performance, with a median error of 4 m, and a long statistical tail out to 12 m. Such large errors would mean, for example, that the locating of a person to a room within a building would not be possible. In contrast, the new method performance, particular with eight base stations has a median error which is smaller than the size of rooms, and thus could be used for reliable location of people within buildings. However, even with the new improved methods there remains a significant tail to the error distributions. This is a consequence of using signal strength measurements in a NLOS radio propagation environment, and these large but infrequent positional errors are significantly worse than those associated with other more accurate methods of indoor radio position determination, such as those based on time-of-arrival measurements. However, given the convenience of using existing radio technology infrastructure such as WiFi and common mobile devices such as smart phones, the new position methods described provide a good compromise between convenience, widespread availability and positional accuracy.

CONCLUDING REMARKS

The idea of using the received radio signal strength for determining the position of a device has been proposed for a long time. With the widespread adoption of both WiFi and smart mobile devices this concept has become more appropriate for indoor situations where GNSS reception is not possible or is lacking the necessary accuracy of a few meters. While the concept is simple and easy to implement, typical existing methods have such poor accuracy that such applications are of limited practical use. Thus the main trust of the methods described in this chapter is to develop new algorithms that are particularly suited to using the RSSI signals from smart mobile devices for moderately accurate positioning indoors. While the raw RSSI data generally has very large apparently random variations, by applying knowledge of the behaviour of radio propagation under non LOS conditions these variations can be minimized. Further, by using a map of the building it is shown that much of this variation can be attributed to the propagation losses through walls, and thus to some degree these variations are deterministic. Thus by adopting a loss excess model and reducing the effects of signal fading, the standard deviation in the signal strength data can be reduced substantially, from the order of 10 dB to 4 dB. Further, it is shown that determining range from the propagation loss data is not only not necessary but a poor strategy for determining positions. Using only the raw RSSI data and a simple power law propagation model the resulting range standard deviation was 11 m; by adopting the proposed methods this was reduced to 3 m.

The algorithms developed are particularly intended for adoption by application programs (Apps) in smart devices to be used in situations where there is minimal knowledge of the characteristics of the access points in a building apart from their location. It is shown with the test data that even with these limitations a positional accuracy of 3 m is possible with communications to as few as five access points, and positional errors as small as 1 m have been demonstrated when more simultaneous communications are available. While these performance results were obtained with a limited set of data, the results indicate that techniques described could form that basis of practical application software. Such software would need to be tested to ascertain the performance in realistic real world situations.

REFERENCES

Chan, Y. T., & Ho, K. C. (1994). A simple and efficient estimator for hyperbolic location. *IEEE Transactions on Signal Processing*, *42*(8), 1905–1915. doi:10.1109/78.301830

Juan, J. M., Hernandez-Pajares, M., Sanz, J., Ramos-Bosch, P., Aragon-Angel, A., Orus, R., & Tossaint, M. et al. (2012). Enhanced precise point positioning for GNSS users. *IEEE Transactions on Geoscience and Remote Sensing*, *50*(10), 4213–4222. doi:10.1109/TGRS.2012.2189888

Lui, G., Gallagher, T., Li, B., Dempster, A. G., & Rizos, C. (2011). Differences in RSSI readings made by different Wi-Fi chipsets: A limitation of WLAN localization. *Proceedings of International Conference on Localization and GNSS*, 53-57. doi:10.1109/ICL-GNSS.2011.5955283

Prasithsangaree, P., Krishnamurthy, P., & Chrysanthis, P. K. (2002). On indoor position location with wireless LANS. *Proceedings of IEEE International Symposium on Personal, Indoor and Mobile Radio Communications*, 720-724. doi:10.1109/PIMRC.2002.1047316

Sathyan, T., Humphrey, D., & Hedley, M. (2011). WASP: A system and algorithms for accurate radio localization using low-cost hardware. *IEEE Transactions on Systems, Man, and Cybernetics. Part C*, *41*(2), 211–222.

Sharp, I., Yu, K., & Guo, Y. J. (2009). GDOP analysis for positioning system design. *IEEE Transactions on Vehicular Technology*, *58*(7), 3371–3382. doi:10.1109/TVT.2009.2017270

Sharp, I., Yu, K., & Hedley, M. (2012). On the GDOP and accuracy for indoor positioning. *IEEE Transactions on Aerospace and Electronic Systems*, *48*(3), 2032–2051. doi:10.1109/TAES.2012.6237577

Teunissen, P. J. G. (2002). *Adjustment Theory – an introduction, Series on Mathematical Geodesy and Positioning*. VSSD.

Yu, K., Sharp, I., & Guo, Y. J. (2009). *Ground-Based Wireless Positioning*. IEEE-Wiley Press. doi:10.1002/9780470747988

ENDNOTES

1 The convention used is that lower case text (*l*) is used for the loss in the standard form, and upper case (*L*) when specified in decibels.

2 Here by definition the *path loss* is a number greater than unity, or in decibels the loss is a positive number. Alternatively, the *path gain* is the reciprocal of the loss, and will be less than unity and a negative number in decibels.

APPENDIX: ANALYSIS OF NOISE CHARACTERISTICS IN SIGNAL STRENGTH DATA

This Appendix analyzes the measurement noise in RSSI data relative to the large scale mean signal level due to Rayleigh amplitude statistics of NLOS signal paths. While large signal fades (greater than 6 dB) can be largely eliminated by pre-processing, and samples averaged to reduce noise effects, some residual noise will remain. The following analysis estimates the standard deviation of this residual noise. For accurate position determination, this random noise must be reduced to as small a value as possible.

Normalized Statistical Distributions

When the propagation paths have no LOS, the received signal in a severe scattering indoor environment will exhibit approximately Rayleigh statistics. However, as determining positions from the signal strength is based on the mean signal, it is convenient to normalize the Rayleigh PDF to the mean signal level for calculations. As the Rayleigh distribution has only one parameter, such normalization results in the normalised PDF and CDF with no parameters, namely

$$\begin{aligned} p_{rayleigh}\left(x\right) &= \frac{\pi}{2} x e^{-\frac{\pi}{4}x^2} \\ cdf_{rayleigh}(x) &= 1 - e^{-\frac{\pi}{4}x^2} \end{aligned} \tag{A-1}$$

The mean of the normalized PDF is (by definition) unity, and the corresponding standard deviation is $\sigma_{ray} = \sqrt{\left(4-\pi\right)/\pi} = 0.523$. However, as the signal strength is measured in decibels, what is really required is the statistical distributions of the associated decibel signal strength. Thus with the signal strength defined as $s = 20\log\left(x\right) = (20 / \ln(10))\ln(x)$, the associated normalized decibel signal PDF is

$$\hat{p}_{rayleigh}\left(s\right) = \frac{\pi \ln(10)}{40} 10^{s/10} \exp\left(-\frac{\pi}{4} 10^{s/10}\right) \tag{A-2}$$

Distribution (A-2) is plotted in Figure 17. What can be clearly observed in Figure 17 is the long negative fading tail, which creates problems when dealing with RSSI measurements defined in dBm. The mean and standard deviations can be derived analytically from the PDF expression (A-2), resulting in

$$\begin{aligned} \mu_{dB} &= \frac{10}{\ln(10)}\left[\ln(4/\pi) - \gamma\right] = -1.46\ dB \\ \sigma_{db} &= \frac{10\pi}{\sqrt{6}\ln\left(10\right)} = 5.57\ dB \end{aligned} \tag{A-3}$$

Figure 17. Normalised Rayleigh distribution with the relative signal strength measured in decibels (mean signal is 0 dB). Also shown is the mean of the decibel signal and the mean minus one standard deviation. The mean is -1.5 dB and the standard deviation is 5.6 dB.

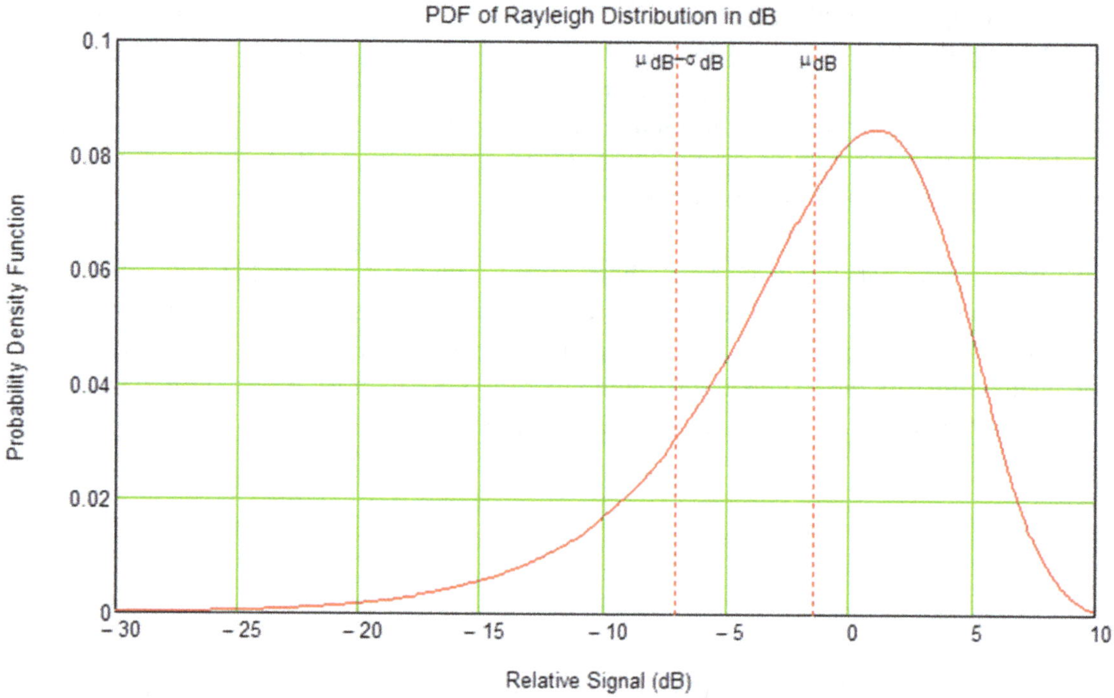

where γ is Euler's constant . Note also that the mean of the decibel signal is not zero. Thus even without the medium-scale log-normal effects, the small-scale signal variation is considerable with a small negative bias, and thus would result in poor positional accuracy.

Two Antenna Performance

One technique for minimizing the small-scale fading effects is to use two antennas on the mobile device. These antennas should be separated by at least one wavelength (12 cm for the Zigbee radios) so that the two received signals are statistically independent to Rayleigh fading. The basic strategy is then to test the signal strength at both antennas (using a switch to direct the signal to a common receiver), and then use the antenna with the strongest signal. Such a procedure will largely remove deep fades, but some variation relative to the nominal mean signal will remain. Note for data communications a signal stronger than the mean signal would be desirable, but for signal strength positioning such strong signals will result in position errors, the same as a signal fade with the same differential. Thus the aim of any measurement is for the mean decibel signal to be as close as possible to the true mean, and the standard deviation to be as small as possible. Based on these desirable characteristics, the performance of a dual antenna measurement system is analysed, using the normalised distribution described in the *Normalised Statistical Distributions* subsection of this Appendix.

The probability density function associated with the above procedure can be determined as follows. The probability that one antenna signal amplitude is x is $p_{rayleigh}\left(x\right)dx$, and the probability the second antenna amplitude is less than x is $cdf_{rayleigh}(x)$, so the probability that both conditions occur is

$$p_{2nat}\left(x\right) = \pi x\left(1 - e^{-\frac{\pi}{4}x^2}\right)e^{-\frac{\pi}{4}x^2} \tag{A-4}$$

where the further condition is that the integral of the PDF must be unity over the domain $\left(0, \infty\right)$. The associated mean and standard deviation are

$$\begin{aligned} \mu_{2ant} &= 2 - \sqrt{2} / 2 = 1.29 \\ \sigma_{2ant} &= \sqrt{6 / \pi + 2\sqrt{2} - 9 / 2} = 0.49 \end{aligned} \tag{A-5}$$

However, again the measured signal is in decibels, so the PDF (A-4) becomes

$$p\_dB_{2nat}\left(s\right) = \frac{\pi \ln(10)}{20} 10^{s/10}\left(1 - e^{-\frac{\pi}{4}s/10}\right)e^{-\frac{\pi}{4}s/10} \tag{A-6}$$

The associated mean and standard deviation can be derived analytically from the PDF (A-6), namely

$$\begin{aligned} \mu\, dB_{2ant} &= \frac{10}{\ln(10)}\left[\ln(8 / \pi) - \gamma\right] = 1.55 \text{ dB} \\ \sigma\, dB_{2ant} &= \frac{10}{\ln(10)}\sqrt{\frac{\pi^2}{6} - 2\ln(2)^2} = 3.58 \text{ dB} \end{aligned} \tag{A-7}$$

Comparing (A-7) with (A-3) shows that the mean signal is about 3 dB greater (fades are suppressed), and the standard deviation is about 2 dB less.

Multiple Sample Performance

Another method to minimize signal variation effects is to use multiple statistically independent signal samples. This approach has been taken for the processing of the Zigbee measurements, with four measurements per location point. The simplest strategy with N samples is to average the data. To avoid the non-linear effects of signals measured in decibels, this averaging should be performed in the standard "x" form. To calculate the PDF of averaging N signals the Rayleigh PDF (A-1) needs to be convolved N times. While this could be performed numerically, a much simpler approximate method is to invoke the Central Limit Theorem, so that the averaged signal distribution is approximately Gaussian of the form

$$\overline{p}_{gauss}\left(x\right)=\sqrt{\frac{N}{2\pi}}\frac{1}{\sigma_{ray}}\exp\left[-\frac{N\left(x-1\right)^{2}}{2\sigma_{ray}^{\ 2}}\right] \tag{A-8}$$

Again, as the signal strength is measured in decibels, the corresponding decibel form is

$$\overline{p}\_dB_{gauss}\left(s\right)=\frac{\ln(10)}{40\sigma_{ray}}\sqrt{\frac{2N}{\pi}}10^{s/20}\exp\left[-\frac{N\left(10^{s/20}-1\right)^{2}}{2\sigma_{ray}^{\ 2}}\right] \tag{A-9}$$

Based on the PDF (A-9), the mean signal strength is -0.3 dB, and the standard deviation 2.6 dB for four samples.

As the number of the Zigbee measurements per location is four, the PDF plot in Figure 18 averages $N = 4$ samples. Comparing it with the Rayleigh distribution in Figure 17, the decibel signal variation is much smaller and the mean of -0.3 dB is close to the true value of 0 dB. The averaging process has reduced the standard deviation by 3 dB from 5.6 dB to 2.6 dB. Averaging more signals would further reduce the signal variability.

Figure 18. Normalised Gaussian distribution with the relative signal strength measured in decibels (mean signal is 0 dB). Also shown is the mean of the decibel signal and the mean minus one standard deviation. The mean is -0.3 dB and the standard deviation is 2.6 dB. The number of signals averaged is four.

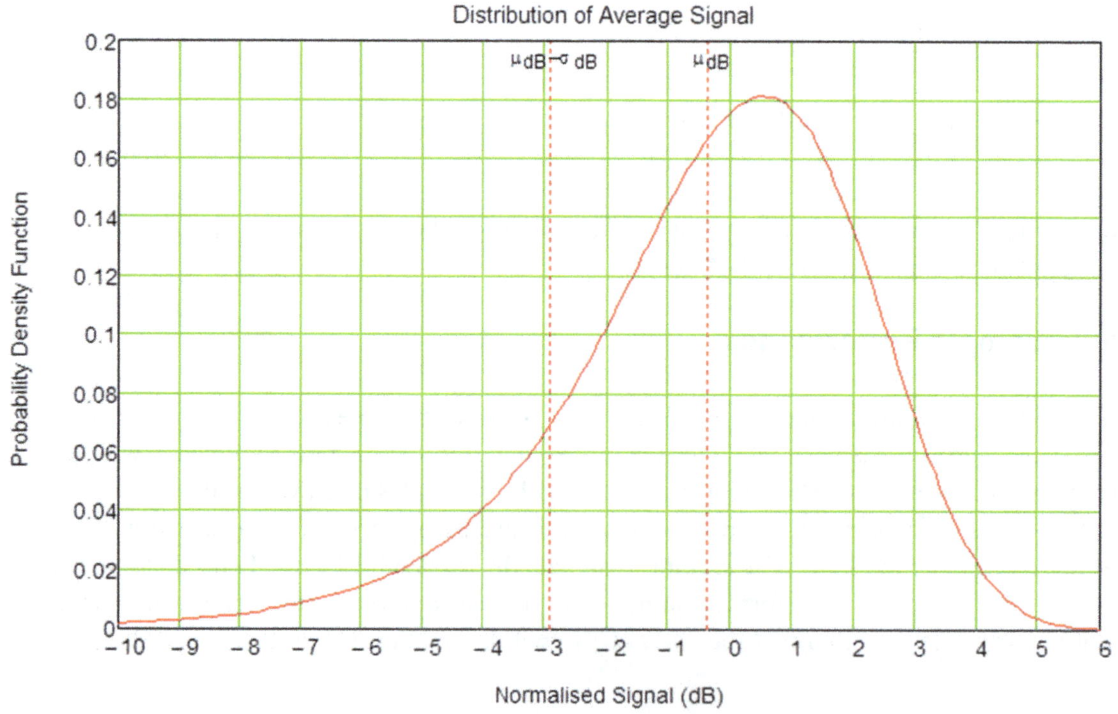

To further analyze the performance, the results of a simulation of the algorithm used to process the raw Zigbee data is shown in Figure 19. As can be observed the vast majority of the processed signals are within ±6 dB of the nominal 0 dB mean signal, with just a few deep fades. The simulation data agrees well with the theoretical expectations. These results show the importance of the pre-processing of raw RSSI data to obtain signal strength data with minimal variation due to Rayleigh fading effects. While averaging more than four samples would somewhat further reduce the variation, in the limit other random processes (particularly medium-scale log-normal variations) will dominate, so that all measurement noise cannot be removed by signal averaging in a small area or over a small time interval for a moving mobile receiver.

Figure 19. Results of a simulation using the "bad" data rejection algorithm used to process the raw Zigbee data. The simulation uses 4000 measurements (4 averaged per sample for 1000 samples) with a normalized Rayleigh distribution. The sample mean is 0.03 dB and the standard deviation is 2.8 dB.

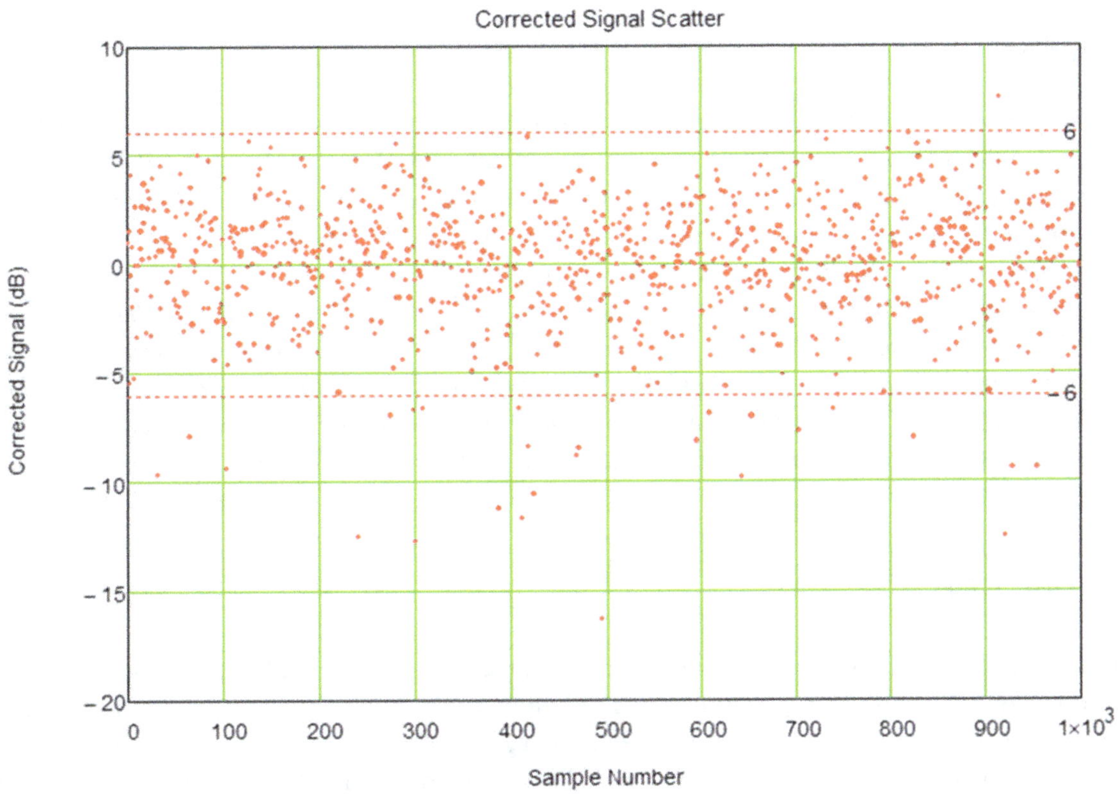

Chapter 2
Received Signal Strength Models for Narrowband Radios

Hüseyin Yiğitler
Aalto University, Finland

Ossi Kaltiokallio
Aalto University, Finland

Riku Jäntti
Aalto University, Finland

ABSTRACT

The advancements in wireless communication technologies have enabled new sensing possibilities where the channel measurements of the radio are used for inferring physical changes in the surrounding environment. Relating the channel measurements to the location and actions of people has been of particular interest due to the wide range of application opportunities enabled by such a sensing capability. As an example, the low-amplitude received signal measurements of low-cost wireless communication systems have been used to detect the presence of a person, to locate and track them, identify gestures and activities of the person, and even monitor their vital signs. This chapter aims to give a deep insight on how people influence radio signals, how these effects are observed at the receiver antenna, and how the measurement system impacts the recorded measurements. These topics are presented to shed light on the relation between the location of people and signal strength measurements of narrowband radios.

INTRODUCTION

Recent advancements in different technologies have triggered a trend in replacing wired communications with wireless technologies, making them ubiquitously available in our everyday environment. Wherever we are, we inevitably interact with electromagnetic (EM) energy emitted by these systems, and alter the characteristics of the EM wave impinging on the receiver antenna. In particular, we alter received signal of the short-range systems, which indicate our presence and actions in close neighborhood of a link-line[1]. This variation has recently been used for non-invasive sensing by associating the

DOI: 10.4018/978-1-5225-3528-7.ch002

received signal variations of different links to the same person in an environment, commonly referred to as *environmental sensing technology*.

The impact of a person on EM waveforms can be related to the person's location and actions even though they do not cooperate with the system by carrying a tag. This property makes the environmental sensing technology an attractive (indoor) device-free localization (Youssef, Mah, & Agrawala, 2007) option for conventional application areas such as ambient assisted living (Bocca, Kaltiokallio, & Patwari, 2013), residential monitoring (Kaltiokallio, Bocca, & Patwari, 2012), security and emergency surveillance (Bjorkbom, et al., 2013), roadside surveillance (Anderson, Martin, Walker, & Thomas, 2014), and forest monitoring (Alippi, Bocca, Boracchi, Patwari, & Roveri, 2015). It has also been shown that the technology can successfully estimate breathing rate of a person (Kaltiokallio O., Yiğitler, Jäntti, & Patwari, 2014), detect the presence of people in an environment (Kosba, Saeed, & Youssef, 2012), (Mrazovac, Todorović, Bjelica, & Kukolj, 2013), or identify when a person falls (Mager, Patwari, & Bocca, 2013). These application areas, however, are constrained by physical factors such as available information, and practical factors such as network management complexity, energy requirements, computational resource requirements and costs. The practical problems are solved by using state-of-the-art methods for existing network deployments since the technology relies on available link-quality measurements of wireless technologies to infer person's location and actions. On the other hand, the information contained in the measurements on the phenomenon of interest has a complex relation with the geometry and electrical properties of the objects[2] in the environment, which is referred to as *communication channel*. As the quality (resolution in terms of both bandwidth and amplitude) of the channel measurements increases, the information about a person increases and the phenomenon of interest can be better inferred. However, higher quality channel measurements also imply a more complex and expensive systems. In this regard, received signal strength (RSS) measurements is very attractive option since it is ubiquitously available also on low-cost devices. In this chapter, the RSS measurements of narrowband wireless communication systems are elaborated in depth, and the variation of the measurements with the person's location and movement are investigated.

The RSS is the received signal power calculated over a certain time duration, and it depends on the propagation channel, interference sources in the environment, transmission power and constant gains of the radio hardware. The information carried by such a compact measure can be used for link quality assessment (Baccour, et al., 2012), adaptive transmission power control (Lin, et al., 2006), and node localization (Patwari, et al., 2005). When the interference sources can be ignored and the radio gains are constant, the RSS varies solely due to propagation channel variations, which also enables RSS-based environmental sensing systems. These systems effectively compare the RSS measurements with the stationary channel case, i.e., base-line RSS, and typically process low-amplitude signals in order to infer the phenomena of interest. Therefore, the validity and accuracy of the RSS models that relate the phenomena of interest with the measurements are very important for successful environmental sensing applications.

This book chapter begins with example real-world RSS measurements to demonstrate how a person affects the RSS and how the related works have modeled these changes. Thereafter, the considered RSS measurement system is introduced, and a general measurement model is presented. The underlying physics of interaction of objects with propagating EM energy is reviewed before, we characterize the temporal variations of the RSS. For this purpose, we elaborate an idealistic deployment scenario, which enables drawing the relation between interacting object's position to the observed RSS variations. Based on these observations, a four-state temporal RSS model is introduced after deriving that the RSS measurements approximately have log-normal distribution under static channel conditions. The deterministic and statistical models are introduced thereafter. We show how the models can be used for regional border crossing, link-line monitoring, and very low complexity radio tomographic imaging. Finally, the chapter is concluded.

MOTIVATION

In wireless communications, the transmitted signal propagates via multiple paths. The associated signal components of all paths sum at the receiver antenna, and produce distorted version of the transmitted signal. In this respect, a person can affect the RSS measured at the receiver either by creating a new multipath component or by altering an existing one (Patwari & Wilson, 2010). In literature, it has been proposed that a person can create a new multipath component by scattering (Norklit & Andersen, 1998) or reflection (Liberti & Rappaport, 1996), (Ghaddar, Talbi, Denidni, & Sebak, 2007). The person may also alter existing multipath components by changing their amplitude and/or phase. For example, the component can diffract around the person (Savazzi, Nicoli, Carminati, & Riva, 2014), (Rampa, Savazzi, Nicoli, & DÁmico, 2015), or transmit through the person (Patwari & Agrawal, 2008), (Wilson & Patwari, 2010). Lastly, it is possible that the person is outside the sensing region of the transmitter (TX) and receiver (RX) pair so that the person does not have an impact on the RSS.

In Figure 1, the RSS measurements of a single TX-RX pair, deployed in an open environment, are shown. The nodes are set on podiums at a height of one meter, they are equipped with Texas Instruments CC2431 radios, and directional antennas are used to assure that the changes in RSS only reflect variations caused by the target. The used directional antennas provide a 8 dBi gain and a horizontal beam width of 75°. A detailed description of the node hardware can be found in (Yiğitler, Jäntti, & Virrankoski, 2014) and the communication protocol is a single transmitter version of the protocol presented in (Yiğitler, Kaltiokallio, & Jäntti, 2013). In the experiments, a person moves along the trajectories shown in Figure 1a and four different link distances are used $d = [3, 5, 7, 9]$ m. In Figure 1, the mean is removed from the RSS since it does not contain information about the person's location, and the measurements are low-pass filtered to enhance the visual quality of the signal.

The first works appeared in the related literature modeled the decrease in RSS when a person obstructs the link-line of the TX-RX pair (Patwari & Agrawal, 2008), (Kanso & Rabbat, 2009). These models are referred to as *shadowing models*, and they assume constant attenuation rate along the link-line. Later, the observed RSS variation in logarithmic scale was modeled with a two-parameter exponential decay when the person is within the first Fresnel zone (Li, Chen, Coates, & Yang, 2011). The aforementioned models assume that the signal is attenuated by human body as it propagates through the person. Since these models fail to include into the model the diffraction phenomena, several diffraction based models have been proposed (Savazzi, Nicoli, Carminati, & Riva, 2014), (Rampa, Savazzi, Nicoli, & DÁmico, 2015), (Wang, Liu, Xu, Bu, & An, 2015). These works simplify the complex diffraction phenomena by assuming cylindrical human body and it is electrically is a perfect absorber. They also assume that knife-edge approximation is valid for the human body.

Just mentioned RSS behavior is visualized in Figure 1. The data is acquired when the person moves on the trajectory A depicted in Figure 1a. When the person obstructs the link-line, the measured RSS decreases on the average as shown in Figure 1b. The shadowing-based models referenced above are capable of explaining the average RSS loss due to effect of the person. They, however, fail to explain the observation that the RSS variation is the largest when the person is close to the TX and RX (Patwari &

Figure 1. In (a), the trajectory of the person in the experiments. In (b), the measured RSS when the person moves along trajectory A and in (c), the RSS when the person moves along trajectory B. In (b) and (c), the thin grey lines depict the RSS on the 16 *different frequency channels and the black thick line is the sample average across all channels. In (d), the box-and-whisker plot of RSS values in shadowing state (black boxes) and fading state (red boxes) with different link distances.*

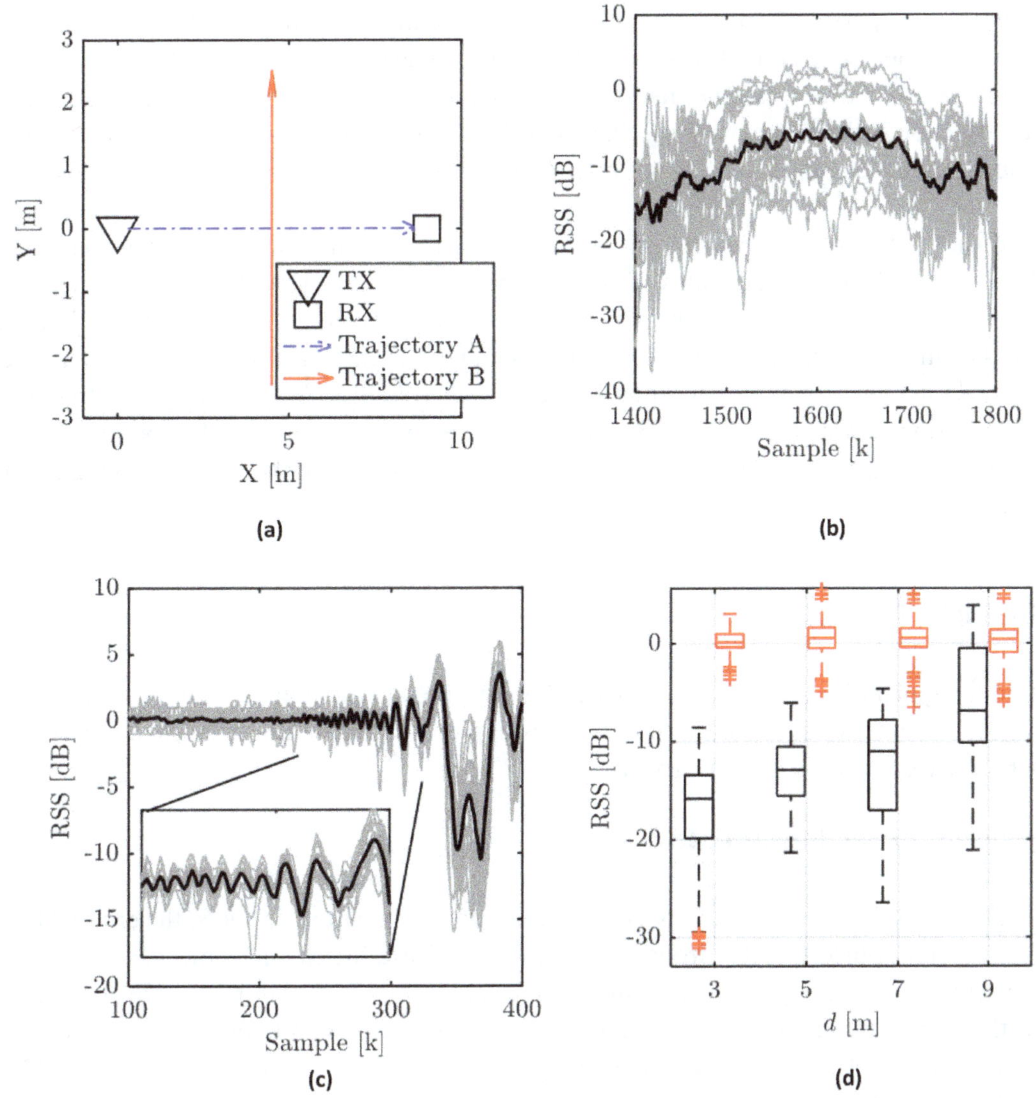

Wilson, 2011). The diffraction-based models depend strongly on the human geometry, and the impact of human body on the RSS increases when the person obstructs larger cross-sectional area of the first Fresnel zone. Since the person obstructs the largest area when the person is close to the transceivers, these models can explain the observed RSS variation. In the remainder of the chapter, this state is referred as *shadowing state*.

Majority of the works have concentrated on modeling the RSS when the person shadows the link, i.e., he/she is in the first Fresnel zone. This region is attractive since the observed RSS variation can be modeled as a linear function of person's occupancy of the link-line (Agrawal & Patwari, 2009). These works disregard the fact that a person can also affect the RSS far from the link-line by scattering and reflection (Patwari & Wilson, 2011). As an example, consider the scenario when a person moves along trajectory B depicted in Figure 1a, which yields the RSS measurements in Figure 1c. In the beginning of the experiment, the person is outside the sensing region of the RX and the RSS are dictated by measurement noise. As the person moves closer to the LoS, the newly created multipath becomes measurable at the RX at $k \approx 240$. After this time instant, the RSS varies increasingly and our recent works have shown that reflection is the dominating mechanism (Yiğitler H., Jäntti, Kaltiokallio, & Patwari, 2017), (Kaltiokallio, Yiğitler, & Jäntti, 2017). We refer to this state as *fading state* in the remainder of the chapter.

Figure 1d shows the box-and-whisker plot of the RSS in the fading and shadowing states with different link distances. As the link distance increases, the person's effect on the signal decreases on the average. This is because the average cross-sectional area of the person, as it appears to the TX and RX, reduces. On the other hand, the RSS changes are larger and the changes are observable over a wider area with longer links in the fading state (Yiğitler H., Jäntti, Kaltiokallio, & Patwari, 2017). In both of the states, the dispersion between the frequency channels increases with longer links and in the literature, channel diversity has been used to increase the probability of capturing the RSS following the models (Kaltiokallio, Bocca, & Patwari, 2012).

MEASUREMENT SYSTEM

In this section, a typical RSS system is introduced. First, the propagation channel models for wireless communication systems[3] are reviewed. Then, the measurement system is introduced, and finally, a general RSS model for narrowband systems is given.

Radio Propagation Channel Model

In a typical single TX – RX pair deployment of wireless communication systems, the emitted EM wave interacts with several objects in the propagation medium. These interacting-objects scatter, reflect or refract the EM wave, generating new waves traversing different paths before impinging on the RX antenna. At the RX, the signals of all the paths sum, resulting in the phenomenon known as *multipath fading*.

The transmitted signal is altered by the interacting objects in the environment according to EM laws of physics. In the received signal perspective, however, the impact of the propagation environment on each multipath component is simply to multiply the transmitted signal with an attenuation factor, add some phase depending on the EM properties of the interacting objects, and delay it due to finite propagation speed. Correspondingly, the complete propagation channel is usually modeled as a time-varying linear filter with impulse response given by

$$h(t,\tau) = \sum_{i=1}^{N(t)} a_i(t) e^{j\phi_i(t)} \delta(t - \tilde{\tau}_i(t)), \tag{1}$$

where $N(t)$ is the number of multipath components at a given time, $\{a_i(t), \phi_i(t), \tilde{\tau}_i(t)\}$ are amplitude, phase, and arrival time of the i th multipath component, and $\delta(\cdot)$ is Dirac delta function.

The RX radio can distinguish different multipath components if it can measure the received signal with a granularity shorter than the smallest arrival time difference. In other cases, the multipath components combine together and form a single tapped-delay line. Therefore, for a practical receiver Eq. (1) reads as

$$h(t,\tau) = \sum_{n=1}^{N} \sum_{m=1}^{M_n(t)} a_{n,k}(t) e^{j\phi_{n,m}(t)} \delta(t - \tilde{\tau}_{n,m}(t)) = \sum_{n=1}^{N} \alpha_n(t) e^{\phi_n(t)} \delta(t - \tau_n), \tag{2}$$

where $M_n(t)$ depends on the time separation between different multipath components, and with slight abuse of notation $\{\alpha_n(t), \phi_n(t)\}$ is the phasor sum of $M_n(t)$ multipath components, τ_n is the time invariant tap delay of the n th tap, and N is the number of taps.

The number of taps N in the tapped delay line model in Eq. (2) is defined by the duration of the transmitted pulses and sampling rate of the receiver. In this regard, the wideband communication systems have a higher number of taps, and can measure impact of different multipath components. In case the sampling rate is very high (e.g. for ultra-wideband communication systems), even densely spaced components can be resolved. On the other hand, for narrowband communication systems, it is customary to take $N = 1$, and assume that all the observed effect of the propagation channel is modeled by single complex quantity, $h(t,\tau) = \alpha(t)$. In the remaining part of the chapter, we study the narrowband communication systems. However, we refer to the channel impulse response in Eq. (1) when more details are needed.

System Overview

Consider the radio wave propagation scenario depicted in Figure 2a, where the propagation environment is completely empty except the interacting object and the TX-RX pair. Suppose that the communication system satisfies the following:

1. The wavelength is much smaller than average geometrical extent of the interacting object of interest (e.g. human body).
2. The antennas have ideal frequency response and gain patterns, and they are vertically polarized.

These assumptions guarantee that the interacting object has a deterministic impact on the received signal and the measurement system does not alter its strength due to polarization or orientation of the antennas. Furthermore, since for phase coherent receivers the reception start instance is explicitly defined (Proakis & Salehi, 2008, pp. 290-326), we also assume that the receiver is phase coherent. For these radios, it is easier to draw the relation between temporal variations in the propagation channel and the changes in RSS, since the reference time ambiguity of the received signals of different multipath components is resolved by the receiver. In sequel, we assume that the shortest path component[4] is assigned to have zero phase, and phases of other components are assigned referencing to the shortest path component.

As mentioned earlier, the RSS is average signal energy in the communication band, measured over a certain duration. In modern digital receivers with quadrature demodulators RSS is calculated using

Figure 2. In (a), an ideal propagation scenario for RSS modeling. The transmitter (TX) and receiver (RX) are separated by dm. In (b), a simplified received signal strength measurement system using digital in-phase $i[k]$ and quadrature $q[k]$ samples, square-sum calculator with round-off error r_n, and a quantized logarithm calculator with quantization level Q.

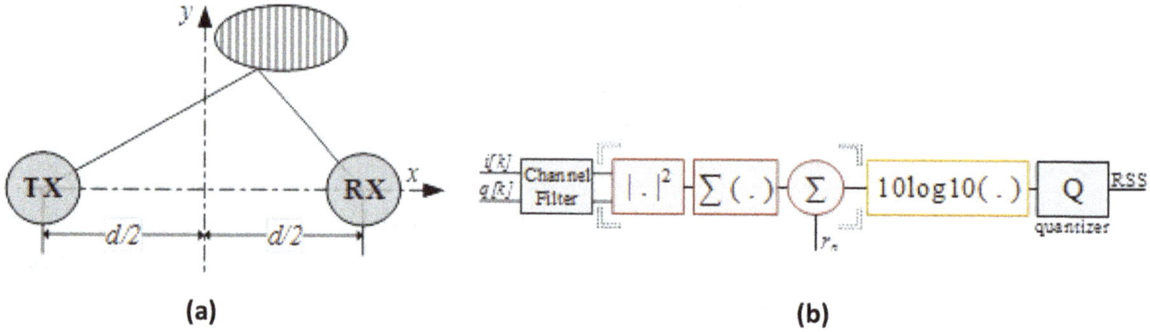

digital samples of in-phase ($i[k]$) and quadrature ($q[k]$) channels, described in Sec. 4.6.2.3 of the book by Gu (Gu, 2005). A simplified RSS measurement system, shown in Figure 2b, outputs RSS indicator (RSSI) values, which are digital RSS in logarithmic scale represented in Q quantization levels. The common receiver components (e.g. filters, low-noise amplifiers and automatic gain control units) can be represented in terms of parameters of the components in the figure. Therefore, the given RSS measurement system is general for the RSS modeling purposes.

In order to capture the variation of $\alpha(t)$ due to interacting objects' actions, the RSSI must be output faster than the bandwidth dictated by the coherence time of the channel. The coherence time of the channel depends on its Doppler spread due to movements of the interacting objects (Rappaport, 2002, pp. 203-205), and within this time interval the channel is usually assumed to be constant. For an interacting object moving with speed v, the maximum Doppler spread f_m (in Hz) and the associated coherence time of the channel T_c are defined as

$$f_m = \frac{v}{\lambda}, \qquad T_C \approx \frac{3}{4\sqrt{\pi} f_m},$$

where λ is the wavelength of the communication system. For example, for $v = 1$ m/s and $\lambda = 0.125$ m, $T_C \approx 52.9$ ms. On the other hand, the output rate of RSSI depends on

1. Sampling rate of the radio and the number of samples used in RSS calculations.
2. Output policy of the RSSI system. For example, the radio may output RSSI only once for each received frame.

For the example system above, the time spacing between two RSSI outputs must be less than 52.9 ms. Therefore, the RSSI output rate of the measurement system must fulfill sampling rate requirement of the fastest speed of the interacting object of interest.

Received Signal Strength Measurements

Suppose that a narrowband signal $s(t)$ is transmitted at time instant t. If the propagation medium is constant for the time period of RSS calculation window, the received baseband signal can be written as (Proakis & Salehi, 2008, p. 846)

$$x(t) = \hat{\alpha}s(t) + \hat{n}(t), \tag{3}$$

where $\hat{\alpha} = \sqrt{g_t g_r}\alpha$ is the channel gain (including constant transmitter side g_t and receiver side g_r gains), $\hat{n}(t)$ is assumed to be a zero-mean circularly symmetric complex white Gaussian noise process with two sided power spectral density $\mathrm{N}_0 / 2$. If round-off error r_n and quantization noise for the RSSI system in Figure 2b are ignored, the RSS calculated using K digital samples of complex signal $x(t)$, $x_k = \hat{\alpha}s_k + \hat{n}_k$, can be written as

$$\mathcal{P} = 10\log_{10}\left(\frac{1}{K}\sum_{k=1}^{K}\left|\hat{\alpha}s_k + \sigma n_k\right|^2\right), \tag{4}$$

where $n_k \triangleq \hat{n}_k / \sigma$ and $\hat{n}_k$ are zero mean circularly symmetric complex Gaussian with equal variance in real and imaginary components σ^2. For a digital receiver, the in-phase and quadrature components of the received signal are sampled with a two-channel analog-to-digital converter, which outputs $i[k]$ and $q[k]$ samples shown in Figure 2b after analog preprocessing and before digital processing starts. Thus, for the purposes of this chapter, we have $\hat{\alpha}s_k + \sigma n_k = i[k] + jq[k]$ for $j = \sqrt{-1}$.

Let us make the following definitions for notational convenience:

$$\bar{X}_K \triangleq \frac{1}{K}\sum_{k=1}^{K}\left|\frac{\hat{\alpha}}{\sigma}s_k + n_k\right|^2, \quad \gamma \triangleq \frac{\mathrm{P}_s}{\sigma^2}, \tag{5}$$

where $\bar{X}_K$ is the noise variance normalized RSS in linear scale, and $\mathrm{P}_s \triangleq \frac{1}{K}\sum_{k=1}^{K} | \hat{\alpha}s_k |^2$ is the signal power in $x(t)$, and γ is the *signal-to-noise ratio* (SNR) of the received signal. Then, the RSS measurement is given by

$$\mathcal{P} = 10\log_{10}(\sigma^2) + \frac{10}{\ln(10)}\ln\left(\bar{X}_K\right). \tag{6}$$

In order to gain further understanding of the RSS output, let us investigate the case where the number of samples K is large. For this purpose, let us expand $\bar{X}_K$ as

$$\bar{X}_K = \gamma + \frac{1}{K}\sum_{k=1}^{K} 2\Re\left\{\frac{\hat{\alpha}}{\sigma} s_k n_k^*\right\} + \frac{1}{K}\sum_{k=1}^{K} \left|n_k\right|^2, \tag{7}$$

where $\Re\{\cdot\}$ is the real part of a complex number, and * denotes complex conjugate. The first sum in Eq. is a weighted (with finite power weights s_k) sum of zero-mean and finite variance independent and identically distributed normal random variables n_k. As $K \to \infty$, this sum approaches to zero almost surely due to *strong law of large numbers* (SLLN) for weighted sequences proposed and proved in Theorem 3.4 of (Baxter, Jones, Lin, & Olsen, 2004) under conditions fulfilled by s_k and n_k. The second sum approaches to the mean of $| n_k |^2$ almost surely due to SLLN. Consequently, RSS $\mathcal{P}$ converges to a scalar given as

$$\lim_{K\to\infty} \mathcal{P}(K) = 10\log_{10}(\sigma^2) + 10\log_{10}\left(2+\gamma\right). \tag{8}$$

In case the propagation channel is stationary, it has been shown in (Yiğitler, Jäntti, & Patwari, 2017) that the logarithmic transformation used in the RSS measurement system makes the RSS $\mathcal{P}$ normally distributed with approximate mean equal to Eq. (8) even for small K. For small K case, therefore, M sample time average of the RSS reads as

$$\frac{1}{M}\sum_{m=1}^{M} \mathcal{P}[m] \approx 10\log_{10}(\sigma^2) + 10\log_{10}\left(2+\gamma\right). \tag{9}$$

The environmental sensing applications in general use the variation of RSS to identify the impact of interacting object of interest compared to the reference measurements, which are acquired when the object is not in the monitored region. Thus, if mean of the reference RSS measurements in logarithmic scale in Eq. (9) are subtracted from all the measurements, the RSS measurement at time sample k is given by

$$r[k] \triangleq \mathcal{P}[k] - \mathcal{P}_r = 10\log_{10}\left(\frac{\bar{X}_K}{2+\gamma_r}\right), \tag{10}$$

where $\mathcal{P}_r$ is the mean of the reference RSS, and γ_r is the SNR value of the received signal for the static propagation channel realization. Then Eq. (7) can be used for approximating $r[k]$ as

$$r[k] \approx 20\log_{10}\left(\frac{\left|\alpha[k]\right|}{\left|\alpha_r + (2/\sqrt{g_t g_r})\right|}\right) + 10\log_{10}\left(\frac{1}{\gamma[k]}\frac{1}{K}\sum_{m=1}^{K}\left|n_m\right|^2\right), \tag{11}$$

Since the actions of the interacting objects merely changes the channel gain $\alpha[k]$ (correspondingly $\gamma[k]$), $r[k]$ is a convenient measurement metric for environmental sensing purposes.

In the above derivations, the impact of round-off error r_n and quantization noise for the RSSI system in Figure 2b are ignored. In general, these errors are not statistically independent of the input, but under certain conditions on the input distribution, they become uncorrelated. Statistical characterization of the round-off errors r_n depends on the implementation of the linear scale RSS estimator. However, when Q is low, the conditions of the first order quantization theorem (approximately) hold (Widrow, Kollar, & Liu, 1996). In this case, well known Shepard's corrections, which increase the second moment, are valid, and its only impact is decreased SNR γ (due to increased σ^2).

The impact of the RSSI quantizer can be included to the RSS model considering the similar arguments as stated for the round-off error. However, by definition of $\bar{X}_K$ in Eq. (5), its variance decreases with increasing K, making the quantization noise neither white nor uncorrelated with the input (Widrow, Kollar, & Liu, 1996). For low-dynamic range quantizer, one can still find mean and variance of the quantizer output using the corrections given by Kollár (Kollár, 1994). The parameters of the resultant log-normally distributed RSSI is given in (Yiğitler, Jäntti, & Patwari, 2017). For high SNR values (higher than 5dB), the mean of the RSSI is approximately equal to the mean given in Eq. (8), and its variance depends on its mean, quantizer free variance and quantization level. Consequently, in the following part of the chapter, we use the processed RSS model given by

$$r[k] = 20\log_{10}\left(\frac{|\alpha[k]|}{|\alpha_r|}\right) + \nu[k], \tag{12}$$

where $\nu[k]$ is the joint noise process of all noise sources including white radio noise, round-off errors, quantization noise, and any source of uncertainty due to modeling errors, and its variance depends also on the current SNR value $\gamma[k]$. In Eq. (12), the term $2/(\sqrt{g_t g_r})$ is ignored since the transmitter and receiver side gains are usually much higher than 4.

PHYSICAL CONSIDERATIONS

In this section, a brief review of the underlying physics for investigating interaction of objects with EM energy is reviewed. First, the antenna radiation characteristics are summarized. Then, the general effect of interacting objects on EM waves is discussed, and thereafter the human body is given.

Transmission and Reception With Passive Antennas

The EM wave propagating on a guided transmission line (e.g. carefully designed copper track on a printed circuit board) is converted to a wave propagating in an unbounded medium (e.g. free space) by antennas. Passive antennas are reciprocal transducer devices; they can convert the EM waves from unbounded medium to the waves in a guided medium. The antennas are modeled in terms of Hertzian dipoles, and they are driven by or they output time-varying conduction current along the z-axis $i(t)$ as shown in Figure 3a.

Figure 3. In (a), radiation of electromagnetic wave from a thin line dipole antenna, propagation spherical coordinate system (d, θ, ϕ), and plane wave approximation at the receiver. In (b), illumination of an interacting object with electromagnetic energy emitted by an antenna. The gray cone is the solid angle subtended by the area of antenna beam, and red cone is the surface area of the object seen by the antenna.

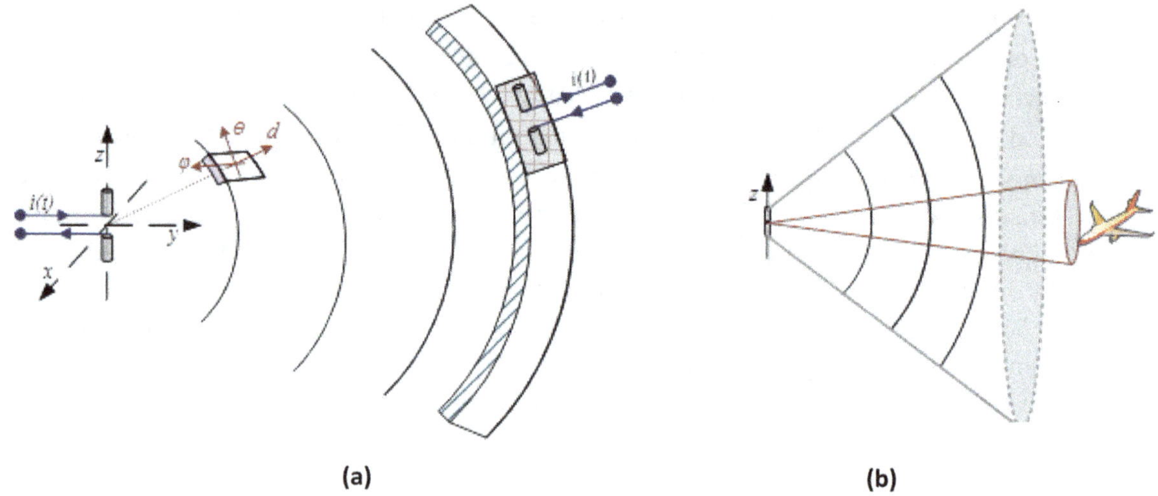

In case the radiated wave is observed near the transmitting antenna, it can be shown that the EM fields are governed by laws of electrostatic and magnetostatic. However, once the observation point is moved far away from the source antenna, electrostatic and magnetostatic effects diminish, and only propagating EM wave is effective. The observed EM field is a spherical wave that has a constant phase on a sphere (a.k.a. spherical wavefront), which is co-centered with the transmitting antenna. Furthermore, it is widely accepted that the spherical wavefront can be approximated by a plane if the plane approximation leads phase error of less than $\pi / 8$ radians. Then, the observation point is in the far field of the antenna, which is defined as points at a distance d satisfying

$$d \geq \frac{2l^2}{\lambda}, \tag{13}$$

where l is the longest linear dimension of the antenna, λ is the wavelength of the wave radiated by the antenna. The plane wave approximation of the transmitted wave is also visualized in Figure 3a for a dipole antenna aperture. The far-field for a full-wave dipole starts after 2λ.

In environmental sensing applications, antennas are operated in their far-field regions. This allows us to approximate the spherical wave as a uniform plane wave, which has uniform properties on a plane, and leads simple solutions to source free wave equations. For uniform plane waves, the vector product of the electric field intensity and the magnetic field is defined as the Poynting vector, half real part of which is the time average power density $\mathcal{S}(\theta, \phi)\hat{\mathbf{d}}$ in the direction of propagation $\hat{\mathbf{d}}$. Then, the differential transmitted power through an elemental area $\delta \mathbf{A} = \delta A \hat{\mathbf{d}}$ is given by

$$\delta \mathrm{P} = \mathcal{S}(\theta,\phi)\delta A.$$

Since the differential area element in spherical coordinates is defined as $\delta A = d^2 \sin(\theta)\delta\theta\delta\phi$, and solid angle $\delta\Omega$ subtended by δA is defined as $\delta\Omega = \delta A / d^2$, the total power through a sphere with radius d is given by

$$\mathrm{P} = d^2 \int_{\Theta}\int_{\Phi} \mathcal{S}(\theta,\phi)\sin(\theta)\delta\theta\delta\phi = d^2 \int_{\Omega} \mathcal{S}(\Omega)\delta\Omega.$$

Finally, since antennas have directivity, the radiated power is not uniform at all solid angles. Thus, for an antenna with $F(\theta,\phi)$ normalized radiation intensity, the total power through the same sphere above reads as

$$\mathrm{P} = d^2 \mathcal{S}_{\max} \int_{\Omega} F(\theta,\phi)\delta\Omega,$$

where $F(\theta,\phi) = \mathcal{S}(\theta,\phi) / \mathcal{S}_{\max}$ and $\mathcal{S}_{\max}$ is the maximum magnitude of the power intensity.

The relation between power intensity and delivered power for antennas dictates affective area of antennas, which is defined as

$$A_e = \frac{\mathrm{P}_{\mathrm{int}}}{\mathcal{S}_i}, \tag{14}$$

where $\mathrm{P}_{\mathrm{int}}$ the intercepted electromagnetic power that can be delivered to a matched load, and $\mathcal{S}_i$ is the incident power intensity. The affective area of an isotropic radiator is given by,

$$A_e = \frac{\lambda^2}{4\pi}.$$

Let us suppose that a transmitting short dipole antenna with gain g_t and normalized radiation intensity $F_t(\theta,\phi)$ radiates P_t Watts of power. The power intensity on a sphere with radius d is $\mathcal{S}$ so that a receiver dipole antenna on the same sphere with gain g_r and normalized radiation intensity $F_r(\theta,\phi)$ intercepts P_r Watts given by

$$\mathcal{S} = g_t \frac{\mathrm{P}_t}{4\pi d^2} F_t(\theta,\phi), \tag{15}$$

$$\mathrm{P}_r = \mathcal{S} g_r \frac{\lambda^2}{4\pi} F_r(\theta,\phi) = g_t g_r \mathrm{P}_t \left(\frac{\lambda}{4\pi d}\right)^2 F_t(\theta,\phi) F_r(\theta,\phi). \tag{16}$$

This last equation is known is Friis Transmission Formula, and it can be used for calculating the received power of direct waves as well as waves reflected from specular surface, i.e. smooth surface, which we will investigate later in this section.

Interaction With Objects

Consider an interacting object illuminated by EM wave emitted by a dipole antenna as in Figure 3a. Let us suppose that the target is d_t m away from the antenna radiating P_t Watts with g_t gain. The power intensity at the object's location is, then, given by Eq.(5). Some of the electromagnetic energy incident on the interacting object is reradiated[5] depending on incident and scattering directions. The total reradiated power is given by

$$\mathrm{P}_{\mathrm{rer}} = g_t \frac{\mathrm{P}_t}{4\pi d_t^2} F_t(\theta,\phi)\sigma(\theta,\phi), \tag{17}$$

where $\sigma(\theta,\phi)$ is the *radar cross section* of a point target, which has units of area (m^2). If the reradiated power is intercepted by an antenna d_r m away from the object, then substituting P_t in Eq. (16) with $\mathrm{P}_{\mathrm{rer}}$, we obtain

$$\mathrm{P}_r = g_t g_r \mathrm{P}_t \frac{\lambda^2}{(4\pi)^3 d_t^2 d_r^2} F_t(\theta,\phi) F_r(\theta,\phi)\sigma(\theta,\phi), \tag{18}$$

which is known as the *bistatic radar equation*.

The radar equation pertains to objects that can be attributed as *point objects*. As visualized in Figure 3b, the point object's surface seen by the antenna leads to a solid angle much smaller than the solid angle of the antenna beams. For bistatic configuration, the solid angles of both receiver and transmitters must satisfy the point target definition. One consequence is any large object (e.g. airplane as in Figure 3b) can be seen as a point object sufficiently far away from both transmitter and receiver antennas, and it is possible to represent them with single scalar field, radar cross section, depending on the antenna angles. The radar cross sections of simple point object geometries are given in (Osipov & Tretyakov, 2017, pp. 87-148).

The three-dimensional problem is represented in two dimensions by considering the wave incidence in terms of direction of propagation and the interacting object's surface. The plane that contains both direction of propagation vector and the surface normal is referred to as plane of incidence. In case the object is close to the antennas, the relation between incident wave and the reradiated wave depends on the wavelength of the radiated wave, object's geometry and fine structures on its surface, and its electrical parameters. The geometry of the smooth (curved) surfaces is approximated with a plane if their radius of curvature is greater than the wavelength (Osipov & Tretyakov, 2017, p. 165). Any fine structures around the mean surface is modeled with a height function. These concepts are visualized in Figure 4.

If the interacting object is in the far-field of the antenna, the EM wave is a planar wave, which has transverse electric field intensity and magnetic field vectors. In other words, the components of these vectors can be defined with respect to their alignment to the incidence plane, which contains the direc-

Figure 4. In (a), radiation of electromagnetic wave from a thin line dipole antenna, and its radiation in direction d *and its interaction with an elliptic cylinder standing in the direction of the antenna alignment. In (b), incidence plane (with normal in* $+z$ *direction) that contains both direction of propagation and surface normal of the interacting object. In (c), plane approximation of the elliptic object due to large radius of curvature.*

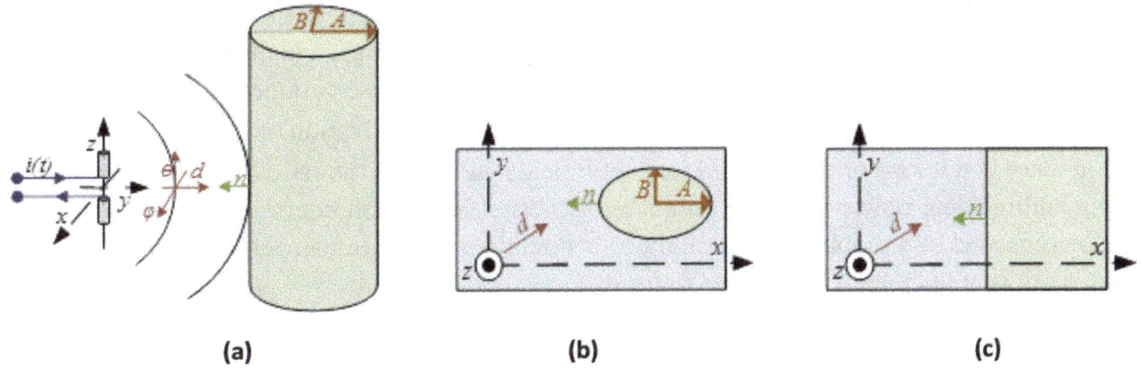

tion propagation vector. It is common to define components parallel to the normal of the incidence plane as *perpendicularly polarized* (perpendicular to the plane) or *horizontally polarized.* The components that are perpendicular to the normal of the incidence plane are defined as *parallel polarized* (parallel to the plane) or *vertically polarized.* Now, for the alignment visualized in Figure 4a, the electric field intensity only has perpendicularly polarized component since the incidence plane has a normal along $+z$ direction.

Large objects (in the sense discussed above) alter the EM wave depending on fine structures on their surfaces. The local variations of the surface are represented as the height function, and the EM interaction mechanisms drastically changes between smooth surfaces and surfaces that have significant (uncorrelated) height variation. A surface of an interacting object is classified with respect to its *degree of roughness*, which is defined as

$$\mathcal{R} = \frac{2\pi}{\lambda} s, \tag{19}$$

for surfaces with root-mean-square (RMS) height of s. The degree of roughness is a measure of how smooth a surface is, and there are different criteria to evaluate the roughness (Pinel, Bourlier, & Saillard, 2010). In radiometric remote sensing literature, Rayleigh criteria and Fraunhofer roughness criteria are checked to assign a surface as smooth (Ulaby, et al., 2014, pp. 421-427):

$$\mathcal{R} < \begin{cases} 0.8 & \text{Rayleigh Criterion} \\ 0.2 & \text{Fraunhofer Criterion} \end{cases} \tag{20}$$

For a large cylinder, the RMS surface height is almost zero so that such a surface is smooth according to both of the criteria.

For perfectly smooth surfaces, the EM wave interaction with an object is governed by the laws of geometrical optics. In particular, the reflection (transmission), and refraction effects can be determined using constitutional parameters of free-space and object's interior, and direction of arrival of propagating incident wave as visualized in Figure 5. Suppose that a plane wave is incident on the object's surface in the direction $\hat{\mathbf{d}}_i$. Some of its energy is reflected in the direction $\hat{\mathbf{d}}_r$, and some of it is transmitted to the second medium and refracted so that it continues to propagate in a new direction $\hat{\mathbf{d}}_t$. The transmission direction depends on the constitutional parameters of the mediums, which are electric permittivity ε, magnetic permeability μ, and conductivity σ. Depending on these parameters and incidence angle θ, Snell's law of reflection ($\theta_i = \theta_r$), Snell's law of refraction, polarization of the incident wave, along with boundary conditions of electric and magnetic fields yields that the ratio between reflected power intensity and incident power intensity is square of Fresnel Reflection coefficient. For the considered scenario composed of a dipole aligned on the $+z$ direction, the wave has perpendicular polarization, and when the first medium is free space (with parameters: permittivity ε_0, permeability μ_0 and conductivity $\sigma_0 = 0$), and the second one is non-magnetic dielectric with zero conductivity, the Fresnel reflection coefficient is given by

$$\Gamma = \frac{\cos(\theta) - \sqrt{\varepsilon_r - \sin^2(\theta)}}{\cos(\theta) + \sqrt{\varepsilon_r - \sin^2(\theta)}}, \tag{21}$$

where $\theta = \theta_i = \theta_r$, and $\varepsilon_1 = \varepsilon_0$, $\varepsilon_2 = \varepsilon_r\varepsilon_0$, $\mu_1 = \mu_2 = \mu_0$ and $\sigma_1 = \sigma_2 = 0$. For more complex dielectric mediums, the conductivity is non-zero making the relative permittivity ε_r a complex quantity; so that Γ is also a complex quantity. Consequently, the reflected power intensity magnitude can be written as

$$\mathcal{S}_r = |\Gamma|^2 \mathcal{S}_i. \tag{22}$$

For smooth surfaces, the reflected EM wave does not change the medium, but the EM field encounters losses since some of its power propagate through the boundary. Furthermore, the boundary conditions of electric and magnetic field imply that some of the components are continuous, and they travel only

Figure 5. Interaction of an electromagnetic wave with smooth objects

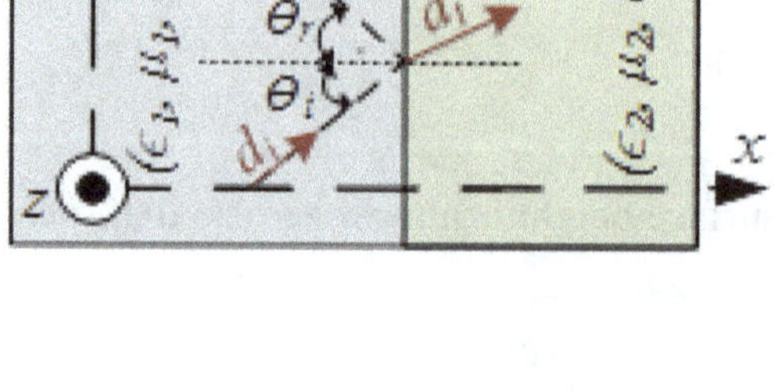

longer distances. In other words, Friis transmission formula in Eq. (16) for wave reflected from a smooth surface can be written as

$$\mathrm{P}_r = g_t g_r \mathrm{P}_t \left(\frac{\lambda \mid \Gamma \mid}{4\pi(d_1 + d_2)} \right)^2 F_t(\theta,\phi) F_r(\theta,\phi),$$

where d_1 is the distance traversed in the first medium before reaching the boundary, and d_2 is the distance traversed after the boundary. In case the object has a non-zero RMS height, the reflected EM energy has two components: coherent and incoherent. The coherent component is the reflected wave due to reflection from the mean surface and non-coherent component is due to scattering of EM energy in different directions (since the surface has height variations). Correspondingly, the energy of the reflected (coherent) component is reduced by an amount given by

$$\Gamma_c = \Gamma \exp\left(-2 \left(\frac{2\pi}{\lambda} s \cos(\theta_i) \right)^2 \right),$$

where θ_i is visualized in Figure 5. The incoherent component has a power intensity magnitude given by

$$\Gamma_{\mathrm{inc}} = \Gamma \sqrt{1 - \exp\left(-4 \left(\frac{2\pi}{\lambda} s \cos(\theta_i) \right)^2 \right)}.$$

Consequently, for large objects having a non-zero roughness, the intensity of the reflected power decreases as the roughness increases. For slightly rough surfaces, however, the coherent component of the reflected EM wave is dominating, and can be observed in RSS measurements.

Human Characterization

The interaction of human body with the EM wave is with respect to interaction mechanisms discussed above. For the EM wave with wavelengths around 10 cm, the human body is almost smooth so that the coherent reflected energy is expected to be dominating. Thus, the amount of reflected power depend on the electrical properties of the human body. Considering humans, clothing is the dielectric boundary, and (dry) human skin is the second reflection layer. Therefore, an accurate model for reflection can be developed by taking into account antenna polarization and gain pattern, electrical properties of multiples of dielectric boundaries, human body geometry and position with respect to link-line.

The relative permittivity of common textile materials are $\varepsilon_r = 1.5 - 2$ at $2.4 - 2.5$ GHz (Sankaralingam & Gupta, 2010), and for human skin $\varepsilon_r = 38$ in the same frequency band (Calculation of the Dielectric Properties of Body Tissues in the frequency range 10 Hz - 100 GHz, 2013). Such values of ε_r implies high reflected RF power for all incidence angles. Furthermore, when the TX antenna is standing in the direction perpendicular to plane of incidence ($+z$ axis in Figure 4), the electric field intensity of the emitted wave is oriented along the same axis so that the reflected waves only have perpendicular polarization, which yields negative Fresnel reflection coefficient for all incidence angles. Consequently, for such configurations, the ratio between reflected signal amplitude and incident signal amplitude Γ is a non-negative real scalar whose sign contributes additional π radians to the phase of the reflected wave.

RSS MODELS

In this section, the available RSS models are reviewed. First, temporal characteristics of human-induced RSS changes are discussed. To aid in the discussion, RSS measurements shown in Figure 6b and Figure 6c are acquired when a robot is moving along a trajectory depicted in Figure 6a. This deployment shows two realizations of the ideal scenario illustrated in Figure 2a. The robot carries a cylindrical container filled with body tissue simulating liquid, and it can accurately localize itself in an indoor environment as described in (Yiğitler & Jäntti, 2016). This experimental setup acquires high quality measurements since geometry and electromagnetic properties of the container and liquid are known, and each RSS measurement can be precisely related to position of the robot. The detailed explanation of the radio hardware can be found in (Yiğitler, Jäntti, & Virrankoski, 2014) and the communication protocol is described in detail in (Yiğitler, Kaltiokallio, & Jäntti, 2013). After giving the temporal characterization, both statistical and deterministic models for each temporal state are given.

Figure 6. In (a), the experimental setup and the robot trajectory (red solid line). The gray areas show the sensitivity regions of the links, and the dashed lines mark the boundary of the shadowing regions. In (b), RSS values of RX_1 and in (c) the RSS measurements of RX_2 are shown. In (b) and (c), the measured $r(k)$ in different communication channels, average $\bar{r}(k)$ over different communication channels, and mean of the modeled RSS $\bar{g}(k)$ are shown. In the figure, reflection and absorption models are evaluated using $A = B = 0.1575\text{m}$, $\varepsilon_r = 2.26$, $\eta = 2.0$, $\rho = 100\text{dB/m}$, $d = 2.8\,m$ for RX_1 and $d = 8.2\,m$ for RX_2.

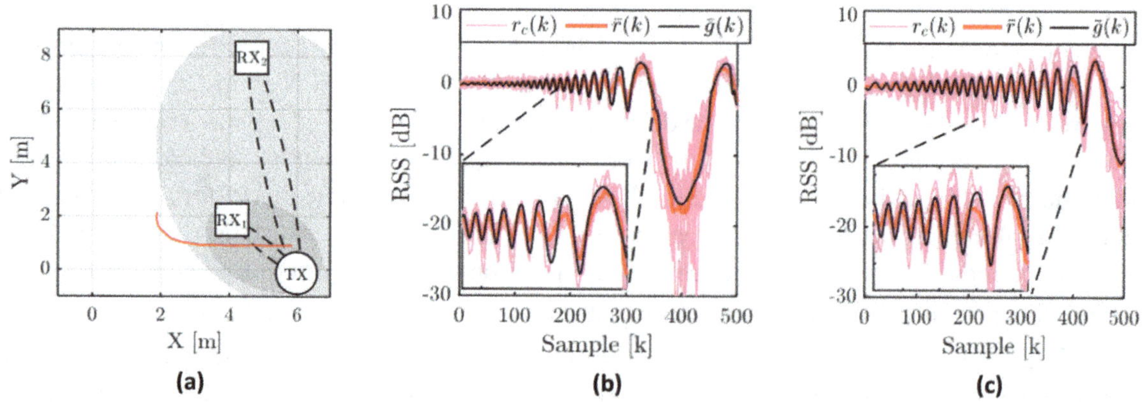

Temporal Characterization

Consider a propagation scenario with a single person in a completely empty environment, as shown in Figure 2a. In addition to the unaltered LoS EM wave, some new waves will be created due to reflection and scattering from the person's surface. Since the propagation speed c_0 is constant, the LoS wave reaches the RX antenna first, and the coherent receiver synchronizes to it. Then, the phases of the signals arriving later are defined with respect to arrival time of the synchronized signal. Therefore, a signal traversing an excess path length Δ m compared to the path length traversed by the LoS signal has a time delay of $\tau = \Delta / c_0$ s. All waves scattered from any surface around an elliptic region having the transceivers located at the foci have the same time delay. Therefore, for fixed node locations, excess path length parameterizes an ellipse, and defines the time-delay of the waves.

A typical radio is only sensitive to the signals with strength above sensitivity threshold of the receiver. Considering the propagation losses only, the radio's sensitivity dictates the maximum excess path length a signal can travel before it falls below the threshold, defining a *sensitivity region* of a TX-RX pair (Chang & Sahai, 2004). If a person is moving outside the sensitivity region, the multipath components affected by the person has very small strength. Thus, the effective propagation channel remains stationary and a single realization of the fading process is observed, and the noise $\nu[k]$ dictates $r[k]$ in Eq.(12). For RX_1, this state is observed in Figure 6b at samples $k = [0, 50]$, and the trajectory of the robot at these time samples is outside the sensitivity region of RX_1 shown in Figure 6b.

If the person starts to move toward the link-line, at some point they enter the sensitivity region of the link. After this, human-induced temporal fading can be observed at the RX due to the interaction of the human body with the EM waves through the mechanisms discussed in the previous section. For large interacting objects such as the human body, both coherent and incoherent scattering[6] (Ulaby, et al., 2014, pp. 194-201) affect the received signal. However, considering average adult human body geometry excited by an Ultra High Frequency (UHF) EM wave, it is expected that the reflection (coherent scattering) is dominating as earlier reported (Ghaddar, Talbi, Denidni, & Sebak, 2007). The reflected waves can cause constructive and destructive fading depending on its excess path length Δ. Moreover, as Δ decreases the amplitude of the reflected signal increases so that the RSS is altered more and more (Rappaport, 2002, pp. 114-125). This state is observed when the robot is within the sensitivity region of the links shown in Figure 6, which corresponds to samples $k = [50, 350]$ in Figure 6b and samples $k = [50, 460]$ in Figure 6c. The results shown in Figure 6 imply that the RSS variations also depend on the link distance, and this should be accounted for in the modeling effort.

The considered propagation process changes significantly as the person gets closer to the link-line. In particular, as the excess path length Δ of the reflected wave gets smaller than a quarter of a wavelength, the RF signal(s) traverse through and around the human as they diffract and refract upon contact to non-homogeneous human tissue. When the person is in the just mentioned region but not obstructing the LoS, and if the human body is assumed to be a perfect absorber, then refraction can be ignored. In this case, the LoS signal experiences significant diffraction due to edges of the human body geometry (Keller, 1962). The observed impact on the electric field can be approximated by knife-edge diffraction, which can be equivalently stated using the excess path length of the diffraction edge as its argument (Durgin, 2009). The accuracy of the model can be improved by considering both vertical and lateral diffraction profile of the human body (Rampa, Savazzi, Nicoli, & DÁmico, 2015), (Mokhtari & Lazaridis, 1999).

As the person finally reaches the link-line and obstructs the LoS, a significant RSS decrease is observed as illustrated in Figure 6b at samples $k = [350, 450]$ and in Figure 6c at samples $k = [460, 500]$. When the person is in this state, his/her body effects the EM waves with various mechanisms – possibly simultaneously. The EM wave may transmit through and may bend around the human body at the same time, if one ignores the complicated mechanisms taking place while the wave traverses through inhomogeneous body tissue. Due to the observed large decrease in signal strength, this temporal state is widely referred to as human-induced *shadowing* (Patwari & Agrawal, 2008), (Li, Chen, Coates, & Yang, 2011), (Guo, et al., 2015).

For the described movements of the person, the RSS measurements experiences transitions among four different states as depicted in Figure 7. Although the transition from non-fading state to fading state (where reflection is the dominating propagation mechanism) is defined by the radio hardware, the transition from reflection dominated state to diffraction dominating state or to shadowing state are not well defined since more than one mechanisms are effective simultaneously. This problem is usually ignored since for a UHF wave, these transitions are within a half of a wavelength distance (e.g. for carrier frequencies around 2.4 GHz this distance is approximately 6 cm).

The temporal state dependence of the RSS measurements in Eq. (12) is solely due to the propagation channel variation, which alters channel gain $\alpha[k]$ and correspondingly SNR $\sigma[k]$. For notational convenience, let us define the RSS measurement models in temporal state s dependent form as follows:

$$r[k] = \nu[k \mid s] + \begin{cases} \mathcal{N}[k] & s = \text{Non}-\text{fading} \\ \mathcal{R}[k] & s = \text{Reflection} \\ \mathcal{D}[k] & s = \text{Diffraction} \\ \mathcal{S}[k] & s = \text{Shadowing} \end{cases} \tag{23}$$

where the notation of $\nu[k \mid s]$ indicates that the noise process may have different characteristics depending on the state. In the remaining part of this section, both statistical and deterministic models for all states are revised.

Figure 7. Temporal states of human-induced electromagnetic wave propagation mechanisms

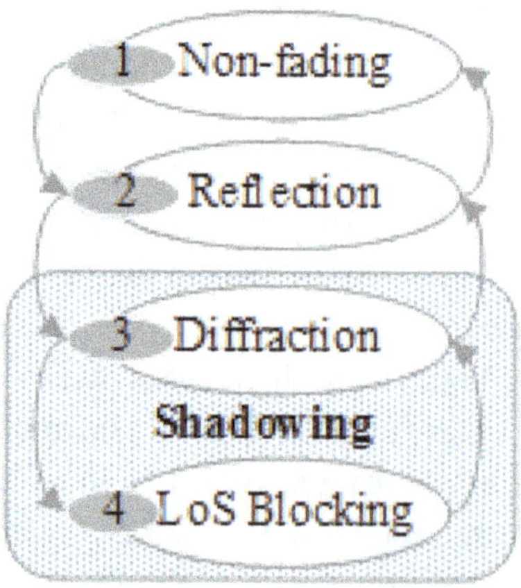

Non-Fading State

For the ideal scenario visualized in Figure 2a, suppose that TX and RX are separated far enough so that their antennas are in their far field regions. If the person is outside of the sensitivity region of the link, and both TX and RX antennas are ideal, the conditions of the well-known Friis transmission formula (Friis, 1946) are satisfied. Thus, the channel gain can be written as

$$\mathcal{P}_r = \mathcal{P}_0 - 10\eta\log_{10}(d / d_0) + \nu, \tag{24}$$

where $\mathcal{P}_0$ is the RSS at reference distance d_0, and η is path-loss exponent generalizing the Friis formula for more complex environments compared to free space. Therefore, the reference measurements in a static environment can be parameterized using link length and path loss exponent η. This model is also known as large-scale propagation model and log-distance path loss model (Rappaport, 2002, pp. 161-162). In the light of Eq. (11) and for $g_t g_r \gg 2$, the channel gain can be written as

$$\alpha_r = \alpha_0(d / d_0)^{-\eta/2}, \tag{25}$$

where α_0 is the channel gain at the reference distance d_0.

In non-fading state, there are no other changes in the propagation medium as compared to the medium when the reference measurements are acquired. In other words, the propagation channel has the same realization; so that we have

$$\gamma[k] = \gamma_r, \qquad \mathcal{N}[k] = 0, \qquad r[k] = \nu[k]. \tag{26}$$

Therefore, the RSS has the same statistics as the noise process $\nu[k]$, which is defined by the statistics of $\bar{X}_K$ defined in Eq. (5).

By definitions, $\bar{X}_K$ is a quadratic form of Gaussian random variables, whose properties have been studied in (Turin, 1960). This quadratic form yields a non-central chi-square random variable with $2K$ degrees of freedom and non-centrality parameter $K\gamma$ (Patnaik, 1949). However, the logarithm of a non-central chi-square random variable has a very complicated density function, as given in (Yiğitler, Jäntti, & Patwari, 2017), leading to investigating approximate distributions with tractable density functions. One option is to approximate $\bar{X}_K$ with a (central) chi-squared (or equivalently Gamma) random variable as proposed earlier (Patnaik, 1949), and used in energy based detector works (Urkowitz, 1967). The central chi-square approximation contains a scale parameter and determines the degrees of freedom, which has later been improved in (Pearson, 1959) and in (Hall, 1983) by including a shift term in the approximation form. The shift term, however, may yield negative values for small arguments, making the distribution undefined. Once Patnaik's central chi-square approximation is accepted, it is a simple matter to show that the analysis of Bartlett and Kendall on logarithmic transformation (Bartlett & Kendall, 1946) is valid, and RSS $\mathcal{P}$ approximately has normal distribution, with mean and variance given as

$$\begin{aligned}
\mathrm{E}\{\gamma\} &\triangleq \mu_{\gamma} = 10\log_{10}\left(\sigma^2 \frac{2}{K}\frac{1+\gamma}{2+\gamma}\right) + \frac{10}{\ln(10)}\psi^{(0)}\left(\frac{K}{2}\frac{\left(2+\gamma\right)^2}{1+\gamma}\right), \\
\mathrm{V}\{\gamma\} &\triangleq \sigma_{\gamma}^2 = \frac{100}{\left(\ln(10)\right)^2}\psi^{(1)}\left(\frac{K}{2}\frac{\left(2+\gamma\right)^2}{1+\gamma}\right),
\end{aligned} \tag{27}$$

where $\psi^{(r)}(\cdot)$ is the polygamma function of order r ($\psi^{(0)}(\cdot)$ is the digamma function) (Abramowitz & Stegun, 1970), and the impact of quantization are ignored. When the impact of quantizer with quantization level Q is included, these two moments become

$$\mu_Q = \mu_{\mathcal{P}} + \frac{Q}{\pi}\sum_{k=1}^{\infty}\frac{(-1)^k}{k}\sin\left(2\pi k\frac{\mu_{\mathcal{P}}}{Q}\right)e^{-\frac{(2\pi k)^2}{2}\frac{\sigma_{\mathcal{P}}^2}{Q^2}}, \tag{28}$$

$$\begin{aligned}
\sigma_Q^2 = \; & \sigma_{\mathcal{P}}^2 - (\mu_Q - \mu_{\mathcal{P}})^2 + \frac{Q^2}{12} \\
& +4\sigma^2\sum_{k=1}^{\infty}(-1)^{k+1}\cos\left(2\pi k\frac{\mu_{\mathcal{P}}}{Q}\right)e^{-\frac{(2\pi k)^2}{2}\frac{\sigma_{\mathcal{P}}^2}{Q^2}} \\
& +\frac{Q^2}{\pi^2}\sum_{k=1}^{\infty}\frac{(-1)^k}{k^2}\cos\left(2\pi k\frac{\mu_{\mathcal{P}}}{Q}\right)e^{-\frac{(2\pi k)^2}{2}\frac{\sigma_{\mathcal{P}}^2}{Q^2}}.
\end{aligned} \tag{29}$$

As a consequence of log-normal RSS, both $\mathcal{P}$ and $\mathcal{P}_r$ are normal. Then, the RSS measurement in Eq. is also a normal random variable.

The data visualized in Figure 6 for experiments of different link lengths ($d = 3$ m and $d = 9$ m) are used for evaluating the log-normality of RSSI measurements in Figure 8. The mean of RSSI measurements of different channels are removed, and the normal pdf is shown along with the data histograms. As expected, the pdf is a quite good fit with a variance value higher than the one imposed by the quantizer ($Q^2 / 12$).

Fading State

In the fading state, the interacting object of interest scatters the incident EM energy using different mechanisms depending on the its distance to the link-line. In fading state, two models for RSS are given: reflection model and diffraction model. The reflection model is used and elaborated in (Yiğitler H., Jäntti, Kaltiokallio, & Patwari, 2017), (Kaltiokallio, Yiğitler, & Jäntti, 2017), and diffraction based model is formulated (in different forms) in (Savazzi, Nicoli, Carminati, & Riva, 2014), (Rampa, Savazzi, Nicoli, & DÁmico, 2015), (Wang, Liu, Xu, Bu, & An, 2015). These two models are parameterized using excess path length Δ defined as

Figure 8. The histogram and normal density fit for non-fading state measurements for the empirical data visualized in Figure 6

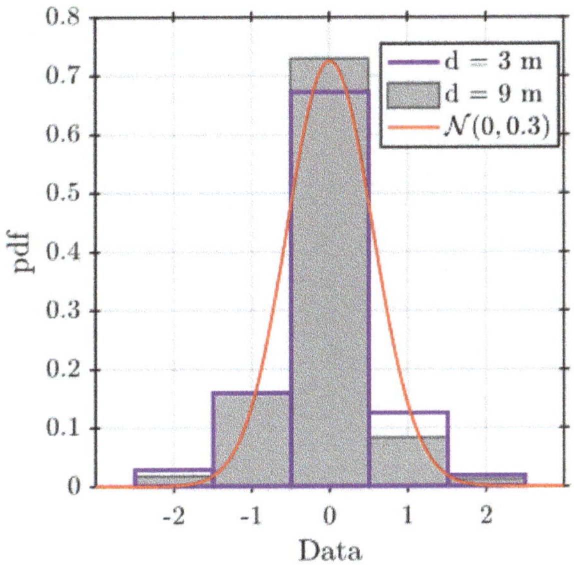

$$\Delta \triangleq d_{TX} + d_{RX} - d, \tag{30}$$

where d_{TX} and d_{RX} distances are visualized in Figure 9a and Figure 9b.

An empirical evaluation of these two models for the data visualized in Figure 6 is shown in Figure 10. Although the reflection model performs better in amplitude for both distances, the diffraction model is continuous within the entire domain whereas reflection model is not valid when the robot obstructs the link-line. Therefore, both models have advantages and disadvantages.

Figure 9. In (a), the reflection model geometry. In (b), the diffraction model geometry. In (c), the absorption model geometry.

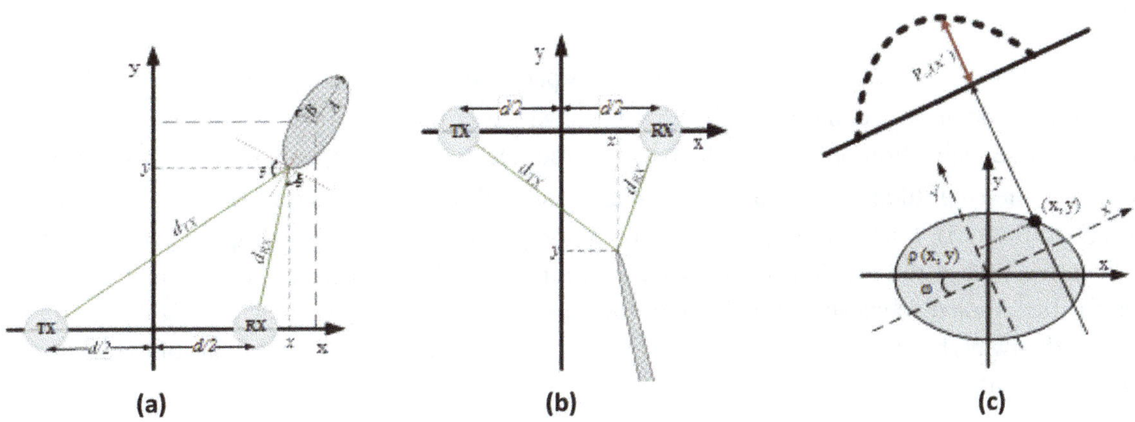

Figure 10. Fading state model comparison using empirical data in Figure 6

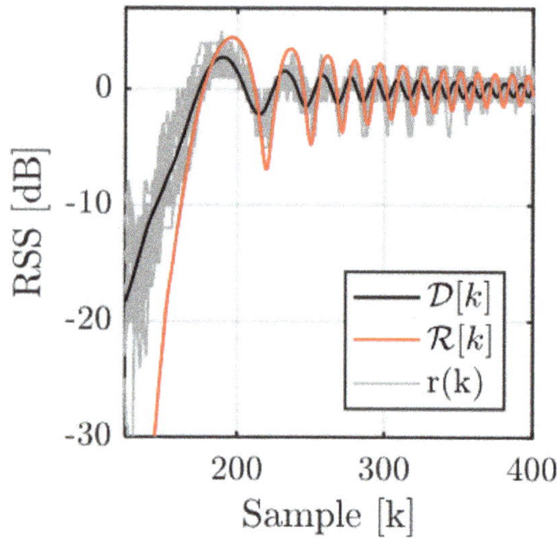

(a) d = 2.8 m

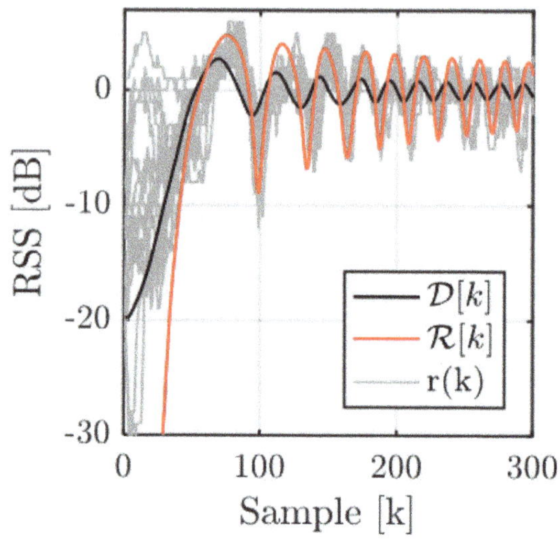

(b) d = 8.2 m

Reflection Model

Consider the radio wave propagation scenario depicted in Figure 2a, where the propagation environment is completely empty except the interacting object and the TX-RX pair. Suppose that the communication system satisfies the assumptions stated in the system overview section. In case the object is within the sensitivity region of the link, some of the electromagnetic energy emitted by the TX is reflected from the surface of the dielectric interacting object and, there are two multipath components impinging on the receiver antenna: the LoS signal and the single-bounce reflected signal. Although a narrowband receiver cannot discriminate between two components, the LoS signal arrives first to the receiver antenna and the delay of this component τ_1 is zero. Then, the received signal for the transmitted narrowband signal $s(t)$ is given by

$$x(t) = \alpha_1 s(t) + \alpha_2 e^{-j\phi} s(t - \tau_2) + \hat{n}(t), \tag{31}$$

where $\phi = 2\pi f_c \tau_2 = 2\pi \Delta f_c / c_0$ and c_0 is the free-space electromagnetic propagation speed.

The amplitude of both components in Eq. (31) have the same decay rate with distance if both experience the same realization of fading, which can be modeled as a site-dependent path-loss exponent η. In this case, the amplitudes of the components decrease with $(-\eta / 2)^{\text{th}}$ exponent of the traversed path-length as in Eq. (25). Furthermore, as discussed in the previous section, the amplitude of the reflected wave is a fraction of the incident wave, which is represented by the Fresnel reflection coefficient Γ. Since for the studied scenario, both TX and RX antennas have perpendicular polarization; so the reflected path always has a π radian additional phase. Consequently, the channel gains of the components can be written as

$$\alpha_1 \approx d^{-\eta/2}, \qquad \alpha_2 \approx -\Gamma\left(d+\Delta\right)^{-\eta/2}, \tag{32}$$

where Δ is the excess path length traversed by the reflected signal. Since the static channel measurements in the non-fading state also has the same channel gain α_1, the RSS model in the reflection state can be written as

$$\mathcal{R}[k] = 10\log_{10}\left(1+\frac{\Gamma^2}{(1+\Delta[k]\,/\,d)^{\eta}}\right)+10\log_{10}\left(1-2\Gamma\frac{(1+\Delta[k]\,/\,d)^{\eta/2}}{\Gamma^2+(1+\Delta[k]\,/\,d)^{\eta}}\cos(2\pi\Delta[k]\,/\,\lambda)\right). \tag{33}$$

Diffraction Model

The diffraction model is based on the physical optics approximation of the complex diffraction phenomenon. In environmental sensing literature, the diffraction based models appeared so far are based on the knife-edge diffraction, which is developed based on Kirchhoff's diffraction theory presented for example in Sec. 8.3.2 of (Born & Wolf, 1975). The theory is valid under the assumption that Kirchhoff's boundary conditions are valid, i.e., the EM energy incident on the interacting object having the edge is completely absorbed and the EM field not-incident on the edge is the same as the field observed in the free space (when the diffracting object is not in the environment).

For the parameter definitions shown in Figure 9b and Eq. (30), the effect of diffraction on the RSS can be written from the knife-edge diffraction formulation (Durgin, 2009) as

$$\mathcal{D}[k] = 10\log_{10}\left(\left|\frac{1+j}{2}\mathrm{F}\left(2\sqrt{\frac{\Delta}{\lambda}}\right)\right|^2\right),$$

where $j=\sqrt{-1}$ and $\mathrm{F}(\cdot)$ is the Fresnel integral

$$\mathrm{F}\left(a\right)=\int_a^{\infty}\exp\left(j\frac{\pi}{2}\tau^2\right)d\tau=\frac{1+j}{2}-\left(C(a)+jS(a)\right),$$

with

$$C(a)=\int_0^a\cos\left(\frac{\pi}{2}\tau^2\right)d\tau, \qquad S(a)=\int_0^a\sin\left(\frac{\pi}{2}\tau^2\right)d\tau.$$

Using these definitions, the diffraction model can be written as

$$\mathcal{D}[k] = 10\log_{10}\left(\frac{1}{2}\left[\left(\frac{1}{2} - C\left(\sqrt{2\pi\frac{\Delta}{\lambda}}\right)\right)^2 + \left(\frac{1}{2} - S\left(\sqrt{2\pi\frac{\Delta}{\lambda}}\right)\right)^2\right]\right). \tag{34}$$

The model in Eq. (34) assumes that the diffraction edge extends from $-\infty$ to a point defining the argument of the Fresnel integral $2\sqrt{\Delta / \lambda}$. For an upside-down configuration, the Fresnel integral must be calculated using $(1 + j) / 2 + (C(a) + jS(a))$, and accordingly the model reads as

$$\mathcal{D}[k] = 10\log_{10}\left(\frac{1}{2}\left[\left(\frac{1}{2} + C\left(\sqrt{2\pi\frac{\Delta}{\lambda}}\right)\right)^2 + \left(\frac{1}{2} + S\left(\sqrt{2\pi\frac{\Delta}{\lambda}}\right)\right)^2\right]\right). \tag{35}$$

The model in Eq. (34) (and Eq. (35)) assumes a single diffraction point and the other edge at the infinity. If the diffracting object is small so that both edges are at comparable distances from link-line, the models above are not valid. The result for this case can be derived as it has been done by Born and Wolf in Sec. 8.7 of their seminal book (Born & Wolf, 1975). However, one crude model can be easily obtained by summing the field intensities due to different edges as follows:

$$\begin{aligned}\mathcal{D}[k] = 10\log_{10}(& \frac{1}{2}\left[\left(\frac{1}{2} \pm C\left(\sqrt{2\pi\frac{\Delta_1}{\lambda}}\right)\right)^2 + \left(\frac{1}{2} \pm S\left(\sqrt{2\pi\frac{\Delta_1}{\lambda}}\right)\right)^2\right] + \\ & \frac{1}{2}\left[\left(\frac{1}{2} \pm C\left(\sqrt{2\pi\frac{\Delta_2}{\lambda}}\right)\right)^2 + \left(\frac{1}{2} \pm S\left(\sqrt{2\pi\frac{\Delta_2}{\lambda}}\right)\right)^2\right]),\end{aligned} \tag{36}$$

where Δ_1 is excess path length of the first edge and Δ_2 is the excess path length of the second edge, the $\pm$ value is defined by the edge configuration with respect to the reference frame as discussed above.

It is to be noted that when the extent of the interacting object along the lateral direction (the z-axis for the reference frame shown in Figure 9) is comparable to wave length, the lateral diffraction effects must be taken into account. For typical deployments of environmental sensing applications, the antennas are placed around chest or abdominal area of the person so that this effect can be ignored.

Statistical Model

The amplitude of the reflected signal depends on the position, geometry, and electrical properties of the person, whereas time-delay τ_2 solely depends on the excess path length traversed by the reflected signal Δ. Thus, in the considered scenario there are two clusters of multipath components with different amplitudes and arrival times. If the region is free of additional scatterers, the amplitude decay of the clusters is the same and therefore, the received signal power is obtained as the modulus sum of the multipath components raised to a certain exponent. The described multipath propagation effect is known to yield Weibull distributed fading (Yacoub, 2002), (Sagias & Karagiannidis, 2005) for which the density function with shape parameter b and scale parameter a is given by

$$f_{r|s}\left(r \mid a, b\right) = \frac{b}{a}\left(\frac{r}{a}\right)^{b-1} \exp\left(-\left(\frac{r}{a}\right)^{b}\right). \tag{37}$$

In case the envelope follows a Weibull distribution, the signal power also follows the Weibull distribution with shape parameter $b / 2$ (Simon & Alouini, 2005, p. 26). Therefore, the RSS in linear scale is expected to be Weibull distributed.

It is to be noted that the RSS in logarithmic scale $\mathcal{P}$ has a distribution of logarithm of Weibull random variable. Due to the domain of Weibull distribution, it is possible to find a closed form of the probability density function, but that effort does not bring any further practical gain.

When the interacting object is not shadowing the link-line but the object is close to it, the diffraction has similar effect as the reflection. In this case, the receiver is in the reflection region of the transmitter (Vaughan, 2003, pp. 123-140), and the statistical arguments above are all valid. However, when the person starts to block the link-line, the process alters and the arguments of the next subsection are valid.

Shadowing State

RF signals can diffract, scatter, reflect and attenuate upon contact with an interacting object, making it a demanding task to accurately model RSS changes in the shadowing state. One option is to ignore transmission of the EM energy through the interior of the interacting object, and consider only diffraction around the body. In this case one can use rigorous theory of diffraction to find correct model, or use approximate models discussed above. Another alternative is to model transmission through the body by using linearity of the losses in logarithmic scale as earlier has been assumed in radio tomographic imaging works, e.g. (Wilson & Patwari, 2010). In this subsection, we take the latter approach and review an interacting object body absorption based RSS model, introduced in (Kaltiokallio, Yiğitler, & Jäntti, 2017).

Absorption Model

The absorption based model can be derived using the basic principle of the Radon transform (Kak & Slaney, 1988, pp. 49-56), which is a *line integral* of the attenuation field $\rho(x, y)$ along a straight line from TX to RX as visualized in Figure 8c. The total attenuation along the line

$$y' = x\cos\omega + y\sin\omega - x' \tag{38}$$

can be written as

$$P_{\omega}(x') = \int_{-\infty}^{\infty}\int_{-\infty}^{\infty} \rho(x, y)\delta(x\cos\omega + y\sin\omega - x')dxdy, \tag{39}$$

where $\delta(\cdot)$ is Dirac delta function. If the interacting object has an elliptic cross section with uniform electrical properties, i.e., $\rho(x,y)=\rho$, the line integral in Eq. (39) yields

$$P_{\omega}(x')=\begin{cases}\dfrac{2\rho AB}{a^2(\omega)}\sqrt{a^2(\omega)-(x')^2} & \text{if } |\, x' \,| \leq a(\omega)\\ 0 & \text{otherwise}\end{cases} \tag{40}$$

where A and B are the semi-minor and semi-major axis and $a^2(\omega)=A^2\cos^2(\omega)+B^2\sin^2(\omega)$.

The total absorption model in Eq. (40) cannot model the empirical observation that the signals are attenuated more when the person is close to the transceivers. This finding can be explained through Fresnel zones which are concentric ellipsoids with radius

$$R_n=\sqrt{\frac{n\lambda d_{TX}d_{RX}}{d_{TX}+d_{RX}}}, \tag{41}$$

where n is the Fresnel zone number, and d_{TX} and d_{RX} are the distances to the TX and RX in respective order. RF signals traversing through space have a greater flux density in W/m^2 the smaller the Fresnel radius is. Thus, the overall attenuation is relative to the area occupied by the person. Since shadowing occurs only inside the first Fresnel zone ($n=1$), this can be approximated by width of the person obstructing the first Fresnel zone divided by the radius of the first Fresnel zone, $\kappa(y)=A/R_1$.

For the considered deployment scenario, the frame of reference is defined with respect to the TX-RX pair resulting that $\omega=0$ and $x'=x$. Now, the measured decrease in RSS is equivalent to summing up the losses along the link-line. As a result, the deterministic *absorption model* can be expressed as

$$\mathcal{S}[k]=-\kappa(y)P_{\omega}(x). \tag{42}$$

Empirical Models

Another modeling option is to define an empirical model. In literature, there are two well-known empirical models: *exponential model* (EM) (Li, Chen, Coates, & Yang, 2011) and *exponential Rayleigh model* (Guo, et al., 2015). In the exponential model, the RSS is modeled to decay exponentially with respect to excess path length Δ as follows

$$\mathcal{S}[k]=\beta_0 e^{-\Delta/\gamma}, \tag{43}$$

where β_0 and γ are model parameters determined using empirical data. The exponential-Rayleigh model consists of two exponential decays given by

$$\mathcal{S}[k]=\beta_0 e^{-\Delta/\gamma_0}+\beta_1\Delta e^{-\Delta^2/\gamma_1}, \tag{44}$$

where the first exponential decay captures the large losses caused by shadowing whereas the second captures the constructive fading effects caused by reflection when a person is in the close proximity of the LoS.

The presented three shadowing state models are compared, using the data presented in the beginning of this section (visualized in Figure 6), in Figure 11. The empirical models are trained with the data of $d = 2.8$ m experiment, and the same parameters are used for evaluating $d = 8.2$ m experiment. The result shows that these models do not adapt to configuration changes. On the other hand, the figure also shows that the absorption model does not have this disadvantage.

Statistical Model

As electromagnetic waves propagate through a person, they are reflected, diffracted and scattered multiples of times due to differences in electrical properties of human tissue. However, most tissues' size are comparable to or less than the wavelength resulting that the path length traveled by different waves is approximately equal. Consequently, the phase variation in the received signal is negligible and it is possible to argue that the phase is deterministic. Thus, the observed effect is solely on the amplitude so that power of the received signal is significantly reduced.

One approach to model amplitude fading in shadowing state is to consider the power transmitted through a person to be mostly scattered by human tissue. In this case, the electromagnetic waves are scattered by a large number of irregular surfaces (tissue). As a consequence, the received power intensity profile is expected to follow a gamma variate (Jakeman, 1982). Similar arguments hold for slow-fading in mobile propagation channels where the gamma distribution has been proposed to characterize shadowing (Abdi & Kaveh, 1999). Therefore, the RSS in linear scale can be assumed to follow the gamma distribution with shape parameter b and scale parameter a with density function

Figure 11. Shadowing state model comparison using empirical data in Figure 6

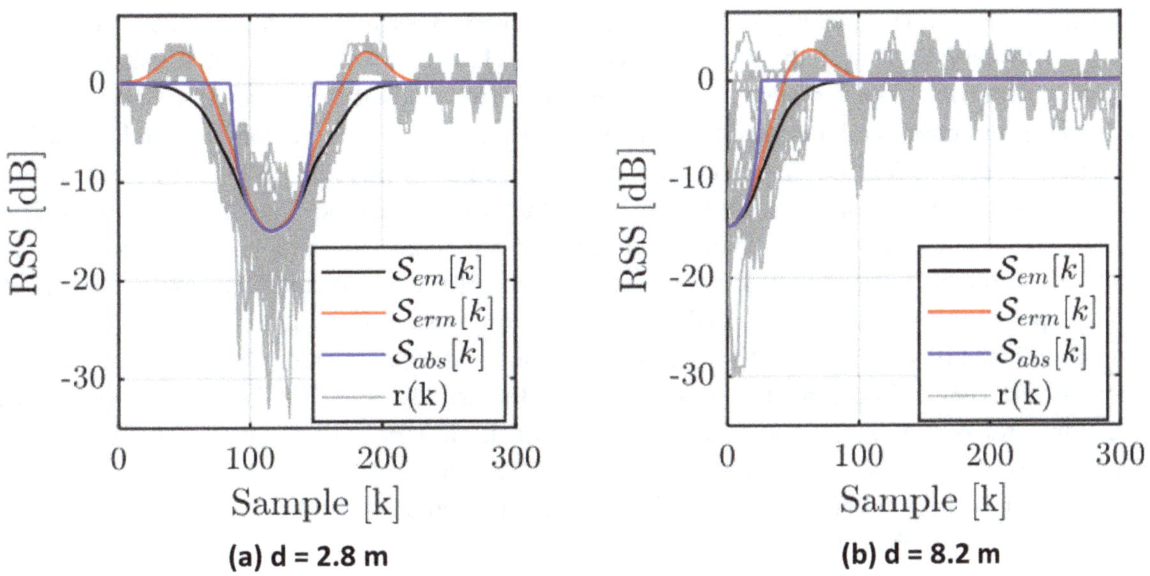

$$f_{r|s}\left(r \mid a,b\right) = \frac{r^{b-1}}{a^{b}\Gamma(b)}\exp\left(-\frac{r}{a}\right), \tag{45}$$

where $\Gamma(\cdot)$ is the gamma function.

It is to be noted that the arguments presented in the non-fading state is based on gamma approximation of signal power in linear scale $\bar{X}_k$ defined in Eq.(5). Since the signal SNR is significantly reduced in shadowing state, the gamma variate approximation improves, making the RSS in logarithmic scale approximately normal. Therefore, it is also possible to use more convenient log-normal distribution for RSS in shadowing state.

APPLICATIONS

In this section, three device-free localization systems are briefly presented and demonstrated. The first system to be introduced is the link line monitoring system presented in (Kaltiokallio, Yiğitler, & Jäntti, 2017) which aims to estimate the temporal state of the propagation channel and track coordinates of the person using as few sensors as possible. The paper models the RSS for three temporal states, and the states are: non-fading, fading and shadowing. For the three states, statistical models are used by a hidden Markov model to estimate the temporal state of the channel. On the other hand, to relate the person's location to the RSS, a single-bounce reflection model is proposed for the fading state, and the theory of computerized tomographic imaging is used for deriving a shadowing model. The state of the person is tracked using a particle filter. The results of the paper demonstrate that a single TX-RX pair can be used for detecting the presence of the person, and probability of detection increases as the person moves closer to the LoS. Once the number of receivers is increased to two, the system can locate and track the person. In Figure 12, two examples of the estimated trajectories using the three-state, exponential and exponential-Rayleigh models are shown. On average, the position estimates are more accurate using the three-state model than with the other two models. As the transceiver distance grows, the enhancement in performance increases. The tracking accuracy increases in every experiment when more receivers are used.

The detector-based RTI system introduced in (Yiğitler H., Jäntti, Kaltiokallio, & Patwari, 2017) is a computationally efficient DFL system that can be implemented on constrained devices. In the paper, the single-bounce reflection model is used to relate RSS variations to a large region around the link-lines. Then, based on the model, the system components are presented including a classifier, a detector, a back-projection based reconstruction algorithm, and a localization method. The detector-based RTI has the following advantages over the other imaging based methods: i.) a simple image reconstruction method that is straight forward to implement; ii.) significantly lower computational complexity such that no floating-point multiplication is required; iii.) data are compressed to a single bit for each link providing improved scalability; iv.) physically significant and repeatable parameters. Despite simplicity of the system, the performance is comparable to other state-of-the-art DFL systems. As an example, the mean localization error is $0.3096\mathrm{m}$ in an apartment experiment, an increase of just 30% with respect to another state-of-the-art system (Kaltiokallio, Bocca, & Patwari, 2014). An example image estimated using detector-based RTI in an apartment experiment is illustrated in Figure 13a and the estimated trajectory is shown in Figure 13b.

Figure 12. Example trajectories using the exponential (black line) (Li, Chen, Coates, & Yang, 2011), the exponential-Rayleigh (grey line) (Guo, et al., 2015) and three-state (dashed red) (Kaltiokallio, Yiğitler, & Jäntti, 2017) models. In the figure, the blue triangles indicate the initial position of the person and the blue dashed line depicts the trajectory.

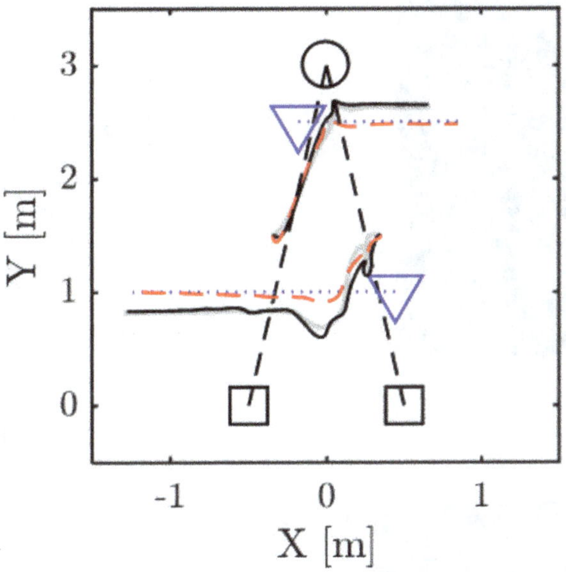

The adaptive radio tomographic imaging (ARTI) system presented in (Kaltiokallio, Jäntti, & Patwari, 2017) aims to maximize the tracking accuracy of the system. It is well known in DFL literature that the RSS can decrease, increase or remain unchanged as the link line is blocked. Moreover, the sensing region of the links vary implying that the model parameters of each link is unique. Spatial models used in DFL are typically set a priori and do not change, which degrades the tracking performance because of the mismatch between the model and the measurements. ARTI provides a means for a RTI system to bootstrap to improve its spatial models using unlabelled data, iteratively improving itself over time. This is done by using a collection of tracking filters to improve the accuracy of image and coordinate estimates. In addition, a smoothing method is presented to fine-tune the model with a given finite latency. The results demonstrate that ARTI achieves high accuracy in various environments, the state estimates can be enhanced using Bayesian smoothing, and improved localization accuracy can be achieved using batch-training. An example propagation field image estimated with ARTI in an apartment experiment is illustrated in Figure 13c and the estimated trajectory is shown in Figure 13d. In this experiment, the mean localization error is 0.1785m .

CONCLUSION

In this chapter, received signal strength models for narrowband receivers are revised. First, the measurement system is introduced, and associated components are discussed to reach a practical received signal strength model for logarithmic scale measurements. Then, underlying radio frequency wave propagation

Figure 13. Example images obtained using detector-based RTI (a) and adaptive RTI (c) in an apartment experiment. The true and estimated trajectories in the experiment with detector-based RTI (b) and adaptive RTI (d).

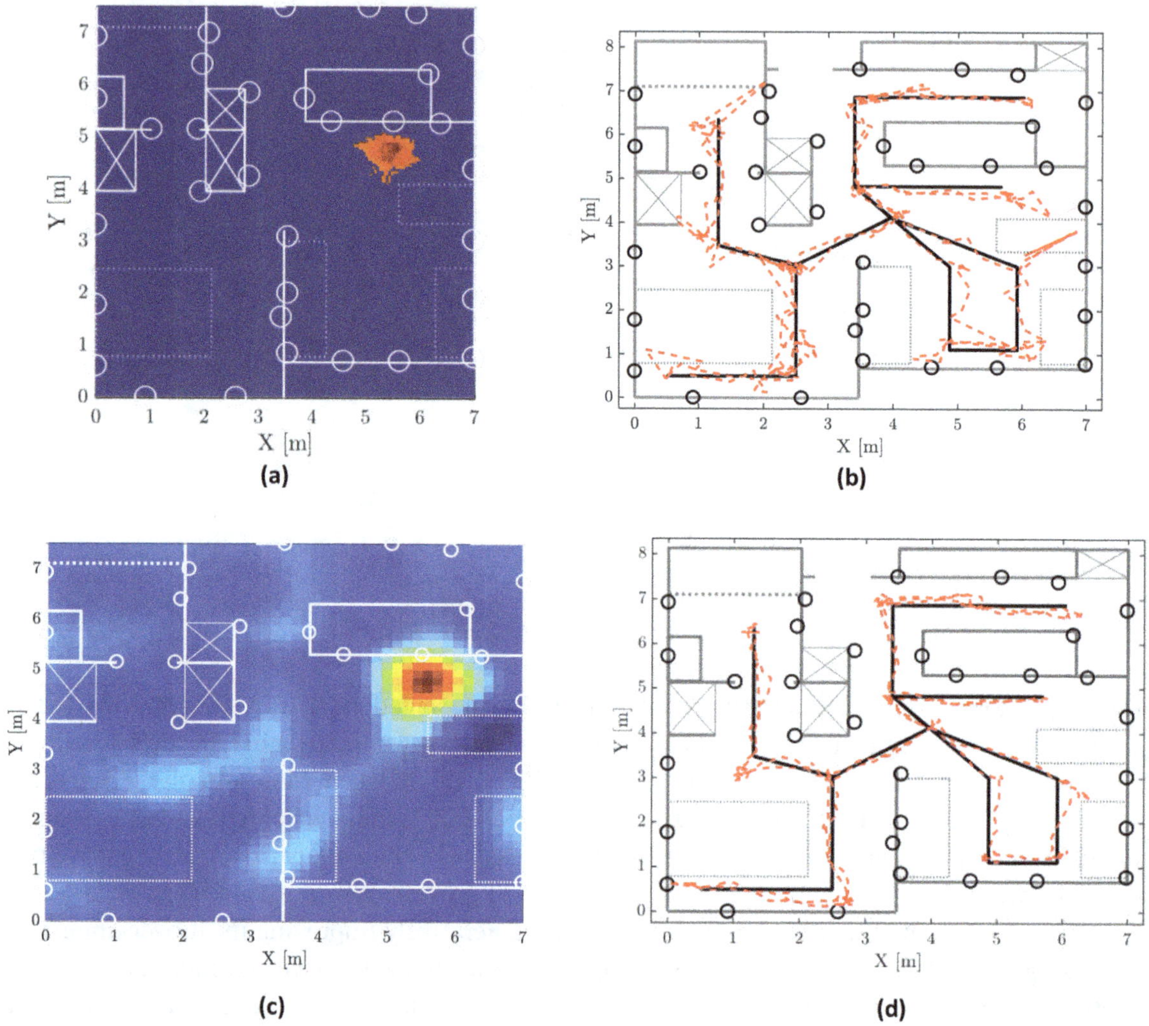

physics are briefly revised to highlight underlying assumptions, and required operating modes of the wireless communication systems. The models are revised after identifying temporal states of the observed measurements. For each state both deterministic models and statistical models are given. Finally, results of some example applications are demonstrated to show the importance of the modeling effort.

The available models have their own advantages and disadvantages. For example, when a continuous model is needed the diffraction based model can be used. However, it must be taken into account that the knife-edge diffraction model makes some assumptions, which cannot generalize for some practical scenarios or make it impossible to reach required level of modeling accuracy. Similarly, relative importance of coherent (reflection) and scattering (rough surface or diffuse scattering) depends on the fine structures of the interacting object's surface as well as the distance of the object to the nodes, making it a challenging task to select a general model to cover both cases. Furthermore, the validity of the statistical arguments of the approximate distributions must be carefully evaluated before using them. Consequently, this chapter can be used as a general guideline to select received signal models for narrowband radios depending on the deployment and considered use cases.

REFERENCES

Abdi, A., & Kaveh, M. (1999). On the utility of gamma PDF in modeling shadow fading (slow fading). *IEEE 49th Vehicular Technology Conference, 3*, 2308-2312.

Abramowitz, M., & Stegun, I. A. (1970). *Handbook of mathematical functions: with formulas, graphs, and mathematical tables*. Dover Publications.

Agrawal, P., & Patwari, N. (2009). Correlated link shadow fading in multi-hop wireless networks. *IEEE Transactions on Wireless Communications, 8*(8), 4024–4036. doi:10.1109/TWC.2009.071293

Alippi, C., Bocca, M., Boracchi, G., Patwari, N., & Roveri, M. (2015). RTI Goes Wild: Radio Tomographic Imaging for Outdoor People Detection and Localization. *IEEE Transactions on Mobile Computing*. doi:10.1109/TMC.2015.2504965

Anderson, C. R., Martin, R. K., Walker, T. O., & Thomas, R. W. (2014, February). Radio Tomography for Roadside Surveillance. *IEEE Journal of Selected Topics in Signal Processing, 8*(1), 66–79. doi:10.1109/JSTSP.2013.2286774

Baccour, N., Koubâa, A., Mottola, L., Zúñiga, M. A., Youssef, H., Boano, C. A., & Alves, M. (2012). Radio link quality estimation in wireless sensor networks: A survey. *ACM Transactions on Sensor Networks, 8*(4), 34. doi:10.1145/2240116.2240123

Bartlett, M., & Kendall, D. G. (1946). The statistical analysis of variance-heterogeneity and the logarithmic transformation. *Supplement to the Journal of the Royal Statistical Society, 8*(1), 128–138. doi:10.2307/2983618

Baxter, J., Jones, R., Lin, M., & Olsen, J. (2004). SLLN for weighted independent identically distributed random variables. *Journal of Theoretical Probability, 17*(1), 165–181. doi:10.1023/B:JOTP.0000020480.84425.8d

Bjorkbom, M., Timonen, J., Yiğitler, H., Kaltiokallio, O., Garcia, J. M., Myrsky, M., & Koivo, H. N. et al. (2013). Localization Services for Online Common Operational Picture and Situation Awareness. *IEEE Access: Practical Innovations, Open Solutions, 1*, 742–757. doi:10.1109/ACCESS.2013.2287302

Bocca, M., Kaltiokallio, O., & Patwari, N. (2013). Radio tomographic imaging for ambient assisted living. In *Evaluating AAL Systems Through Competitive Benchmarking* (pp. 108–130). Springer. doi:10.1007/978-3-642-37419-7_9

Born, M., & Wolf, E. (1975). *Principles of optics: electromagnetic theory of propagation, interference and diffraction of light* (5th ed.). Pergamon Press.

Calculation of the Dielectric Properties of Body Tissues in the frequency range 10 Hz - 100 GHz. (2013). Academic Press. http://niremf.ifac.cnr.it/tissprop/htmlclie/htmlclie.php

Chang, C., & Sahai, A. (2004). Object tracking in a 2D UWB sensor network. *Signals, Systems and Computers, 2004. Conference Record of the Thirty-Eighth Asilomar Conference on*, 1, 1252-1256.

Durgin, G. D. (2009, June). The Practical Behavior of Various Edge-Diffraction Formulas. *IEEE Antennas & Propagation Magazine, 51*(3), 24–35. doi:10.1109/MAP.2009.5251189

Friis, H. T. (1946). A note on a simple transmission formula. *Proceedings of the IRE, 34*(5), 254–256. doi:10.1109/JRPROC.1946.234568

Ghaddar, M., Talbi, L., Denidni, T. A., & Sebak, A. (2007). A conducting cylinder for modeling human body presence in indoor propagation channel. *IEEE Transactions on Antennas and Propagation, 55*(11), 3099–3103. doi:10.1109/TAP.2007.908563

Gu, Q. (2005). *RF system design of transceivers for wireless communications*. Springer Science & Business Media.

Guo, Y., Huang, K., Jiang, N., Guo, X., Li, Y., & Wang, G. (2015, March). An Exponential-Rayleigh Model for RSS-Based Device-Free Localization and Tracking. *IEEE Transactions on Mobile Computing, 14*(3), 484–494. doi:10.1109/TMC.2014.2329007

Hall, P. (1983). Chi squared approximations to the distribution of a sum of independent random variables. *Annals of Probability, 11*(4), 1028–1036. doi:10.1214/aop/1176993451

Hashemi, H. (1993). The indoor radio propagation channel. *Proceedings of the IEEE, 81*(7), 943–968. doi:10.1109/5.231342

Jakeman, E. (1982). Fresnel scattering by a corrugated random surface with fractal slope. *JOSA, 72*(8), 1034–1041. doi:10.1364/JOSA.72.001034

Kak, A. C., & Slaney, M. (1988). *Principles of computerized tomographic imaging*. New York: IEEE.

Kaltiokallio, O., Bocca, M., & Patwari, N. (2012). Enhancing the Accuracy of Radio Tomographic Imaging Using Channel Diversity. *The 9th IEEE International Conference on Mobile Ad hoc and Sensor Systems (MASS), 2012 (pp. 254-262). IEEE.*

Kaltiokallio, O., Bocca, M., & Patwari, N. (2012, Oct). Follow @grandma: Long-term device-free localization for residential monitoring. *IEEE 37th Conference on Local Computer Networks Workshops (LCN Workshops), 2012*, 991-998. doi:10.1109/LCNW.2012.6424092

Kaltiokallio, O., Bocca, M., & Patwari, N. (2014). A Fade Level-based Spatial Model for Radio Tomographic Imaging. *IEEE Transactions on Mobile Computing, 13*(6), 1159–1172. doi:10.1109/TMC.2013.158

Kaltiokallio, O., Jäntti, R., & Patwari, N. (2017, August). ARTI: An Adaptive Radio Tomographic Imaging System. *IEEE Transactions on Vehicular Technology*, *66*(8), 7302–7316. doi:10.1109/TVT.2017.2664938

Kaltiokallio, O., Yiğitler, H., & Jäntti, R. (2017). A Three-State Received Signal Strength Model for Device-Free Localization. *IEEE Transaction on Vehicular Technology*.

Kaltiokallio, O., Yiğitler, H., Jäntti, R., & Patwari, N. (2014, April). Non-invasive respiration rate monitoring using a single COTS TX-RX pair. *Proceedings of the 13th International Symposium on Information Processing in Sensor Networks, IPSN-14*, 59-69. doi:10.1109/IPSN.2014.6846741

Kanso, M. A., & Rabbat, M. G. (2009). Compressed RF Tomography for Wireless Sensor Networks: Centralized and Decentralized Approaches. *5th IEEE Intl. Conf. on Distributed Computing in Sensor Systems (DCOSS-09)*. doi:10.1007/978-3-642-02085-8_13

Keller, J. B. (1962, February). Geometrical Theory of Diffraction. *Journal of the Optical Society of America*, *52*(2), 116–130. doi:10.1364/JOSA.52.000116 PMID:14454948

Kollár, I. (1994). Bias of mean value and mean square value measurements based on quantized data. *IEEE Transactions on Instrumentation and Measurement*, *43*(5), 733–739. doi:10.1109/19.328894

Kosba, A. E., Saeed, A., & Youssef, M. (2012). Rasid: A robust WLAN device-free passive motion detection system. *2012 IEEE International Conference on Pervasive Computing and Communications (PerCom)*, Lugano 180-189.

Li, Y., Chen, X., Coates, M., & Yang, B. (2011). Sequential Monte Carlo Radio-Frequency tomographic tracking. *IEEE International Conference on Acoustics, Speech and Signal Processing (ICASSP)*, 3976-3979. doi:10.1109/ICASSP.2011.5947223

Liberti, J. C., & Rappaport, T. S. (1996, Apr). A geometrically based model for line-of-sight multipath radio channels. *Proceedings of Vehicular Technology Conference - VTC, 2*, 844-848. doi:10.1109/VETEC.1996.501430

Lin, S., Zhang, J., Zhou, G., Gu, L., Stankovic, J. A., & He, T. (2006). ATPC: Adaptive Transmission Power Control for Wireless Sensor Networks. In *Proceedings of the 4th International Conference on Embedded Networked Sensor Systems* (pp. 223-236). New York: ACM. doi:10.1145/1182807.1182830

Mager, B., Patwari, N., & Bocca, M. (2013). Fall detection using RF sensor networks. *IEEE 24th International Symposium on Personal Indoor and Mobile Radio Communications (PIMRC)*, London, 2013, pp. 3472-3476. doi: 10.1109/PIMRC.2013.6666749.

Mokhtari, H., & Lazaridis, P. (1999, January). Comparative study of lateral profile knife-edge diffraction and ray tracing technique using GTD in urban environment. *IEEE Transactions on Vehicular Technology*, *48*(1), 255–261. doi:10.1109/25.740101

Molisch, A. F. (2011). *Wireless communications* (2nd ed.; Vol. 15). John Wiley & Sons.

Mrazovac, B., Todorović, B. M., Bjelica, M. Z., & Kukolj, D. (2013). Device-free indoor human presence detection method based on the information entropy of RSSI variations. *Electronics Letters*, *49*(22), 1386–1388. doi:10.1049/el.2013.1041

Norklit, O., & Andersen, J. B. (1998, June). Diffuse channel model and experimental results for array antennas in mobile environments. *IEEE Transactions on Antennas and Propagation*, *46*(6), 834–840. doi:10.1109/8.686770

Osipov, A. V., & Tretyakov, S. A. (2017). *Modern Electromagnetic Scattering Theory with Applications*. John Wiley & Sons, Ltd. doi:10.1002/9781119004639

Patnaik, P. B. (1949). The non-central and F-distribution and their applications. *Biometrika*, *36*, 202–232. PMID:18132092

Patwari, N., & Agrawal, P. (2008). Effects of Correlated Shadowing: Connectivity, Localization, and RF Tomography. *IEEE/ACM Int'l Conf. on Information Processing in Sensor Networks (IPSN'08)*, 82-93. doi:10.1109/IPSN.2008.7

Patwari, N., Ash, J. N., Kyperountas, S., Hero, I. I., Moses, R. L., & Correal, N. S. (2005). Locating the nodes: Cooperative localization in wireless sensor networks. *IEEE Signal Processing Magazine*, *22*(4), 54–69. doi:10.1109/MSP.2005.1458287

Patwari, N., & Wilson, J. (2010). RF sensor networks for device-free localization: Measurements, models, and algorithms. *Proceedings of the IEEE*, *98*(11), 1961–1973. doi:10.1109/JPROC.2010.2052010

Patwari, N., & Wilson, J. (2011). Spatial models for human motion-induced signal strength variance on static links. *IEEE Transactions on Information Forensics and Security*, *6*(3), 791–802. doi:10.1109/TIFS.2011.2146774

Pearson, E. S. (1959). Note on an Approximation to the Distribution of Non-Central. *Biometrika*, *46*(3/4), 364–364. doi:10.2307/2333533

Pinel, N., Bourlier, C., & Saillard, J. (2010). Degree of roughness of rough layers: Extensions of the Rayleigh roughness criterion and some applications. *Progress In Electromagnetics Research B*, *19*, 41–63. doi:10.2528/PIERB09110907

Proakis, J. G., & Salehi, M. (2008). *Digital Communications* (5th ed.). New York, NY: McGraw-Hill.

Rampa, V., Savazzi, S., Nicoli, M., & D'Amico, M. (2015, November). Physical Modeling and Performance Bounds for Device-free Localization Systems. *IEEE Signal Processing Letters*, *22*(11), 1864–1868. doi:10.1109/LSP.2015.2438176

Rappaport, T. S. (2002). Wireless Communications: Principles and Practice (2nd ed.). Prentice Hall.

Sagias, N. C., & Karagiannidis, G. K. (2005). Gaussian class multivariate Weibull distributions: Theory and applications in fading channels. *IEEE Transactions on Information Theory*, *51*, 3608–3619. doi:10.1109/TIT.2005.855598

Sankaralingam, S., & Gupta, B. (2010). Determination of Dielectric Constant of Fabric Materials and Their Use as Substrates for Design and Development of Antennas for Wearable Applications. *IEEE Transactions on Instrumentation and Measurement*, *59*(12), 3122–3130. doi:10.1109/TIM.2010.2063090

Savazzi, S., Nicoli, M., Carminati, F., & Riva, M. (2014, February). A Bayesian Approach to Device-Free Localization: Modeling and Experimental Assessment. *IEEE Journal of Selected Topics in Signal Processing*, *8*(1), 16–29. doi:10.1109/JSTSP.2013.2286772

Simon, M. K., & Alouini, M.-S. (2005). *Digital communication over fading channels* (Vol. 95). Wiley.

Turin, G. L. (1960). The characteristic function of Hermitian quadratic forms in complex normal variables. *Biometrika*, *47*(1-2), 199–201. doi:10.1093/biomet/47.1-2.199

Ulaby, F. T., Long, D. G., Blackwell, W. J., Elachi, C., Fung, A. K., Ruf, C., & Van Zyl, J. (2014). *Microwave radar and radiometric remote sensing* (Vol. 4). University of Michigan Press Ann Arbor.

Urkowitz, H. (1967). Energy detection of unknown deterministic signals. *Proceedings of the IEEE*, *55*(4), 523–531. doi:10.1109/PROC.1967.5573

Vaughan, R. (2003). *Channels, propagation and antennas for mobile communications.* Iet.

Wang, Z., Liu, H., Xu, S., Bu, X., & An, J. (2015, November). A Diffraction Measurement Model and Particle Filter Tracking Method for RSS-Based DFL. *IEEE Journal on Selected Areas in Communications*, *33*(11), 2391–2403. doi:10.1109/JSAC.2015.2430517

Widrow, B., Kollar, I., & Liu, M.-C. (1996). Statistical theory of quantization. *IEEE Transactions on Instrumentation and Measurement*, *45*(2), 353–361. doi:10.1109/19.492748

Wilson, J., & Patwari, N. (2010). Radio Tomographic Imaging with Wireless Networks. *IEEE Transactions on Mobile Computing*, *9*(5), 621–632. doi:10.1109/TMC.2009.174

Yacoub, M. D. (2002). The α - μ distribution: a general fading distribution. *13th IEEE International Symposium on Personal, Indoor and Mobile Radio Communications, 2*, 629-633. doi:10.1109/PIMRC.2002.1047298

Yiğitler, H., & Jäntti, R. (2016, March). Experimental Accuracy Assessment of Radio Tomographic Imaging Methods. *IEEE International Conference on Pervasive Computing and Communication Workshops (PerCom Workshops)*, 1-6. doi:10.1109/PERCOMW.2016.7457117

Yiğitler, H., Jäntti, R., Kaltiokallio, O., & Patwari, N. (2017). Detector based Radio Tomographic Imaging. *IEEE Transaction on Mobile Computing.*

Yiğitler, H., Jäntti, R., & Patwari, N. (2017, April). On Log-Normality of RSSI in Narrowband Receivers Under Static Conditions. *IEEE Signal Processing Letters*, *24*(4), 367–371. doi:10.1109/LSP.2017.2657332

Yiğitler, H., Jäntti, R., & Virrankoski, R. (2014). pRoot: An Adaptable Wireless Sensor-Actuator Hardware Platform. *2014 12th IEEE International Conference on Embedded and Ubiquitous Computing, Milano*, 281-286.

Yiğitler, H., Kaltiokallio, O., & Jäntti, R. (2013, Aug). A management framework for device-free localization. *International Joint Conference on Neural Networks (IJCNN)*, 1-8. doi:10.1109/IJCNN.2013.6707131

Youssef, M., Mah, M., & Agrawala, A. (2007). Challenges: device-free passive localization for wireless environments. *ACM Int'l Conf. Mobile Computing and Networking, 2007. MobiCom'07.*, 222-229. doi:10.1145/1287853.1287880

KEY TERMS AND DEFINITIONS

Absorption Model: The shadowing state model aiming to represent absorption of electromagnetic energy while propagating through the human tissue.

Device-Free Localization: Localization technology that estimates position of people or assets using intrinsic measurements of local/personal area wireless networking devices deployed in an area of interest.

Diffraction Model: The fading/shadowing state model that is based on diffraction theory of electromagnetic waves. A simple model is obtained by using the knife-edge diffraction model.

Diffraction State: The temporal state that represents the interacting object's locations that yield dominating diffraction affect.

Environmental Sensing: Recent sensing applications where intrinsic radio measurements of low-range radios are used to infer some physical phenomenon of interest.

Fading State: The temporal state that represents the interacting object's locations that yield dominating reflection/diffraction affects. This is the state between non-fading state and shadowing state, and it is parameterized with respect to a second electromagnetic wave that travels an excess path length compared to the shortest distance path.

Interacting Object: Any object in the propagation medium altering the characteristics of electromagnetic waves impinging on the receiver antenna.

Non-Fading State: The temporal state that represents the interacting object's locations that yield nonsignificant variations in the measurements.

Propagation Channel: A linear time-varying filter that alters the transmitted signal. It is characterized by discrete components with specific amplitude, phase, and delay.

Received Signal Strength: The strength of a received signal by a radio. It is averaged received signal energy over a certain time window.

Reflection Model: The fading state model that is based on coherent scattering of electromagnetic waves. The reflection model is developed based on geometrical optics models when the surface roughness is smooth.

Temporal State: The variation of effect of interacting object with time as the person moves within the sensitivity region of a link. There are four temporal states and for each of these states the observed signal strength variations can be modeled using different models.

ENDNOTES

1. We refer to the artificial line connecting transmitter and receiver as link-line.
2. We refer to any objects altering the characteristics electromagnetic wave on the receiver antenna as interacting object. When the context is clear, we use human body or person to indicate that the considered interacting object is a human body.

3 In this subsection, a brief review of the propagation channel model is given. The interested reader is referred to Part II in (Molisch, 2011) and (Hashemi, 1993).

4 The shortest path multipath component is also referred to as line-of-sight (LoS) component. We use these terms interchangeably.

5 The reradiation is commonly referred to as scattering in the Radar literature.

6 In the remaining part of this section, we refer to coherent component from specular human body surface as reflected component, and incoherent (or diffuse) component as scattered component following convention used in the environmental sensing literature, e.g. (Patwari & Wilson, 2011).

Chapter 3
Robustness in Fingerprinting-Based Indoor Positioning Systems

Shih-Hau Fang
Yuan-Ze University, Taiwan

ABSTRACT

Indoor positioning systems have received increasing attention for supporting location-based services in indoor environments. Received signal strength (RSS), mostly utilized in Wi-Fi fingerprinting systems, is known to be unreliable due to two reasons: orientation mismatch and variations in hardware. This chapter introduces an approach based on histogram equalization to compensate for orientation mismatch in robust Wi-Fi localization. The proposed method involves converting the temporal-spatial radio signal strength into a reference function (i.e., equalizing the histogram). This chapter also introduces an enhanced positioning feature, which is called delta-fused principal strength, to enhance the robustness of Wi-Fi localization against the problem of heterogeneous hardware. This algorithm computes the pairwise delta RSS and then integrates with RSS using principal component analysis. The proposed methods effectively and efficiently improve the robustness of location estimation in the presence of mismatch orientation and hardware variations, respectively.

INTRODUCTION

Location awareness has become a crucial concern for various mobile applications (Gu, Lo & Niemegeers, 2009; Perez-Ramirez, Borah & Voelz, 2013; Sun, Zhu, Zhang & Fang, 2011). Recently, numerous studies have addressed location estimation by using existing Wi-Fi infrastructure to compensate for the weakness of the Global Positioning System (GPS) in urban areas and indoor environments (Bahl & Padmanabhan, 2000; Youssef & Agrawala, 2008; Fang, Wang, Huang, Yang & Chen, 2012). Because of the frequent deployment of access points (APs), Wi-Fi-based localization has gained considerable attention over the last several years, and the received signal strength (RSS) is commonly adopted as a positioning characteristic (Bshara, Orguner, Gustafsson & Van Biesen, 2010; Figuera, *et al.* 2011; Liao,

DOI: 10.4018/978-1-5225-3528-7.ch003

et al. 2011). Among the various Wi-Fi positioning systems, the fingerprinting-based approach, is one of the most feasible solutions (Youssef, Agrawala & Udaya Shankar, 2003; Tsui, Lin, Chen, Huang & Chu, 2010; Fang & Wang, 2011). This chapter focues on the location fingerprinting systems, which is based on a RSS-mapping approach. Based on the pre-recorded RSS from different reference locations, denoted as fingerprints or radio map, the user's location is estimated by mapping the currently measurement with the pre-stored fingerprints. Because this approach provides a high positioning accuracy in a GPS-less environment, researchers have recently proposed various fingerprinting-based localization algorithms (Kuo & Tseng, 2008; Shin, Chon & Cha, 2012). Although Wi-Fi-based localization shows great promise, a key challenge in real-time location estimation is managing the robustness issues (Bernardos, Casar & Tarrio, 2010; Mazuelas, *et al.* 2009). This chapter will discuss two robustness issues of the fingerprinting localization systems, including the orientation mismatch and heterogeneous hardware.

The first robustness issue discussed in this chapter is the orientation mismatch. The performance of fingerprinting based localization systems degrades severely when radio environments between training and testing RSSs differ from each other (Fang, Wang, Chiou & Lin, 2012). In fact, radio irregularity is a common occurrence in wireless environments (Gezici, 2008; Liu, Darabi, Banerjee & Liu, 2007; Roberts & Pahlavan, 2009). One type of radio mismatch is caused by the different user orientations, referred to as "orientation mismatch". The human body consists of more than 50% water and, hence, might block the transmission of 2.4-GHz Wi-Fi radios (Ladd, *et al.* 2002). In a fingerprinting-based system, the mismatch between training and locating orientations makes it difficult to accurately determine the location based on RSS patterns. Researchers have performed experiments regarding the impact of user/device orientation on RSS. Previous studies have acknowledged this orientation problem and have indicated that RSS varies substantially depending on the user's orientation, even at a fixed location (Liu, *et al.* 2012; Kaemarungsi & Krishnamurthy, 2004).

To overcome this problem, (Papapostolou & Chaouchi, 2009) proposed a compass-assisted approach, in which the testing RSS is matched according to the orientation-matched radio map only, rather than the entire database. A similar approach was presented in (Sanchez, Quinteiro, Hernandez-Morera & Martel-Jordan, 2012; Papapostolou & Chaouchi, 2009), in which fingerprinting is extended with orientation information. These approaches effectively reduce the orientation mismatch effect; however, they require prior orientation data to enrich the radio map and real-time feedback from a digital compass. Although current high-level smartphones can be equipped with compass sensing functions that provide real-time user orientation, reconstructing a radio map for every possible orientation is extremely time-consuming and infeasible.

The second robustness issue discussed in this chapter is the heterogeneous hardware. Many studies have indicated that RSS variation between diverse Wi-Fi devices may exceed 25 dBm, even in the same location (Kaemarungsi & Krishnamurthy, 2004; Kaemarungsi, 2006). Possible reasons include a lack of consistency in the various industry standards and the use of different chipsets and antenna. Based on IEEE 802.11 standard (IEEE standard for information technology, 2007), RSS is not restrictively defined, and it is left as implementer's choice, but must indicate energy in the channel. Variation in RSS resulting from heterogeneous devices causes large bias error in localization systems (Figuera, *et al.* 2011; Fang, Wang, Chiou & Lin, 2012). Such variation in hardware is inevitable in the real world due to the tremendous growth in recent years of new Wi-Fi devices, such as iPhones, iPads, and Android devices, which is expected to continue.

To overcome this difficulty, researchers have devoted considerable effort to designing robust localization systems that provide valid location information despite the use of heterogeneous devices. Common approaches for handling variations in hardware fall into two categories: device mapping (Haeberlen, *et al.* 2004; Tsui, Chuang & Chu, 2009; Bernardos, Casar & Tarrio, 2010) and robust location features (Kjærgaard & Munk, 2008; Chan, Baciu & Mak, 2010; Mahtab Hossain, Van, Jin & Soh, 2007). The first method attempts to transform the measurement from user devices to training devices through a mapping function. The advantage of this approach is a significant improvement in positioning accuracy. However, determining mapping functions for every possible device is time consuming and unfeasible due to the enormous number of products on the market. The second method extracts robust location features to mitigate the effects of heterogeneous devices, such as HLF (hyperbolic location fingerprinting) (Kjærgaard, 2011) and SSD (signal strength difference) (Mahtab Hossain, Jin, Soh & Van, 2013). HLF uses the ratio of RSSs instead of RSS. However, due to the extended pairwise dimensions, this approach consumes more memory and requires more computational time (Figuera, *et al.* 2011). SSD presented an independent subset approach to reduce the overhead under the assumption of independence between APs; however, this assumption could be erroneous (Fang, Lin & Lin, 2008). An enhanced positioning feature, called delta fused principal strength (DFPS), was proposed in this chapter to improve the robustness of Wi-Fi localization in heterogeneous hardware.

BACKGROUND

This section contains two subsections. First, this section introduces the background of the indoor location fingerprinting systems. Then, the physical property of received signal strength, the most widely used features in fingerprinting systems, is illustrated.

Indoor Location Fingerprinting Systems

Indoor location based service is not simply an extension of outdoor localization. It can work for specific scenario and improve life quality as well as boost business market. Bahl et al. (2000) proposed an in-building user location and tracking system, called the RADAR system, used the k-nearest-neighbors (KNN) search. The KNN search applies different deterministic mathematical models which need a searching space constructed by a radio propagation model to estimate the location of the target device. The RADAR system can provide 2-D absolute position information and the accuracy is about 2-3 meters (Bahl & Padmanabhan, 2000). Seshadri et al. used multiple probabilistic models based on the RSSs known as Bayesian inference to enhance the performance of the original system (Seshadri, Zaruba & Huber, 2005). Youssef et al. proposed the Horus system approach which utilized the strongest APs to decrease the computational complexity of the positioning algorithm. The Horus system treats RSSs at each location as a statistical random variable. Each candidate location is regarded as a cluster. The position is estimated while its likelihood is the highest (Youssef, Agrawala & Udaya Shankar, 2003; Youssef & Agrawala, 2004).

Fingerprints are utilized in many literatures to assist indoor positioning, and the most widely used one is Wi-Fi signal strength. Wi-Fi is one of the most used wireless technologies. It follows a series of standards in IEEE 802.11 and uses two license-exempt bands: 2.4 GHz, and 5 GHz. A lot of Wi-Fi access points (APs) have been deployed in public areas such as super malls, office buildings, hospitals, shopping malls, and train stations. Most commercial products with the sensing function are available on all Wi-Fi-enabled devices (e.g., smart phones, tablets and laptops). Wi-Fi based positioning systems can be constructed with a small deployment cost by utilizing the deployment of Wi-Fi APs as an infrastructure. With all the advantages, Wi-Fi is a mainstream technology in literatures for indoor localization. The fingerprinting is one of the important technologies used in location-based system (LBS). The position estimations usually use RSS values on indoor localization system. The RSS signal from the multiple APs creates a specific signal pattern. Based on the pre-recorded signal pattern from different locations, the user's position is estimated by mapping the current measurements with the prestored signal pattern.

Generally speaking, the localization system can be divided into two phases. First, in the offline phase (training phase), the signal patterns, donoted as fingerprints, are recorded from reference locations in database. Youssef et al. position based on the current RSSs and the previously stored radio map is called on-line phase (position phase). Different kinds of methods have been used to determine position by fingerprints such as the mapping, geometric and statistic approaches (Kjærgaard, 2007; Roos, Myllymaki & Tirri, 2002; Gezici, 2008). The thesis is called location fingerprinting, which focuses on the mapping approach which uses RSSs signal to provide a distance and location information depending on a radio map.

Received Signal Strength

Based on the radio propagation model, the free space loss of the signal power (Saunders & Simon, 1999, Vaseghi, 2006) can be given by

$$L = \left(\frac{4\pi d}{\lambda}\right)^2 \quad (1)$$

where L is the free-space path loss, λ is the wavelength of an electromagnetic wave, and d is the distance between the sender and the receiver. Such property shows the relationship between the receiving power and the distance. However, the relationship between RSS and distance in a real world is difficult to model. For example, the log-distance path loss model (Saunders & Simon, 1999) is

$$PL(d) = PL(d_0) + 10n\ log(d / d_0) + X_\sigma \quad (2)$$

where d_0 is a reference distance, X is the noise, and n is path loss exponent. This model is in logarithmic scale. In fact, estimating the real value of n is difficult because it depends on the propagation environment. In some environments, such as buildings, stadiums and other indoor environments, the path loss exponent can vary in a certain range. The theoretical value of the path loss exponent is 2 as implied in Eq. (1), but in indoor environments it is typical to observe values 4 to 6. On the other hand, a tunnel may act as a waveguide, resulting in a path loss exponent less than 2. In Eq. (2), X represents small-scale fading and shadowing effects. These indeterministic effects, usually modeled as a Gaussian random variable, results in the temporal variation of RSS even at a fixed location. Several approaches such as the temporal filter or singular value decomposition are utilized to remove the noisy effect (Fang & Lin, 2008). In reality, the propagation of a RF wave is attenuated when it hits the surface of an obstacle. Eq. (2) shows that a longer distance results in a higher power loss, thus making a lower RSS. Nevertheless, in addition to the line-of-sight (LOS) signal, there are possibly multiple signals arriving at the receiver through different paths. This multipath effect is even more severe in indoor environments where a ceiling, floor, and walls are present. As a result, it is possible for a closer receiver to have a lower RSS than a more distant one. Consequently, a simple relationship between received power and separating distance cannot be established. Therefore, this time varying and vulnerable RSSs value creates undesirable localization errors.

OREIENTATION MISMATCH AND HETEROGENEOUS HARDWARE ISSUES

This section briefly discuss the robustness and some related issues in indoor fingerprinting system. Although Wi-Fi-based localization shows great promise, a key challenge in real-time location estimation is managing the robustness from various perspectives (Mazuelas, Bahillo, Lorenzo, Fernandez, Lago, Garcia, Blas & Abril, 2009; Bernardos, Casar & Tarrio, 2010). For example, researchers have addressed heterogeneity in the hardware of Wi-Fi devices (Mahtab Hossain, Jin, Soh & Van, 2010; Kjærgaard, 2011; Dong, Chen, Liu, Ning & Piao, 2009). Furthermore, previous studies have designed secure location systems that can identify malicious network attacks and reject them from the positioning system (Li, Jiang & Guibas, 2011; Fang, Chuang & Wang, 2012; Chen, Yang, Trappe & Martin, 2010). Several studies have investigated noise and multipath distortion (Medawar, Handel & Zetterberg, 2013; Yu & Dutkiewicz, 2013; Bialer, Raphaeli & Weiss, 2013). Robust pattern matching, outliers removement, and data-overfitting problems have also been discussed in the literature (Roos, Myllymaki & Tirri, 2002; Gezici, 2008; Garcia-Valverde, Garcia-Sola, Botia & Gomez-Skarmeta, 2012).

The device mapping changes the measurements of user's RSS values to minimize the different devices from training device. Haeberlen et al. treats signal variation as a Gaussian mean-value shift and uses a linear model to fit the RSS values on user's and training devices (Haeberlen, Flannery, Ladd, Rudys, Wallach & Kavraki, 2004). Tsui et al. propose a transformation function which uses online RSS measurements to solve the hardware variance problem (Tsui, Chuang & Chu, 2009). Bernardos et al. propose a hyperbolic trangulation positioning algorithm called the Least Mean Squares. This technique calibrates RSS samples based on approximating the true gradient of the squared error in real time (Bernardos, Casar & Tarrio, 2010).

The robust location fingerprints extract the robust location features to mitigate the effect of heterogeneous devices. Kjaergaard et al. used signal strength ratios between a pair of access points as fingerprints to overcome the hardware variance problem (Kjærgaard, 2011; Kjærgaard & Munk, 2008). Della Rosa et al. propose a method called Collaborative Mapping (CM) based on the exploitation of the RSS from ad-hoc connections among neighboring base stations (Della Rosa, Leppa, Biancullo & Nurmi, 2010). The estimation of the spatial proximity among them is the metric of comparison for the database calibrations and final positioning estimation. Chan et al. transforms the signal strength into the Fourier descriptors which is frequency domain curve (Chan, Baciu & Mak, 2010). The curves are used in KNN location estimation. The signal strength difference (SSD) are more robust than RSS when we use it as location fingerprints (Mahtab Hossain, Van, Jin & Soh, 2007; Hossain, Van & Soh, 2010). SSDoA which used SSD is similar to the Time Difference of Arrival technique. Instead of mapping methods, this technique defined the hyperbolic curves solution for determining location. These approaches do not require manually collecting measurements from various devices (Papadakis & Traganitis, 2010).

The first robustness issue discussed in this chapter is the orientation mismatch. In fact, radio irregularity is a common occurrence in wireless environments. Although Wi-Fi-based localization shows great promise, a key challenge in real-time location estimation is managing the robustness from various perspectives (Bernardos, Casar & Tarrio, 2010; Mazuelas, *et al.* 2009).

Radio irregularity is a common occurrence in wireless environments (Gezici, 2008; Liu, Darabi, Banerjee & Liu, 2007; Roberts & Pahlavan, 2009). The performance of fingerprinting based localization systems degrades severely when radio environments between training and testing RSSs differ from each other (Fang, Wang, Chiou & Lin, 2012). One type of radio mismatch is caused by the different user orientations. Users tend to carry mobile devices in front of them so that they can view the displays and use the touchscreen. However, the human body consists of more than 50% water and, hence, might block the transmission of 2.4-GHz Wi-Fi radios (Ladd, *et al.* 2002). Researchers have performed experiments regarding the impact of user/device orientation on RSS. Previous studies have acknowledged this orientation problem and have indicated that RSS varies substantially depending on the user's orientation, even at a fixed location (Liu, *et al.* 2012; Kaemarungsi & Krishnamurthy, 2004). However, in a fingerprinting-based system, the mismatch between training and locating orientations makes it difficult to accurately determine the location based on RSS patterns. This severe problem is referred to as "orientation mismatch."

To overcome this problem, in (Papapostolou & Chaouchi, 2009) a compass-assisted approach was proposed, in which the testing RSS is matched according to the orientation-matched radio map only, rather than the entire database. A similar approach was presented in (Sanchez, Quinteiro, Hernandez-Morera & Martel-Jordan, 2012; Papapostolou & Chaouchi, 2009), in which fingerprinting is extended with orientation information. These approaches effectively reduce the orientation mismatch effect; however, they require prior orientation data to enrich the radio map and real-time feedback from a digital compass. Although current high-level smartphones can be equipped with compass sensing functions that provide real-time user orientation, reconstructing a radio map for every possible orientation is extremely time-consuming and infeasible. Liao et al. proposed the orientation classifier approach, which is used to predict the user's orientation based on the moving trajectory and RSS variations (Liao & Kao, 2008). However, the effectiveness of this method depends on the assumption that a user moves at a constant speed in a fixed direction. This assumption could be erroneous, and modeling the orientation process is difficult due to the random user behavior.

The second robustness issue discussed in this chapter is the heterogeneous hardware. The RSSs are not always stable even if at the same place with the same devices (Kaemarungsi & Krishnamurthy, 2004; Kaemarungsi, 2006). Nevertheless, the heterogeneity of the WLAN cards embedded into various mobile devices, such as laptops, desktop PCs, mobile phones, and Internet tablets is a huge problem for researchers and application developers. Such heterogeneity may result from the usage of different chipsets and the lack of standard between industries. Hardware variance can impact the accuracy of the system because different devices provide different RSS-patterns.

To overcome this difficulty, researchers have devoted considerable effort to designing robust localization systems that provide valid location information despite the use of heterogeneous devices. Common approaches for handling variations in hardware fall into two categories: device mapping (Haeberlen, *et al.* 2004; Tsui, Chuang & Chu, 2009; Bernardos, Casar & Tarrio, 2010) and robust location features (Kjærgaard & Munk, 2008; Chan, Baciu & Mak, 2010; Mahtab Hossain, Van, Jin & Soh, 2007). The first method attempts to transform the measurement from user devices to training devices through a mapping function. The advantage of this approach is a significant improvement in positioning accuracy. However, determining mapping functions for every possible device is time consuming and unfeasible due to the enormous number of products on the market. The second method extracts robust location features to mitigate the effects of heterogeneous devices, such as HLF (hyperbolic location fingerprinting) (Kjærgaard, 2011) and SSD (signal strength difference) (Mahtab Hossain, Jin, Soh & Van, 2013). HLF uses the ratio of RSSs instead of RSS. However, due to the extended pairwise dimensions, this approach consumes more memory and requires more computational time (Figuera, *et al.* 2011). SSD presented an independent subset approach to reduce the overhead under the assumption of independence between APs; however, this assumption could be erroneous (Fang, Lin & Lin, 2008). The following two sections present solutions to deal with the orientation mismatch and heterogeneous hardware problems mentioned above.

THE SOLUTION FOR ORIENTATION MISMATCH PROBLEM

This section contains of four parts including "Problem Statement", "Theoretical Foundation of Histogram Equalization", "Proposed Indoor Positioning Algorithm" and "Experimental Result and Analysis". In the first two subsections, we formulate the mismatch problems of the traditional Wi-Fi location fingerprinting systems, and introduce the theoretical foundation of histogram equalization. In the next two subsections, we will show how to incorporate the concept of histogram equalization into a positioning system design, and then report the experimental results.

Problem Statement

A probabilistic location algorithm estimates the likelihood of observations for all candidate locations and selects the one having maximum likelihood as the result. This can be formulated as

$$j^* = \arg\max_j P\left(\Lambda_j \mid S\right)$$

$$= \arg\max_j \frac{P\left(S \mid \Lambda_j\right)}{P\left(S\right)}$$

$$= \arg\max_j P\left(S \mid \Lambda_j\right)$$

$$= \arg\max_j \prod\nolimits_{i=1,\dots,M} P\left(S_i \mid \Lambda_j\right), \tag{3}$$

where j is the index of the reference location, j^* is the maximum-likelihood estimation, Λ_j is the model parameters of the jth reference location, *S* is the RSS observation, and s_i is the *i*th element of *S*. In (3), $P(\Lambda_j)$ is assumed to be equal $\forall j$, and *P*(*S*) does not affect the decision result. This approach is widely used for indoor wireless local area network (WLAN) location fingerprinting systems. However, this positioning algorithm is sensitive to orientation mismatch. Assuming Y is an orientation-mismatched observation, this process becomes

$$j^{**} = \arg\max_j P\left(\Lambda_j \middle| Y\right)$$

$$= \arg\max_j \int P(\Lambda_j \mid Y, S) P(S \mid Y) dS$$

$$= \arg\max_j \int P(\Lambda_j \mid S) P(S \mid Y) dS$$

$$= \arg\max_j \int \frac{P(S \mid \Lambda_j) P(\Lambda_j)}{P(S)} P(S \mid Y) dS$$

$$= \arg\max_j \prod_{i=1,..,M} \int \frac{P(S_i \mid \Lambda_j)}{P(s_i)} P(s_i \mid Y) ds_i \,, \tag{4}$$

where j^{**} is the new estimation in this mismatched scenario. We assume that the RSS vector is only dependent on the mismatched RSS of current vector. In this way, $P(s_i \mid Y) = P(s_i \mid y_i)$. Based on the point estimate, we assume $P(s_i \mid y_i) = \delta(s_i - \hat{s}_i(y_i))$; thus, the new estimation in (4) can be rewritten as

$$j^{**} = \arg\max_j \prod_{i=1,..,M} P(\hat{s}_i(y_i) \mid \Lambda_j), \tag{5}$$

Comparing (5) with (3), we can observe the difference between the likelihood formulations, i.e., $j^{**} \neq j^*$. The goal of the robust location estimation is to make j^{**} approaches j^*. That is, $\hat{s}_i(y_i) = f(y_i) \to s_i$. However, designing the mapping function that achieves the aforementioned goal is difficult without a set of paired training and testing data.

Now, consider a rather naive assumption that the distributions of training data and observation data are both Gaussians and parameterized by $\{\mu_S, \sigma_S\}$ and $\{\mu_Y, \sigma_Y\}$, respectively. Then, the goal is to estimate a mapping function so that the two distributions become close. The simplest way is to match the mean of these two distributions, which can be done in two types of implementation: direct and indirect approaches. For the direct approach, testing RSSs are transformed to match training data directly, and the mapping function is $f(y_i) = y_i - \mu_Y + \mu_S$, such that the distribution of the testing data has the same mean as that of the training data. For the indirect approach, both training and testing data are transformed to a same reference distribution, and we perform $f(y_i) = y_i - \mu_Y$ and $f(s_i) = s_i - \mu_S$; Therefore, both the distributions of the testing data and training data have zero mean. If we consider to normalize both mean and variance, the direct approach applies $f(y_i) = (y_i - \mu_Y)(\sigma_S / \sigma_Y) + \mu_S$, and the indirect approach performs $f(y_i) = (y_i - \mu_Y) / \sigma_Y$ and $f(s_i) = (s_i - \mu_S) / \sigma_S$. The direct approach transforms the testing data to share the same mean and variance with the training data, whereas the indirect approach transforms both training and testing data to the same zero mean and unit variance Gaussian distribution.

However, in realistic situations, the impact of user orientation mismatch on the RSS values may not be linear, and the training and testing distributions are not necessary Gaussians. Moreover, more accurate matching is to normalize higher statistical moments apart from the mean and variance. This motivates us to adopt histogram equalization to overcome this problem. The histogram equalization described here equalizes all the moments of the probability distribution to those of the reference probability distribution. Therefore, this procedure can be considered an extension of mean and variance normalization.

Traditional works assume the availability of orientation information during the offline and online phases while dealing with the mismatch problem (King, Kopf, Haenselmann, Lubberger & Effelsberg, 2006; Sanchez, Quinteiro, Hernandez-Morera & Martel-Jordan, 2012; Papapostolou & Chaouchi, 2009). In other words, these methods are based on the orientation-aware radio map and the likelihood becomes

$$j^{***} = \arg\max_{j} \prod_{i=1,...,M} P(s_i^{(0)} \mid \Lambda_j^{(0)}), \tag{6}$$

where o indicates the current user's orientation, and Λ_j^0 represents the model parameters estimated from the data with the same orientation o. Due to the same orientation population, j^{***} approaches j^*, and such compass-assisted approach can be regarded as a performance upper bound. (Feng et al & Liao & Kao, 2008; Russ, 1995) combines affinity propagation (De la Torre, *et al.* 2005) and compressive sensing to improve the positioning accuracy, where the current user orientation is ignored. These approaches effectively reduce the orientation mismatch effect; however, they require prior orientation data to enrich the radio map and real-time feedback from a digital compass.

Although current high-level smartphones can be equipped with compass sensing functions that provide real-time user orientation, reconstructing a radio map for every possible orientation is extremely time-consuming and infeasible. The goal is to alleviating the impact of orientation mismatch without the need of orientation information from either radio map or embedded sensors. Liao *et al.* (Dharanipragada & Padmanabhan, 2000) proposed the orientation classifier approach, which is used to predict the user's orientation based on the moving trajectory and RSS variations. However, the effectiveness of this method depends on the assumption that a user moves at a constant speed in a fixed direction. This assumption could be erroneous, and modeling the orientation process is difficult due to the random user behavior.

Theoretical Foundation of Histogram Equalization

HEQ has been recognized as an effective digital image processing method for optimizing the dynamic gray-level ranges and improving the brightness and contrast (Kjærgaard & Munk, 2008). The purpose of HEQ is to provide a transformation, i.e., $x_1 = F(x_0)$, that converts the probability density function $p_0 = (x_0)$ of an original variable into a reference probability density function $p_1 = (x_1)$. In other words, the transformation equalizes the histograms (Mahtab Hossain, Van, Jin & Soh, 2007). The formulation of HEQ is described as follows.

Let x_0 and x_1 be the original and reference random variables, respectively, with distributions $p_0(x_0)$ and $p_1(x_1)$. A transformation, i.e., $x_1 = F(x_0)$, modifies the probability distribution according to the following expression (Kjærgaard & Munk, 2008; Mahtab Hossain, Van, Jin & Soh, 2007):

$$p_1(x_1) = p_0(\mathrm{G}(x_1))\frac{\partial G(x_1)}{\partial x_1}, \tag{7}$$

where $G(x_1)$ is the inverse function of $F(x_0)$. Note that we make $F(\cdot)$ invertible by using probabilistic one-to-one mapping functions $C_1(\cdot)$ and $C_0(\cdot)$. The relationship between the cumulative probability density functions (cdfs) associated with these probability distributions is given by

$$\mathrm{C}_0(x_0) = \int_{-\infty}^{x_0} p_0(x_0')dx_0'$$

$$= \int_{-\infty}^{F(x_0)} p(G(x_1'))\frac{\partial G(x_1)}{\partial x_1'}dx_1'$$

$$= \int_{-\infty}^{F(x_0)} p_1(x_1')dx_1'$$

$$= C_1(F(x_0)), \tag{8}$$

where $C_1(F(x_0))$ and $\mathrm{C}_0(x_0)$ are cdfs of the reference and original distribution, respectively. The transformation $x_1 = F(x_0)$, which converts distribution $p_0(x_0)$ into reference distribution $p_1(x_1)$, also converts the cumulative probability $\mathrm{C}_0(x_0)$ into $\mathrm{C}_1(x_1)$. Hence, the transformation function $F(x_0)$ converting $p_0(x_0)$ into $p_1(x_1)$ is expressed as

$$x_1 = F(x_0) = C_1^{-1}[C_0(x_0)] \tag{9}$$

where C_1^{-1} denotes the inverse function of the cdf $\mathrm{C}_1(x_1)$, specifying the value x_1 that corresponds to a certain cumulative probability.

Considering practical implementations, the cumulative histograms are typically used because the number of observations is finite. Therefore, the procedure is referred to as HEQ in the literature. The main concept of HEQ is shown in Figure 1, where the green and orange integral areas indicate the cdfs of x_0 and x_1, respectively. The black line represents the transformation by which the algorithm equalizes the histograms $\mathrm{C}_0(x_0) = \mathrm{C}_1(x_1)$, as indicated in (9). In addition to digital image processing, HEQ has also been introduced to the field of speech processing (Mahtab Hossain, Van, Jin & Soh, 2007; Kjærgaard, 2011).

Figure 1. Main idea of the histogram equalization procedure

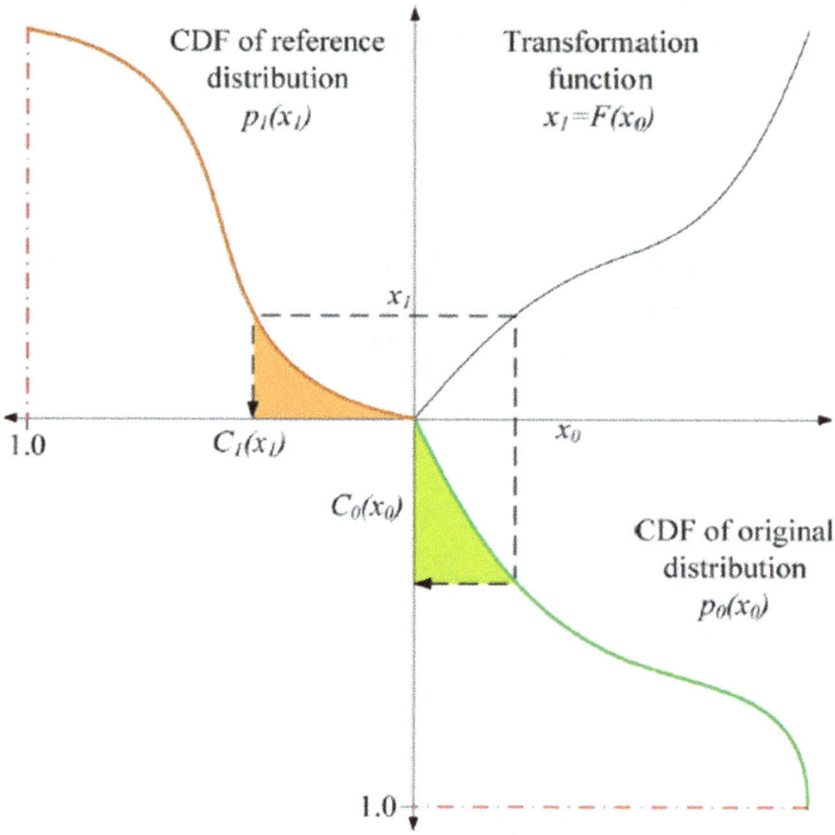

In this chapter, the HEQ technique converts the temporal–spatial radio signal strength into a reference (i.e., equalizing the histogram). Now, we assume $z_i = f(x_i)$, where z_i and x_i denote the *i*th transformed and original data. Letting p_z and p_r denote the density of transformed and reference distributions, respectively, HEQ has been recognized by minimizing the Kullback–Liebler distance between two probability densities as

$$L(f) = D(p_z \mid p_r)$$

$$= \int_z p_z(z) \log p_z(Z) dz - \int_z p_z(z) \log p_r(Z) dz \quad (10)$$

The goal of HEQ is to obtain f^* that minimizes $L(f)$, i.e., $f^* = \operatorname{argmin} L(f)$, so that p_z approaches p_r. Assume that $C_x(x) = \int_{-\infty}^{x} p_x(w)dw$ and $C_r(r) = \int_{-\infty}^{r} p_r(w)dw$ where C_x and C_r being cdfs, are single valued and monotonically increasing in the interval [0 1]. If we define the transform $f(\cdot)$ 5= $C_r^{-1}C_x(\cdot)$ and apply $z_i = f(x_i)$, p_z equals p_r. HEQ can be implemented in two manners: direct and indirect. In the former, we assume $x_i = y_i$ and $p_r = p_s$, where p_s is the distribution of training data. Thus, HEQ performs $z_i = f(y_i)$ with the aim of $p_z = p_s$. In the latter, we transform both y_i and s_i to a reference distribution p_r, where p_r is a predefined distribution. The following describes how it can be adapted for Wi-Fi indoor localization with details.

Proposed Indoor Positioning Algorithm

Offline Phase: The proposed indoor WLAN positioning system consists of two phases: offline and online. In the offline phase, RSS readings from Wi-Fi APs at reference positions are collected to build a database, which is called a radio map. The raw RSS data are denoted by $\psi_{i,j}^{(o)}[\tau]$ and indicate the τ th RSS value measured from the *i*th AP at the *j*th reference location, with orientation *o*. The offline-constructed radio map can then be spanned and represented by $\boldsymbol{\Psi}$ as

$$\boldsymbol{\Psi}^{(o)} = \begin{pmatrix} \psi_{1,1}^{(o)}[\tau] & \psi_{1,2}^{(o)}[\tau] & \cdots & \psi_{1,N}^{(o)}[\tau] \\ \psi_{2,1}^{(o)}[\tau] & \psi_{2,2}^{(o)}[\tau] & \cdots & \psi_{2,N}^{(o)}[\tau] \\ \vdots & \vdots & \ddots & \vdots \\ \psi_{M,1}^{(o)}[\tau] & \psi_{M,2}^{(o)}[\tau] & \cdots & \psi_{M,N}^{(o)}[\tau] \end{pmatrix}, \tag{11}$$

where *M* is the number of APs, *N* is the number of reference locations, *o* is the measured orientations from an embedded digital compass, $\tau = \{1, \ldots, Q\}$ is the index of temporal RSS samples, and *Q* is the number of samples for each raw RSS data. The rows of $\boldsymbol{\Psi}$ indicates the spatial RSS vectors over *N* reference locations with *Q* temporal length in the target environment. The columns of $\boldsymbol{\Psi}$, representing the *M*-dimensional RSS training samples for the *j*th reference location, can be represented as

$$\boldsymbol{\Psi}_j = [\psi_{1,j}^{(o)}, \psi_{2,j}^{(o)}, \ldots, \psi_{M,j}^{(o)}]^T, \tag{12}$$

Incorporating the orientation parameter is a generalization of the RSS collection procedure. This value may be missing for certain indoor positioning systems. It represents the difficulty of determining the location based on a history of RSS patterns $\boldsymbol{\Psi}$. The goal of the proposed algorithm is to achieve robust location estimation in the presence of orientation mismatch without assistance from a digital compass. That is, the proposed algorithm allows the attribute *o* to be ignored when the orientation is unavailable because of the lack of a digital compass.

Online Phase: During the online phase, a mobile device requests positioning services by using the online RSSs from *M* APs. Using similar notations, the real-time measurements can be represented as

$$\boldsymbol{\Psi}_r^{(o)} = \begin{pmatrix} \psi_{1,r}[1] & \psi_{1,r}[2] & \cdots & \psi_{1,r}[q] \\ \psi_{2,r}[1] & \psi_{2,r}[2] & \cdots & \psi_{2,r}[q] \\ & \vdots & & \\ \psi_{M,r}[1] & \psi_{M,r}[2] & \cdots & \psi_{M,r}[q] \end{pmatrix}_{M*q}, \tag{13}$$

where the location index r and the orientation attribute are unknown, representing that a digital compass is unnecessary in the proposed algorithm. In this chapter, the real-time measurements $\Psi_r^{(o)}$ is called the temporal–spatial radio, which consists of q temporal RSSs from M APs. HEQ is then applied to each component of this temporal–spatial measurement matrix. The first step is to reshape $\Psi_r^{(o)}$ into a vector form as follows:

$$\Phi = [\phi_1, \phi_2, \cdots, \phi_q, \phi_{q+1}, \cdots, \phi_{M\cdot q}], \tag{14}$$

Equation (14) is used to reshape the array in (13) into a row vector Φ with $M \cdot q$ elements, where $\phi_1 = \psi_{1,r}[1]$, $\phi_q = \psi_{1,r}[q]$, $\phi_{q+1} = \psi_{2,r}[1]$, and $\phi_{M\cdot q} = \psi_{m,r}[q]$.

For simplification, a reorganized sequence S consisting of $M{\cdot}q$ elements of a test RSS component in ϕ_r is then defined as

$$S = [s_1, s_2, \ldots, s_{M.q}] \tag{15}$$

The difference between (15) and (14) lies in the order statistics, which can be represented as

$$\phi_{T(1)} \leq \phi_{T(2)} \leq \cdots \leq \phi_{T(a)} \leq \cdots \leq \phi_{T(M.q)}, \tag{16}$$

where $T(\cdot)$ denotes the original index of the test RSS components in such a way that $s_a = \phi_{T(a)}$. Given S, an estimate of the equalized RSS (ERSS) component based on (9) is obtained using the following:

$$\hat{S}_i = F(s_i) = C_{ref}^{-1}[C_s(s_i)], \tag{17}$$

where C_{ref}^{-1} is the inverse of the reference cdf, C_s represents the cdf of the sequence S, F is the equalized transformation, and $\hat{S}_i$ is the ith ERSS. One can compute $C_s(s_i)$ by the empirical distribution function.

The selection of reference distribution C_{ref} for equalization discussed in (Russ, 1995; De la Torre, *et al.* 2005), and two widely used methods are adopted in this chapter: Gaussian distribution (indirect approach) and training distribution (direct approach). First, while adopting a typical Gaussian distribution as a reference (called Gaussianization) in the indirect approach, the inverse cdf can be expressed as

$$c_{ref}^{-1}(z) = [\Phi(\frac{z-\mu}{\sigma})]^{-1} = [\Phi(z')]^{-1}, \tag{18}$$

where z is a reference Gaussian variable with mean μ and standard deviation (STD) σ. The variable z' is the normalized Gaussian variable, and Φ is an error function expressed as

$$\Phi(z') = \frac{1}{\sqrt{2\pi}} \int_{-\infty}^{z'} e^{-\frac{u^2}{2}} d_u \tag{19}$$

Next, the training distribution means estimating cdf from the training radio map as reference for equalization. This direct approach can be viewed as a nonparametric version of the Gaussian distribution in which the shape is estimated empirically. The only requirement is sufficient temporal–spatial radio data to calculate the training distribution. Note that the training distribution can be stored in the offline stage such that the equalization requires no additional real-time estimation process.

Localization Using the Equalized RSS: After equalizing the online temporal–spatial RSSs, the equalized components are adopted for positioning using a probabilistic approach. This method involves mapping the ERSSs to a physical location by interpolation using the normalized likelihood value, i.e.,

$$\hat{L} = \sum\nolimits_{j=1}^{N} L_j \cdot \hat{P}_j, \tag{20}$$

where L_j represents the coordinates of the *j*th reference location, and $\hat{P}_j$ is a normalized likelihood with respect to all posterior probabilities as

$$\hat{P} = \left(\frac{p(\hat{s} \mid L_j)}{\sum\nolimits_{j=1}^{N} p(\hat{s} \mid L_j)} \right), \tag{21}$$

In (21), $p(\hat{s} \mid L_j)$ represents the likelihood of measuring $\hat{s}$ at the *j*th reference location, where $\hat{s} = \left[\hat{s}_1, \hat{s}_2, \ldots, \hat{s}_M\right]^T$ are the ERSSs from *M* APs. Assuming the equalized signals are modeled as Gaussian distributions, the statistical parameters, including a mean vector $\hat{u}_j \in \Re^{M*1}$ and a covariance matrix $\Omega_j \in \Re^{M*M}$ are calculated and stored for each L_j during the offline phase using

$$\hat{u}_j = \frac{1}{q} \sum\nolimits_{\tau=1}^{q} \hat{\Psi}_j[\tau] \tag{22}$$

$$\Omega_j = E\left[\left(\hat{\Psi}_J - \hat{u}_j\right) \left(\hat{\Psi}_J - \hat{u}_J\right)^T \right], \tag{23}$$

where $\hat{\Psi}_j[\tau]$ indicates the τ th equalized training RSS sample at the *j*th reference location, and the notation *E* represents the average over *q* temporal samples. HEQ was applied as the Wi-Fi radio representation for both online measurements and offline training RSSs. This procedure is required for the likelihood computation. The equalized training RSS samples can be obtained using (17) as follows

$$\hat{\Psi}_J = c_{ref}^{-1}\left[C_{\hat{\Psi}_J}\left(\hat{\Psi}_J\right)\right], \quad (24)$$

where the reference probabilistic distribution is the same as that used in the online phase. If one selects a typical Gaussian distribution as reference probabilistic distributions, C_{ref}^{-1} can be obtained using (18). If one selects the training distribution as reference, estimating cdf from the training radio map is required to compute C_{ref}^{-1} for equalization. $C_{\hat{\Psi}_J}\left(\hat{\Psi}_J\right)$ can be empirically estimated by dividing the number of observations less than a certain value by the total numbers. After estimating the statistical parameters based on the ERSSs, the likelihood weights $P(\hat{S} \mid L)$ in (21) can be calculated as follows:

$$P\left(\hat{s} \middle| L_j\right) = \prod_{m=1}^{M} \frac{1}{\sqrt{2\pi \cdot \Omega_j\left(m,m\right)}} \exp\left(-\frac{\left(\hat{s}_m - \hat{u}_j\left(m\right)\right)^2}{2\Omega_j\left(m,m\right)}\right), \quad (25)$$

where $\hat{u}_j$ *(m)* is the *m*th element of the mean vector $\hat{u}_j$, and the APs are assumed uncorrelated. Although assuming the diagonal matrix could be erroneous, this is widely used for implementing the probabilistic approach due to some practical constraints, such as insufficient training data. In this chapter, the amount of data is not sufficient to train an accurate full matrix; thus, a diagonal matrix is used for the location estimation. Note that, although this algorithm requires histogram estimation, it is easier than traditional methods, which require either a digital compass or an assumption of user behavior. The effect of orientation mismatch is assumed to cause a certain transformation of the RSS representation space. Therefore, the proposed algorithm equalizes RSSs to prevent the performance degradation caused by orientation mismatch. During positioning, HEQ is implemented for training data and real-time measurements in parallel, and the device coordinates are estimated by mapping the ERSSs to the physical location using the normalized likelihood values. While performing the nonlinear transformation, this approach preserves the ranking of RSSs instead of the numerical RSS values. When RSS values are reliable, i.e., perfectly following the radio propagation model, the transformation may lose some location information, and the side effect is possible in some clean environments.

Experimental Results and Analysis

Experimental Setup

This section shows the experimental setup. We conducted experiments on the fourth floor of the telecommunications building at Yuan Ze University, Taoyuan, Taiwan. Figure 2 shows where the experiments were performed. The size of the test field was 63.5 (m) × 36.5 (m), and 30 detectable APs were present in the environment (M = 30). An Asus laptop with Windows XP and NetStumbler network software were used to gather Wi-Fi RSSs from the APs. The RSS readings were recorded at 70 reference locations (N = 70), separated by a distance of 1.2–2 m and arranged in four orientations, to build the radio map for the indoor positioning system. The testing samples were measured at 30 distinct locations in this test field. To evaluate the orientation effects, users orientated in four directions, i.e., north, east, south, and west, and the corresponding RSS values were recorded. At each reference location, 50 RSS samples (Q

Figure 2. Part of the fourth floor plane of the test-bed building, where we performed the experiments

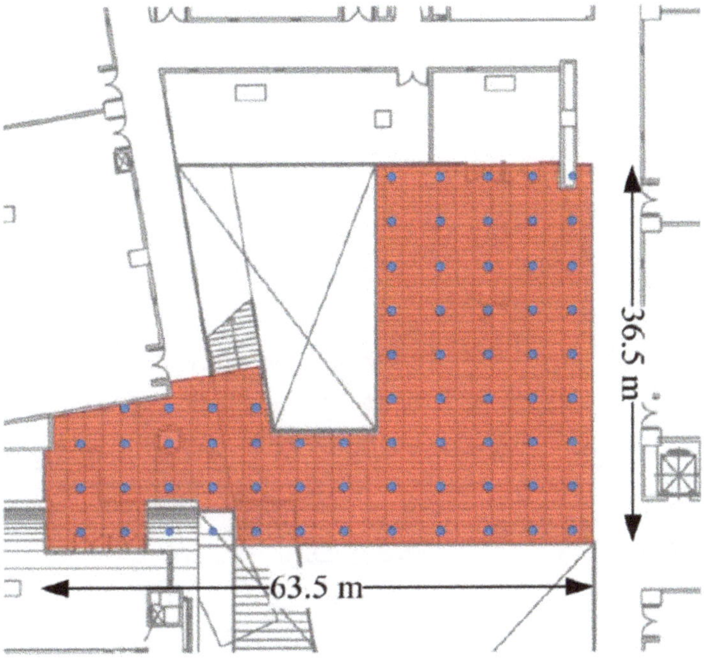

= 50) were recorded for each orientation. A Gaussian distribution with zero mean and unity variance was used as the reference distribution. HEQ was applied to each component of the temporal–spatial Wi-Fi radio map for both offline and online phases. To obtain the transformation of each component, the cumulative histogram was estimated by considering 100 uniform intervals between $u_i \pm 4\sigma_i$ where u_i and σ_i are the mean and STD of the *i*th component of the reshaped RSS vector, respectively.

Performance Evaluation

First, Figure 3 shows the distribution of RSS measurements from four orientations, at the same location, from the same AP. The lines indicate that the average RSS values from north, east, south, and west were −57.7, −56.6, −63.5, and −61.1 dBm, respectively. This figure clearly reveals the changing RSS patterns problem caused by the user orientation variation. It indicates the difficulty of accurately determining the location based on RSSs without the assistance of orientation information.

Figure 4 shows a box plot of the statistical positioning errors in orientation match and mismatch conditions in the fingerprinting system. Each box red line represents the median, the blue boxes represent the 25th and 75th percentile values, and the dashed lines represent the maximum and minimum value. In this experiment, the term "match" means that both testing and training data are measured in the same user orientation. By contrast, "mismatch" means that the testing data are measured from orientations different from which the training data were measured. For example, the first box for the mismatch condition indicates that the training data are measured from the north, while the testing data are measured from three orientations: east, south, and west. Figure 4 shows that a match between the testing and training

Figure 3. RSS distributions from the same AP with Different orientations

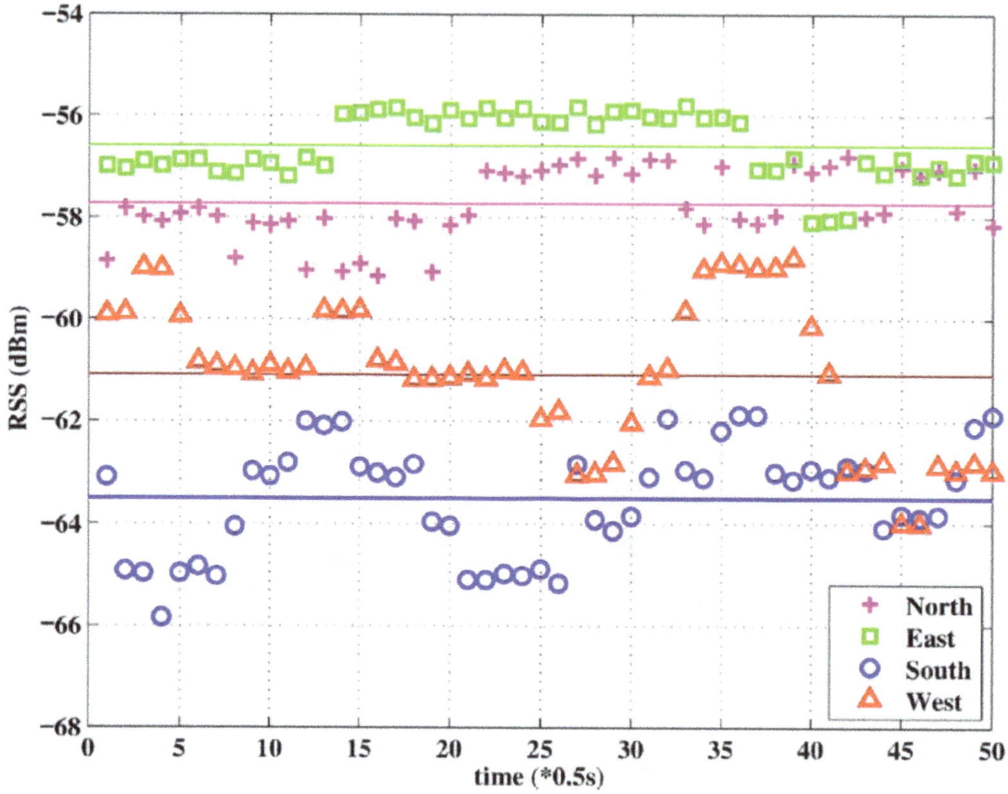

orientation is desirable for minimizing the positioning error in a fingerprinting-based location system. By contrast, the performance degrades when the orientation differs between the testing phase and the training phase. In the match condition, the median error was approximately 2.1–3 m, whereas that of the mismatch condition increased to 4–5.3 m. Numerical results showed that, in mismatched conditions, the performance of the median positioning error degraded 43.91% on average. These results verify that the mismatched orientation easily influences the positioning accuracy because the RSS patterns did not match in this scenario.

Figure 5 shows the cumulative positioning error distribution of the RSS and the Gaussian ERSS in the condition of orientation mismatch. In this experiment, the training data were measured from the west, whereas the testing data were measured from four distinct directions: north, east, south, and random. This figure shows that the proposed approach outperformed the original method under various mismatch conditions. In this mismatched orientation, the 67th percentile localization error of the proposed algorithm remained below 4.5 m, whereas that of the RSS increased to 7 m. The proposed algorithm, based on the ERSS, improved the performance of the raw RSS by approximately 28.84% in all cases. Figure 6 shows the cumulative distribution of positioning error under an alternative orientation mismatched condition, where the training data were measured from the north, and the testing data was measured from east, south, west, and random. According to Figure 6, ERSS evidently enhanced the robustness, and the accuracy within 4 m is improved from 48.75% to 79.12%. The results generally agree well with Figure

Figure 4. Box plot of positioning errors under orientation matched and mismatched conditions

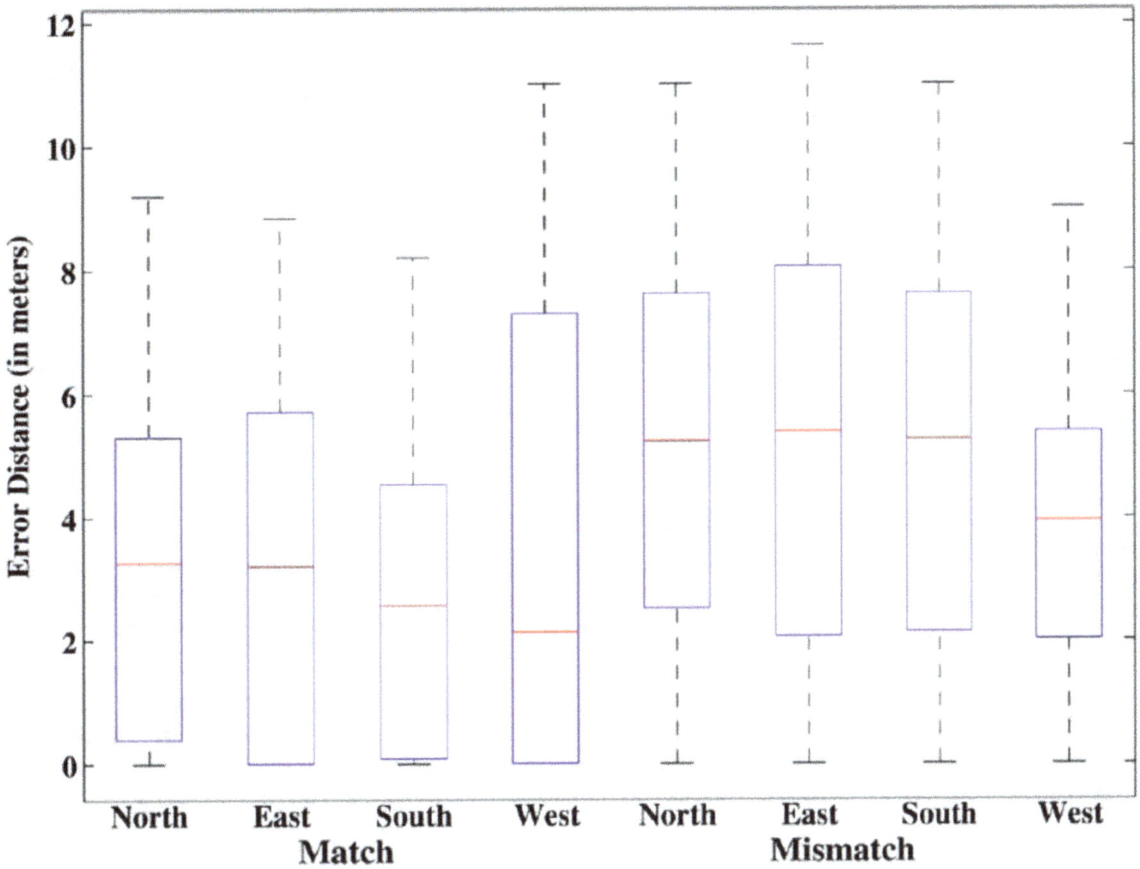

5, indicating that the proposed algorithm is capable of achieving robust location estimation, even when orientation is mismatched, by equalizing the temporal–spatial radio power.

Finally, the proposed algorithm is compared with four traditional approaches for addressing the problem of orientation variation: the compass-assisted method (King, Kopf, Haenselmann, Lubberger & Effelsberg, 2006; Sanchez, Quinteiro, Hernandez-Morera & Martel-Jordan, 2012), the orientation classifier (Liao & Kao, 2008), the compressive sensing (Feng, Au, Valaee & Tan, 2012), and the radio map filtering methods (Papapostolou & Chaouchi, 2009). The compass-assisted approach involves the assumption that both the device and radio map contain real-time orientation information and one estimates the testing measurement according to the orientation matched radio map only. Thus, this approach can be regarded as a perfect match case. The orientation classifier estimates the user's orientation according to the moving trajectory and varying RSSs based on the assumption that a user moves in a constant direction at a constant speed. When implementing the orientation classifier, this chapter assumed 75% accuracy of estimated direction. The compressive sensing approach combines affinity propagation to improve the positioning accuracy, where the real-time orientation information can be ignored. In this method, we set the number of clusters to 5, and the reduced dimension to 10. The radio map filtering approach removes incompatible RSS samples during the offline and online phases. We

Figure 5. Cumulative positioning error distribution under the orientation mismatched condition (the orientation of training data is west)

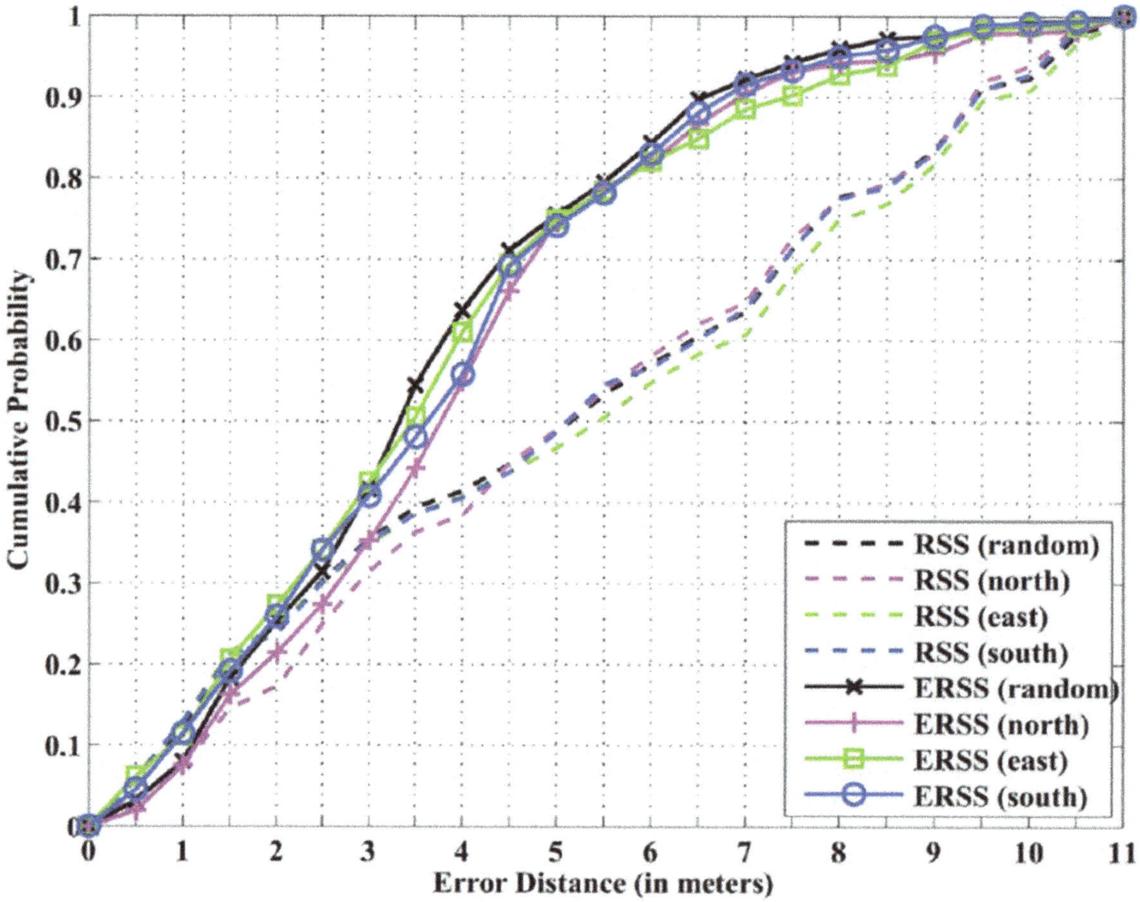

have implemented these methods and compared the positioning accuracy in a mismatch scenario. Note that, although compressive sensing and radio map filtering are robust to orientation mismatch, they are still based on an extended orientation-aware radio map (i.e., assume the availability of digital compasses during the offline and online phase).

Figure 7 shows a comparison of the positioning accuracy of various algorithms in a mismatch scenario, where the errors are cumulated from all mismatch conditions, and the orientation information of radio map is not given. This figure indicates the effectiveness of the orientation classifier approach. Compared with the traditional RSS method, which is easily influenced by the user orientation. The orientation classifier improved the median error by 38.34%. The compass-assisted approach achieved the most favorable performance, reducing the median error by 51.43% compared with the RSS. This is because the location is accurately estimated according to the orientation-matched radio map. This figure shows that the radio map filtering does not provide significant improvement, as compared with the RSS-based method. This is because probabilistic schemes are based on the statistics of the samples, and a sufficient number of samples are required to estimate the model parameters. This result agrees well with that in (Papapostolou & Chaouchi, 2009). and the result indicates the effectiveness of the

Figure 6. Cumulative positioning error distribution under an alternative orientation mismatched condition (the orientation of training data is north)

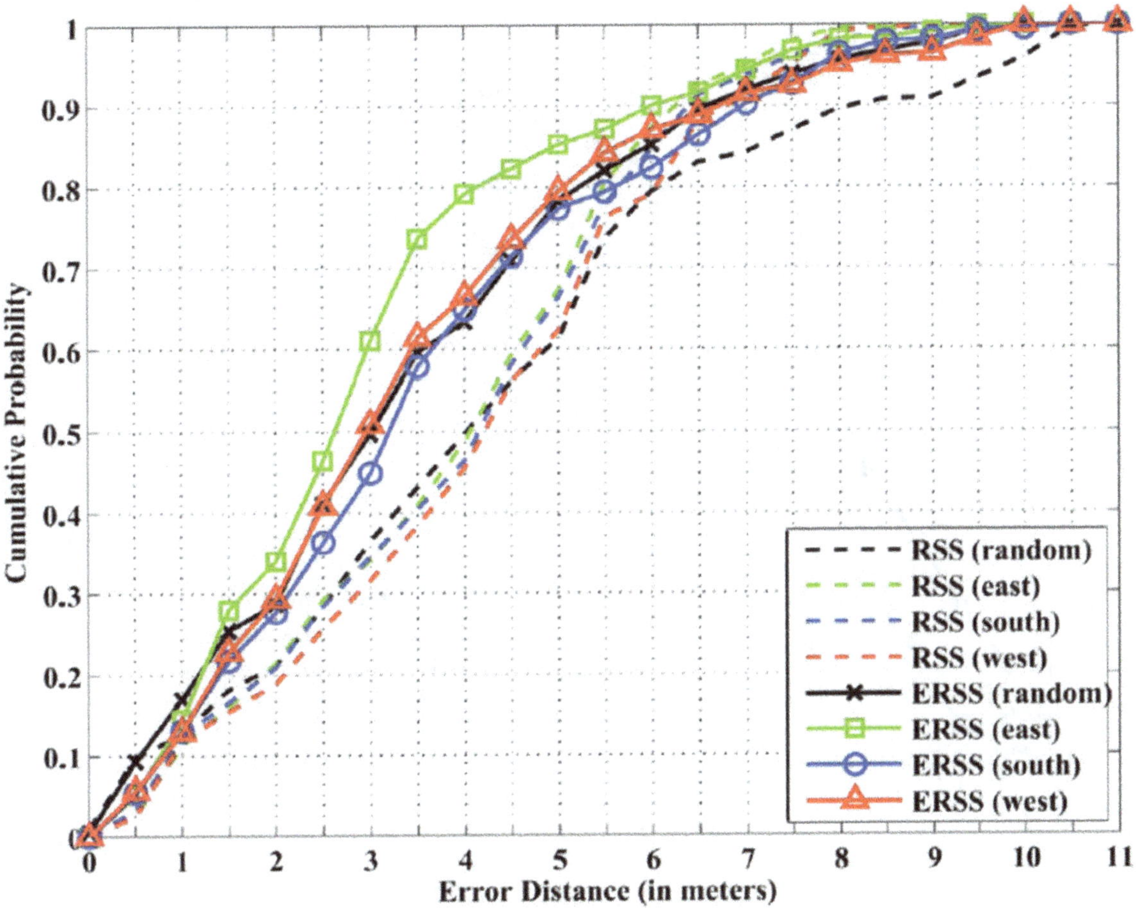

compressive sensing approach. The reason may be that the clustering method reduces the large error, as fewer reference locations are considered. More importantly, Figure 7 shows that the proposed algorithm provided better performance compared with the orientation classifier and even comparable accuracy to the compass-assisted method. The result again verified that the ERSS is a robust positioning characteristic. The advantage of the proposed algorithm is that no assumptions are made regarding the orientation process, which is difficult to model because of the random user behavior. Moreover, unlike traditional methods, which require assistance from a digital compass, the proposed algorithm does not require additional hardware and orientation information from the radio map.

Variation of Reference Distributions

Here, we performed extra experiments to evaluate the improvement of the proposed algorithm using different reference distributions. We estimated the reference distribution through the training temporal–spatial radio map instead of using Gaussian to perform HEQ. Compared with Gaussianization, the

Figure 7. Performance comparison between different positioning algorithms

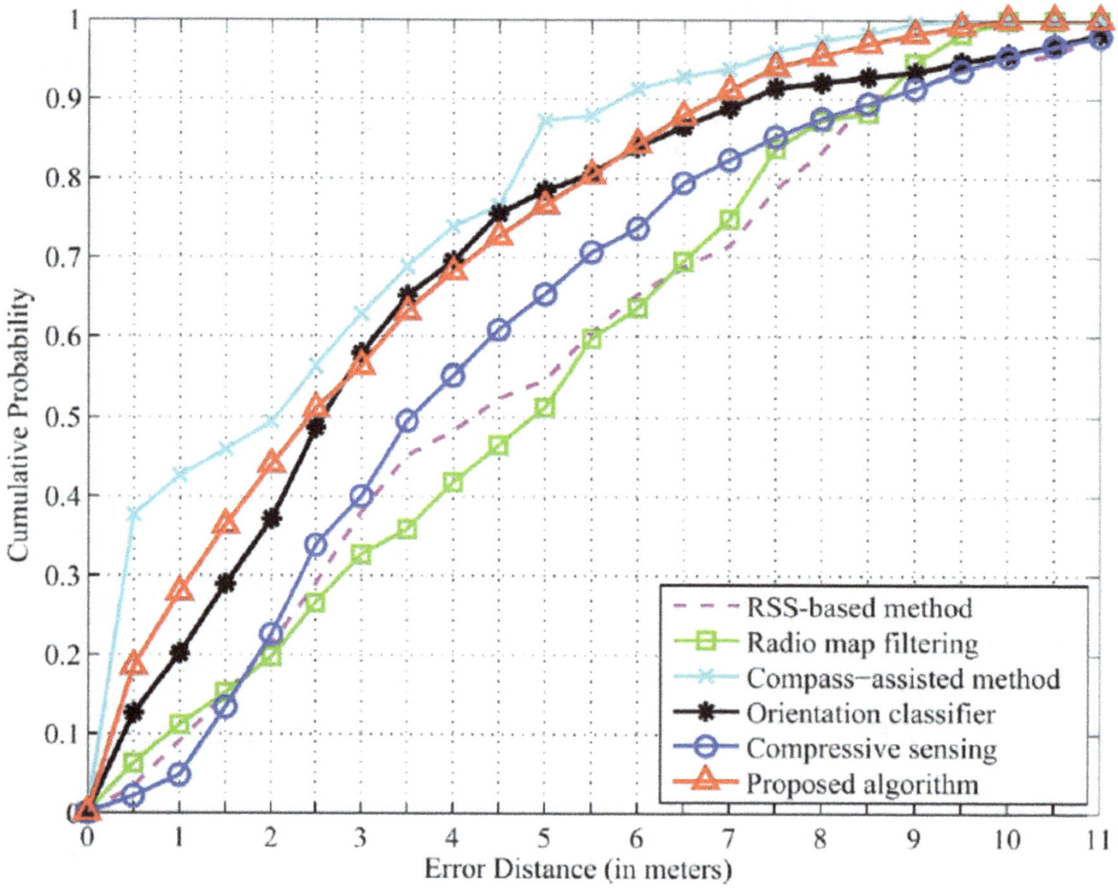

advantage of this approach is that only the testing RSSs are required to be transformed. Table 1 and Table 2 report five error measures numerically, including average, 50%, 67%, and 90% percentile errors, as well as STD in two mismatch conditions, where ref-G and ref-T indicate that the reference distribution is obtained, respectively, by Gaussianization and training data. The experimental setup in Table 1 and Table 2 is the same as that in Figure 5 and Figure 6, respectively.

Both tables show that no matter what reference distributions the proposed algorithm uses, ERSS still maintained robustness superior to that of RSS. This again demonstrates the effectiveness of the proposed mechanism. Nevertheless, we noticed that the performance slightly differs in different scenarios. For example, Table 2 shows that the selection of training data as reference gives better results than Gaussian for localization. It seems intuitive because the distribution is empirically calculated using additional information from radio map. However, in Table 1, Gaussianization outperforms the training data reference when the testing orientations are east and south because the estimated probabilistic distribution is susceptible to data variation. Changing the orientation of training data may introduce bias or errors in the global statistic, making precise distribution estimation difficult. Another critical issue is the amount of training data from the radio map. When the target indoor environment is large, the amount of estimation samples may be insufficient for a reliable histogram computation. As stated previously, the data-driven

Table 1. Five positioning error measures (in meters) for different orientation (training: West)

Testing: North					
Method/Performance	**Average**	50th	67th	90th	**STD**
RSS-based method	5.22	5.22	7.15	9.41	3.04
ERSS-based (Ref-G)	3.95	3.69	4.52	6.89	2.27
ERSS-based (Ref-T)	3.75	3.58	4.50	6.72	2.13
Testing: East					
Method/Performance	**Average**	50th	67th	90th	**STD**
RSS-based method	5.23	5.37	7.38	9.50	3.35
ERSS-based (Ref-G)	3.73	3.46	4.27	7.39	2.42
ERSS-based (Ref-T)	3.88	3.72	4.69	6.67	2.13
Testing: South					
Method/Performance	**Average**	50th	67th	90th	**STD**
RSS-based method	5.10	5.25	7.28	9.48	3.23
ERSS-based (Ref-G)	3.74	3.60	4.29	6.71	2.26
ERSS-based (Ref-T)	3.86	3.72	4.77	7.05	2.20
Testing: Random					
Method/Performance	**Average**	50th	67th	90th	**STD**
RSS-based method	5.09	5.19	7.20	9.42	3.24
ERSS-based (Ref-G)	3.64	3.26	4.18	6.51	2.15
ERSS-based (Ref-T)	3.79	3.60	4.38	7.43	2.44

reference distribution could enhance the robustness against orientation mismatch. Nevertheless, we note that Gaussianization is a safer approach to accommodate as much data variation as possible in the face of limited training data, unknown indoor environments, and training orientation.

Analysis of Results

To provide insight into the proposed algorithm, the mechanism that the ERSS compensates for the effect of orientation mismatch in location estimation is analyzed here. Figure 8 shows the original Wi-Fi RSS probability distributions, which were estimated according to the measurements from 1500 temporal–spatial signals (50 temporal samples from 30 APs) at a specific reference location. This figure shows that the diverse orientations severely affect the probability distributions of the Wi-Fi radio power. In the proposed algorithm, the effect of the orientation is assumed to be a certain transformation of the RSS representation space, incurring a considerable mismatch problem when training and locating orientations differ, which is verified by Figure 8. This observation is generally consistent for all reference locations. Figure 9 shows the absolute RSS values of the reshaped vector in (12). This figure shows how the multidimensional RSSs from four orientations change over time. Similarly, this figure shows that a

Table 2. Five positioning error measures (in meters) for different orientation (training: north)

Testing: North					
Method/Performance	**Average**	50th	67th	90th	**STD**
RSS-based method	3.83	4.03	5.00	6.15	1.90
ERSS-based (Ref-G)	2.93	2.59	3.23	5.99	1.90
ERSS-based (Ref-T)	2.59	2.31	3.23	6.27	2.27
Testing: East					
Method/Performance	**Average**	50th	67th	90th	**STD**
RSS-based method	3.90	4.10	5.04	6.39	2.02
ERSS-based (Ref-G)	3.53	3.17	4.11	6.99	2.24
ERSS-based (Ref-T)	2.49	2.31	3.12	5.82	2.34
Testing: South					
Method/Performance	**Average**	50th	67th	90th	**STD**
RSS-based method	4.04	4.12	5.09	6.67	2.02
ERSS-based (Ref-G)	3.37	2.95	4.00	6.67	2.24
ERSS-based (Ref-T)	2.41	2.23	3.00	6.09	2.41
Testing: Random					
Method/Performance	**Average**	50th	67th	90th	**STD**
RSS-based method	4.18	4.03	5.20	8.03	2.72
ERSS-based (Ref-G)	3.34	3.00	4.21	6.59	2.28
ERSS-based (Ref-T)	2.42	2.24	3.05	6.14	2.37

substantial influence resulted from the different orientations. This verifies the effect of orientation on measuring Wi-Fi radio signal power, and the mismatch makes it difficult to determine the location based on training RSS patterns when using the fingerprinting-based system.

Figure 10 shows an equalization case, where the testing RSS measured from the *E* are transformed such that their cumulative histogram matches that of the reference training data measured from the *W*. For example, the proposed algorithm changes the RSS values from −85 into −75 dBm in Figure 10 because the same cumulative probability of 50% is presented in the original and equalized domains, respectively. Note that, in this case, the positioning error using RSS is 6.84 m on average under orientation mismatch conditions, whereas it was reduced to 2.84 m using the ERSS. This means that the impact of orientation was significantly alleviated through the equalization process. In other words, the proposed algorithm compensates for variation in user orientations, which distorts the RSS space. Similar plots can be observed for testing data at different locations and orientations, and only a typical case is shown in Figure 10 because of space limitations.

Figure 11 shows the transformations obtained in Gaussianization case and used to convert the original RSS histograms into the reference histogram, by using the proposed algorithm. This figure shows that, in this case, the RSS values ranging from −50 to −100 dBm were transformed into an equalized version rang-

Figure 8. Impact of diverse orientations on the probability density of temporal–spatial RSS samples at a fixed reference location

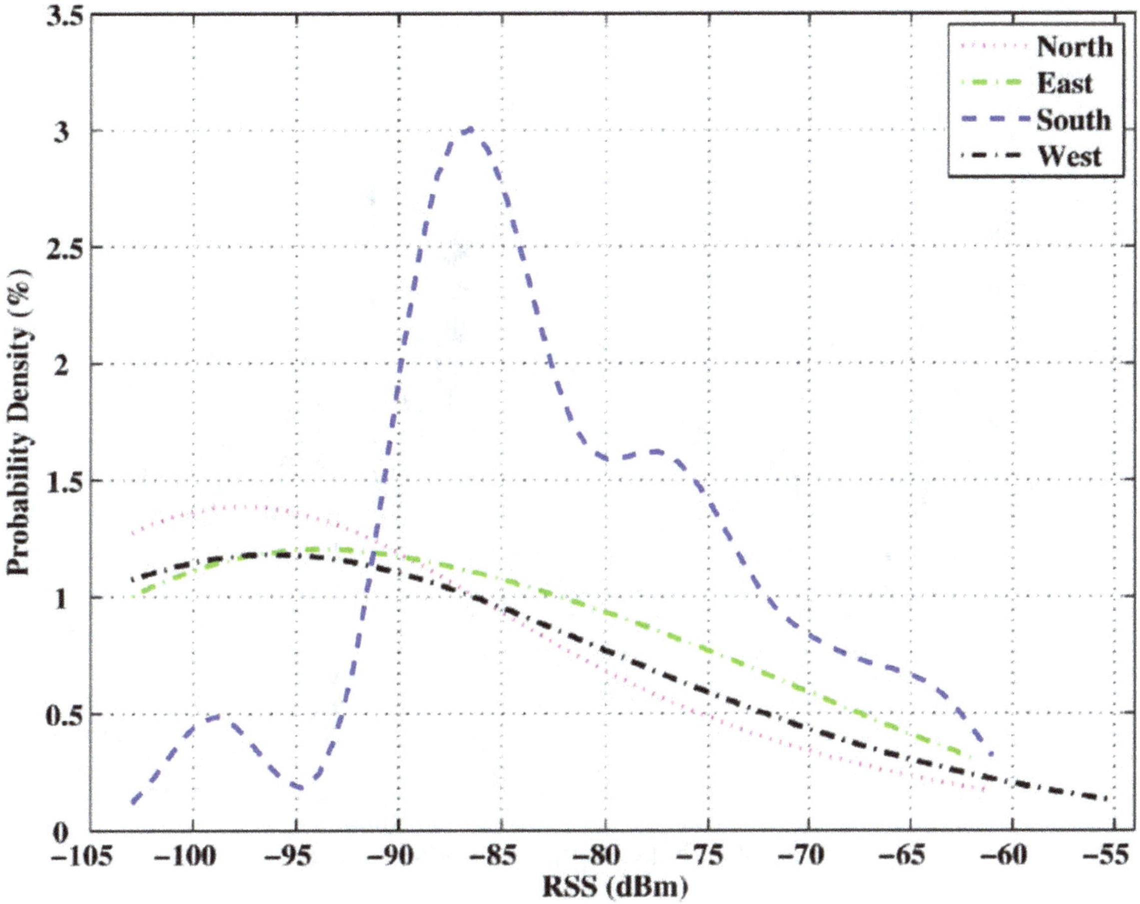

ing from −4 to +4. When a Gaussian was used as the reference distribution, the corresponding reference histogram was easily obtained and is also depicted in Figure 11. After the equalization, the histograms of the transformed RSS representations are shown in Figure 12. This figure shows that the transformed RSSs approximate the reference probability distribution and indicates that the impact of orientation on the distribution was considerably alleviated, as indicated in Figure 8 and Figure 9. When an orientation mismatch occurs between training and locating conditions, the proposed algorithm normalizes all moments of RSS distribution to those of the reference distributions. Note that, although the absolute RSS vales were equalized, the relative information between the multidimensional RSSs was still preserved for localization. Finally, Figure 13 shows the temporal–spatial trajectory of the ERSSs. Compared with Figure 9, Figure 13 shows that the impact of orientation mismatch was substantially reduced.

The experiments and analysis reveal the effectiveness of the proposed algorithm in providing robust indoor Wi-Fi localization in the presence of orientation mismatch.

Figure 9. Impact of diverse orientations on the absolute RSS values of the reshaped temporal-spatial vector (10) at a fixed reference location

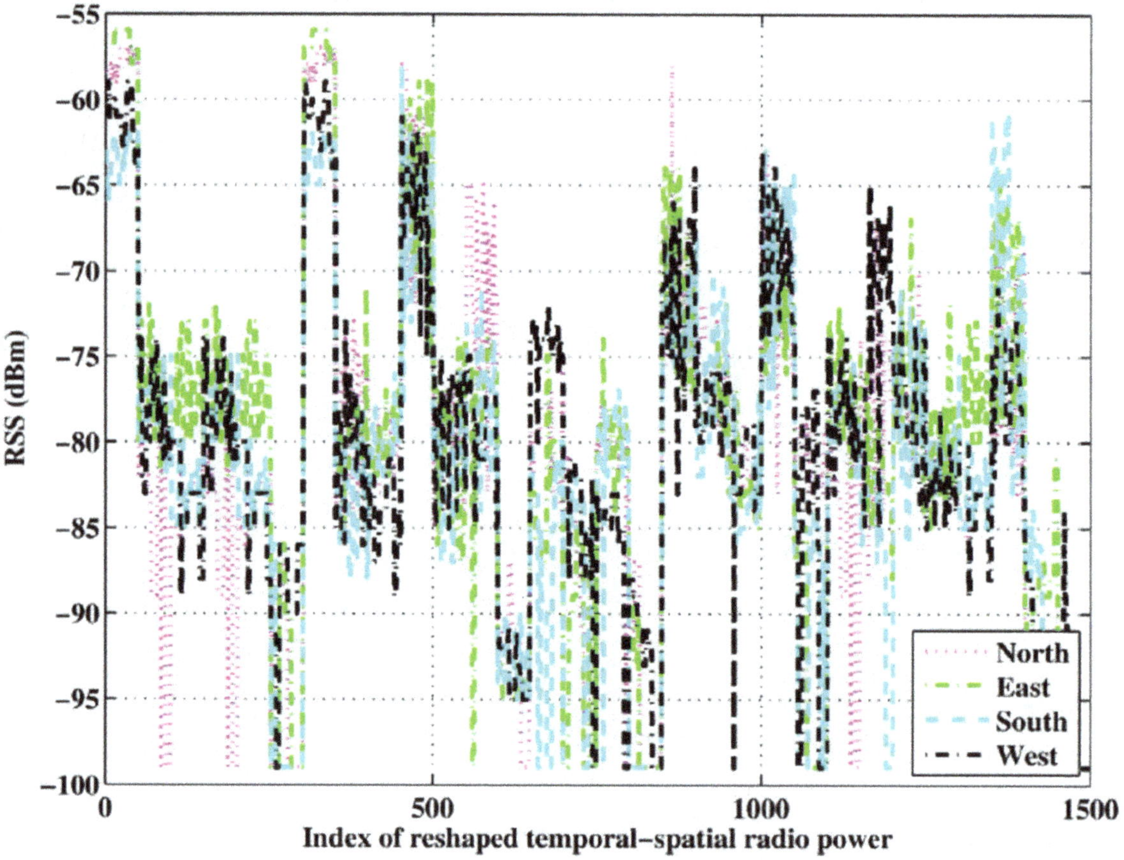

THE SOLUTION FOR HETEROGENEOUS HARDWARE PROBLEM

In this section, we are going to discuss the solution for heterogeneous hardware problem. It contains of three parts, "Delta Signal Strength", "Delta-Fused Principal Strength" and "Experimental Result and Analysis".

In the first and second parts, we present a novel positioning feature, called delta-fused principal strength (DFPS) to improve the robustness of Wi-Fi localization against the problem of heterogeneous hardware. In "Delta Signal Strength", we provide a series of subtraction process to reduce problem associated with hardware variation. The analysis explains why differential RSS is a more robust positioning feature than RSS with heterogeneous devices. In "Delta-Fused Principal Strength", we explain the side effects of using ΔRSS, which is the increased dimensions and the loss of discriminative information, and we provide ways to solve them.

Figure 10. Equalization case. The testing RSSs measured from the E are transformed such that their cumulative histogram matches that of the reference training data measured from the W.

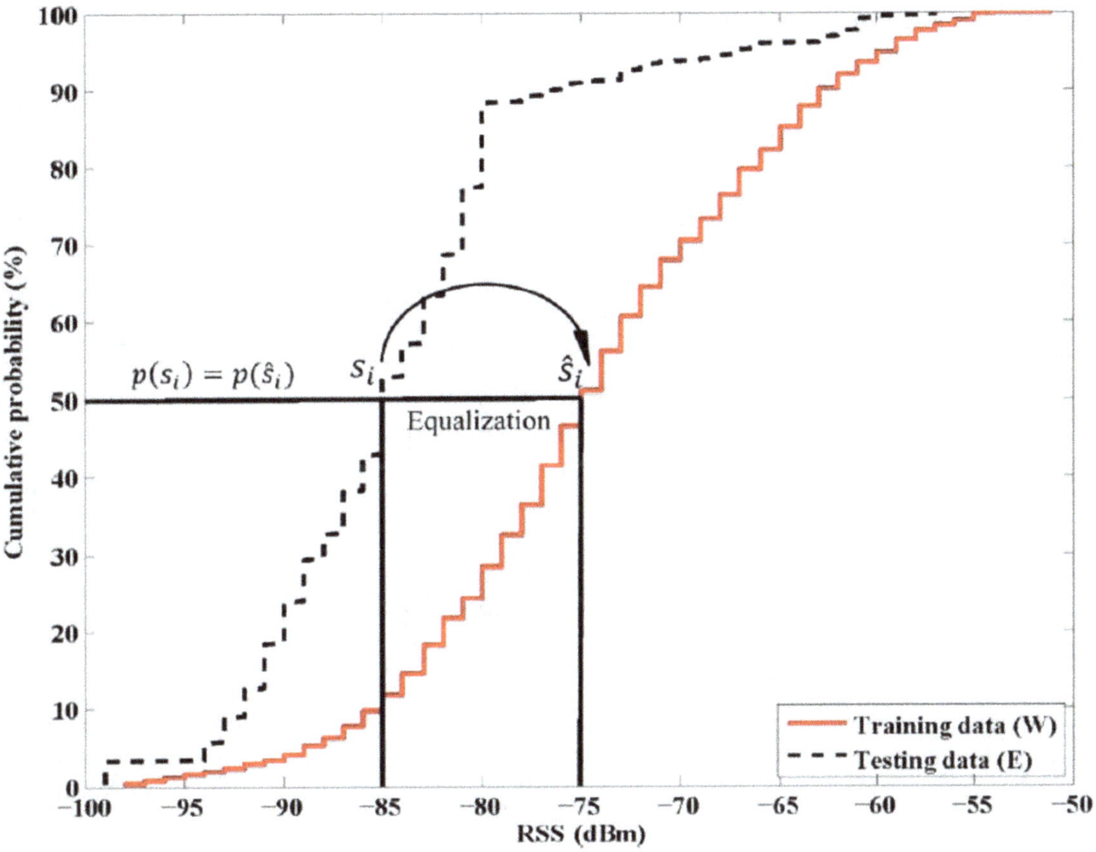

In "Experimental Result and Analysis", we additionally divide this section into six parts including "Experimental Setup", "Performance Evaluation", "Analysis of Delta-Fused Principal Strength", "Alternative Positioning Algorithm"," Multiple Devices Simulation" and "Different Test-Bed Environments and Devices". In "Experimental Setup", we discuss how and where to set up the experiment in details. In "Performance Evaluation", we state the result of the experiment. In "Analysis of Delta-Fused Principal Strength", we analyze the proposed DFPS algorithm from different perspectives. In "Alternative Positioning Algorithm", extra experimental results are used to evaluate the improvement of DFPS using an alternative positioning algorithm, support vector machines (SVM) regression. In "Multiple Devices Simulation", we consider a scenario in which the testing devices have an alternative receiver gain. Finally, in "Different Test-Bed Environments and Devices", results are provided from conducting an additional experiment in a largescale indoor environment.

Delta Signal Strength

The first step in DFPS is to compute the pairwise delta received signal strength (Δ RSS) to deal with the problem of hardware variation in RSS. It is assumed that the effect of heterogeneity can be cancelled

Figure 11. Example of the transformation function to equalize RSS and the reference histogram function

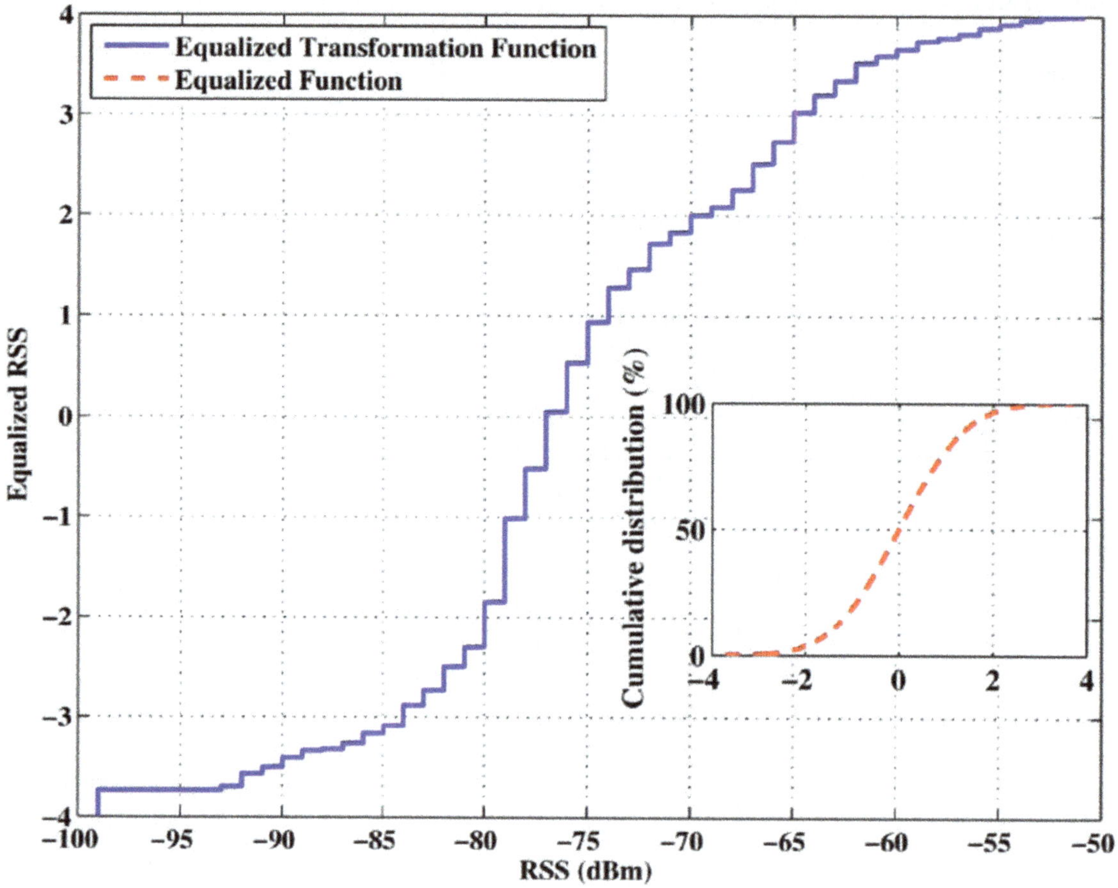

each other out through a subtraction process. In general, the log-normal RSS model can be expressed as (Rappaport, 2001):

$$P_r(d) = P_t - PL(d), \tag{26}$$

where $P_r(d)$ refers to the received power at a particular distance d, P_t is the transmission power, and $PL(d_0)$ is the path loss at distance d as:

$$PL(d) = PL(d_0) + 10n\log\left(\frac{d}{d_0}\right) + X_\sigma \tag{27}$$

In Eq. (27), $PL(d_0)$ represents the path loss at reference distance d_0, n is an exponent that depends on the propagation environment, and X_σ is a random variable representing uncertainty in the model.

Figure 12. Impact of diverse orientation on the probability density of equalize Gaussian RSSs at a fix reference location

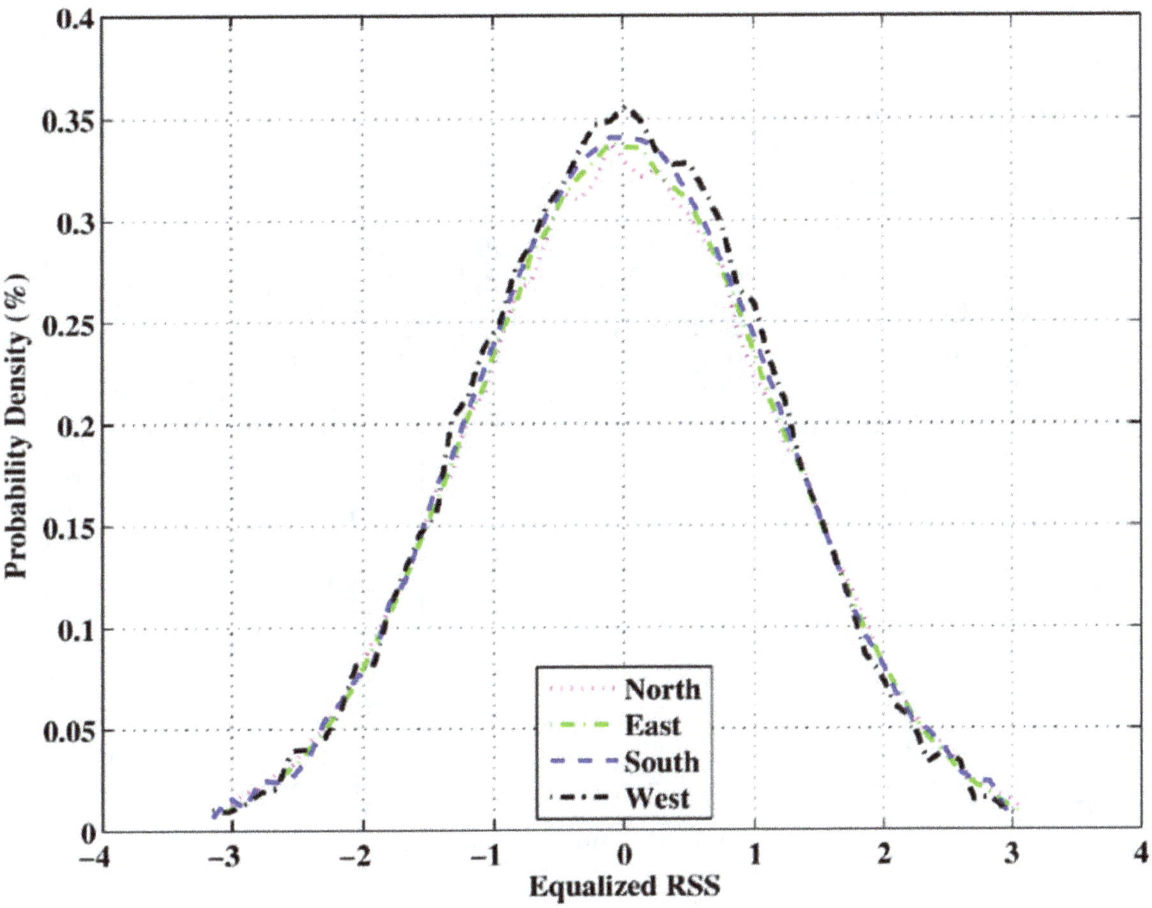

Path loss $PL(d_0)$ is defined as the difference between the transmitted power and received power, which may include the effect of antenna gain as:

$$PL(d) = -10\log\left(\frac{G_t G_r \lambda_t^2}{(4\pi)^2 d_0^2}\right) \tag{28}$$

where G_t is the transmitter antenna gain, G_r is the receiver antenna gain, and λ is the wavelength of the transmitted radio signal. We note that, since d_0 is a short reference distance, assuming a propagation exponent of two is appropriate and such an assumption widely used. One may use different propagation exponents here as long as the propagation exponent remains a constant. Substituting Eq. (27) and Eq. (28) into Eq. (26), we obtain

Figure 13. ERSSs of the reshaped temporal–spatial vector at a fixed reference location

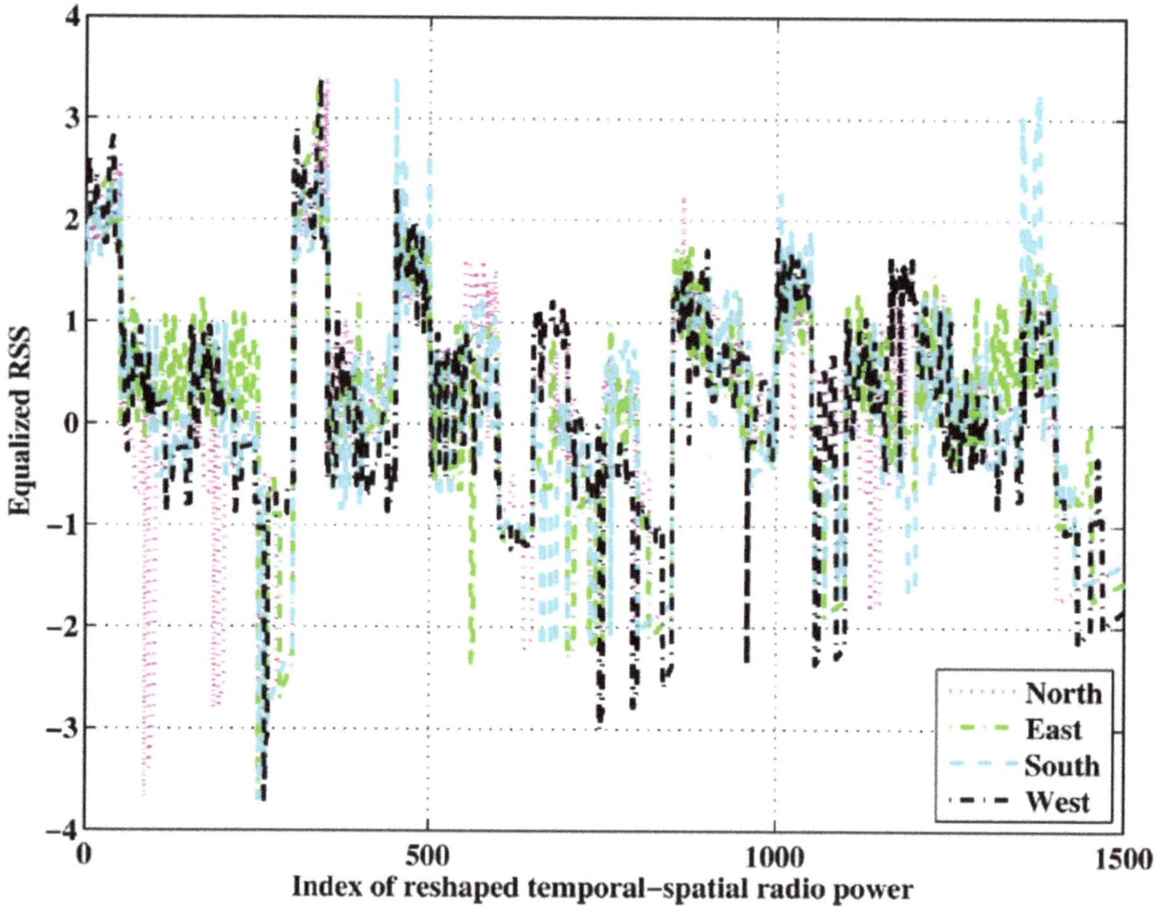

$$P_r(d) = P_t + 10\log\left(\frac{G_t G_r \lambda_t^2}{(4\pi)^2 d_0^2}\right) - 10n\log\left(\frac{d}{d_0}\right) - X_\sigma \tag{29}$$

Eq. (29) shows that two factors introduce variation into the measurements. The first factor is the random variable X_σ, which causes temporal variation in RSS due to the nature of radio propagation. The next factor is the receiver antenna gain G_r Under heterogeneous conditions, variations in G_r lead to entirely different RSS values, even over a fixed distance. Thus, because the measurement process for obtaining the RSS values is quite different from one device to another, Δ RSS is proposed to mitigate the effects of such variations in hardware. Assuming that two RSSs (x_1 and x_2) are measured from two different Aps (AP_1 and AP_2) at distances d_1 and d_2, we write

$$x_1 = P_1 + 10\log\left(\frac{G_1 G_r \lambda_1^2}{\left(4\pi\right)^2 d_0^2}\right) - 10n\log\left(\frac{d_1}{d_0}\right) - X_{\sigma 1} \tag{30}$$

$$x_2 = P_2 + 10\log\left(\frac{G_2 G_r \lambda_2^2}{\left(4\pi\right)^2 d_0^2}\right) - 10n\log\left(\frac{d_2}{d_0}\right) - X_{\sigma 2} \tag{31}$$

The delta signal strength can then be computed and simplified by subtracting Eq. (31) from Eq. (30):

$$x_1 - x_2 = -10n\log\left(\frac{d_1}{d_2}\right) - \left(X_{\sigma 1} - X_{\sigma 2}\right) \tag{32}$$

where we assume that AP_1 and AP_2 have the same transmission power, antenna gain, and the wavelength ($P_1 = P_2 = P_t$, $G_1 = G_2 = G_t$ and $\lambda_1 = \lambda_2 = \lambda_t$). Equation (32) shows that delta signal strength $\Delta X_{12} = X_1 - X_2$) is independent of mobile device antenna gain G_r, which reduces impact associated with hardware variation.

Delta-Fused Principal Strength

The above analysis explains why Δ RSS is a more robust positioning feature than RSS with heterogeneous devices. However, the side effects of using Δ RSS are the increased dimensions and the loss of discriminative information. First, due to the extended pairwise computation, it suffers from having to deal with a high-dimension space comprising input variables from *N* to C_2^N, where *N* is the number of RSSs from the sensed APs and the notation C_2^N refers to the combinations of *N* RSSs taken 2 at a time. That is, using Δ RSS requires approximately N^2 times more memory space and computation overhead than conventional methods. Moreover, discriminative power is reduced using Δ RSS, compared to RSS, because distance information becomes a distance ratio. In other words, the robustness is gained at the expense of some discriminative information. Thus, this subsection investigates how to exploit the complementary advantages of both Δ RSS and RSS, while overcoming the drawbacks simultaneously.

The proposed algorithm enhances discriminative power by fusing pairwise Δ RSS with traditional RSS. As mentioned above, we have *N* RSSs $\{X_1, X_2, \ldots, X_N\}$ and C_2^N Δ RSSs as

$$\Delta X_{ij} = X_i - X_j \; i, j \in \left(1, N\right), i < j \tag{33}$$

Unlike traditional methods, which select an individual subset from either Δ RSS or RSS, we combine Δ RSS and RSS into a fused vector X:

$$X = \left[\Delta x_{12}, \cdots, \Delta x_{ij}, x_1, \cdots, x_N\right]^T \tag{34}$$

The main objective of this step is to keep whole discriminative power as much as possible by retaining all features. Note that the dimension of the fused vector increases to $S_{\Sigma} = \sum_{k=1}^{k} \left(\widehat{X}_k - \widehat{X}\right)\left(\widehat{X}_k - \widehat{X}\right)^T$, . Thus, the next step in implementing DFPS is the design of the dimensionality reduction. The proposed algorithm extracts the useful location-related information through the transformation A:

$$y_d = A_{(d)}^T X, \tag{35}$$

where y_d represents the d-th component of DFPS and $A \in \Re^{\left(N+C_2^N\right)*N}$ represents a projection by which the Δ RSS and RSS are combined to determine DFPS. In Eq. (35), $A_{(d)}^T$ is the transpose of the d-th row of the projection. The coefficients in $A_{(d)}^T$ indicate how we fuse Δ RSS and RSS to produce the d-th DFPS y_d. Note that the dimension of DFPS is reduced from $N + C_2^N$ to N due to the multiplication of A.

If the projection A is suitably chosen, the transformed DFPS can exhibit high location-related information packing properties compared with the fused vector. This clearly differs from the traditional method, in which the subset of X is preserved to reduce the dimensions. Obviously, Δ RSS, RSS, SSD, and DIFF are special cases of DFPS, in which the corresponding coefficients are set to zero, respectively. Because the coefficients can take any real values in DFPS, it is possible to obtain different fused positioning features with different parameters of $A_{(d)}^T$. Moreover, we can easily control the required components by changing the number of columns in A such that the dimension of DFPS ranges from one to N. Here we selected N columns to maintain the same dimensions as RSS and avoid computational overhead.

Next, this study adopted principal component analysis (PCA) to determine the transformation A. This is the reason that the generated positioning feature is named delta-fused principal strength. Previous works have shown that PCA more efficiently utilizes location information in a fingerprinting-based system (Fang, Lin & Lin, 2008; Fang & Lin, 2012). However, because RSS and delta RSS represent different physical meanings and show significant scale difference, a standardization is required to rescale the fused features and make them more comparable. Here, standardizing means centering and then dividing each variable by its standard deviation (SD), so that all variables have zero mean and unit variance. This can be seen as a convenient change of units/scales, to make all variables comparable and remove outliers. Standardizing has been recognized a typical approach for handling the scale issues in PCA (Flury, 1997; Baxter, 1995; Arnold & Collins, 1993). In this study, we apply this standardizing procedure to prevent RSS's variability swamping all the other delta RSSs. Before performing PCA, we standardize all the fused components of X to ensure that delta RSS and RSS are treated on the same scale and the standardized vector is denoted by $\widehat{X}$. Given $\widehat{X}$, PCA attempts to find the projection that best represents the data set by selecting the maximum-variance directions as follows:

$$A_{opt} = Arg\max_{A} \left|A^T S_{\Sigma} A\right|, \tag{36}$$

where $S_{\Sigma} \in \Re^{\left(N+C_2^N\right)*\left(N+C_2^N\right)}$ is the covariance matrix of all normalized Δ RSS and RSS combined vectors $\widehat{X}$ from Eq. (34) as:

$$S_{\Sigma} = \sum_{k=1}^{k} \left(\widehat{X}_k - \bar{X}\right)\left(\widehat{X}_k - \bar{X}\right)^T, \tag{37}$$

where $\widehat{X}_k$ is the k-th normalized vector, k is the number of all training samples, and $\bar{X}$ is the global mean of all $\widehat{X}_k$.

Computing the projection in Eq. (36) reduces the problem of an eigenvalue-eigenvector. The solution A_{opt} is in fact a subset of eigenvectors of the covariance matrix S_{Σ}, given as:

$$A = \left[e_1, e_2, \cdots e_N\right], \tag{38}$$

where e_d is the d-th eigenvector obtained from the eigen-decomposition of the matrix in Eq. (37). The eigenvalue corresponding to e_d is λ_i, which quantifies the information contributed by the d-th DFPS component y_d. Because the covariance matrix is a positive-semidefinite symmetric matrix, the eigenvectors are geometrically orthonormal and statistically uncorrelated. This nice property guarantees that the generated DFPS components are uncorrelated to each other.

Note that standardizing all the variables prior to the analysis is equivalent to performing PCA on the correlation matrix instead of covariance matrix. While performing PCA, it is recommended to use the correlation matrix when within-variable range and scale widely differ, and use covariance matrix to preserve variance if the range and scale of variables are similar or in the same units of measure. Second, it preserves the correlation between the original data. If we divide data by the corresponding SDs, the correlation between the standardized data equals the correlation between the original data, which is a desirable property. However, (Flury, 1997) described this issue as an anti-property of principal components. The argument against automatically using correlation matrices is that standardizing is quite a brutal way of normalizing the data (Jolliffe, 1986).

The computation of $\boldsymbol{A}$ requires only the RSS values taken with a particular device in the training phase. The same transformation $\boldsymbol{A}$ is then applied to both the fingerprints and the online observation such that the pattern matching is feasible. Although we directly assign N DFPS components to perform a fair comparison with RSS, the proposed approach provides a mechanism with which to objectively determine the required dimensions under various constraints. A measure of how well the retained DFPS provides information is given by the relative proportion of eigenvalues:

$$\frac{\sum_{d=1}^{L} \lambda_d}{\sum_{d=1}^{c_2^N+N} \lambda_d} \geq \eta^*, \tag{39}$$

where L is the desired number of DFPS components. We can set a cutoff threshold η^* to represent the percentage of total information contributed by the selected DFPS components. The value of L is thus the smallest integer satisfying Eq. (39), in which the specified threshold η^* is exceeded. The value of η^* can sometimes be higher or lower depending on practical applications. If computational overhead is a major concern, a smaller η^* can be selected to reduce the number of parameters involved. On the other hand, selecting η^* to be approximately 90% is a practical approach, which preserves most of the information. It can be noted that by definition of combined vector in Eq. (34), the correlation among measurements due to gains can be mitigated by explicitly including them into the model. Since the difference variables and measurements are correlated, the covariance estimate must provide a robust estimate for uncorrelated measurement vector if the back-end fingerprinting system uses a traditional probabilistic approach. Alternatively, the fingerprinting systems may use some popular machine learning algorithms to handle the incurred correlation among the extended dimensions in Eq. (34).

Experimental Results and Analysis

Experimental Setup

To evaluate the effectiveness of our DFPS algorithm, experiments were conducted in the FarEasTone Telecommunications Building at Yuan-Ze University, as shown in Figure 14. We used an Asus laptop and HTC smart phones as heterogeneous devices to gather RSSs. Figure 15 shows the RSS measurements over time at the same location and from the same AP, using the laptop and smart phone, respectively. The lines represent that the average RSS value from laptop is −47.6 dBm whereas that from the smart phone is −66.2 dBm. This figure clearly reveals the hardware variation problem since the RSS patterns could not match with heterogeneous devices. In the experiments, RSS data of seven APs ($N = 7$) were collected 200 times for each reference location. This means that each RSS measurement contained seven numerical values. We collected Wi-Fi RSS at 25 different reference locations in the area on different days and at four different user orientations. We then selected 17 locations, separated by a distance of 1.2 to 2 meters, as the training data to build the baseline positioning system. Through the developed software, the mobile devices record the instantaneous RSS values in dBm. We take the average of 5 RSS samples as inputs and feed to the location system.

Using the experimental results, the DFPS is compared with four different approaches which are, RSS, HLF (Kjærgaard & Munk, 2008; Kjærgaard, 2011), SSD (Mahtab Hossain, Van, Jin & Soh, 2007; Fang, Lin & Lin, 2008), and DIFF (Dong, Chen, Liu, Ning & Piao, 2009) methods. Note that DFPS maintains the same dimensions as RSS by setting the cutoff threshold η^* as 98%. By contrast, the dimensions of HLF and DIFF increase from 7 to 21 C_2^7 due to the extended pairwise strength. Unlike HLF, SSD reduces the dimensions to 6 based on an independent subset selection mechanism. In fact, it reduces the

Figure 14. Part of the sixth floor plane of the FarEasTone Telecommunications Building, where the makers indicate the locations of APs

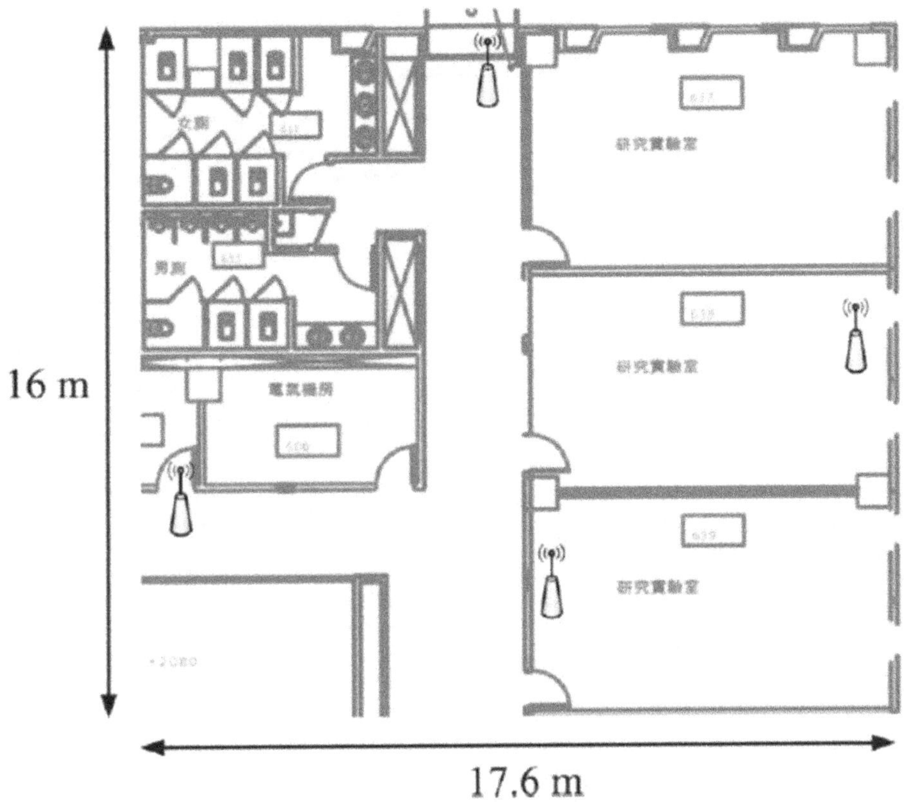

dimensions from N to $N-1$ under the assumption of independence between APs. Nevertheless, this assumption could be erroneous. Our positioning system adopts a fingerprinting approach via Multi-layer Perceptron (MLP) (Battiti, Nhat & Villani, 2002; Fang & Lin, 2008; Stella, Russo & Begusic, 2007). This method consists of a number of interconnected neurons in parallel to nonlinearly map RSS to coordinates. The adopted MLP contained an input layer, a hidden layer (a TANSIG transfer function with 10 neurons), and an output layer. MLP is trained using a typical gradient-based algorithm to minimize the positioning errors. This study defines the positioning error as the Euclidean distance between the estimated results and the true positions.

Before performing DFPS, we standardize the RSS and Δ RSS values into the same scale. Indeed, there are many normalizing methods which can make variables comparable. For example, one can indeed rescale variables by dividing different measures of spread such as the median absolute deviation. However, it does not preserve the correlation information between original data. Applying the standardizing procedure before PCA has shown effectiveness when the data are measured in different units/scales. Since standardization has been recognized as a typical approach for handling the scale issues in PCA, we use standardization to generate DFPS in the following experiments and believe that it will not mislead the contribution of this study.

Figure 15. The RSS values over time using diverse device from the same AP and at the fixed location indicate the f APs

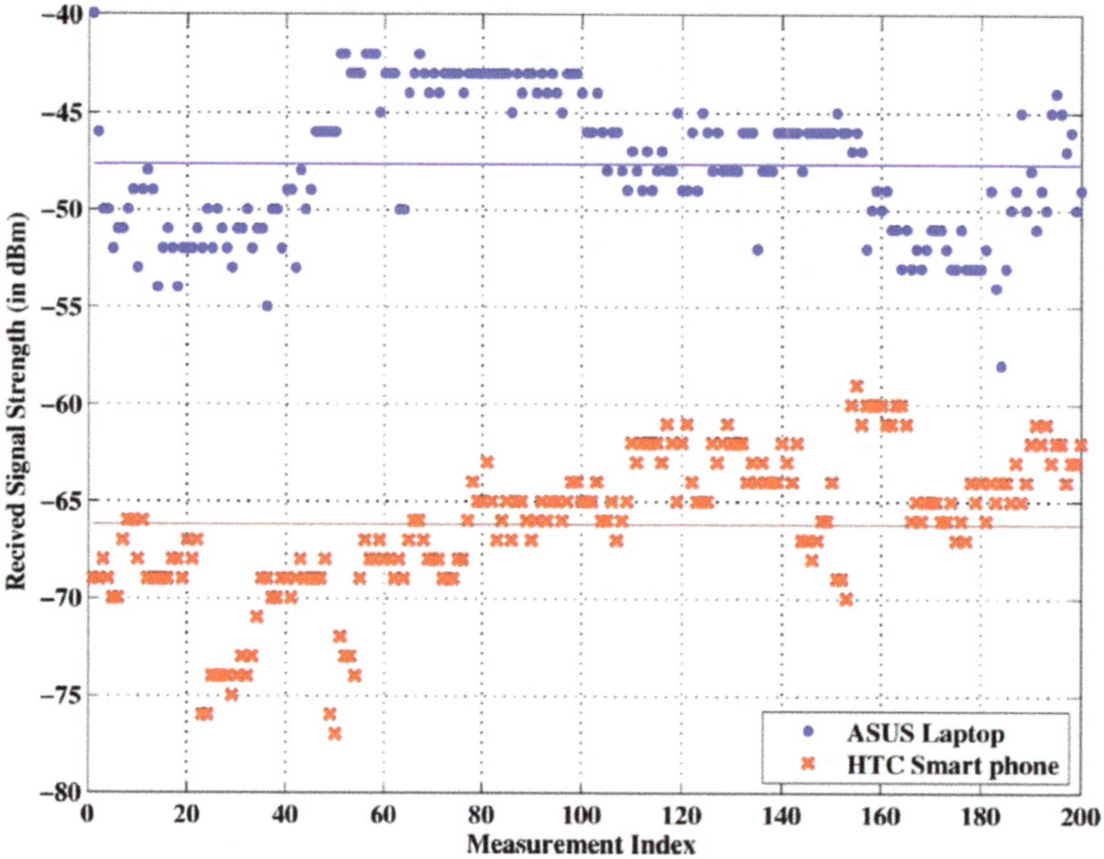

Performance Evaluation

Figure 16 and Figure 17 display the cumulative positioning error distribution of five approaches: RSS, SSD, HLF, DIFF and DFPS using heterogeneous and homogeneous devices. Here, heterogeneous refers to the measurement of testing data using the smart phones while training data is obtained using a laptop. Figure 16 compares the performance, showing that delta-based positioning features, including SSD, HLF, and DIFF, outperformed the original RSS with heterogeneous devices. This verifies that Δ RSS improves robustness when dealing with the problem of hardware variation by extracting hardware-independent features. Figure 18 shows the average variation of RSS and Δ RSS using a range of diverse devices, in which Δ RSS clearly demonstrates a less variation than RSS for all training locations. This explains why Δ RSS outperforms RSS under heterogeneous conditions. More importantly, Figure 16 shows that the proposed DFPS outperforms traditional methods, reducing the 67-th percentile localization error by 12.8%, 12.9%, 34.8% and 96%, compared to DIFF, SSD, HLF and RSS, respectively. These results clearly demonstrate the ability of DFPS to overcome the effects of heterogeneous hardware.

Figure 16. Cumulative positioning error distribution with heterogeneous devices

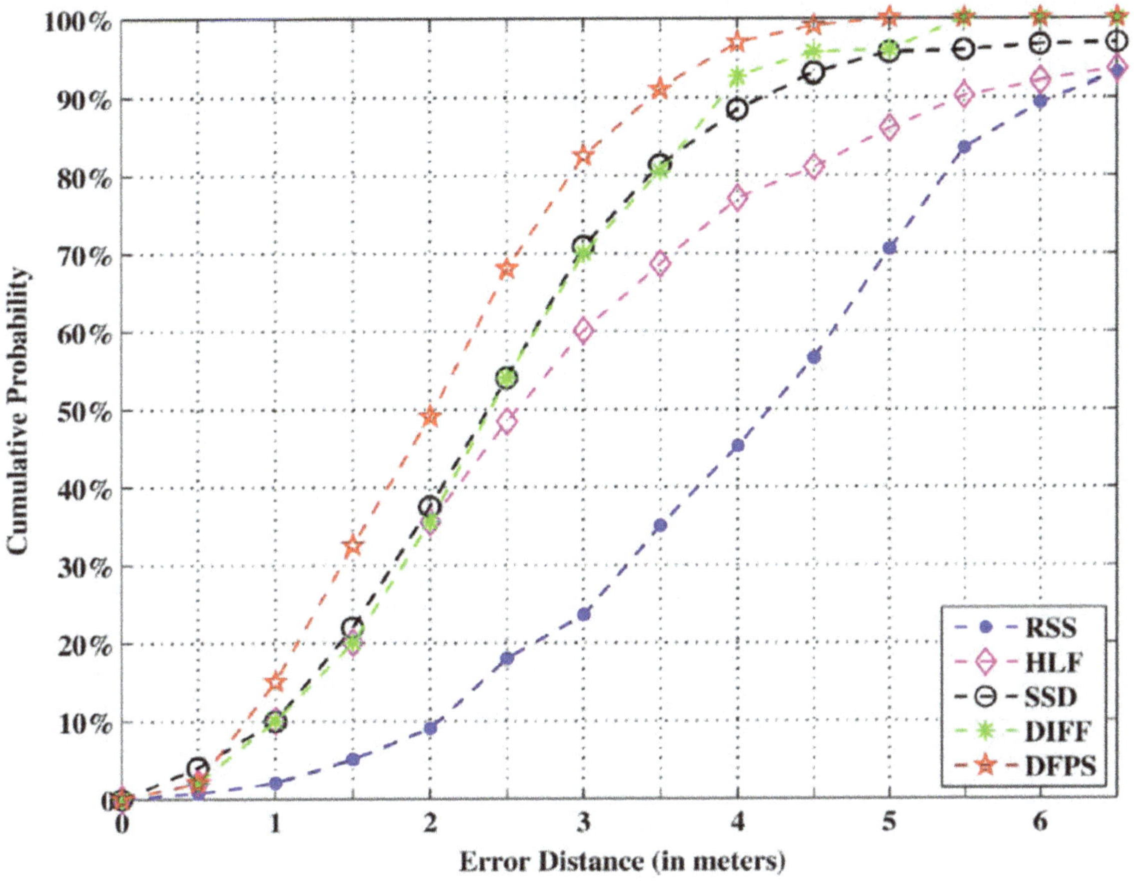

(a)

Figure 17 compares positioning results using homogeneous devices, in which the testing and training data are both measured from a laptop. DFPS achieved a three-meter accuracy of 92.85%, while DIFF, SSD, HLF and RSS were 88.31%, 83.52%, 79.83% and 85.87%, respectively. From Figure 17, the performance of HLF and SSD is clearly inferior to that of RSS in this scenario because with the same hardware, the effective positioning features are better characterized by RSS values than by Δ RSS. These results shows that, although the delta-based features perform well with heterogeneous devices, the improvement may be gained at the expense of discriminative information. Fortunately, the proposed DFPS overcomes such limitations by embedding discriminative power within Δ RSS, while compensating for the problems RSS has in dealing with variations in hardware. Figure 16 and Figure 17 show that only the proposed DFPS is capable of achieving superior positioning performance with both heterogeneous and homogeneous devices. This is because the complementary advantages of RSSs and Δ RSS are intelligently fused through principal component analysis. Thus, it can be concluded that DFPS overcomes the main drawbacks inherent in traditional robust location fingerprinting methods.

Figure 17. Cumulative positioning error distribution with homogeneous devices

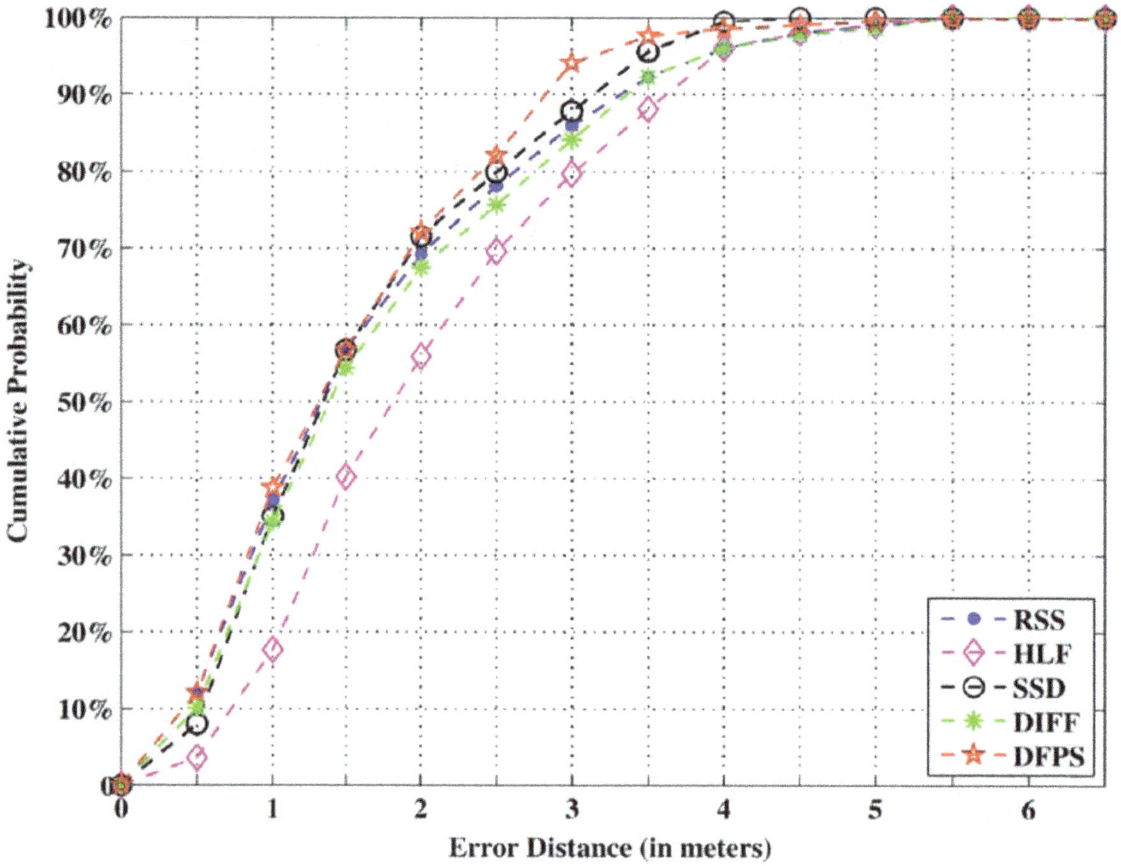

(b)

Figure 17 shows that DIFF delivers performance similar to that of RSS; however, the dimensions of DIFF (21) far exceed those of RSS (7). That means that DIFF consumes a great deal of memory space and requires a longer processing time due to the extension of Δ RSS. By contrast, DFPS maintains the same dimensions as RSSs by squeezing the useful information into relatively fewer components. Therefore, it does not require any more memory or computational overhead than conventional methods. As indicated in Figure 16 and Figure 17, the performance of DFPS using homogeneous devices is generally superior to that using heterogeneous devices. The two-meter accuracy of RSS and DFPS improved from 10% to 70% and 50% to 73%, respectively. Finally, five error measures were numerically reported in Table 3 including STD (standard deviation), average, 50%, 67%, and 90% percentile errors. This table shows that DFPS provides the best positioning performance in terms of the five error measures among the five compared approaches in two different scenarios.

Finally, we perform a simple experiment to study how the rescaling operations affect the positioning performance. Figure 19 compares the positioning accuracy between DFPS and RSSs with PCA under heterogeneous conditions, where r_1 denotes a typical standardization procedure, PCA and PCA$\left(r_1\right)$ indicate the PCA results without and with the standardization on RSSs, respectively.

Figure 18. Variations in RSS and Δ RSS resulting from heterogeneous devices in different locations. Here the Δ RSS is obtained by $x_1 - x_2$.

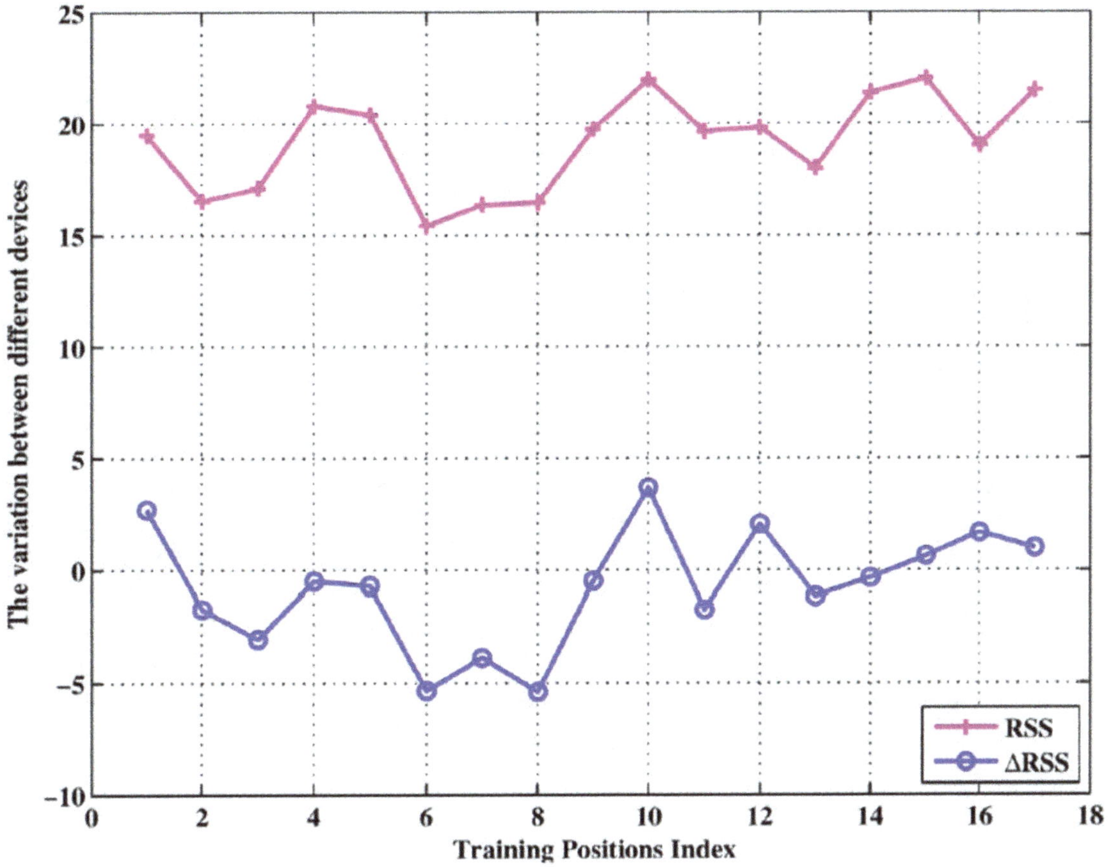

Figure 19 shows that, DFPS (r_1) offers robustness improvement compared to PCA (r_1) even based on the same standardization method. This figure also shows a slight difference between DFPS (r_1) and DFPS (r_1), indicating normalization methods achieve diverse performance in different positioning error regions. More importantly, this indicates that, although different normalization schemes may generate different results, the proposed algorithm outperforms previous robust positioning features, whether the normalization is standardization or not. The best way to normalize data is still an open issue in applied statistics (Flury, 1997; Bieber, Brooks, Harbottle & Sayre, 1976; Wright, 1989). The empirical results in (Baxter, 1995) show that standardization and log-transformation lead to equivalent results in many cases. The literature summarized that the best normalization method may depend on a subjective choice, experience and cross-validation (Flury, 1997; Baxter, 1995). The optimization of normalization will be an interesting topic to further investigations. In the following experiments, we uses standardization (r_1) to generate DFPS.

Figure 19. The impact of the rescaling operation on RSSs with PAC and DFPSs under heterogeneous Devices conditions

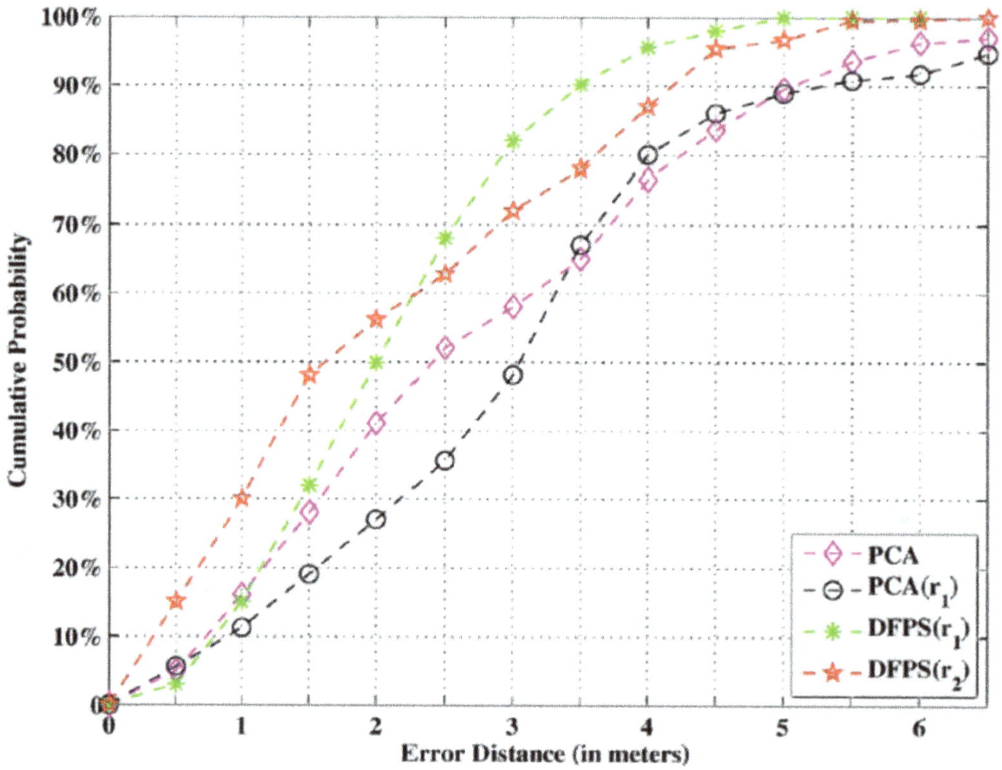

Analysis of Delta-Fused Principal Strength

In this section, we analyze the proposed DFPS algorithm from different perspectives. Figure 20 illustrates the first two DFPS components (y_1 and y_2) and the principal component scores (the coefficients in matrix ***A***) for each observation (x_i and Δx_{ij}). In Figure 20, some important variables are represented by contributed vectors. The direction and length of the vectors indicate how each observation contributes to the first two DFPS components.

For example, the first DFPS component y_1 , represented by the horizontal axis, has positive coefficients for the vectors, including Δx_{36} and Δx_{35} , which are directed into the right half of this figure. It can be observed that x_3 , x_6 , Δx_{36} and Δx_{67} are significant in the production of the first DFPS component y_1 . From Figure 20, we also see that the important observations in the generation of the second DFPS component y_2 are x_2 and Δx_{23} This figure illustrates how the proposed DFPS combines RSS and ΔRSS in an intelligent manner, such that the complementary advantages of different positioning characteristics can be simultaneously exploited.

Figure 20 also shows that the cross correlation between AP exists because the weights of Δx are not zero. This is because many factors may introduce the dependence between RSSs and the inherent spatial

Figure 20. Contribution from RSSs(x_i) and Δ RSS(Δx_{ij}) to the first two DEPS components (y_1 and y_2)

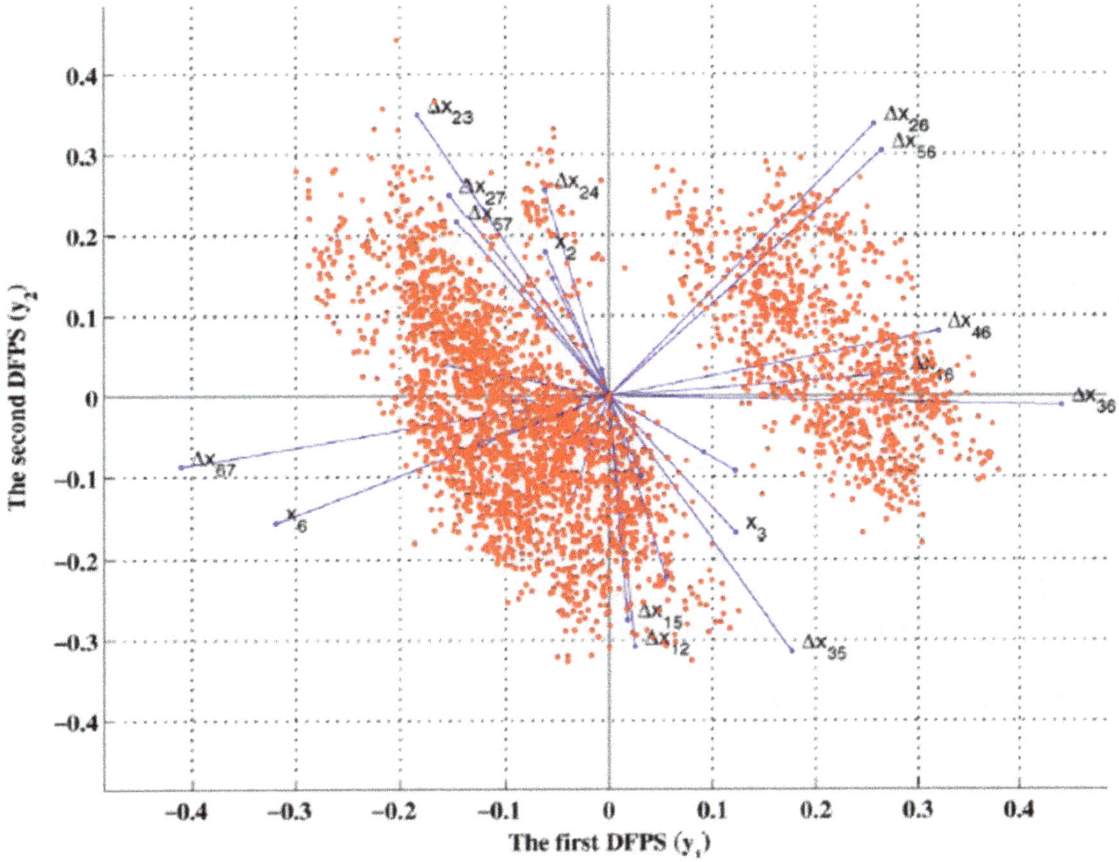

dependence is definitely one such example. For example, let us assume a long corridor with two APs placed at the opposite sides of the corridor. If the RSS from one AP is very high (i.e. the device is close to it) then with very high probability the RSS from the other AP will be low.

Moreover, according to the standard, there are only three non-overlapping channels in the IEEE 802.11b/g 2.4 GHz frequency band (channels 1, 6, and 11). That means, the interference between the RSSs from Wi-Fi APs exists when there are more than three APs in the environment. This also explains the dependency between RSSs from different APs in some sense. Some studies have reported that the positioning performance of a fingerprinting-based system is not monotonically improved as the number of AP increases (Fang & Lin, 2012; Kushki, Plataniotis & Venetsanopoulos, 2007; Yin, Yang & Ni, 2008). This also justifies the dependence of APs from an experimental perspective. Although fusing the delta measurements may introduce extra correlation, the proposed algorithm compresses the useful information into relatively smaller DFPS dimensions. The novelty of DFPS lies in the combination of RSS with delta RSS in a unified framework, and the independence can be viewed as a special case in the proposed algorithm. When Aps are independent, DFPS reduces the dimensions by removing all depending components through PCA.

Table 3. Five positioning error measures (in meters) for different algorithms based on mlp

Method	STD	50th Percentile	67th Percentile	90th Percentile	Average
Heterogeneous Case (Training: Laptop, Testing: Mobile Phone)					
RSS	1.52	4.33	4.89	6.70	6.14
HLF	1.46	2.55	3.43	5.51	4.50
DIFF	1.35	2.41	2.75	4.19	3.27
SSD	0.98	2.47	2.87	3.83	3.30
DFPS	0.90	2.05	2.57	3.51	2.92
Homogeneous Case (Training: Laptop, Testing: Laptop)					
RSS	0.88	1.25	1.90	3.31	3.72
HLF	0.86	1.74	2.35	3.67	4.45
DIFF	0.92	1.32	1.83	3.13	3.67
SSD	0.78	1.40	2.00	3.39	3.82
DFPS	0.87	1.20	1.80	2.76	2.50

Next, we discuss the number of required components, which is a critical element in the DFPS approach. In DFPS, a measurement of how well the projected dimensions provide information is given by the relative proportion of the eigenvalues in Eq. (39). This measure indicates the quantity of information explained by the corresponding DFPS. Figure 21 displays the individual and the cumulative percentages of information explained by DFPS. The first three DFPS components explain more than two-thirds

Figure 21. Individual and the cumulative percentages of Information explained by DFPS

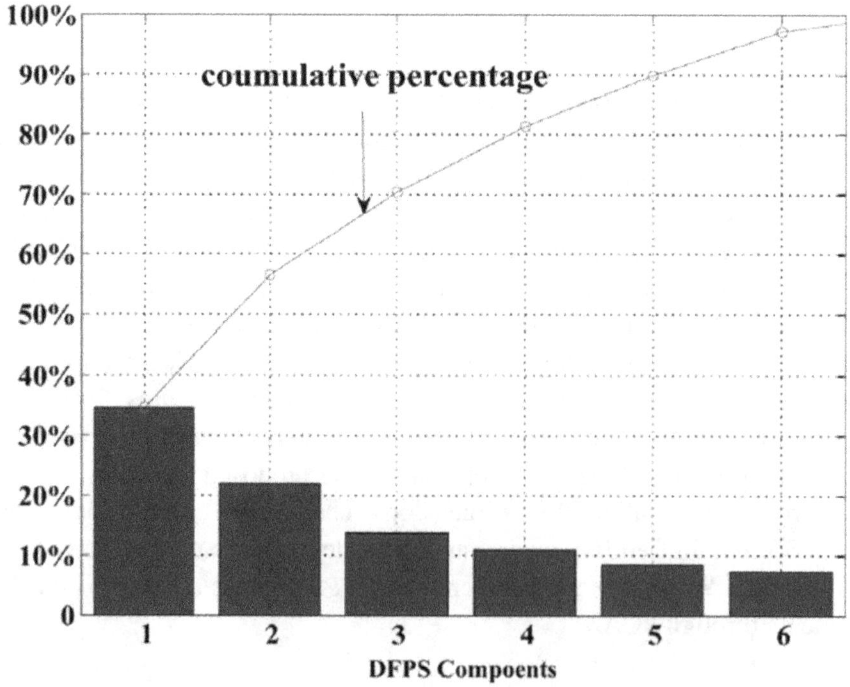

of the total information provided by RSSs and ΔRSSs. In other words, when the cut-off threshold is set as 70%, only three DFPS components are required. In particular, Figure 22 displays the eigenvectors and the cumulative percentage of the eigenvalues, in which the computed eigenvalues approach zero after the eighth component. Because 7 DFPS components explain 98% information in the original variables, reducing the dimensions to 7 is a reasonable approach to avoid unnecessary computation. Note that this analysis is data dependent. The distribution of computed eigenvalues differs according to the environments, as a result of geometric diversity and different AP placements. In general, selecting the threshold of approximately 90% is a safer approach to preserve most of the information.

Alternative Positioning Algorithm

In this section, we performed extra experiments to evaluate the improvement of DFPS using an alternative positioning algorithm, support vector machines (SVM) regression. The basic idea of SVM is to find an optimal separating hyperplane which simultaneously maximizes the margin between groups and minimizes the misclassification rate. We utilized the open software LIBSVM tools (Chang & Lin, 2011) and selected linear kernel with regression to implement the back-end localization algorithm. Table 4

Figure 22. (a) Obtained eigenvalues; (b) Cumulative percentage of eigenvalues

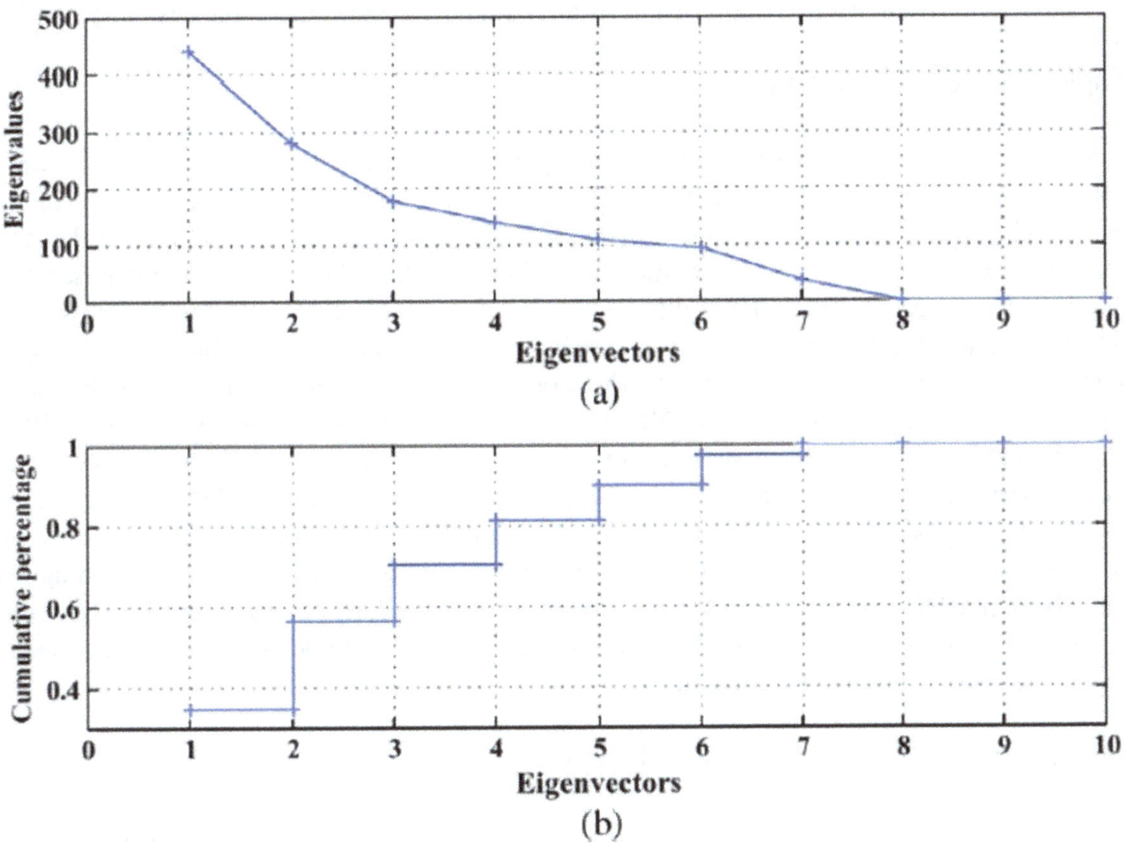

Table 4. Five positioning error measures (in meters) for different algorithms based on svm

Heterogeneous Case (Training: Laptop, Testing: Mobile Phone)					
Method	STD	50th Percentile	67th Percentile	90th Percentile	Average
RSS	0.92	2.35	2.59	3.93	2.25
HLF	1.12	2.38	2.40	4.03	2.01
DIFF	1.31	0.75	2.14	3.89	1.73
SSD	1.39	0.53	2.33	4.05	1.69
DFPS	1.34	0.42	2.07	3.95	1.39
Homogeneous Case (Training: Laptop, Testing: Laptop)					
RSS	1.31	0.33	2.05	3.77	1.15
HLF	1.39	0.50	2.38	4.03	1.27
DIFF	1.42	0.53	2.00	3.84	1.31
SSD	1.34	0.42	2.07	3.95	1.19
DFPS	1.11	0.29	0.92	3.32	0.97

provides the numerical values of different error measures. The experimental results indicate that the proposed algorithm still provides higher position accuracy based on the SVM-based positioning system. The results show a consistent tendency as previous results using MLP and confirm the advantages of DFPS.

Multiple Devices Simulation

This section considers a scenario in which the testing devices have an alternative receiver gain. Because performing measurements for all devices is infeasible, we assume heterogeneous devices mainly introduce a constant bias in RSSs in this experiment. In addition to real measurements from mobile devices, we simulate the multiple-testing devices condition by adding a constant bias τ and a random noise v to each testing RSS. Figure 23 illustrates the results in this scenario at test-bed 1, where the bias τ ranges from -10 to 10 dBm and the noise v is a uniformly distributed noise variable ranging between ± 2. Figure 23 demonstrates that the proposed approach still offers consistent accuracy improvement under the simulated multiple-testing devices while SSD and DIFF performed similarly. As the bias increases to 10 dBm, the positioning errors clearly decreases because the scenario is similar to the homogeneous case. These results support that, DFPS can improve positioning accuracy even when testing devices have alternative gains. This finding is of great practical importance because different devices inevitably produce biased measurements. Although the bias is unknown, the proposed DFPS is a robust fused positioning feature that can be applied to various RSS-based fingerprinting location systems. These results, based on real measurements from six heterogeneous devices and simulations of multiple unseen devices, again verify the advantages of the proposed algorithm.

Different Test-Bed Environments and Devices

This study conducted an additional experiment in a largescale indoor environment (test-bed 2). Compared with the previous test-bed, the area of this test-bed is increased to 57×40 square meters with 20

Figure 23. Multiple-testing devices simulation with varying biases on testing RSSs

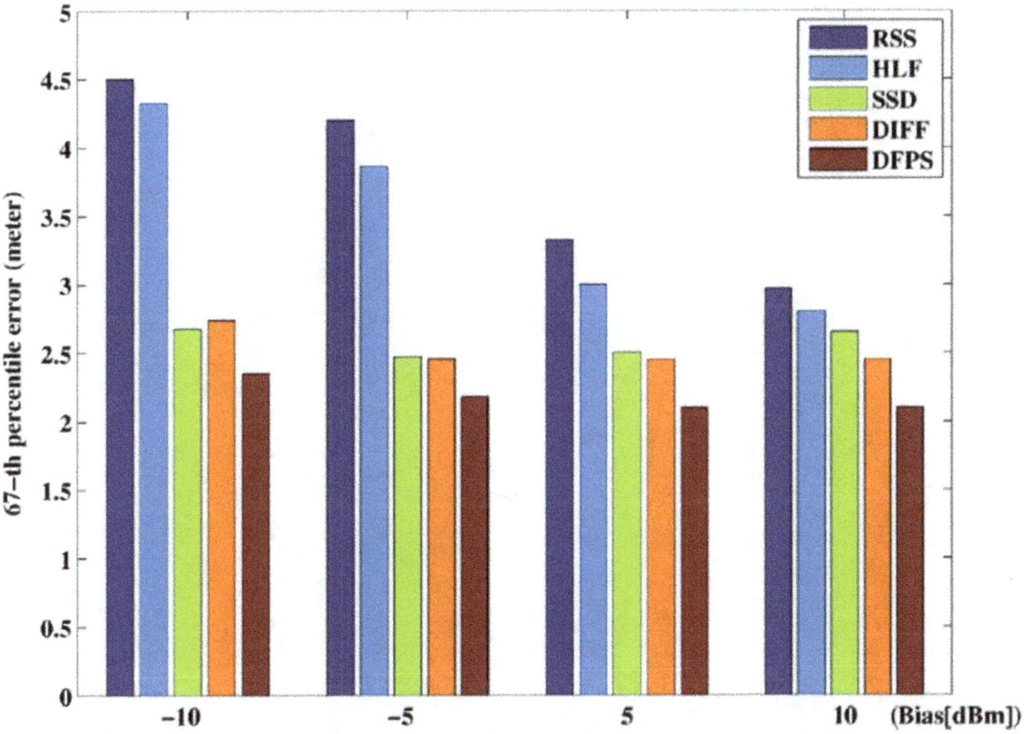

detectable APs, as shown in Figure 24. In this experiment, we developed various software tools for six devices from different brands and operating systems.

Specifically, we used one tablet (Samsung), two laptops (Acer and Asus), and three smart phones (Sony and HTC) as heterogeneous devices to gather RSSs and to evaluate the performance of the proposed DFPS algorithm. We collected 200 Wi-Fi RSS samples at 50 different testing locations for each device in this area, and again used an Asus laptop to gather RSS at 100 distinct reference locations as training data. Figure 25 shows the probability density functions of RSS samples from six different devices at test-bed 2. In this figure, the effect of heterogeneous devices shows a certain transformation of the RSS representation space, thus incurring a considerable mismatch problem when training and testing devices differ.

Table 5 provides the numerical values of median positioning errors based on MLP, in which DFPS achieved a 1.2–2.8 m, while RSS, HLF, DIFF, and SSD were 1.3–4.8 m, 1.7–3.6 m, 1.3–3 m and 1.4–3 m, respectively. Table 6 provides the results using SVM. Both tables show that, with different back-end positioning algorithms, the proposed DFPS feature is capable of achieving superior positioning performance with heterogeneous devices than traditional methods. The performance of using Acer laptop is generally superior to that using smartphones and tablet. This is because the distributions of smartphones incur a more significant difference (a larger mean shift) than that of laptops, and thus resulting in a greater influence. This observation agrees well with Figure 25. The experiments again reveal the effectiveness of the proposed algorithm in providing robust indoor Wi-Fi localization using heterogeneous devices.

Figure 24. The fourth floor plane of Far Eas Tone Telecommunications Building, where we performed the experiment

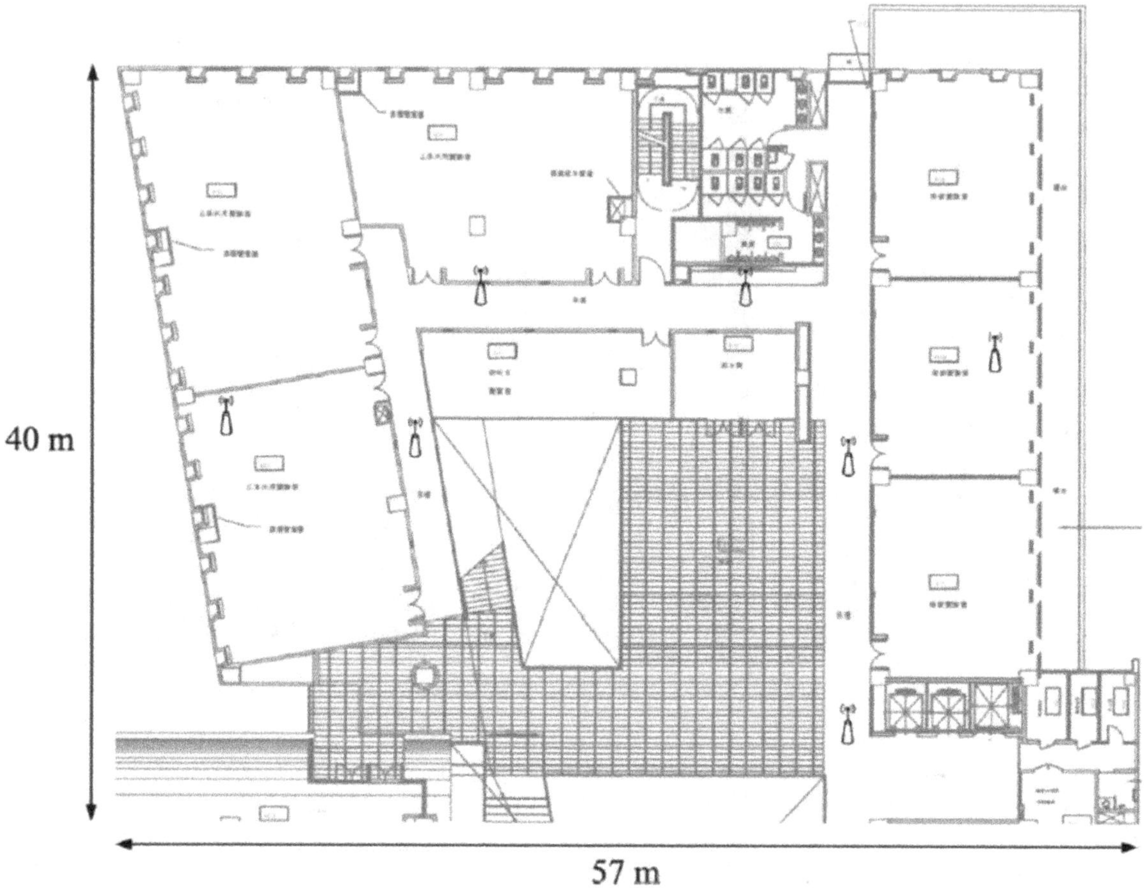

Table 5. Performance comparison (50-th percentile errors in meters) at test-bed 2 using heterogeneous devices based on mlp

Method/ Device	Sony Xperia	HTC Tattoo	HTC Desire	Samsung Galaxy	Acer M5-581	ASUS M5200
RSS	4.79	4.33	3.70	3.24	3.69	1.25
HLF	3.63	2.55	2.51	2.63	3.14	1.74
DIFF	2.95	2.41	2.19	2.98	2.55	1.32
SSD	3.07	2.47	2.83	2.70	2.29	1.40
DFPS	2.77	2.05	1.81	2.06	1.96	1.20

Figure 25. The impact of different devices on the probability density of RSS samples at a fixed reference location

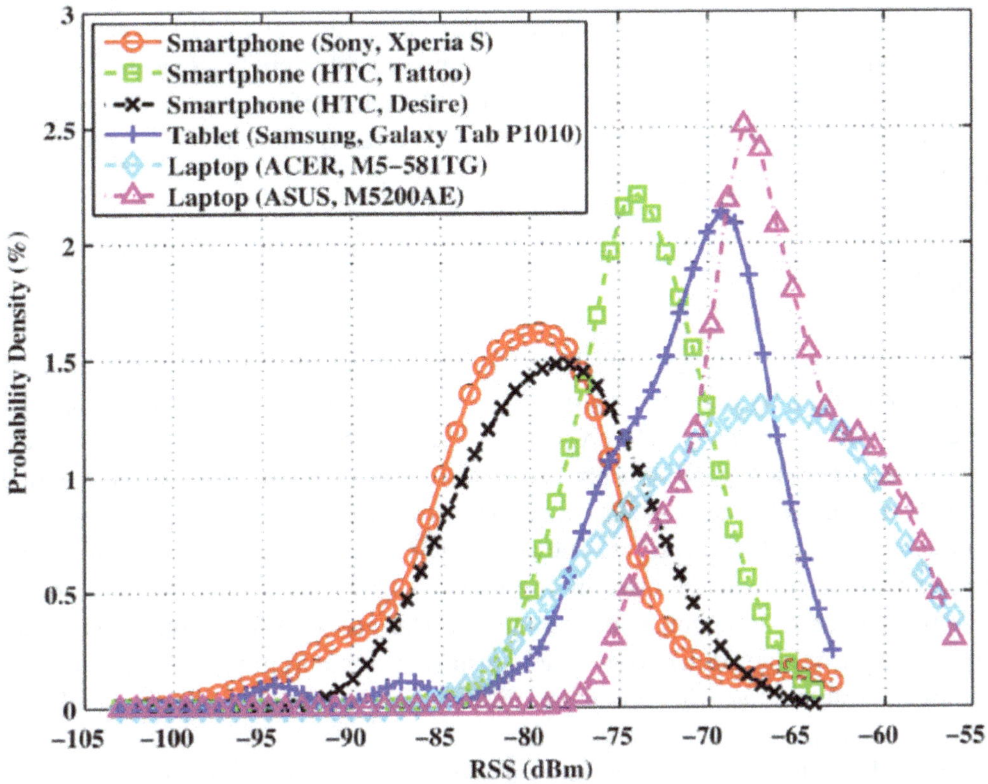

Table 6. Performance comparison (50-th percentile errors in meters) at test-bed 2 using heterogeneous devices based on svm

Method/ Device	Sony Xperia	HTC Tattoo	HTC Desire	Samsung Galaxy	Acer M5-581	ASUS M5200
RSS	3.69	3.78	3.69	3.69	2.79	1.89
HLF	3.49	3.13	3.13	3.03	2.24	1.75
DIFF	3.47	3.47	3.01	3.47	2.49	1.51
SSD	2.83	2.83	3.47	2.96	2.45	1.64
DFPS	2.72	2.69	2.39	1.98	1.71	1.43

FUTURE RESEARCH DIRECTIONS

Future wireless applications create insatiability demand for "high data rate" and "high link quality" wireless access. The main idea is that multiple antennas were used to provide diversity gain and increase the reliability of wireless links. It utilize the idea of using more spectrum for increasing wireless capacity, data rate, spectrum utilization and spectral efficiency. MIMO offers significant increases in data throughput and link range without additional bandwidth or transmit power (Halperin, Hu, Sheth & Wetherall, 2010). Channel state information (CSI) describes the properties of a wireless communication link between the transmitters and receivers. Although CSI provides amount of location information for precise position, it is excessively complex, such as having too many parameters. Recently, CSI has been adopted as an enhanced wireless channel measurement instead of received signal strength (RSS) for indoor WiFi positioning systems. However, although CSI contains richer location information, a challenging problem is the severe dynamic range and fluctuation among the high-dimensional channels, which may degrade accuracy and cause overfitting problems.

CONCLUSION

In this chapter, we proposed a novel approach based on HEQ to compensate for an orientation mismatch in robust Wi-Fi localization. By transforming the temporal–spatial radio signal strength into a reference histogram, the proposed algorithm effectively prevents the performance degradation under orientation mismatch conditions. Specifically, HEQ is applied to each component of temporal–spatial radio to improve the robustness of indoor localization systems. The proposed approach is conceptually simple and easy to implement in practical applications. The advantages of the proposed algorithm are that it does not involve assumptions regarding the random user behavior and does not require embedding a digital compass in mobile devices. The experiments, conducted in an indoor Wi-Fi environment, confirm the superiority of the proposed algorithm in orientation mismatch compensation. Results show that the proposed algorithm outperforms the orientation classifier method and provides a comparable positioning accuracy to that of the compass-assisted approach.

Another key challenge for WiFi-RSS-based localization is how to handle the heterogeneous devices problems. In this chapter, we also propose an enhanced positioning feature, named delta-fused principal strength (DFPS), to defend against such hardware variation effects. The main idea behind DFPS is to fuse RSS and ΔRSS intelligently, compensating for the heterogeneous devices problems of the former and embedding the discriminatory power into the latter. In addition, DFPS offers greater flexibility in determining the required number of components, thereby achieving better computational efficiency. The proposed algorithm was evaluated using empirical data acquired in two indoor Wi-Fi environments, where realistic RSS measurements were collected from six different W-Fi equipped mobile devices. Experimental results demonstrate the effectiveness of DFPS. Compared with other positioning features, such as RSS, HLF, SSD, and DIFF, DFPS performs the best under both heterogeneous and homogeneous conditions.

REFERENCES

Arnold, G. M., & Collins, A. J. (1993). Interpretation of transformed axes in multivariate analysis. *Journal of the Royal Statistical Society. Series C, Applied Statistics*, *42*(2), 381–400.

Bahl, P., & Padmanabhan, V. (2000). RADAR: An in-building RF-based user location and tracking system. *Proceedings IEEE INFOCOM 2000. Conference on Computer Communications. Nineteenth Annual Joint Conference of the IEEE Computer and Communications Societies*, *2*, 775-784.

Battiti, R., Nhat, T. L., & Villani, A. (2002). *Location-aware computing: A neural network model for determining location in wireless LANs.* University of Trento, Trento, Italy, Tech. Rep. DIT-02-0083. Retrieved from http://citeseerx.ist.psu.edu/viewdoc/download?doi=10.1.1.20.2366&rep=rep1&type=pdf

Baxter, M. J. (1995). Standardization and transformation in principal component analysis, with applications to archaeometry. *Applied Statistics-Journal of the Royal Statistical Society Series C*, *44*(4), 513–527.

Bernardos, A., Casar, J., & Tarrio, P. (2010). Real time calibration for RSS indoor positioning systems. *2010 International Conference on Indoor Positioning and Indoor Navigation*, 1-7. doi:10.1109/IPIN.2010.5648231

Bialer, O., Raphaeli, D., & Weiss, A. (2013, June). Maximum-likelihood direct position estimation in dense multipath. *IEEE Transactions on Vehicular Technology*, *62*(5), 2069–2079. doi:10.1109/TVT.2012.2236895

Bieber, A. M., Brooks, D. W., Harbottle, G., & Sayre, E. V. (1976, February). Application of multivariate techniques to analytical data on aegean ceramics. *Archaeometrys*, *18*(1), 59–74. doi:10.1111/j.1475-4754.1976.tb00145.x

Bshara, M., Orguner, U., Gustafsson, F., & Van Biesen, L. (2010, January). Fingerprinting localization in wireless networks based on received-signal-strength measurements: A case study on WiMAX networks. *IEEE Transactions on Vehicular Technology*, *59*(1), 283–294. doi:10.1109/TVT.2009.2030504

Chan, E., Baciu, G., & Mak, S. (2010). Wi-Fi positioning based on Fourier descriptors. *International Conference on Communications and Mobile Computing*, *3*, 545–551.

Chang, C.-C., & Lin, C.-J. (2011, Apr). LIBSVM: A library for support vector machines. *ACM Transactions on Intelligent Systems and Technology*, *2*(3), 27:1–27:27. Retrieved from http://www.csie.ntu.edu.tw/~cjlin/libsvm

Chen, Y., Kleisouris, K., Li, X., Trappe, W., & Martin, R. P. (2009, Feb). A security and robustness performance analysis of localization algorithms to signal strength attacks. *ACM Transactions on Sensor Networks*, *5*, 2:1-2:37.

Chen, Y., Yang, J., Trappe, W., & Martin, R. (2010). Detecting and localizing identity-based attacks in wireless and sensor networks. *IEEE Transactions on Vehicular Technology*, *59*(5), 1–1. doi:10.1109/TVT.2010.2044904

Coleri Ergen, S., Tetikol, H., Kontik, M., Sevlian, R., Rajagopal, R., & Varaiya, P. (2014, January). RSSI-fingerprinting-based mobile phone localization with route constraints. *IEEE Transactions on Vehicular Technology*, *63*(1), 423–428. doi:10.1109/TVT.2013.2274646

De la Torre, A., Peinado, A. M., Segura, J. C., Perez-Cordoba, J. L., Benitez, M. C., & Rubio, A. J. (2005, May). Histogram equalization of speech representation for robust speech recognition. *IEEE Transactions on Speech and Audio Processing*, *13*(3), 355–366. doi:10.1109/TSA.2005.845805

Della Rosa, F., Leppa, H., Biancullo, S., & Nurmi, J. (2010). Ad-hoc networks aiding indoor calibrations of heterogeneous devices for fingerprinting applications. *International Conference on Indoor Positioning and Indoor Navigation*, 1-6. doi:10.1109/IPIN.2010.5647827

Dharanipragada, S., & Padmanabhan, M. (2000). A nonlinear unsupervised adaptation technique for speech recognition. *Proceedings of INTERSPEECH*, 556–559.

Dong, F., Chen, Y., Liu, J., Ning, Q., & Piao, S. (2009). A calibration-free localization solution for handling signal strength variance. In *MELT* (pp. 79–90). Berlin: Springer-Verlag. doi:10.1007/978-3-642-04385-7_6

Fang, S.-H., Chuang, C.-C., & Wang, C. (2012). Attack-resistant wireless localization using an inclusive disjunction model. *IEEE Transactions on Communications*, *60*(5), 1209–1214. doi:10.1109/TCOMM.2012.040212.100291

Fang, S.-H., & Lin, T.-N. (2008). Robust wireless LAN location fingerprinting by SVD-based noise reduction. *2008 3rd International Symposium on Communications, Control and Signal Processing*, 295-298.

Fang, S.-H., & Lin, T.-N. (2008, November). Indoor location system based on discriminant-adaptive neural network in IEEE 802.11 environments. *IEEE Transactions on Neural Networks*, *19*(11), 1973–1978. doi:10.1109/TNN.2008.2005494 PMID:19000967

Fang, S.-H., & Lin, T.-N. (2012, January). Principal component localization in indoor WLAN environments. *IEEE Transactions on Mobile Computing*, *11*(1), 100–110. doi:10.1109/TMC.2011.30

Fang, S.-H., Lin, T.-N., & Lin, P.-C. (2008, May). Location fingerprinting in a decorrelated space. *IEEE Transactions on Knowledge and Data Engineering*, *20*(5), 685–691. doi:10.1109/TKDE.2007.190731

Fang, S.-H., & Wang, C.-H. (2011, March). A dynamic hybrid projection approach for improved Wi-Fi location fingerprinting. *IEEE Transactions on Vehicular Technology*, *60*(3), 1037–1044. doi:10.1109/TVT.2011.2107757

Fang, S.-H., Wang, C.-H., Chiou, S.-M., & Lin, P. (2012). Calibration-free approaches for robust Wi-Fi positioning against device diversity: A performance comparison. *Proc. IEEE Transactions on Vehicular Technology Conference Spring*, 1–5. doi:10.1109/VETECS.2012.6240088

Fang, S.-H., Wang, C.-H., Huang, T.-Y., Yang, C.-H., & Chen, Y.-S. (2012, April). An enhanced ZigBee indoor positioning system with an ensemble approach. *IEEE Communications Letters*, *16*(4), 564–567. doi:10.1109/LCOMM.2012.022112.120131

Feng, C., Au, W., Valaee, S., & Tan, Z. (2009). Orientation-aware indoor localization using affinity propagation and compressive sensing. *IEEE International Workshop on Computational Advances in Multi-Sensor Adaptive Processing (CAMSAP)*, 261–264. doi:10.1109/CAMSAP.2009.5413285

Feng, C., Au, W., Valaee, S., & Tan, Z. (2012, December). Received-signal-strengthbased indoor positioning using compressive sensing. *IEEE Transactions on Mobile Computing*, *11*(12), 1983–1993. doi:10.1109/TMC.2011.216

Figuera, C., Rojo-Álvarez, J. L., Mora-Jiménez, I., Guerrero-Curieses, A., Wilby, M., & Ramos-López, J. (2011, July). Time-space sampling and mobile device calibration for WiFi indoor location systems. *IEEE Transactions on Mobile Computing*, *10*(7), 913–926. doi:10.1109/TMC.2011.84

Flury, B. (1997). *A First Course in Multivariate Statistics*. New York: Springer-Verlag. doi:10.1007/978-1-4757-2765-4

Frey, B. J., & Dueck, D. (2007, February). Clustering by passing messages between data points. *Science*, *315*(5814), 2007. doi:10.1126/science.1136800 PMID:17218491

Garcia-Valverde, T., Garcia-Sola, A., Botia, J., & Gomez-Skarmeta, A. (2012). Automatic design of an indoor user location infrastructure using a memetic multi-objective approach. *IEEE Transactions on Systems, Man and Cybernetics. Part C, Applications and Reviews*, *42*(5), 704–709. doi:10.1109/TSMCC.2012.2187443

Gezici, S. (2008). A survey on wireless position estimation. *Wireless Personal Communications*, *44*(3), 263–282. doi:10.1007/s11277-007-9375-z

Ghaboosi, N., & Jamalipour, A. (2012). *Location estimation using geometry of overhearing*. Academic Press.

Gu, Y., Lo, A., & Niemegeers, I. (2009, January). A survey of indoor positioning systems for wireless personal networks. *EEE Communications Surveys & Tutorials*, *11*(1), 13–32. doi:10.1109/SURV.2009.090103

Haeberlen, A., Flannery, E., Ladd, A. M., Rudys, A., Wallach, D. S., & Kavraki, L. E. (2004). Practical robust localization over large-scale 802.11 wireless networks. In MobiCom (pp. 70-84). ACM.

Halperin, D., Hu, W., Sheth, A., & Wetherall, D. (2010, August). Predictable 802.11 packet delivery from wireless channel measurements. *Computer Communication Review*, 41.

Hossain, A. K. M. M., Van, H. N., & Soh, W.-S. (2010). Utilization of user feedback in indoor positioning system. *Pervasive and Mobile Computing*, *6*(4), 467–481. doi:10.1016/j.pmcj.2010.04.003

IEEE Standard for Information Technology. (2007, June). *Telecommunications and information exchange between systems - local and metropolitan area networks - specific requirements - part 11: Wireless LAN medium access control (MAC) and physical layer (PHY) specifications," IEEE Std 802.11-2007 (Revision of IEEE Std 802.11-1999)*. IEEE.

Jolliffe, I. (1986). *Principal Component Analysis*. New York: Springer-Verlag. doi:10.1007/978-1-4757-1904-8

Kaemarungsi, K. (2006). Distribution of WLAN received signal strength indication for indoor location determination. *Proc. ISWPC*, 1–6. doi:10.1109/ISWPC.2006.1613601

Kaemarungsi, K., & Krishnamurthy, P. (2004). Modeling of indoor positioning systems based on location fingerprinting. *Proc. IEEE 23rd Annual Joint Conference on Computing and Communication. Soc.*, *2*, 1012–1022. doi:10.1109/INFCOM.2004.1356988

Kaemarungsi, K., & Krishnamurthy, P. (2004). Properties of indoor received signal strength for WLAN location fingerprinting. *Proc. 1st Annual International Conference on Mobile Ubiquitous Systems Network Servers*, 14–23. doi:10.1109/MOBIQ.2004.1331706

King, T., Kopf, S., Haenselmann, T., Lubberger, C., & Effelsberg, W. (2006). COMPASS: A probabilistic indoor positioning system based on 802.11 and digital compasses. *Proc. WiNTECH*, 34–40. doi:10.1145/1160987.1160995

Kjærgaard, M., & Munk, C. (2008). Hyperbolic location fingerprinting: A calibration-free solution for handling differences in signal strength. *Proc. IEEE PerCom*, 110–116.

Kjærgaard, M. B. (2007). A taxonomy for radio location fingerprinting. in Proceedings of the 3rd international conference on Location-and context-awareness. *LoCA'07,* 139-156.

Kjærgaard, M. B. (2011, February). Indoor location fingerprinting with heterogeneous clients. *Pervasive and Mobile Computing*, *7*(1), 31–43. doi:10.1016/j.pmcj.2010.04.005

Kuo, S.-P., & Tseng, Y.-C. (2008, May). A scrambling method for fingerprint positioning based on temporal diversity and spatial dependency. *IEEE Transactions on Knowledge and Data Engineering*, *20*(5), 678–684. doi:10.1109/TKDE.2007.190730

Kuo, S.-P., & Tseng, Y.-C. (2011, February). Discriminant minimization search for largescale RF-based localization systems. *IEEE Transactions on Mobile Computing*, *10*(2), 291–304. doi:10.1109/TMC.2010.67

Kushki, A., Plataniotis, K., & Venetsanopoulos, A. (2007, June). Kernel-based positioning in wireless local area networks. *IEEE Transactions on Mobile Computing*, *6*(6), 689–705. doi:10.1109/TMC.2007.1017

Ladd, A. M. (2002). *Robotics-based location sensing using wireless ethernet*. Proc. Mobile Computing Network.

Li, M., Jiang, X., and Guibas, L. (2011). Fingerprinting mobile user positions in sensor networks: Attacks and countermeasures. *IEEE Transactions on Parallel and Distributed Systems*.

Li, W., Jia, Y., Du, J., & Zhang, J. (2013, July). Distributed multiple-model estimation for simultaneous localization and tracking with NLOS mitigation. *IEEE Transactions on Vehicular Technology*, *62*(6), 2824–2830. doi:10.1109/TVT.2013.2247073

Liao, I.-E., & Kao, K.-F. (2008). Enhancing the accuracy of WLAN-based location determination systems using predicted orientation information. *Information Sciences*, *178*(4), 1049–1068. doi:10.1016/j.ins.2007.09.025

Liao, L., Chen, W., Zhang, C., Zhang, L., Xuan, D., & Jia, W. (2011, June). Two birds with one stone: Wireless access point deployment for both coverage and localization. *IEEE Transactions on Vehicular Technology*, *60*(5), 2239–2252. doi:10.1109/TVT.2011.2109405

Lin, T., Fang, W. S., Tseng, W.-H., Lee, C.-W., & Hsieh, J.-W. (2014, October). A group-discrimination-based access point selection for WLAN fingerprinting localization. *IEEE Transactions on Vehicular Technology*, *63*(8), 3967–3976. doi:10.1109/TVT.2014.2303141

Liu, H., Darabi, H., Banerjee, P., & Liu, J. (2007, November). Survey of wireless indoor positioning techniques and systems. *IEEE Transactions on System, Management, Cybernetics, Part C, Applications and Reviews*, *37*(6), 1067–1080. doi:10.1109/TSMCC.2007.905750

Liu, H., Gan, Y., Yang, J., Sidhom, S., Wang, Y., Chen, Y., & Ye, F. (2012). *Push the limit of WiFi based localization for smartphones*. Mobicom. doi:10.1145/2348543.2348581

Mahtab Hossain, A., Jin, Y., Soh, W.-S., & Van, H. N. (2010). SSD: A robust RF location fingerprint addressing mobile devices' heterogeneity. *IEEE Transactions on Mobile Computing, 12*(1), 65–77. doi:10.1109/TMC.2011.243

Mahtab Hossain, A., Van, H. N., Jin, Y., & Soh, W.-S. (2007). Indoor localization using multiple wireless technologies. *IEEE International Conference on Mobile Adhoc and Sensor Systems*, 1-8.

Mazuelas, S., Bahillo, A., Lorenzo, R., Fernandez, P., Lago, F., Garcia, E., & Abril, E. et al. (2009). Robust indoor positioning provided by real-time RSSI values in unmodified WLAN networks. *IEEE Journal of Selected Topics in Signal Processing, 3*(5), 821–831. doi:10.1109/JSTSP.2009.2029191

Medawar, S., Handel, P., & Zetterberg, P. (2013). Approximate maximum likelihood estimation of rician K-Factor and investigation of urban wireless measurements. *IEEE Transactions on Wireless Communications, 12*(6), 2545–2555. doi:10.1109/TWC.2013.042413.111734

Mihaylova, L., Angelova, D., Bull, D., & Canagarajah, N. (2011). Localization of mobile nodes in wireless networks with correlated in time measurement noise. *IEEE Transactions on Mobile Computing, 10*(1), 44–53. doi:10.1109/TMC.2010.132

Papadakis, S., & Traganitis, A. (2010). *Wireless positioning using the signal strength difference on arrival*. MASS. doi:10.1109/MASS.2010.5663794

Papapostolou, A., & Chaouchi, H. (2009). Orientation-based radio map extensions for improving positioning system accuracy. *Proc. International Conference on Wireless Communication and. Mobile Computing, Connecting the World Wirelessly*, 947–951. doi:10.1145/1582379.1582586

Perez-Ramirez, J., Borah, D., & Voelz, D. (2013, September). Optimal 3-D landmark placement for vehicle localization using heterogeneous sensors. *IEEE Transactions on Vehicular Technology, 62*(7), 2987–2999. doi:10.1109/TVT.2013.2255072

Rappaport, T. (2001). *Wireless Communications: Principles and Practice* (2nd ed.). Upper Saddle River, NJ: Prentice-Hall.

Roberts, B., & Pahlavan, K. (2009). Site-specific RSS signature modeling for WiFi localization. *Proc. GLOBECOM*, 1–6. doi:10.1109/GLOCOM.2009.5425740

Roos, T., Myllymaki, P., & Tirri, H. (2002). A statistical modeling approach to location estimation. *IEEE Transactions on Mobile Computing, 1*(1), 59–69. doi:10.1109/TMC.2002.1011059

Russ, J. C. (1995). *Image Processing Handbook*. Boca Raton, FL: CRC.

Sanchez, D., Quinteiro, J., Hernandez-Morera, P., & Martel-Jordan, E. (2012). Using data mining and fingerprinting extension with device orientation information for WLAN efficient indoor location estimation. *Proc. Conference on Wireless Mobile Computing, Networking, and Communications*, 77–83. doi:10.1109/WiMOB.2012.6379164

Saunders, S. R., & Simon, S. R. (1999). *Antennas and Propagation for Wireless Communication Systems* (1st ed.). New York: John Wiley & Sons, Inc.

Seshadri, V., Zaruba, G., & Huber, M. (2005). A bayesian sampling approach to indoor localization of wireless devices using received signal strength indication. *Pervasive Computing and* Communications, 75–84.

Sharp, I., & Yu, K. (2013, August). Enhanced least-squares positioning algorithm for indoor positioning. *IEEE Transactions on Mobile Computing*, *12*(8), 1640–1650. doi:10.1109/TMC.2012.124

Shin, H., Chon, Y., & Cha, H. (2012, November). Unsupervised construction of an indoor floor plan using a smartphone. *IEEE Transactions on Systems, Man and Cybernetics. Part C, Applications and Reviews*, *42*(6), 889–898. doi:10.1109/TSMCC.2011.2169403

Stella, M., Russo, M., & Begusic, D. (2007). *Location determination in indoor environment based on RSS fingerprinting and artificial neural network*. Proc. ConTel. doi:10.1109/CONTEL.2007.381886

Sun, J., Zhu, X., Zhang, C., & Fang, Y. (2011, March). Rescueme: Location-based secure and dependable VANETs for disaster rescue. *IEEE Journal on Selected Areas in Communications*, *29*(3), 659–669. doi:10.1109/JSAC.2011.110314

Teng, J., Snoussi, H., Richard, C., & Zhou, R. (2012, June). Distributed variational filtering for simultaneous sensor localization and target tracking in wireless sensor networks. *IEEE Transactions on Vehicular Technology*, *61*(5), 2305–2318. doi:10.1109/TVT.2012.2190631

Tsui, A., Lin, W.-C., Chen, W.-J., Huang, P., & Chu, H.-H. (2010, November). Accuracy performance analysis between war driving and war walking in metropolitan Wi-Fi localization. *IEEE Transactions on Mobile Computing*, *9*(11), 1551–1562. doi:10.1109/TMC.2010.121

Tsui, A. W., Chuang, Y.-H., & Chu, H.-H. (2009, October). Unsupervised learning for solving RSS hardware variance problem in WiFi localization. *Mobile Networks and Applications*, *14*(5), 677–691. doi:10.1007/s11036-008-0139-0

Vaseghi, S. V. (2006). *Advanced Digital Signal Processing and Noise Reduction*. John Wiley & Sons.

Wright, R. V. S. (1989). *Doing Multivariate Analysis and Prehistory*. Sydney, Australia: University of Sydney Press.

Yin, J., Yang, Q., & Ni, L. (2008, July). Learning adaptive temporal radio maps for signal-strength-based location estimation. *IEEE Transactions on Mobile Computing*, *7*(7), 869–883. doi:10.1109/TMC.2007.70764

Youssef, M., & Agrawala, A. (2004, March). Handling samples correlation in the horus system. *INFOCOM 2004. Twenty-third Annual Joint Conference of the IEEE Computer and Communications Societies*, *2*, 1023-1031.

Youssef, M., & Agrawala, A. (2008, June). The Horus location determination system. *Wireless Networks*, *14*(3), 357–374. doi:10.1007/s11276-006-0725-7

Youssef, V., Agrawala, A., & Udaya Shankar, A. (2003). WLAN location determination via clustering and probability distributions. Pervasive Computing and Communications, 143-150.

Yu, K., & Dutkiewicz, E. (2012, February). Geometry and motion-based positioning algorithms for mobile tracking in NLOS environments. *IEEE Transactions on Mobile Computing*, *11*(2), 254–263. doi:10.1109/TMC.2011.24

Yu, K., & Dutkiewicz, E. (2013, July). NLOS identication and mitigation for mobile tracking. *IEEE Transactions on Aerospace and Electronic Systems*, *49*(3), 1438–1452. doi:10.1109/TAES.2013.6557997

Yu, K., & Guo, Y. (2009, January). Statistical NLOS identication based on AOA, TOA, and signal strength. *IEEE Transactions on Vehicular Technology*, *58*(1), 274–286. doi:10.1109/TVT.2008.924975

Zhang, Y., Liu, W., Fang, Y., & Wu, D. (2006). Secure localization and authentication in ultra-wideband sensor networks. *IEEE Journal on Selected Areas in Communications*, *24*(4), 829–835. doi:10.1109/JSAC.2005.863855

KEY TERMS AND DEFINITIONS

Hardware Variation Location Fingerprinting: Hardware variance can impact the accuracy of the system because different devices provide different RSS-patterns.

Heterogeneity: Substance consisting of dissimilar or diverse ingredients or constituents.

Histogram Equalization (HEQ): The received signal is projected through the transfer function to the specified distribution without additional positional information, while solving the problem of different hardware, and the lack of positional information.

Mobile Positioning: The ability to pinpoint the location of a vehicle or mobile caller in transit.

Principle Component Analysis (PCA): Principal component analysis (PCA) is a statistical procedure that uses an orthogonal transformation to convert a set of observations of possibly correlated variables into a set of values of linearly uncorrelated variables called principal components.

Received Signal Strength (RSS): The RSS sometimes referred as RSSI (received signal strength indicator) is a measurement of the power present in a received radio signal.

User Orientation Wi-Fi: The variation of RSSs is causing by changing of direction of devices.

Chapter 4
RFID Indoor Localization Techniques

Yongtao Ma
Tianjin University, China

Zheng Gao
Tianjin University, China

Yang Zhao
Tianjin University, China

ABSTRACT

Radio frequency identification (RFID) is a technique using two-way radio transmission pattern to transmit information through the device of interrogator (also called reader) and tag. It is considered to be one of the most popular techniques for internet of things (IOT). In this chapter, the authors study indoor localization techniques based on passive UHF RFID, which works around the frequency of 900MHz. Passive RFID has the advantage of reasonable reading distance, non-contact, easy deployment, and low cost. The tags do not need battery and it can harvest power through wireless charging. Due to those advantages, passive UHF RFID positioning has always been an active research area in the past few decades. This chapter discusses the key techniques in passive UHF RFID positioning, which include range-based, range-free, tag-based (device-based), tag-free (device-free), and improved positioning methods. All the techniques studied are suited to be implemented in RFID systems, each of which can be accommodated to a specific application scenario.

INTRODUCTION

With the rapid development of Internet of Things (IOT) technology, radio frequency identification (RFID) localization has attracted wide attention. RFID positioning system consists of reader, antenna and tag. RFID localization technology has many advantages such as non-contact, easy deployment, and low cost, hence it has been widely used in logistics, medicine, library management, and so on. RFID tags are divided into active tags and passive tags. The primary difference between an active tag and a

DOI: 10.4018/978-1-5225-3528-7.ch004

passive tag is whether or not a power supply (i.e. battery) is included. The active tag has a power supply so that it has a long read distance, but it requires periodic battery replacement. The passive tag does not have a power supply so it harvests power typically through electromagnetic induction when a reader is within the radio range of the tag. Comparatively, a passive tag is cheaper and has a smaller size, however, it has a shorter read distance.

This chapter introduces the main positioning techniques based on passive ultra-high frequency (UHF) RFID. After briefly reviewing the various RFID positioning approaches in the introduction section, five different techniques for passive RFID positioning are described in detail in five separate sections, which are LANDMARC, range-based method, improved range-based method, optimization based localization, and device free localization.

BACKGROUND

The signal transmission characteristics for wireless indoor localization have been investigated for many years, resulting in some useful conclusions and experiences which have a positive effect on the advances in UHF RFID localization. The UHF RFID has a backscattering channel, so one research focus is on the effect of channel characteristics, link budgets, and modeling methods on positioning performance. Griffin and Durgin (2009) studied the link budgets for RFID backscattering channel. They analyzed the budgets for passive tag's power-up link, backscatter radio link, and antenna configuration. Lázaro *et al.* presented a method to estimate the read range of a tag in multipath environments and analyzed the different effects on read range. A RFID system simulator was also presented using the ray tracing method (Lázaro, Girbau & Salinas, 2009). Arnitz *et al.* proposed a method to calculate wideband channel parameters of backscatter channels based on the parameters of the constituent one-way channels and studied characteristics that are vital for narrowband and wideband ranging, such as the K-factor w.r.t. the direct (line-of-sight) path and the root mean square (RMS) delay spread (Arnitz, Muehlmann & Witrisal, 2012a). They also conducted a comprehensive investigation on ranging and localization with UHF RFID, using measurements made inside a warehouse portal (Arnitz, Muehlmann & Witrisal, 2012b). The influence of reflecting/absorbing backplanes as well as the effects of a reflective pallet on the channel was discussed in detail. Furthermore, the developed channel model was analyzed for deterministic components and compared to a number of channel models. Faseth *et al.* studied the effect of multipath propagation on phase-based positioning methods based on UHF RFID and a channel model was derived (Faseth, Winkler & Arthaber, 2011). Simulations were conducted to understand the effect of the channel parameters on the narrowband phase-based localization methods.

In recent years, new algorithms especially related to optimization have been proposed. Many are based on the LANDMARC (Ni et al. 2004) and VIRE (Zhao, Liu & Ni, 2007). LANDMARC is a kind of range-free localization which uses reference tags to position. LANDMARC will be discussed in detail later. VIRE is a positioning method using virtual reference elimination. Zhang *et al.* analyzed how tag interaction affects a tag antenna radiation pattern and RSSI to improve the RSSI-based localization algorithms such as the k-nearest neighbor (k-NN) algorithm and the Simplex algorithm (Zhang et al. 2014). Experimental results show that the performance of the modfied k-NN and the modified Simplex algorithms is improved significantly and they are more robust against variations in the number, distribution density, and material type of target objects. DiGiampaolo and Martinelli (2012) studied a global localization problem for a mobile robot that was able to detect the presence of passive RFID tags. Kalman filtering was used to fuse RFID readings with odometer data. Experimental results showed that an average position error of about 0.1 m can be obtained with a tag density of less than 1 tag/m^2 .

The phase information can also be used for UHF RFID localization. Nikitin *et al.* presented an overview of how to obtain spatial identification (position and velocity) of modulated backscatter UHF RFID tags using RF phase information. Three main phase difference of arrival (PDOA) based techniques, namely time domain, frequency domain, and spatial domain technique, are studied and evaluated through with both a free space model and a multi-ray channel model in a deterministic environment (Nikitin *et al.* 2010). Scherhäufl *et al.* introduced a two-dimensional localization system based on phase-of-arrival evaluation of the backscattered tag signal. To improve localization accuracy and reduce multipath effect, a frequency-stepped continuous-waveform approach was employed (Scherhäufl *et al.* 2013). DiGiampaolo and Martinelli (2014) proposed a global robot localization approach based on the phase of the signals coming from a set of passive RFID tags located in known positions on the ceiling of the room. A multi-hypothesis extended Kalman filter (EKF) is used to fuse the RFID data and the odometer data obtained with the encoders on the wheels of the vehicle. Results demonstrate that in the case of two tags, the location precision is about 4cm, outperforming other methods in both accuracy and computation time. Särkkä *et al.* presented a location tracking system, which is based on measuring the phases of backscattered signals of single carrier frequency from RFID tag using multiple spatially distributed antennas. The wavelength ambiguity of the phase measurements was resolved by using the EKF and the Rauch-Tung-Striebel (RTS) smoother, where the state variables include the positions, velocity, and phase offsets of antennas (Särkkä, Viikari & Huusko et al. 2012).

In the following sections, five UHF RFID positioning techniques are presented. They are either widely studied algorithms or recently proposed methods, which are suited to different scenarios and can achieve nice positioning accuracy.

BASIC RFID SYSTEM AND LANDMARC

LANDMARC is a famous indoor RFID positioning system originally proposed in 2003. LANDMARC uses reference tags with known positions to assist in positioning. Because the selected reference tags are close to the target tag and hence their propagation environments are very similar, the NLOS effect is eliminated and multipath effect can be significantly reduced(Ni, Liu, Lau & Patil, 2003). More details of LANDMARC principle and performance are provided after briefing introducing the components of a basic RFID system.

Basic RFID System Structure

A typical passive RFID localization system consists of readers, readers' antennas and tags as shown in Figure 1.

Tag

The tag consists of an internal chip and an antenna, which exchanges data with a reader through communication when they are within the radio range of each other. RFID tags are divided into active tag and passive tag. The active tag has a power supply and can actively send signals to the reader. The passive

Figure 1. The composition of RFID localization system

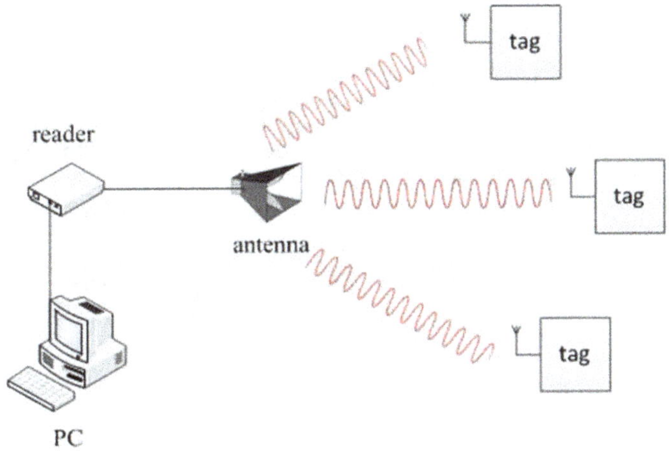

tag doesn't have a power supply, but uses electromagnetic induction by the signal sent by a reader to generate power to transmit a signal to the readers. A passive RFID localization system uses passive tags.

Antenna

The two types of antennas can be used in an RFID system, which are directional antenna and omni-directional antenna. The former can only transmit and receive radio signals in a particular direction, while the latter transmits and receives wireless signals in every direction.

Reader

The RFID reader consists of power module, RF module and control module, which controls signal transmission, reception and processing. The main functions of the RFID reader include the selection of signal frequency and the adjustment of transmit power and receiver sensitivity.

Basic Principles of LANDMARC

The LANDMARC system locates the target tag by deploying reference tags in advance at fixed locations (Han & Cho, 2010), as illustrated in Figure 2.

Assume that there are N readers in the localization environment and let S_{ik} represent the RSS of the signal transmitted from the i-th reference tag and received at the k-th reader. Thus $[S_{i1}, S_{i2}, \cdots, S_{iN}]$ denotes the RSS vector of the ith reference tag and let $[S_1, S_2, \cdots, S_N]$ denote the RSS vector of the target tag, The RSS Euclidian distance E_i between the target tag and the reference tags can be expressed as

Figure 2. The schematic diagram of LANDMARC

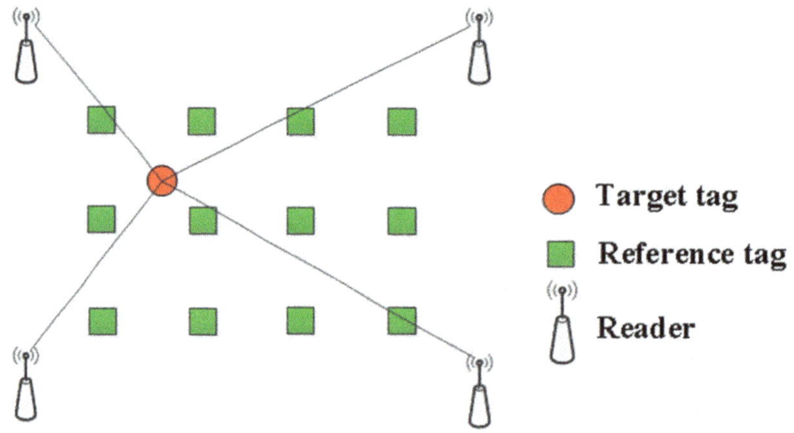

$$E_i = \sqrt{\sum_{q=1}^{N} (S_{iq} - S_q)^2} \tag{1}$$

As mentioned earlier, it is usually true that the closer a reference tag is to the target tag, the smaller the RSS Euclidian distance is. The four reference tags with the smallest RSS Euclidian distances are selected, which are likely closest to the target tag since they have very similar propagation environments.

The coordinate of the target tag (x, y) can be calculated by the weighted sum of the nearest four reference tag coordinates:

$$(x, y) = \sum_{i=1}^{k} w_i \cdot (x_i, y_i) \tag{2}$$

where (x_i, y_i) are the coordinates of i th reference tag, w_i is the weighting coefficient for the i -th reference tag, which can be defined as:

$$w_i = \frac{1}{E_i^2} \Big/ \sum_{i=1}^{k} \frac{1}{E_i^2} \tag{3}$$

It can be seen that the weighting coefficient is inversely proportional to the squared RSS Euclidian distance. This is based on the consideration that a smaller RSS Euclidian distance roughly corresponds to a smaller geometric distance, which indicates closer to the target tag. A closer reference tag should be weighted more in the calculation of the target tag position.

Factors Affecting LANDMARC Performance

LANDMARC system produces high-accuracy tag location estimates by use of a large number of reference tags and a small number of readers in the location scenario because a reader is much more expensive than a tag, LANDMARC system is widely used due to its high price-performance ratio. Nevertheless, the performance of LANDMARC system is still affected by a number of factors as described below.

Impact of Environment

The indoor environment is complex and dynamic, where the multipath and NLOS phenomenon is serious especially due to the placement of various items and personnel movements. Because of multipath and NLOS effects as well as interference, the positioning accuracy of some traditional range-based positioning algorithms may not be satisfactory. However, LANDMARC system uses reference tags to mitigate the impact of environmental factors on positioning accuracy. This is because the target tag and selected reference tags are within a small area, the propagation conditions of their signals are similar.

Number of Readers

As mentioned previously, LANDMARC system assumes that the RSS Euclidian distance is proportional to the true geometric distance. In theory, if the number of readers is larger, the relationship between the RSS Euclidian distance and the true geometric distance is more consistent, so that the positioning accuracy is better. However, since the reader is very expensive, it is useful to make a trade-off between accuracy and cost.

Reference Tag Density

For a given location environment, a higher density of the reference tags means that more reference tags are required. In theory, the higher the reference tag density is, the higher the positioning accuracy is, as is illustrated in Figure 3. Because a reader is much more expensive than a tag, it is better to increase the number of reference tags instead of readers to improve positioning accuracy. But, as the number of reference tags increases, the interference between the tags increases. Therefore, it is necessary to determine the optimal reference tag density to achieve the best positioning accuracy for the environment of (Jin, Lu & Park, 2006).

Figure 3. Two different reference tag densities of LANDMARC

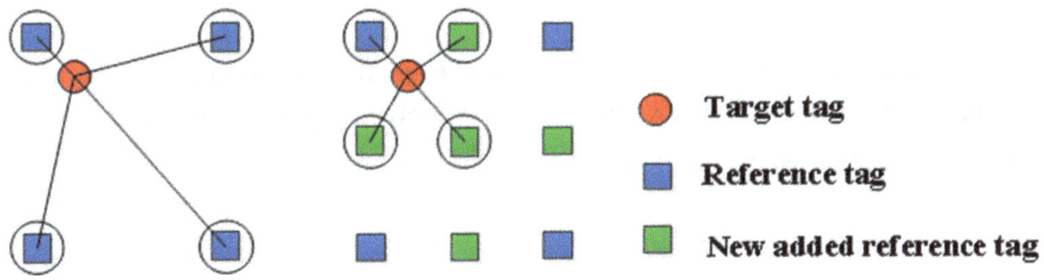

RANGE BASED LOCALIZATION

Introduction

Indoor localization methods are generally divided into two categories, range-based methods and range-free methods. The common range-free method is fingerprinting such as LANDMARC mentioned earlier. The commonly used range-based methods include RSS (Wang, Balasingham, Zhang & Huang, 2011), phase of arrival (POA) (Ma, Zhou, Liu & Wang, 2014), time of arrival (TOA) (Geng, He, Deng & Kaveh, 2013), and time difference of arrival (TDOA) (Hara, Anzai, Yabu, Lee, Derham & Zemek, 2013).

In range-based methods, ranging accuracy plays an important role in positioning accuracy. The performance of a range-based method is affected considerably by the application scenario. For example, TOA has a very high ranging accuracy in LOS scenario such as using ultra-wideband (UWB) technology, while in NLOS condition, the ranging and hence positioning performance of TOA is significantly degraded. Another problem of TOA is that it requires expensive clock synchronization (Gao, Zhang, Wang & Li, 2016). POA has a very high ranging accuracy in a short distance, but when the distance exceeds a critical value, POA method fails due to the issue of integer phase ambiguity. RSS is widely used in ranging due to the low cost and easy access. However, due to multipath and NLOS effects, the accuracy of RSS ranging is rather low. In RFID indoor localization, the common ranging methods are RSS and PDOA, which will be discussed in detail later.

Wireless Channel Model

The wireless channel model is divided into statistical model and deterministic model. Statistical models are produced based on statistical analysis of a large number of measurement results. Statistical modeling has a mature theoretical basis and does not require detailed location information for the various obstacles in the actual environment. Therefore, the statistical model is characterized by a simple model but a relatively low accuracy. The deterministic model is a radio wave propagation prediction model based on the Maxwell equation in a specific real environment, which is characterized by a complex model but with high accuracy. However, an indoor radio wave propagation environment is very complex, accurate solution Maxwell equation is difficult to obtain, so the deterministic model is basically a Maxwell equation to approximate the electromagnetic field distribution. The ray tracing is one of the commonly used approaches to produce deterministic models.

Statistical Channel Model

Assume that the reader utilizes the same antenna to receive and transmit signals and the uplink channel is identical to the downlink one. Therefore, the power of received signal in the reader, P_{r_reader}, is given by the modified Friis transmission equation:

$$P_{\mathrm{r\_reader}} = \frac{P_{t\_reader} G_{reader}^2 G_{tag}^2 \Delta G^2 \rho^2 L_p^2}{L_{sys}^2} \tag{4}$$

where P_{t_reader} is the power transmitted by the reader, G_{reader} and G_{tag} are the gains of the reader and tag antennas, respectively. The term ΔG includes the gain penalty due to detuning and the gain reduction when the tag is on contact with materials. L_{sys} is the cable loss, L_p is the path loss and ρ is the differential reflection coefficient of the tag (where $\rho = \rho_1 - \rho_2$, ρ_1 and ρ_2 are the 0 and 1 states of the chip reflection coefficient, which depend on the chip load).

The signal received by RFID readers is heavily influenced by reflections in the environment. Path loss L_p can be modeled as the sum of LOS path component and several multipath components due to reflection by ground, walls, ceiling or other objects:

$$L_p = \frac{\lambda^2}{(4\pi)^2}\left|H\right|^2 \tag{5}$$

$$H = \frac{1}{d_0}e^{-jkd_0} + \sum_{i=1}^{N}\Gamma_i\frac{1}{d_i}e^{-jkd_i} \tag{6}$$

where d_0 is the length of the direct path, d_i is the length of the ith reflection path, k is the wavenumber, N is the total number of reflection paths, and Γ_i is the complex reflection coefficient for the i-th reflection path.

In statistical channel modeling, the Rice distribution is used to generate amplitude attenuation of reflection paths. It can be assumed that the arrival time sequence of all paths is a Poisson process and the arrival time interval sequence is exponentially distributed (Xia, Bertoni, Maciel, Lindsay-Stewart & Rowe, 1993).

Ray Tracing

Ray tracing is based on the principle of geometrical optics (Kanatas, Kountouris, Kostaras & Constantinou, 1997). Under the high frequency condition, the propagation of the wireless signal is approximated as a linear propagation. By searching the propagation path of direct, reflection and diffraction of the wireless signal, the total received signal is synthesized by vector superposition at the receiving end.

The ray tracing method models the actual environment of the wireless signal propagation and finds out all the ray paths which may reach the receiver's antenna when the positions of transmitter and receiver are known.

The schematic diagram of ray tracing model can be seen in Figure 4. In this scene, there are one direct path, one reflection path and one diffraction path. The electric field of each path is calculated as follows.

1. Direct path

The direct path represents the ray path from the transmitter to the receiver when the line of sight between the transmitting end and the receiving end is not blocked by any obstacle. The electric field strength of direct path is calculated as

Figure 4. Ray tracing model

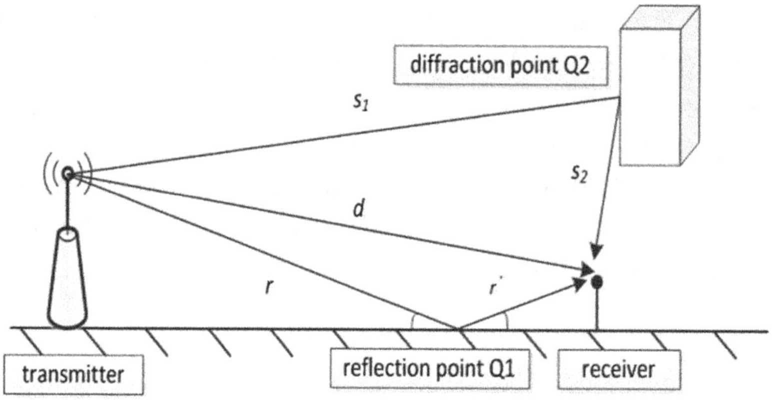

$$E_{LOS} = E_0 \frac{e^{-jkd}}{d} \tag{7}$$

where E_0 is the emitted electric field and d is the distance from the receiver to the transmitter of the LOS path.

2. Reflection path

The signal is reflected when a plane is encountered during the propagation of the ray. According to the geometric optics theory, the angle of incidence is equal to the angle of reflection. The electric field strength of reflection path is calculated as

$$E_R = E_0 R \frac{e^{-jk(r+r')}}{r+r'} \tag{8}$$

where E_0 is the emitted electric field, r is the distance between the transmitter and reflection point Q1, r' is the distance between reflection point Q1 and receiver, and R is the reflection coefficient.

3. Diffraction path

When an edge of an obstacle is encountered during propagation, the electromagnetic wave is diffracted. According to the uniform theory of diffraction (UTD), the electric field strength of diffraction path is calculated as

$$E_D = E^i(Q2) \cdot \bar{D} \cdot A_s(s) \cdot e^{-jks_2} \tag{9}$$

where $E^{i}(Q2)$ is the electric field at the diffraction point Q2, $A_s(s)$ is the diffusion factor, and $\bar{D}$ is the diffraction coefficient.

RSS Ranging

Ranging with RSS requires a RSS-distance relational model (Shi & Wong, 2011). In free space without any obstacles, there is only one direct path between the transmitter and the receiver. The received power P_r in free space can be expressed by Friis formula:

$$P_r = P_t \frac{\lambda^2}{(4\pi d)^2} \tag{10}$$

where P_t is the transmitted power, λ is the wavelength and d represents the distance between the transmitter and receiver.

The log-normal path loss model is commonly used to describe the power loss in indoor environment:

$$PL(d) = PL(d_0) + 10n\log(\frac{d}{d_0}) + z \tag{11}$$

where d_0 is the reference distance, satisfying d_0<d, n is the path loss exponent, and z is the noise modeled as a zero-mean Gaussian random variable.

In RFID system, as illustrated in Figure 5, suppose that the transmitted power at tag is P_t (in dBm), the transmitter antenna gain of tag is G_t, and the receiver antenna gain of reader is G_r. The RSS P on reader can be represented by:

Figure 5. Signal travels from a tag to a reader

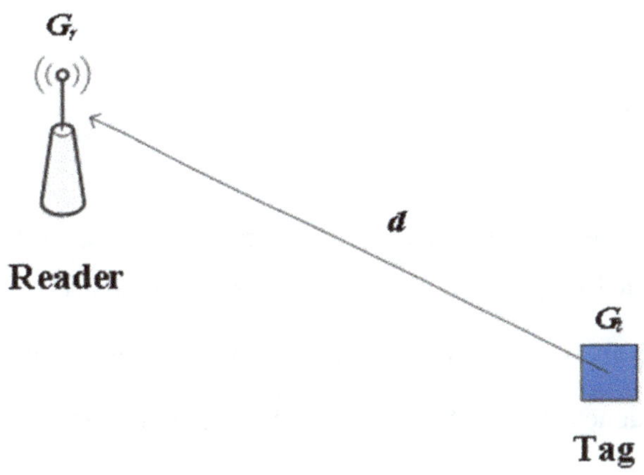

$$P = P_t + G_t + G_r - PL(d_0) - 10n\log(\frac{d}{d_0}) - z \tag{12}$$

Where $PL(d_0)$ is the path loss over the reference distance d_0, which is determined in advance. When the transmitted power and antenna gain are determined and the noise is ignored, the RSS is only determined by the distance between the tag and the reader. Therefore, given the RSS P, one can calculate the distance d as:

$$\begin{aligned} d &= 10^{\frac{c-p}{10n}} \\ c &= P_t + G_t + G_r \text{-} PL(d_0) + 10n\log(d_0) \end{aligned} \tag{13}$$

PDOA Ranging

PDOA is a phase-based ranging technique which allows coherent signal processing. The signal received by readers can be utilized to estimate the distance between readers and tags (Li, Zhang & Amin, 2009).

Suppose that a reader transmits dual-frequency continuous wave (CW) with frequencies f_1 and f_2 to provide power for tags to send uplink signals. The signal transmitted by a reader can be expressed as

$$s_i = \rho_{s_i}\cos(2\pi f_i\,\mathrm{t} + \varphi_{s_i})\ \ (i = 1,2) \tag{14}$$

where ρ_{s_i} and φ_{s_i} are the amplitude and phase of the transmitted sinusoidal signal, respectively. Once triggered by the reader signal through induction, the tag emits back the sinusoidal signal which is received by the reader and the received signal can be expressed as

$$r_i = \rho_{r_i}'\cos(2\pi f_i\,\mathrm{t} + \varphi_{r_i}')(i = 1,2) \tag{15}$$

where ρ_{r_i}' and φ_{r_i}' are the amplitude and phase of the received tag signal, respectively. Therefore, it can be written as

$$\varphi_{r_i}' + 2N_i\pi = 2\pi f_i(t_i - \Delta\tau) + \varphi_{s_i}\ (i = 1,2) \tag{16}$$

where t_i is the time interval between the reader transmits the signal and the reader receive the signal, and $t_i = \tau + \Delta\tau$. $\Delta\tau$ is the time interval between the reader signal arrives at the tag and the tag transmits a signal. $\tau = \frac{2d}{c}$, c is the speed of light, and N_i is the integer number of complete cycles of the sinusoidal signal with frequency f_i over the distance $2d$. Equation (16) can be rewritten as

$$\varphi_{r_1}^{'} + 2N_1\pi = 2\pi f_1(t_1 - \Delta\tau) + \varphi_{s_1} \tag{17}$$

$$\varphi_{r_2}^{'} + 2N_2\pi = 2\pi f_2(t_2 - \Delta\tau) + \varphi_{s_2} \tag{18}$$

Subtracting (18) by (17), produces

$$\Delta\varphi = \varphi_{r_2}^{'} - \varphi_{s_2} - \varphi_{r_1}^{'} + \varphi_{s_1} + 2N\pi = \frac{4\pi(f_2 - f_1)d}{c} \tag{19}$$

where $\Delta\varphi$ is the phase difference, and $N = N_2 - N_1$ which is unknown.

To generate a unique solution for the distance, the value of N must satisfy $|\Delta\varphi| \leq \pi$; otherwise an ambiguity occurs. The maximum estimated distance which can be estimated depends on the frequency offset. Consider the case where $f_2 > f_1$. When $\Delta\varphi > 0$, the estimated distance can be expressed as

$$d = \frac{c\Delta\varphi}{4\pi(f_2 - f_1)} \tag{20}$$

When $\Delta\varphi < 0$, the wave of f_2 went through one more period than the wave of f_1, the estimated distance can be obtained as

$$d = \frac{c(\Delta\varphi + 2\pi)}{4\pi(f_2 - f_1)} \tag{21}$$

Figure 6 shows the relationship between the frequency offset and the maximum distance which can be estimated. As expected from (20) and (21), a smaller frequency offset means that a longer distance can be estimated using the PDOA method. It is worth noting that in the presence of noise and when the phase difference is very close to zero, it is likely to use the wrong formula to calculate the distance, which should be avoided.

LOCALIZATION BASED ON IMPROVED RANGING METHOD

In the indoor environment, due to the influence of multipath, NLOS and noise, the RSS technique that utilizes the attenuation of transmitted signal strength to estimate the distance between the transmitter and the receiver produces large ranging errors. In recent years, pattern matching algorithms are drawing significant attention, which uses low-cost passive reference tags to establish database on-line such as LANDMARC described earlier. Pattern matching algorithms can reduce multipath effect significantly since the reference tags and target tags are in the very similar propagation conditions. This section presents an improved ranging method to improve localization accuracy.

Figure 6. Maximum distance which can be estimated versus frequency separation

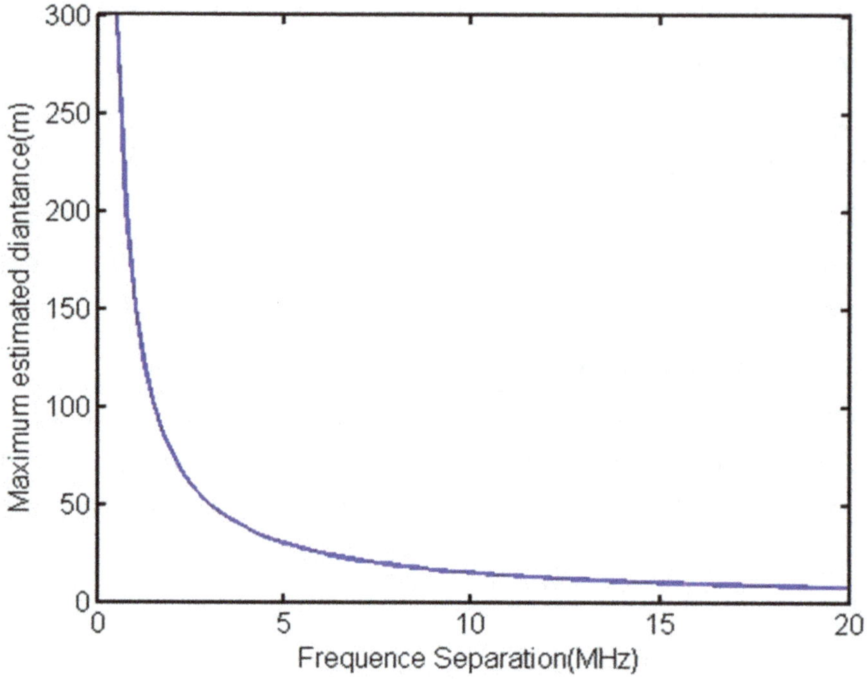

One of the most popular localization systems is LANDMARC which uses a kind of non-ranging algorithm with active reference tags to estimate the locations of target tags. Another way of using reference tags is range estimation and a passive scheme is proposed in (Wang, Wu & Tzeng, 2007). Here the RFID reader is assumed to have K transmission power levels, and the readers start with the lowest power level. When the transmission power is gradually increased each reader receives the responses from tags. The increase of transmission power is done until they receive the response from the target tag. At the same time, the set of responded tags is recorded, for example let θ_w be the set of tags that responds when the transmission power level is w. Considering a reader antenna *i* and a target tag *j*, the distance between antenna *i* and target tag *j* is denoted by L_{ij}. If $j \notin \theta_w$ and $j \in \theta_{w+1}$, L_{ij} is calculated by using the average distance from the *i*th reader antenna to all references that are in θ_{w+1} but not in θ_w. After a set of approximate distances from the *j*th target tag to all the reader antennas are calculated, the location is estimated by following range based localization algorithms.

Similarity Analysis Based Ranging Algorithm

As described earlier, in the k-NN algorithm, the coordinates are estimated using the known coordinates of the selected k candidate reference tags which are selected by comparing the similarity of RSS information. But it is still an important problem to effectively set the K value. Clustering algorithms partition patterns into a certain number of clusters (groups, subsets, or categories) with certain similarity (Xu & Wunsch, 2005). In passive RFID technology, each pattern consists of two kinds of information, i.e., RSS and phase of signal. The choosing of information utilized as a pattern characteristic is also a key factor. RSS is commonly used in pattern matching, which means only one feature is utilized. With the increasing interest of phase information, phase based localization algorithms are getting more and more attention (Ma, Zhou, Liu & Wang, 2014; Yang, Chen & Li et al. 2014; Liu, Yang & Lin et al. 2014). In traditional pattern recognition and other related fields, a pattern object should be described with a set of features, and usually represented in a multidimensional vector. So combination of RSSI and phase as a two-dimension vector is reasonable in clustering as shown in Figure 7 (Zhao, Liu & Ma et al. 2017). To avoid the ambiguity problem of phase, we use the phase difference of two tones. The phase value is in the range of $[0, 2\pi]$, theoretically the corresponding RSS is in the range of $[0, \mathrm{RSS}_m]$. RSS_m denotes the RSS value when the propagation distance is $\mathrm{d}_{\max}$.

Traditional layout of reference tags is grid-based pattern as shown in Figure 8 (Zhao, Liu & Ma et al. 2017). RSS perceived by readers can be utilized in distance estimation and location estimation respectively. For instance, some algorithms utilize the coordinates of reference tags to directly estimate the location of a target tag, like LANDMARC.

In Figure 8, four antennas are connected to one reader. The red dotted lines are arc lines (denoted as $A1, A2, \ldots, A10$) whose center is located at antenna1. The radius of each arc line is increased equally with reference spacing, denoted as *d*. The reference tags which are surrounded by two adjacent arc lines show high similarity in transmission distances.

Consider another range based indoor localization method which is termed SAIL (Zhao, Liu & Ma et al. 2017). Let N and M be the number of reader antennas and the number of reference tags, respec-

Figure 7. Clustering analysis

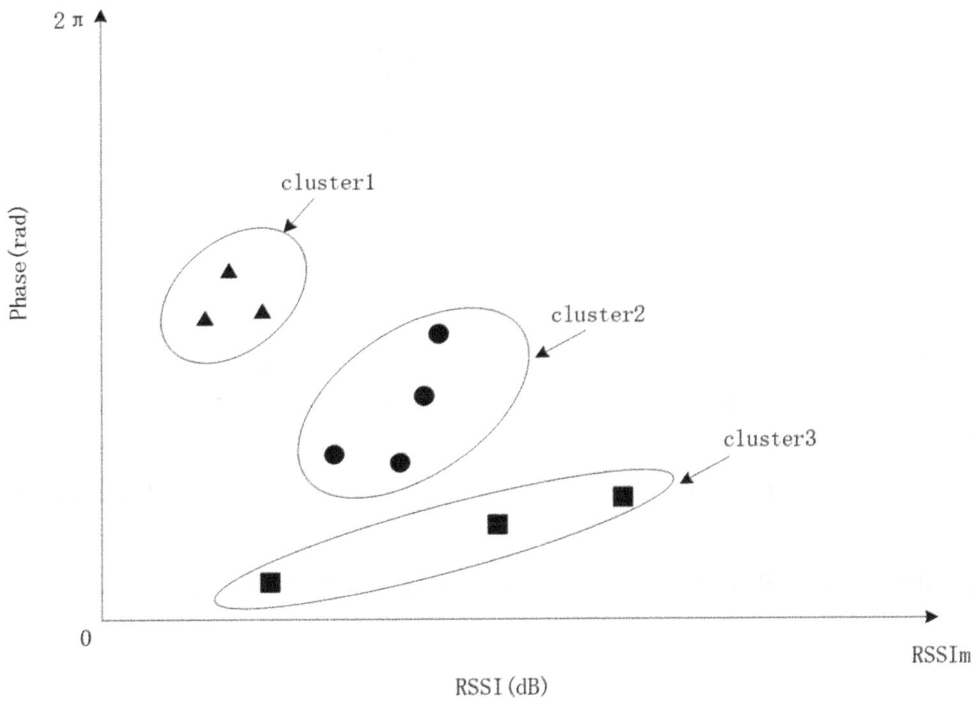

Figure 8. Traditional layout of reference tags

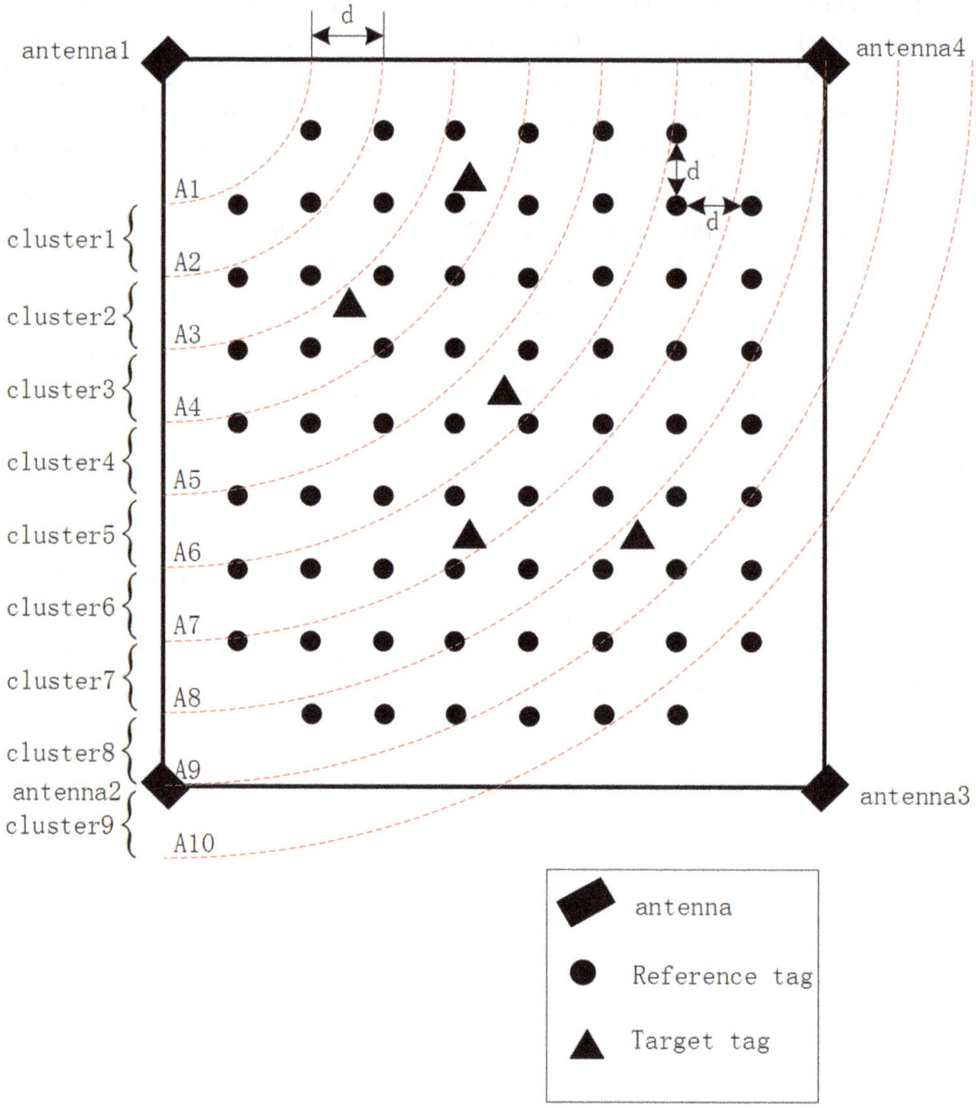

tively. The RSS vector of the *i*th target tag is defined as $TR_i = (tr_{i,1}, tr_{i,2}, tr_{i,3}, \ldots, tr_{i,N})$, where $tr_{i,j}$ denotes the RSS of the *i*th target tag perceived by the *j*th reader antenna, where $j \in (1, N)$. The phase vector of the *i*th target tag is denoted as $TP_i = (tp_{i,1}, tp_{i,2}, tp_{i,3}, \ldots, tp_{i,N})$. As for the *m*th reference tag, the corresponding RSS vector is denoted as $RR_m = (rr_{m,1}, rr_{m,2}, rr_{m,3}, \ldots, rr_{m,N})$ and the phase vector is $RP_m = (rp_{m,1}, rp_{m,2}, rp_{m,3}, \ldots, rp_{m,N})$.

Correspondingly, the RSS vector observed at the *j*th reader antenna is denoted as

$$\mathrm{Re}\,aderR_j = (tr_{i,j}, rr_{1,j}, rr_{2,j}, rr_{3,j}, \ldots, rr_{M,j}),$$

and the phase vector is

$$\mathrm{Re}\,aderP_j = (tp_{i,j}, rp_{1,j}, rp_{2,j}, rp_{3,j}, \ldots, rp_{M,j}).$$

By performing normalization for $\mathrm{Re}\,aderR_j$ and $\mathrm{Re}\,aderP_j$ according to

$$X_{norm} = \frac{X - X_{\min}}{X_{\max} - X_{\min}} \tag{22}$$

where X denotes information vector and $X_{\min}$ and $X_{\max}$ respectively denote the minimum and maximum value of X, the RSS and phase vector are normalized and the new vectors are represented by $\mathrm{Re}\,aderR_j^{'} = (tr_{i,j}^{'}, rr_{1,j}^{'}, rr_{2,j}^{'}, rr_{3,j}^{'}, \ldots, rr_{M,j}^{'})$ and $\mathrm{Re}\,aderP_j^{'} = (tp_{i,j}^{'}, rp_{1,j}^{'}, rp_{2,j}^{'}, rp_{3,j}^{'}, \ldots, rp_{M,j}^{'})$. It can be seen that the elements in these two vectors are within [0, 1]. The RSS and phase vectors can be combined to form a new vector as

$$\mathrm{Re}\,ader_j^{'} = ((tr_{i,j}^{'}, tp_{i,j}^{'}), (rr_{1,j}^{'}, rp_{1,j}^{'}), (rr_{2,j}^{'}, rp_{2,j}^{'}), \ldots, (rr_{M,j}^{'}, rp_{M,j}^{'})) \tag{23}$$

where can be further rewritten as $\mathrm{Re}\,ader_j^{'} = (x_1, x_2, \ldots, x_{M+1})$ where each component is a two-dimensional real-valued vector.

Here *k*-means algorithm is utilized for clustering. *k*-means clustering partitions the observations into *k* clusters by minimizing the distance from a point to the cluster centroid. The algorithm involves the following steps:

Step 1: Initialize cluster centroids $\mu_1, \mu_2, \ldots, \mu_k \in R^2$ randomly.
Step 2: Define the cluster number S_i for x_i

$$S_i := \arg\min_j dis(x_i, \mu_j) \tag{24}$$

where $dis(o, p)$ is a function for calculating the distance or measuring the similarity between variable *o* and variable *p*. Here function $dis(o, p)$ is defined as cosine similarity:

$$dis(o, p) := \frac{o^T p}{\|o\| \|p\|} \tag{25}$$

Step 3: Update the *j*th cluster centroid by

$$\mu_j := \frac{\sum_{i=1}^{M+1} 1\{S_i = j\} x_i}{\sum_{i=1}^{M+1} 1\{S_i = j\}} \tag{26}$$

where $1\{S_i = j\}$ is 1 when the cluster number S_i is the *j*th cluster.

Repeat step 2 and step 3 until convergence is reached. The information vector of the cluster which the *i*th target tag is assigned is denoted as $x_i, X_1, \ldots, X_Q$, *Q* is the number of reference tags belong to the same cluster.

According to information clustering, the *i*th target is assigned to one cluster with several reference tags for each reader antenna. The reference tags within the same cluster share high distance similarity with the target tag. This distance similarity is used to estimate the distance between the *j*th reader antenna and the *i*th target tag as shown in (27).

$$\tilde{d}_{i,j} = \sum_{q=1}^{Q} \omega_q d_{q,j} \tag{27}$$

where $d_{q,j}$ is the distance between *q*th reference tag and the *j*th reader antenna, and ω_q is the weighting factor. The weighting factor is proposed and calculated by

$$\omega_q = \frac{1 / \left\| x_1 - X_q \right\|^2}{\sum_{q=1}^{Q} 1 / \left\| x_1 - X_q \right\|^2} \tag{28}$$

After the distances of the target tag to all the reader antennas are estimated, the tag position is determined by the linear-least-squares algorithm.

The performance of *k*-NN algorithm is considerably affected by the selection of parameter k. Unlike *k*-NN algorithm, the clustering in SAIL algorithm is automatic; the number of reference tags within the same cluster is also adaptively selected by the algorithm. To some degree, SAIL algorithm is more flexible than *k*-NN algorithm.

Experimental Results

An experiment was conducted in a room of dimensions of $6.5 \times 10.0m^2$. Four directional antennas were fixed on four tripods which were deployed in the four corners of an area of $2.4 \times 3.0m^2$. The RFID tags were all attached to white paper boxes with dimensions of $5cm \times 5cm \times 5cm$, which were placed on the ground as shown in Figure 9. The target tag was marked by red on the top, roughly located in the center of the reference tags. To evaluate the performance of the different algorithms, a total of 128 target tag positions were tested and the distance between two adjacent positions is 10 cm. Only when the tag was activated by all the four reader antennas, the measurements were used for tag position determi-

Figure 9. Experiment scene

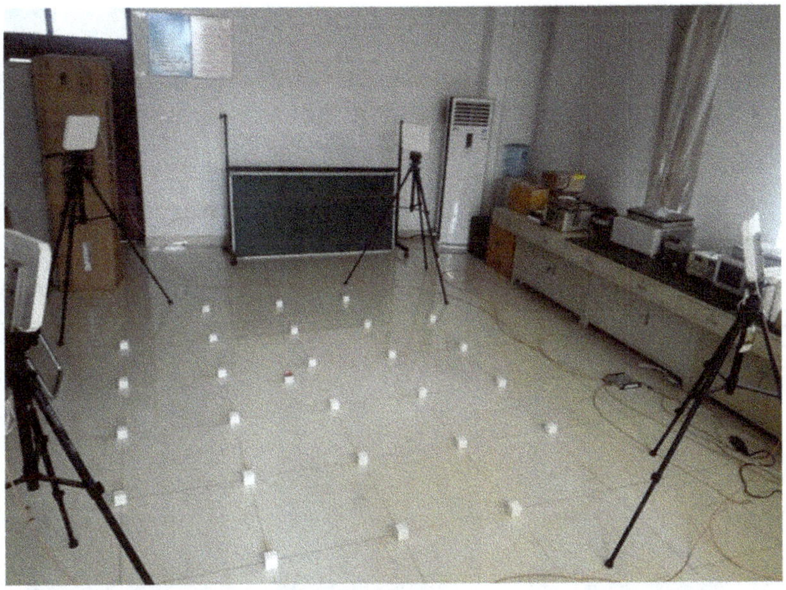

nation. The same experiment was repeated five times and all the measurements were used for performance evaluation. One clustering result per target tag position point was recorded in each experiment test.

Several localization schemes are compared with SAIL in terms of the ranging and localization accuracy. These schemes include the following:

1. SAIL-RSS: The calculation is the same as SAIL just except that only RSS information is utilized and the similarity is calculated with RSS Euclidian distance.
2. SAIL-phase: It is similar to SAIL-RSS except that the phase Euclidian distance is utilized as similarity measure.
3. KNN-k: This is the localization algorithm used in LANDMARC system with RSS. The common k values in k-NN algorithm are considered as 4 and 5.

Firstly, ranging performance of the three schemes (SAIL, SAIL-RSS, and SAIL-phase) are shown in Figure 10. The KNN-k method is not involved with ranging, so it is not applicable. It can be seen that SAIL slightly outperforms SAIL-RSS, while the performance of SAIL-phase is the worst. So the combination of RSS and phase only improves ranging accuracy rather marginally. Figure 11 shows the position errors of the four algorithms. SAIL still outperforms the other three algorithms and SAIL-phase is the worst, indicating that SAIL-phase should not be used alone. KNN-5 is a little better than KNN-4 but KNN-5 requires one more reference tag. Figure 12 shows the position accuracy in terms of error cumulative distribution function (CDF). Again, SAIL is the best and SAIL-phase is the worst, while SAIL-RSS, KNN-4, and KNN-5 achieve very similar performance. The CDF results are in consistent with RMSE result shown in Figure 11. It is also observed that the maximum error of SAIL is 1.96m, while those of other algorithms are greater than 2.23m.

Figure 10. Ranging error comparison between three methods

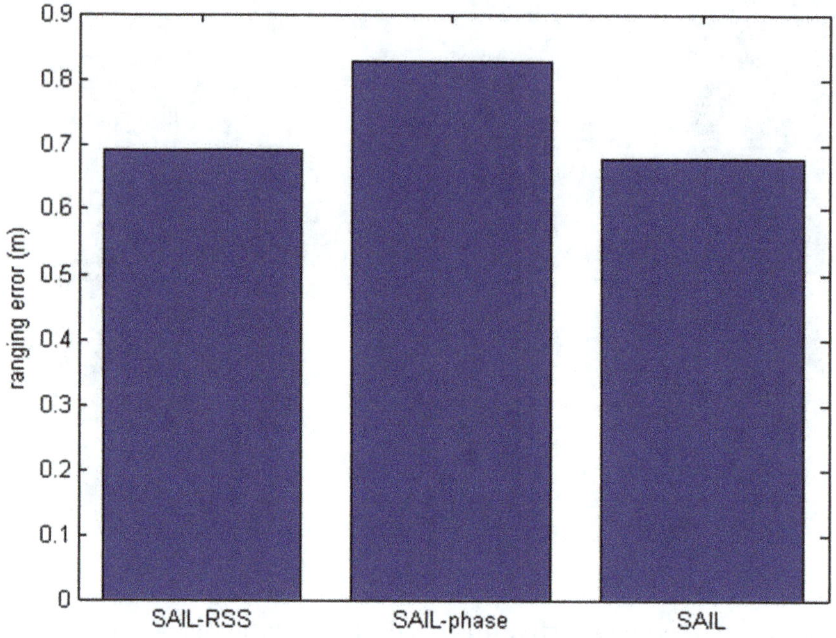

Figure 11. Localization error comparison between four methods

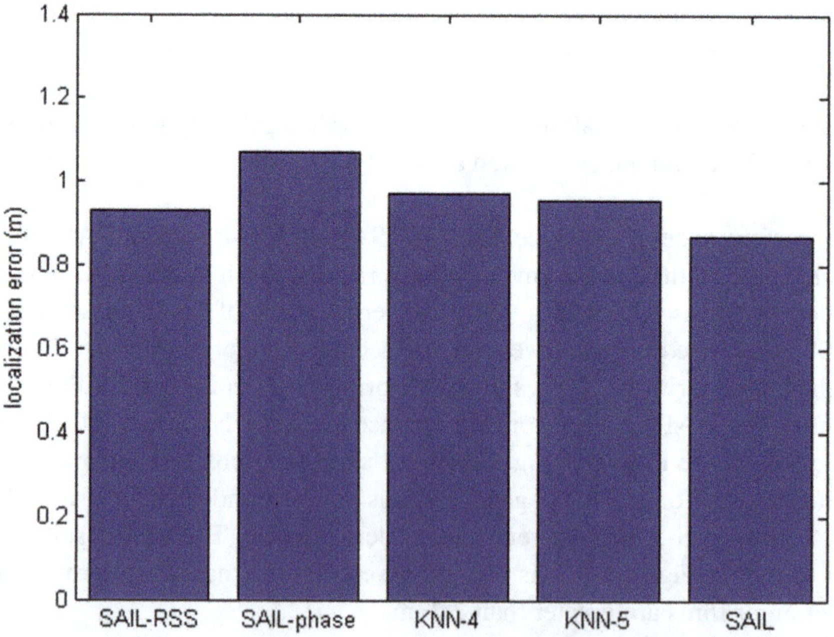

Figure 12. Position error CDF of four algorithms

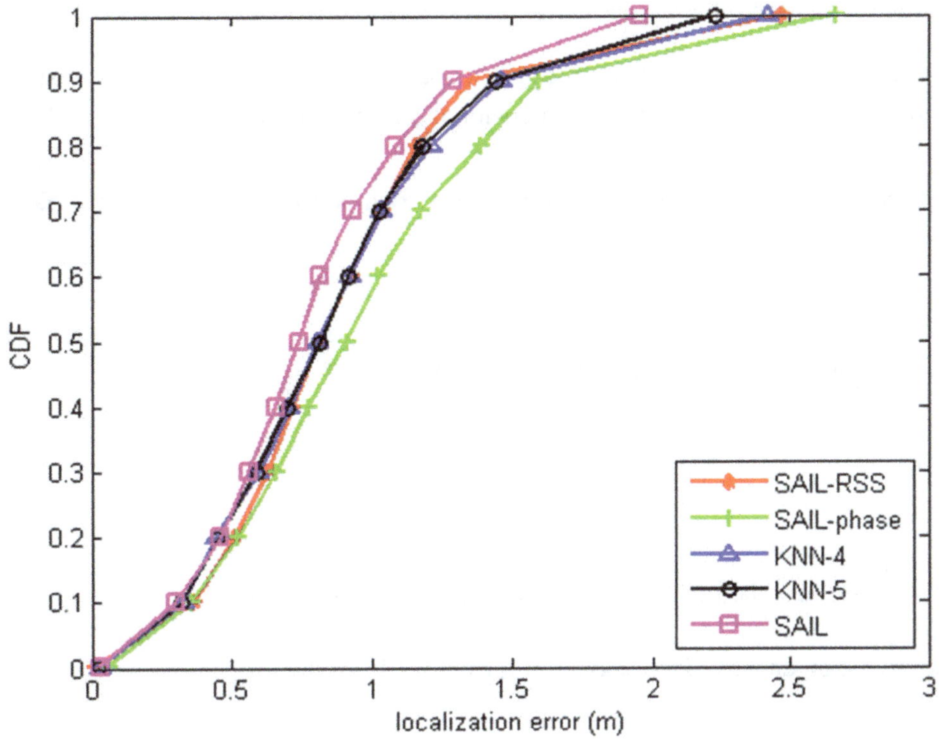

OPTIMIZATION BASED LOCALIZATION ALGORITHMS

Multidimensional Scaling (MDS)

MDS is a statistical method, which can be used to obtain the coordinates of all involved nodes in multidimensional space by the similarity relation between nodes. This method is widely used in psychology, biology, sociology, and so on. Shang *et al.* first introduced MDS into node localization in wireless sensor networks and proposed an algorithm named MDS-MAP based on classical MDS (Shang, Ruml, Zhang & Fromherz, 2003). Classical MDS uses Euclidian distance as the similarity between nodes, and the coordinates of the nodes are obtained by a series of matrix operations. In this section, the principle of classical MDS is first described. Then, the residual weighted MDS (RWMDS) is used for RFID tag localization in multipath channel environment.

Classical MDS

The relative distance map of all position points in multidimensional space can be constructed by the matrix of dissimilarity between entities. The goal of MDS is to make the dissimilarity between entities as close as possible to the distance between the position points in the multidimensional space. In the positioning application, the Euclidian distance between the nodes is often used to construct the dissimilarity matrix (Cheung & So, 2005).

Suppose that there are *n* nodes whose positions are to be determined. Classical MDS algorithm can be divided into the following steps:

1. The distance measurement of each node pair is obtained by ranging, and all the distance measurements are combined to form the two dimensional dissimilarity matrixes $[P_{ij}]$ of dimensions *n* by *n*.
2. Each of the components (i.e. distances) in $[P_{ij}]$ is squared to produce the matrix **D** as:

$$D = [(P_{ij})^2] \in R^{n*n} \tag{29}$$

3. Applying double centering to the matrix D yields:

$$B = -\frac{1}{2} JDJ^T \tag{30}$$

where $J = I - \frac{1}{n} ee^T$, **I** is the identity matrix of order *n*, **e** is a column vector of ones.

4. Performing the eigenvalue decomposition for matrix **B** produces:

$$B = V \wedge V^T \tag{31}$$

where **V** the corresponding eigenvectors and $\wedge$ is the diagonal matrix In the case of two-dimensional location, taking the largest two eigenvalues and the corresponding eigenvectors which are normalized to be unit eigenvalue vectors. The relative coordinates of nodes can be calculated using the following formula:

$$X = V_2 \wedge_2^{\frac{1}{2}} \tag{32}$$

Classical MDS can only produce the estimates of the relative coordinates of the nodes in an unknown coordinate system. If a number of nodes are anchors whose positions are also known in a reference coordinate system, then coordinate transformation can be realized through rotation and translation to obtain the position estimates of all nodes in the reference coordinate system.

Residual Weighted Multidimensional Scaling

As discussed above positioning accuracy can be improved by deploying reference tags in the positioning environment where the wireless signal propagation is affected by multipath as illustrated in Figure 13. Therefore, the estimated distance between reader and tag is affected by multipath, which can be expressed as

Figure 13. The schematic diagram of multipath environment

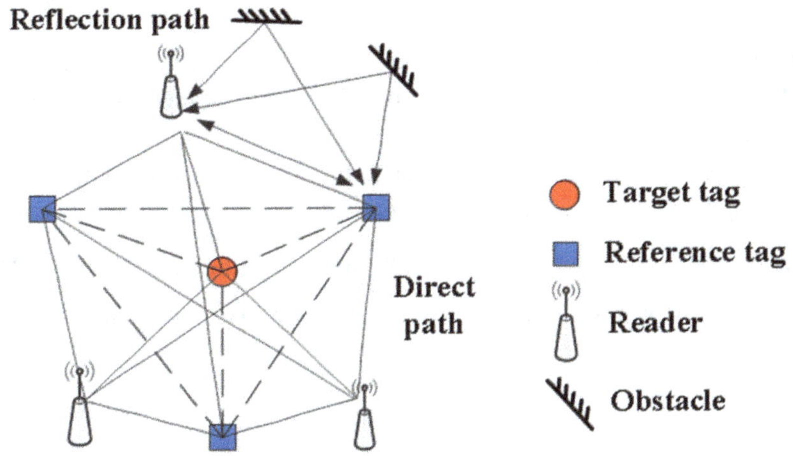

$$\hat{d} = d + e = d + e_{noise} + e_{multipath} \tag{33}$$

where d is the true Euclidian distance, e is the total error, e_{noise} is the noise, and $e_{multipath}$ is the error caused by multipath.

The distance from the tag to the reader can be obtained by a ranging method such as PDOA. However, the distances between tags are missing, because there is no communication between tags. The distance between tags can be obtained by means of data fitting (Wang, Ma, Zhao & Liu, 2015).

Suppose that N_1 reference tags and N_2 readers are deployed in the localization environment. Let $\widehat{\boldsymbol{\theta}} = \begin{bmatrix} \hat{\theta}_1 & \cdots & \hat{\theta}_{N_2} \end{bmatrix}^T$ be the phase vector of the target tag and $\widehat{\boldsymbol{\varphi}}_i = \begin{bmatrix} \hat{\varphi}_{i,1} & \cdots & \hat{\varphi}_{i,N_2} \end{bmatrix}^T$ be the phase vector of the i th reference tag.

Then, the phase difference Euclidian distance between reference tags *i* and *j* can be defined as

$$E_{\varphi_{ij}} = \| \widehat{\boldsymbol{\varphi}}_i - \widehat{\boldsymbol{\varphi}}_j \|_2 \quad i, j = 1, 2, \cdots N_1 \tag{34}$$

where $\|\cdot\|_2$ is the norm operation.

Assume that the phase difference Euclidian distance $E_{\varphi_{ij}}$ and the geometric distance $d_{\varphi_{ij}}$ between reference tags have a linear relationship:

$$d_{\varphi_{ij}} = a \cdot E_{\varphi_{ij}} + b \quad i, j = 1, 2, \cdots N_1 \tag{35}$$

where a and b are the coefficients which can be obtained by data fitting. Therefore, given the phase difference Euclidian distance, the geometric distance can be calculated. Similarly, the distance between the target tag and the i th reference tag can be calculated by

$$d_{\theta,\varphi_i} = a \cdot E_{\theta,\varphi_i} + b \quad i = 1,2,\cdots N_2 \tag{36}$$

Because the target tag and reference tags are in the same localization environment, the multipath effects on the target tag and reference tags are the same. Then the distance estimation between the target tag and reference tag is influenced by Gaussian noise.

The basic principle of the residual weighted multidimensional scaling (RWMDS) is described as follows. Let $x_0 = [x_0\ y_0]^T$ be the location of target tag, $x_{r_i} = [x_{r_i}\ y_{r_i}]^T, i = 1,2\cdots N_1$, be the ith reference tag location, and $x_{R_j} = [x_{R_j}\ y_{R_j}]^T, j = 1,2\cdots N_2$, be the jth reader location. There $N_1 + N_2 + 1$ are tags and readers in total. Without considering the ranging error, a squared distance matrix can be constructed as

$$\mathbf{D} = \begin{bmatrix} 0 & d_{r_1}^2 & \cdots & d_{r_{N_1}}^2 & d_{R_1}^2 & \cdots & d_{R_{N_2}}^2 \\ d_{r_1}^2 & 0 & d_{Rr_{21}}^2 & \cdots & \cdots & \cdots & d_{Rr_{1M}}^2 \\ \vdots & d_{Rr_{21}}^2 & 0 & \vdots & \vdots & \vdots & \vdots \\ d_{r_{N_1}}^2 & \vdots & \vdots & 0 & \vdots & \vdots & \vdots \\ d_{R_1}^2 & \vdots & \vdots & \vdots & 0 & \vdots & \vdots \\ \vdots & \vdots & \vdots & \vdots & \vdots & 0 & \vdots \\ d_{R_{N_2}}^2 & d_{Rr_{2M}}^2 & \cdots & \cdots & \cdots & \cdots & 0 \end{bmatrix}_{(M+1)\times(M+1)} \tag{37}$$

where $M = N_1 + N_2$, d_{r_i} is the distance between the target tag and the ith reference tag, d_{R_j} is the distance between the target tag and the jth reader and $d_{Rr_{mn}}$ is the distance between the mth reader and the nth reference tag. Assume that the origin of the coordinate system is located at the centroid of all the reference tags and readers with known locations. The position of the centroid is calculated as: $x_c = \frac{1}{M}\left(\sum_{i=1}^{N_1} x_{r_i} + \sum_{j=1}^{N_2} x_{R_j}\right)$. Then, the coordinates of all nodes are translated and a coordinate matrix can be constructed as

$$\begin{aligned} \mathbf{X}_s &= \begin{bmatrix} \mathbf{x}_0 - \mathbf{x}_c & \mathbf{x}_{r_1} - \mathbf{x}_c & \cdots & \mathbf{x}_{r_{N_1}} - \mathbf{x}_c & \mathbf{x}_{R_1} - \mathbf{x}_c & \cdots & \mathbf{x}_{R_{N_2}} - \mathbf{x}_c \end{bmatrix}^T \\ &= \mathbf{P} - \mathbf{Q}\mathbf{x}_0^T \end{aligned} \tag{38}$$

where $\mathbf{P} = \begin{bmatrix} -\mathbf{x}_c & \mathbf{x}_{r_1} - \mathbf{x}_c & \cdots & \mathbf{x}_{r_{N_1}} - \mathbf{x}_c & \mathbf{x}_{R_1} - \mathbf{x}_c & \cdots & \mathbf{x}_{R_{N_2}} - \mathbf{x}_c \end{bmatrix}^T$ and $\mathrm{Q} = [-1,\ 0,\ \cdots 0]^T$.

Alternatively, (38) can be written as

$$\mathbf{X}_s = \mathbf{X} - 1_{M+1}\mathbf{w}^T\mathbf{X} = \mathbf{P}_w\mathbf{X} \tag{39}$$

where $\mathbf{X} = \begin{bmatrix} \mathbf{x}_0 & \mathbf{x}_{r_1} & \cdots & \mathbf{x}_{r_{N_1}} & \mathbf{x}_{R_1} & \cdots & \mathbf{x}_{R_{N_2}} \end{bmatrix}^T$, $\mathbf{w} = \begin{bmatrix} 0 & 1/M & \cdots & 1/M \end{bmatrix}^T$, $\mathrm{P}_w = \mathrm{I}_{M+1} - 1_{M+1}\mathrm{w}^T$, I_{M+1} and 1_{M+1} are the identity matrix with dimensions $(M+1)\times(M+1)$ and $(M+1)\times 1$ column vector of all ones, respectively.

Define matrix B_s as the scalar product of matrix X_s and its transpose, which can be described as

$$\mathbf{B}_s = \mathbf{X}_s\mathbf{X}_s^T = \mathbf{P}_w\mathbf{X}\mathbf{X}^T\mathbf{P}_w^T = -0.5(\mathbf{P}_w\mathbf{J})\mathbf{D}(\mathbf{P}_w\mathbf{J})^T = -0.5\mathbf{P}_w\mathbf{D}\mathbf{P}_w^T \tag{40}$$

where $\mathrm{J} = \mathrm{I}_{M+1} - \frac{1}{M+1}1_{M+1}1_{M+1}^T$, $\mathrm{P}_w\mathrm{J} = \mathrm{P}_w$ is used.

It can be approved that the scalar product matrix $\mathbf{B}_s$ is symmetric and positive semi-definite. Decomposing $\mathbf{B}_s$ by eigenvalue decomposition yields $\mathbf{B}_s = \mathbf{V}\wedge\mathbf{V}^T$, where $\Lambda = diag\{\lambda_1 \ \cdots \ \lambda_{M+1}\}$, $\lambda_1 \geq \lambda_2 \geq \cdots \geq \lambda_{M+1}$, and V is a unitary matrix. For two dimensional localization, $\mathrm{rank}(\mathbf{B}_s) = 2$, hence the matrix $\mathbf{B}_s$ can be also represented as

$$\mathbf{B}_s = \mathbf{V}_s \wedge_s \mathbf{V}_s^T \tag{41}$$

and

$$\mathbf{V}_n^T\mathbf{B}_s\mathbf{V}_n = \mathbf{O}_{(M-1)\times(M-1)} \tag{42}$$

where $\mathbf{V}_s = [\mathbf{V}_1, \mathbf{V}_2]$ and $\mathbf{V}_n = [\mathbf{V}_3, \mathbf{V}_4, \cdots \mathbf{V}_{M+1}]$ represent the signal subspace and noise subspace of the scalar product matrix, respectively, and $\mathrm{O}_{(M-1)\times(M-1)}$ is a matrix of all zeroes with dimensions $(M-1)\times(M-1)$.

Because we use data fitting method to generate the distances between reference tags and target tag and the distances between target tag and readers are obtained through PDOA ranging. The ranging error in these two modes is different.

The distance measurements between the target tag and ith reference tag are

$$\hat{d}_{r_i} = d_{r_i} + e_i, i = 1, 2, \cdots N_1 \tag{43}$$

And the distance measurements between the target tag and the jth reader are

$$\hat{d}_{R_j} = d_{R_j} + q_j, j = 1, 2, \cdots N_2 \tag{44}$$

Where e_i and q_j represents the ranging error, respectively. The errors are independent of each other.

Therefore, we use different weights to weight the RWMDS algorithm. The RWMDS algorithm can outperform through a few of iterations. The detailed derivation of the RWMDS algorithm is shown in the paper (Wang, Ma, Zhao & Liu, 2015).

The number of multipath reflects the complexity of the environment. From Figure 14 we can see that the RWMDS algorithm is superior to the traditional localization methods. With the increase in the number of multipath, compared with the traditional positioning method, the advantages of the RWMDS algorithm become more and more obvious. As the environment becomes more complex, the robustness of the RWMDS algorithm is optimal.

Convex Optimization

Convex relaxation can be used to handle the optimization problem. When the formulation of the localization problem modeling is non-convex, semi-definite relaxation or second-order cone relaxation can be applied to obtain the convex semi-definite program (SDP) or second-order cone program (SOCP). Interior-point method is usually utilized to solve convex optimization problem, which is effective to solve small-scale convex optimization problem.

Fundamental Theory

1. Convex sets

A set C is convex if the line segment between any two points in C lies in C, i.e., if for any $x_1, x_2 \in C$ and any α with $0 \le \alpha \le 1$, we have $\alpha x_1 + (1-\alpha)x_2 \in C$.

Figure 14. Algorithm comparisons in different multipath environments

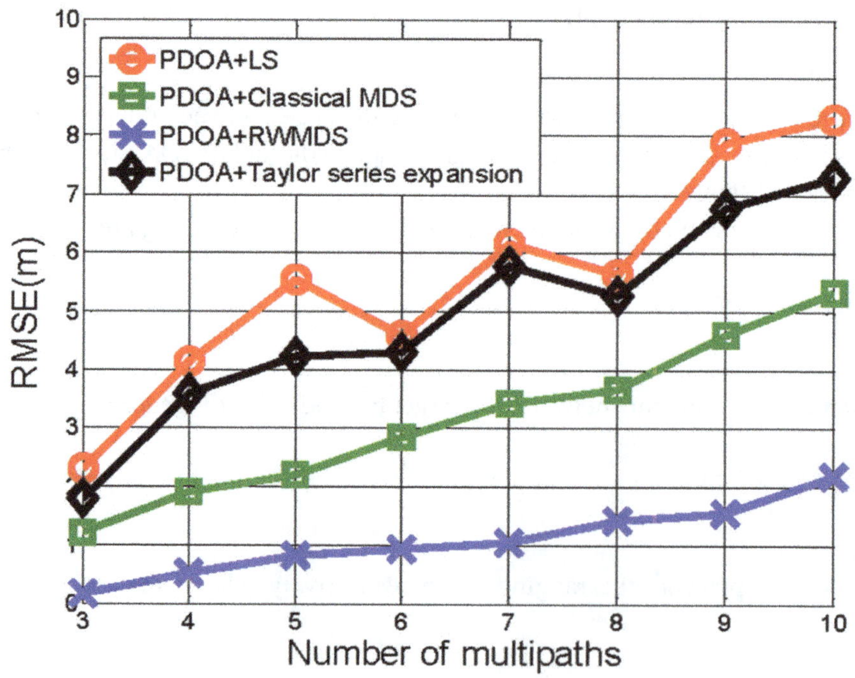

Common convex set contains single point set, empty set and the entire Euclidean space. In one-dimensional space, the convex set is a single point or an uninterrupted line. In two-dimensional or three-dimensional space, the convex set is an intuitive convex pattern. Convex set and non-convex set are illustrated in Figure 15.

2. Convex function

A function $f : R^n \rightarrow R$ is convex if $\text{domain}f$ is a convex set and if for all $x, y \in \text{dom}f$, we have $f(\alpha x_1 + (1-\alpha)x_2) \leq \alpha f(x_1) + (1-\alpha)f(x_1)$, where, $0 \leq \alpha \leq 1$. Geometrically, this inequality indicates that the line segment between $(x, f(x))$ and $(y, f(y))$ is above or on the curve of the function f as shown in Figure 16.

The common convex functions include exponential function, screen function, and negative entropy function. In general, the method to determine whether a function is a convex function is to determine whether its Hessian matrix is semi-definite, and if so, the function is a convex function, and if the matrix is positive definite, then the function is a strictly convex function (Wang & Wu, 2017).

The convex optimization problem can be expressed as

$$\begin{aligned} &\min\text{imize} \quad f_0(x) \\ &\text{subject} \quad to \;\; f_i(\mathrm{x}) \leq \mathrm{b}_i, i = 1, \cdots, m, \end{aligned} \tag{45}$$

Figure 15. Convex and non-convex graphs

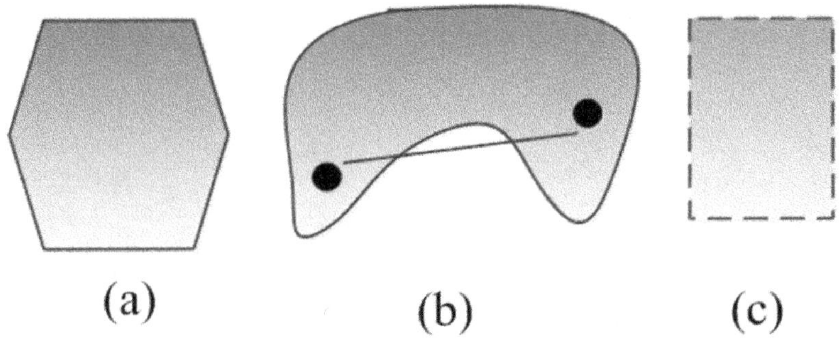

Figure 16. Graph of a convex function

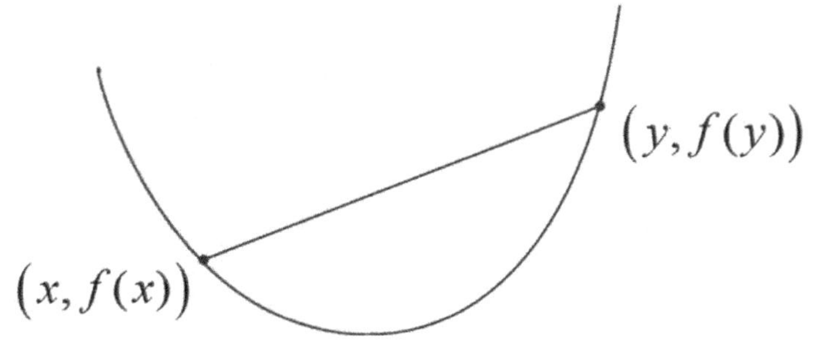

where the function $f_0(x),\cdots,f_m : R^n \to R$ are all convex.

Localization via Convex Optimization

1. Localization via least squares

Assume that that there are N readers in the environment, the position of the target tag is $\mathbf{s} = \begin{bmatrix} x & y \end{bmatrix}^T$, and the position of the ith reader is $\mathbf{R}_i = \begin{bmatrix} x_i & y_i \end{bmatrix}^T, i = 1,\cdots,N_1$. The distance between the ith reader and the target tag $\hat{d}_{R_i}$ can be expressed as

$$\hat{d}_{R_i} = \left\| \mathbf{s} - \mathbf{R}_i \right\| + \lambda_{R_i} + n_{R_i} \quad i = 1,2,\cdots,N_1 \tag{46}$$

where, λ_{R_i} is the multipath error, n_{R_i} is the measurement noise which is assumed to have a zero-mean Gaussian distribution $\mathcal{N}\left(0,\sigma_{R_i}^2\right)$.

By ignoring the multipath effect and measurement noise, (46) becomes

$$\hat{d}_{R_i} = \left\| \mathbf{s} - \mathbf{R}_i \right\| = \sqrt{(x-x_i)^2 + (y-y_i)^2}, i = 1,\cdots,N_1 \tag{47}$$

Taking the square of both sides of eq.(64) leads to

$$\hat{d}_{R_i}^2 = x^2 - 2xx_i + \mathrm{x}_i^2 + y^2 - 2yy_i + y_i^2, i = 1,\cdots,N_1 \tag{48}$$

Sorting (48) yields

$$2(x_i - x_1)x + 2(y_i - y_1)y = (x_i^2 + y_i^2) - (x_1^2 + y_1^2) - d_i^2 + d_1^2, i = 1,\cdots,N_1 \tag{49}$$

The equation can be expressed in a matrix form as

$$\mathbf{As} = \mathbf{B} \tag{50}$$

Where $\mathbf{A} = 2\begin{bmatrix} x_2 - x_1 & y_2 - y_1 \\ x_3 - x_1 & y_3 - y_1 \\ \vdots & \vdots \\ x_{N_1} - x_1 & y_{N_1} - y_1 \end{bmatrix}_{(N_1-1)\times 2}$, $\mathbf{B} = \begin{bmatrix} (x_2^2 + y_2^2) - (x_1^2 + y_1^2) - d_2^2 + d_1^2 \\ (x_3^2 + y_3^2) - (x_1^2 + y_1^2) - d_3^2 + d_1^2 \\ \vdots \\ (x_{N_1}^2 + y_{N_1}^2) - (x_1^2 + y_1^2) - d_{N_1}^2 + d_1^2 \end{bmatrix}_{(N_1-1)\times 1}$.

The least squares solution to (50) is given by

$$\hat{\mathbf{s}} = (\mathbf{A}^T\mathbf{A})^{-1}\mathbf{A}^T\mathbf{B} \tag{51}$$

2. Localization via semi-definite program

As discussed in (36), the distance between the target tag and the ith reference tag is

$$\hat{d}_{r_i} = \left\|\mathbf{s} - \mathbf{r}_i\right\| + \lambda_{r_i} + n_{r_i} \quad i = 1, 2, \cdots, N_2 \tag{52}$$

where $\mathbf{r}_i$ is the location of the ith reference tag. λ_{r_i} is the mean ranging error of the ith reference tag, $n_{r_i} \sim \mathcal{N}\left(0, \sigma_{r_i}^2\right)$, both of which could be estimated by reference tags except for the ith reference tag (Vaghefi & Buehrer, 2015).

In order to estimate the location of the target tag, (46) and (52) can be equally written as

$$\hat{d}_{R_i} - \lambda_{R_i} - \left\|\mathbf{s} - \mathbf{R}_i\right\| = n_{R_i} \tag{53}$$

$$\hat{d}_{r_i} - \lambda_{r_i} - \left\|\mathbf{s} - \mathbf{r}_i\right\| = n_{r_i} \tag{54}$$

Based on (53) and (54), the maximum likelihood (ML) estimation of can be formulated as

$$\min_s \sum_{i=1}^{N_1} \frac{\left(\hat{d}_{R_i} - \lambda_{R_i} - \left\|\mathbf{s} - \mathbf{R}_i\right\|\right)^2}{\sigma_{R_i}^2} + \sum_{i=1}^{N_2} \frac{\left(\hat{d}_{r_i} - \lambda_{r_i} - \left\|\mathbf{s} - \mathbf{r}_i\right\|\right)^2}{\sigma_{r_i}^2} \tag{55}$$

which can be rewritten as

$$\begin{aligned} \min_s \quad & \sum_{i=1}^{N_1} \frac{\left(\hat{d}_{R_i} - \lambda_{R_i}\right)^2 - 2\left(\hat{d}_{R_i} - \lambda_{R_i}\right)\left\|\mathbf{s} - \mathbf{R}_i\right\| + \left\|\mathbf{s} - \mathbf{R}_i\right\|^2}{\sigma_{R_i}^2} \\ & + \sum_{i=1}^{N_2} \frac{\left(\hat{d}_{r_i} - \lambda_{r_i}\right)^2 - 2\left(\hat{d}_{r_i} - \lambda_{r_i}\right)\left\|\mathbf{s} - \mathbf{r}_i\right\| + \left\|\mathbf{s} - \mathbf{r}_i\right\|^2}{\sigma_{r_i}^2} \end{aligned} \tag{56}$$

which implies

$$\begin{aligned}\min_{\substack{s,v,\{g_{R_i}\},\{g_{r_i}\},\\ \{h_{R_i}\},\{h_{r_i}\}}} \quad & \sum_{i=1}^{N_1}\frac{1}{\sigma_{R_i}^2}\left(\left(\hat{d}_{R_i}-\lambda_{R_i}\right)^2-2\left(\hat{d}_{R_i}-\lambda_{R_i}\right)h_{R_i}+g_{R_i}\right)\\ & +\sum_{i=1}^{N_2}\frac{1}{\sigma_{r_i}^2}\left(\left(\hat{d}_{r_i}-\lambda_{r_i}\right)^2-2\left(\hat{d}_{r_i}-\lambda_{r_i}\right)h_{r_i}+g_{r_i}\right)\\ s.t \quad & g_{R_i}=v\text{ - }2s^T\mathbf{R}_i+\left\|\mathbf{R}_i\right\|^2, i=1,2,\cdots,N_1\\ & g_{r_i}=v\text{ - }2\mathbf{s}^{\mathrm{T}}\mathbf{r}_i+\left\|\mathbf{r}_i\right\|^2,, i=1,2,\cdots,N_2\\ & g_{R_i}=h_{R_i}^2, i=1,2,\cdots,N_1\\ & g_{r_i}=h_{r_i}^2, i=1,2,\cdots,N_2\\ & v=\left\|\mathbf{s}\right\|^2\end{aligned} \tag{57}$$

where $g_{R_i}=\left\|s-R_i\right\|^2$, $h_{R_i}=\left\|s-R_i\right\|$, $g_{r_i}=\left\|s-r_i\right\|^2$, $h_{r_i}=\left\|s-r_i\right\|$.

Eq.(57) is a non-convex problem. However, the constraint of of $v=\left\|\mathbf{s}\right\|^2$ can be relaxed as $v\geq\left\|\mathbf{s}\right\|^2$. And hence Eq.(57) is relaxed to become an SDP:

$$\begin{aligned}\min_{\substack{s,v,\{g_{R_i}\},\{g_{r_i}\},\\ \{h_{R_i}\},\{h_{r_i}\}}} \quad & \sum_{i=1}^{N_1}\frac{1}{\sigma_{R_i}^2}\left(\left(\hat{d}_{R_i}-\lambda_{R_i}\right)^2-2\left(\hat{d}_{R_i}-\lambda_{R_i}\right)h_{R_i}+g_{R_i}\right)\\ & +\sum_{i=1}^{N_2}\frac{1}{\sigma_{r_i}^2}\left(\left(\hat{d}_{r_i}-\lambda_{r_i}\right)^2-2\left(\hat{d}_{r_i}-\lambda_{r_i}\right)h_{r_i}+g_{r_i}\right)\\ s.t \quad & g_{R_i}=v\text{ - }2\mathbf{s}^{\mathrm{T}}\mathbf{R}_i+\left\|\mathbf{R}_i\right\|^2, i=1,2,\cdots,N_1\\ & g_{r_i}=v\text{ - }2\mathbf{s}^{\mathrm{T}}\mathbf{r}_i+\left\|\mathbf{r}_i\right\|^2,, i=1,2,\cdots,N_2\\ & g_{R_i}\geq h_{R_i}^2, i=1,2,\cdots,N_1\\ & g_{r_i}\geq h_{r_i}^2, i=1,2,\cdots,N_2\\ & v\geq\left\|\mathbf{s}\right\|^2\end{aligned} \tag{58}$$

which can be rewritten as

$$\begin{aligned}
\min_{\substack{s,v,\{g_{R_i}\},\{g_{r_i}\},\\ \{h_{R_i}\},\{h_{r_i}\}}} \quad & \sum_{i=1}^{N_1} \frac{1}{\sigma_{R_i}^2}\left(\left(\hat{d}_{R_i}-\lambda_{R_i}\right)^2-2\left(\hat{d}_{R_i}-\lambda_{R_i}\right)h_{R_i}+g_{R_i}\right) \\
& +\sum_{i=1}^{N_2} \frac{1}{\sigma_{r_i}^2}\left(\left(\hat{d}_{r_i}-\lambda_{r_i}\right)^2-2\left(\hat{d}_{r_i}-\lambda_{r_i}\right)h_{r_i}+g_{r_i}\right) \\
s.t \quad & g_{R_i}=v-2\mathbf{s}^{\mathrm{T}}\mathbf{R}_i+\left\|\mathbf{R}_i\right\|^2, i=1,2,\cdots,N_1 \\
& g_{r_i}=v-2\mathbf{s}^{T}\mathbf{r}_i+\left\|\mathbf{r}_i\right\|_i^2,, i=1,2,\cdots,N_2 \\
& \begin{bmatrix} 1 & h_{R_i} \\ h_{R_i} & g_{R_i} \end{bmatrix} \succ 0, i=1,2,\cdots,N_1 \\
& \begin{bmatrix} 1 & h_{r_i} \\ h_{r_i} & g_{r_i} \end{bmatrix} \succ 0, i=1,2,\cdots,N_2 \\
& \begin{bmatrix} \mathbf{I}_2 & \mathbf{s} \\ \mathbf{s}^{\mathrm{T}} & v \end{bmatrix} \succ 0
\end{aligned} \tag{59}$$

where $\mathbf{I}_2$ is a two-dimensional identity matrix.

3. Localization via second-order cone program

Eq.(46) and eq.(52) can be equivalently written as

$$\hat{d}_{R_i}-\lambda_{R_i}=\left\|\mathbf{s}-\mathbf{R}_i\right\|+n_{R_i} \tag{60}$$

$$\hat{d}_{r_i}-\lambda_{r_i}=\left\|\mathbf{s}-\mathbf{r}_i\right\|+n_{r_i} \tag{61}$$

Squaring both sides of (60) and (61) yields

$$\left(\hat{d}_{R_i}-\lambda_{R_i}\right)^2=\left\|\mathbf{s}-\mathbf{R}_i\right\|^2+2\left\|\mathbf{s}-\mathbf{R}_i\right\|n_{R_i}+n_{R_i}^2 \tag{62}$$

$$\left(\hat{d}_{r_i}-\lambda_{r_i}\right)^2=\left\|\mathbf{s}-\mathbf{r}_i\right\|^2+2\left\|\mathbf{s}-\mathbf{r}_i\right\|n_{r_i}+n_{r_i}^2 \tag{63}$$

which can be simplified as

$$\frac{\left(\hat{d}_{R_i}-\lambda_{R_i}\right)^2-\left\|\mathbf{s}-\mathbf{R}_i\right\|^2}{2\left\|\mathbf{s}-\mathbf{R}_i\right\|}=n_{R_i} \tag{64}$$

$$\frac{\left(\hat{d}_{r_i}-\lambda_{r_i}\right)^2-\left\|\mathbf{s}-\mathbf{r}_i\right\|^2}{2\left\|\mathbf{s}-\mathbf{r}_i\right\|}=n_{r_i} \tag{65}$$

Owing to $n_{R_i}^2 << \left\|\mathbf{s}-\mathbf{R}_i\right\| n_{R_i}$, $n_{r_i}^2 << \left\|\mathbf{s}-\mathbf{r}_i\right\| n_{r_i}$, $n_{R_i}^2$ and $n_{r_i}^2$ are negligible (Wang, Chen, Li & Ansari, 2014).

Based on (64) and (65), the ML estimation of s can be formulated as

$$\min_s \sum_{i=1}^{N_1} \frac{\left(\left(\hat{d}_{R_i}-\lambda_{R_i}\right)^2-\left\|\mathbf{s}-\mathbf{R}_i\right\|^2\right)^2}{4\sigma_{R_i}^2\left\|\mathbf{s}-\mathbf{R}_i\right\|^2} + \sum_{i=1}^{N_2} \frac{\left(\left(\hat{d}_{r_i}-\lambda_{r_i}\right)^2-\left\|\mathbf{s}-\mathbf{r}_i\right\|^2\right)^2}{4\sigma_{r_i}^2\left\|\mathbf{s}-\mathbf{r}_i\right\|^2} \tag{66}$$

which can be rewritten as

$$\begin{aligned} \min_{\substack{s,v,\\ \{g_{R_i}\},\{g_{r_i}\}}} \; & \sum_{i=1}^{N_1} \frac{\left(\left(\hat{d}_{R_i}-\lambda_{R_i}\right)^2-g_{R_i}\right)^2}{4\sigma_{R_i}^2 g_{R_i}} + \sum_{i=1}^{N_2} \frac{\left(\left(\hat{d}_{r_i}-\lambda_{r_i}\right)^2-g_{r_i}\right)^2}{4\sigma_{r_i}^2 g_{r_i}} \\ s.t \quad & g_{R_i}=v-2\mathbf{s}^{\mathrm{T}}\mathbf{R}_i+\left\|\mathbf{R}_i\right\|^2, i=1,2,\cdots,N_1 \\ & g_{r_i}=v-2\mathbf{s}^{\mathrm{T}}\mathbf{r}_i+\left\|\mathbf{r}_i\right\|^2, i=1,2,\cdots,N_2 \\ & v=\left\|\mathbf{s}\right\|^2 \end{aligned} \tag{67}$$

where $g_{R_i}=\left\|\mathbf{s}-\mathbf{R}_i\right\|^2$, $g_{r_i}=\left\|\mathbf{s}-\mathbf{r}_i\right\|^2$.

Eq.(67) is also a non-convex problem. Similarly, by relaxing the constraint $v=\left\|\mathbf{s}\right\|^2$ can be $v \geq \left\|\mathbf{s}\right\|^2$. Eq.(67) is relaxed to be:

$$\begin{aligned}
\min_{\substack{s,v,\\ \{g_{R_i}\},\{g_{r_i}\}}} \quad & \sum_{i=1}^{N_1} \frac{\left(\left(\hat{d}_{R_i} - \lambda_{R_i}\right)^2 - g_{R_i}\right)^2}{4\sigma_{R_i}^2 g_{R_i}} + \sum_{i=1}^{N_2} \frac{\left(\left(\hat{d}_{r_i} - \lambda_{r_i}\right)^2 - g_{r_i}\right)^2}{4\sigma_{r_i}^2 g_{r_i}} \\
s.t \quad & g_{R_i} = v - 2\mathbf{s}^{\mathrm{T}}\mathbf{R}_i + \left\|\mathbf{R}_i\right\|^2, i = 1, 2, \cdots, N_1 \\
& g_{r_i} = v - 2\mathbf{s}^{\mathrm{T}}\mathbf{r}_i + \left\|\mathbf{r}_i\right\|^2, i = 1, 2, \cdots, N_2 \\
& v \geq \left\|\mathbf{s}\right\|^2
\end{aligned} \tag{68}$$

which can be rewritten as

$$\begin{aligned}
\min_{s,v,\{g_{R_i}\},\{g_{r_i}\}} \quad & \sum_{i=1}^{N_1} t_{R_i} + \sum_{i=1}^{N_2} t_{r_i} \\
s.t \quad & \frac{\left(\left(\hat{d}_{R_i} - \lambda_{R_i}\right)^2 - g_{R_i}\right)^2}{4\sigma_{R_i}^2 g_{R_i}} \leq t_{R_i}, i = 1, 2, \cdots, N_1 \\
& \frac{\left(\left(\hat{d}_{r_i} - \lambda_{r_i}\right)^2 - g_{r_i}\right)^2}{4\sigma_{r_i}^2 g_{r_i}} \leq t_{r_i}, i = 1, 2, \cdots, N_2 \\
& g_{R_i} = v - 2\mathbf{s}^{\mathrm{T}}\mathbf{R}_i + \left\|\mathbf{R}_i\right\|^2, i = 1, 2, \cdots, N_1 \\
& g_{r_i} = v - 2\mathbf{s}^{\mathrm{T}}\mathbf{r}_i + \left\|\mathbf{r}_i\right\|^2, i = 1, 2, \cdots, N_2 \\
& v \geq \left\|\mathbf{s}\right\|^2
\end{aligned} \tag{69}$$

which can be further rewritten as an SOCP form:

$$\begin{aligned}
\min_{s,v,\{g_{R_i}\},\{g_{r_i}\}} \quad & \sum_{i=1}^{N_1} t_{R_i} + \sum_{i=1}^{N_2} t_{r_i} \\
s.t \quad & \left\|\left[\left(\hat{d}_{R_i} - \lambda_{R_i}\right)^2 - g_{R_i}; g_{R_i}\sigma_{R_i}^2 - t_{R_i}\right]\right\| \leq g_{R_i}\sigma_{R_i}^2 + t_{R_i}, \\
& \qquad i = 1, 2, \cdots, N_1 \\
& \left\|\left[\left(\hat{d}_{r_i} - \lambda_{r_i}\right)^2 - g_{r_i}; g_{r_i}\sigma_{r_i}^2 - t_{r_i}\right]\right\| \leq g_{r_i}\sigma_{r_i}^2 + t_{r_i}, \\
& \qquad i = 1, 2, \cdots, N_2 \\
& g_{R_i} = v - 2\mathbf{s}^{\mathrm{T}}\mathbf{R}_i + \left\|\mathbf{R}_i\right\|^2, i = 1, 2, \cdots, N_1 \\
& g_{r_i} = v - 2\mathbf{s}^{\mathrm{T}}\mathbf{r}_i + \left\|\mathbf{r}_i\right\|^2, i = 1, 2, \cdots, N_2 \\
& \left\|\left[\mathbf{s}; v - \frac{1}{4}\right]\right\| \leq v + \frac{1}{4}
\end{aligned} \tag{70}$$

DEVICE FREE LOCALIZATION

Device-free localization (DFL) (Youssef et al., 2007) is a promising localization technique. The traditional active positioning technologies, either GNSS-based or non-GNSS-based, require a target to carry an electronic device for signal reception. However, such a collaborative localization way is impractical in many scenarios. For instance, a thief who breaks into a house to steal properties would not turn on a smartphone for localization purpose. This also applies to the scenario where a spy tries to steal secret documents in an office room. Another example is that a person is trying to cross a border illegally. Thus, new kind of positioning solutions are needed and one is device-free localization (DFL) (Youssef et al., 2007). DFL do not need the target to carry any auxiliary equipment, so it is also called the non-cooperative localization. It is a promising technique for applications in defend and public security, intrusion detection, and prevention of burglary.

There are many popular algorithms in DFL systems, including: (1) the fingerprint (FP) (Zhou, Yang & Wu, 2014; Hong & Ohtsuki, 2015); (2) the support vector machine (SVM) (Deak, Curran & Condell et al., 2013; Yin, Cui & Wu et al., 2015); (3) radio tomographic imaging (RTI) (Wilson & Patwari, 2010); (4) the particle filter (PF) (Wilson & Patwari, 2012; Wang, Gao & Yu et al., 2013); (5) compressed sensing (CS) (Wang, Chen & Fang et al., 2015; Wang, Gao & Wang et al., 2015); and (6) the deep learning(DP) (Wang, Zhang & Gao et al., 2016). Among these algorithms, the RTI-based and the fingerprint-based DFL methods have received more attention in recent years, which are the focus of this section.

Radio Tomographic Imaging

RTI is an emerging technology for imaging the attenuation caused by physical objects in wireless networks (Wilson & Patwari, 2010), as depicted in Figure 17. The algorithm has low computational complexity, high positioning accuracy, no training data, and good real-time performance. Due to the prominent properties, the algorithm has been extensively investigated and different techniques have been proposed to improve the performance. Wilson *et al.* presented a Variance-Based RTI (VRTI) for localizing and tracking motion behind walls in real time. The method takes advantage of the motion-induced variance of RSS measurements made in a wireless peer-to-peer network. Using a multipath channel model, they show that the signal strength on a wireless link is largely dependent on the power contained in multipath components that travel through space containing moving objects. A statistical model relating variance to spatial locations of movement is presented and used as a framework for the estimation of a motion image. From the motion image, the Kalman filter is applied to recursively track the coordinates of a moving target (Wilson & Patwari, 2011). Zhao *et al.* proposed the subspace decomposition method variance-based radio tomography (SubVRT) and least squares variance-based radio tomography (LS-VRT) to reduce the impact of the variations caused by intrinsic motion. They discussed how intrinsic motion, such as leaves moving in the wind, increases measured RSS variance in a way that is "noise" to a localization system. LSVRT outperforms SubVRT in localization accuracy, and it requires tuning of fewer parameters compared to SubVRT. They also proposed an online calibration method so that both SubVRT and LSVRT can use real-time online measurements to perform calibration instead of using "empty-area" offline calibration measurements. Experiments were conducted to investigate the effect of intrinsic motion and the performance of SubVRT and LSVRT (Zhao & Patwari, 2015). Hamilton *et al.* examined two existing shadowing models for RF tomography, the NeSH model and Normalized Ellipse model, and presented Inverse Area Elliptical (IAE) model. The experimental results show that

Figure 17. An illustration of an RTI network. Each node broadcasts to the others, creating many projections that can be used to reconstruct an image of objects inside the network area.

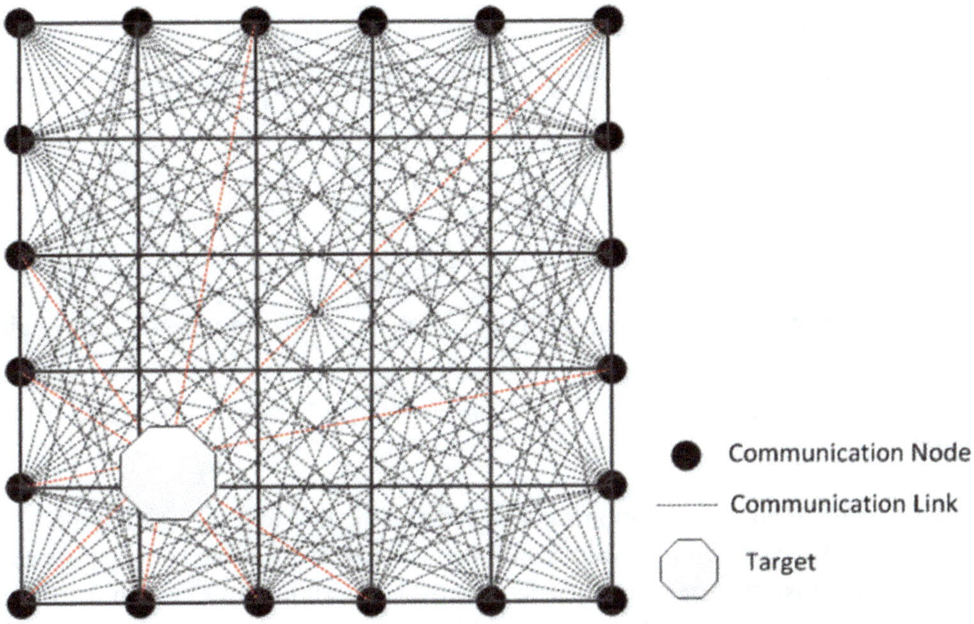

the positioning accuracy of the IAE model is higher than the Elliptical weighting model. They also described the design of their RF tomography testbed which includes a complicated building-like structure and was designed to collect high fidelity signal strength measurements. Measurements collected from a field test using the testbed were then used to analyze both how well the three shadowing models matched the observed radio propagation environment and their utility for RF tomography. The experimental results show that the positioning accuracy of the IAE model is higher than the Elliptical weighting model (Hamilton et al., 2014).

The following is the specific implementation of the RTI algorithm.

Linear Formulation

When wireless nodes communicate with each other, the radio signals pass through the physical area of the network. Objects within the area absorb, reflect, diffract, or scatter some of the transmitted power. The goal of an RTI system is to determine an image vector of dimension R^N that describes the amount of radio power attenuation occurring due to physical objects within *N* voxels of a network region. Since voxel locations are known, RTI allows one to know where attenuation in a network is occurring, and therefore, where objects are located.

We set $y_i(t)$ as the received signal strength at time t for link i. So we have:

$$y_i(t) = p_i(t) - l_i(t) - s_i(t) - f_i(t) - v_i(t) \tag{71}$$

where $p_i(t)$ represents the transmitted power, $s_i(t)$ indicates the shadow loss, $f_i(t)$ represents multipath fading, $l_i(t)$ represents static loss caused by distance, $v_i(t)$ represents measurement noise.

Here $s_i(t)$ is equivalent to the sum of all pixels attenuation. Considering the contribution of pixels in the area to the ith link is different, we need to distribute the weights for each pixel. The mathematical form is:

$$s_i(t) = \sum_{j=1}^{N} w_{ij} \cdot x_j(t) \tag{72}$$

where $x_j(t)$ represents attenuation values at time t for pixel j, w_{ij} represents the weight value of pixel j for link i.

Weight Model

An ellipsoid with foci at each node location can be used as a method for determining the weighting for each link in the network (Patwari & Agrawal, 2008). If a particular voxel falls outside the ellipsoid, the weighting for that voxel is set to zero. If a particular voxel is within the LOS path determined by the ellipsoid, its weight is set to be inversely proportional to the square root of the link distance. Intuitively, longer links will provide less information about the attenuation in voxels that they cross. When link distances are very long, the signals reflect and diffract around the obstructions. A link with a distance of only a few feet will experience more change in RSS when an obstruction occurs than a link with a length of hundreds of feet.

The elliptical weighting model can be described as:

$$w_{ij} = \frac{1}{\sqrt{d_i}} \begin{cases} 1, & if \quad d_{ij}(1) + d_{ij}(2) < d_i + \lambda \\ 0, & otherwise \end{cases} \tag{73}$$

where d_i indicates the length of link i, $d_{ij}(1)$ and $d_{ij}(2)$ respectively represent the distance from the center of pixel j to the two endpoints of link i and λ is an adjustable parameter to control the size of the ellipse. Figure 18 shows the elliptical weighting model in a link of wireless sensor network (WSN).

Equation Solving

We assume that the transmission power and the static loss do not change over time, so the variation of the received power at the ith link is described as follows:

$$\Delta y_i = y_i(t) - y_i(0) = s_i(t) - s_i(0) + f_i(t) - f_i(0) + v_i(t) - v_i(0) \tag{74}$$

that is:

Figure 18. An illustration of the elliptical weighting model

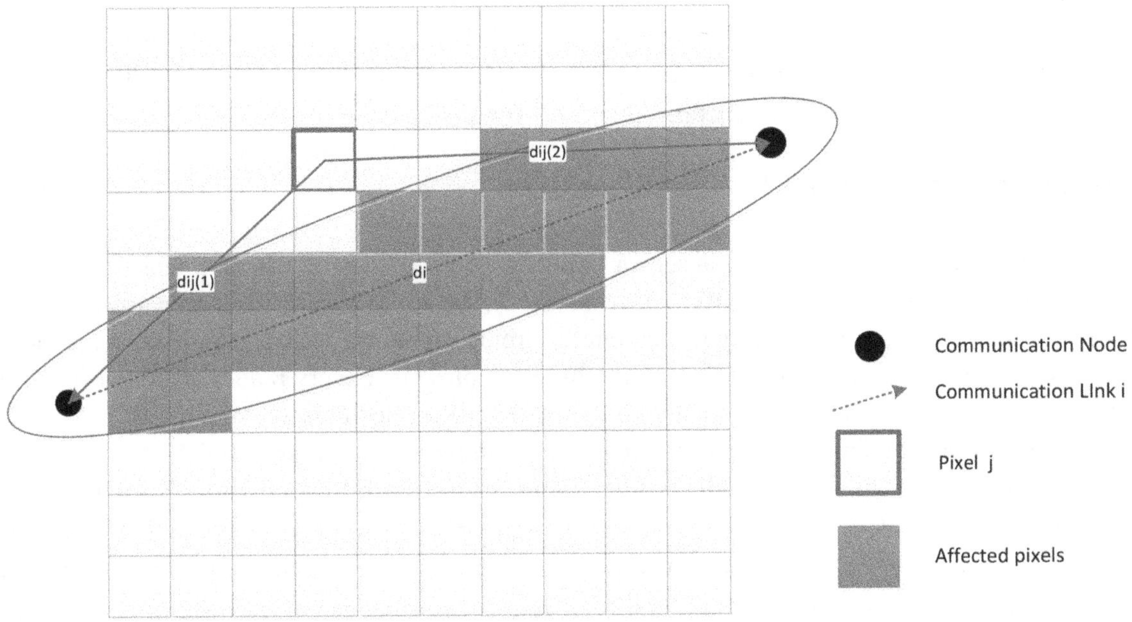

$$\Delta y_i = \sum_{j=1}^{N} w_{ij} \cdot \Delta x_j + n_i \tag{75}$$

where N is the number of pixels, Δx_j represents the attenuation values at pixel j, n_i is the overall system noise, including the multipath noise and measurement noise. We assume that the network has q sensor nodes, so there are $C_q^2 = q(q-1)/2 = m$ links. These equations are expressed in a matrix form as:

$$\Delta \mathbf{y} = \mathbf{W} \cdot \Delta \mathbf{x} + \mathbf{n} \tag{76}$$

where $\Delta \mathbf{y}$ is a column vector of length m, containing the RSS variations of all links. $\mathbf{W}$ is the weighting matrix of dimensions $m \times n$, with each column representing a single voxel, and each row describing the weighting of each voxel for that particular link. $\Delta \mathbf{x}$ is a column vector of length n, containing the RSS change per pixel. $\mathbf{n}$ is a noise vector. Using the least squares method and the method of regularization to solve the equation (76) (Wilson & Patwari, 2010), we can have:

$$\Delta \hat{\mathbf{x}} = (\mathbf{W}^T \mathbf{W} + \sigma_N^2 \mathbf{C}_x^{-1})^{-1} \mathbf{W}^T \Delta \mathbf{y} \tag{77}$$

where σ_N^2 is the regularization parameter. The priori covariance matrix $\mathbf{C}_x$ is calculated by the following index space attenuation model:

$$[\mathbf{C}_x]_{jk} = \sigma_x^2 e^{-d_{jk}/\delta_c} \tag{78}$$

where d_{jk} represents the distance between pixel j and pixel k, σ_x^2 is the variance at each pixel, and δ_c is a "space constant" correlation parameter.

Radio Tomographic Imaging for Passive RFID

Referring to the arrangement of wireless sensor network, and considering the link density of wireless signal coverage (Wagner, Patwari, & Timmermann, 2012), the RFID localization scenario is shown in Figure 19. Reference tags are uniformly placed on the circumference of a square, and four reader antennas are placed outside the four sides of the square. In this deployment every reader antenna is powering the tags on the line in front of it. This ensures a maximum power transmission to the transponders. Once triggered, the tags are sending their data to each of the other reader antennas.

One main difference between the RFID field in Figure 19 and the WSN in Figure 17 is in the number of communication links. In the WSN, each node is both the sender and the receiver. If there are n sensor nodes, the network has C_n^2 different links. But in RFID network, there is only forward links from each of the four readers to tags in front of it and backward links from each tag to three other readers. As a consequence, the number of communication links is decreased dramatically. Another advantage of the setup in Figure 19 is that the reader transmits power only to a row of tags in the front of the antenna, so these tags can gain more energy to send out signals. In this case, we assume that the target only affects the backward link, and we can put the decay on backward link as the decay in this link, which means we only need to research the backward link. At this time, the number of links for positioning is $n_{tag} \times n_{read}$,

Figure 19. RFID localization scenario

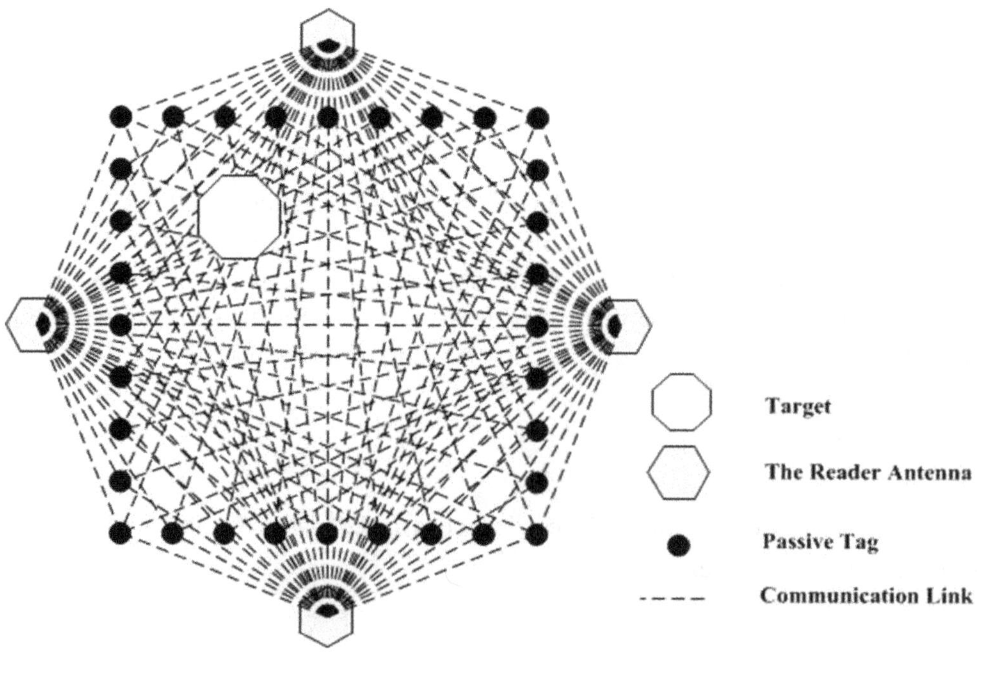

where n_{tag} represents the number of tags, n_{read} represents the number of tags, represents the number of reader antennas. From now on, the sender is the tag, while the receiver is the reader antenna.

The impact of the human body on the link has been widely studied and different models have been proposed. One of these models is (Lieckfeldt et al., 2009):

$$\Delta\hat{p}(d_{exc}) = A d_{exc}^{B} \cos(2\pi \frac{d_{exc}}{\hat{\lambda}} + \phi_{refl}) \tag{79}$$

$$d_{exc} = d_{nlos}^{t} + d_{nlos}^{r} - d_{los} \tag{80}$$

$\Delta\hat{p}$ indicates the RSS change for a link, d_{exc} is path length difference, d_{nlos}^{t} represents the distance between sender to the target, d_{nlos}^{r} represents the distance between target to the receiver, and d_{los} is the distance between sender and receiver. $\hat{\lambda}$ is the signal wavelength, ϕ_{refl} is the initial reference phase, and *A* and *B* are environment constants. According to the RTI algorithm model (76) and the solution formula (77), use $\Delta\mathbf{\hat{p}}$ instead of $\Delta\mathbf{y}$, which can be obtained:

$$\Delta\mathbf{\hat{x}} = (\mathbf{W}^{T}\mathbf{W} + \sigma_{N}^{2}\mathbf{C}_{x}^{-1})^{-1}\mathbf{W}^{T}\Delta\mathbf{\hat{p}} \tag{81}$$

The significance of the parameters in the upper equation is the same as in (77). It is clear that the pixel value of pixels ($\Delta\mathbf{\hat{x}}$) is mainly determined by the weight of the pixel ($\mathbf{W}$)and the attenuation value of each link($\Delta\mathbf{\hat{p}}$).

The last step of the RTI method is imaging the $\Delta\mathbf{\hat{x}}$ and an example of the imaging result is shown in Figure 20. The experimental area is square and the length is 4.8m. Each tag is 0.3m apart with a total of 40 labels. We divide the location region into 40 * 40 square grids, and the grid length is 0.1m. And the coordinate axis in the graph represents the number of grids. Usually the point with the maximum pixel value is regarded as the target location.

The RTI Multi-objective positioning is an important research topic, but only a few references are available in the literature, especially in the RFID scenarios. In the future, it is useful to combine the RTI with other positioning methods to realize multi-target recognition in the complex environment, by identifying the size, shape and posture of multiple targets.

Fingerprint Based DFL Method in RFID Environment

Fingerprinting is one of the primary methods of DFL, involving two main stages (Youssef, Mah & Agrawala, 2007; Mager, Lundrigan & Patwari, 2015; Zhou, Yang & Wu, 2014; Hong & Ohtsuki, 2015). During the offline stage, we need to collect some information for fingerprint matching. We first place tags at some fixed positions. These tags are not attached to moving objects. Each tag forms a link by communicating with the reader. Then we choose several reference points. The target will be placed at one of these points at a time. At each time, the system records all the links' status which is regarded as the fingerprint of the reference point. After collecting all the reference points' fingerprints, a database

Figure 20. The RTI localization result

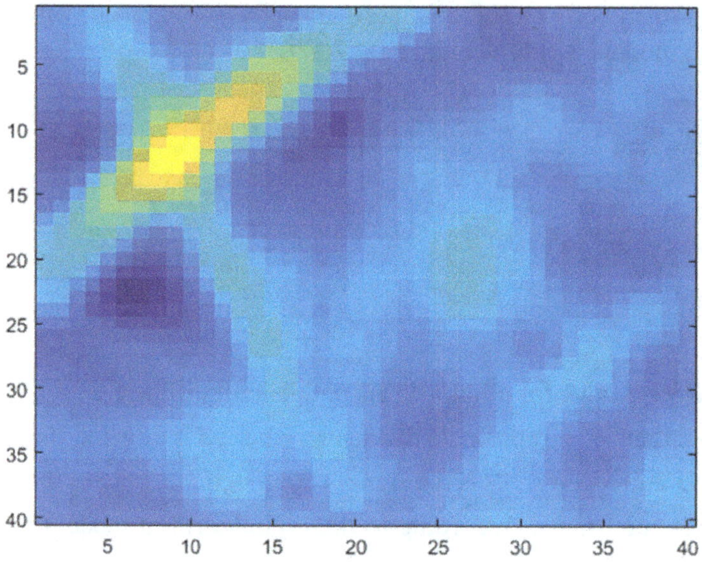

named as radio map is constructed. During the online phase, a real time link status is collected by system. This real time link status is compared with each fingerprint in the radio map to estimate the target's position. Since fingerprinting is a nonparametric location method, it does not use channel parameters. That means it does not need to deal with the parameter errors caused by multipath, diffraction and reflection. Thus, it can achieve high positioning accuracy in complex indoor environment. Fingerprint's localization accuracy mainly depends on the radio map information. More fingerprints and link status means that a higher accuracy can be achieved. To get more link information, it needs to deploy more wireless devices. Passive UHF RFID tag has advantages of small size, long life, low cost, and low energy consumption. With large amount of tags deployed in the environment, the RFID system can afford abundant link information. So fingerprint based method has a pervasive usage in DFL for RFID (Lieckfeldt, You & Timmermann, 2009; Han, Qian & Wang, 2016).

RFID Localization Environment

One advantage of passive RFID tag is low cost. That means we can deploy a large amount of tags in the environment without considering the total cost. One example of the basic deployment of RFID tags and readers in 3D environment is shown in Figure 21.

The readers are deployed on the middle point of each side of the rectangular environment. RFID tags are uniformly deployed on the floor of the environment. Tags hanged in the air or attached to some objects may change their position in some specific scenario like warehouse. This position variation may lead to position error. Before localizing target, the environment is first divided into a number of small equal grids. Each grid's central position stands for the reference point. RFID tags are deployed on the corners of each grid.

Figure 21. Deployment of RFID tags and readers in 3D environment

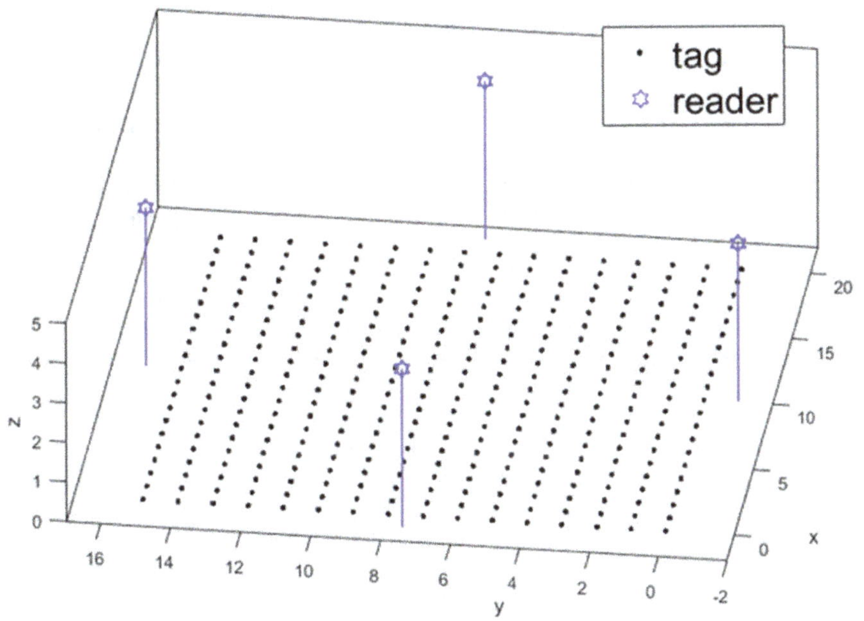

Link Characteristics

Since RFID system adopts a two-way communication technology. Firstly a reader sends a message to all tags. When tags detect the message, they will respond to the reader by adding their own identity information in their response message. Reader can distinguish responses from different tags. So in a RFID scenario, each tag can form a unique link with reader. When a target moves in the scenario, some links will be disturbed as shown in Figure 22.

The most common used RFID signal parameter is RSS. The measured RSS for link j at time t can be described as (Wang, Gao & Yu, 2013):

$$R_j(t) = P_j(t) + G_j(t) - F_j(t) - 10\gamma \log_{10}^{d_j} - T_j(t) + n_j(t) \tag{82}$$

where $P_j(t)$ is the transmitting power, $G_j(t)$ is the receiver antenna gain, $F_j(t)$ is the reference attenuation power when the distance is one meter, γ is the path loss parameter, d_j is the link distance, $T_j(t)$ is the attenuation caused by target, and $n_j(t)$ is the additive noise. $G_j(t)$, $F_j(t)$ and $P_j(t)$ are considered as constants.

For a specific link, the difference between RSS at t_1 and that at time t_0 is:

Figure 22. Deployment of RFID tags and readers in 2D environment

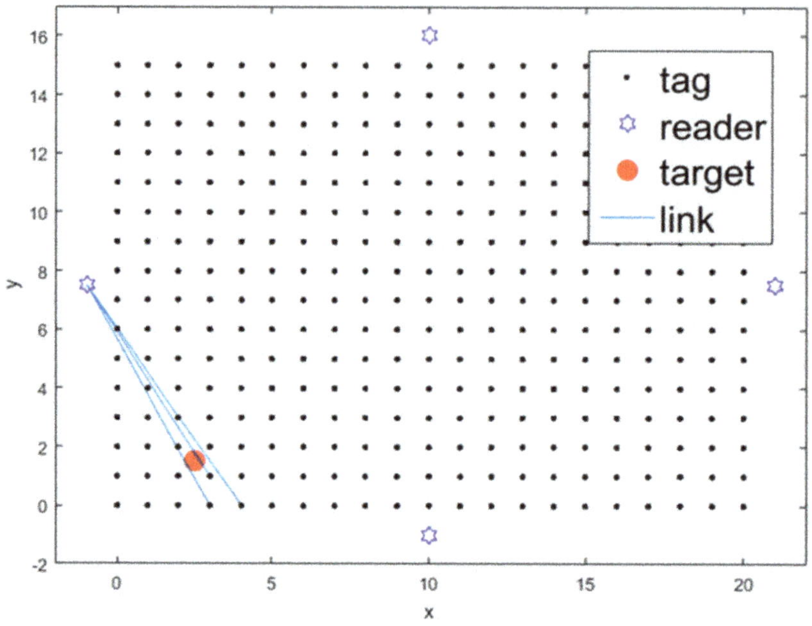

$$\begin{aligned}\Delta R_j(t_1) &= R_j(t_1) - R_j(t_0) \\ &= T_j(t_0) - T_j(t_1) + n_j(t_1) - n_j(t_0) \\ &\approx T_j(t_0) - T_j(t_1)\end{aligned} \tag{83}$$

So the main variation is caused by the presence of the target. As mentioned previously, at the offline phase, the link variation has been recorded when target stays at each reference point. The major work of the online phase is matching the real time RSS measurement with the most similar fingerprint in the database. Next we will introduce some basic fingerprint matching algorithms for position determination.

Matching Algorithm

1. Random forest

Random forest or random decision forest is a learning method based on decision tree which is a pervasively used machine learning algorithm. The block diagram of random forest is given in Figure 23. In a classification process, there are several attributes of a sample which could be used to decide the class the sample belongs to. Decision tree is the process that determines the attributes' sequence and each attribute's degree. Random forest is a further improvement of decision tree. When a traditional decision tree decides to classify the sample, it just chooses a best fitted attribute in the current available attribute set. But in random forest process, K attributes are randomly chosen to form a subset. Each subset can form a sub decision tree which is termed as base learning device. Then each base learning device will do

Figure 23. The processing of random forest

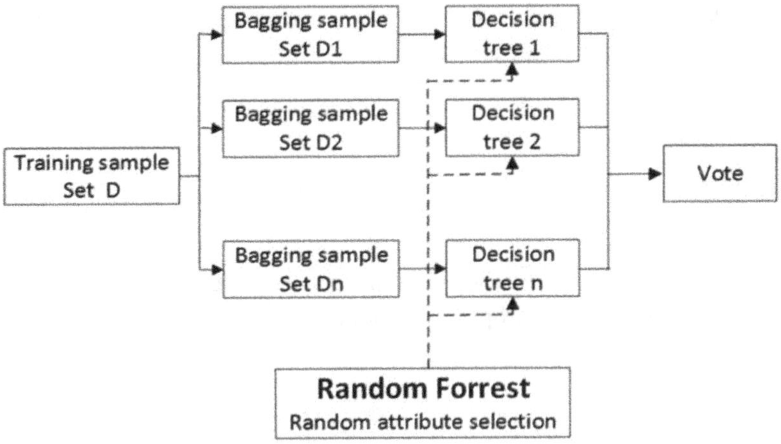

the decision tree process. Because of random selection of attributes, each base learning device produces a distinct result. Although the performance of each base learning device may not be as good as that of decision tree, combining the results from all the base learning devices will likely produce better performance than decision tree. Random forest has advantages of simplicity, low computational complexity, and easy implementation.

3. Support vector machine

To deal with our pre-collected fingerprint samples, one of our ideas is to classify these samples to find the class that is most similar to the data we get at the online stage. For a given training set $D = \{x_1, x_2, \ldots, x_n\}$ where x_n is a m-dimensional vector (suppose there are m links), these points are distributed in a m-dimensional space. A hyperplane can divide this space into two parts. Each part contains several training samples. To classify the training sample set D, we just need to find a proper hyperplane. A hyperplane can be described by a linear equation:

$$w^T x + a = 0 \tag{84}$$

$w = (w_1, w_2, \ldots, w_m)$ is the normal vector of the hyperplane. a is the displacement item of the hyperplane. We can record a hyperplane as (w, a). The distance between one point in this m-dimensional place and the hyperplane is:

$$d = \frac{\left| w^T x_i + a \right|}{\left\| w \right\|} \tag{85}$$

Suppose the hyperplane (w, a) can correctly divide the training set. Let r_i be the classification result of point x_i. Since a hyperplane can divide a place into two part. The result r has two possible results. We set the two possible results as +1 and -1. Then, we can get the equation:

$$\begin{cases} w^T x_i + a \geq +1 & r_i = +1 \\ w^T x_i + a \leq -1 & r_i = -1 \end{cases} \tag{86}$$

The m-dimensional point that has the shortest distance to hyperplane is called support vector. The sum of distances of two heterogeneous support vectors to the hyperplane is:

$$d = \frac{2}{\|w\|} \tag{87}$$

d is defined as margin.

To find the hyperplane which can get the maximum margin γ, we have:

$$\max_{w,a} \frac{2}{\|w\|} \quad s.t. \;\; r_i(w^T x_i + a \geq 1), \quad i = 1, 2, ..., n \tag{88}$$

In order to maximize the margin, we only need to maximize $\|w\|^{-1}$ which is equal to the minimum of $\|w\|^2$. So we can rewrite the above formula as:

$$\min_{w,a} \frac{1}{2}\|w\|^2 \qquad s.t. \;\; r_i(w^T x_i + a) \geq 1, \qquad i = 1, 2, \ldots, n \tag{89}$$

That is the basic principle of support vector machine as illustrated in figure 24.

To solve equation (89), we can use Lagrange method. Adding Lagrange factor χ_i to constrain each equation in (89), we can get:

$$G(w, a, \chi) = \frac{1}{2}\|w\|^2 + \sum_{i=1}^{n} \chi_i (1 - r_i(w^T x_i + a)) \tag{90}$$

By resolving partial derivatives of and, we get:

$$w = \sum_{i=1}^{n} \chi_i r_i x_i \tag{91}$$

$$0 = \sum_{i=1}^{n} \chi_i r_i \tag{92}$$

Substituting equation 91 and 92 into equation 89, we can get:

$$\max_{\chi} \ \sum_{i=1}^{n} \chi_i - \frac{1}{2}\sum_{i=1}^{n}\sum_{j=1}^{n} \chi_i \chi_j r_i r_j x_i^T x_j \quad s.t. \ \sum_{i=1}^{n} \chi_i r_i = 0, \quad \chi \geq 0, \quad i = 1, 2, \ldots, n \tag{93}$$

This equation is the dual problem of equation 89. To solve this problem, many different methods can be applied. When using the SMO (Sequential Minimal Optimization) method, two main steps are involved:

First, we choose two Lagrange factors χ_i and χ_j;

Second, we fix all the other factors except χ_i and χ_j. Using equation 93 to get the new value of χ_i and χ_j;

Repeat these two steps until convergence is reached.

Simulation Performance of Matching Algorithms

Except for the two matching algorithms described above, there are many other methods. Five matching algorithms (KNN, random forest, SVM, naïve Bayes, DA) are chosen for performance comparison by

Figure 24. Support vector machine

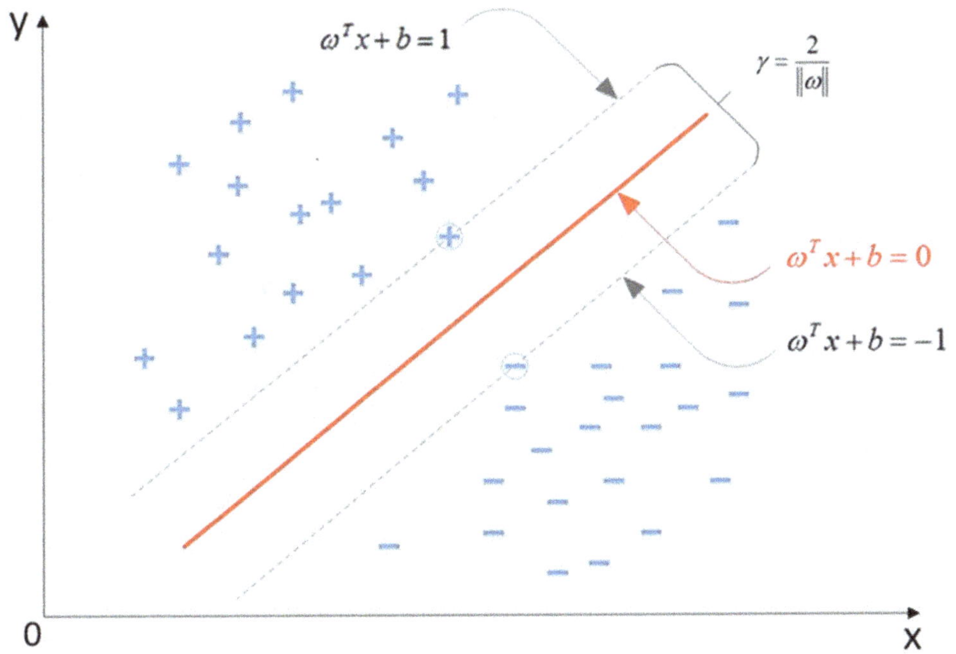

simulation. The simulation is conducted in the environment as shown in Figure 21. The environment is set as a 20m×15m room space. The simulated deployment of the readers and tags are exactly the same as that shown in Figure 21.. The simulation results in terms of error CDF are shown in Figure 25. From the figure, we can see that these five algorithms have the similar performance.

CONCLUSION AND FUTURE RESEARCH DIRECTIONS

Nowadays, internet of things (IOT) has become one hot topic in technology. Due to the fact that RFID is among the key techniques of IOT, UHF RFID has been attracting a lot attention from both industry and academia, and is envisioned as one of the leading techniques of localization in the future. The main methods of UHF RFID positioning include range-based and range-free. The range-based method mainly focuses on two aspects: one is how to acquire accurate distance between reader and tag, and the other is the localization algorithm. As for range-free method, there are even more issues to be resolved. We think machine learning will play a great role in future range-free localization not only in RFID, but also in other positioning schemes. There is another classification criterion, that is, RFID localization can be divided into tag-based (device-based) and tag-free (device-free). Tag-based localization is common, and will be implemented everywhere in future, because there will be tags everywhere. The research for tag-based localization is plentiful, involving every aspect of classical localization, such as range-based or rang-free methods. Tag-free localization is relatively a new research area, suited for scenarios where tag-based methods are not applicable.

Figure 25. The error cumulative distribution function of five matching algorithms

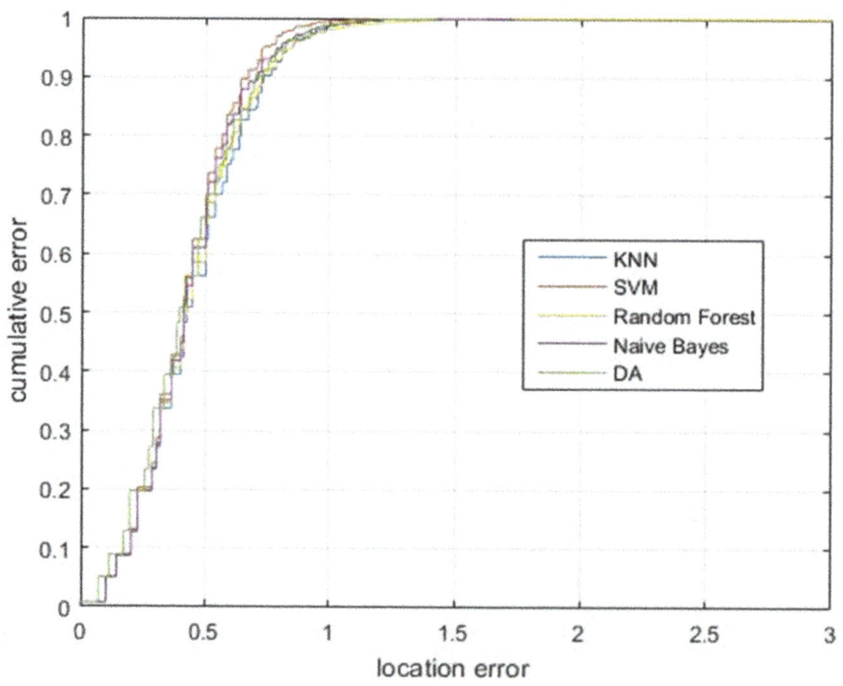

This chapter has discussed the key techniques for passive UHF RFID positioning, which are essential to deliver location information for IOT. These positioning techniques include range-based, range-free, tag-based (device-based), and tag-free (device-free), and improved range-based methods. The fundamental principles of each positioning technique were described in detail. Simulation and experimental results were provided to evaluate some of the positioning techniques. Each positioning technique owns advantages and disadvantages, which may be desired for some specific application scenarios.

One challenging issue in tag-free localization is the diversity in a person's body size and clothes, which make the online measurement rather different from the offline measurement. In addition, the propagation environment varies with the object's position. So how to characterize a precise channel model is a future research problem. Furthermore, it is a useful research topic to detect and track multiple tag-free objects in a complex indoor environment. Modifications and new ideas are still desired to enhance passive RFID positioning performance to better serve IOT.

REFERENCES

Arnitz, D., Muehlmann, U., & Witrisal, K. (2012a). Wideband characterization of backscatter channels: Derivations and theoretical background. *IEEE Transactions on Antennas and Propagation*, *60*(1), 257–266. doi:10.1109/TAP.2011.2167923

Arnitz, D., Muehlmann, U., & Witrisal, K. (2012b). Characterization and modeling of UHF RFID channels for ranging and localization. *IEEE Transactions on Antennas and Propagation*, *60*(5), 2491–2501. doi:10.1109/TAP.2012.2189705

Cheung, K. W., & So, H. C. (2005). A multidimensional scaling framework for mobile location using time-of-arrival measurements. *IEEE Transactions on Signal Processing*, *53*(2), 460–470. doi:10.1109/TSP.2004.840721

Deak, G., Curran, K., Condell, J., Deak, D., & Kiedrowski, P. (2013). Support vector machine classification in a device-free passive localisation (DFPL) scenario. In *Image Processing and Communications Challenges 4* (Vol. 184, pp. 253–260). Berlin: Springer. doi:10.1007/978-3-642-32384-3_31

DiGiampaolo, E., & Martinelli, F. (2012). A passive UHF-RFID system for the localization of an indoor autonomous vehicle. *IEEE Transactions on Industrial Electronics*, *59*(10), 3961–3970. doi:10.1109/TIE.2011.2173091

DiGiampaolo, E., & Martinelli, F. (2014). Mobile robot localization using the phase of passive UHF RFID signals. *IEEE Transactions on Industrial Electronics*, *61*(1), 365–376. doi:10.1109/TIE.2013.2248333

Faseth, T., Winkler, M., & Arthaber, H. (2011, Sept.). *The influence of multipath propagation on phase-based narrowband positioning principles in UHF RFID*. Paper presented at Antennas and Propagation in Wireless Communications (APWC), IEEE-APS Topical Conference on, Torino, Italy. doi:10.1109/APWC.2011.6046829

Gao, S., Zhang, S., Wang, G., & Li, Y. (2016). Robust second-order cone relaxation for TW-TOA-based localization with clock imperfection. *IEEE Signal Processing Letters*, *23*(8), 1047–1051. doi:10.1109/LSP.2016.2580743

Geng, Y., He, J., & Deng, H., & Kaveh. (2013, June). *Modeling the effect of human body on TOA ranging for indoor human tracking with wrist mounted sensor*. Paper presented at the International Symposium on Wireless Personal Multimedia Communications (WPMC), Atlantic City, NJ.

Griffin, J., & Durgin, G. (2009). Complete link budgets for backscatter-radio and RFID systems. *IEEE Antennas & Propagation Magazine*, *15*(2), 11–25. doi:10.1109/MAP.2009.5162013

Hamilton, B. R., Ma, X., Baxley, R. J., & Matechik, S. M. (2014). Propagation modeling for radio frequency tomography in wireless networks. *IEEE Journal of Selected Topics in Signal Processing*, *8*(1), 55–65. doi:10.1109/JSTSP.2013.2287471

Han, J., Qian, C., Wang, X., Ma, D., Zhao, J., Xi, W., & Wang, Z. et al. (2016). Twins: Device-free object tracking using passive tags. *IEEE/ACM Transactions on Networking*, *24*(3), 1605–1617. doi:10.1109/TNET.2015.2429657

Han, K., & Cho, S. (2010, December). *Advanced LANDMARC with adaptive k-nearest algorithm for RFID location system*. Paper presented at the 2010 2nd IEEE International Conference on Network Infrastructure and Digital Content, Beijing, China. doi:10.1109/ICNIDC.2010.5657852

Hara, S., Anzai, D., Yabu, T., Lee, K., Derham, T., & Zemek, R. (2013). A perturbation analysis on the performance of TOA and TDOA localization in mixed LOS/NLOS environments. *IEEE Transactions on Communications*, *61*(2), 679–689. doi:10.1109/TCOMM.2013.012313.110509

Hong, J., & Ohtsuki, T. (2015). Signal eigenvector-based device-free passive localization using array sensor. *IEEE Transactions on Vehicular Technology*, *64*(4), 1354–1363. doi:10.1109/TVT.2015.2397436

Jin, G., Lu, X., & Park, M. (2006, June). *An indoor localization mechanism using active RFID tag*. Paper presented at the IEEE International Conference on Sensor Networks, Ubiquitous, and Trustworthy Computing (SUTC'06), Taichung.

Kanatas, A., Kountouris, I., Kostaras, G., & Constantinou, P. (1997). A UTD propagation model in urban microcellular environments. *IEEE Transactions on Vehicular Technology*, *46*(1), 185–193. doi:10.1109/25.554751

Lázaro, A., Girbau, D., & Salinas, D. (2009). Radio link budgets for UHF RFID on multipath environments. *IEEE Transactions on Antennas and Propagation*, *57*(4), 1241–1251. doi:10.1109/TAP.2009.2015818

Li, X., Zhang, Y., & Amin, M. (2009 May). *Multifrequency-based range estimation of RFID Tags*. Paper presented at the 2009 IEEE International Conference on RFID, Orlando, FL.

Lieckfeldt, D., You, J., & Timmermann, D. (2009, October). *Exploiting RF-scatter: human localization with bistatic passive UHF RFID-systems*. Paper presented at the meeting of 2009 IEEE International Conference on Wireless and Mobile Computing, Networking and Communications, Beijing, China. doi:10.1109/WiMob.2009.39

Lieckfeldt, D., You, J., & Timmermann, D. (2009, Oct.). *Characterizing the influence of human presence on bistatic passive RFID-system*. Paper presented at 2009 IEEE International Conference on Wireless and Mobile Computing, Networking and Communications, Marrakech, Morocco. doi:10.1109/WiMob.2009.64

Liu, T., Yang, L., Lin, Q., Guo, Y., & Liu, Y. (2014, April). *Anchor-free backscatter positioning for RFID tags with high accuracy*. Paper presented at IEEE Conference on Computer Communications (INFOCOM), Toronto, Canada. doi:10.1109/INFOCOM.2014.6847960

Ma, Y., Zhou, L., Liu, K., & Wang, J. (2014). Iterative phase reconstruction and weighted localization algorithm for indoor RFID-based localization in NLOS environment. *IEEE Sensors Journal*, *14*(2), 597–611. doi:10.1109/JSEN.2013.2286220

Mager, B., Lundrigan, P., & Patwari, N. (2015). Fingerprint-based device-free localization performance in changing environments. *IEEE Journal on Selected Areas in Communications*, *33*(11), 2429–2438. doi:10.1109/JSAC.2015.2430515

Ni, L., Liu, Y., Lau, Y., & Patil, A. (2003, April). *LANDMARC: indoor location sensing using active RFID*. Paper presented at the First IEEE International Conference on Pervasive Computing and Communications (PerCom 2003), Fort Worth, TX. doi:10.1109/PERCOM.2003.1192765

Ni, L. M., Liu, Y., Lau, Y. C., & Patil, A. P. (2004). LANDMARC: Indoor location sensing using active RFID. *Wireless Networks*, *10*(6), 701–710. doi:10.1023/B:WINE.0000044029.06344.dd

Nikitin, P., Martinez, R., Ramamurthy, S., ... (2010, April). *Phase based spatial identification of UHF RFID tags*. Paper presented at the 2010 IEEE International Conference on RFID, Orlando, FL. doi:10.1109/RFID.2010.5467253

Patwari, N., & Agrawal, P. (2008, Apr.). *Effects of correlated shadowing: connectivity, localization, and RF tomography*. Paper presented in Int'l Conf. Information Processing in Sensor Networks (IPSN '08), St. Louis, MO. doi:10.1109/IPSN.2008.7

Särkkä, S., Viikari, V., Huusko, M., & Jaakkola, K. (2012). Phase-based UHF RFID tracking with nonlinear kalman filtering and smoothing. *IEEE Sensors Journal*, *12*(5), 904–910. doi:10.1109/JSEN.2011.2164062

Scherhäufl, M., Pichler, M., Schimbäck, E., Muller, D. J., Ziroff, A., & Stelzer, A. (2013). Indoor localization of passive UHF RFID tags based on phase-of-arrival evaluation. *IEEE Transactions on Microwave and Techniques*, *61*(12), 4724–4729. doi:10.1109/TMTT.2013.2287183

Shang, Y., Ruml, W., Zhang, Y., & Fromherz, M. P. (2003, Jun.). *Localization from mere connectivity*. Paper presented in the 4th international symposium on Mobile ad hoc networking & computing, Annapolis, MD. doi:10.1145/778415.778439

Shi, W., & Wong, V. (2011 July). *MDS-based localization algorithm for RFID systems*. Paper presented at the 2011 IEEE International Conference on Communications (ICC), Kyoto, Japan. doi:10.1109/icc.2011.5962800

Vaghefi, R. M., & Buehrer, R. M. (2015). Cooperative localization in NLOS environments using semidefinite programming. *IEEE Communications Letters*, *19*(8), 1382–1385. doi:10.1109/LCOMM.2015.2442580

Wagner, B., Patwari, N., & Timmermann, D. (2012, Mar.). *Passive RFID tomographic imaging for device-free user localization*, Paper presented at 2012 9th Workshop on Positioning Navigation and Communication (WPNC), Dresden, Germany. doi:10.1109/WPNC.2012.6268750

Wang, C., Wu, H., & Tzeng, N. F. (2007, May). *RFID-based 3-D positioning schemes*. Paper presented at the 26th IEEE International Conference on Computer Communications, Anchorage, AK. doi:10.1109/INFCOM.2007.147

Wang, G., Chen, H., Li, Y., & Ansari, N. (2014). NLOS error mitigation for TOA-based localization via convex relaxation. *IEEE Transactions on Wireless Communications*, *13*(8), 4119–4131. doi:10.1109/TWC.2014.2314640

Wang, J., Chen, X., Fang, D., Wu, C., Yang, Z., & Xing, T. (2015). Transferring- compressive-sensing-based device-free localization across target diversity. *IEEE Transactions on Industrial Electronics*, *62*(4), 2397–2409. doi:10.1109/TIE.2014.2360140

Wang, J., Gao, Q., Wang, H., Cheng, P., & Xin, K. (2015). Device-free localization with multidimensional wireless link information. *IEEE Transactions on Vehicular Technology*, *64*(1), 356–366. doi:10.1109/TVT.2014.2318084

Wang, J., Gao, Q., Yu, Y., Cheng, P., Wu, L., & Wang, H. (2013). Robust device-free wireless localization based on differential RSS measurements. *IEEE Transactions on Industrial Electronics*, *60*(12), 5943–5952. doi:10.1109/TIE.2012.2228145

Wang, J., Ma, Y., Zhao, Y., & Liu, K. (2015). A multipath mitigation localization algorithm based on MDS for passive UHF RFID. *IEEE Communications Letters*, *19*(9), 1652–1655. doi:10.1109/LCOMM.2015.2450217

Wang, J., Zhang, X., Gao, Q., Yue, H., & Wang, H. (2016). Device-free wireless localization and activity recognition: A deep learning approach. *IEEE Transactions on Vehicular Technology*. DOI: 10.1109/TVT.2016.2635161

Wang, Q., Balasingham, I., Zhang, M., & Huang, X. (2011). Improving RSS-based ranging in LOS-NLOS scenario using GMMs. *IEEE Communications Letters*, *15*(10), 1065–1067. doi:10.1109/LCOMM.2011.080811.111087

Wang, Y., & Wu, Y. (2017). An efficient semidefinite relaxation algorithm for moving source localization using TDOA and FDOA measurements. *IEEE Communications Letters*, *21*(1), 80–83. doi:10.1109/LCOMM.2016.2614936

Wilson, J., & Patwari, N. (2010). Radio tomographic imaging with wireless networks. *IEEE Transactions on Mobile Computing*, *9*(5), 621–632. doi:10.1109/TMC.2009.174

Wilson, J., & Patwari, N. (2011). See through walls: Motion tracking using variance-based radio tomography networks. *IEEE Transactions on Mobile Computing*, *10*(5), 612–621. doi:10.1109/TMC.2010.175

Wilson, J., & Patwari, N. (2012). A fade-level skew-laplace signal strength model for device-free localization with wireless networks. *IEEE Transactions on Mobile Computing*, *11*(6), 947–958. doi:10.1109/TMC.2011.102

Xia, H., Bertoni, H., Maciel, L., Lindsay-Stewart, A., & Rowe, R. (1993). Radio propagation characteristics for line-of-sight microcellular and personal communications. *IEEE Transactions on Antennas and Propagation*, *41*(10), 1439–1447. doi:10.1109/8.247785

Xu, R., & Wunsch, D. II. (2005). Survey of clustering algorithms. *IEEE Transactions on Neural Networks*, *16*(3), 645–678. doi:10.1109/TNN.2005.845141 PMID:15940994

Yang, L., Chen, Y., Li, X., Xiao, C., Li, M., & Liu, Y. (2014, September). *Tagoram: Real-time tracking of mobile RFID tags to high precision using COTS devices.* Paper presented at the 20th annual international conference on Mobile computing and networking, Maui, HI. doi:10.1145/2639108.2639111

Yin, Z., Cui, K., Wu, Z., & Yin, L. (2015). Entropy-based TOA estimation and SVM-based ranging error mitigation in UWB ranging systems. *Sensors (Basel)*, *15*(3), 11701–11724. doi:10.3390/s150511701 PMID:26007726

Youssef, M., Mah, M., & Agrawala, A. (2007, September). *Challenges: device-free passive localization for wireless environments.* Paper presented at the meeting of the 13th annual ACM international conference on Mobile computing and networking, Montreal, Quebec, Canada. doi:10.1145/1287853.1287880

Zhang, Z., Lu, Z., Saakian, V., Qin, X., Chen, Q., & Zheng, L.-R. (2014). Item-level indoor localization with passive UHF RFID based on tag interaction analysis. *IEEE Transactions on Industrial Electronics*, *61*(4), 2122–2135. doi:10.1109/TIE.2013.2264785

Zhao, Y., Liu, K., Ma, Y., Gao, Z., Zang, Y., & Teng, J. (2017). Similarity analysis based indoor localization algorithm with backscatter information of Passive UHF RFID tags. *IEEE Sensors Journal*, *17*(1), 185–193. doi:10.1109/JSEN.2016.2624314

Zhao, Y., Liu, Y., & Ni, L. (2007, Oct.). VIRE: active RFID-based localization using virtual reference elimination. Paper presented at the 2007 International Conference on Parallel Processing, Xi'an, China. doi:10.1109/ICPP.2007.84

Zhao, Y., & Patwari, N. (2015). Robust estimators for variance-based device free localization and tracking. *IEEE Transactions on Mobile Computing*, *14*(10), 2116–2129. doi:10.1109/TMC.2014.2385710

Zhou, Z., Yang, Z., Wu, C., Shangguan, L., & Liu, Y. (2014). Omnidirectional coverage for device-free passive human detection. *IEEE Transactions on Parallel and Distributed Systems*, *25*(7), 1819–1829. doi:10.1109/TPDS.2013.274

KEY TERMS AND DEFINITIONS

Convex Optimization: A mathematical method of optimization that studies the problem of minimizing convex functions over convex sets. The convexity property can make optimization in some sense "easier" than the general case – for example, any local minimum of a convex function must be a global minimum.

Device Free Localization (DFL): DFL is a kind of localization method that the target does not carry or switch on any compatible equipment. The goal of localization can be achieved by analyzing the influence of target on the surrounding electro-magnetic environments.

Fingerprinting Algorithm: A fingerprinting algorithm in localization is a procedure that maps a string of RSS data to a specific position; it is an analogue of human fingerprints which can be uniquely identified. A RSS fingerprint associated with a position is usually different from that related to another position, enabling unique localization.

K-Nearest Neighbors (K-NN): K-NN is a type of instance-based learning, or lazy learning, where the function is only approximated locally and all computation is deferred until classification. The k-NN algorithm is among the simplest of all machine learning algorithms.

Localization: It is to determine the position of an object. There are many kinds of localization methods and categories, for example, satellite localization, radio localization, wireless localization, indoor localization, robot localization, active localization, and passive localization.

Multidimensional Scaling (MDS): MDS is a means of visualizing the level of similarity of individual cases of a dataset. It refers to a set of related ordination techniques used in information visualization, in particular to display the information contained in a distance matrix. It is a form of non-linear dimensionality reduction.

Radio Frequency Identification (RFID): RFID uses electromagnetic fields to automatically identify and track tags attached to objects. The tags contain electronically stored information. Passive tags collect energy from a nearby RFID reader's interrogating radio waves.

Chapter 5
TDOA-Based Acoustic Direction Finding

Xunxue Cui
Anhui University of Science and Technology, China

Kegen Yu
Wuhan University, China

Songsheng Lu
New Star Institute of Applied Technology, China

ABSTRACT

This chapter focuses on estimating the azimuth and elevation angles of a sound emitter based on time-difference-of-arrival (TDOA) measurements using an array of acoustic sensors. The TDOA-based direction-finding problem is appropriate because in a range of scenarios the source only emits a transient signal and TDOA measurements provide a simple method of finding the direction of the received signal. Given the measurement of TDOA, three methods for calculating the actual bearing of an acoustic source are considered—algebraic calculations based on trigonometric functions, linear least squares, and nonlinear least squares—and these results are also compared with the Cramer-Rao lower bound (CRLB). In this chapter, a comprehensive analysis of TDOA-based direction-finding methods is presented with regard to different application conditions, while their estimation performances are analysed with both simulation and field experimental results produced by 3-D microphone array.

INTRODUCTION

Source localization and Direction of Arrival (DOA) estimation using sensor arrays has drawn significant attention from both academia and industry in the array signal processing field for decades. DOA estimation is usually simply called direction finding. If the source is in the far field of a receiving array, the source signal wavefront arriving at the array can be assumed to be plane wave. Then the DOA can be accurately estimated, while the range estimation error and hence position estimation error would be large especially for the high range-to-baseline ratio case. Thus multiple subarrays are needed for accurate

DOI: 10.4018/978-1-5225-3528-7.ch005

position estimation. For instance, in the two-dimensional case, the crossing of bearing lines produced by two subarrays can be used to estimate the source location.

Many high resolution algorithms such as Multiple Signal Classification (MUSIC) and Estimating Signal Parameters via Rotational Invariance Techniques (ESPRIT) have been considered as the popular beamforming methods for DOA estimation. In contrast to the beamforming approach where the data are processed in the frequency domain, the Time Difference of Arrival (TDOA) based DOA estimator performs the data processing in the time domain. In the case where the acoustic signal is transient and does not have a pre-defined waveform pattern, the signal characteristics are unknown such that only the relative time delays can be measured and utilized. The azimuth and elevation can be directly calculated from the estimated time delays so that it does not need a time-consuming two dimensional search over the array manifold as the beamformer does.

The TDOA technique is one of the widely used localization schemes in cellular networks and wireless sensor networks including the use of acoustic arrays. Some indoor localization systems use ultrasonic transceivers to deliver the localization solution with a high accuracy based on the TDOA technique (Yu *et al* 2009). In a typical scenario where the source is surrounded by a number of sensors (or receivers), a single acoustic target (source) emits a signal, and the sensors detect the signal and measure the arrival time of the source signal. By calculating the signal TDOA at a number of receivers with respect to a reference receiver, the source location can be determined.

Military operations consider the accurate location of weapon system as a vital factor on the battlefield by using relative time of arrival (TOA) of the sound impulse at several accurately located microphones. The relative TOA is commonly referred to as TDOA or simply time delay. The firing of a gun or mortar generates an acoustic impulse, which propagates radially outwards from the source. The TDOA measurements by an array are used to determine the source bearing, and the battlefield target location is cross-fixed by triangulation using the bearing estimates from multiple widely separated arrays (Sallai *et al* 2011; Brian *et al* 2002).

Figure 1 illustrates the scenario where an acoustic array is employed to detect and locate the acoustic signal from a single explosion source. The sound source generates a far field signal traveling as an approximate plane wave. In the past the long baseline scheme has been used dominantly to localize a transient explosion source, because the precision of data acquisition systems was not high enough to distinguish the tiny TDOA measurements associated with small distance differences. However, the performance of the present-day equipment has been improved significantly to enable the measurement of the small TDOA values, which allows practical estimation of DOA of far field sources with a number of TDOA measurements.

Such an estimation can be realized either in the time or frequency domains. The frequency domain method involves a Fourier transform, but with this method there may be an issue related to the signal wavelength and the antenna aperture size. On the other hand, the time domain method requires a larger amount of processor memory and a higher sampling rate, and may be limited to direction finding of a single sound source. However, the time domain method has been widely applied to locate acoustic sources in many fields. For example it is especially suited for direction finding and localization of the source of a transient signal such as gunshot, cannon or bomb explosion. Furthermore, in the case where the signal is transient, the arrival difference of pulse time can be directly measured by the localization system hardware. In these scenarios, beamforming in general is not applicable since the signal characteristics are unknown; only the relative time delays can be measured. There are several techniques to estimate

Figure 1. Simple schematic of detection of DOA with an acoustic signal by a sensor array

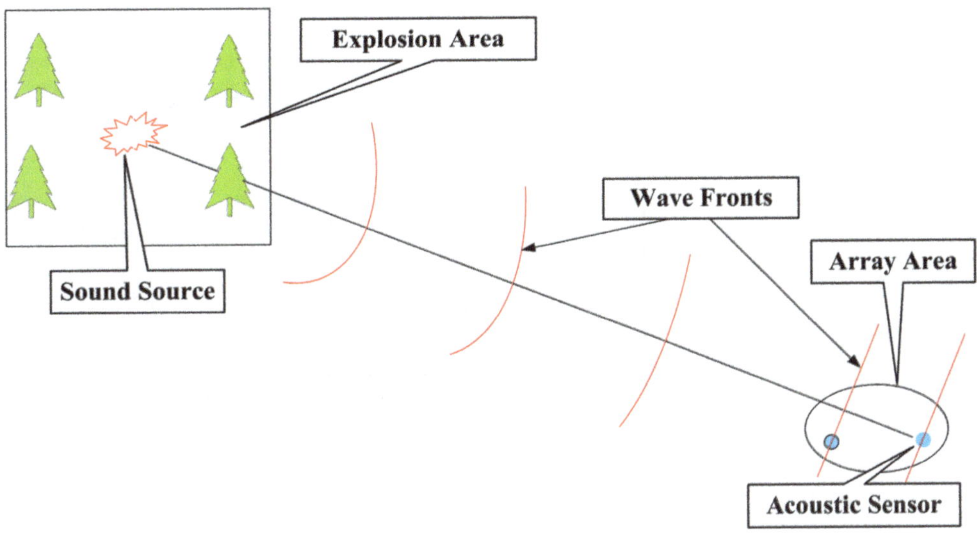

TDOA from the measured signal at multiple sensors, including the generalized cross-correlation method and the maximum peak detector method (Vakulya & Simon 2011).

This chapter introduces the TDOA-based direction finding model and a number of DOA estimation methods including a few recently developed methods. The Cramer-Rao Lower Bound (CRLB) is also provided in the background section of the chapter. One advantage of an isometric array with a fixed structure is that it is easy to set up, and the source bearing can be readily determined by algebraic operation. The case of a four-microphone planar array is addressed to illustrate how such an array can be used for direction finding. The chapter also includes a description and the theoretical performance analysis of the classical TDOA-based direction finding method, i.e. the linear least squares (LLS) estimator for the 1-D, 2-D, and 3-D arrays. Two evolutionary TDOA-based direction finding methods are also presented especially for an asymmetrical 3-D array and their performance are verified by simulation and field experimental results.

BACKGROUNDS INFORMATION

TDOA-Based Direction Finding Model

The following material provides information and conventions used in the mathematical analyses in this chapter. Here the cap '$\wedge$' upon a symbol represents an evaluated or measurement value. All vectors and matrixes are boldfaced. Consider first direction finding in the 3-D space, as illustrated in Figure 2. The angle vector involving the azimuth and elevation of an emitter is defined as

$$\boldsymbol{\gamma} = \left[\phi, \theta\right]^{\mathrm{T}} \tag{1}$$

Figure 2. Diagram illustrating the notation and geometric relations of direction finding

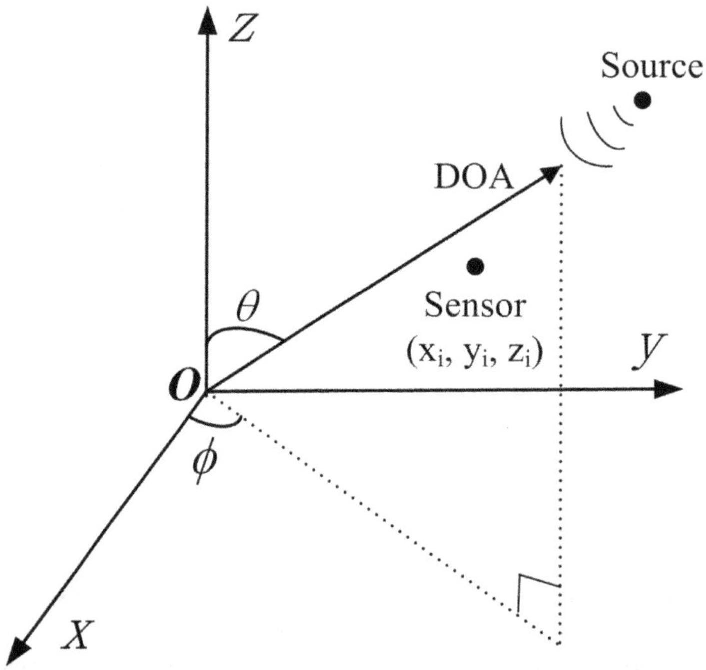

where ϕ denotes the source azimuth (clockwise angle between the positive x-axis and a line from the origin to a source perpendicularly projected onto the horizontal plane), θ denotes the source elevation (clockwise angle between the positive z-axis and a source position vector), and T is the transpose operator. Then the unit DOA vector $\mathbf{k}$ of the emitter is represented by

$$\mathbf{k} = \begin{bmatrix} k_x \\ k_y \\ k_z \end{bmatrix} = \begin{bmatrix} \sin\theta\cos\phi \\ \sin\theta\sin\phi \\ \cos\theta \end{bmatrix} \tag{2}$$

where the three components k_x, k_y and k_z are the corresponding projection of the DOA vector onto the three coordinate axes.

Without loss of generality, the coordinate origin of the sensor array is located at the sensor closest to the array geometric centre, which is labelled as the zeroth sensor and the TDOA measurement reference point. The aim is to estimate the DOA vector of the emitter from TDOA measurements. If the array size is N, the spatial coordinates of the i-th sensor is denoted as $\mathbf{s}_i = \left[x_i, y_i, z_i\right]^{\mathrm{T}}$ $(i = 1, \ldots, N-1)$, then the corresponding TDOA measurement equation becomes

$$\hat{c}\hat{\tau}_i = x_i \sin\hat{\theta}\cos\hat{\phi} + y_i \sin\hat{\theta}\sin\hat{\phi} + z_i \cos\hat{\theta} + n_i \tag{3}$$

where $\hat{\tau}_i$ is the TDOA measurement of the i-th sensor relative to the zeroth sensor whose TOA is set as zero, $\hat{c}$ is the in-situ sound speed measurement taken as a scalar quantity, and n_i is the measurement error which takes into account both TDOA and sound speed error. The TDOA related to the i-th sensor and the zeroth sensor is the difference between the TOA values of the two sensors.

The DOA vector of the zeroth sensor and that of the i-th sensor are parallel for a plane wave, as illustrated in Figure 3. If the projection point of the i-th sensor position $\mathrm{P}_i\left(x_i, y_i, z_i\right)$ onto the DOA line of the zeroth sensor is denoted P_i', then the line segment length $\left|\overline{OP_i'}\right|$ represents the absolute value of the difference between the two ranges from the two sensors to the source, namely $\left|c\tau_i\right|$. Alternatively, the range difference $c\tau_i$ (either positive or negative) can be determined by the inner product of the DOA vector and the vector $\overline{OP_i}$, and represents the projection of the vector $\overline{OP_i}$ along the DOA vector. Then

$$c\tau_i = \mathbf{s}_i\mathbf{k} \tag{4}$$

which is the basic measurement equation of a TDOA-based direction finding system.

Figure 3. Schematic relationship between DOA and TDOA for an individual sensor and a plane wave

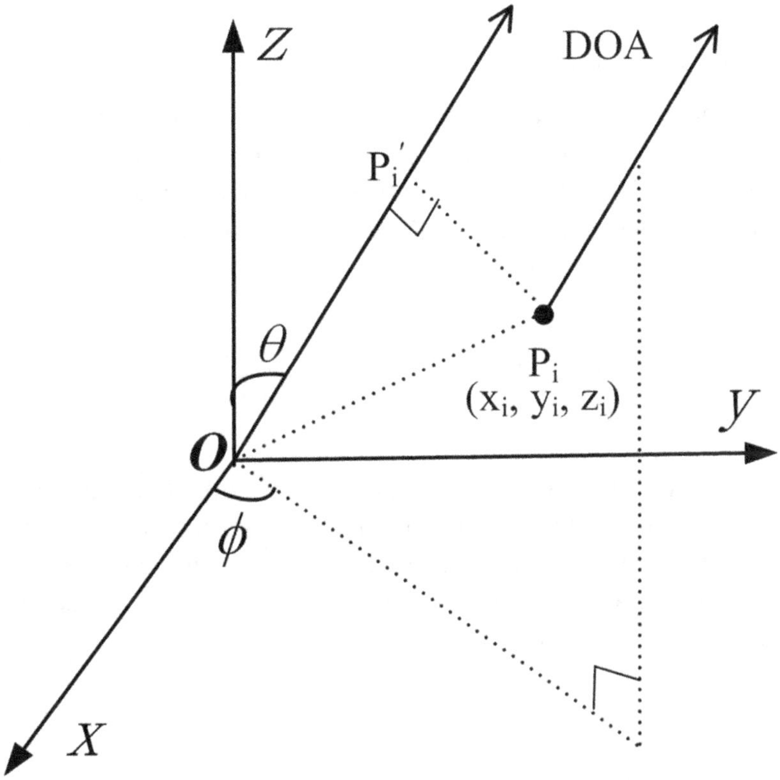

If the array coordinates except for the zeroth sensor are constructed as a matrix $\mathbf{S} = \left[\mathbf{s}_1, \mathbf{s}_2, \ldots, \mathbf{s}_{N-1}\right]^{\mathrm{T}} \in \mathbb{R}^{(N-1)\times 3}$, and the corresponding set of all TDOA measurements is $\hat{\boldsymbol{\tau}} = \left[\hat{\tau}_1, \hat{\tau}_2, \ldots, \hat{\tau}_{N-1}\right]^{\mathrm{T}} \in \mathbb{R}^{(N-1)\times 1}$, then the direction finding problem to determine a DOA estimate can be modelled as the following matrix form

$$\hat{c}\hat{\boldsymbol{\tau}} = \mathbf{S}\hat{\mathbf{k}} + \mathbf{n} \tag{5}$$

where $\mathbf{n}$ is the measurement error vector with components $\{n_i\}$, $i = 1, 2, \ldots, N-1$.

CRLB for Azimuth and Elevation Angles

The CRLB has been widely used in signal processing and parameter estimation to set a lower bound on any unbiased estimators in the presence of unbiased random noise. It is also a key performance evaluation benchmark in target positioning. If a vector of TDOA error has an asymptotically Normal distribution with covariance matrix given by Σ, the conditional probability density function (PDF) of TDOA measurement error in (5) is described by

$$p(\boldsymbol{\tau} \mid \boldsymbol{\gamma}) = \frac{1}{\left(2\pi\right)^{(N-1)/2} \left|\Sigma\right|^{1/2}} \exp\left\{-\frac{1}{2}\left(\boldsymbol{\tau} - \frac{\mathbf{Sk}}{c}\right)^{T} \Sigma^{-1} \left(\boldsymbol{\tau} - \frac{\mathbf{Sk}}{c}\right)\right\} \tag{6}$$

where $\left|\Sigma\right|$ is the determinant of the covariance matrix.

The Fisher information matrix (FIM) is known to calculate the covariance matrices associated with maximum likelihood estimates. The CRLB states that the inverse of the Fisher information matrix is a lower bound on the variance of any unbiased estimator of an unknown parameter. If $F(\boldsymbol{\gamma})$ denotes the FIM, and γ_1, γ_2 respectively denote ϕ and θ as defined in (1), then the elements of the FIM are defined by

$$[F(\boldsymbol{\gamma})]_{k,k} = -E\left[\frac{\partial^2 \ln p(\boldsymbol{\tau} \mid \boldsymbol{\gamma})}{\partial \gamma_k \partial \gamma_l}\right] k,\ l = 1, 2 \tag{7}$$

Nielsen (1994) has derived the CRLB expressions for the azimuth and elevation estimation error when using an arbitrary 3-D array of identical omnidirectional sensors. However, these formulated expressions are based on a frequency domain analysis, so a Fourier transform is required to transform the received time domain signal data. To conveniently analyse a TDOA-based direction finding system, the following time domain CRLB expressions are required:

$$\begin{cases} CRLB(\phi) = F_{\theta\theta} \,/\, (F_{\phi\phi}F_{\theta\theta} - F_{\theta\phi}^2) \\ CRLB(\theta) = F_{\phi\phi} \,/\, (F_{\phi\phi}F_{\theta\theta} - F_{\theta\phi}^2) \end{cases} \tag{8}$$

The entries of the corresponding FIM can be determined by

$$\begin{cases} F_{\theta\theta} = \dfrac{\partial f^T(\gamma)}{\partial \theta} \Sigma^{-1} \dfrac{\partial f(\gamma)}{\partial \theta} \\ F_{\phi\phi} = \dfrac{\partial f^T(\gamma)}{\partial \phi} \Sigma^{-1} \dfrac{\partial f(\gamma)}{\partial \phi} \\ F_{\phi\theta} = F_{\theta\phi} = \dfrac{\partial f^T(\gamma)}{\partial \phi} \Sigma^{-1} \dfrac{\partial f(\gamma)}{\partial \theta} \end{cases} \tag{9}$$

where

$$f_i(\gamma) = \mathbf{s}_i\mathbf{k} \,/\, \mathbf{c} \ (i = 1, \ldots, N-1) \tag{10}$$

$$f(\gamma) = \left[f_1(\gamma), \ldots, f_{N-1}(\gamma) \right]^T \tag{11}$$

$$\begin{cases} \dfrac{\partial f_i(\gamma)}{\partial \phi} = \dfrac{\sin\theta}{c}(y_i \cos\phi - x_i \sin\phi) \\ \dfrac{\partial f_i(\gamma)}{\partial \theta} = \dfrac{1}{c}(x_i \cos\theta\cos\phi + y_i \cos\theta\sin\phi - z_i \cos\theta) \end{cases} \tag{12}$$

When the TDOA measurement errors are mutually independent Gaussian random variables of mean zero and variances given as $\sigma_{n_i}^2$, the conditional PDF $p(\tau \mid \gamma)$ can be rewritten as

$$p(\tau \mid \gamma) = \frac{1}{(2\pi)^{(N-1)/2} \prod_{i=1}^{N-1} \sigma_{n_i}} \exp\left\{ -\sum_{i=1}^{N-1} \frac{1}{2\sigma_{n_i}^2} \left(\tau_i - \frac{\mathbf{s}_i\mathbf{k}}{c} \right)^2 \right\} \tag{13}$$

Taking natural logarithms on both sides of (13) and ignoring the irrelevant constants, we can obtain its log-likelihood function. Based on the definition of the FIM in (7), the entries of the FIM can be derived as

$$\begin{cases} F_{\theta\theta} = \sum_{i=1}^{N-1} \dfrac{1}{\sigma_{n_i}^2} \left[\dfrac{\partial f_i(\gamma)}{\partial \theta} \right]^2 \\ F_{\phi\phi} = \sum_{i=1}^{N-1} \dfrac{1}{\sigma_{n_i}^2} \left[\dfrac{\partial f_i(\gamma)}{\partial \phi} \right]^2 \\ F_{\phi\theta} = F_{\theta\phi} = \sum_{i=1}^{N-1} \dfrac{1}{\sigma_{n_i}^2} \dfrac{\partial f_i(\gamma)}{\partial \phi} \dfrac{\partial f_i(\gamma)}{\partial \theta} \end{cases} \tag{14}$$

FOUR-MICROPHONE PLANAR ARRAY FOR DIRECTION FINDING

The TDOA-based direction finding problem has been of considerable interest, particularly in the public security domain for determining the location of a gunshot using an isometric array. If the distances from the sensors to the array centre are equal for the fixed configuration, a set of TDOA measurement equations can be constructed through trigonometric functions and directly resolved to obtain the azimuth and elevation of an emitter (James 2008). In the case of a four-microphone planar array, the basic direction finding method is as follows.

The diagram of a four-microphone planar array is illustrated in Figure 4. Let *P* denote the position of an acoustic and point source, and let P_i $(i = 0, 1, 2, 3)$ denote the positions of four array sensors. Suppose that the four sensors are placed at the four corners of a square with side length *L* on the *xoy* plane of the 3-D Cartesian coordinate system and the centre of the square is the origin of the coordinate system. Then the positions of the four sensors are $P_0(L / 2, L / 2, 0)$, $P_1(-L / 2, L / 2, 0)$, $P_2(-L / 2, -L / 2, 0)$ and $P_3(L / 2, -L / 2, 0)$, respectively. The position of the source is denoted $P(x, y, z)$ whose distance to the origin is denoted *r*.

Assume that the zeroth sensor is regarded as the TDOA measurement reference point, and the relative time delay measurements are denoted by $\hat{\tau}_i$ $(i = 1, 2, 3)$ for the *i*-th sensor. Denote the range between the source and the *i*-th sensor as r_i, and the range difference between the *i*-th and *j*-th sensors as d_{ij}. Thus the corresponding TDOA is $\tau_i - \tau_j$. If the sound speed is denoted *c*, then

Figure 4. Diagram illustrating four-microphone planar array for direction finding

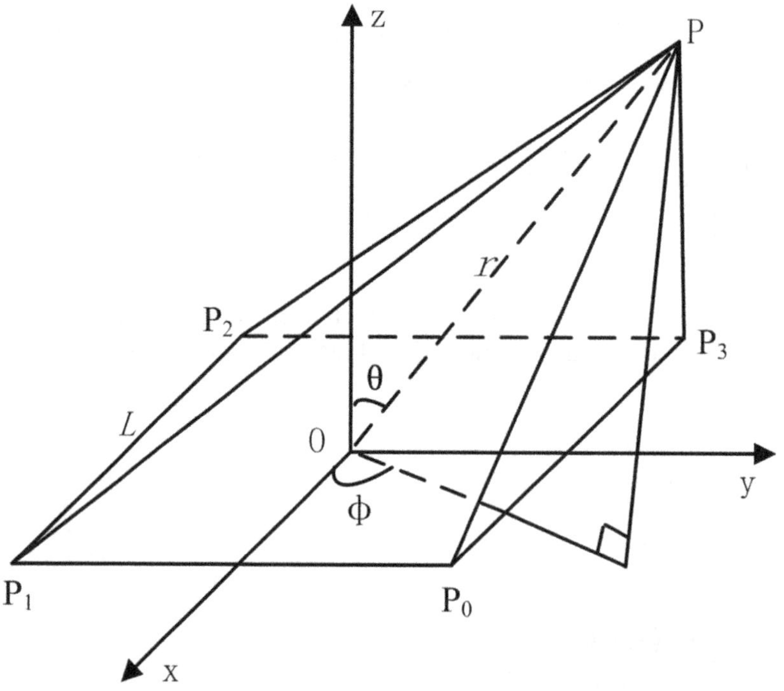

$$d_{ij} = r_i - r_j = c\left(\tau_i - \tau_j\right) \tag{15}$$

The five squared distance equations are

$$\begin{cases} x^2 + y^2 + z^2 = r^2 \\ (x - \frac{L}{2})^2 + (y - \frac{L}{2})^2 + z^2 = r_0^{\,2} \\ (x + \frac{L}{2})^2 + (y - \frac{L}{2})^2 + z^2 = r_1^{\,2} \\ (x + \frac{L}{2})^2 + (y + \frac{L}{2})^2 + z^2 = r_2^{\,2} \\ (x - \frac{L}{2})^2 + (y + \frac{L}{2})^2 + z^2 = r_3^{\,2} \end{cases} \tag{16}$$

The relationship between the Cartesian coordinate system and the spherical coordinate system is

$$\begin{cases} x = r\sin\theta\cos\phi \\ y = r\sin\theta\sin\phi \\ z = r\cos\theta \end{cases} \tag{17}$$

Substituting (17) into (16) yields

$$\begin{cases} r_0^{\,2} = r^2 + \frac{L^2}{2} - Lr\sin\theta(\cos\phi + \sin\phi) \\ r_1^{\,2} = r^2 + \frac{L^2}{2} + Lr\sin\theta(\cos\phi - \sin\phi) \\ r_2^{\,2} = r^2 + \frac{L^2}{2} + Lr\sin\theta(\cos\phi + \sin\phi) \\ r_3^{\,2} = r^2 + \frac{L^2}{2} - Lr\sin\theta(\cos\phi - \sin\phi) \end{cases} \tag{18}$$

Subtracting (18d) from (18c), and (18b) from (18c), respectively, we obtain

$$\begin{cases} r_2^{\,2} - r_3^{\,2} = 2Lr\sin\theta\cos\phi \\ r_2^{\,2} - r_1^{\,2} = 2Lr\sin\theta\sin\phi \end{cases} \tag{19}$$

Substituting (17) into (19a) and (19b) produces

$$\begin{cases} x = \dfrac{r_2^{\,2} - r_3^{\,2}}{2L} \\ y = \dfrac{r_2^{\,2} - r_1^{\,2}}{2L} \end{cases} \tag{20}$$

Then

$$\tan\phi = \frac{y}{x} = \frac{(r_2 - r_1)(r_2 + r_1)}{(r_2 - r_3)(r_2 + r_3)} \tag{21}$$

When the source is in the far-field of an array, we can obtain

$$r_2 + r_1 \approx r_2 + r_3 \approx 2r \tag{22}$$

When the source azimuth ranges between $[0°, 90°]$, the estimated azimuth angle can be determined from (21) and (22) as

$$\hat{\phi} = \tan^{-1}\left(\frac{r_2 - r_1}{r_2 - r_3}\right) = \tan^{-1}\left(\frac{\hat{\tau}_2 - \hat{\tau}_1}{\hat{\tau}_2 - \hat{\tau}_3}\right) \tag{23}$$

where the last expression in (23) is based on measured values that will include measurement errors. Note also that (23) only uses TDOA measurements. According to (20a), (20b) and (22), we can get

$$\sin\theta = \frac{\sqrt{x^2 + y^2}}{r} = \frac{\sqrt{(r_2 - r_3)^2(r_2 + r_3)^2 + (r_2 - r_1)^2(r_2 + r_1)^2}}{2Lr} = \frac{\sqrt{(r_2 - r_3)^2 + (r_2 - r_1)^2}}{L} \tag{24}$$

Assume that the source elevation satisfies $0° \leq \theta \leq 90°$, we obtain

$$\hat{\theta} = \sin^{-1}\left(\frac{c\sqrt{(\hat{\tau}_2 - \hat{\tau}_3)^2 + (\hat{\tau}_2 - \hat{\tau}_1)^2}}{L}\right) \tag{25}$$

which again only requires TDOA measurements, but a knowledge of the sound speed is also required.

LINEAR LEAST SQUARES DIRECTIONF FINDING

Linear Least Squares Estimator

Baruch et al. first investigated this problem and developed a linear least squares (LLS) method to resolve the direction finding problem (Baruch et al 1999). The LLS estimator regards the three components of an estimated DOA vector as three unknown parameters, and the azimuth and elevation angles are determined by trigonometric function calculation. This procedure adds one unknown variable, degrading estimation accuracy.

Let $\Delta\tau$ represent the TDOA measurement noise vector and hence the TDOA measurement equation becomes

$$\hat{\tau} = \tau + \Delta\tau \tag{26}$$

If $\Delta\mathbf{k}$ is the differential vector between the real DOA vector and the corresponding estimated DOA vector, i.e. the error vector, the estimated DOA vector can be written as

$$\hat{\mathbf{k}} = \mathbf{k} + \Delta\mathbf{k} \tag{27}$$

It is desirable to estimate the source bearing as the one which, in some sense, best fits the measured TDOAs in a set of over-determined equations. The LLS estimator utilizes the linear relation between the TDOA vector and a hypothesized DOA vector, producing

$$\hat{\mathbf{k}} = \hat{c}\left(\mathbf{S}^{\mathrm{T}}\mathbf{W}\mathbf{S}\right)^{-1}\mathbf{S}^{\mathrm{T}}\mathbf{W}\hat{\tau} \tag{28}$$

Herein $\mathbf{W}$ is the weighting matrix. If the covariance matrix of $\Delta\tau$ is given by $\Sigma = \mathrm{cov}(\Delta\tau)$, then $\mathbf{W} = \Sigma^{-1}$. In the case of unknown covariance matrix, $\mathbf{W}$ may be determined based on signal-to-noise ratio (SNR). Otherwise, when the measurement clock errors of TOA for each sensor are independent and identically distributed (i.i.d.) Gaussian random variables with an identical variance σ_t, $\mathbf{W}$ may be simply assigned as (Chan and Ho 1994)

$$\mathbf{W} = \left(\mathbf{I} + \mathbf{1}\times\mathbf{1}^{\mathrm{T}}\right)^{-1} \tag{29}$$

where $\mathbf{I}$ is $(N-1)\times(N-1)$ identity matrix, and $\mathbf{1}$ is $(N-1)\times 1$ vector of ones. According to the matrix inversion lemma, (29) becomes

$$\mathbf{W} = \mathbf{I} - (\mathbf{1}\times\mathbf{1}^{\mathrm{T}})/N \tag{30}$$

Given the DOA vector estimate of the 3-D array, a closed-form but suboptimal estimate for the azimuth and elevation is obtained as

$$\begin{cases} \hat{\phi} = \tan^{-1}(\hat{k}_y / \hat{k}_x) \\ \hat{\theta} = \tan^{-1}(\sqrt{\hat{k}_x^{\,2} + \hat{k}_y^{\,2}} / \hat{k}_z) \end{cases} \tag{31}$$

With regard to the 2-D array where $\mathbf{s}_i = \left[x_i, y_i\right]^{\mathrm{T}}$ $(i = 1,\ldots, N-1)$, the azimuth and elevation estimates are calculated by

$$\begin{cases} \hat{\phi} = \tan^{-1}(\hat{k}_y / \hat{k}_x) \\ \hat{\theta} = \cos^{-1}(\hat{k}_z) = \cos^{-1}\left(\sqrt{1 - (\hat{k}_x^{\,2} + \hat{k}_y^{\,2})}\right) \end{cases} \tag{32}$$

For the 1-D array configuration where $\mathbf{s}_i = x_i$ $(i = 1,\ldots, N-1)$, we can only estimate $\hat{k}_x$ without elevation angle. Then the azimuth of signal source is estimated by

$$\hat{\phi} = \cos^{-1}(\hat{k}_x) \tag{33}$$

We can see that the azimuth estimate from the 2-D array configuration and both the azimuth and the elevation estimates from the 3-D one, are independent of the sound speed. The elevation estimation with the 2-D array involves the sound speed parameter as does the four-microphone planar array solution.

Theoretical Performance of the LLS Estimator

In this section analytic expressions for the accuracy of the source bearing estimation are given for 1-D, 2-D, and 3-D array cases when using the LLS estimator. The theoretical accuracy of the parameter estimation is also compared with the expression of the related CRLB.

When the TOA measurement errors of all sensors are i.i.d. Gaussian random variables with an identical variance σ_t^2, the associated TDOA error variance is given by $\sigma_\tau^2 = 2\sigma_t^2$. For the linear 1-D array configuration, $\gamma = \phi$, $\mathbf{s}_i = x_i$, and $\mathbf{k} = k_x = \cos\phi$, the variance of the azimuth estimation is given as

$$\sigma_\phi^2 = \frac{c^2\sigma_t^2}{\sin^2\phi \sum_{i=1}^{N-1} x_i^2} \tag{34}$$

The CRLB for the 1-D case is determined by $CRLB(\phi) = 1 / F_{\phi\phi}$. Substituting $y_i = 0$ and $\theta = \pi / 2$ in (14b) and (12a) yields

$$CRLB(\phi) = \frac{c^2\sigma_t^2}{\sin^2\phi \sum_{i=1}^{N-1} x_i^2} \tag{35}$$

As observed the CRLB is exactly the same as the derived variance given by (34), indicating that the LLS estimator achieves the CRLB in this case.

It is known that the source elevation angle in the 3-D space can also be estimated by using a planar 2-D array. Thus, for the 2-D array case where $\gamma = \left[\phi, \theta\right]^{\mathrm{T}}$, $\mathbf{k} = \left[k_x, k_y\right]^{\mathrm{T}} = \left[\sin\theta\cos\phi, \sin\theta\sin\phi\right]^{\mathrm{T}}$ and $\mathbf{s}_i = \left[x_i, y_i\right]^{\mathrm{T}}$, the variances of the azimuth and elevation estimation can be respectively derived as

$$\sigma_\phi^2 = \frac{c^2\sigma_t^2\left(\cos^2\phi\sum_{i=1}^{N-1}x_i^2 + 2\sin\phi\cos\phi\sum_{i=1}^{N-1}x_iy_i + \sin^2\phi\sum_{i=1}^{N-1}y_i^2\right)}{\sin^2\theta\left[\sum_{i=1}^{N-1}x_i^2\sum_{i=1}^{N-1}y_i^2 - \left(\sum_{i=1}^{N-1}x_iy_i\right)^2\right]} \tag{36}$$

and

$$\sigma_\theta^2 = \frac{c^2\sigma_t^2\left(\sin^2\phi\sum_{i=1}^{N-1}x_i^2 - 2\sin\phi\cos\phi\sum_{i=1}^{N-1}x_iy_i + \cos^2\phi\sum_{i=1}^{N-1}y_i^2\right)}{\cos^2\theta\left[\sum_{i=1}^{N-1}x_i^2\sum_{i=1}^{N-1}y_i^2 - \left(\sum_{i=1}^{N-1}x_iy_i\right)^2\right]} \tag{37}$$

Applying the constraint $z_i = 0$ to the CRLB for the 3-D array case, it can be seen that the resulting CRLB expressions are identical to (36) and (37). In other words, the LLS estimator is a statistically efficient method for the 2-D array with an arbitrary configuration structure.

For the spatial 3-D array configuration where $\gamma = \left[\phi, \theta\right]^{\mathrm{T}}$, $\mathbf{s}_i = \left[x_i, y_i, z_i\right]^{\mathrm{T}}$ and $\mathbf{k} = \left[k_x, k_y, k_z\right]^{\mathrm{T}} = \left[\sin\theta\cos\phi, \sin\theta\sin\phi, \cos\theta\right]^{\mathrm{T}}$, the variances of the azimuth and elevation estimation are given respectively by

$$\sigma_\phi^2 = \frac{c^2\sigma_t^2}{\sin^2\theta}\left[\sin\phi \quad \cos\phi \quad 0\right]\left(\mathbf{S}^{\mathrm{T}}\mathbf{S}\right)^{-1}\left[\sin\phi \quad \cos\phi \quad 0\right]^{\mathrm{T}} \tag{38}$$

and

$$\sigma_\theta^2 = c^2\sigma_t^2 \times \left[\cos\theta\cos\phi \quad \cos\theta\sin\phi \quad -\sin\theta\right]\left(\mathbf{S}^{\mathrm{T}}\mathbf{S}\right)^{-1}\left[\cos\theta\cos\phi \quad \cos\theta\sin\phi \quad -\sin\theta\right]^{\mathrm{T}} \tag{39}$$

In general this estimator is not necessarily the same as the CRLB. However, if the array obeys the following geometrical constraints,

$$\begin{cases} \sum_{i=1}^{N-1}x_iy_i = \sum_{i=1}^{N-1}x_iz_i = \sum_{i=1}^{N-1}y_iz_i = 0 \\ \sum_{i=1}^{N-1}x_i^2 = \sum_{i=1}^{N-1}y_i^2 \end{cases} \tag{40}$$

then the variance of the azimuth estimation becomes

$$\sigma_{\phi}^{2} = \frac{c^{2}\sigma_{t}^{2}}{\sin^{2}\theta\sum_{i=1}^{N-1} x_{i}^{2}} \tag{41}$$

It can be shown that the CRLB for the azimuth estimation under the same constraints has the same expression. Thus under the condition given by (40), the LLS estimator achieves the optimal solution for the azimuth estimation. If the array geometry has a fully balanced array structure, which obeys the following conditions

$$\begin{cases} \sum_{i=1}^{N-1} x_{i} y_{i} = \sum_{i=1}^{N-1} x_{i} z_{i} = \sum_{i=1}^{N-1} y_{i} z_{i} = 0 \\ \sum_{i=1}^{N-1} x_{i}^{2} = \sum_{i=1}^{N-1} y_{i}^{2} = \sum_{i=1}^{N-1} z_{i}^{2} \end{cases} \tag{42}$$

Then the variance of elevation estimation is obtained as

$$\sigma_{\theta}^{2} = \frac{c^{2}\sigma_{t}^{2}}{2\sum_{i=1}^{N-1} x_{i}^{2}} \tag{43}$$

which exactly equals the CRLB. Therefore, under the condition of a fully balanced array configuration, the LLS estimator achieves the CRLB for both the azimuth and the elevation estimation.

The LLS estimator utilizes the linear relationship between the TDOA measurement vector and the source DOA vector in Cartesian coordinate system. As this linear relationship allows a closed form estimation of the DOA vector, the LLS estimator can determine the azimuth and elevation angles directly from the TDOA measurements. Thus it is considerably more efficient in terms of computational load and does not need the prior knowledge of the sound speed to estimate the azimuth for the 2-D and 3-D array configuration and the elevation for the 3-D one.

EVOLUTIONARY TDOA-BASED DIRECTION FINDING METHODS

This section studies a direction finding approach based on evolutionary computation methods in artificial intelligence science. The intelligent optimization process is adopted to search for the optimal source azimuth and elevation estimation solutions. Two evolutionary computation estimators are presented, which are the genetic algorithm (GA) and the particle swarm optimization (PSO) algorithm (Cui *et al* 2015).

Idea of Evolutionary Estimator

The primary idea in the development of the evolutionary algorithm is three-fold: (i) the in-situ sound speed is estimated using real-time measured temperature, wind speed, and wind direction, as well as the LLS based DOA estimation results; (ii) the azimuth and elevation estimates from the LLS estimator are used as the initial azimuth and elevation values for GA and PSO; (iii) the cost function is defined as the squared-error sum of 3-D array direction finding.

The estimated sound speed is required to calculate the nonlinear cost function during the search process of GA and PSO. In an outdoor environment, the sound propagation speed is affected by a number of weather parameters, mainly including air temperature, wind speed and wind direction. Assume that the available real-time in-situ weather parameters are the air temperature ($\hat{T}_0$), wind speed ($\hat{V}_W$), and wind direction ($\hat{\phi}_W$). The estimate of the sound propagation direction denoted by $\hat{\phi}_0$ can be obtained by the LLS method; hence one obtains the intersection angle between the estimated DOA and the measured wind direction as $\left|\hat{\phi}_W - \hat{\phi}_0\right|$. The sound speed estimate is then computed according to

$$\hat{c} = 20.06\sqrt{273.15 + \hat{T}_0} + \hat{V}_W \cos(\hat{\phi}_W - \hat{\phi}_0) \tag{44}$$

Although the LLS method may not produce a very accurate estimate of the sound azimuth angle, usually the estimate does not deviate too much from the true value. Therefore, it can be seen from (44) that the sound speed estimation is not very sensitive to the angle estimation error after cosine operation is included.

To speed up the convergence of the evolutionary process and to avoid a possible local minimum, the initial direction is set to be the LLS based estimates, namely $\hat{\phi}_{LLS}$ and $\hat{\theta}_{LLS}$, which would not deviate from the actual direction too much. The two DOA parameters are also constrained by

$$\begin{cases} \hat{\phi}_{LLS} + \mathrm{B}_\phi \geq \phi \geq \hat{\phi}_{LLS} - \mathrm{B}_\phi \\ \hat{\theta}_{LLS} + \mathrm{B}_\theta \geq \theta \geq \hat{\theta}_{LLS} - \mathrm{B}_\theta \end{cases} \tag{45}$$

where B_ϕ and B_θ are constants to constrain the two DOA parameters to a rectangular area of length $2B_\phi$ and width $2B_\theta$ and centered at $(\hat{\phi}_{LLS}, \hat{\theta}_{LLS})$.

The two bound parameters are defined as follows:

$$\begin{cases} \mathrm{B}_\phi = k_\phi c\sigma_\tau \\ \mathrm{B}_\theta = k_\theta c\sigma_\tau \end{cases} \tag{46}$$

where the unit of B_ϕ and B_θ is degree, k_ϕ and k_θ are scaling factors which are selected as 100 in the simulation to be discussed later, and σ_τ is the standard deviation (STD) of TDOA measurement error whose unit is second. In the case where $\left\{\hat{\tau}_{ij}\right\}$ have different STDs, their average value may be used.

In a practical localization system the DOA estimation error would be constrained to 30 degrees. A too small bounded space would not enable a normal evolutionary search, while a too large bounded space would result in a slow convergence for the optimization process. Typically GA and PSO estimators would be constrained to search in a small area, e.g. several degrees wide, after some evolutionary generations.

For convenience of algorithm implementation, the PSO uses real number (decimal) coding, while GA uses binary coding. However, the binary system can be transformed to the decimal system or vice versa. Also, the azimuth and elevation values can adopt the degree instead of the radian unit during the evolutionary process. In fact the two units would be easily converted to each other as required.

If the zeroth sensor is chosen as reference node, then the cost function is defined as follows:

$$f(\phi,\theta) = \sum_{j=1}^{N-1}\left[x_j \sin\theta\cos\phi + y_j \sin\theta\sin\phi + z_j \cos\theta - \hat{c}\hat{\tau}_j\right]^2 \tag{47}$$

The aim of the evolutionary DOA estimators is to minimize $f(\phi,\theta)$ with respect to the azimuth and elevation, and find the global optimal solution.

The GA direction finding method adopts the best solution preservation strategy of the evolutionary computation theory, that is, the solution with the optimal fitness would not be replaced in the next generation. Thus the global convergence of the evolutionary process can be reached theoretically according to the swarm intelligence theory. In practice, the population size and evolutionary generation can be tuned within the simulation, so that the global convergence can still be ensured in general. On the other hand, the PSO direction finding method cannot always guarantee a global convergence theoretically. Since the DOA estimation of LLS estimator is exploited as the initial direction guess and constraints are applied to the two parameters, the divergence problem related to the traditional PSO can be usually avoided.

Evolutionary Estimator Description

Genetic Algorithm

The GA procedure can be described in the following steps:

Step 1: Determine an appropriate population size, maximum genetic generation, float-point precision of the azimuth and elevation as two real number variables, crossover probability (P_c) and mutation probability (P_m).

Step 2: Input the total number of sensors, 3-D coordinates of each sensor, TDOA measurement between each pair of sensors, measured air temperature, wind speed, and wind direction.

Step 3: Determine the variable bounds of initial candidate solutions (that is the individuals in the field of the swarm intelligence discipline): (i) if the STD of TDOA (σ_τ) is known, it is decided by (45) after the scaling factor is set; (ii) if the knowledge about σ_τ cannot be acquired in advance, the bounds are selected by user experience.

Step 4: Estimate the DOA using the LLS estimator.

Step 5: Initialize population: i) the initial values of the two real-valued DOA variables are randomly selected from the bound given by (45); ii) the DOA variables are changed into two segment chromosomes of binary coding with the variable float-point precision.

Step 6: Compute the fitness function of each individual using (47).

Step 7: Use the roulette wheel selection strategy to select the male and female individuals by the individual fitness.

Step 8: Randomly and respectively select two crossover points on the male and female chromosomes with two segments, and then execute crossover and mutation operations for each segment of two chromosomes by crossover probability and mutation probability.

Crossover probability (P_c) is the probability that a pair of chromosomes will be crossed. Mutation probability (P_m) is the probability that a gene on a chromosome will be mutated randomly. Usually P_c and P_m can be set as 0.75 and 0.05, respectively. It should be pointed out that the generated new DOA individual still is a feasible solution after crossover and mutation, as the decoding result of a binary individual is limited within the DOA parameter bound.

Step 9: Preserve the best solution of current generation, and then repeat *Step 6*, *Step 7* and *Step 8* to replace other individuals in the old population until the new population has been generated.
Step 10: Stop the evolutionary procedure by two termination criteria: if the maximum genetic generation has been reached, or else if the elite solution has not changed for a certain number of generations, and then output the two variables of the elite as DOA estimate.

Particle Swarm Optimization

The following steps describe the basic principles of the PSO algorithm.

Step 1: Determine an appropriate number of total particles, maximum iteration number, learning factor c_1 and c_2, particle inertia weight ω and velocity update maximum (vMax).
Step 2: Input the total number of sensors, 3-D coordinates of each sensor, TDOA measurement between any two sensors, measured air temperature, wind speed, and wind direction.
Step 3: Determine the variable bounds and set the initial particle in the same manner as used in *Step 3* of GA procedure.
Step 4: Estimate DOA with the LLS estimator.
Step 5: Initialize particle swarm: (i) the first and the second dimension fly positions of each particle are randomly selected from the DOA parameter bound respectively; (ii) the initial fly velocities of two dimensions are set to be zero.

It should be explained that when using PSO, a possible solution is represented by the particle fly position vector. Here the vector consists of two-dimensional parameters, i.e. azimuth and elevation. Additionally, each particle has a current fly velocity vector for the associated dimensional fly positions, which represents a magnitude and direction toward a new solution.

Step 6: Compute the cost function of each particle using (47).
Step 7: Determine the optimal solution of current generation (G^{best}) and the global optimal solution (T^{best}) with all particle fitness functions.
Step 8: Assume that the velocity $v_{i,j}(t)$ and position $p_{i,j}(t)$ of *i*-th particle in *t*-th generation are known; here *j* is the space dimensions (*j*=1, 2). If r_1 and r_2 are random numbers uniformly distributed within [0, 1], then iteratively compute the following equations:

$$v_{i,j}(t+1) = \omega v_{i,j}(t) + c_1 r_1 \left[G^{best}(t) - p_{i,j}(t)\right] + c_2 r_2 \left[T^{best}(t) - p_{i,j}(t)\right] \tag{48}$$

$$p_{i,j}(t+1) = p_{i,j}(t) + v_{i,j}(t) \tag{49}$$

Step 9: If the updated absolute value of particle velocity is larger than a predefined velocity threshold vMax, then renew the velocity by

$$v_{i,j}(t+1) = sign(v_{i,j}(t+1)) \bullet vMax \tag{50}$$

where $sign(u) = 1$ for $u > 0$, otherwise $sign(u) = -1$.

Step 10: Stop the evolutionary procedure by two termination criteria: if the maximum iteration generation has been reached, or else if the global optimal solution has not changed for a certain number of generations, and then output the first and the second particle fly position parameters of global optimal solution as DOA estimate.

Simulation of Evolutionary Estimator

Some illustrative simulation examples are provided below to show the performance of the evolutionary estimators described above.

Simulation Setup

The weather parameter measurements are fixed as follows: air temperature is $20\,°\mathrm{C}$; horizontal wind speed is 10 m/s, which equals to 5–6 level gales according to the Beaufort scale; horizontal wind direction is $\pi / 3$. Usually the wind velocity near the ground means the horizontal wind speed, as the vertical speed is usually unknown. Under the ideal steady weather conditions, air temperature, wind speed and wind direction are assumed to remain the same during the period from the transient sound generation to the completion of measurement data sampling, thus the sound speed is constant.

In an outdoor environment, weather parameters are usually not constant, resulting in a time-varying sound speed. According to meteorology theory (Holton 2004), their typical variation ranges are given by

$$\begin{cases} \hat{T}_0 + 5 \geq T_0 \geq \hat{T}_0 - 5 \\ \hat{V}_W + 6 \geq V_W \geq \hat{V}_W - 6 \\ \hat{\phi}_W + 22.5 \geq \phi_W \geq \hat{\phi}_W - 22.5 \end{cases} \tag{51}$$

where T_0, V_W, and ϕ_W are the time-varying temperature ($°\mathrm{C}$), wind speed (m/s) and direction (degree), respectively, while $\hat{T}_0$, $\hat{V}_W$, and $\hat{\phi}_W$ are respectively their means. In simulations these real weather parameters are set to vary around these means within the above ranges under the non-steady weather condition. The array and source locations are known in advance, and the real azimuth of the sound wave can be obtained from these locations.

Assume that the sound source location (x, y, z) is (500 m, 300 m, 100 m). If the sensor array is not symmetrical, the *x*-axis and *y*-axis positions are randomly selected on a ring of 30m radius. As the actual altitude difference among employed sensors would not be large for typical practical cases, the *z*-axis positions are uniformly randomly selected in the small range [–10m, 10m]. Although it is convenient to deploy such a 3-D array with small and random altitude variations, the consequence is that considerable elevation estimation performance degradation would be expected.

The asymmetrical array size (number of elements/sensors) is set as eight, which is in accordance with the experimental hardware setup to be discussed in the following section. The position coordinates of the eight sensors are selected in units of meter as: (4.00, 29.73, 0.38), (26.21, –14.58, –9.35), (–29.86, –2.83, 0.65), (–25.84, –15.23, –0.18), (–1.86, 29.94, 9.11), (–6.32, –29.32, 0.81), (17.43, –24.41, –2.94), and (–29.62, –4.72, 4.20). The zeroth sensor can be set as the coordinate origin through coordinate translation.

In this case the range-to-baseline ratio is only about eight, while in practice the ratio can be as high as many dozens. For instance, the gunfire locator namely Boomerang, was developed by DARPA to locate snipers by finding the direction of gunfire and the system was tested during the Iraq War. The device is actually an array consisting of seven microphones, which can be mounted on mobile vehicles. A single Boomerang device would not be used for position determination, since the position error would be extremely large when the distance between the sniper and the device is around 100 meters.

When *P* denotes the position of an acoustic source, the ratio of array aperture to the distance between array and source is given by

$$\rho_i = \|\mathbf{s}_i\| / \|\mathbf{p}\| \tag{52}$$

where the definition is under the assumption that the center of the array is the origin of the coordinate system. For far-field sources $\rho_i \ll 1$, the signal is treated as a plane wave. We can take any element of the array as the reference point of the DOA estimation, as the plane wave has the same direction for any point on the array. However, if ρ_i is not much less than 1, the DOA estimation performance would be affected considerably, depending on the value of ρ_i and the selection of the benchmark point.

To calculate the DOA estimation performance conveniently, the TDOA measurement errors are assumed as Gaussian random variables of a specific STD without systematic measurement bias. The TDOA noise power ($10\log\sigma_\tau^2$) ranges from –70dB to –40dB, so that σ_τ ranges from 0.3 ms to 10 ms. This TDOA noise bound would be basically within the range of realistic measurement noise/errors. The simulated TDOA measurement is the signal arrival distance difference divided by the simulated sound speed plus Gaussian noise, namely

$$\hat{\tau}_i = \frac{\|\mathbf{s}_i\|}{c} + \varepsilon_{\tau_i} \tag{53}$$

Additionally, the simulation parameters of GA and PSO are selected as follows. The GA float-point precisions of azimuth and elevation are set as 0.01, cross probability is 0.75, and mutation probability is 0.05. The PSO particle inertia weight is 0.08, the learning factor $c_1 = 2$ and $c_2 = 2$, and the maximum renewal velocity is 5 degrees. The population sizes of both GA and PSO algorithms are set as 100. When the global optimal solution has not changed in recent 20 generations, GA and PSO stop searching according to the evolutionary criterion described above.

DOA Estimation Results

For the random array case, 2000 simulations are run for each noise power under steady weather conditions with a constant sound speed at first. The root mean squared error (RMSE) criterion is used to evaluate DOA estimation performance, while the square root of the CRLB is computed as a reference. Figure 5 and Figure 6 show the RMSE of the azimuth and elevation estimation results, respectively for the case of a random array. As observed, the elevation estimation performance is much inferior to the azimuth estimation, and the GA and PSO algorithms outperform the LLS considerably when the noise power is low.

The azimuth and elevation simulation results within 2000 runs for the random array and non-steady weather conditions are respectively given in Figure 7 and Figure 8. As expected, the DOA estimation error increases as the TDOA measurement noise increases. It can be seen that the results are similar to those in the steady weather case. From these results it can be observed that the LLS estimator does not need any information about the sound speed. Although knowledge of sound speed is required for

Figure 5. RMSE of azimuth estimation with a random array under steady weather condition

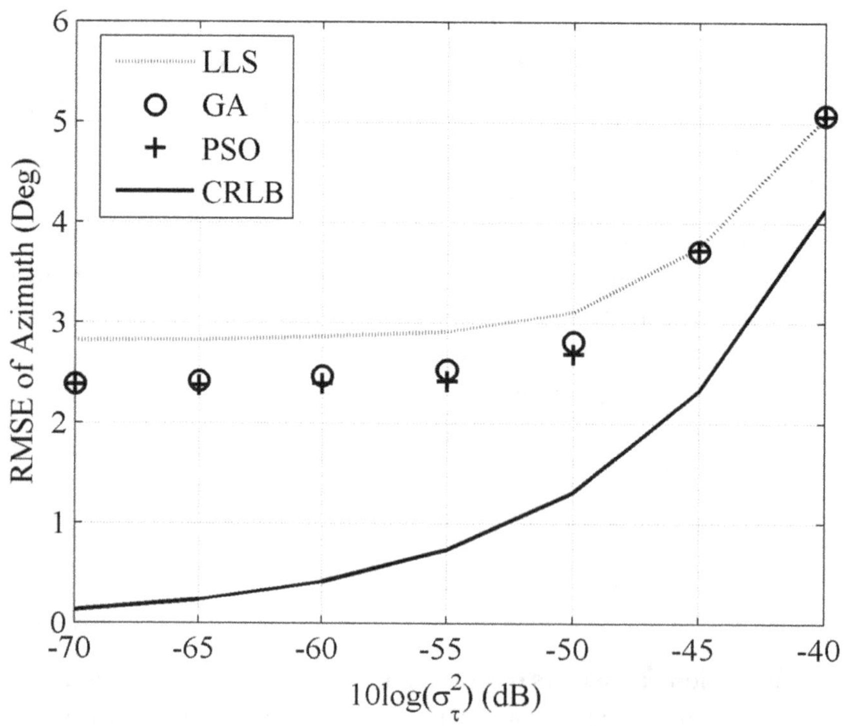

Figure 6. RMSE of elevation estimation with a random array under steady weather condition

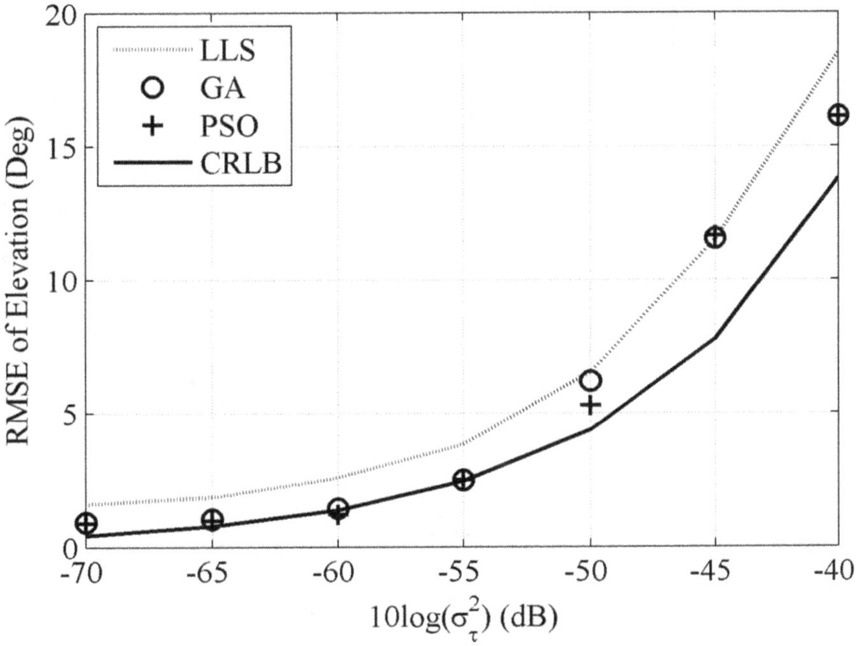

Figure 7. RMSE of azimuth estimation with a random array under non-steady weather condition

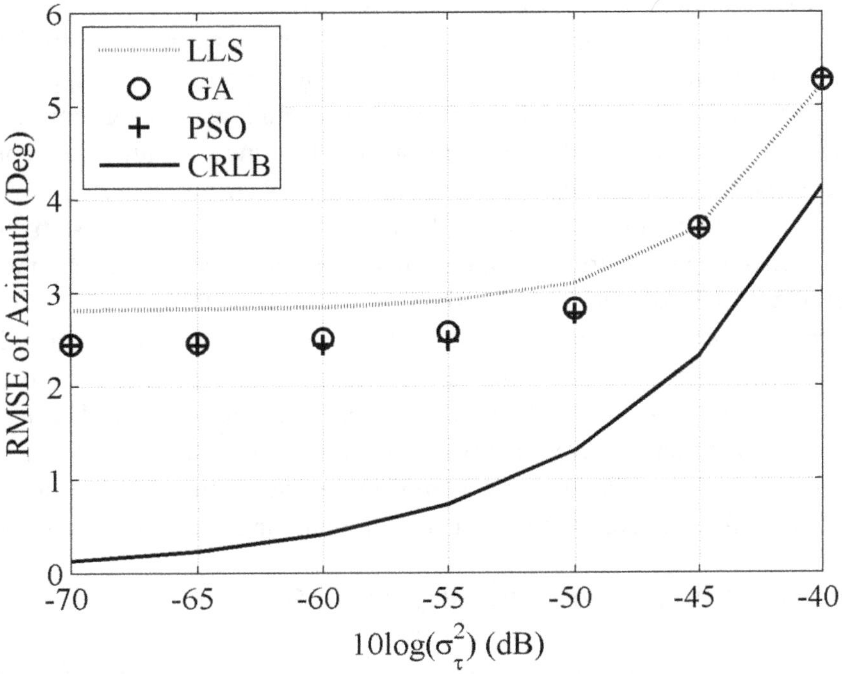

Figure 8. RMSE of elevation estimation with a random array under non-steady weather condition

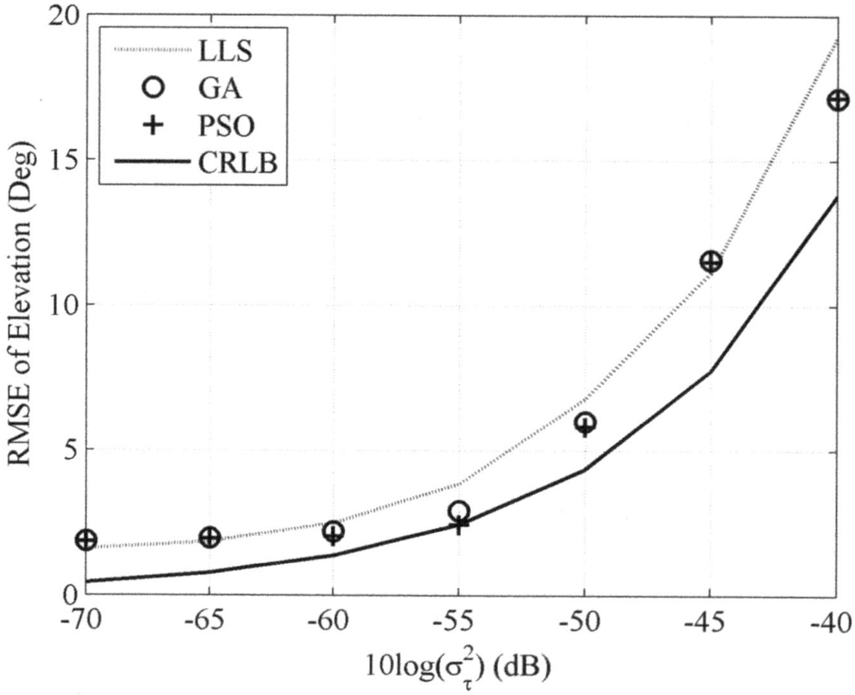

the GA-based and PSO-based methods, these two methods appear to be almost insensitive to the sound speed variations typical in outdoor non-steady weather conditions.

When the cost function does not change, or the change is very small in the recent 20 generations, it is considered that the evolutionary process of the GA and PSO algorithms has approached the global optimum solution from a statistical probability perspective. Evolutionary generations are counted until the algorithm approaches the steady state, and the results of GA and PSO are correspondingly plotted using "boxplot" as shown in Figure 9 and Figure 10. The central stripe on each box is the generation median, and the box top and bottom edge are the 25th and 75th percentiles, respectively. The box top and bottom whisker extending to the most extreme data points respectively represent the maximum and minimum number of generations experienced in the simulation with 2000 runs.

From these boxplots we can see the approximate generation ranges and their probability distribution. The GA-based and PSO-based methods converge after about 30 and 90 generations, respectively. Notably their maximum generations are around 120 and 350, respectively. It is shown that the GA-based and PSO-based DOA estimation methods are effectively capable of obtaining their global solutions within a finite evolutionary generation. Thus they can complete the evolutionary search process rapidly and always converge, despite their random statistical nature in searching for the best solution.

Field Experiment Verification of Evolutionary Estimator

To verify the feasibility of evolutionary TDOA-based direction finding methods with three-dimensional acoustic arrays, extensive experiments in the field were conducted. The acoustic signal sources are bomb

Figure 9. Graphical representations of evolutionary generation number of GA

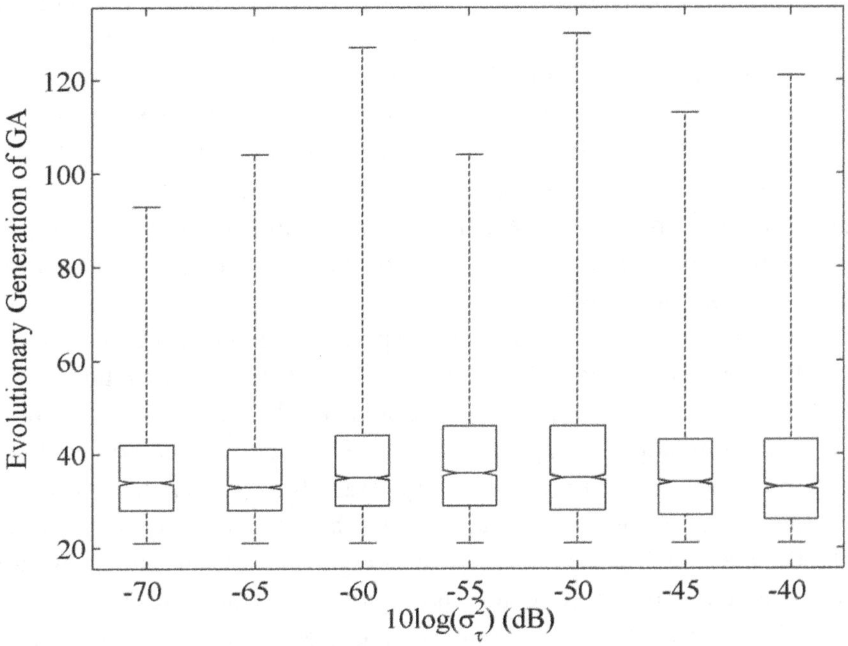

Figure 10. Graphical representations of evolutionary generation number of PSO

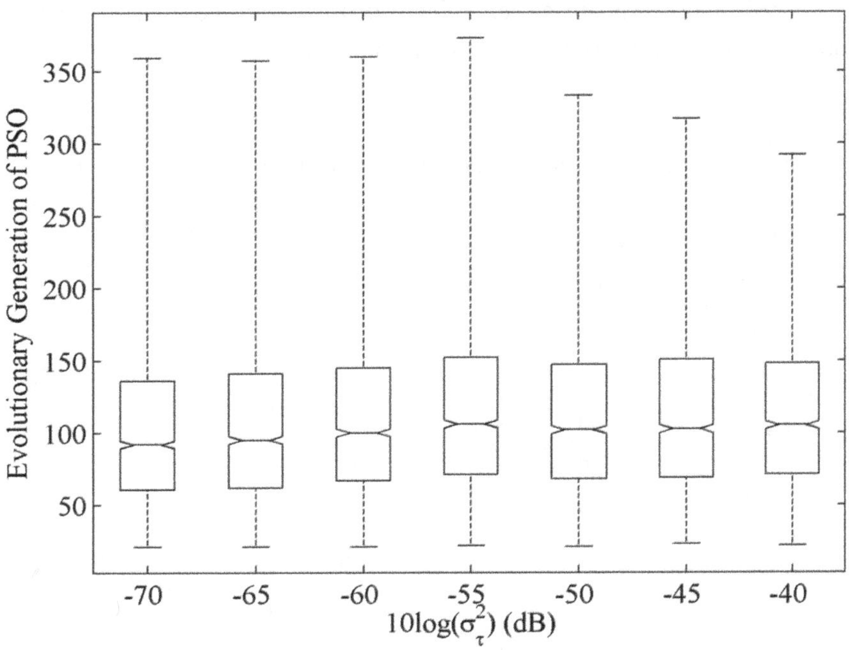

explosions. A total station was used to precisely measure the locations of sensors and sound source, which can achieve centimetre accuracy. The total station is an electronic/optical instrument used in modern surveying and building construction, which actually is an electronic theodolite integrated with an electronic distance meter to read slope distances from the instrument to a particular point.

The eight sensor locations are given in units of meter as follows: (6.06, –26.44, –0.48), (–5.36, –25.13, 1.70), (–20.92, –9.99, 5.33), (–24.35, –3.50, 5.45), (–3.67, 29.42, –1.50), (8.86, 28.13, –3.21), (25.86, 20.88, –5.43), and (13.54, –13.36, –1.85). The sound source location is (251.18, –361.50, 21.20). The longest and the shortest distances between sound source and sensors are 467m and 415m, respectively, while the largest distance between a pair of sensors is 57m. Figure 11 shows the 3-D view schematic of geometric arrangement of deployed sensors and sound source, where the digit near sensor location indicates the sequential number of related sensor.

The resulting transient pressure pulse was measured using a National Instruments Compact RIO-9234 data acquisition board to synchronously collect the acoustic signals. The sampling frequency of data acquisition is 51200 Hz. The sensitivity of acoustic sensor is 50 mV/Pa, whose responding frequency is 6.3 Hz – 20 kHz and maximum output voltage is 4.5 Vrms. The experimental process lasted two hours. The transient explosions occurred when the temperature was 19 °C (Celsius), the average wind speed was 6 m/s, and the wind direction was approximately along the direction from sensor#4 to sensor#1.

The scenes of the field test environment are shown in Figure 12 and Figure 13, where Figure 12 is the data acquisition instrument at the array center, and Figure 13 demonstrates the acoustic sensor with a windshield on a tripod. Figure 14 shows an example of sound pressure curves recorded by two sensors with the waveforms having obvious primary peaks followed by decaying ringing signals. The peak of each leading pulse can be used to determine the TDOA of a pair of sensors.

Table 1 shows a summary of the measured peak time points of all the eight sensors for the ten experiments/explosions. In this table the peak time point is decimal number and not an integer, as these data are outputted by the NI LabVIEW software with a higher data resolution precision. The TDOA in units of second between any two sensors is simply calculated as the difference between the two peak time points (unit in number of samples) divided by the sampling frequency.

Figure 11. Geometric arrangement of deployed sensors and sound source

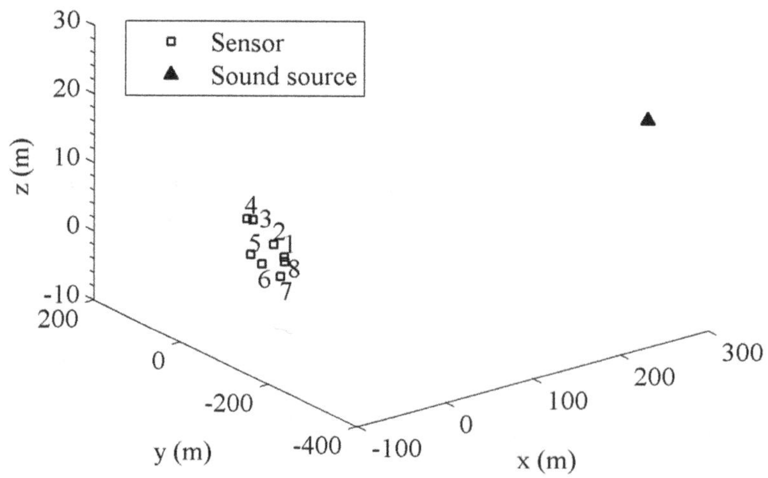

Figure 12. Data acquisition instrument in the field test environment

Figure 13. Acoustic sensor in the field test environment

The individual DOA estimation errors of the ten field experiments and the RMSE of the different methods are given in Table 2. The RMSE of field data based estimation is calculated by $RMSE = \sqrt{STD^2 + Mean^2}$, where STD and Mean represent the standard deviation and mean error, respectively. For comparison the RMSE values related to $10 \log \sigma_{\tau}^2 = -40dB$ in Figure 7 and Figure 8 are also listed in Table 2 with "RMSE1" and "RMSE2" denoting the experimental results and the simulation results, respectively.

From Table 2 it can be seen that there is good consistency between the field experimental results and the simulation results shown in Figure 7 and Figure 8. The elevation estimation does suffer performance degradation due to the small difference among sensor altitudes, but the GA-based and PSO-based methods can have much better performance in azimuth estimation.

Figure 14. Typical waveforms recorded at two acoustic sensors during a field experiment

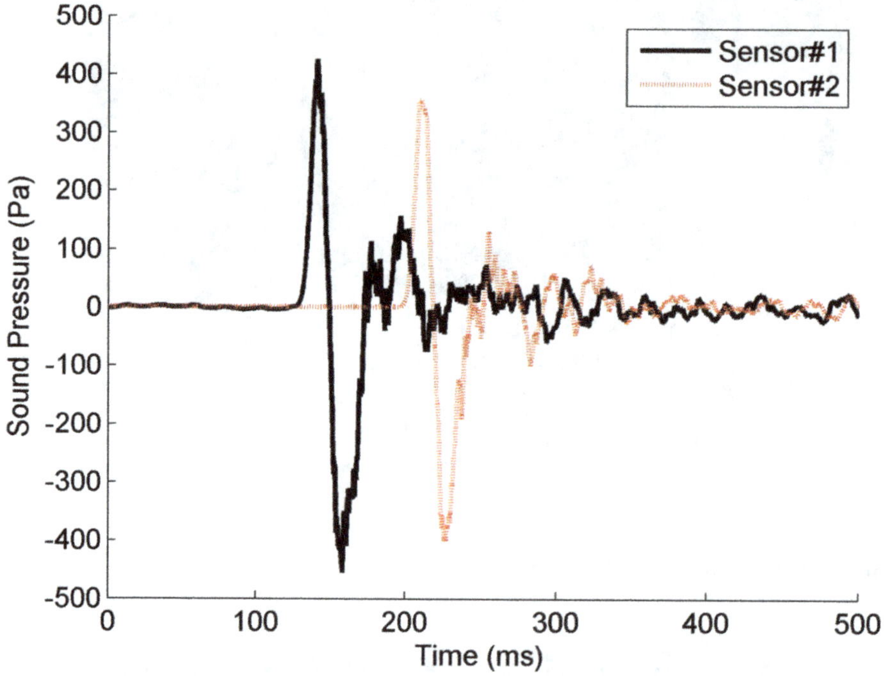

Table 1. Peak time points (number of samples) of received sound waveforms at eight sensors.

Sensor	Peak time point									
S_1	0	0	0	0	0	0	0	0	0	0
S_2	1091.3	1084.6	1194.7	1121.2	1113.7	1148.4	1145.0	1133.0	1094.0	1093.7
S_3	4104.7	4134.5	4206.7	4162.0	4140.5	4169.3	4135.0	4179.0	4136.0	4125.8
S_4	5174.6	5195.7	5384.9	5241.2	5217.3	5231.4	5221.0	5235.0	5216.0	5194.3
S_5	7588.2	7331.1	7325.1	7853.6	7907.2	7962.4	7946.0	7264.0	7839.0	7746.5
S_6	6502.3	6286.7	6263	6770.9	6917.6	6932.5	6909.0	6172.0	6729.0	6635.6
S_7	4380.5	4122.4	4127.5	4593.8	4721.9	4741.7	4749.0	4032.0	4552.0	4513.0
S_8	994.0	682.3	688.6	1270.1	1337.1	1353.8	1327.0	646.0	1203.0	1132.2

DIRECTION FINDING BASED ON BIASED TDOA MEASUREMENTS

TDOA bias is a systematic error of TDOA measurement between any two sensors in the presence of biased measurement errors. Its physical quantity can be considered as the extent of systematic outlier in a measurement instrument system. In this section, an efficient expectation-maximization (EM) algorithm is used for accurately estimating the DOA of the signal from a far field source in presence of biased TDOA measurements.

TDOA Error Distribution Problem

Commonly TDOA errors without bias are modeled as a zero-mean Gaussian random variable. However, experimentation allows the measurement data to be compared with other models to ascertain the representation. The following analysis investigates four different distributions (Gaussian, Laplace, Gaussian-Laplace, and generalized Normal distribution (GND)) that have been widely used in modeling various physical phenomena. Note that Gaussian and Laplace distributions are two special cases of GND.

Several factors can significantly affect TDOA measurement noise, which include SNR, non-line-of-sight (NLOS) propagation, multipath distortion, phase mismatch between circuitries of each pair of sensors, and atmospheric turbulence effect on sound propagation (Chen 1999). For mobile localization in wireless networks especially in cellular networks, the NLOS induced TOA bias is often modeled as an exponential variable. The resulting bias variable in the TDOA error is the difference between two exponential variables, which has a Laplace distribution or double exponential distribution.

In the case of using an acoustic array for determining the direction of a far field source, the TDOA noise would mostly result from the following three sources:

1. NLOS and multipath propagation induced error, which can be modeled by a Laplace distribution. The scenario considered is that the source is in the far field and the propagation distance from the sound source to a sensor is much longer (e.g. many tens of times) than the distance between a pair of sensors, and hence a symmetric Laplace distribution may be assumed for analytical simplicity.
2. Systematic bias errors such as the synchronization errors at the different instrument ports, which can be considered as an unknown constant for each port; but can be difficult to calibrate.
3. Receiver noise, which is usually assumed to be a zero mean Gaussian random variable.

The PDF of the Gaussian noise combined with the systematic bias is given by

$$f_G(u) = \frac{1}{\sqrt{2\pi}\sigma}\exp\left(-\frac{(u-\mu)^2}{2\sigma^2}\right) \tag{54}$$

Table 2. DOA estimation results of measured data.

Methods		DOA estimation error (°)										RMSE1 (°)	RMSE2 (°)
		1	2	3	4	5	6	7	8	9	10		
Azimuth	LLS	-6.28	-2.10	-4.33	-10.00	-12.18	-11.80	-11.44	-1.92	-7.80	-8.00	8.51	5.29
	GA	-2.76	0.04	0.39	-7.35	-8.43	-8.18	-8.26	1.04	-7.46	-6.75	6.21	5.22
	PSO	-2.75	-0.01	0.37	-7.32	-8.27	-8.16	-8.21	1.05	-7.44	-4.38	5.95	5.27
Elevation	LLS	-13.06	-3.81	-20.9	-23.61	-27.91	-26.64	-24.62	-8.70	-14.01	-16.77	19.72	19.26
	GA	4.50	6.99	2.89	-10.63	-9.67	-8.22	-8.36	6.78	-12.13	-12.76	9.19	17.16
	PSO	4.57	6.94	2.80	-10.58	-9.12	-7.91	-8.21	6.63	-12.24	1.03	8.09	17.17

* RMSE1 denotes the field experimental results; RMSE2 denotes the simulation results in Figure 7 and Figure 8 when $10\log\sigma_{\tilde{A}}^2$ equals –40dB.

where μ is the constant systematic bias and σ is the noise STD. The symmetric Laplace distribution of the NLOS and multipath induced error is described by

$$f_L(v) = \frac{1}{2\alpha}\exp\left(-\frac{|v|}{\alpha}\right) \tag{55}$$

where α is the positive scale parameter. As shown in (Cui *et al* 2016), the TDOA error, if modeled as the sum of a Gaussian random variable and a Laplace random variable ($\varepsilon = \mu + v$), has a PDF given by the convolution of the two PDFs resulting in the complex expression

$$f_{GL}(\varepsilon) = \frac{1}{4\alpha}\exp\left(\frac{1}{2\alpha}\left(\frac{\sigma^2}{\alpha} + 2\mu - 2\varepsilon\right)\right) \left\{erfc\left(\frac{1}{\sqrt{2}\alpha}\left(\frac{\sigma^2}{\alpha} + \mu - \varepsilon\right)\right) + \exp\left(\frac{2}{\alpha}(\varepsilon - \mu)\right) erfc\left(\frac{1}{\sqrt{2}\alpha}\left(\frac{\sigma^2}{\alpha} + \varepsilon - \mu\right)\right)\right\} \tag{56}$$

where erfc(x) is the complementary error function. For simplicity, this Gaussian-Laplace distribution is called GL distribution. Clearly, the TDOA errors may be modeled as Gaussian random variables, if the NLOS and multipath induced errors are negligible. On the other hand, the errors can be modeled to have Laplace distribution, if the noise is negligible.

Alternatively, the TDOA measurement errors can be modeled to have a GND with the PDF

$$f_{GND}(\varepsilon) = \frac{1}{2\alpha\Gamma(1/\beta)}\exp\left(-\left(\frac{|\varepsilon - \mu|}{\alpha}\right)^{\beta}\right) \tag{57}$$

where μ and α respectively denote the mean and the scale parameter, β denotes the shape parameter, and Γ is the Gamma function. The GND can be applied to many applications, as its flexible parametric form can be adapted to a large family of symmetric distributions including Laplace ($\beta = 1$) and Gaussian ($\beta = 2$). Although a GND is more complicated than a Gaussian distribution or a Laplace distribution, it is more compacted and has a simpler form than the PDF given by (56).

The TDOA measurement equation associated with receiving signals from a common far-field transmitter source at the i-th and j-th sensors can be described by

$$\hat{\tau}_{j,i} = \tau_{j,i} + \varepsilon_{j,i} \tag{58}$$

where $\tau_{j,i}$ is the error free TDOA given by

$$\tau_{j,i} = \frac{1}{c}\left[(x_j - x_i)\sin\theta\cos\phi + (y_j - y_i)\sin\theta\sin\phi + (z_j - z_i)\cos\theta\right] \tag{59}$$

where (x_i, y_i, z_i) indicates the position coordinates of the i-th sensor. The TDOA measurement error $\varepsilon_{j,i}$ can be described as

$$\varepsilon_{j,i} = \xi_{j,i} + n_{j,i} \tag{60}$$

where $\xi_{j,i}$ represents the TDOA bias error and $n_{j,i}$ denotes the zero mean random component of the TDOA error.

Given the above statistical models, it is useful to compare the TDOA error model distributions discussed above with the measured data. The experimental TDOA measurements are obtained by the same field experiments as described in the former section. As shown in Figure 15 and Figure 16, the PDFs of the TDOA errors of two pairs of sensors are approximated by histograms. The two histograms are respectively produced by 70 TDOA error samples, where the width of each bin is 1.0 ms. The errors are fitted to the Normal, Laplace with a nonzero bias, GND and GL distributions, respectively, and the four least-squares fitted curves are also displayed in Figure 15 and Figure 16. It can be observed that the measured distributions are highly peaked at the mean TDOA, which is indicative that the Laplace distribution is a better model than the Gaussian distribution; GND and GL distribution have a moderate peak between Laplace and Gaussian distribution. Both GND and GL distribution have three distribution hyper-parameters to be estimated in the presence of biases. With three hyper-parameters to control their shapes, GND and GL distribution can provide a more flexible representation of kurtosis.

The corresponding cumulative distribution functions (CDFs) are shown in Figure 17 and Figure 18. It is observed that all the four model distributions are close to the CDF of measured data distribu-

Figure 15. The first example of biased TDOA error PDFs with field measurements

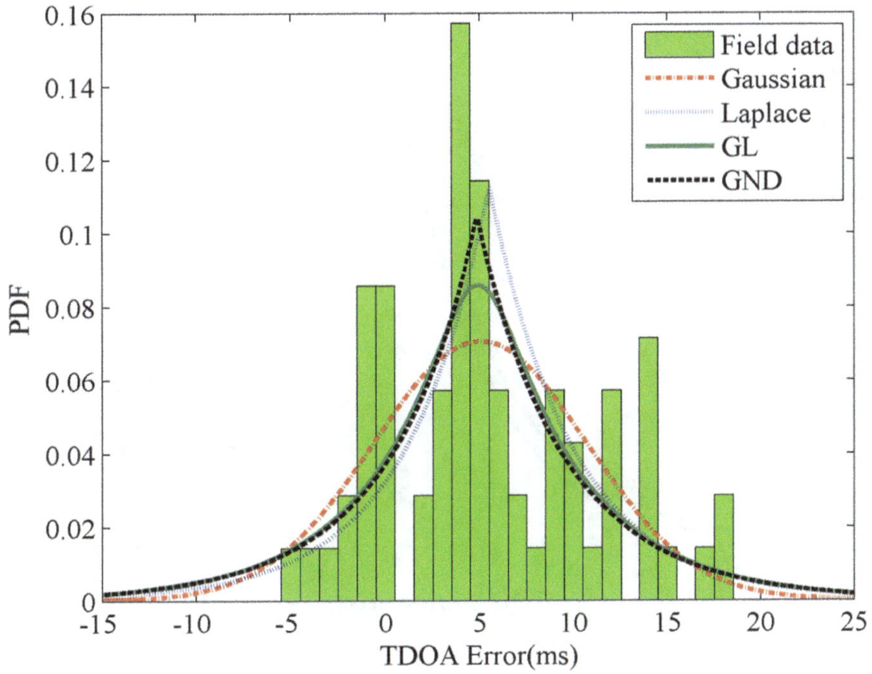

Figure 16. The second example of biased TDOA error PDFs with field measurements

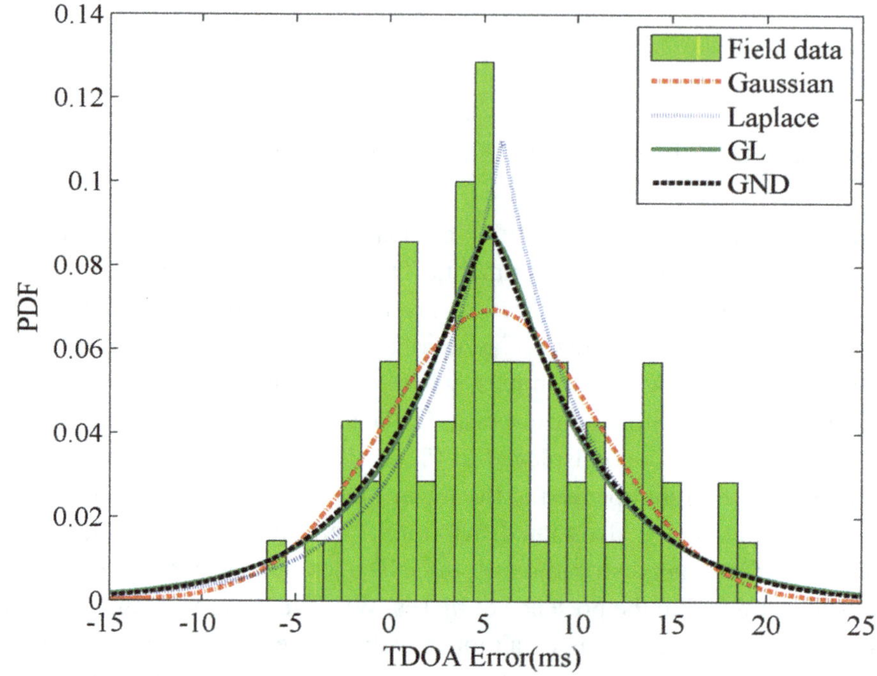

Figure 17. The first example of biased TDOA error CDFs with field measurements

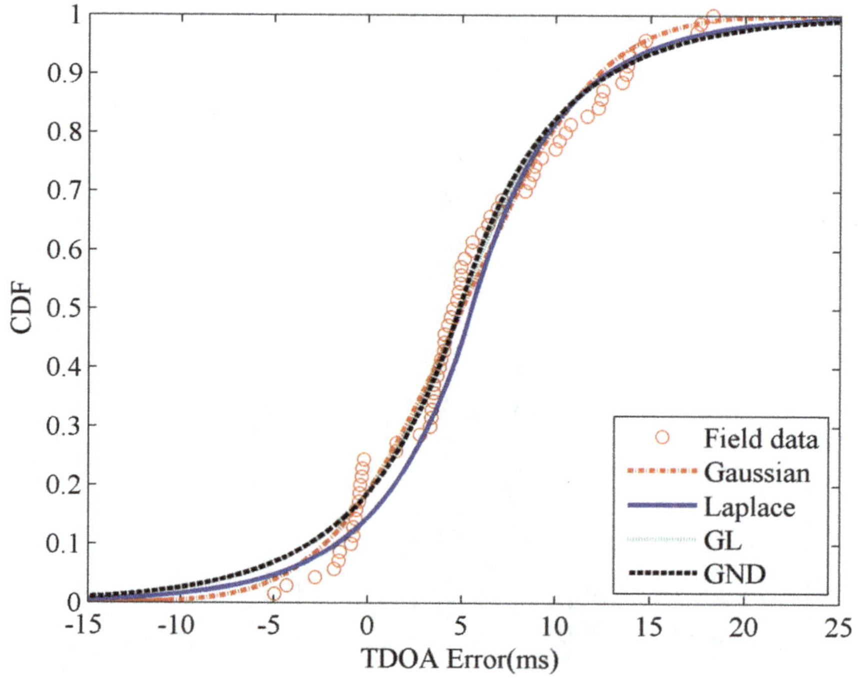

Figure 18. The second example of biased TDOA error CDFs with field measurements

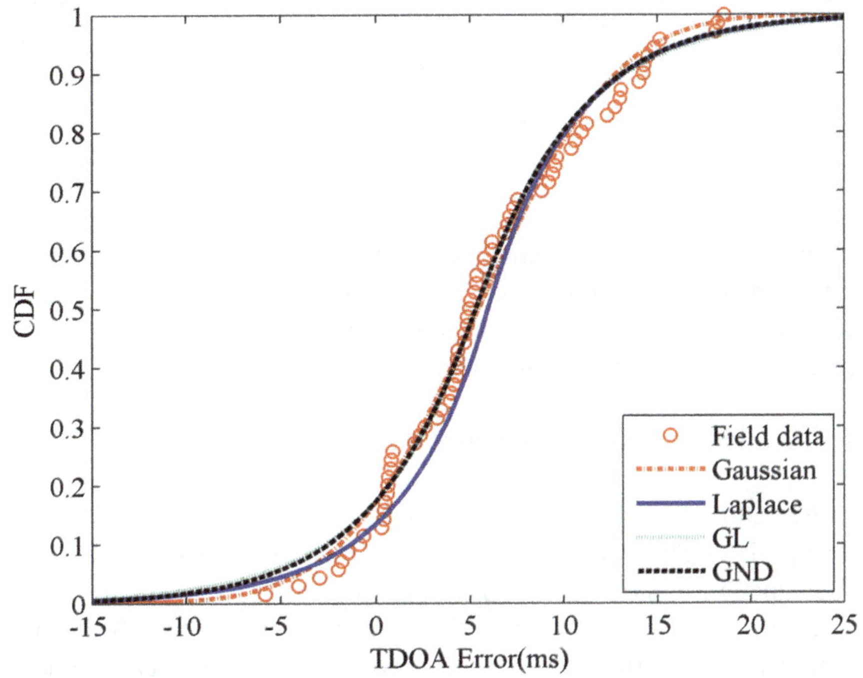

tion. The fitted distribution parameters and the RMSE of CDF fitting are listed in Table 3 and Table 4, respectively. It can be seen that GND has the best match with the smallest RMSE, while the Gaussian distribution has the worst match, but these differences are not statistically significant.

In addition, the shape parameter of GND in Table 3 is close to one for both cases, so that the NLOS and multipath propagation is the dominant source of the TDOA error. That is, it may be appropriate to model the TDOA error simply as a Laplace random variable for scenarios where a far source direction finding or localization is concerned. This simplification would usually not result in a significant degradation in performance, since the Laplace fitting error is very similar to the GND fitting error. Whether a Laplace distribution is suitable to describe the TDOA error mainly depends on the particular environment. In practical applications, it would be useful to conduct field experiments to collect data

Table 3. Fitted TDOA Error Distribution Parameters

	Case a	Case b
Gaussian	μ=5.00, σ=5.65	μ=5.42, σ=5.74
Laplace	μ=5.50, α=4.44	μ=5.92, α=4.51
GL	σ=1.15, μ=4.89, α=4.87	σ=1.02, μ=5.31, α=4.91
GND	β=0.96, μ=4.85, α=4.67	β=1.17, μ=5.29, α=5.90

Table 4. RMSE of Fitted CDF Error

	Case a	Case b
Gaussian	0.0415	0.0348
Laplace	0.0384	0.0334
GL	0.0392	0.0335
GND	0.0383	0.0330

and then perform analysis. In any case, it is an option to use GND to fit the field data to determine the shape parameter. Furthermore, as shown in Table 3, the mean of the TDOA errors is approximately 5.0 ms for both cases, indicating that the systematic bias is significant. Therefore, it is vital to consider the bias when developing the direction finding algorithms.

DOA Estimator for Biased TDOA Measurements

As the TDOA measurements of sound pulses are subjected to a number of sources of measurement errors, the statistical analysis of TDOA errors is important in the overall DOA estimation. This section considers the DOA estimator for biased TDOA measurements.

The main idea considered is to jointly estimate the DOA and the biases to mitigate the bias effects by using the EM method (Cui *et al* 2016). Here the bias is considered as a *hidden* parameter. In statistics, hidden parameters are not directly observed, but instead are inferred through a mathematical model from other directly measured variables. The EM algorithm is an iterative method for finding maximum likelihood or maximum *a posteriori* estimates of parameters in statistical models that depend on unobserved hidden variables. The standard EM algorithm was proposed by Dempster et al. for the maximum likelihood estimation with incomplete data (Dempster *et al* 1977).

Cost Function

Suppose that b_i denotes the TOA bias at the i-th sensor, $i = 0,\dots, N-1$, $\mathbf{b} = \left[b_0, b_1,\dots, b_{N-1}\right]^{\mathrm{T}} \in \mathbb{R}^{N\times 1}$ denotes the TOA bias vector, ξ_i denotes the TDOA bias related to the i-th and the zeroth reference sensor, and $\xi = \left[0, \xi_1,\dots, \xi_{N-1}\right]^{\mathrm{T}} \in \mathbb{R}^{N\times 1}$ denotes the TDOA bias vector with the zeroth sensor as reference. If the TDOA measurement error vector is denoted by $\mathbf{n}$ whose i-th item is n_i for the i-th sensor, then from (5) the corresponding TDOA measurement equation becomes

$$\hat{\tau}_i = \frac{1}{c}\mathbf{s}_i\hat{\mathbf{k}} + \hat{\xi}_i + \mathrm{n}_i \tag{61}$$

If $\mathbf{e}_{\mathrm{N}}$ is the column vector of ones with size N, then the TDOA bias vector $\hat{\xi}$ and the TOA bias vector $\hat{\mathbf{b}}$ have a relationship given by

$$\hat{\xi} = \hat{b}_0 \mathbf{e}_N - \hat{\mathbf{b}} \tag{62}$$

Then (61) can be rewritten in the matrix form as

$$\hat{\tau} = \frac{1}{c}\mathbf{S}\hat{\mathbf{k}} + \hat{b}_0 \mathbf{e}_N - \hat{\mathbf{b}} + \mathbf{n} \tag{63}$$

This is a nonlinear equation in terms of the unknown direction vector and the bias vector, so that there is no closed-form solution to the direction finding problem. The number of unknown parameters is equal to *N+3* (*N* TOA bias variables and three DOA components), while there are only *N* independent original TOA measurement equations. However, it is still possible to solve the underdetermined system problem by developing a minimization or optimization algorithm and using initial parameter estimates which are not too far from the true parameter values. Define the cost function in a quadratic form as

$$f(\hat{\mathbf{k}}) = \left\| \hat{\tau} - \frac{1}{c}\mathbf{S}\hat{\mathbf{k}} - (\hat{b}_0 \mathbf{e}_N - \hat{\mathbf{b}}) \right\|^2 \tag{64}$$

The parameter estimates are obtained by minimizing the cost function $f(\hat{\mathbf{k}})$. If the measurement bias is not considered, the cost function is simplified as

$$f(\hat{\mathbf{k}}) = \left\| \hat{\tau} - \frac{1}{c}\mathbf{S}\hat{\mathbf{k}} \right\|^2 \tag{65}$$

As defined earlier, $\hat{\mathbf{k}}$ is a three-component vector. Given $\hat{\mathbf{k}}$, the estimates of the azimuth and elevation angles can be readily obtained. Alternatively, a two-component parameter vector such as γ may be considered. Although the two-component scheme can directly produce the estimate of the two angles, the cost function is more complex in terms of the unknown angle parameters, resulting in a higher computational complexity. The two-component parameter vector has a smaller estimation error than the three-component parameter vector, but the three-parameter scheme is more suitable in real-time implementation. A modified Gauss-Newton algorithm can be used to perform the nonlinear least squares (NLS) estimation problem with the cost function given by (64) and (65).

Initialization and Convergence

The initial DOA estimate used in the EM-based method should not be too far from the true DOA values. As mentioned earlier, the angle estimates provided by the LLS method can be used as the initial angle parameter values. If the sensor locations, the initial estimate of the azimuth and elevation angles, the estimated speed of sound, and the STD of TDOA errors are all available, then we can calculate the STD of azimuth and elevation estimation error of the LLS method by (38) and (39). The DOA estimation error of the LLS method may be approximated as a Gaussian random variable. Thus based on the Gaussian PDF, the probability of an estimate deviating from the mean by more than three STDs is approximately 0.3%. A reasonable initial angle estimate of EM-based method can be randomly chosen within the ranges $\left[-3\sigma_{\hat{\phi}_0}+\hat{\phi}_0, 3\sigma_{\hat{\phi}_0}+\hat{\phi}_0\right]$ and $\left[-3\sigma_{\hat{\theta}_0}+\hat{\theta}_0, 3\sigma_{\hat{\theta}_0}+\hat{\theta}_0\right]$, where $\hat{\phi}_0$ and $\hat{\theta}_0$ are the DOA estimates by the LLS method.

One of the popular methods in nonlinear minimization is the modified Gauss-Newton algorithm which is suited especially when the cost function is the sum of squares of nonlinear functions. Nevertheless, similar to other nonlinear minimization algorithms, global convergence cannot always be guaranteed. However, this limitation can be overcome if a set of different initial parameter vectors are randomly selected within the azimuth and elevation ranges. The minimization process is performed with each of the initial parameter vectors, and if convergence occurs it is added to a list of potential solutions. The DOA estimate producing the minimum cost function is taken as the final direction estimate. The number of initial parameter vectors may be determined through simulation. In the simulation to be discussed later, the number is set to be ten.

In the case where the TDOA measurement errors are large, the finally computed cost function might still be large or the DOA estimate may not be reasonable, such as the elevation exceeding the maximum range $[0°, 180°]$. This case typically would occur with a tiny probability. To avoid the abnormal cases, a threshold is used to determine whether the cost function should be accepted. If it is smaller than the threshold, the corresponding DOA estimate with the minimum cost function is accepted as the final estimate; otherwise, the initial DOA estimate by the LLS method is treated as the final output.

DOA Estimation Procedure

One of the prominent properties of the EM-based estimator is that the TDOA bias parameters and hence the associated TOA bias parameters are considered as unknown variables whose vector value is estimated as part of the process. If the estimate $\hat{\mathbf{k}}$ is available, then the LLS solution of the TOA bias vector is given by

$$\hat{\mathbf{b}} = (\mathbf{A}^{\mathrm{T}}\mathbf{A})^{-1}\mathbf{A}^{\mathrm{T}}(\hat{\boldsymbol{\tau}} - \frac{1}{\hat{c}}\mathbf{S}\hat{\mathbf{k}}) \tag{66}$$

where $\mathbf{A} \in \mathbb{R}^{N\times N}$ is a coefficient matrix whose item in the k-th row and the j-the column $(j = 0,\ldots,N-1;\ k = 0,\ldots,N-1)$ is defined as

$$a(k,j) = \begin{cases} 1 & j = 0 \; and \; k \neq j \\ -1 & k = j \; and \; k \neq 0 \\ 0 & other \end{cases} \tag{67}$$

The above procedure to determine TOA biases is the expectation step (E-step). The currently estimated bias and TDOA measurements are used as the known parameters to calculate the cost function. The modified Gauss-Newton algorithm is used to perform the maximization step (M-step), providing a new DOA estimate. E-step and M-step are alternated until that the DOA estimate variation is sufficiently small.

To summarize, the operation of the EM-based estimator is described in Table 5. In particular, in Step 8, the E-step is performed to estimate the biases, while the M-step is carried out from Step 9 to Step 11. If Step 8 is omitted and the measurement bias is not considered, then the EM procedure in Table 5 becomes the simplex NLS optimization procedure, which ignores the measurement bias.

One issue related to the EM-based estimators is whether it is necessary to determine if there are biases in the TDOA measurements. In the absence of measurement bias, the simplex NLS methods might be preferable. However, when performing direction finding of a far field source in an outdoor environment, the measurements would usually be corrupted by multipath propagation and perhaps NLOS propagation. Moreover, time synchronization errors typically exist among different receiving ports of the signal-collecting instrument. Therefore, typically the raw EM-based estimators can be directly used to estimate DOA as long as an outdoor environment and a far field source are involved especially when the range-to-baseline ratio is very high.

Simulation for Biased TDOA Measurements

This section performs a computer simulation of the DOA estimation problem for biased TDOA measurements based on the theory and concepts described in previous sections. The performance evaluation metrics are again the RMSE of azimuth and elevation angle estimation and CDF of the angle estimation errors. The angle error is simply the difference between the estimated and the true source bearing. Simulation parameters are selected as follows.

Table 5. Expectation maximum direction finding procedure.

Input: $\mathbf{S}$, $\hat{\tau}$, $\hat{T}_0$, $\hat{V}_W$, $\hat{\phi}_W$, InitN: initial DOA number, iteration precision, cost function threshold. Output: $\hat{\gamma} = \left[\hat{\phi}, \hat{\theta}\right]^T$
1: Apply the LLS estimator to estimate initial DOA as $\hat{\gamma}_0$ 2: Calculate $\hat{c}$ using (44) 3: Calculate cost function using (64), and average it as STD of TDOA error 4: Calculate $\sigma_{\hat{\phi}_0}$ and $\sigma_{\hat{\theta}_0}$ using (38) and (39) 5: Determine the initial bearing range using $\sigma_{\hat{\phi}_0}$, $\sigma_{\hat{\theta}_0}$ and $\hat{\gamma}_0$ 6: for t:= 1 to InitN do 7: Randomly select a set of estimates within the initial bearing range 8: Estimate a new speed of sound and calculate the TOA biases by using (66) 9: Calculate current cost function, its gradient ∇f and the increment of DOA estimate $\Delta\hat{\gamma}$ 10: Set modification coefficient $k_1 = 1$, $k_2 = 10^{-5}$, if $f(\hat{\gamma} + k_1\Delta\hat{\gamma}) > f(\hat{\gamma}) + 2k_1k_2(\Delta\hat{\gamma})^T\nabla f$, then $k_1 \leftarrow k_1 / 2$, and return this step; otherwise, $\hat{\gamma} \leftarrow \hat{\gamma} + k_1\Delta\hat{\gamma}$ 11: If iteration precision does not satisfy termination qualification, go to Step 8 12: If $\hat{\gamma}$ is not rational, the cost function is assigned as a large number 13: end for 14: If the cost function of each initialization exceeds threshold, take the estimate from Step 1 as output; otherwise, select the DOA estimate corresponding to the minimum cost function as output

1. Each simulation is conducted over 10,000 Monte Carlo computer program experiments. Iteration precision is 10^{-5}; and cost function threshold is 10^{4}.
2. The asymmetric array size is set as eight, which is in accordance with the actual field hardware setup. Sensor locations are identical to the data in the simulation section of the evolutionary estimators described above.
3. The sound source location is (2000 m, 2000 m, 2000 m) whose distance to array center is denoted by r. Herein we define SNR as $SNR = 10\log\left(\frac{1}{2r^2\sigma_\tau^2}\right)$ where σ_τ represents the TDOA STD. SNR ranges from –35 dB to –15 dB because this SNR range can provide a rational and practical estimation result of source bearing. For the given SNR range and the distance between the source and array, TDOA STD is determined. TDOA random error has a Laplace distribution and TDOA error bias is proportional to the TDOA error STD with a scaling factor randomly selected in the range [–5, 5].
4. Ten initial DOA parameter vectors are randomly selected and each is used in turn to start the search process and potentially generate a solution. The weather parameters are fixed as follows: air temperature is $20\,^{\circ}\mathrm{C}$, wind speed is 10 m/s, and wind direction is $\pi / 3$. In an outdoor environment, weather parameters are usually time-varying. Their rational variation ranges are given as follows: air temperature variation is $\pm 5\,^{\circ}\mathrm{C}$, wind speed variation is ±6 m/s, and wind direction variation is $\pm\pi / 8$. The in-situ measurement of each parameter is used as its mean and the actual real time parameter value varies around its mean within the variation range.

Figure 19 and Figure 20 respectively compare the azimuth and elevation performance of the three methods in terms of RMSE, which include the LLS estimator, the simplex NLS estimator, and the EM estimator. It can be seen from Figure 19 and Figure 20 that EM outperforms the other two methods significantly for both azimuth and elevation estimation.

Field Experiment Verification for Biased TDOA Measurements

The same experimental data used for the evaluation of the evolutionary estimators are employed in the following analysis. The individual DOA estimation errors of the ten field experiments and the RMSE of the different methods are given in Table 6. For comparison the simulated RMSE values related to $SNR = -35dB$ in Figure 19 and Figure 20 are also listed in this table. It can be seen that EM achieves the best performance in azimuth estimation, while the simplex NLS estimator is the best in elevation estimation. Regarding elevation estimation performance comparison between NLS and EM, the experimental results are rather different from the simulation results. But it is observed that EM performs as almost well as NLS in estimating the source elevation angle based on the experimental data.

SUMMARY

This chapter introduced the TDOA-based direction finding techniques for a single far-field acoustic source using a multi-sensor receiving array. A general model was developed for the TDOA-based direction finding problem and the analytic expressions of the CRLB were presented to benchmark the estimation

Figure 19. RMSE of azimuth estimation

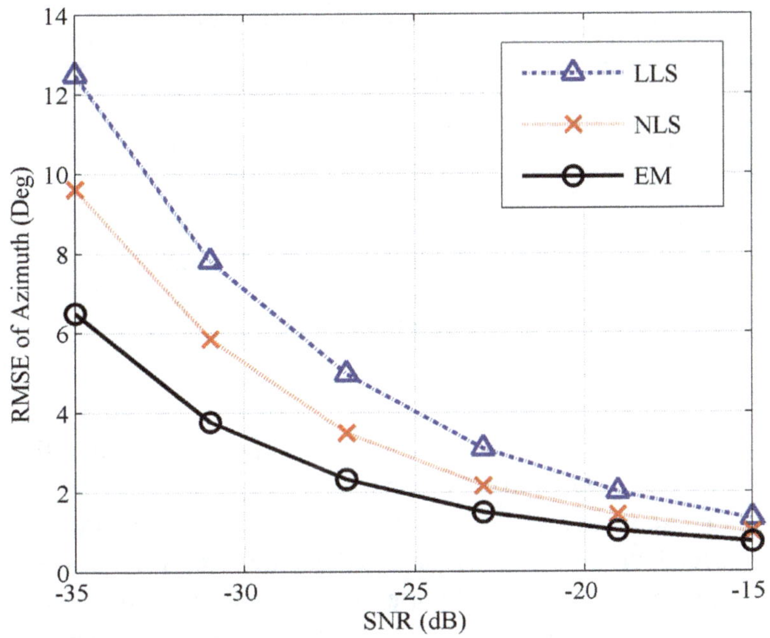

Figure 20. RMSE of elevation estimation

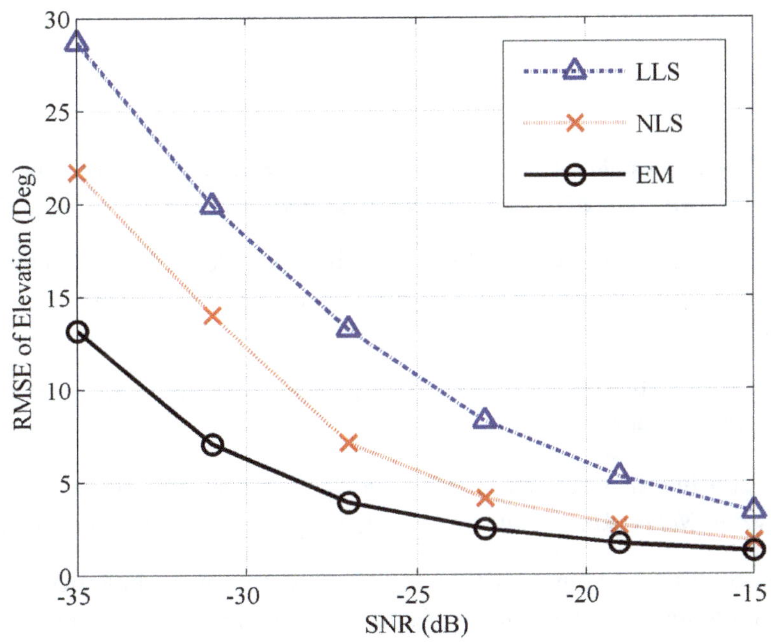

Table 6. DOA estimation results of measured data

	Methods	Experimental DOA estimation error (°)										Experimental	Simulation
		1	2	3	4	5	6	7	8	9	10	RMSE (°)	RMSE (°)
Azimuth	EM	0.12	1.93	2.83	-1.46	-2.49	-2.44	-2.74	3.11	-1.63	-1.12	2.40	6.49
	NLS	-0.27	2.35	2.90	-2.70	-4.01	-4.27	-4.16	3.39	-2.35	-1.64	3.19	9.60
	LLS	-4.88	-0.71	-2.94	-8.61	-10.78	-10.40	-10.05	-0.52	-6.41	-6.61	7.30	12.47
Elevation	EM	12.44	9.62	8.54	12.42	13.25	13.35	12.08	10.15	10.50	11.08	10.56	13.17
	NLS	10.18	11.90	8.92	5.15	4.19	2.50	3.67	11.64	6.35	8.02	8.00	21.68
	LLS	-12.83	-3.58	-20.70	-23.38	-27.68	-26.40	-24.39	-8.47	-13.78	-16.54	19.52	28.66

* *Simulation RMSE results come from Fig. 4 when SNR equals -35 dB*

errors of the azimuth and elevation angle when using a 3-D array. Four direction finding methods were described in detail for different application conditions.

First, the algebraic expression was derived to estimate the bearing angle for the four-microphone array with a fixed structure. Second, the DOA estimation and theoretical variances of the LLS estimator were provided for any 1-D, 2-D, and 3-D array configurations. Furthermore, two evolutionary computation methods were presented, which are GA and PSO algorithm. Their performance was evaluated through simulation and field experimental results using a 3-D acoustic sensor array. Finally, when sensor measurements contain non-negligible consistent systematic biases as observed in experimental data, the accuracies of three different algorithms (EM, NLS and LLS) were examined and compared using both simulation and experimental data. On average, the EM algorithm achieved the best performance, while the NLS algorithm had the highest accuracy in elevation estimation with experimental data.

REFERENCES

Baruch, B., Doron, M. A., Rosenhouse, J., & Azhari, H. (1999). On direction finding of an emitting source from time delays. *The Journal of the Acoustical Society of America*, *106*(6), 3355–3363.

Brian, G. F., Lionel, G. C., & Kam, W. L. (2002). Locating far-field impulsive sound sources in air by triangulation. *The Journal of the Acoustical Society of America*, *111*(1), 104–116. doi:10.1121/1.1402618 PMID:11831786

Chan, Y., & Ho, K. (1994). A simple and efficient estimator for hyperbolic location. *IEEE Transactions on Signal Processing*, *42*(8), 1905–1915. doi:10.1109/78.301830

Chen, P. C. (1999). A non-line-of-sight error mitigation algorithm in location estimation. In *Proc. Wireless Communications and Networking Conference* (vol. 1, pp. 316-320). Piscataway, NJ: IEEE Press.

Cui, X., Yu, K., & Lu, S. (2015). Evolutionary TDOA-based direction finding methods with 3-D acoustic array. *IEEE Transactions on Instrumentation and Measurement*, *64*(9), 2347–2359. doi:10.1109/TIM.2015.2415051

Cui, X., Yu, K., & Lu, S. (2016). Direction finding for transient acoustic source based on biased TDOA measurement. *IEEE Transactions on Instrumentation and Measurement*, *65*(11), 2442–2453. doi:10.1109/TIM.2016.2583224

Dempster, A. P., Laird, N. M., & Rubin, D. B. (1977). Maximum likelihood from incomplete data via the EM algorithm. *Journal of the Royal Statistical Society. Series B. Methodological*, *39*(1), 1–38.

Holton, J. R. (2004). *An Introduction to Dynamic Meteorology* (4th ed.). Burlington, MD: Elsevier Inc.

James, M. D. (2008). *Multiarray passive acoustic localization and tracking* (Ph.D. dissertation). Dept. Mech. Eng., Virginia Polytechnic Institute and State Univ., Blacksburg, VA.

Nielsen, R. O. (1994). Estimation of azimuth and elevation angles for a plane wave sine wave with a 3-D array. *IEEE Transactions on Signal Processing*, *42*(11), 3274–3276. doi:10.1109/78.330396

Sallai, J., Hedgecock, W., Volgyesi, P., Nadas, A., Balogh, G., & Ledeczi, A. (2011). Weapon classification and shooter localization using distributed multichannel acoustic sensors. *Journal of Systems Architecture*, *57*(10), 869–885. doi:10.1016/j.sysarc.2011.04.003

Vakulya, G., & Simon, G. (2011). Fast adaptive acoustic localization for sensor networks. *IEEE Transactions on Instrumentation and Measurement*, *60*(5), 1820–1829. doi:10.1109/TIM.2011.2108074

Yu, K., Sharp, I., & Guo, Y. (2009). *Ground-Based Wireless Positioning*. Chichester, UK: Wiley Press. doi:10.1002/9780470747988

Chapter 6
Cooperative Localization in Wireless Networks

Yunlong Wang
Tsinghua University, China

Yang Cai
Tsinghua University, China

Yuan Shen
Tsinghua University, China

ABSTRACT

This chapter describes cooperative localization in wireless networks, where mobile nodes with unknown positions jointly infer their positions through measuring and exchanging information with each other. The technique of cooperation localization, efficiently even in harsh propagation environment, enables amounts of location-based services that rely on high-accuracy position information of mobile nodes. After a brief introduction of cooperative localization, the Cramer-Rao lower bound is given as a standard metric for performance. Then the information in the temporal and spatial domain is illustrated with geometrical interpretations. Two classes of cooperative localization algorithms, namely, centralized and distributed algorithms, are presented to show the implementation of the cooperative localization in a wireless network. Then the performance of cooperative localization under non-line-of-sight condition is analyzed. Lastly, numerical results are given to illustrate the performance of cooperative localization algorithms.

INTRODUCTION

The availability of absolute or relative position information is essential in many applications, such as localization services in cellular networks, search-and-rescue operations, asset tracking, blue force tracking, traffic monitoring and autonomous vehicles (Win et al., 2011). Automatic localization of the nodes in a wireless network is a key enabling technology which, mainly because the availability of the node position is highly relevant to the value of the data it collects. With the integration of global positioning system (GPS) into cell phones, in conjunction with other common commercial signals, such as global system

DOI: 10.4018/978-1-5225-3528-7.ch006

for mobile (GSM), WiFi, Bluetooth and LED, the ubiquitous location-aware services with sub-meter accuracy are envisioned in the near future, bringing new opportunities for many wireless applications in commercial, public safety, and military sectors. The wide range of potential applications will also motivate an increasing research interest in localization and tracking technologies for wireless networks.

Location-aware networks generally consist of two kinds of nodes: anchor nodes and agent nodes, where anchors have known positions while agents have unknown positions. Two common examples of anchor-based techniques include the GPS and beacon localization. In GPS, an agent determines its location based on the signals received from a constellation of GPS satellites. However, GPS does not operate well in harsh propagation environments, such as indoors or in urban canyons, since the signals suffer from severe attenuation when across obstacles and non-line-of-sight (NLOS) signals will degrade the localization accuracy severely. Beacon localization, on the other hand, relies on terrestrial anchors, such as WiFi access points or GSM base stations. In areas where network coverage is sparse, e.g., in emergency situations, localization errors can be unacceptably large. Therefore, there is an urgent need for localization systems that can achieve high accuracy in harsh propagation environments with limited infrastructure requirements.

Figure 1 illustrates an example of comparison between traditional and cooperative localization/navigation in a simple network with two mobile agents and four static anchors. Conventionally, each agent determines its position based only on the range measurements to at least three anchors, e.g., Figure 1-(a), which is also known as triangulation. These techniques often fail to localize an agent (e.g., Agent 1) when there are insufficient neighboring anchors within the communication range, or fail to provide satisfactory localization accuracy (e.g., Agent 2) when the range measurements are subject to large errors due to NLOS propagation. However, when the agents cooperate in both spatial and temporal domains, they are able to not only localize themselves with enough line-of-sight (LOS) signals, but also significantly improve the localization accuracy with more valid constraints e.g., Figure 1-(b).

If the agents with unknown locations can make measurements among themselves, such as relative ranges and relative angles, the additional information gained from these measurements between pairs of agents can enhance the accuracy and robustness of the localization system. We call this type of coopera-

Figure 1. A network with two mobile agents (blue rings) and four static anchors (red rings) in two time steps, in which light blue and dark blue rings denote the agents in the previous and current time step. Inter- and intra-node measurements are denoted by red and black arrows, respectively. Anchor 4 provides NLOS signals due to harsh propagation conditions while other three anchors provide LOS signals. (a) Before spatiotemporal cooperation. (b) After spatiotemporal cooperation.

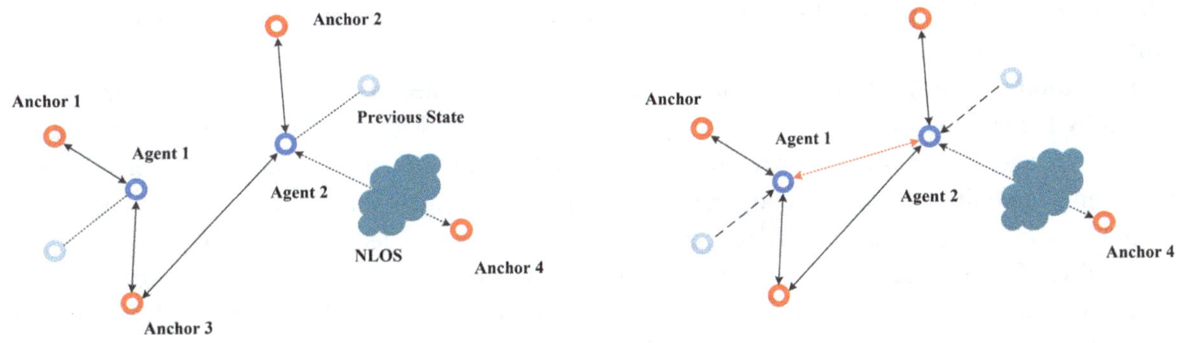

tion as spatial cooperation since the relative information in the space is utilized (Patwari et al., 2005). Moreover, if an agent can also incorporate its previous states and then the performance can be further improved. This type of cooperation can be viewed as temporal cooperation since the nodes' states at different time instants are utilized. The combination of these two types of cooperation in the wireless network enables the joint spatial and temporal cooperation, leading to a new paradigm for real-time localization and navigation in wireless networks, which is referred to as network localization (Win et al., 2011). The joint spatial and temporal cooperation among network also improve the robustness of localization in harsh propagation environments by adding more information.

This chapter will introduce the cooperative localization techniques in wireless networks. After a brief review of both theory and algorithm of cooperative localization, the cooperative information in spatial and temporal domain is given with geometric interpretations. Then two classes of cooperative localization algorithms, namely, centralized and distributed algorithms are presented to show the implementation of the cooperative localization in a wireless network. Finally, numerical results are conducted to illustrate the performance of cooperative localization algorithms.

BACKGROUND

Although the GPS can provide localization accuracy on the order of meters in open outdoor environment, which has been widely used in practice, it may fail in several harsh propagation environments, such as in building and urban canyons (Shen, Wymeersch & Win, 2010; Wang & Wu, 2017; Cheng, Wang & Shen, 2017). The blocking and NLOS signal will degrade the localization accuracy severely. Therefore, the terrestrial localization system is important in these GPS-challenged or even GPS-denied areas. Generally, these systems are based on cellular networks, radio frequency identification (RFID) networks, wireless local area networks (WLANs) and wireless networks and each of them has unique infrastructure requirements. In indoor environments, the cellular networks experience signal attenuation and cannot be used for localization in practice. WLANs typically suffer from the problems of obstruction, scattering and reflection from obstacles, furniture and people. Although RFIDs can provide the desired accuracy, it requires high node density to guarantee the stability. Ultra wideband (UWB) technology offers high ranging accuracy in harsh propagation environments due to its ability to resolve multipath and penetrate obstacles (Gezici et al., 2005; Dardari et al., 2009). It is envisioned that UWB-based localization will play an important role in future high-definition situation-aware and RFID networks. In particular, the IEEE 802.15.4a standard is the first to contemplate both communication and localization with high levels of availability and sub-meter accuracy. Moreover, since generally a wireless network is off-battery or difficult to recharge, the problem of energy consumption and allocation should be considered in network localization (Dai, Shen & Win, 2017).

Cooperation among agents for localization has attracted great interests due to its superior performance obtained by the sharing of information among spatial neighbors. Moreover, enabling the communications between agents circumvents the requirements of high-power transmission and, high-density anchor deployment. In addition to spatial cooperation with neighbors, agents in the network can also benefit from the information obtained in different time instants, which is called temporal cooperation and can be utilized via filtering techniques in a process commonly known as tracking (Foxlin, 2005).

The errors sources for cooperative localization and the Cramér-Rao bound (CRB) analysis were first presented in (Patwari et al., 2005). In a view of graph theory, the grounded graphs were constructed in

(Aspnes et al., 2006) and applied graph rigidity theory to test the conditions for unique localizability and to construct uniquely localizable networks. Then by studying the computational complexity of network localization, a subclass of grounded graphs was investigated to compute the localization efficiently. The fundamental limits of localization accuracy for wideband wireless cooperative networks in terms of a performance measure named square position error bound was derived in (Shen, Wymeersch & Win, 2010). It was shown that the network localization information can be decomposed into building blocks associated with every pair of the nodes, called the ranging information. Moreover, the contribution of the a priori knowledge of the channel parameters and the agent's positions were discussed and showed that when an infinite a priori knowledge is available, agents could be treated as anchors. By using the eigen-decomposition, both of the non-cooperative and cooperative localization were further presented with geometric interpretation, showing the enhancement of cooperation on the information ellipse. For a dynamic network, the problem of network navigation was analyzed in (Shen, 2014) and the corresponding theoretical framework was established. It is shown that the cooperation information in a dynamic network includes: spatial cooperation and temporal cooperation (Win et al., 2011). The former characterizes the navigation information corresponding to the cooperation via inter-node measurements and the later illustrates the cooperation via intra-node measurements.

Although the relative information between nodes was utilized earlier, especially in the field of robotic control and monitoring, the cooperative localization in wireless network was first mentioned in (Savarese, Rabaey & Beutel, 2001) and proposed in (Patwari et al., 2005). Based on different processing procedures, the cooperative localization algorithms can be categorized into two sets: centralized algorithms and distributed algorithms (Patwari et al., 2005). The former requires a central processor to collect all the information of the network before calculating the agents' positions, while the distributed algorithms only allow nodes to share information only with their neighbors.

The centralized methods are generally based on the non-Bayesian estimators. Least square (LS) and maximum likelihood (ML) are the two common criteria for implementing algorithms. A centralized maximum likelihood estimator (MLE) is proposed in (Patwari et al., 2003) to solve the relative location estimation in wireless networks. However, the algorithm involves a multivariable nonlinear optimization problem that makes it difficult to implement in practice. To alleviate this problem, the ML estimation was being relaxed to a centralized semi-definite programming (SDP) problem (Biswas et al., 2006) and a distributed second-order cone programming problem, respectively. With semidefinite relaxation (SDR), the network localization problem can be converted into a convex optimization program (Boyd & Vandenberghe, 2004), which can be solved efficiently. Another popular centralized algorithm is multidimensional scaling (MDS), which utilizes pairwise distances available from the received signal strength or time-of-arrival measurements (Leeuw, 1988; Shang et al., 2004). While the MDS cost function is non-convex, it can still be solved approximately (Ramsay, 1982; Zinnes, 1983). A generalized MDS method was proposed in (Chan, 2009) which utilizing the matrix transformation and subspace decomposition. It showed by replacing the transformation matrix, the classical MDS is obtained. Based on the connectivity of the network, MDS algorithms can also be developed to a distributed manner with partially connected network (Chan & So, 2009).

Distributed cooperative localization algorithms generally belong to the Bayesian inference. In (Caceres, Sottile & Garello, 2010; Sottile et al., 2011), Bayesian estimators are utilized to localize the agents by treating the nodes' positions as random variables. The filtering techniques, such as the Kalman filter (KF), extended Kalman filter (EKF) and unscented Kalman filter (UKF) are proposed for cooperative localization in (Caceres, 2010). Both the EKF and UKF interpret the effects of neighbors' position uncertainties similarly to the range measurement errors, which lead to performance losses. A particle filter based cooperative localization is proposed in (Sottile, 2011) for hybrid GPS-terrestrial environments, which is suitable for the nonlinear systems. However, it often suffers from a large computational complexity.

A class of novel methods based on the graph theory, which is referred to as message passing algorithms, were studied for the cooperative localization (Ihler et al., 2005; Wymeersch, Lien & Win, 2009; Savic & Zazo, 2012). Particularly, a nonparametric belief propagation (NBP) method was proposed for self-localization in wireless networks (Ihler et al., 2005), which approximated the messages by particles with the corresponding weights. However, the large numbers of particles lead to a huge computational complexity and communication overhead. To reduce the communication overhead in NBP, a method using the approximations on the position beliefs by Gaussian mixtures and information censoring was proposed in (Savic & Zazo, 2012). A cooperative localization method based on the sum-product algorithm and factor graph was proposed in (Wymeersch, Lien & Win, 2009), which was known as the sum-product algorithm over a wireless network (SPAWN) algorithm. Unfortunately, the direct implementation of SPAWN still experiences the computational complexity and further approximation is necessary. In the literature of (Lien et al., 2012), the parametric and sample-based message passing algorithms of cooperative localization were compared. It indicated that parametric message representation combined with simple censoring can yield excellent performance at a relatively low complexity.

In practical applications, the performance of network localization mainly determined by the following two factors: i) the geometric configuration of the system (i.e., the deployment of the anchors and agents), and ii) the quality of the measurements (waveform and ranges). Therefore, the design of such location-aware networks requires a clear understanding of these factors and their influences. For example, the obstacles may block partial or complete LOS, which leads to positively biased range estimates for time based ranging techniques. Discarding or refining unreliable range measurements can improve the localization performance. Particularly, regardless of the specific range-based localization algorithm used, the following three-step procedure can be adopted: (i) calculate the preliminary position estimate $\hat{\mathbf{x}}^{(1)}$ from initial range estimates based on the measurements; (ii) correct range estimates based on $\hat{\mathbf{x}}^{(1)}$ by removing biases according to bias models; and (iii) estimate a refined position $\hat{\mathbf{x}}^{(2)}$ with the correct values.

In harsh propagation environments with the presence of obstacles, the NLOS biases will result in large localization errors if the a priori knowledge of NLOS distribution is unavailable (Shen, Wymeersch & Win, 2010). An effective way to alleviate this problem is to identify the NLOS signals and then discard the corresponding measurements (if there are sufficient LOS paths) or reweight the contributions of different signals in the localization process (Marano et al., 2010; Wang & Zekavat, 2012). Amounts of techniques were proposed to identify NLOS conditions and mitigate their influence. The identification of signal obstruction is mainly accomplished by performing a likelihood ratio test on binary hypotheses. This requires the extraction of features that are significantly affected by the propagation conditions from the received waveform. For example, the NLOS identification technique in (Güvenc et al., 2008) is based on signal features such as root mean square delay-spread, Kurtosis, and mean excess delay. Other approaches include regression with support vector machines and Gaussian processes. The characterization of each possible link, by means of ranging and waveform measurements, enables the system designer to understand how to harness environmental knowledge, identify LOS and NLOS conditions, take advantage of cooperation, and choose cooperating nodes.

However, the paths may be identified incorrectly: i) the LOS signals are misidentified as NLOS (false alarm) and ii) NLOS signals are misidentified as LOS (missed detection) and both of them can significantly impact the localization accuracy (Wang, So & Li, 2016). To cope with the NLOS condition, an alternative way is to measure the propagation statistics of the channel information and then localize the agent from a scattering model (Lv et al., 2015). The difficulty is to obtain an accurate model, and the model may change in a time varying environment which requires high computational resources. Some previous works were attempting to localize agents with all NLOS and LOS measurements, and provided weights to minimize the effects of the NLOS contributions (Abu, Zhou & Abhayapala, 2016). The map information can be also utilized for improving the localization accuracy in the indoor environment (Wang et al., 2017). Unlike non-cooperative localization, the NLOS mitigation is more complex in cooperative localization since the communication range and link between agents should be considered.

System Model and Localization Methods

The network is based on intra- and inter-node measurements and the types of measurements available depend on the technologies utilized. The intra-node measurements include acceleration, angular velocity, Doppler shift and orientation etc., while inter-node measurements are waveforms, ranges and directions. For example, an inertial measurement unit (IMU) can support measures of distance and direction of the node itself while an agent with transceivers can infer the distance and direction to its neighbors based on signal metrics extracted from exchanged radio frequency waveforms. The implementation of localization methods generally includes non-Bayesian and Bayesian methods. The former treats the unknown parameters as deterministic variables and the common estimators in non-Bayesian methods are based on LS and ML. When the unknowns are considered to have a certain distribution, Bayesian methods are suitable for these cases and the common estimators are generally based on minimum mean square error (MMSE) or maximum *a posteriori* (MAP).

Network Model

The network considered in this chapter is with N anchors and M agents, where each of the agents is equipped with multiple sensors that can perform intra- and inter-node measurements for the purpose of localization and navigation. The position of agents and anchors are denoted by $\mathbf{X} = [\mathbf{x}_1, \mathbf{x}_2, \ldots, \mathbf{x}_M]^{\mathrm{T}}$ and $\mathbf{X}_{\mathrm{b}} = [\mathbf{x}_{M+1}, \mathbf{x}_{M+2}, \ldots, \mathbf{x}_{M+N}]^{\mathrm{T}}$. Using these intra- and inter-node measurements, represented by $\mathbf{z} = [\mathbf{z}_{\mathrm{s}}, \mathbf{z}_{\mathrm{r}}]$, the agents can infer their positions $\mathbf{X}$. Let $\mathcal{M} = \{1, 2, \ldots, M\}$ and $\mathcal{N} = \{M+1, M+2, \ldots, M+N\}$ be the sets of agents and anchors, respectively. The distance between node pairs (anchor-anchor, anchor-agent and agent-agent) is denoted by $\{d_{i,j}\}$, $i, j \in \mathcal{M} \cup \mathcal{N} \times \mathcal{M} \cup \mathcal{N}$. Since the nodes are with the constraint of communication range, we use the set $\mathcal{N}(i)$ to denote the nodes set that node *i* can make measurements with i $\notin \mathcal{N}(i)$ and $\mathcal{N}(i) \subset \{1, \ldots, M+N\}$.

Signal Metrics

The classification of network localization can be categorized by the type of measurement metrics, such as range-based, direction-based and proximity-based. Among these metrics, the range-based methods are more suitable due to its high accuracy and lower complexity. Moreover, two common techniques are used to obtain range measurements: delay-based and received signal strength (RSS)-based systems.

In a delay-based system, the range can be measured from the time-of-arrival (TOA) which is the time that a transmitted signal first arrives at a receiver in LOS environments. Suppose the transmitter and the receiver are synchronized, the receiver can determine the TOA, which is the time of transmission plus the time delay of propagation. The propagation induced time-delay between node i and node j is denoted by $\tau_{i,j}$, is equal to the distance $d_{i,j}$, divided by the velocity of propagation v. The propagation velocity for RF signals is the speed of light. The process of time delay can be accomplished by one-way TOA, two-way TOA or time difference of arrival (TDOA). The factors that influence the time delay estimation are mainly noises, multipath effects, obstacles, interferences, and clock drifts. Generally, in a scenario with the dense multipath channel, the first path is often not the strongest which makes the TOA estimation challenging. The TDOA metric can be used to reduce the clock bias or in a non-cooperative localization scenario where the TOA cannot be measured directly. The method of ML-based time delay estimation can achieve the CRB at high signal-to-noise ratio (SNR) regimes, but the implementation of ML estimators usually requires sampling at the Nyquist sampling rate or higher, which is a challenging task for wideband and UWB signals.

The estimation accuracy of the propagation time at the receiver determines the localization performance directly in the time based approaches. It is mainly affected by additive noises and the multipath effects. Typically, the time delay can be estimated from the highest peak of cross-correlation coefficients, but in the multipath channel, the first-arriving peak should be used for the delay estimation because the LOS signal may not be the strongest of the arriving signals. It is shown in (Gezici et al., 2005) that the accuracy of TOA of first LOS signal estimation satisfies

$$\mathrm{var}(\hat{\tau}) \geq \frac{1}{8\pi^2\beta^2 \cdot \mathrm{SNR}} \tag{1}$$

where SNR is the signal to noise ratio, and β is the effective signal bandwidth defined by

$$\beta^2 = \frac{\int_{-\infty}^{\infty} f^2 \left|S(f)\right|^2 \mathrm{d}f}{\int_{-\infty}^{\infty} \left|S(f)\right|^2 \mathrm{d}f}. \tag{2}$$

By increasing the SNR and the effective signal bandwidth, the accuracy of TOA can be improved. For example, the time based techniques can achieve extremely accurate ranging estimates using UWB radios (Cassioli, Win & Molisch, 2002), which can transmit and receive signals with a very large bandwidth. Compared to the additive noise, the range error caused by the multipath effect is much larger. Since there is no guarantee that the LOS signal will be the strongest among all the arrival signals, the TOA is usually estimated by finding the first-arriving peak of the cross correlation. The higher temporal resolution can be achieved via increasing the signal bandwidth, since the peak width is inversely proportional to the signal bandwidth.

The distance is measured by the strength of the received signal, which follows the physical law that the signal experiences path loss due to the propagation in the medium. The measure of RSS can be based on theoretical or empirical models and the choice of mode strongly affects the ranging accuracy. A widely used mode is given by $P_r(d) = P_0 - 10n_p \log_{10}(d) + S$, where $P_r(d)$ is the received power (dBm), P_0 is the receiver power (dBm) at 1m, the distance from the transmitter and the receiver is denoted by d, n_p is the path loss exponent and S is the large-scale fading (shadowing) which is commonly modeled as a Gaussian random variable with zero mean and standard deviation σ_S. RSS is a signal metric that is easy to measure and generally cannot provide high-accuracy ranging, because it may suffer from the mismatch errors between distance and signal attenuation. In (Dardari et al., 2009), two kinds of TOA estimators based on energy detection and the effects of impairments are analyzed.

The angle-of-arrival (AOA) measurements provide information about the directions from neighboring nodes. There are two common ways to measure the AOA. The first method uses a node with antenna arrays whose locations are known, and estimates the AOA from the phase delays on different elements. The second approach uses two antennas with different directions equipped on the node, and estimates the AOA from the ratio of their individual RSS values.

Graphic Model

Before introducing the cooperative localization, the following assumptions are given for clarity. These mild assumptions include movement model and measurement model which are common in many practical scenarios (Wymeersch, Lien & Win, 2009).

Before we introduce the assumptions, several notations are defined first. The state of agent i at time t is written as $\mathbf{x}_i^{(t)}$ and the state from time 0 to T is denoted by $\mathbf{x}^{(0:T)}$. Denote by $\mathcal{S}_{\to i}^{(t)}$ the set of nodes from which agent i may receive signals during time slot t, and $\mathcal{S}_{i\to}^{(t)}$ the set of agents that may receive a signal from agent i. At time slot t, the agent i may estimate self-metrics $y_{i,\mathrm{s}}^{(t)}$ (e.g., orientation or acceleration measurements), by intra-node measurements and signal metrics $y_{j\to i}^{(t)}$ (e.g., TOA or AOA measurements) by received signals. The collection of all relative signal metrics $\mathbf{y}_\mathrm{r}^{(t)}$ and local self-metrics $\mathbf{y}_\mathrm{s}^{(t)}$ of all agents is denoted by $\mathbf{y}^{(t)}$ at time t.

For the movement model, we assume that the agents move independently according to a memoryless walk:

$$p\left(\mathbf{x}^{(0:T)}\right) = p\left(\mathbf{x}^{(0)}\right)\prod_{t=1}^{T} p\left(\mathbf{x}^{(t)} \mid \mathbf{x}^{(t-1)}\right) = p\left(\mathbf{x}^{(0)}\right)\prod_{t=1}^{T}\left\{\prod_{i=1}^{M} p\left(\mathbf{x}_i^{(t)} \mid \mathbf{x}_i^{(t-1)}\right)\right\}. \tag{3}$$

For the measurement model, we first assume that the intra-node measurements and the inter-node measurements are independent conditioned on the states

$$p\left(\mathbf{y}^{(1:T)} \mid \mathbf{x}^{(0:T)}\right) = p\left(\mathbf{y}_\mathrm{s}^{(1:T)} \mid \mathbf{x}^{(0:T)}\right) p\left(\mathbf{y}_\mathrm{r}^{(1:T)} \mid \mathbf{x}^{(0:T)}\right). \tag{4}$$

The intra-node measurements are mutually independent for different nodes and different time slots, and depend only on the current and previous location of the agent

$$p\left(\mathbf{y}_{\mathrm{s}}^{(1:T)} \mid \mathbf{x}^{(0:T)}\right)=\prod_{t=1}^{T} p\left(\mathbf{y}_{\mathrm{s}}^{(t)} \mid \mathbf{x}^{(t-1)}, \mathbf{x}^{(t)}\right)=\prod_{t=1}^{T}\left\{\prod_{i=1}^{M} p\left(z_{i,\mathrm{s}}^{(t)} \mid \mathbf{x}_{i}^{(t-1)}, \mathbf{x}_{i}^{(t)}\right)\right\}. \tag{5}$$

The inter-node measurements at different time slots are independent, conditioned on the nodes' locations, and each of them depends only on the current locations of the two nodes involved:

$$p\left(\mathbf{z}_{\mathrm{r}}^{(1:T)} \mid \mathbf{x}^{(0:T)}\right)=\prod_{t=1}^{T} p\left(\mathbf{z}_{\mathrm{r}}^{(t)} \mid \mathbf{x}^{(t)}\right)=\prod_{t=1}^{T}\left\{\prod_{i=1}^{M} \prod_{j\in\mathcal{S}_{\to i}^{(t)}} p\left(z_{j\to i}^{(t)} \mid \mathbf{x}_{i}^{(t)}, \mathbf{x}_{j}^{(t)}\right)\right\}. \tag{6}$$

Non-Bayesian Methods

In the non-Bayesian framework, the parameters to be estimated are treated as unknown, but deterministic variables. Two common estimators are LS estimator (Nguyen et al., 2015) and the ML estimator. The LS approach minimizes the unfitness between the true observations and the model induced results, while the ML approach maximizes the likelihood function of the observations given the concerned parameter (Poor, 2013). Both methods find their estimators by minimizing cost functions $\mathrm{C}^{(t)}(\mathbf{X})$ with respect to $\mathbf{X}=[\mathbf{x}_1,\cdots,\mathbf{x}_M]$ at time t.

For LS estimator, the cost function is given by

$$\mathrm{C}_{\mathrm{LS}}^{(t)}(\mathbf{X})=\sum_{i=1}^{M}\sum_{j\in\mathcal{S}_{\to i}^{(t)}}\left[y_{j\to i}^{(t)}-m(\mathbf{x}_i,\mathbf{x}_j)\right]^2 \tag{7}$$

where $m(\mathbf{x}_i,\mathbf{x}_j)$ is a suitable function based on the signal metrics. For example, in the case of ranging measurements, the function is the Euclid distance between the two nodes, which is $m(\mathbf{x}_i,\mathbf{x}_j)=\|\mathbf{x}_i-\mathbf{x}_j\|$.

For ML estimator, the cost function becomes

$$\mathrm{C}_{\mathrm{ML}}^{(t)}(\mathbf{X})=-\log p(\mathbf{y}_{\mathrm{r}}^{(t)}\mid\mathbf{X})=-\sum_{i=1}^{M}\sum_{j\in\mathcal{S}_{\to i}^{(t)}}\log p(y_{j\to i}^{(t)}\mid\mathbf{x}_i,\mathbf{x}_j) \tag{8}$$

where p is determined by the statistical model of the measurements. A common model is additive Gaussian noise $\mathcal{N}(0,\sigma^2)$, and in this case

$$p(y_{j\to i}^{(t)}\mid\mathbf{x}_i,\mathbf{x}_j)=\frac{1}{\sqrt{2\pi}\sigma}\cdot\exp\left\{-\frac{[y_{j\to i}^{(t)}-m(\mathbf{x}_i,\mathbf{x}_j)]^2}{2\sigma^2}\right\}. \tag{9}$$

Note that the cost function can be summarized as $\mathrm{C}^{(t)}(\mathbf{X})=\sum_{i=1}^{M}\sum_{j\in\mathcal{S}_{\to i}^{(t)}}c_{j\to i}(y_{j\to i}^{(t)};\mathbf{x}_i,\mathbf{x}_j)$ for both LS estimator and the ML estimator aforementioned. To minimize this cost function, the gradient of the function should be set as a zero vector. This is equivalent to setting its partial derivatives with respect to all $\mathbf{x}_i$ $(i = 1, 2, \cdots, M)$ as zeros simultaneously. Specially, the derivative with respect to $\mathbf{x}_i$ is given by

$$\frac{\partial C^{(t)}(\mathbf{X})}{\partial \mathbf{x}_i} = \sum_{j \in \mathcal{S}_{\to i}^{(t)}} \frac{\partial c_{j\to i}(y_{j\to i}^{(t)}; \mathbf{x}_i, \mathbf{x}_j)}{\partial \mathbf{x}_i} + \sum_{k \in \mathcal{S}_{i\to}^{(t)}} \frac{\partial c_{i\to k}(y_{i\to k}^{(t)}; \mathbf{x}_k, \mathbf{x}_i)}{\partial \mathbf{x}_i} = 0. \tag{10}$$

A closed-form solution for the above problem has not been found yet. In practice, the gradient decent based algorithms are usually applied to iteratively minimize the cost function $C^{(t)}(\mathbf{x})$.

Bayesian Methods

In the Bayesian framework, the parameter $\mathbf{x}$ is treated as a realization of a random variable $\mathbf{X}$ with an *a priori* distribution $p_{\mathbf{X}}(\mathbf{x})$. Two common estimators are MMSE estimator and MAP estimator. The MMSE approach minimizes the mean squared error of all possible realizations of the parameter, and the MAP approach finds the peak of the conditional probability of the concerned parameters given the observations (Poor, 1984). The key step of both methods is to find the posterior distribution.

The MMSE estimator has been proved to be the mean of the *a posteriori* distribution

$$\hat{\mathbf{x}}_{\text{MMSE}} = \int \mathbf{x} \cdot p_{\mathbf{X}|\mathbf{Z}}\left(\mathbf{x} \mid \mathbf{y}\right) \mathrm{d}\mathbf{x}. \tag{11}$$

The MAP estimator is the mode of the *a posteriori* distribution

$$\hat{\mathbf{x}}_{\text{MAP}} = \arg\max_{\mathbf{x}} \mathrm{p}_{\mathbf{X}|\mathbf{Z}}\left(\mathbf{x} \mid \mathbf{y}\right). \tag{12}$$

Both estimators can be calculated by each component rather than the entire vector, which means that the estimation of the k-th component can be performed by the marginal distribution of the variable $\mathbf{X}_k$. With a little abuse of definition, the word marginal is used. For MMSE estimation, the marginal distribution of the k-th component is given by

$$p_{\mathbf{X}_k}\left(\mathbf{x}_k\right) = \int_{\sim\{\mathbf{x}_k\}} p_{\mathbf{X}}\left(\mathbf{x}\right) \tag{13}$$

and for MAP estimation, it is

$p_{\mathbf{X}_k}\left(\mathbf{x}_k\right) = \max_{\sim\{\mathbf{x}_k\}} p_{\mathbf{X}}\left(\mathbf{x}\right)$ (14)where $\sim\{\mathbf{x}_k\}$ means applying a certain operation with respect to all variables but $\mathbf{x}_k$. The fast calculation of the marginal distribution of each component is essential to the realization of Bayesian methods.

SPATIAL AND TEMPORAL COOPERATION

The concept of cooperation has been applied to wireless sensor networks (WSNs), where distributed nodes work together to draw a consensus about the environment or to make measurements with each other via spatial and temporal manner (Akyildiz et al., 2002). Similarly, network localization allows agents to communicate with each other in estimating their positions, offering additional benefits such as improved localization accuracy, resilience to system failure, increase in coverage, and reduction of cost per node (Wymeersch, Lien & Win, 2009; Patwari et al., 2005; & Shen, 2014). Understanding the fundamentals of wireless cooperative localization is important, which not only provides a performance benchmark, but also guides algorithm and network design. The accuracy of location estimation is inherently limited due to random phenomena such as noise, fading, shadowing, and multipath propagation. The fundamental limits of such accuracy have been derived by using the information inequality (Shen, Wymeersch & Win, 2010).

From the view of the theoretical analysis, the CRB provides a lower bound on the variance achievable by any unbiased location estimator (Kay, 2013). It provides a guideline for how a best estimator can possibly do and a measure of the estimator that we designed. Given an unbiased estimator $\hat{\mathbf{x}}$, it must satisfy that

$$\text{var}(\hat{\mathbf{x}}) \geq \mathbf{J}(\mathbf{x})^{-1} \tag{15}$$

and

$$\mathbf{J}(\mathbf{x}) = \mathbb{E}\left\{-\frac{\partial^2 \ln f(\mathbf{y}\,|\,\mathbf{x})}{\partial\mathbf{x}\partial\mathbf{x}^{\mathrm{T}}}\right\} \tag{16}$$

where $\mathbf{J}(\mathbf{x})$ is the Fisher information matrix (FIM) of $\mathbf{x}$, $\text{var}(\hat{\mathbf{x}})$ is the variance of the estimator and $f(\mathbf{y}\,|\,\mathbf{x})$ is the likelihood function of $\mathbf{y}$ condition on $\mathbf{x}$. Then the CRB can be obtained by taking inversion of FIM, that is

$$\text{CRB}_{.} = \text{trace}\left(\mathbf{J}^{-1}(\mathbf{x})\right). \tag{17}$$

The bound is also similar to the sensitivity analysis, applied to random measurements since the CRB is based on the curvature of the log-likelihood function. Generally, a sharp log-likelihood function performs superior to the broad one, resulting a lower bound. It should be noted that the CRB is limited to unbiased estimators. Such estimators provide coordinate estimates that, if averaged over enough realizations, are equal to the true coordinates. Although unbiased estimation is a desirable property, some bias might be tolerated to reduce variance; in such cases, the bound can be adapted (Hero III, Fessler & Usman, 1996).

Since the CRB is typically used in the high SNRs region and it is not accurate at low and moderate SNRs for TOA estimation. TOA estimation, like many other estimation problems, also can be characterized by the threshold effect with distinct SNR regions corresponding two different modes of operation. Particularly, at low SNRs (*a priori region*), measurements do not provide significant additional information, and the mean square error (MSE) of the estimator is close to that obtained solely from the *a priori* information about TOA. At high SNRs (*asymptotic region*), the mean square error of the estimator can be accurately described by the CRB. Furthermore, the region between these two extremes is called *transition* or *ambiguity* region where the performance is subject to ambiguities. Therefore, many other bounds like Ziv-Zakai bound (Chazan, Zakai & Ziv, 1975) were proposed to characterize the performance by accounting both ambiguity effects and *a priori* information.

Spatial Cooperation

In a static network, or a dynamic network at a given time instant, only spatial cooperation among agents can be exploited. By using the notion of the FIM, the squared position error for agent k is bounded by the following inequality

$$\mathbb{E}\left\{\left\|\mathbf{x}_k - \hat{\mathbf{x}}_k\right\|^2\right\} \geq \mathrm{tr}\left\{\left[\mathbf{J}(\mathbf{x})\right]_{\mathbf{x}_k}^{-1}\right\} \tag{18}$$

where $\mathbf{J}(\mathbf{x})$ is the $2M \times 2M$ FIM for $\mathbf{x}$; $\mathbb{E}\{\}$ and $\mathrm{tr}\{\}$ are the expectation and trace operators, respectively, and $[\ ]_{\mathbf{x}_k}$ denotes the square submatrix on the diagonal corresponding to $\mathbf{x}_k$. It can be shown that the FIM $\mathbf{J}(\mathbf{x})$ is a sum of two parts, the localization information from anchors (shown as the block-diagonal matrices consisting of $\mathbf{K}_i, i = 1,2,\ldots,M$) and that from agents' spatial cooperation (shown as the structured matrix consisting of $\mathbf{C}_{ij},\ i, j = 1,2,\ldots,M$), shown as follows

$$\mathbf{J}(\mathbf{x}) = \begin{bmatrix} \mathbf{K}_1 + \sum\limits_{j\in\mathcal{M}\backslash\{1\}} \mathbf{C}_{1j} & -\mathbf{C}_{12} & \cdots & -\mathbf{C}_{1M} \\ -\mathbf{C}_{12} & \mathbf{K}_2 + \sum\limits_{j\in\mathcal{M}\backslash\{2\}} \mathbf{C}_{2j} & & -\mathbf{C}_{2M} \\ \vdots & & \ddots & \\ -\mathbf{C}_{1M} & -\mathbf{C}_{2M} & & \mathbf{K}_{N_a} + \sum\limits_{j\in\mathcal{M}\backslash\{M\}} \mathbf{C}_{Mj} \end{bmatrix}. \tag{19}$$

The diagonal matrix in $\mathbf{J}(\mathbf{x})$ is the summation of information from anchor $\mathbf{K}_i$ and cooperation from other agents $\mathbf{C}_{ij}\ (i \neq j)$. The basic building blocks of the FIM $\mathbf{J}(\mathbf{x})$ represent the localization information between pairs of nodes in the form $\lambda\mathbf{u}(\phi)\mathbf{u}^{\mathrm{T}}(\phi)$ where λ is a positive scalar that characterizes the ranging information intensity (RII) (Shen, Wymeersch & Win, 2010). The value of λ depends on the ranging technique as well as the power and bandwidth of the received waveform, multipath propagation, and prior statistical channel knowledge. In particular, the RII is proportional to the square of the effective bandwidth (Shen, Wymeersch & Win, 2010). Moreover, the localization information from anchors can be expressed in a canonical form as a weighted sum of "one-dimensional" information from individual anchors; while cooperation always improves the localization accuracy since it adds a positive semi-definite matrix to the FIM corresponding to non-cooperative localization. It was shown that the accuracy of localization is affected by two factors: the quality of point-to-point measurements, reflected by the expression of RII, and the network topology, reflected by the block structure of the matrix $\mathbf{J}(\mathbf{x})$.

In the absence of cooperation, every $\mathbf{C}_{ij}$ corresponding to cooperation between nodes *i* and *j* becomes 0, resulting in a total FIM $\mathbf{J}(\mathbf{x})$ with a block-diagonal structure. In such cases, localization data for different agents are uncorrelated, and the SPEB can be calculated using only local information at the agents.

As aforementioned analyses, the measurements between node pairs can be TOA, AOA, RSS or connectivity. To derive the CRB of the cooperative localization, the FIM $\mathbf{J}(\mathbf{x})$ is given as follows

$$\mathbf{J}_{\mathrm{TR}} = \begin{cases} \dfrac{\gamma}{d_{k,i}^{s}} \sum_{i \in \mathcal{N}(k)} \mathbf{u}(\phi_{k,i})\mathbf{u}^{\mathrm{T}}(\phi_{k,i}), & k = l \\ -\dfrac{\gamma}{d_{k,i}^{s}} I_{\mathcal{N}(k)}(l)\mathbf{u}(\phi_{k,i})\mathbf{u}^{\mathrm{T}}(\phi_{k,i}), & k \neq l \end{cases} \tag{20}$$

$$\mathbf{J}_{\mathrm{A}} = \begin{cases} \dfrac{\gamma}{d_{k,i}^{s}} \sum_{i \in \mathcal{N}(k)} \mathbf{u}(\phi_{k,i} + \dfrac{\pi}{2})\mathbf{u}^{\mathrm{T}}(\phi_{k,i} + \dfrac{\pi}{2}), & k = l \\ -\dfrac{\gamma}{d_{k,i}^{s}} I_{\mathcal{N}(k)}(l)\mathbf{u}(\phi_{k,i} + \dfrac{\pi}{2})\mathbf{u}^{\mathrm{T}}(\phi_{k,i} + \dfrac{\pi}{2}), & k \neq l \end{cases} \tag{21}$$

where $\mathbf{u}(\phi) = [\cos(\phi),\ \sin(\phi)]^{\mathrm{T}}$ is the unit vector with direction given by ϕ denoting the angle from one node to the other, $\mathbf{J} = \mathbf{J}_{\mathrm{TR}}$ denotes the FIM of TOA and RSS metrics, $\mathbf{J} = \mathbf{J}_{\mathrm{A}}$ denotes the FIM of AOA metric, γ is a channel constant and s is an exponent, both of which are functions of the measurements type and are given in the Table 1. σ_{τ}^{2} and σ_{α}^{2} are the variance of the TOA estimates and AOA estimates, respectively. σ_{dB} is the deviation of the received power. The indicator function $I_{\mathcal{N}(k)}(l)$ represents the available measurements set with node *l*, i.e., $I_{\mathcal{N}(k)}(l) = 1$ when node $l \in \mathcal{N}(k)$ can make measurements with node *k*, or 0 if not.

Spatial-Temporal Cooperation

Building on the understanding of cooperative localization, we now discuss the case of cooperative navigation where agents in a dynamic network cooperate in both the space and time domains. For each time instant, the contribution of cooperation in space is similar to what we have seen in cooperative localization. In addition, another layer of cooperation in time, exploiting intra-node measurements and

Table 1. Differences in CRB by measurements type

	Channel constant	Exponent	FIM
TOA	$\gamma = 1/(v\sigma_{\tau})^{2}$	s=0	$\mathbf{J}$=$\mathbf{J}_{\mathrm{TR}}$
RSS	$\gamma = (10n_{p}/\sigma_{\mathrm{dB}}\log 10)$	s=2	$\mathbf{J}$=$\mathbf{J}_{\mathrm{TR}}$
AOA	$\gamma = 1/\sigma_{\alpha}^{2}$	s=2	$\mathbf{J}$=$\mathbf{J}_{\mathrm{A}}$

mobility (dynamic) models, yields new information for navigation. Such information at time t is characterized by $\mathbf{T}^{(t)}$ matrices in the total FIM $\mathbf{J}(\mathbf{x}^{(1:T)})$ where $\mathbf{x}^{(1:T)}$ consists of the positions of all agents from time 1 to T. Consequently, the FIM for each agent at a given time instant can be given by

$$\mathbf{J}(\mathbf{x}^{(1:T)}) = \begin{bmatrix} \mathbf{S}^{(1)} + \mathbf{T}^{(2)} & -\mathbf{T}^{(2)} & & & \\ -\mathbf{T}^{(2)} & \mathbf{S}^{(2)} + \mathbf{T}^{(2)} + \mathbf{T}^{(3)} & & & \\ & & \ddots & & \\ & & & \mathbf{S}^{(T-1)} + \mathbf{T}^{(T-1)} + \mathbf{T}^{(T)} & -\mathbf{T}^{(T)} \\ & & & -\mathbf{T}^{(T)} & \mathbf{S}^{(T)} + \mathbf{T}^{(T)} \end{bmatrix} \tag{22}$$

where $\mathbf{S}^{(t)}$ represents the spatial cooperation information at time t and $\mathbf{S}^{(t)}$ has the same form of and $\mathbf{T}^{(t)} = \operatorname{diag}\left\{\mathbf{T}_1^{(t)}, \mathbf{T}_2^{(t)}, \ldots, \mathbf{T}_M^{(t)}\right\}$ represents the temporal information.

Some observations can be drawn from the structure of the FIM for cooperative navigation (Marano et al., 2010). First, the total FIM consists of two major components: cooperation in space as well as in time. The former characterizes the localization information from inter-node measurements in the entire network at each time instant (shown as the diagonal $2M \times 2M$ block matrices), and the latter characterizes the information from intra-node measurements and mobility models at each individual agent (shown as components $\mathbf{T}^{(t)}$ outside the main block-diagonal). Second, since intra-node measurements and the mobility models for different agents are independent, the corresponding matrices $\mathbf{T}^{(t)}$ form a block diagonal matrix in the upper-right and lower-left quarter of the total FIM. Third, the $\mathbf{T}^{(t)}$ components can be viewed as the temporal link that connects localization information from spatial cooperation of the previous time instant to the current one. If the temporal link is not available (e.g., $\mathbf{T}^{(t)}$ matrices are zero), the total FIM is block diagonal, implying that position inference is independent from time to time. The structure of the FIM for cooperative navigation allows a recursive method to calculate the FIM at each time instant. This view also provides insights into the information evolution of spatial-temporal cooperation in cooperative navigation.

In spatial cooperation, one agent will incorporate the beliefs from neighbor agents to determine its position and a centralized processor is required to update the joint belief of all agents by using all the measurements as well as the prior joint belief. Unfortunately, the position belief update and interference is a high dimensional optimization that is difficult to handle. Since the position belief of all agents will transmit to the central processor and then communicate back after update, the large communication bandwidth and the resources are required, which is inefficient in practice. Moreover, this kind of centralized solutions relies on the stability of the central processor and this kind of network is not robust to failure. Therefore, the distributed algorithms in cooperative networks are attractive to tackle these problems. From Figure 2-(b) we can observe that the FIM does not have a block-diagonal structure due to the correlation between the inferred agents' positions. Thus, the distributed algorithms for cooperative localization cannot achieve the CRB.

A common method to implement distributed cooperative localization is by the loopy belief propagation (Wymeersch, 2009) which assumes the position beliefs are independent and propagate the beliefs of each agent through the cooperative network in an iterative manner by the sum-product algorithm. The Bayes' updates and marginalization are corresponding to the products and sums (integrals) respectively.

Figure 2. Localization information provided by spatial cooperation. The red and blue ellipses respectively represent the localization information and error.

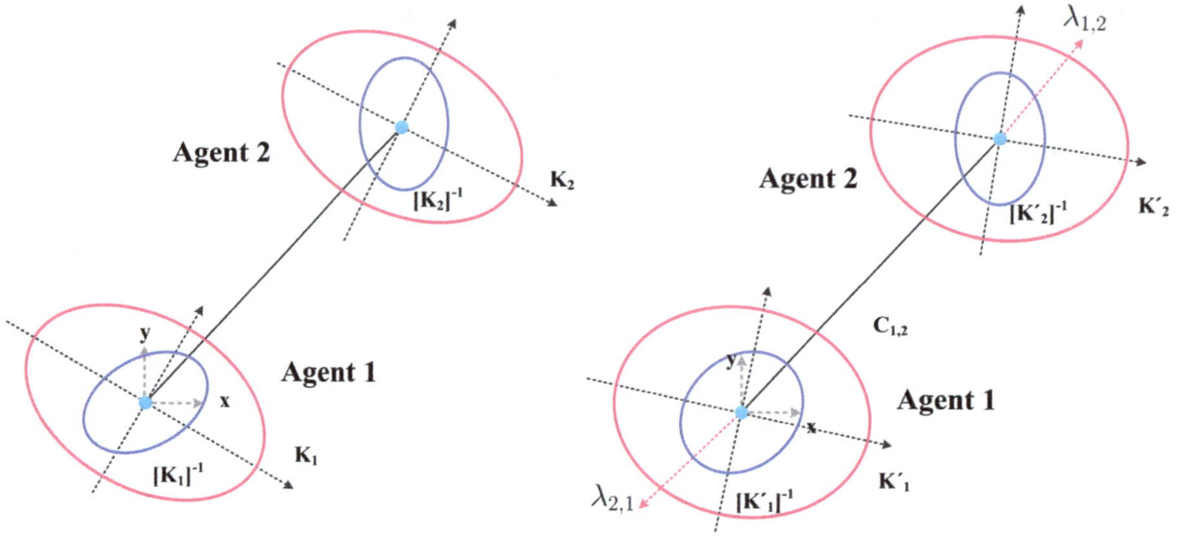

To illustrate the benefits of the cooperative localization, an example of two agents with spatial cooperation is shown in Figure 2, describing the position errors before and after cooperation. The red and blue ellipses in Figure 2-(a) and Figure 2-(b) represent the agents' localization information and errors, respectively. As the two agents cooperate shown in the figures, the information ellipses enlarged after cooperation while the error ellipses shrunk, as predicted by the theoretical results. Correspondingly, the position beliefs of the two agents become more concentrated after cooperation.

In spatiotemporal cooperation, the agent's belief not only obtains from the neighbor agents, but also from the intra-node measurements and mobility models. The block-diagonal matrix $\mathbf{T}^{(t)}$ in is the additional information to improve the network localization performance. Given $\mathbf{x}$ denotes the position states like positions and orientations, etc. The temporal cooperation can be accomplished using mobility and intra-node measurement models. The former statistically describes the evolution of the position states in time by $p(\mathbf{x}^{(t)} \mid \mathbf{x}^{(t-1)})$ and the relationship between intra-node measurements and the position states is $p(\mathbf{y}_s \mid \mathbf{x}^{(t)})$. Thus, the belief updates based on Bayes' rule and marginalization can be performed in the following two steps:

1. Prediction

$$p\left(\mathbf{x}^{(t)} \mid \mathbf{y}^{(1:t-1)}\right) = \int p\left(\mathbf{x}^{(t-1)} \mid \mathbf{y}^{(1:t-1)}\right) p\left(\mathbf{x}^{(t)} \mid \mathbf{x}^{(t-1)}\right) \mathrm{d}\mathbf{x}^{(t-1)} \tag{23}$$

2. Correction:

$$p\left(\mathbf{x}^{(t)} \mid \mathbf{y}^{(1:t)}\right) \propto p\left(\mathbf{x}^{(t)} \mid \mathbf{y}^{(1:t-1)}\right) p\left(\mathbf{y}^{(t)} \mid \mathbf{x}^{(t)}\right) \tag{24}$$

where $p\left(\mathbf{y}^{(t)} \mid \mathbf{x}^{(t)}\right) = p\left(\mathbf{y}_{\mathrm{s}}^{(t)} \mid \mathbf{x}^{(t)}\right) p\left(\mathbf{y}_{\mathrm{r}}^{(t)} \mid \mathbf{x}^{(t)}\right)$.

If both measurements and mobility models are linear-Gaussian and the previous beliefs are also Gaussian, the updated beliefs at each time instant can be evaluated exactly by means of the Kalman filter. However, for nonlinear or non-Gaussian models in many scenarios, several approximation methods, such as, extended KF or unscented KF, to particle filters (PFs) based on sequential Monte Carlo techniques can be taken to improve the localization performance.

LOCALIZATION TECHNIQUES

In this section, two classes of cooperative localization algorithms: centralized and distributed algorithms will be introduced. The centralized algorithms include the subspace-based MDS method and SDP-based optimization method. After that, a distributed algorithm called SPAWN is presented.

Localization and Navigation Algorithms

From the perspective of Bayesian, the localization and navigation algorithm aims to determine the posterior distribution $p(\mathbf{x} \mid \mathbf{y})$, also referred to its position belief. Once the posterior distribution is determined, point estimates can be accomplished with the MMSE or MAP estimates, respectively. The primary tools to obtain the posterior distribution are Bayes' rule and marginalization, updating beliefs according to new measurements and reducing the dimension of inference problem, respectively.

In the traditional non-cooperative localization network, each agent can only determine its position from the neighbor anchors independently. The belief update and MMSE or MAP estimation with Bayes' rule can be calculated from the ML function modeled by a tractable statistical distribution. For a Gaussian distribution, the MMSE and MAP estimators for the agent's position are the solution of weighted least squares (WLS), which is

$$\hat{\mathbf{x}}_{\mathrm{MMSE}} = \hat{\mathbf{x}}_{\mathrm{MAP}} = \arg\min_{\mathbf{x}} \sum_{j \in \mathcal{N}} \frac{1}{\sigma_{\mathbf{y}_j}^2} \| \mathbf{y}_j - h\left(\mathbf{x}, \mathbf{x}_j\right) \|^2 + \frac{1}{\sigma_{\mathbf{x}}^2} \| \mathbf{x} - \overline{\mathbf{x}} \|^2 \tag{25}$$

where $\sigma_{\mathbf{y}_j}^2$ and $\sigma_{\mathbf{x}}^2$ are the variance of the noise in each measurement and variance of the agent's distribution, respectively, and $h(\mathbf{x}, \mathbf{x}_j)$ is a functional of the distance from anchor j to agent, $\overline{\mathbf{x}}$ is the mean of the positon. When the likelihood function is a non-Gaussian distribution, the optimization process for MMSE and MAP may be complicated, in which case the WLS solution can be used as a tractable suboptimal solution. While each agent determines its belief based on the information obtained from its local measurements of anchors, the localization performance can be improved if agents incorporate the beliefs of neighbors (spatial cooperation) and the beliefs obtained in the past (temporal cooperation). The following parts will introduce the method of the spatial cooperation and temporal cooperation in algorithm designs.

CENTRALIZED ALGORITHM

If the measurements can be described well by a particular statistical model (e.g., Gaussian or log-normal), then the MLE can be derived and implemented. One reason that these estimators are used is that their variance asymptotically (as the SNR goes large) approaches the lower bound given by the CRB. However, there are two difficulties with this approach. i) Local maxima: an initial estimate close to the correct solution is required; ii) model dependency: If measurements deviate from the assumed model (or model parameters), the results are no longer guaranteed to be optimal. The method to tackle the local maxima problem is to formulate the localization as a convex optimization problem. Convex constraints are utilized to estimate nodes' location in (Doherty, Pister & Ghaoui, 2001).

More general constraints can be considered if SDP techniques are used (Biswas & Ye, 2003; Luo et al., 2010). One difficulty that must be overcome in both techniques is their high computational complexity. Toward this end, a distributed SDP-based localization algorithm was presented in (Biswas & Ye, 2003; Luo et al., 2010). MDS-based algorithm is a well-known network localization algorithm by formulating localization from range measurements as an LS problem (Shang et al., 2003; Shang & Ruml, 2004) and the location estimation can be solved by eigen-decomposition in classical MDS. The MDS do not suffer the local maxima problem and can achieve the global convergence, which enables the MDS as an attracting method. The square distance matrix is the kernel of MDS algorithms which contains the relative distance between all nodes in the network. According to the connectivity of the network (fully connected or partially connected), the MDS algorithms can be further divided into centralized or distributed version. By considering the sensor node uncertainties, a cooperative localization based on WLS was proposed in an asynchronous network (Rui & Ho, 2014). The unknown locations and clock parameters are jointly estimated through WLS minimization. The least square cooperative localization was further investigated in a NLOS propagation environment and the theoretical bounds were derived to show that the Gaussian NOS bias is the worst case among all biases (Nguyen et al., 2015).

Subspace-Based Algorithm

MDS is a field of study related to the search for a low-dimensional space (Cheung & So, 2005; Tinsley & Brown, 2000) in which points in the space represent the objects such that the distances, or dissimilarities, between the points in the space matching.

MDS was first used in psychometrics and psychophysics and it can be seen as a set of data analysis techniques that display the structure of distance-like data as a geometrical picture (Borg & Groenen, 1997). Generally, the MDS starts with one or more distance matrices (or similarity matrices) that are presumed to have been derived from points in a multidimensional space. It is usually used to find a placement of the points in a low dimensional space, usually two or three-dimensional, where the distances between points resemble the original similarities. MDS is often used as part of exploratory data analysis or information visualization. By visualizing objects as points in a low-dimensional space, the complexity in the original data matrix can often be reduced while preserving the essential information. MDS is related to principal component analysis, factor analysis, and cluster analysis. The approaches based on MDS have been applied in many fields, such as physical, biological, machine learning and computational chemistry, etc.

MDS-based methods belong to deterministic localization and by applying the MDS to a fully connected network, the centralized MDS is developed. For a partially connected network, the distributed

MDS can be developed. In this section, we will introduce the MDS localization method (Chan & So, 2009) in a cooperative network.

Consider a fully connected network with $M+N$ nodes in a plane (three-dimensional case can be extended straightforward). Let $\mathbf{x}_i = [x_i, y_i]^{\mathrm{T}}$ denotes the i-th node (i=1, 2, …, $M+N$). We repeat the positions of agents and anchors for completeness, $\mathbf{X} = [\mathbf{x}_1, \mathbf{x}_2, \ldots, \mathbf{x}_M]^{\mathrm{T}}$ are agents while $\mathbf{X}_{\mathrm{b}} = [\mathbf{x}_{M+1}, \mathbf{x}_{M+2}, \ldots, \mathbf{x}_{M+N}]^{\mathrm{T}}$ are anchor nodes and our mission is to estimate $\mathbf{X}$ with inter-measurements from anchors and intra-measurements from agents.

Let $\mathbf{X}_{\mathrm{fs}} \in \mathbb{R}^{(\mathrm{C}_{M+N}^2 - \mathrm{C}_N^2)\times 2}$ where $\mathrm{C}_i^j = i!/\,j!(i-j)!$ includes all the differences of $x_i - x_j$ and $y_i - y_j$, $i, j = 1, 2, \ldots, M+N$ and $i > j$ with at least one of the $\mathbf{x}_i$ and $\mathbf{x}_j$ being unknown. The matrix $\mathbf{X}_{\mathrm{fs}}$ has the following form

$$\mathbf{X}_{\mathrm{fs}} = \begin{bmatrix} x_1\mathbf{1}_N - \mathbf{s}_1^{\mathrm{x}} & y_1\mathbf{1}_N - \mathbf{s}_1^{\mathrm{y}} \\ x_2\mathbf{1}_{N+1} - \mathbf{s}_2^{\mathrm{x}} & y_2\mathbf{1}_{N+1} - \mathbf{s}_2^{\mathrm{y}} \\ \vdots & \vdots \\ x_M\mathbf{1}_{M+N-1} - \mathbf{s}_{N-1}^{\mathrm{x}} & y_M\mathbf{1}_{M+N-1} - \mathbf{s}_{N-1}^{\mathrm{y}} \end{bmatrix} \tag{26}$$

where $\mathbf{s}_i^{\mathrm{x}} = [x_{M+1}\ x_{M+2}\ \cdots\ x_{M+i}]^{\mathrm{T}}$, $\mathbf{s}_i^{\mathrm{y}} = [y_{M+1}\ y_{M+2}\ \cdots\ y_{M+i}]^{\mathrm{T}}$ and $\mathbf{1}_i$ denotes the i-dimensional vector with all 1. The matrix $\mathbf{X}_{\mathrm{fs}}$ can be expressed as an affine combination of matrices with coefficients x_i and y_i, as

$$\mathbf{X}_{\mathrm{fs}} = \sum_{i=1}^{2M} \varphi_i \mathbf{X}_i + \mathbf{X}_0 \tag{27}$$

for $i = 1, 2, \ldots, M$, and φ_i is the element of the vector $\mathbf{\Phi}_{\mathrm{fs}}$ which is defined as

$$\mathbf{\Phi}_{\mathrm{fs}} = \left[x_1, y_1, x_2, y_2, \cdots x_M, y_M\right]^{\mathrm{T}}. \tag{28}$$

The matrix $\mathbf{X}_0$ can be yielded by removing the coordinates of agents and the matrix $\mathbf{X}_i$ contains the signs of coordinates of anchors, i.e., $\mathbf{X}_{2i-1}$ and $\mathbf{X}_{2i}$ contain the signs of x_i and y_i for $i = 1, 2, \ldots, M$, respectively. The other entries of $\mathbf{X}_i$ are all zero when the corresponding element is absent.

$$\mathbf{F}_{\mathrm{fs}} = \mathbf{X}_{\mathrm{fs}} \mathbf{X}_{\mathrm{fs}}^{\mathrm{T}}. \tag{29}$$

The matrix $\mathbf{X}_{\mathrm{fs}}$ can be seen as full of column rank which is usually true in practice. Therefore, we have $\mathrm{rank}(\mathbf{F}_{\mathrm{fs}}) = \mathrm{rank}(\mathbf{X}_{\mathrm{fs}}) = 2$.

The matrix $\mathbf{F}_{\mathrm{fs}}$ is constructed from the distance between node pairs (anchor-anchor, anchor-agent and agent-agent) $\{d_{i,j}\}$ where $d_{i,j}$ denotes the distance between the i-th node and j-th node. With the use of scalar dot product, we have

$$\left(\mathbf{x}_i - \mathbf{x}_j\right)^{\mathrm{T}}\left(\mathbf{x}_m - \mathbf{x}_j\right) = \frac{1}{2}\left(d_{i,j}^2 + d_{j,m}^2 - d_{i,m}^2\right). \tag{30}$$

Extending the formula above, yields

$$\begin{aligned}\left(\mathbf{x}_i - \mathbf{x}_j\right)^{\mathrm{T}}\left(\mathbf{x}_m - \mathbf{x}_n\right) &= \left(\mathbf{x}_i - \mathbf{x}_j\right)^{\mathrm{T}}\left(\mathbf{x}_m - \mathbf{x}_j - \left(\mathbf{x}_n - \mathbf{x}_j\right)\right)\\ &= \left(\mathbf{e}_{M+N,i}^{\mathrm{T}}\mathbf{R}\mathbf{e}_{M+N,m} - \mathbf{e}_{M+N,i}^{\mathrm{T}}\mathbf{R}\mathbf{e}_{M+N,\mathrm{n}} + \mathbf{e}_{M+N,j}^{\mathrm{T}}\mathbf{R}\mathbf{e}_{M+N,n} - \mathbf{e}_{M+N,j}^{\mathrm{T}}\mathbf{R}\mathbf{e}_{M+N,n}\right)\\ &= -\frac{1}{2}\left(\mathbf{e}_{M+N,i} - \mathbf{e}_{M+N,j}\right)^{\mathrm{T}}\mathbf{R}\left(\mathbf{e}_{M+N,m} - \mathbf{e}_{M+N,n}\right)\end{aligned} \tag{31}$$

where $[\mathbf{R}]_{i,j} = d_{i,j}^2$ and $\mathbf{e}_{N,i}$ stands for the i-th column of the identity matrix with $M+N$ dimension. With the above formula, the matrix $\mathbf{F}_{\mathrm{fs}}$ can be expressed as

$$\mathbf{F}_{\mathrm{fs}} = -\frac{1}{2}\mathbf{K}\mathbf{R}\mathbf{K}^{\mathrm{T}} \tag{32}$$

where $\mathbf{K} = [\mathbf{H}_1 \;\; \mathbf{H}_2 \;\; \cdots \;\; \mathbf{H}_M]^{\mathrm{T}}$ and $\mathbf{H}_i = [-\mathbf{I}_{i-1} \;\; \mathbf{1}_{i-1} \;\; \mathbf{O}_{(i-1)\times(M+N-i)}]^{\mathrm{T}}$ for $i = 1, 2, \cdots, M$. Square matrix $\mathbf{I}_K$ is an identity matrix with K dimension and $\mathbf{O}_{(i-1)\times(M+N-i)}$ denotes the $(i-1)\times(M+N-i)$-dimensional matrix with all zero.

Note that if matrix $\mathbf{K}$ is replaced by $\mathbf{J} = \mathbf{I}_{M+N} - \mathbf{1}_{M+N}\mathbf{1}_{M+N}^{\mathrm{T}} / (M+N)$, then the matrix of $\mathbf{F}_{\mathrm{fs}}$ is corresponding to the classical MDS (Shang et al., 2004; Cheung & So, 2005). Since $\mathbf{F}_{\mathrm{fs}}$ is a symmetric positive semidefinite and of rank-2, it can be decomposed with eigenvalue factorization as

$$\mathbf{F}_{\mathrm{fs}} = \mathbf{U}_{\mathrm{s}}\Lambda_{\mathrm{s}}\mathbf{U}_{\mathrm{s}}^{\mathrm{T}} + \mathbf{U}_{\mathrm{n}}\Lambda_{\mathrm{n}}\mathbf{U}_{\mathrm{n}}^{\mathrm{T}} \tag{33}$$

where $\mathbf{U}_{\mathrm{s}} \in \mathbb{R}^{(C_{M+N}^2 - C_N^2)\times 2}$ and $\mathbf{U}_{\mathrm{n}} \in \mathbb{R}^{(C_{M+N}^2 - C_N^2)\times(C_{M+N}^2 - C_N^2 - 2)}$ are the signal and noise eigenvectors, respectively. $\Lambda_{\mathrm{s}} = \mathrm{diag}(\lambda_1, \lambda_2)$ with eigenvalues $\lambda_1 \geq \lambda_2 \geq 0$, and $\Lambda_{\mathrm{n}} = \mathrm{diag}(\lambda_3, \lambda_4, \ldots, \lambda_{C_{M+N}^2 - C_N^2})$ where all entries are zeros. Note that $\mathbf{X}_{\mathrm{fs}}$ is orthogonal to the subspace spanned by $\mathbf{U}_{\mathrm{n}}$, then we have

$$\mathbf{U}_{\mathrm{n}}^{\mathrm{T}}\mathbf{X}_{\mathrm{fs}} = \mathbf{0}_{C_{M+N}^2 - C_N^2 - 2} \tag{34}$$

or

$$\sum_{i=1}^{2M} \varphi_i \mathbf{U}_{\mathrm{n}}^{\mathrm{T}} \mathbf{X}_i = -\mathbf{U}_{\mathrm{n}}^{\mathrm{T}} \mathbf{X}_0. \tag{35}$$

When the noise is present, the measured distance $\left\{r_{i,j}\right\}$ can be expressed by

$$r_{i,j} = d_{i,j} + q_{i,j} \tag{36}$$

where $q_{i,j} = q_{j,i}$ is the Gaussian noise with zero mean and variance $\sigma_{i,j}^2 > 0$. Then $\{d_{i,j}\}$ will be replaced with $\{r_{i,j}\}$ in (32) and (33) and the orthogonality between $\mathbf{X}_{\mathrm{fs}}$ and $\mathbf{U}_{\mathrm{n}}$ established approximately.

Therefore, we have

$$\mathbf{A}_{\mathrm{fs}} \mathbf{\Phi}_{\mathrm{fs}} \approx \mathbf{b}_{\mathrm{fs}} \tag{37}$$

where $\mathbf{A}_{\mathrm{fs}} = [\mathrm{vec}(\mathbf{U}_{\mathrm{n}}^{\mathrm{T}}\mathbf{X}_1)\ \ \mathrm{vec}(\mathbf{U}_{\mathrm{n}}^{\mathrm{T}}\mathbf{X}_2)\ \ \cdots\ \ \mathrm{vec}(\mathbf{U}_{\mathrm{n}}^{\mathrm{T}}\mathbf{X}_{2M})]$ and $\mathbf{b}_{\mathrm{fs}} = -\mathrm{vec}(\mathbf{U}_{\mathrm{n}}^{\mathrm{T}}\mathbf{X}_0)$ with vec() being the vectorization operator. Obviously, the equation above can be solved by the least square

$$\mathbf{\Phi}_{\mathrm{fs}} \approx \left(\mathbf{A}_{\mathrm{fs}}^{\mathrm{T}} \mathbf{A}_{\mathrm{fs}}\right)^{-1} \mathbf{A}_{\mathrm{fs}}^{\mathrm{T}} \mathbf{b}_{\mathrm{fs}}. \tag{38}$$

Note that the above algorithm is formulated based on the fully connected network and if the network is partially connected, the distributed version can be deduced. More details can be found in (Chan & So, 2009).

After constructing the square distance matrix, the resolving process of MDS is based on the subspace principle which is in essence the same as the multiple signal classification (MUSIC) (Schmidt, 1986) algorithm for AOA estimation. However, the AOA estimation of MUSIC from the noise subspace requires the search of the spatial area of interest, while the MDS algorithm for localization based on the full-set subspace support a closed-form solution of .

By solving the equation of, all unknown coordinates of agents can be obtained and these results can be used to refine the measurements between agents, and therefore, the above algorithm can behave in an iterative manner to achieve better performance.

SDP-Based Algorithm

Let $\mathcal{M} \times \mathcal{M}$ and $\mathcal{M} \times \mathcal{N}$ denote the set of agent-agent and agent-anchor edges, respectively. The measurements between two possible nodes are denoted by $\{d_{i,k} : (i,k) \in \left(\mathcal{M} \cup \mathcal{N}\right) \times \mathcal{M}\}$ (agent-agent and agent-anchor), then we have

$$\left\|\mathbf{x}_i - \mathbf{x}_k\right\|^2 = d_{i,k}^2, \quad \left(i,k\right) \in \left(\mathcal{M} \cup \mathcal{N}\right) \times \mathcal{M}. \tag{39}$$

In general, the equations above are difficult to solve, as the quadratic constraints in them are nonconvex. The problem of determining the feasibility of (40) is non-deterministic polynomial hard (NP-hard). However, one can derive a computationally efficient semidefinite relaxation of (40) as follows. First, we observe that

$$\left\|\mathbf{x}_i-\mathbf{x}_k\right\|^2=\mathbf{x}_i^{\mathrm{T}}\mathbf{x}_i-2\mathbf{x}_i^{\mathrm{T}}\mathbf{x}_k+\mathbf{x}_k^{\mathrm{T}}\mathbf{x}_k,\quad (i,k)\in\mathcal{M}\times\mathcal{M}. \tag{40}$$

In particular, we can find that $\|\mathbf{x}_i-\mathbf{x}_k\|^2$ is linear in the inner products $\mathbf{x}_i^{\mathrm{T}}\mathbf{x}_i$, $\mathbf{x}_i^{\mathrm{T}}\mathbf{x}_k$ and $\mathbf{x}_k^{\mathrm{T}}\mathbf{x}_k$. Hence, we can rewrite the equation above as

$$\left\|\mathbf{x}_i-\mathbf{x}_k\right\|^2=\left(\mathbf{e}_{M,i}-\mathbf{e}_{N_a,k}\right)^{\mathrm{T}}\mathbf{X}^{\mathrm{T}}\mathbf{X}\left(\mathbf{e}_{M,i}-\mathbf{e}_{M,k}\right)=\mathrm{trace}\left(\mathbf{E}_{ik}\mathbf{X}^{\mathrm{T}}\mathbf{X}\right) \tag{41}$$

where $\mathbf{E}_{ik}=(\mathbf{e}_{M,i}-\mathbf{e}_{M,k})(\mathbf{e}_{M,i}-\mathbf{e}_{M,k})^{\mathrm{T}}$. Similarly, for the anchor-agent node pairs, we have

$$\left\|\mathbf{x}_i-\mathbf{x}_k\right\|^2=\mathbf{x}_i^{\mathrm{T}}\mathbf{x}_i-2\mathbf{x}_i^{\mathrm{T}}\mathbf{x}_k+\mathbf{x}_k^{\mathrm{T}}\mathbf{x}_k,\quad (i,k)\in\mathcal{N}\times\mathcal{M}. \tag{42}$$

and it can be transformed as the following form (anchor-agent link) (Luo, 2010)

$$\left\|\mathbf{x}_i-\mathbf{x}_k\right\|^2=\begin{bmatrix}\mathbf{x}_i^{\mathrm{T}} & -\mathbf{e}_{M,k}^{\mathrm{T}}\end{bmatrix}\begin{bmatrix}\mathbf{I}_2 & \mathbf{X}\\ \mathbf{X}^{\mathrm{T}} & \mathbf{X}^{\mathrm{T}}\mathbf{X}\end{bmatrix}\begin{bmatrix}\mathbf{x}_i\\ \mathbf{e}_{M,k}\end{bmatrix}=\mathrm{trace}\left(\mathbf{M}_{ik}\mathbf{Z}\right),\quad (i,k)\in\mathcal{N}\times\mathcal{M} \tag{43}$$

where

$$\mathbf{M}_{ik}=\begin{bmatrix}\mathbf{x}_i\\ \mathbf{e}_{M,k}\end{bmatrix}\begin{bmatrix}\mathbf{x}_i^{\mathrm{T}} & \mathbf{e}_{M,k}^{\mathrm{T}}\end{bmatrix},\quad (i,k)\in\mathcal{N}\times\mathcal{M}$$

and

$$\mathbf{Z}=\begin{bmatrix}\mathbf{I}_2 & \mathbf{X}\\ \mathbf{X}^{\mathrm{T}} & \mathbf{X}^{\mathrm{T}}\mathbf{X}\end{bmatrix}. \tag{44}$$

Note that the matrix $\mathbf{Z}$ is a rank-2 positive semidefinite matrix with the upper left submatrix constrained to be an identity matrix. Moreover, using the Schur complement, it is not difficult to show that any rank-2 positive semidefinite matrix $\mathbf{Z}$ whose upper left sub-block is an identity matrix must have the form given by for some $\mathbf{X}$. Moreover, by defining the matrix (agent-agent link)

$$\mathbf{M}_{ik} = \begin{bmatrix} \mathbf{O}_{2\times 2} & \mathbf{O}_{2\times n} \\ \mathbf{O}_{2\times n}^{\mathrm{T}} & \mathbf{E}_{ik} \end{bmatrix}, \quad (i,k) \in \mathcal{M} \times \mathcal{M} \tag{45}$$

Then the cooperative localization can be formulated as

$$\begin{aligned} \text{Find} \quad & \mathbf{Z} \\ \text{s.t.} \quad & \operatorname{trace}\left(\mathbf{M}_{ik}\mathbf{Z}\right) = d_{i,k}^2 \quad \left(i,k\right) \in \left(\mathcal{M} \cup \mathcal{N}\right) \times \mathcal{M}, \\ & \mathbf{Z}_{1:2,1:2} = \mathbf{I}_2, \\ & \mathbf{Z} \succeq \mathbf{O}, \\ & \operatorname{rank}\left(\mathbf{Z}\right) = 2. \end{aligned} \tag{46}$$

By omitting the rank constraint that causes non-convexity, the above formulation gives the SDR of cooperative localization. If we solve the SDR of, and obtain a rank-*r* solution $\mathbf{Z}$, then we can extract from a set of *r*-dimensional coordinates for the nodes such that those coordinates satisfy the distance constraints (So & Ye, 2007).

Note that the above discussions only focused on the case where the measured distances are noise-free. However, in practice, the measured distances are usually corrupted by noise (say, by an additive Gaussian noise). In this case, there is no solution to satisfy the constraints in the above SDR, and a more general optimization problem should be set up. A MLE based method is introduced as follows (Biswas & Ye, 2006; Liang, Wang & Ye, 2004).

Suppose there are some measurement errors between agent-agent and agent-anchor denoted by $w_{i,k}$ and $\bar{w}_{i,k}$, respectively

$$\hat{d}_{i,k} = \left\|\mathbf{x}_i - \mathbf{x}_k\right\| + w_{i,k}, \quad (i,k) \in \left(\mathcal{M} \cup \mathcal{N}\right) \times \mathcal{M} \tag{47}$$

where we assume the noise $w_{i,k} \sim \mathcal{N}\left(0, \sigma_{i,k}^2\right)$, and they are independent. Let the ML function *p* to estimate **X**, using all distance measurements, be

$$p\left(\left(\hat{d}_{i,k}, \left(i,k\right) \in \left(\mathcal{M} \cup \mathcal{N}\right) \times \mathcal{M}\right), \mathbf{X}\right) = \prod_{(i,k)\in\left(\mathcal{M} \cup \mathcal{N}\right)\times\mathcal{M}} \frac{1}{2\pi^{\frac{1}{2}}\sigma_{i,k}} \exp\left(-\frac{1}{2\sigma_{i,k}^2}\left(\hat{d}_{i,k} - d\left(\mathbf{x}_i, \mathbf{x}_k\right)\right)^2\right) \tag{48}$$

where $d\left(\mathbf{x}_i, \mathbf{x}_k\right)$ denote the true distance between $\mathbf{x}_i$ and $\mathbf{x}_k$ for $\left(i,k\right) \in \left(\mathcal{M} \cup \mathcal{N}\right) \times \mathcal{M}$. Then the MLE of the locations can be given by

$$\mathbf{X}_{\mathrm{ML}} = \arg\max_{\mathbf{X}} p\left(\left(\hat{d}_{i,k}, (i,k) \in \left(\mathcal{M} \cup \mathcal{N}\right) \times \mathcal{M}\right), \mathbf{X}\right). \tag{49}$$

Then $\mathbf{X}_{\mathrm{ML}}$ can be written explicitly as

$$\mathbf{X}_{\mathrm{ML}} = \arg\min_{\mathbf{X}} \sum_{(i,k)\in(\mathcal{M}\cup\mathcal{N})\times\mathcal{M}} \frac{1}{2\sigma_{i,k}^2}\left(\hat{d}_{i,k} - d\left(\mathbf{x}_i,\mathbf{x}_k\right)\right)^2. \tag{50}$$

Note that since the variances of the distance measurements are unknown, the following optimization formulation can solve the MLE problem

$$\begin{aligned} \min \quad & \sum_{(i,k)\in(\mathcal{M}\cup\mathcal{N})\times\mathcal{M}} \varepsilon_{i,k}^2 \\ \text{s.t.} \quad & \left(\hat{d}_{i,k} - \left\|\mathbf{x}_i - \mathbf{x}_k\right\|\right) = \varepsilon_{i,k}, \quad \forall(i,k)\in\left(\mathcal{M}\cup\mathcal{N}\right)\times\mathcal{M} \end{aligned} \tag{51}$$

and then can be relaxed as

$$\begin{aligned} \min \quad & \alpha \\ \text{s.t.} \quad & \operatorname{trace}\left(\mathbf{M}_{ik}\mathbf{Z}\right) - \hat{d}_{i,k}^2 = \varepsilon_{i,k}, \quad \forall(i,k)\in\left(\mathcal{M}\cup\mathcal{N}\right)\times\mathcal{M}, \\ & \left(\sum_{(i,k)\in(\mathcal{M}\cup\mathcal{N})\times\mathcal{M}} \varepsilon_{i,k}^2\right)^{1/2} \le \alpha, \\ & \mathbf{Z}_{1:2,1:2} = \mathbf{I}_2, \\ & \mathbf{Z} \succeq \mathbf{O}. \end{aligned} \tag{52}$$

Solving the SDR of, the coordinates of all agents can be obtained. If we restrict the communication range of the node, the optimization problem of and can be implemented by modifying the available link set.

DISTRIBUTED ALGORITHMS

Although the centralized algorithm can achieve a better accuracy and reach the CRB asymptotically, a central processor is required to process the measurements (Zhou & Chen, 2016). When the central processor is unavailable, the distributed algorithm is necessary. Particularly, when a large network of nodes should be processed, the communication bandwidth and computational resources in the central processor become the bottleneck of the performance of the whole network. Moreover, the selection of the node pairs for cooperation, referred to as network scheduling, is a vital part in a large network scenario (Wang et al., 2017). The network scheduling strategies affect the localization errors directly and it is crucial to schedule the nodes with an appropriate distributed manner. Generally, the distributed algorithms for cooperative localization generally fall into one of two categories:

1. Network multilateration: Each node estimates its multi-hop range to the nearest reference nodes. Note that finding the shortest path is readily distributed across the network. When each node has multiple range estimates to known positions, its coordinates are calculated locally via multilateration (Torrieri, 1984; Caffery, 1999).

2. Successive refinement: These algorithms try to find the optimum of a global cost function, e.g., LS, WLS, or ML criteria. After the self-localization process, each node will send the related information to its neighbors and neighbors will refine their locations again until convergence.

Typically, better statistical performance is achieved by successive refinement compared to network multilateration, but convergence issues must be addressed. The outage probability of localization in a wireless network can be referred to (Zhou & Shen, 2017).

In this section, distributed localization algorithms are introduced. The study of distributed cooperative localization algorithms is essential, since they are more scalable than the centralized ones and can be applied to large localization networks. Four algorithms will be introduced, namely the cooperative LS and ML localization, and the SPAWN with MMSE and MAP.

Cooperative LS/ML Localization

A centralized method for LS/ML based localization has been introduced in (Mao, Fidan & Anderson, 2007). However, note that agent i is not aware of the measurements $z_{i\to k}^{(t)}$ $(k \in \mathcal{S}_{i\to}^{(t)})$. Thus in a distributed realization, the second term of the is omitted, and only the first term is considered

$$\frac{\partial C_{\text{dis}}^{(t)}(\mathbf{x})}{\partial \mathbf{x}_i} = \sum_{j\in\mathcal{S}_{\to i}^{(t)}} \frac{\partial c_{j\to i}(z_{j\to i}^{(t)};\mathbf{x}_i,\mathbf{x}_j)}{\partial \mathbf{x}_i}. \tag{53}$$

Similarly, a gradient decent algorithm can be applied to achieve suboptimal solution.

The initial state can be chosen as the result of location estimation in time slot $(t-1)$

$$\hat{\mathbf{x}}_i^{(t,0)} = \hat{\mathbf{x}}_i^{(t-1,N_{\text{iter}})} \tag{54}$$

where N_{iter} represents the number of iterations at a given time instant. In each iteration, every agent broadcasts its current location estimation $\hat{\mathbf{x}}_i^{(t,l-1)}$ and receives the location estimations from its neighbors $\hat{\mathbf{x}}_j^{(t,l-1)}$ $(j \in \mathcal{S}_{\to i}^{(t)})$. The agent then updates it estimation as

$$\hat{\mathbf{x}}_i^{(t,l)} = \hat{\mathbf{x}}_i^{(t,l-1)} + \delta_i^{(t,l)} \sum_{j\in\mathcal{S}_{\to i}^{(t)}} \psi_{j\to i}^{(t,l-1)} \tag{55}$$

where for notation convenience, we introduce

$$\psi_{j\to i}^{(t,l-1)} \triangleq \left.\frac{\partial c_{j\to i}\left(z_{j\to i}^{(t)};\mathbf{x}_i,\hat{\mathbf{x}}_j^{(t,l-1)}\right)}{\partial \mathbf{x}_i}\right|_{\mathbf{x}_i=\hat{\mathbf{x}}_i^{(t,l-1)}}. \tag{56}$$

Step size $\delta_i^{(t,l)}$ controls the convergence speed and can be chosen by methods such as backtracking line search (Boyd & Vandenberghe, 2004). The updated result in the maximum number of iteration is set as the location estimation of each agent $\hat{\mathbf{x}}_i^{(t)} = \hat{\mathbf{x}}_i^{(t,N_{\text{iter}})}$.

Factor Graph

A factor graph $\mathcal{G} = (\mathcal{V}, \mathcal{E})$ is a bipartite graph that describes the structure of function factorization (Kschischang, 2001), where $\mathcal{V}$ and $\mathcal{E}$ are respectively the sets of vertices and edges. Consider a high dimensional function g, with variables collected in the set $\mathcal{X} = \{\mathbf{x}_1, \cdots, \mathbf{x}_M\}$. Suppose it can be written as a product of several local functions, each having some subset as arguments

$$g(\mathbf{x}_1, \cdots, \mathbf{x}_M) = \prod_{j \in \mathcal{J}} f_j(\mathcal{X}_j) \tag{57}$$

where $\mathcal{J}$ is a discrete index set, $\mathcal{X}_j$ is a subset of $\mathcal{X}$, and $f_j(\mathcal{X}_j)$ is a function having $\mathcal{X}_j$ as the argument. Then the factor graph corresponds to the factorization can be constructed by the following rules

In set $\mathcal{V}$, there is a node for each variable $\mathbf{x}_i$ and a vertex for each local function f_j.
In set $\mathcal{E}$, there is an edge connecting node $\mathbf{x}_i$ with vertex f_j if and only if $\mathbf{x}_i \in \mathcal{X}_j$.

Based on the factor graph, the calculation of marginal distributions for each variable can be efficiently completed via the sum-product algorithm (Liu, Dai & Win, 2017).

Sum Product Algorithm

Let $\mu_{a \to b}$ denotes the message passed from node a to node b in the operation of the sum-product algorithm, and let $\mathcal{N}_v$ be the set of neighbors of a given node v. It can be that the neighbors of a variable node are function vertices, and the neighbors of a function vertex are variable nodes in a factor graph. Thus there is one and only one variable node for each edge, and it is called the variable associated with the edge.

Before we introduce the information update rule, an operation named summary should be defined. Summary over variable x means doing certain operation without x. In the context of MMSE estimation, the operation refers to integral; and it refers to maximization for MAP estimation. Without ambiguity, the nonstandard notation $\sum_{\sim\{x\}} f$ is used to denote summary over x.

The Sum-Product Update Rule: the message sent from a node v on an edge e is the product of the local function at v (or the unit function if v is a variable node) with all messages received at v on edges other than e, summarized for the variables associated with e.

By applying the above results, an explicit presentation of the message is given by

Variable to local function:

$$\mu_{x\to f}\left(x\right)=\prod_{h\in\mathcal{N}_x\setminus\{f\}}\mu_{h\to x}\left(x\right). \tag{58}$$

Local function to variable:

$$\mu_{f\to x}(x)=\sum_{\sim\{x\}}\left(f(\mathcal{N}_f)\prod_{y\in\mathcal{N}_f\setminus\{x\}}\mu_{y\to f}(y)\right). \tag{59}$$

By the definition of the factor graph, the marginal distribution g_i of the variable x_i is the product of all messages directed toward node x_i and thus can be calculated. Equivalently, it can be calculated as the product of the messages that are passed in opposite directions over any single edge associated with the variable x_i.

The calculation of marginal distributions for more than one variable is of much interest in many problems. Such a task can be accomplished by applying the procedure above for all indices separately. However, many intermediate values can be reused. In a cycle-free finite graph, the computation for all indices simultaneously can be efficiently accomplished by starting from any leaf node, passing through all possible edges in both directions, and calculating the message. For factor graph with cycles, iterative algorithms are developed but the convergence is not guaranteed. For more details, we refer readers to (Kschischang, Frey & Loeliger, 2004).

SPAWN Algorithm

As previously mentioned, the factorization of $p(\mathbf{x}^{(0:T)}\mid\mathbf{y}^{(1:T)})$ is of great concern in Bayesian methods. By the previous assumptions, the conditional probability function $p(\mathbf{x}^{(0:T)}\mid\mathbf{y}^{(1:T)})$ can be factorized as

$$p\left(\mathbf{x}^{(0:T)}\mid\mathbf{y}^{(1:T)}\right)\propto p\left(\mathbf{x}^{(0)}\right)p\left(\mathbf{y}_{\mathrm{s}}^{(1:T)}\mid\mathbf{x}^{(0:T)}\right)p\left(\mathbf{y}_{\mathrm{r}}^{(1:T)}\mid\mathbf{x}^{(0:T)}\right). \tag{60}$$

Then we have

$$p\left(\mathbf{x}^{(0:T)}\mid\mathbf{y}^{(1:T)}\right)\propto p\left(\mathbf{x}^{(0)}\right)\prod_{t=1}^{T}\left\{\left[\prod_{i=1}^{M}p\left(\mathbf{x}_i^{(t)}\mid\mathbf{x}_i^{(t-1)}\right)p\left(y_{i,\mathrm{s}}^{(t)}\mid\mathbf{x}_i^{(t-1)},\mathbf{x}_i^{(t)}\right)\right]\times p\left(\mathbf{y}_{\mathrm{r}}^{(t)}\mid\mathbf{x}^{(t)}\right)\right\} \tag{61}$$

where the likelihood function of the relative measurements at time t is given by

$$p\left(\mathbf{y}_{\mathrm{r}}^{(t)}\mid\mathbf{x}^{(t)}\right)=\prod_{i=1}^{M}\prod_{j\in\mathcal{S}_{\to i}^{(t)}}p\left(y_{j\to i}^{(t)}\mid\mathbf{x}_i^{(t)},\mathbf{x}_j^{(t)}\right). \tag{62}$$

An example network and its factor graph are show in Figure 3.

Figure 3. Each agent links to all anchors, and there is a cooperation link between Agent 1 and Agent 3. The red box represents the factor graph of the communication topology.

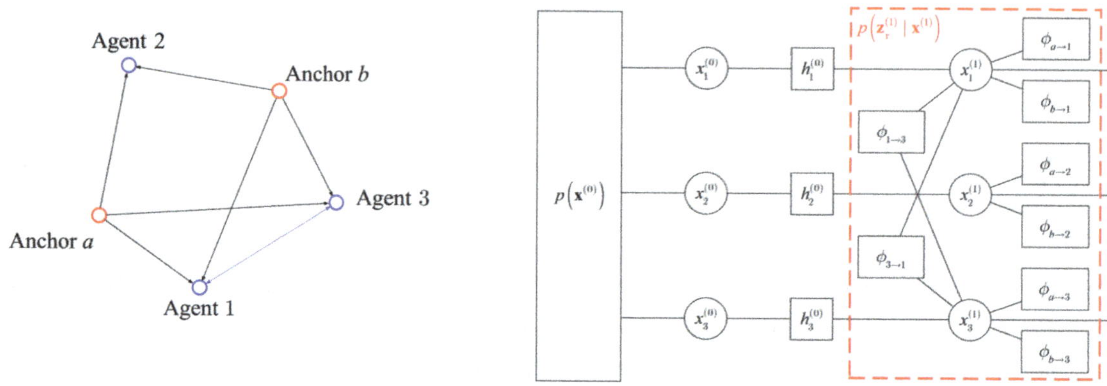

In mobile networks, the topology of the network varies from time to time. We can break the messages into intra-node messages and inter-messages. The former type is computed locally at each node, while the latter is passed via communication links between node pairs (Wymeersch, Lien & Win, 2009).

Since the SPAWN algorithm is based on Bayesian filtering, the description of the SPAWN algorithm can be departed into the prediction phase and the correction phase.

In the prediction phase, each node predicts its current position based on the estimation in the last time slot and the intra-node measurement, which is

$$\mu_{h_i^{(t-1)} \to X_i^{(t)}}\left(\mathbf{x}_i^{(t)}\right) \propto \int \underbrace{p\left(\mathbf{x}_i^{(t)} \mid \mathbf{x}_i^{(t-1)}\right) p\left(z_{i,s}^{(t)} \mid \mathbf{x}_i^{(t-1)}, \mathbf{x}_i^{(t)}\right)}_{=h_i^{(t-1)}\left(\mathbf{x}_i^{(t-1)}, \mathbf{x}_i^{(t)}\right)} \times \mu_{X_i^{(t-1)} \to h_i^{(t-1)}}\left(\mathbf{x}_i^{(t-1)}\right) \mathrm{d}\mathbf{x}_i^{(t-1)}. \tag{63}$$

In the correction phase, we define message $b_{X_i^{(t)}}^{(l)}(\cdot)$ as the belief of agent i at the l-th iteration in time slot t, which is initialized by the estimation in the prediction phase

$$b_{X_i^{(t)}}^{(0)}(\cdot) = \mu_{h_i^{(t-1)} \to X_i^{(t)}}(\cdot). \tag{64}$$

In the following iteration, each node broadcast its belief, and receives the ones from its neighbors. The message passing from node j to agent i is given by

$$\mu_{\phi_{j\to i} \to X_i^{(t)}}^{(l)}\left(\mathbf{x}_i^{(t)}\right) \propto \int p\left(z_{j\to i}^{(t)} \mid \mathbf{x}_i^{(t)}, \mathbf{x}_j^{(t)}\right) b_{X_j^{(t)}}^{(l-1)}\left(\mathbf{x}_j^{(t)}\right) \mathrm{d}\mathbf{x}_j^{(t)} \tag{65}$$

and the belief of agent i after this iteration is updated by

$$b_{X_i^{(t)}}^{(l)}\left(\mathbf{x}_i^{(t)}\right) \propto \mu_{h_i^{(t-1)} \to X_i^{(t)}}\left(\mathbf{x}_i^{(t)}\right) \prod_{j \in \mathcal{S}_{\to i}^{(t)}} \mu_{\phi_{j \to i} \to X_i^{(t)}}^{(l)}\left(\mathbf{x}_i^{(t)}\right). \tag{66}$$

At time t, the MMSE/MAP estimations can be determined at each agent by taking the mean or the mode of the final messages $b_{X_i^{(t)}}^{(N_{\text{iter}})}(\cdot)$.

COOPERATIVE LOCALIZATION IN COMPLEX ENVIRONMENTS

As aforementioned, localization suffers accuracy degradation in harsh propagation environments, i.e., in the presence of obstacles, where the NLOS errors will result in large localization errors when *a priori* knowledge of NLOS distribution is not available (Shen, Wymeersch & Win, 2010). To understand the theoretical performance of cooperative localization in wireless network is essential to the system design and algorithm development. The theoretical bound of cooperative localization under NLOS conditions was analyzed in (Nguyen et al., 2015) and a distributed cooperative algorithm based on LS-Gaussian variational message passing was proposed. The localization problem based on a geometric interpretation was formulated as an implicit convex feasibility problem in which some of the sets depend on the agents' positions and apply a parallel projection onto convex sets approach to estimate the unknown agents' positions (Gholami et al., 2013). By taking the NLOS biases as nuisance parameters along with agents' locations, a simple and efficient SDP-based localization was given in (Vaghefi & Buehrer, 2015).

In this subsection, the fundamental limits of cooperative localization under NLOS conditions will be briefly discussed. Note that the localization under NLOS conditions in a cooperative network is still not fully investigated and many open problems require further investigation.

Cooperative Localization Model Under NLOS Conditions

The network settings are repeated here for completeness. There are M agents and N anchors with positions $\mathbf{X} = [\mathbf{x}_1, \mathbf{x}_2, \ldots, \mathbf{x}_M]^{\mathrm{T}}$ and $\mathbf{X}_{\mathrm{b}} = [\mathbf{x}_{M+1}, \mathbf{x}_{M+2}, \ldots, \mathbf{x}_{M+N}]^{\mathrm{T}}$, respectively. The distance estimate between two node i and j at n-th measurement under NLOS conditions can be model as (Van, 2016)

$$\hat{d}_{i,j}[k] = \left\|\mathbf{x}_i - \mathbf{x}_j\right\| + z_{i,j}[k] + b_{i,j}[k], i, j \in \mathcal{M} \cup \mathcal{N} \times \mathcal{M} \cup \mathcal{N} \tag{67}$$

where $z_{i,j}[k] \sim \mathcal{N}\left(0, \sigma_{i,j}^2\right)$ is Gaussian measurement noise; $b_{i,j}[k]$ is the bias term accounting for the NLOS effect with $\mathbb{E}\{b_{i,j}\} = \mu_{i,j}$ and $\mathbb{V}\mathrm{ar}\{b_{i,j}\} = \varsigma_{i,j}^2$. NLOS errors $b_{i,j}[k]$ are always positive and generally assumed to be much larger than the measurement noise. The sample mean of K (k=1, 2, …, K) ranging measurements is given as follows

$$\hat{d}_{i,j} = \frac{1}{K}\sum_{k=1}^{K} \hat{d}_{i,j}[k] = \left\|\mathbf{x}_i - \mathbf{x}_j\right\| + z_{i,j} + b_{i,j} \tag{68}$$

where $z_{i,j} = (1/K)\sum_{k=1}^{K} z_{i,j}[k]$ and $b_{i,j} = (1/K)\sum_{k=1}^{K} b_{i,j}[k]$. Note that the LOS biases $b_{i,j}[k]$ are independent and identical distributed random variables underlying the same known PDF with a set of unknown parameters.

The LS solution of agents' positions for cooperative localization can be written as

$$\mathbf{X} = \arg\min_{\mathbf{x}} f_{\mathrm{LS}}\left(\mathbf{x}\right) \tag{69}$$

with

$$f_{\mathrm{LS}}\left(\mathbf{x}\right) = \sum_{i \in \mathcal{M}} \sum_{\substack{j>i \\ j \in \mathcal{M} \cup \mathcal{M}}} \frac{c_{i,j}}{\sigma_{i,j}^2 + \varsigma_{i,j}^2} \left(\hat{d}_{i,j} - \hat{b}_{i,j} - \left\|\mathbf{x}_i - \mathbf{x}_j\right\|\right)^2$$

where $\hat{b}_{i,j}$ is the estimate of $b_{i,j}$, $c_{i,j}$ is an indicator, such that $c_{i,j} = 1$ if nodes *i* and *j* can make measurements and $c_{i,j} = 0$ otherwise. When the bias estimate $\hat{b}_{i,j}$ is assumed to be unbiased and this LS solution achieves the CRB as $K \to \infty$ (Larsson & Danev, 2010).

Moreover, if we square the range estimate, the square-range LS (SR-LS) estimator is given by

$$\mathbf{X} = \arg\min_{\mathbf{x}} f_{\text{SR-LS}}\left(\mathbf{x}\right) \tag{70}$$

where

$$f_{\text{SR-LS}}\left(\mathbf{x}\right) = \sum_{i \in \mathcal{M}} \sum_{\substack{j>i \\ j \in \mathcal{M} \cup \mathcal{M}}} \frac{c_{i,j}}{\sigma_{i,j}^2 + \varsigma_{i,j}^2} \left[(\hat{d}_{i,j} - \hat{b}_{i,j})^2 - \left\|\mathbf{x}_i - \mathbf{x}_j\right\|^2\right]^2. \tag{71}$$

Note that the LS and SR-LS formulations are non-convex problem and it is hard to find an exact solution. The SR-LS localization of a non-cooperative case was investigated in (Beck, Stoica & Li, 2008) and its globally optimal solution can be obtained by generalized trust region sub-problem (GTRS) technique (Moré, 1993).

By introducing the weighting factors, the square-range WLS (SR-WLS) for cooperative localization can be obtained by

$$\mathbf{X} = \arg\min_{\mathbf{x}} f_{\text{SR-WLS}}\left(\mathbf{x}\right) \tag{72}$$

where

$$f_{\text{SR-WLS}}(\mathbf{x}) = \sum_{i\in\mathcal{M}} \sum_{\substack{j>i \\ j\in\mathcal{M}\cup\mathcal{M}}} \left\{ \frac{c_{i,j}}{\sigma_{i,j}^2 + \varsigma_{i,j}^2} \frac{1}{(\hat{d}_{i,j} - \hat{b}_{i,j})^2} \times \left[(\hat{d}_{i,j} - \hat{b}_{i,j})^2 - \left\|\mathbf{x}_i - \mathbf{x}_j\right\|^2 \right]^2 \right\}.$$

It is shown that the LS and SR-WLS cooperative solutions are asymptotically efficient as $K \to \infty$ even in the presence of NLOS biases.

FIM of Cooperative Localization Under NLOS Conditions

Let $\hat{\mathbf{d}}$ be the vector representing all ranging observations $\hat{d}_{i,j}$, the FIM can be calculated by as

$$\mathbf{J}(\mathbf{x}) = \mathbb{E}_{\hat{\mathbf{d}}} \left\{ - \begin{bmatrix} \dfrac{\partial^2 \ln f(\hat{\mathbf{d}};\mathbf{x})}{\partial \mathbf{x}_1 \partial \mathbf{x}_1^{\mathrm{T}}} & \cdots & \dfrac{\partial^2 \ln f(\hat{\mathbf{d}};\mathbf{x})}{\partial \mathbf{x}_M \partial \mathbf{x}_1^{\mathrm{T}}} \\ \vdots & \ddots & \vdots \\ \dfrac{\partial^2 \ln f(\hat{\mathbf{d}};\mathbf{x})}{\partial \mathbf{x}_1 \partial \mathbf{x}_M^{\mathrm{T}}} & \cdots & \dfrac{\partial^2 \ln f(\hat{\mathbf{d}};\mathbf{x})}{\partial \mathbf{x}_M \partial \mathbf{x}_M^{\mathrm{T}}} \end{bmatrix} \right\} \tag{73}$$

with

$$f(\hat{\mathbf{d}};\mathbf{x}) = \prod_{i\in\mathcal{M}} \prod_{\substack{j>i \\ j\in\mathcal{M}\cup\mathcal{M}}} \left[p_{\hat{d}_{i,j}}(\hat{d}_{i,j}) \right]^{c_{i,j}}. \tag{74}$$

Note that $\hat{d}_{i,j} \mid b_{i,j} \sim \mathcal{N}_1(b_{i,j} + \left\|\mathbf{x}_i - \mathbf{x}_j\right\|, \sigma_{i,j}^2 / K)$, then we have

$$\frac{\partial p_{\hat{d}_{i,j}}(x)}{\partial \mathbf{p}_i} = \frac{K}{\sigma_{i,j}^2} \frac{\mathbf{x}_i - \mathbf{x}_j}{\left\|\mathbf{x}_i - \mathbf{x}_j\right\|} \left[(x - \left\|\mathbf{x}_i - \mathbf{x}_j\right\|) p_{\hat{d}_{i,j}}(x) - \sqrt{\frac{K}{2\pi\sigma_{i,j}^2}} \mathbb{E}\left\{ b_{i,j} \mathrm{e}^{-\frac{K}{2\sigma_{i,j}^2}(x - b_{i,j} - \left\|\mathbf{x}_i - \mathbf{x}_j\right\|)^2} \right\} \right]. \tag{75}$$

Let

$$V_{i,j} = \frac{\sqrt{K}}{\sigma_{i,j}^2}(z_{i,j} + b_{i,j}), \quad U_{i,j} = \frac{\sqrt{K}}{\sigma_{i,j}^2} b_{i,j}$$

be the normalized noise-plus-bias variable and the normalized NLOS bias of ranging $\hat{d}_{i,j}$, respectively. Then it follows the form of that

$$\mathbb{E}_{\hat{d}_{i,j}}\left\{-\frac{\partial^2 \ln p_{\hat{d}_{i,j}}(\hat{d}_{i,j})}{\partial \mathbf{p}_i \partial \mathbf{p}_j^{\mathrm{T}}}\right\} = \frac{N}{\sigma_{i,j}^2}\frac{\left(\mathbf{x}_i - \mathbf{x}_j\right)\left(\mathbf{x}_i - \mathbf{x}_j\right)^{\mathrm{T}}}{\left\|\mathbf{x}_i - \mathbf{x}_j\right\|}\chi_{i,j}, \ \forall j \in \mathcal{M} \cup \mathcal{M}, j > i$$

where $\chi_{i,j}$ is called the LOS figure for the link between nodes *i* and *j*, is given by

$$\chi_{i,j} = 1 - \frac{\varsigma_{i,j}^2 + K\mu_{i,j}^2}{\sigma_{i,j}^2} + \int_{-\infty}^{+\infty} \frac{\left(\mathbb{E}_{U_{i,j}}\{U_{i,j} e^{-\left(x - U_{i,j}\right)^2/2}\}\right)^2}{\sqrt{2\pi}\mathbb{E}_{U_{i,j}}\{e^{-\left(x - U_{i,j}\right)^2/2}\}} dx. \tag{76}$$

Then we obtain the (i, j) 2×2 submatrix of the FIM $\mathbf{J}\left(\mathbf{x}\right)$ as

$$\mathbb{E}_{\hat{\mathbf{d}}}\left\{-\frac{\partial^2 \ln f\left(\hat{\mathbf{d}};\mathbf{x}\right)}{\partial \mathbf{x}_i \partial \mathbf{x}_j^{\mathrm{T}}}\right\} = \begin{cases} \dfrac{Kc_{i,j}}{\sigma_{i,j}^2}\dfrac{\left(\mathbf{x}_i - \mathbf{x}_j\right)\left(\mathbf{x}_i - \mathbf{x}_j\right)^{\mathrm{T}}}{\left\|\mathbf{x}_i - \mathbf{x}_j\right\|}\chi_{i,j} & j = i, \\ -\sum_{\substack{l \in \mathcal{M} \cup \mathcal{M} \\ l \neq i}} \dfrac{Kc_{i,l}}{\sigma_{i,l}^2}\dfrac{\left(\mathbf{x}_i - \mathbf{x}_l\right)\left(\mathbf{x}_i - \mathbf{x}_l\right)^{\mathrm{T}}}{\left\|\mathbf{x}_i - \mathbf{x}_l\right\|}\chi_{i,j} & j \neq i. \end{cases} \tag{77}$$

Note that when all LOS figures are equal to 1, reduces to the FIM expression for the LOS-only environment. The best and worst case FIMs can be given as follows.

Let $\chi_{i,j}^{\mathrm{G}}$ and $\chi_{i,j}^{\mathrm{LOS}}$ be the LOS figures for a Gaussian bias $b_{i,j} \sim \mathcal{N}_1(\mu_{i,j}, \varsigma_{i,j}^2 / K)$ and a constant bias $b_{i,j}$ with $\varsigma_{i,j}^2 = 0$ (i.e., $b_{i,j} = \mu_{i,j}$ with probability 1), respectively. Then, we have

$$\frac{\sigma_{i,j}^2}{\sigma_{i,j}^2 + \varsigma_{i,j}^2} = \chi_{i,j}^{\mathrm{G}} \leq \chi_{i,j} \leq \chi_{i,j}^{\mathrm{LOS}} = 1 \tag{78}$$

and the FIM $\mathbf{J}\left(\mathbf{x}\right)$ is bounded as

$$\mathbf{G}\left(\mathbf{x}\right) \preceq \mathbf{J}\left(\mathbf{x}\right) \preceq \mathbf{L}\left(\mathbf{x}\right) \tag{79}$$

where $\mathbf{G}\left(\mathbf{x}\right)$ represents the FIM corresponding to Gaussian biases for all links, i.e., $\chi_{i,j} = \chi_{i,j}^{\mathrm{G}}$ whereas $\mathbf{L}\left(\mathbf{x}\right)$ is the FIM for full LOS links or equivalently constant biases for all links ($\chi_{i,j} = 1$).

The proof of above conclusion in can be referred to (Nguyen et al., 2016) and it discloses that the NLOS biases always degrade the localization accuracy, and the Gaussian bias is the worst case among those with the same mean and variance.

NUMERICAL RESULTS

In this part, the simulations are presented for evaluating the performance gain of the cooperative localization algorithm. Both of centralized methods and distributed methods will be illustrated and the NLOS errors are also considered to test the effectiveness of cooperation among agent nodes.

Centralized Methods

Consider a two-dimensional dynamic network within a square of $20 \times 20\ \mathrm{m}^2$, consisting five anchors and four agents. The location of anchors are [20, 20]m, [20, -20]m, [-20, 20]m, [-20, -20]m and [0 0]m while agents are randomly placed in a rectangle area bounded by $[\pm 15, \pm 15]$ m. The MDS and SDP based algorithms with cooperative and non-cooperative version are illustrated in Figure 4. The agents in the network are assumed to be fully connected with all anchors and agents.

As shown in the Figure 4, the root mean squared errors (RMSE) of algorithms (cooperative and non-cooperative) show the superior performance of cooperative algorithm due to the additional measurements of unknown agents. The performance comparison also shows that the cooperative SDP performs superior

Figure 4. The RMSE comparison of MDS and SDP based non-cooperative and cooperative algorithms. The standard deviation of range measurement noise σ between anchor and agent is ranged from 0 to 1 meter, and the measurement noise between agent and agent is 0.1σ.

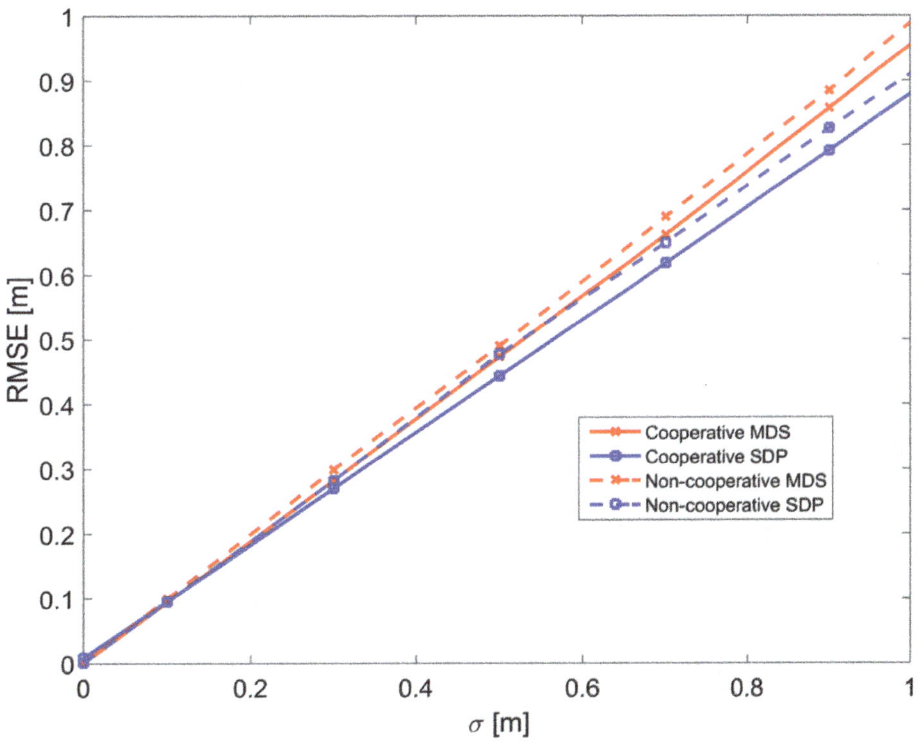

to the MDS counterparts. Note that the trend of these curves is linear with the noise standard deviation which is consistent with our previous analyses.

Furthermore, we consider the localization performance comparison under NLOS propagation environments. The standard deviation of measurement noise is $\sigma = 0.3$ m. From Figure 5, we can observe that, with NLOS propagation, the cooperation between agents brings more information in localization which performs superior to the noncooperation methods. The performance gain is larger than the result shown in Figure 4, especially when the NLOS errors are large. It is shown in Figure 4 that the SDP-based method is more robust than the MDS-based method and even the non-cooperative SDP holds lower RMSE in NLOS conditions.

The cooperation gain with the number of agents is shown in Figure 6 and we can observe that the both of two cooperative methods have significant performance gain when the number of agent increases. It illustrates that more agent cooperation will bring more performance gain. Moreover, the curve of cooperative SDP method experiences a steeper slope which means the cooperation constrains are tighter than cooperative MDS method. On the other hand, the RMSE of non-cooperative methods almost keep constant during the all range since no more additional information is available.

Figure 5. The RMSE comparison of MDS and SDP based non-cooperative and cooperative algorithms. The standard deviation of range measurement noise σ is 0.3 m, and the upper bound of NLOS errors are varying from zero to one m. We randomly choose one to three anchors provide the NLOS measurements to each agent and the NLOS errors obey the uniform distribution between 0 to b meters.

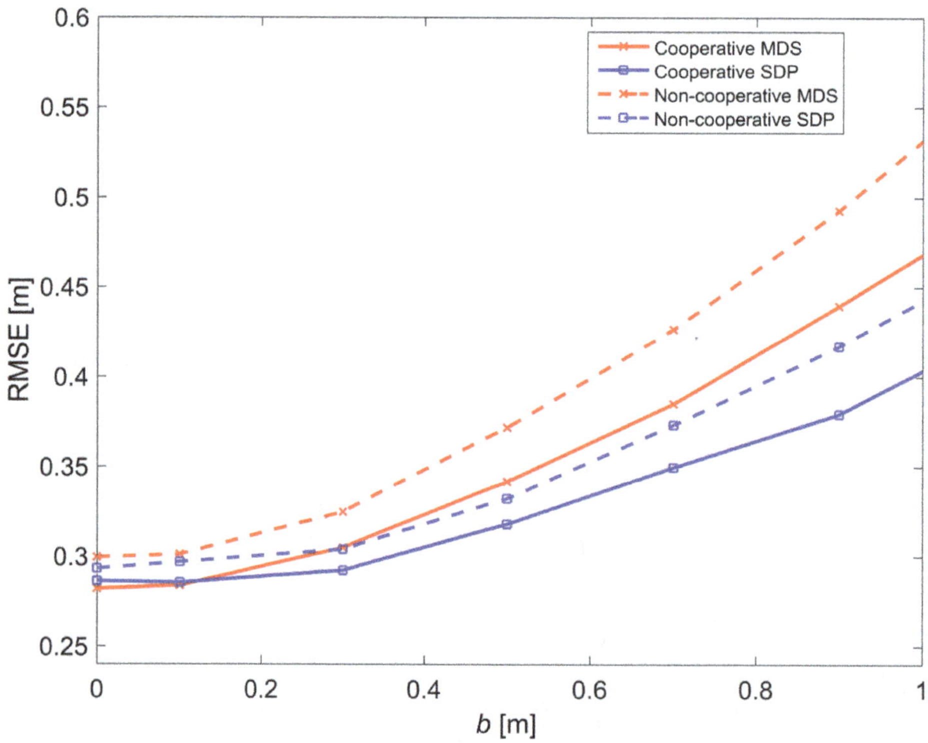

Figure 6. The cooperation gain in centralized methods. The standard deviation of measurement noise σ is 0.3 meters and the number of agents varying from two to eight.

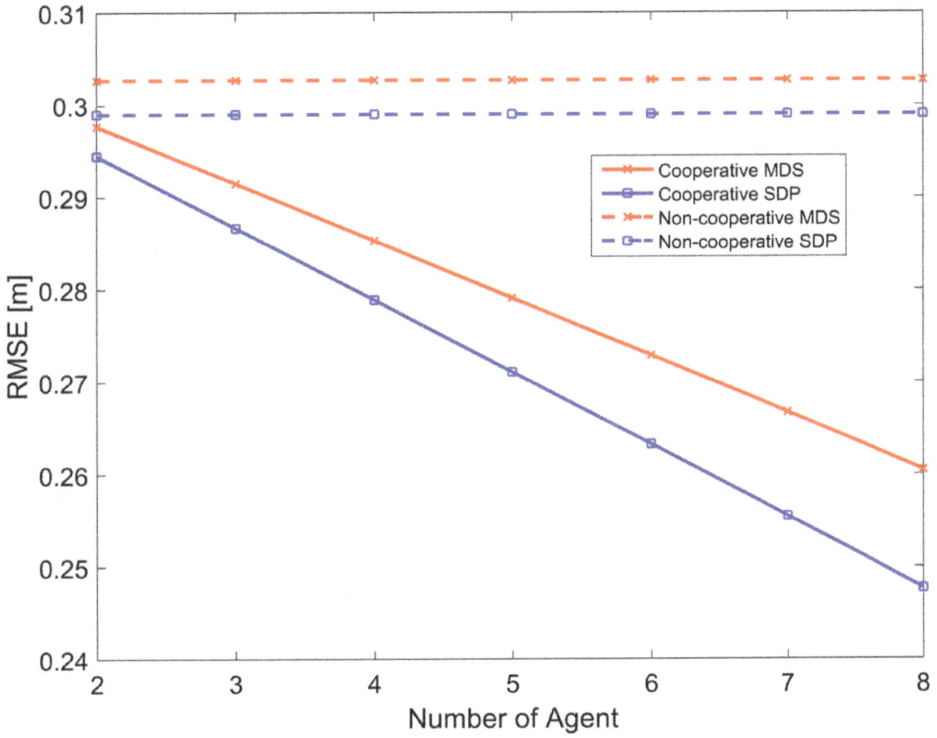

Distributed Methods

Consider a two-dimensional dynamic network within a square of $20 \times 20\ \mathrm{m}^2$, consisting five agents and four anchors placed at its corners. The agents move independently according to Gaussian random walk $\mathcal{N}(0\ \mathrm{m}, \mathbf{I}_{2\times2}\ \mathrm{m}^2)$ reflecting in the area, and are equipped with IMUs whose measure errors are modeled as Gaussian variables, $\mathcal{N}(0\ \mathrm{m}, 0.01 \times \mathbf{I}_{2\times2}\ \mathrm{m}^2)$. Assume that each agent is within the communication range of all nodes, and the maximum iteration number is set as $N_{\mathrm{iter}} = 20$.

The RMSE of location estimation increases as the variance of ranging measurements grows. The comparison between non-Bayesian and Bayesian methods under both non-cooperative and cooperative schemes are presented in Figure 7. It can be seen that both of two cooperative methods performs better than their non-cooperative counterparts and the cooperative gain is larger when the noise standard deviation is higher.

The curves in Figure 8 show the superior performance of distributed cooperative localization under NLOS propagation environment since the cooperation between agents is more robust to the NLOS er-

Figure 7. The comparison between Bayesian method (MMSE) and the non-Bayesian method (LS). The standard deviation of range measurement noise σ is ranged from zero to one meter. It is shown that cooperation between agents helps to improve the localization accuracy.

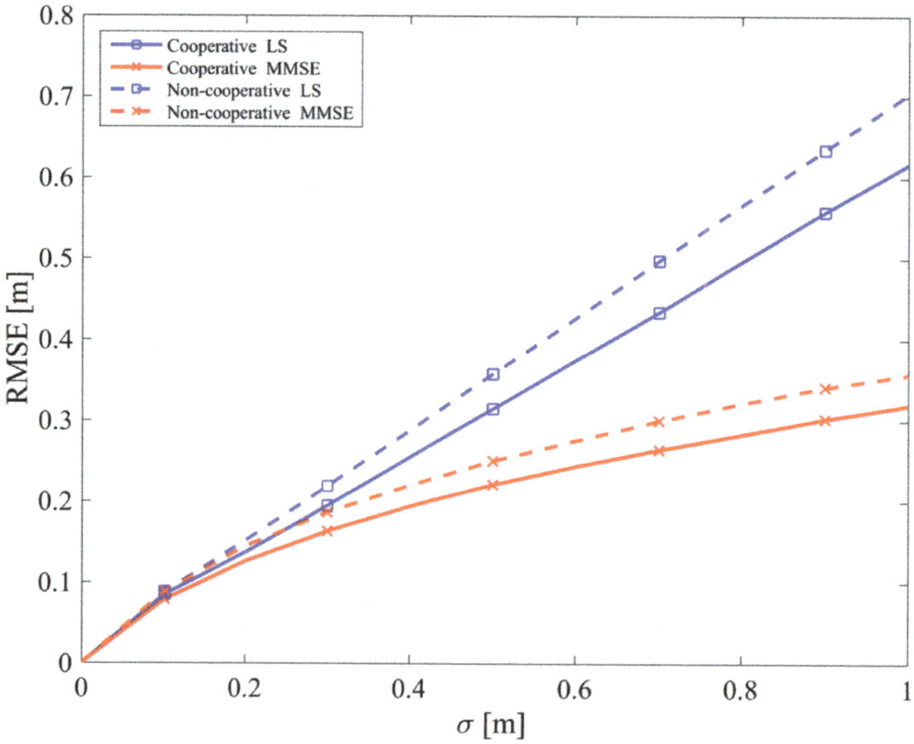

rors. In Figure 8, one to three anchors are randomly chosen that suffer from the NLOS errors which is uniformly distributed between 0 to b meters. From Figure 7 and Figure 8, it is illustrated the cooperative LS is better than non-cooperative MMSE when NLOS errors becomes large. Although the MMSE based methods hold lower RMSE, LS based methods have higher cooperative gain when the NLOS error is large. It is shown that cooperative LS method is more robust to the NLOS errors by utilizing the constraints more efficiently.

In Figure 9, we can see that when the number of agents increases, the cooperative methods achieve higher localization accuracy due to more cooperation constraints between these agents. The improvement of cooperative gain of the LS method is larger than MMSE, but the MMSE performs much better than LS. The cooperative LS is approximately equivalent to the non-cooperative MMSE when the number of agents is 7. As we know, non-cooperative methods do not benefit from this additional information and the localization accuracy remains constant in the figure.

Figure 8. The comparison between Bayesian method (MMSE) and the non-Bayesian method (LS) under NLOS propagation environment. The standard deviation of measurement noise σ is 0.3 meters and the upper bound of NLOS errors are varying from zero to one meter.

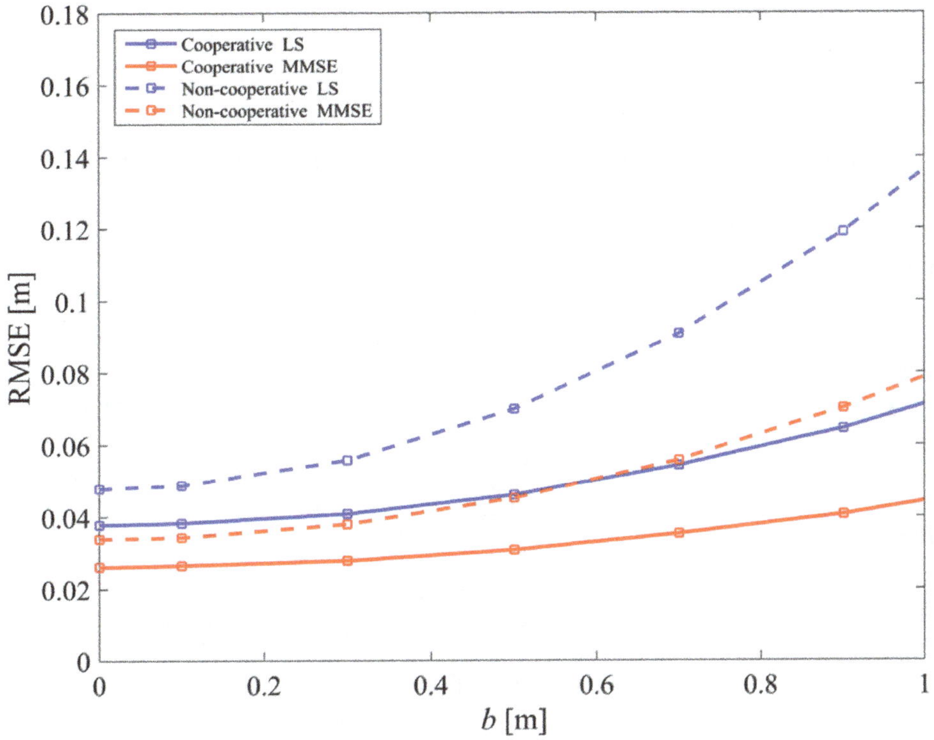

CONCLUSION

This chapter introduces both the theory and algorithms of cooperative localization in wireless networks. There are two kinds of cooperation among nodes in a network, i.e., spatial cooperation between neighboring nodes and temporal cooperation from the previous states of a node itself. With the relative measurements between agents, cooperative localization can provide high-accuracy location estimates of mobile agents, which is important for many coming location-based services. The analysis based on the FIM demonstrates the superior performance of cooperative localization compared to their non-cooperative counterparts. In terms of algorithms, cooperative localization algorithms can be categorized into centralized and distributed ones, depending on whether centralized computation is required. The centralized algorithms can generally achieve higher localization accuracy, while distributed ones require lower communication overheads, which is more suitable for practical usage. Then the analyses of cooperative localization under NLOS conditions show that the NLOS biases always degrade the localization accu-

Figure 9. The cooperation gain in Bayesian and non-Bayesian cooperative algorithms. The standard deviation of measurement noise σ is 0.3 meters and the number of agents varying from two to eight.

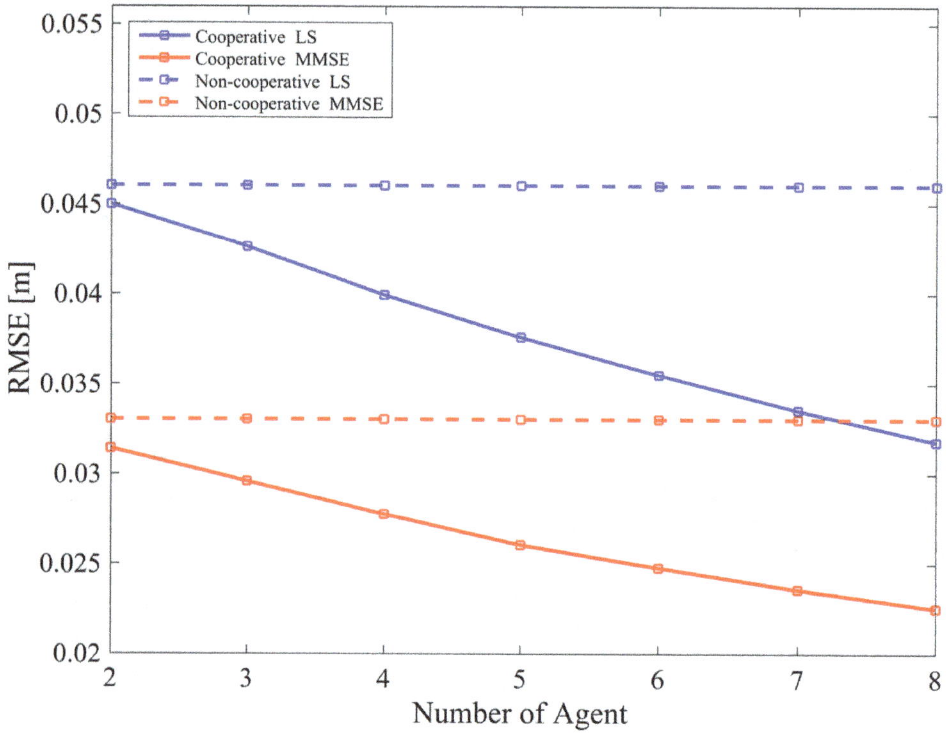

racy, and the Gaussian bias is the worst case among all of the same mean and variance. Finally, the two types of localization algorithms are compared with numerical results, and the advantages of cooperative localization are also verified through these results.

ACKNOWLEDGMENT

This work is supported by National Science Foundation of China under Grant No. 61501279 and 91638204.

REFERENCES

Abu-Shaban, Z., Zhou, X., & Abhayapala, T. D. (2016). A novel TOA-based mobile localization Technique under mixed LOS/NLOS conditions for cellular networks. *IEEE Transactions on Vehicular Technology*, *65*(11), 8841–8853. doi:10.1109/TVT.2016.2517151

Akyildiz, I. F., Su, W., Sankarasubramaniam, Y., & Cayirci, E. (2002). A survey on sensor networks. *IEEE Communications Magazine*, *40*(8), 102–114. doi:10.1109/MCOM.2002.1024422

Aspnes, J., Eren, T., Goldenberg, D. K., Morse, A. S., Whiteley, W., Yang, Y. R., & Belhumeur, P. N. et al. (2006). A Theory of Network Localization. *IEEE Transactions on Mobile Computing*, *5*(12), 1663–1678. doi:10.1109/TMC.2006.174

Beck, A., Stoica, P., & Li, J. (2008). Exact and approximate solutions for source localization problems. *IEEE Transactions on Signal Processing*, *56*(5), 1770–1778. doi:10.1109/TSP.2007.909342

Biswas, P., Lian, T. C., Wang, T. C., & Ye, Y. (2006). Semidefinite programming based algorithms for sensor network localization. *ACM Transactions on Sensor Networks*, *2*(2), 188–220. doi:10.1145/1149283.1149286

Biswas, P., Liang, T. C., Toh, K. C., Wang, T. C., & Ye, Y. (2006). Semidefinite programming approaches for sensor network localization with noisy distance measurements. *IEEE Transactions on Automation Science and Engineering*, *3*(4), 360–371. doi:10.1109/TASE.2006.877401

Biswas, P., & Ye, Y. (2003). *Semidefinite programming for ad hoc wireless sensor network localization*. Stanford, CA: Tech. Rep.

Biswas, P., & Ye, Y. (2003). *A distributed method for solving semidefinite programs arising from ad hoc wireless sensor network localization*. Department of Computer Science, Stanford Univ., Stanford, CA, Tech. Rep., Oct.

Borg, I., & Groenen, P. (1997). *Modern multidimensional scaling, Theory and Applications New York*. Springer-Verlag. doi:10.1007/978-1-4757-2711-1

Boyd, S., & Vandenberghe, L. (2004). *Convex optimization*. New York: Cambridge University Press. doi:10.1017/CBO9780511804441

Caceres, M. A., Sottile, F., & Garello, R. (2010). Hybrid GNSS-TOA localization and tracking via cooperative unscented Kalman filter. *IEEE International Symposium on Personal, Indoor and Mobile Radio Communications Workshops*, 272-276. doi:10.1109/PIMRCW.2010.5670378

Caffery, J. J. (1999). *Wireless Location in CDMA Cellular Radio Systems*. Norwell, MA: Kluwer.

Capkun, S., Hamdi, M., & Hubaux, J. P. (2001). GPS-free positioning in mobile ad hoc networks. *Proc. 34th IEEE Hawaii Int. Conf. System Sciences (HICSS-34)*, *34*, 1-10.

Cassioli, D., Win, M. Z., & Molisch, A. F. (2002). The ultra-wide bandwidth indoor channel: From statistical model to simulations. *IEEE Journal on Selected Areas in Communications*, *20*(6), 1247–1257. doi:10.1109/JSAC.2002.801228

Chan, F. K. W., & So, H. C. (2009). Efficient weighted multidimensional scaling for wireless sensor network localization. *IEEE Transactions on Signal Processing*, *57*(11), 4548–4553. doi:10.1109/TSP.2009.2024869

Chan, F. K. W., So, H. C., & Ma, W. K. (2009). A novel subspace approach for cooperative localization in wireless sensor networks using range measurements. *IEEE Transactions on Signal Processing*, *57*(1), 260–269. doi:10.1109/TSP.2008.2005870

Chazan, D., Zakai, M., & Ziv, J. (1975). Improved lower bounds on signal parameter estimation. *IEEE Transactions on Information Theory*, *21*(1), 90–93. doi:10.1109/TIT.1975.1055325

Cheng, Z., Wang, Y., & Shen, Y. (2017). Direct position determination of multiple targets via reduced-dimension beamspace. *IEEE International Conference on Communications Workshops*, 1030-1035. doi:10.1109/ICCW.2017.7962794

Cheung, K. W., & So, H. C. (2005). A Multidimensional scaling framework for mobile location using time-of-arrival measurements. *IEEE Truncations Signal Processing*, *53*(2), 460–470. doi:10.1109/TSP.2004.840721

Dai, W., Shen, Y., & Win, M. Z. (2017). A computational geometry framework for efficient network localization. *IEEE Transactions Information Theory*.

Dardari, D., Conti, A., Ferner, U., Giorgetti, A., & Win, M. Z. (2009). Ranging with ultra wide bandwidth signals in multipath environments. *Proceedings of the IEEE*, *97*(2), 404–426. doi:10.1109/JPROC.2008.2008846

Doherty, L., Pister, K. S. J., & Ghaoui, L. E. (2001). Convex position estimation in wireless sensor networks. *Proceedings - IEEE INFOCOM*, *3*, 1655–1663.

Foxlin, E. (2005). Pedestrian tracking with shoe-mounted inertial sensors. *IEEE Computer Graphics and Applications*, *25*(6), 38–46. doi:10.1109/MCG.2005.140 PMID:16315476

Gezici, S., Tian, Z., Biannakis, B. G., Kobayashi, H., Molisch, A. F., Poor, H. V., & Sahinoglu, Z. (2005). Localization via ultra-wideband radios: A look at positioning aspects for future sensor networks. *IEEE Transactions on Signal Processing*, *22*(4), 70–84. doi:10.1109/MSP.2005.1458289

Gholami, M. R., Tetruashvili, L., Strom, E. G., & Censor, Y. (2013). Cooperative wireless sensor network positioning via implicit convex feasibility. *IEEE Transactions on Signal Processing*, *61*(23), 5830–5840. doi:10.1109/TSP.2013.2279770

Güvenç, I., Chong, C. C., Watanabe, F., & Inamura, H. (2007). NLOS identification and weighted least-squares localization for UWB systems using multipath channel statistics. *EURASIP Journal on Advances in Signal Processing*, *36*, 1–14.

Hero-III, A. O., Fessler, J. A., & Usman, M. (1996). Exploring estimator bias-variance tradeoffs using the uniform CR bound. *IEEE Transactions on Signal Processing*, *44*(8), 2026–2041. doi:10.1109/78.533723

Ihler, A. T., Fisher, J. W., Moses, R. L., & Willsky, A. S. (2005). Nonparametric belief propagation for self-localization of sensor networks. *IEEE Journal on Selected Areas in Communications*, *23*(4), 809–819. doi:10.1109/JSAC.2005.843548

Kay, S. M. (2013). *Fundamentals of statistical signal processing: Practical algorithm development*. Pearson Education.

Kschischang, F. R., Frey, B. J., & Loeliger, H. A. (2001). Factor graphs and the sum-product algorithm. *IEEE Transactions on Information Theory*, *47*(2), 498–519. doi:10.1109/18.910572

Larsson, E. G., & Danev, D. (2010). Accuracy comparison of LS and squared range LS for source localization. *IEEE Transactions on Signal Processing*, *58*(2), 916–923. doi:10.1109/TSP.2009.2032036

Leeuw, J. D. (1988). Convergence of the majorization method for multidimensional scaling. *Journal of Classification*, *5*(2), 163–180. doi:10.1007/BF01897162

Liang, T. C., Wang, T. C., & Ye, Y. (2004). *A gradient search method to round the semidefinite programming relaxation solution for ad hoc wireless sensor network localization*. Sanford University.

Lien, J., Ferner, U. J., Srichavengsup, W., Wymeersch, H., & Win, M. Z. (2012). A Comparison of Parametric and Sample-Based Message Representation in Cooperative Localization. *International Journal of Navigation and Observation*, *2012*, 1–10. doi:10.1155/2012/281592

Liu, Z., Dai, W., & Win, M. Z. (2017). Mercury: An infrastructure-free system for network localization and Navigation. *IEEE Transactions Mobile Computing*.

Loeliger, H. A. (2004). An introduction to factor graphs. *IEEE Signal Processing Magazine*, *21*(1), 28–41. doi:10.1109/MSP.2004.1267047

Luo, Z.-Q., Ma, W. K., So, A. M.-C., Ye, Y., & Zhang, S. (2010). Semidefinite relaxation of quadratic optimization problems. *IEEE Signal Processing Magazine*, *27*(3), 20–34. doi:10.1109/MSP.2010.936019

Lv, T., Gao, H., Li, X., Yang, S., & Hanzo, L. (2016). Space-time hierarchical-graph based cooperative localization in wireless sensor networks. *IEEE Transactions on Signal Processing*, *64*(2), 322–334. doi:10.1109/TSP.2015.2480038

Mao, G., Fidan, B., & Anderson, B. D. (2007). Wireless sensor network localization techniques. *Computer Networks*, *51*(10), 2529–2553. doi:10.1016/j.comnet.2006.11.018

Marano, S., Gifford, W. M., Wymeersch, H., & Win, M. Z. (2010). NLOS identification and mitigation for localization based on UWB experimental data. *IEEE Journal on Selected Areas in Communications*, *28*(7), 1026–1035. doi:10.1109/JSAC.2010.100907

Moré, J. J. (1993). Generalizations of the trust region problem, Optimum. *Methods Software*, *2*(3-4), 189–209. doi:10.1080/10556789308805542

Nguyen, T. V., Jeong, Y., Shin, H., & Win, M. Z. (2015). Least square cooperative localization. *IEEE Transactions on Vehicular Technology*, *64*(4), 1318–1330. doi:10.1109/TVT.2015.2398874

Patwari, N., Ash, J. N., Kyperountas, S., Hero, A. O., Moses, R. L., & Correal, N. S. (2005). Locating the nodes: Cooperative localization in wireless sensor networks. *IEEE Signal Processing Magazine*, *22*(4), 54–69. doi:10.1109/MSP.2005.1458287

Patwari, N., Hero, A. O., Perkins, M., Correal, N. S., & O'dea, R. J. (2003). Relative location estimation in wireless sensor networks. *IEEE Transactions on Signal Processing*, *51*(8), 2137–2148. doi:10.1109/TSP.2003.814469

Poor, H. V. (2013). *An introduction to signal detection and estimation*. Springer Science & Business Media.

Ramsay, J. O. (1982). Some statistical approaches to multidimensional scaling data. *Journal of the Royal Statistical Society. Series A (General)*, *145*(3), 285–312. doi:10.2307/2981865

Rui, L., & Ho, K. C. (2014). Algebraic solution for joint localization and synchronization of multiple sensor nodes in the presence of beacon uncertainties. *IEEE Transactions on Wireless Communications, 13*(9), 5196–5210. doi:10.1109/TWC.2014.2322053

Savarese, C., Rabaey, J. M., & Beutel, J. (2001). Location in distributed ad-hoc wireless sensor networks. *IEEE International Conference on Acoustics, Speech, and Signal Processing, 4*, 2037-2040. doi:10.1109/ICASSP.2001.940391

Savic, V., & Zazo, S. (2012). Reducing communication overhead for cooperative localization using nonparametric belief propagation. *IEEE Wireless Communications Letters, 1*(4), 308–311. doi:10.1109/WCL.2012.042512.120172

Saxe, J. B. (1980). *Embeddability of weighted graphs in k-space is strongly NP-hard. Carnegie-Mellon University*. Department of Computer Science.

Schmidt, R. (1986). Multiple emitter location and signal parameter estimation. *IEEE Transactions on Antennas and Propagation, 34*(3), 276–280. doi:10.1109/TAP.1986.1143830

Shang, Y., Rumi, W., Zhang, Y., & Fromherz, M. (2004). Localization from connectivity in sensor networks. *IEEE Transactions on Parallel and Distributed Systems, 15*(11), 961–974. doi:10.1109/TPDS.2004.67

Shang, Y., & Ruml, W. (2004). Improved MDS-based localization. *INFOCOM 2004. Twenty-third Annual Joint Conference of the IEEE Computer and Communications Societies, 4*, 2640-2651.

Shang, Y., Ruml, W., Zhang, Y., & Fromherz, M. P. (2003). Localization from mere connectivity. *Proceedings of the 4th ACM International Symposium Mobile Ad Hoc Networking & Computing*, 201-212. doi:10.1145/778415.778439

Shen, Y. (2014). *Network localization and navigation: Theoretical framework, efficient operation, and security assurance*. Cambridge, MA: Massachusetts Institute of Technology.

Shen, Y., Wymeersch, H., & Win, M. Z. (2010). Fundamental limits of wideband localization - Part II: Cooperative networks. *IEEE Transactions on Information Theory, 56*(10), 4981–5000. doi:10.1109/TIT.2010.2059720

So, A. M. C., & Ye, Y. (2007). Theory of semidefinite programming for sensor network localization. *Mathematical Programming, 109*(2-3), 367–384. doi:10.1007/s10107-006-0040-1

Sottile, F., Wymeersch, H., Caceres, M. A., & Spirito, M. A. (2011). Hybrid GNSS-terrestrial cooperative positioning based on particle filter. *IEEE Global Telecommunications Conference*, 1-5. doi:10.1109/GLOCOM.2011.6134002

Tenenbaum, J. B., De Silva, V., & Langford, J. C. (2000). A global geometric framework for nonlinear dimensionality reduction. *Science, 290*(5500), 2319–2323. doi:10.1126/science.290.5500.2319 PMID:11125149

Tinsley, H. E., & Brown, S. D. (2000). *Handbook of applied multivariate statistics and mathematical modeling*. Academic Press.

Torrieri, D. J. (1984). Statistical theory of passive location systems. *IEEE Transactions on Aerospace and Electronic Systems, AES-20*(2), 183–198. doi:10.1109/TAES.1984.310439

Vaghefi, R. M., & Buehrer, R. M. (2015). Cooperative localization in NLOS environments using semidefinite programming. *IEEE Communications Letters, 19*(8), 1382–1385. doi:10.1109/LCOMM.2015.2442580

Wang, G., So, M. C., & Li, Y. (2016). Robust convex approximation methods for TDOA-based localization under NLOS conditions. *IEEE Transactions on Signal Processing, 64*(13), 3281–3296. doi:10.1109/TSP.2016.2539139

Wang, J., & Shen, Y. (2017). On the Discretization Schemes in Map-aided Indoor Localization. *IEEE Communications Letters.*

Wang, T., Shen, Y., Conti, A., & Win, M. Z. (2017). Network navigation with scheduling: error evolution, *IEEE Transactions Information Theory.*

Wang, Y., & Wu, Y. (2017). An efficient semidefinite relaxation algorithm for moving source localization using TDOA and FDOA measurements. *IEEE Communications Letters, 21*(1), 80–83. doi:10.1109/LCOMM.2016.2614936

Wang, Z., & Zekavat, S. A. (2012). Omni-directional mobile NLOS identification and localization via multiple cooperative nodes. *IEEE Transactions on Mobile Computing, 11*(12), 2047–2059. doi:10.1109/TMC.2011.232

Win, M. Z., Conti, A., Mazuelas, S., Shen, Y., Gifford, W. M., Dardari, D., & Chiani, M. (2011). Network localization and navigation via cooperation. *IEEE Communications Magazine, 49*(5), 56–62. doi:10.1109/MCOM.2011.5762798

Wymeersch, H., Lien, J., & Win, M. Z. (2009). Cooperative localization in wireless networks. *Proceedings of the IEEE, 97*(2), 427–450. doi:10.1109/JPROC.2008.2008853

Zhou, B., & Chen, Q. (2016). On the particle-assisted stochastic search mechanism in wireless cooperative localization. *IEEE Transactions on Wireless Communications, 15*(7), 4765–4777.

Zhou, F., & Shen, Y. (2017). On the outage probability of localization in randomly deployed wireless networks. *IEEE Communications Letters, 21*(4), 901–904. doi:10.1109/LCOMM.2017.2647919

Zinnes, J. L., & MacKay, D. B. (1983). Probabilistic multidimensional scaling: Complete and incomplete data. *Psychometrika, 48*(1), 27–48. doi:10.1007/BF02314675

Chapter 7
Position Estimation for IR-UWB Systems:
DPE, Multi-TOA Approach, and Their Compressive Sensing-Based Modeling

Eva Lagunas
University of Luxembourg, Luxembourg

Monica Navarro
Centre Tecnologic de Telecomunicacions de Catalunya (CTTC), Spain

Pau Closas
Northeastern University, USA

Montse Najar
Universitat Politecnica de Catalunya, Spain

Ricardo Garcia-Gutierrez
Universitat Politecnica de Catalunya, Spain

Genís Floriach
Universitat Politecnica de Catalunya, Spain

ABSTRACT

IR-UWB has emerged as a promising candidate for positioning passive nodes in wireless networks due to its extremely short time domain transmitted pulses. The two-step approaches in which first different TOAs are estimated and then fed into a triangulation procedure are suboptimal in general. This is because in the first stage of these methods, the measurements at distinct anchors are independent and ignore the constraint that all measurements must be consistent with a single emitter location. In this chapter, the authors investigate two techniques to overcome this issue. First, a two-step procedure based on multi-TOA estimation is proposed. Second, a positioning approach omitting the intermediate known as DPE is presented. Complementarily, the authors explore the CS-based modeling of both approaches so that the temporal sparsity of the UWB received signal and the consequent sparseness of the discrete spatial domain are exploited to select the most significant TOAs and to reduce the amount of information to be sent to a central fusion unit in the DPE approach.

DOI: 10.4018/978-1-5225-3528-7.ch007

INTRODUCTION

Localization and positioning represent one of the main themes that are being developed in the field of wireless communications. One of the existing technologies that offers better features related to the location applications is Impulse-Radio Ultra-Wideband (IR-UWB) (Gezici et al., 2005). This radio technology can potentially enable very accurate ranging and location applications, given the extremely short duration of the pulses.

To provide accurate position information of nodes, a number of reference nodes with known position (anchors) collect ranging information from radio signals emanating from the node with unknown position.

Range accuracy depends on the narrowness of the pulse so that a pulse having a very short duration will result in the highest time resolution, which implies the highest range accuracy. The large signal bandwidth of IR-UWB not only offers outstanding ranging capabilities but also provide a means for resolving multipath indoor components. The latter is important since dense multipath environments, which are usually the propagation conditions in indoor positioning, make the shortest path delay (containing information about the distance) difficult to be accurately estimated.

The conventional approach to solve the localization problem consists of a two-step procedure. First, some parameters of the transmitted signal are measured, and next, these parameters are combined using a simple trilateration or multi-lateration algorithm yielding the location of the target nodes. According to (Dardari, Conti, Ferner, Giorgetti, & Win, 2009), positioning systems can be divided into three main categories: Time-of-Arrival (TOA), Direction-of-Arrival (DOA) and Signal-Strength based systems. Signal-Strength method does not exploit the advantages of large bandwidth and, thus, they are not the best option for IR-UWB systems. The Angle-of-Arrival (AOA) or DOA based positioning technique involve the use of antenna arrays. Moreover, and due to the large bandwidth of the UWB signal, the number of paths may be very large, and therefore, accurate angle estimation becomes very challenging. Clearly, TOA-based techniques are better motivated for IR-UWB, due to its enhanced precision.

In this chapter, we focus on TOA-based IR-UWB positioning techniques. We first focus on two-step procedures, where proper first path detection acquires special relevance. Discrimination of the first arrival is extremely challenging in indoor environments since the first path may not necessarily be the one bearing the highest power. The performance of two-step TOA-based procedures has been shown to be very sensitive to the threshold selection used to select the first arriving path. Moreover, the high bandwidth nature of UWB systems imposes very high sampling rate requirement, which translate in high computational time, higher storage capabilities and strong constraints into the Analog-to-Digital Converter (ADC).In this context, we will introduce a TOA-based scheme that avoids the use of the threshold selection while relaxing the sampling requirements by introducing concepts from Compressive Sensing (CS) technology. CS (Candes & Wakin, 2008) provides a solution to the computational complexity issue by allowing to sample signals that are sparse or compressible under a certain basis, at a rate significantly less than the Nyquist sampling limit. Herein, exploiting the time-domain sparsity inherent in the IR-UWB signals, we consider CS techniques for compression and detection applied to TOA estimation for IR-UWB, where the basic strategy is to extract the most significant energy paths present in a frequency-domain Power Delay Profile (PDP) via CS-based reconstruction techniques.

An alternative to two-step procedure is the positioning approach omitting the intermediate step, which is known in the literature as Direct Position Estimation (DPE). DPE focuses on the direct estimation of position coordinates, which are indeed the parameters of interest to the end-user. Furthermore,

DPE does not require a threshold selection upon which accuracy of two-step approaches depend on. In this chapter, we will propose a frequency domain receiver architecture to accomplish DPE in IR-UWB systems for high resolution estimators based on Power-Position Profile (PPP).

However, the DPE comes at the cost of higher computational complexity, since the signals gathered at the anchors have to be combined. This represents a major challenge in IR-UWB systems, given its extremely high sampling rates thus generating huge amount of sampled data. As a solution to handle the issues associated with two-step and DPE methods, this chapter presents a novel technique which, unlike the DPE approach, does not send the complete received signals to the fusion center but instead it sends a set of TOAs estimates. All the measurements received at the fusion center are used to determine the target node position. Consequently, it reduces the ambiguity in the selection of the first path detection while reducing the amount of data to be transferred to the fusion center.

In addition, this chapter proposes CS techniques in the context of DPE-based localization for IR-UWB technology, in this case, exploiting the inherent spatial sparsity of the target 2D scene. In particular, CS is used to compress the signals to be transferred to the fusion center and to improve the detection accuracy on the PPP by using conventional CS reconstruction algorithms. In so doing, we avoid the PPP grid-search-based where the multiple location hypotheses are evaluated independently, thus allowing a reduction in computational time.

Finally, this chapter concludes with some interesting recommendations for future research directions and the conclusions.

RESEARCH TRENDS IN POSITIONING WITH IR-UWB

In this section, we describe the recent research trends in IR-UWB positioning technologies from the existing state of the art focusing on the three key components tackled in this chapter: Two-step approaches, DPE and CS.

Two-Step Approaches

In general, the positioning problem is divided into two sub-problems, in the first stage certain parameters are estimated from the signal and, in a second step, they are used to estimate the target node position. This procedure is known in the literature as two-step positioning. In this section, we give an overview of the possible positioning techniques based on the two-step approach. The two-step approaches are mainly based on the estimation of the received signal strength (RSS), DOA, TOA or time difference of arrival (TDOA) (Dardari, Closas & Djuric 2015). It is also important to note that is possible to estimate multiple of these parameters per signal in order to improve positioning accuracy.

Received Signal Strength (RSS)

In RSS-based positioning techniques, the distance between two nodes is calculated by measuring the energy of the received signal. The distance between two nodes provides a circle of uncertainty for the positioning of the target node. Therefore, a minimum of three distance measurements from three different anchor nodes are necessary to determine the unknown position. The target position is given by the intersection of these three circles, as indicated in the trilateration procedure illustrated in Figure 1. However, due to inaccuracies the three circles might not intersect. To overcome this issue, the least squares solution is typically used as estimation.

The signal strength dependency with distance comes from the path-loss which can be modeled as follows:

$$P(d) = P_0 - 10n \log_{10}(d) + S \tag{1}$$

where n is the path-loss exponent, $P(d)$ is the received power in dB at distance d, P_0 is the transmitted power and S represents the large-scale fading variations (i.e. shadowing). The RSS is affected by several propagation impairments, such as shadowing effects and small-scale fading. As RSS measurements depend on the channel characteristics, RSS-based positioning algorithms are very sensitive to the estimation of channel parameters. An advantage of the RSS method is that no synchronization between the target and the anchor nodes is required. By taking into account log-normal shadowing effects only, with a shadowing standard deviation defined as σ_{sh}, the square root of the Cramer-Rao Lower Bound (CRLB) for the unbiased estimated distance is:

$$\sigma_{\hat{d}} \geq \frac{(\ln 10) \cdot \sigma_{sh} \cdot d}{10n} \tag{2}$$

As we can see from (1.2), the accuracy of the distance estimation depends on the channel parameters and on the distance between the two nodes but not on the signal bandwidth. Therefore, the unique characteristic of UWB signals which is its large bandwidth is not exploited by the RSS approach. Moreover,

Figure 1. Position estimation via trilateration

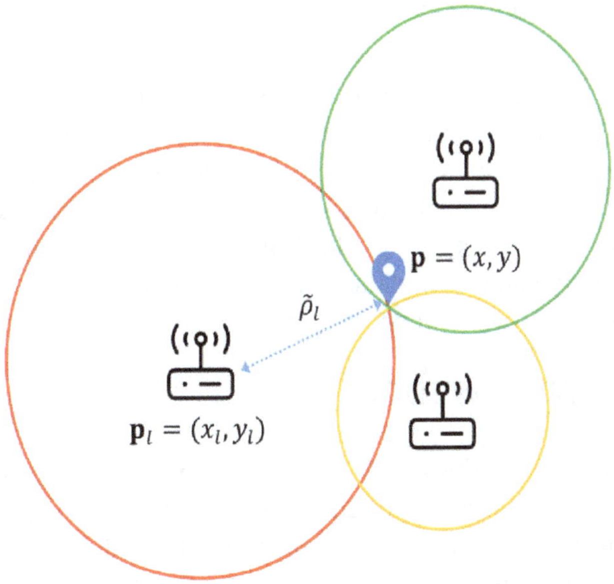

Figure 2. DOA-based position estimation via triangulation

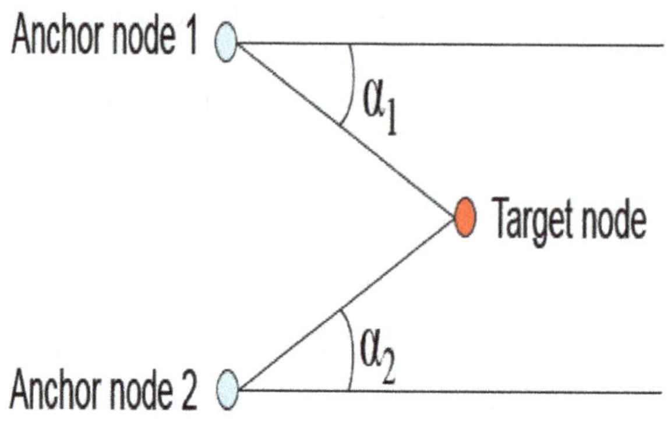

the RSS-based localization techniques tend to perform worse than their TOA-based counterparts. As a consequence, the RSS method is barely applicable for UWB positioning.

Direction of Arrival (DOA)

In DOA-based positioning, the position of the target node is obtained by measuring the angles of the straight lines that connect the target node and at least 2 anchor nodes (triangulation), as shown in Figure 2.

Antenna arrays are needed to measure the DOA of a signal. In particular, the angle information is obtained by measuring the differences in arrival times of an incoming signal at different antenna elements. Under a single-path additive white Gaussian noise (AWGN) channel, the square root of the CRLB for the standard deviation of the unbiased estimated angle for a Uniform Linear Array (ULA) with N elements is:

$$\sigma_{\hat{\theta}} \geq \frac{\sqrt{3} \cdot c}{\sqrt{2} \cdot \pi \cdot \sqrt{SNR} \cdot \beta \cdot \sqrt{N(N^2 - 1)} \cdot d \cdot \cos(\theta)} \tag{3}$$

where θ is the DOA, *SNR* represents the Signal-to-Noise Ratio, β is the effective bandwidth and c is the speed of the light. It is noted that, an increase in the SNR, in the effective bandwidth, in the inter-element spacing or in the number of antenna elements improves the accuracy of DOA estimation. Therefore, unlike RSS methods, DOA methods take advantage of the large bandwidth of UWB signal. However, this approach is not the most suitable for UWB systems as it requires antenna arrays which increase the system complexity and the implementation cost, annulling one of the main advantages of UWB systems which is low-cost transceivers.

Time-Based Approach (TOA and TDOA)

Time-based measurements, such as TOA and TDOA, provide information about the distances between two nodes by estimating the time of flight of a signal that travels from the target node to the anchor node. As in the RSS approach, by using the measurement of at least three distances the position can be trilaterated. In general, this approach requires the nodes to be synchronized. To this end, nodes must either have a common clock, or exchange timing information by certain protocols such as two-way ranging protocol. This last protocol consists of measuring the Round Trip Time (RTT), which is the time from the transmission of a signal to the reception of an acknowledgment of that signal, defined as,

$$RTT = 2TOF + T_{ack} \tag{4}$$

Where TOF is the time of flight that provides distance information and T_{ack} is the time the node takes to process the received signal and start to transmit back. When there is no synchronization between the target node and the anchor nodes, but there is synchronization among the anchors nodes, then Time Difference of Arrival (TDOA) techniques can be employed. In this case, the TDOA of two signals traveling between the target and two references nodes is estimated, which determines a hyperbola, with foci at the two reference nodes. Then, the position can be calculated as the intersection of at least two hyperboles, for which at least three anchor nodes are required.

Under a single-path AWGN channel, the square root of the CRLB for the standard deviation of the unbiased estimated TOA is:

$$\sigma_{\hat{\tau}} \geq \frac{1}{2\sqrt{2}\pi\sqrt{SNR}\beta} \tag{5}$$

As we can see in the expression above the accuracy of the TOA estimation can be improved by increasing the SNR and/or the effective bandwidth. Therefore, large bandwidths of UWB signals can provide very precise TOA measurements. Moreover, they are less complex and costly than DOA based approach. That is the reason why this technique has attracted the interest of the scientific community and the first commercially available UWB positioning systems is based on this approach (Decawave, 2017) (Ubisense 2017) (Alereon 2017).

TOA Estimation Algorithms

In this section, we will give a brief overview of the state of the art algorithms for TOA estimation in IR-UWB systems.

Matched-Filter

TOA estimation can be accomplished by the classical correlation estimator (or, equivalently, using a matched filter), where the TOA is obtained by correlating the received signal with a signal template and choosing the time shift of the signal template that produces a maximum. The signal template should be the transmitted signal convolved with the channel impulse response. Therefore, all the parameters regarding the channel response (multipath, delays and attenuations) should be estimated. Since there are too many parameters to estimate, the conventional approach uses the transmitted signal (assuming undistorted received pulses) as a template, which makes it sub-optimal. However, thanks to the high time resolution of UWB signals, the multipath components do not overlap and then TOA can be accurately estimated by selecting the first peak above a certain threshold. The reason for selecting the first peak that surpasses a threshold and not the maximum absolute peak is because in indoor channels the first path may not be the strongest one.

Unfortunately, the matched-filter approach has strong practical limitations as it requires sampling the output matched filter at the Nyquist rate. In this line, there are several proposals (Gezici et al 2005) to achieve a reliable TOA estimation sampling the output matched filter at symbol-rate or frame-rate, which is far lower than the Nyquist sampling rate. When symbol-rate samples are employed, the TOA estimation can take a very long time because the receiver needs to process many symbols to accurately estimate the TOA. To address this problem, two-step TOA estimation algorithm, with an initial coarse estimation followed by a higher resolution stage, has been proposed (Dardari, Chong & Win 2008).The later integrates the output of a square-law device over intervals comparable to the pulse width and, thus, the TOA estimation is given by the first interval where the energy overcomes a suitable threshold The ranging accuracy of these type of schemes depends on the duration of the integration window which shall be selected to satisfy a good trade-off between accuracy and low sampling constraints (the sampling rate is given by the length of the integration window).

Maximum-Likelihood and Generalized Maximum-Likelihood

The conventional Maximum Likelihood (ML) solution for the TOA estimation in UWB systems has strong practical limitations due to the requirements of very high sampling rates and complexity. To address this problem, there are several approaches that try to reduce the complexity (Yang & Giannakis 2005). An alternative approach is based on the Generalized Maximum Likelihood (GML) criterion used to detect the TOA of the first path (Lee & Scholtz 2002). This method relies on the assumption that the number of multipath components is finite and known, and all the multipath coefficients and their arrival times are jointly estimated. In a typical indoor environment, the multipath is dense, the number of paths is large and unknown, and therefore, the GML is required to estimate too many parameters which include all the multipath arrival instants, amplitudes and phases. Thus, in dense multipath, the GML is not a practical solution.

Frequency Domain Approach

TOA estimation has also been addressed applying frequency domain processing. These techniques are known as frequency-domain super-resolution TOA estimation (Pallas & Jourdain 1991). In this context, high resolution spectral techniques such as Multiple Signal Classification (MUSIC) or root-MUSIC can be used as TOA estimators (Li& Pahlavan 2004). These methods are based on the separation of the signal and the noise subspaces. This characterization is associated with high computational load as it requires eigenvalue computation. An alternative approach based on frequency domain processing, consists on using the conventional periodogram spectral estimator to obtain a Power Delay Profile (PDP) of the received signal (Navarro & Najar 2011). Thanks to the large signal bandwidth, periodogram spectral estimator can achieve high resolution TOA estimation at significantly lower complexity. Moreover, sampling requirements of frequency domain estimation methods can be significantly reduced when applying a channelization approach to the receiver architecture.

Channel Impulse Response Estimation Algorithms

Algorithms to estimate the Channel Impulse Response (CIR) can also be used to estimate the TOA, since the TOA is given by the first component of the channel impulse response. A classical approach to estimate the CIR is to deconvolve the received signal by dividing the spectrum of the received signal by the spectrum of the template and then taking the inverse Fourier transform (Riad1986). Better results can be obtained by using more sophisticated approaches such as the CLEAN algorithm which is a time domain deconvolution technique (Vaughan & Scott 1999).

Trilateration Algorithm

In Figure 1 an example illustrates the trilateration approach for positioning in a two-dimensional space. The position estimate is obtained from the intersection of the circles, centered at each anchor node, which have a radius equal to the ranging measurement value $\tilde{\rho}_\ell$. The ranging measurement is a distance measurement. The target position in Cartesian coordinates is denoted by $\boldsymbol{p} = (x, y)$, while $\boldsymbol{p}_\ell = (x_\ell, y_\ell)$ denotes the known anchor position for anchor ℓ. By defining position vectors in their matrix form,

$$\boldsymbol{p}_\ell = [x_\ell, y_\ell]^T, \quad \ell = 1, \ldots, N_A \quad \boldsymbol{p} = [x, y]^T \tag{6}$$

the geometrical relation between the target and the ℓ-th anchor node is defined as the Euclidean norm in $\mathbb{R}^2$,

$$\xi_\ell(\boldsymbol{p}) \triangleq \| \boldsymbol{p} - \boldsymbol{p}_\ell \| \tag{7}$$

In general, the trilateration problem is posed as an overdetermined system of equations with the set of noisy measurements (therefore, proportional to the number of anchor nodes $N_A \geq 3$). The ranging measurements are assumed to follow a Gaussian distribution $\tilde{\rho}_\ell \sim \mathcal{N}\left(\xi_\ell(\boldsymbol{p}), \sigma_\ell^2\right)$. Such assumption is supported by the fact that in the case of obtaining distance measurements by means of an unbiased estimator, it is known that the expected value approximates the true geometrical distance $\xi_\ell(\boldsymbol{p})$.

Let us consider the following model for the distance measurements:

$$\hat{d}_\ell = d_\ell(\boldsymbol{x}) + w_\ell \quad \ell = 1, \ldots, N_A \tag{8}$$

where $\hat{d}_\ell$ is the estimated distance between the target node, located in position $\boldsymbol{x}$, and the ℓ-th anchor node obtained during the TOA estimation step, $d_\ell(\boldsymbol{x})$ is the true distance and w_ℓ is the measurement error. Trilateration algorithms can be classified into two groups: Statistical estimators and Geometric estimators.

Statistical Estimators

The Nonlinear Least Squares (NLS) position estimate is the solution of the following non-convex optimization problem:

$$x = \arg\min_x \left\{ \sum_{\ell=1}^{N_A} \frac{\left(\hat{d}_\ell - d_\ell(x)\right)^2}{\sigma_\ell^2} \right\} \tag{9}$$

This is also the ML estimate when the errors w_i are independent, identically distributed and zero-mean Gaussian random variables. To recast the trilateration problem into a linear least squares problem, we need to define a model that is linear in the unknown parameter. In order to do so, there exist several linearization techniques (Gholami et al 2010).

Geometric Estimators

In this case, the problem is formulated as a convex feasibility instead of nonlinear least squares. Each term in the sum of (1.9) obtains its minimum on the circumference of disc:

$$D_\ell = \left\{ x \in \mathbb{R}^2 : d_\ell \leq \hat{d}_\ell \right\} \tag{10}$$

An estimator of the target node position would be any point in the intersection D of the convex sets D_ℓ:

$$\hat{x} \in D = \bigcap_{\ell=1}^{N_A} D_\ell \tag{11}$$

Due to noise in the measurements the intersection D might be empty. The Projection On to Convex Sets (POCS) algorithm (Rydstorm, Strom & Svensson 2006) (Gholami, Reza, Rydstrom & Strom 2010) solves this problem minimizing the following cost function:

$$\hat{x} = \arg\min_x \sum_{\ell=1}^{N_A} || x - \mathcal{P}_{D_\ell}(x) || \tag{12}$$

where $\mathcal{P}_{D_\ell}(x)$ is the projection of the point x onto the convex set D_ℓ .Then, what POCS algorithm does is to project onto the different convex sets iteratively. The steps taken are the following:

Initialize x^0 arbritarly

Iteratively: $x^{k+1} = x^k + \lambda_k \left[\mathcal{P}_{D_{K(k)}}\left(x^k\right) - x^k \right]$

Where $K(k) = k \bmod N_A$ (iterate cyclically onto the different convex sets) and λ_k are relaxation parameters. Normally, relaxation parameters are initially set to 1 and after a certain number of iterations (k_0) is decreased at a rate 1/(k- k_0). Figure 3 shows the path that the algorithm takes to find the solution. The black lines represent the path taken by the algorithm and the colored circles are the contours of the disks defined by each anchor node, in this case we have considered a squared scenario. The purpose of Figure 4 and Figure 5 is to show the effect of the relaxation parameter (the image has been zoomed around the optimal point), if we set λ_k to 1 the algorithm gets stuck and do not attains the minimum of the cost function.

Direct Position Estimation

As discussed in the previous sections, wireless tracking systems basically involve the presence of a number of reference wireless nodes (anchors) deployed at fixed locations (if available) and of one or more mobile nodes (Dardari, Closas & Djuric 2015). In general, the problem of localization or tracking involves the determination of the location of the mobile nodes given a set of position-related measurements. These measurements are typically related to physical quantities affected by the geometric and inertial configuration of the nodes. They could provide information between the mobile node and some anchor node, the mobile node to other mobile nodes in the scene, and self-measurements of the mobile node (e.g., on-board inertial sensors). The nature of the measurements can be very broad, depending on the geometrical information they provide. As mentioned in the previous section, the main ones RSS, TOA and DOA. In geometric trilateration based positioning, RSS and TOA are widely used. RSS is relatively easy to measure in any RF front-end, although its range estimates are in general not very accurate. On the other hand, TOA is more involved to measure but range estimates are typically better than those obtained from RSS observations.

Figure 3. Path for POCS solving the intersection of 3 discs

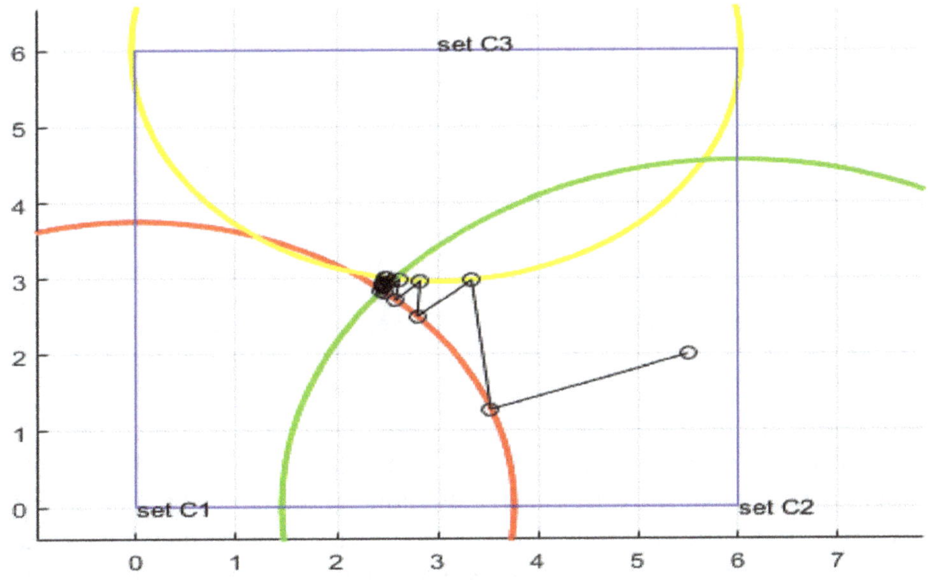

Figure 4. λ_k *set to 1*

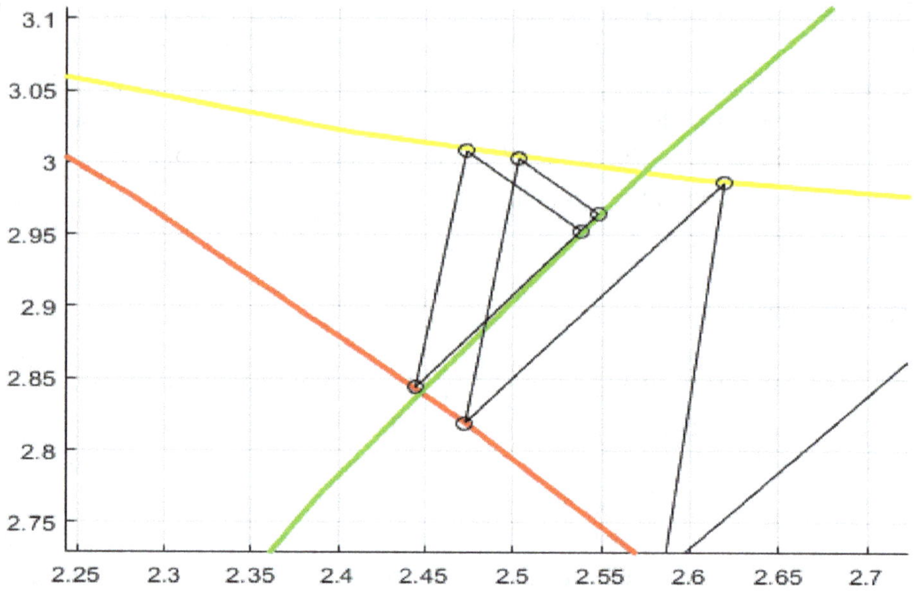

Figure 5. Effect of decreasing λ

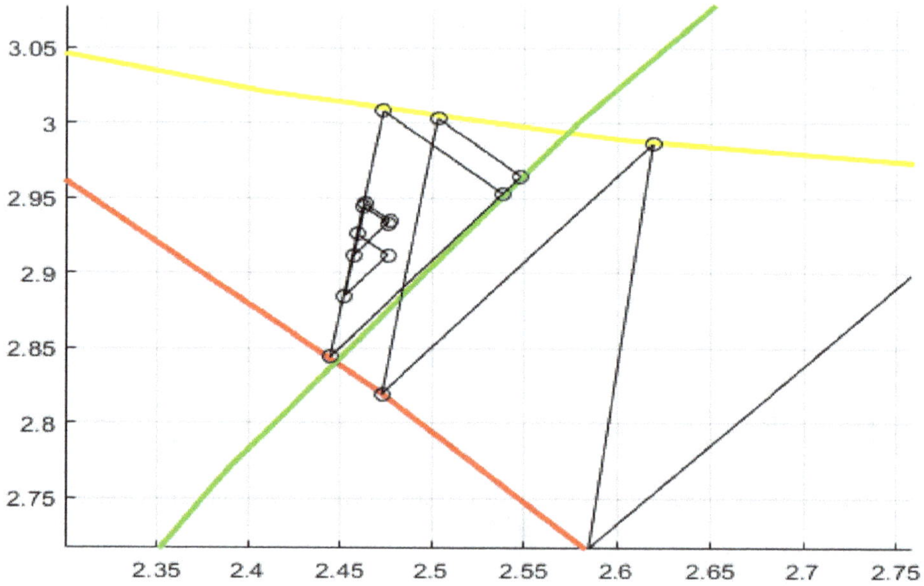

In general, the two-step positioning approach has established itself as the de facto technique in positioning. This is due to its modularity, re-use of well-known receiver blocks, and, importantly, notable performance over the years. The proposal of new techniques is blooming due to the advances in digital signal processing devices, which allow increased computational complexity at faster rates (Dampf et al. 2015). Recently, fundamental modifications to the conventional receiver architecture were proposed.

Here, we refer to those approaches that not only substitute certain parts of the receiver by more sophisticated algorithms, but those which entail an essential modification of the receiver's operation. One of the most promising alternatives is the so-called DPE concept introduced in (Closas, Fernandez-Prades & Fernandez-Rubio 2007) for global navigation satellite systems (GNSS) (including GPS, Glonass, Galileo, etc). DPE is a completely different approach to the receiver design, where a position solution is estimated in one step. Rather than estimating a set of TOA (RSS, etc.) to infer the associated mobile's position, a DPE-enabled receiver directly estimates position from the sampled data (therefore the name). The method was first introduced in the context of localization of narrowband radio frequency transmitters (Weiss 2004) and for multiple radio signals (Weiss & Amar 2005). In the context of self-localization, (Closas, Fernandez-Prades & Fernandez-Rubio 2007) presented the approach for GNSS receivers.

The main advantage of DPE with respect to two-steps is twofold: 1) usage of more information in the synchronization process thanks to the (geometrical) coupling among. The key idea is to account that all RSSs or TOAs are referred to the same receiver location, fact not used in the two-steps approach; and 2) use of prior/side information appears naturally, since tracked parameters are related to receiver's motion for which a number of sensors are available (as is the case considered in this proposal). As consequence, receivers based on DPE's approach are able to enhance tracking of signals from individual anchor nodes and deliver position estimates under challenging scenarios such as multipath propagation or weak signal under fading conditions. Contrary to two-steps architectures, DPE approach has the ability to use (and extract useful information from) weak signals, cope with signal blockages, allowing fast recovery and even operate with low signal coverage (Closas et al 2015). Current challenges involve developing efficient, lightweight position estimators under DPE's paradigm, either in open or closed loop architectures. From a theoretical perspective, it has been shown in (Amar & Weiss 2008) that direct localization outperforms two-steps solutions in terms of achievable mean square error (MSE). In the context of positioning (i.e., self-localization), more results exist where the Cramér-Rao lower bounds on position estimation are derived for both positioning approaches, highlighting the potential benefits of adopting DPE (Closas, Fernandez-Prades & Fernandez-Rubio 2009) (Gusi-Amigo, Closas, Mallat & Vandendorpe 2014). Further research is required in terms of developing novel MSE estimation bounds yielding to a better understanding of DPE and its potential. These bounds, based on the Ziv-Zakai methodology, would allow determining, for instance, the signal to noise ratio at which both DPE and two-steps approaches are able to operate before breaking down. It turns out from this analysis that, under certain scenarios, there is an increase in sensitivity for DPE in the order of 10log(M) dB, M being the total number of anchor nodes. This highlights the theoretical sensitivity increase of a DPE receiver.

In the context of IR-UWB, DPE concept was applied in (Navarro, Closas & Najar 2013) (Lagunas, Navarro, Closas & Najar 2014) in relevant IEEE 802.15.4a channels. Although the potential performance benefits brought by DPE are promising, its application to UWB-based localization has particular challenges that needs careful consideration. Mainly, this is due to the fact of the extremely large number of samples available in this type of technology, and which needs to be processed in a reasonable time and within bounded computational complexity. This is due to the high data bandwidth in UWB, which on the other side makes it particularly suitable for high-precision ranging and positioning applications. As a consequence, UWB is able to resolve extremely close multi path components or propagation paths. Interestingly, this requires additional algorithmic modifications when computing DPE's solution, since multiple peaks appear in the cost function that has to be optimized. This chapter provides additional insights and results on the use of direct-positioning techniques with UWB signals.

Compressive Sensing

Signal acquisition is a main topic in signal processing. Sampling theorems provide the bridge between the continuous and the discrete-time worlds. The most famous theorem is often attributed to Shannon (but usually called Nyquist theorem) and states that the sampling rate must be twice the maximum frequency present in the signal in order to perfectly recover the signal. In the field of data conversion, standard ADC technology implements the usual quantized Shannon representation: the signal is uniformly sampled at or above the Nyquist rate. For many signals the Nyquist rate can be very high, thus, very large number of samples need to be acquired, leading to impractical and costly designs. However, sampling at the Nyquist rate is shown to be inefficient when the signals of interest contain only a small number of significant frequencies relative to the band-limit. To alleviate the strong hardware requirements of the ADCs as well as the subsequent storage and processing bottlenecks, a promising alternative for this type of sparse signals is the use of sub-Nyquist sampling techniques.

In this context, in 2006 appeared the Compressed Sensing (CS) concept (Candes & Wakin 2008, Donoho 2006), a novel sampling paradigm that goes further than Shannon-Nyquist's theorem. CS allows to compress signals with sparse or compressible representation while they are sampled. It originates from the idea that it is not necessary to invest a lot of power into observing the entries of a sparse signal because most of them will be zero. Considering a sparse signal, it should be possible to collect only a small number of measurements that still allow for reconstruction.

A vastly growing research literature on some of the foundational developments in CS is available in http://www.compressedsensing.com. In this Section, we review the basic concepts in CS theory. CS allows to compress signals with sparse or compressible representation while they are sampled. Let us consider the following finite length, discrete signal $\mathbf{x} \in \mathbb{C}^L$ that has K non-zero coefficients for some $K \leq L$. The framework of CS aims at recovering the unknown vector $\boldsymbol{x}$ from an underdetermined system of linear equations,

$$\mathbf{y} = \mathbf{\Phi x} \tag{13}$$

where $\mathbf{y} \in \mathbb{C}^{\kappa}$ is the received data vector and matrix $\mathbf{\Phi} \in \mathbb{C}^{\kappa \times L}$ is the sub-sampling matrix or sensing matrix since the number of rows is less than the number of columns. Since $\kappa \leq L$, this system has more unknowns than equations, and thus the system is not invertible. In particular, equation (13) has infinitely many solutions. Among the infinitely many solutions of (13), we are only interested in the sparsest one. CS reconstruction can be performed via l_1-norm minimization. l_1-norm is very interesting and popular due to its tendency to sparsify the solution. The vector $\mathbf{x}$ can be recovered from $K \log_2(L / K)$ number of linear non-adaptive measurements (Donoho 2006, Candes & Wakin 2008) by solving,

$$\min_{\theta} \left\|\boldsymbol{\theta}\right\|_{l_1} \text{ subject to } \mathbf{y} \approx \mathbf{\Phi x} \tag{14}$$

where $\left\|\boldsymbol{\theta}\right\|_{l_1} = \sum_i \left|\theta_i\right|$. Several methods are available in the literature to solve the optimization problem in (14). The l_1 minimization is a convex problem and can be recast as a linear program (LP) (Boyd & Vandenberghe 2004). This is the foundation for the Basis Pursuit (BP) techniques (Candes, Romberg &

Tao 2006, Candes & Tao 2006, Chen, Donoho & Saunders 1999). Alternatively, greedy methods, known as Matching Pursuit (MP), can be used to solve (14) iteratively (Mallat & Zhang 1993, Tropp 2004).

The temporal sparsity property of UWB together with CS have been applied to localization in (Yang, Li & Peterson 2013, Lingwen & Zhenhui 2010, Cevher, Duarte & Baraniuk 2008, Ping, Huailin & Fuhua 2012, Weis 2011). In (Yang, Li & Peterson 2013, Lingwen & Zhenhui 2010, Cevher, Duarte & Baraniuk 2008), high resolution UWB signals at different base stations are reconstructed from reduced volume of data. Although the signal is not precisely recovered, the accuracy of the TOA estimation obtained from the reconstructed signal is shown to be good enough for two-step positioning. The same idea was also applied for channel impulse response estimation from compressive measurements in (Ping, Huailin & Fuhua 2012), where direct path detection is used to estimate the TOA. In (Weiss 2011), a source localization method based on Time Difference of Arrival (TDOA) in a multipath channel is suggested by exploiting the sparsity of the multipath channel for estimation of the LOS component.

All the above methods employed CS for channel and/or signal reconstruction from compressive measurements. Recently, CS ideas have also been exploited to achieve substantial improvements on localization accuracy given the implicit sparse representation of the discrete spatial domain (Ke & Wu 2012, Pourhomayoun, Fowler & Wu 2012). Generally, potential targets cover a small part of the total discrete spatial domain, which allow the localization problem to be linearized over a set of possible locations and solved using sparsity-based recovery techniques. Spatial sparsity-based is exploited for DPE methods in (Ke & Wu 2012, Pourhomayoun, Fowler & Wu 2012). These approaches were originally introduced to work on the full data volume and did not account for nor were they ever examined under reduced data measurements, UWB technology and/or indoor propagation.

In this context, the work in (Lagunas, Navarro, Closas & Najar 2014) deserves special attention. In (Lagunas et al. 2014), and based on the inherent spatial sparsity of the scene, a multiple location hypotheses is introduced through the use of an overcomplete basis, which highlights the sparseness of the spatial domain. Therefore, (Lagunas et al. 2014) makes use of CS-based sampling and sparsity-based reconstruction techniques to jointly evaluate all possible hypotheses.

IR-UWB Signal Model

We consider the tag to periodically transmit IR-UWB pulses of very short duration. The transmitted signal is modeled as,

$$s(t) = \sum_{k=-\infty}^{\infty} p(t - kT_f) \tag{13}$$

where T_f is the pulse repetition period also referred to as frame period and $p(t)$ refers to the signal pulse waveform, being typically a Gaussian monocycle or one of its derivatives of duration T_p. Signal $s(t)$ propagates through an M-path fading channel whose response to $p(t)$ is $\sum_{m=0}^{M-1} h_m p(t-\tau_m)$, with $\tau_0 < \cdots < \tau_{M-1}$, being τ_0 the TOA that has to be estimated as it corresponds to the first arriving path. The received signal is then the summation of multiple delayed and attenuated replicas of the received pulse waveform $\tilde{p}(t)$, which includes the antenna and filters distortion,

$$y(t) = \sum_{m=0}^{M-1} \sum_{k=-\infty}^{\infty} h_m \tilde{p}(t - kT_f - \tau_m) + w(t) \tag{14}$$

where $w(t)$ is an AWGN and is modeled as $N(0, \sigma_w^2)$. We assume that the received pulse from each m-th path exhibits the same waveform but experiences a different fading coefficient, h_m, and time delay, τ_m.

The signal associated to the k-th transmitted pulse, in the frequency domain yields,

$$Y_k(\omega) = h_0 S_k(\omega) e^{-j\omega\tau_0} + \sum_{m=1}^{M-1} h_m S_k(\omega) e^{-j\omega\tau_m} + V_k(\omega) \tag{15}$$

where the Line-of-Sight (LOS) contribution is explicitly separated in the first term from the signal replicas associated to multipath. The frequency component associated to the shifted pulse is given by,

$$S_k(\omega) = \tilde{\mathbf{P}}(\omega) e^{-j\omega T_f} \tag{16}$$

where $\tilde{\mathbf{P}}(\omega)$ denotes the Fourier Transform of the pulse $\tilde{p}(t)$ and $V_k(\omega)$ is the noise in the frequency domain associated to the k-th symbol. Sampling (18) at $\omega_n = \omega_0 n$ for $n = 0, 1, \ldots, N-1$ where $\omega_0 = \dfrac{2\pi}{N}$ and rearranging the frequency domain samples $Y_k(\omega)$ into the vector $\mathbf{Y}_k \in \mathbb{C}^N$ yields,

$$\mathbf{Y}_k = h_0 \mathbf{S}_k \mathbf{e}_{\tau_0} + \tilde{\mathbf{V}}_k \tag{17}$$

with $\tilde{\mathbf{V}}_k = \mathbf{S}_k \mathbf{E} \mathbf{h} + \mathbf{V}_k$ and $\mathbf{S}_k \in \mathbb{C}^{N \times N}$ is a diagonal matrix whose components are the frequency samples of $S_k(\omega)$. The matrix $\mathbf{E} \in \mathbb{C}^{N \times (M-1)}$ contains the delay-signature vectors associated to each arriving delayed signal (multipath),

$$\mathbf{E} = [\mathbf{e}_{\tau_1} \cdots \mathbf{e}_{\tau_m} \cdots \mathbf{e}_{\tau_{M-1}}] \tag{18}$$

with $\mathbf{e}_{\tau_m} = [1, e^{-j\omega_0\tau_m}, \ldots, e^{-j\omega_0(N-1)\tau_m}]^T$. The channel fading coefficients, except for h_0, are arranged in the vector $\mathbf{h} = [h_1, \ldots, h_{M-1}]^T$,and the noise samples in vector $\mathbf{V}_k \in \mathbb{C}^N$.

TWO-STEP MULTI-TOA-BASED POSITIONING WITH SPARSITY EXPLOITATION

In this section we present a novel technique that combines multiple TOA estimates from multiple anchors in order to skip the threshold-based step of selecting the first arriving path.

The proposed two step multi-TOA positioning scheme for IR-UWB systems is based on the classical TOA estimator and decomposes the positioning problem, just like in TOA estimation algorithms, into two sub-problems: ranging to the anchor nodes at known locations and trilateration to obtain a position estimate.

In the first stage, the high-resolution TOA estimate is based on calculating the pseudo-periodogram or PDP, defined as the signal energy distribution with respect to propagation delays. Namely,

$$P_\ell\left(\tau\right) = \boldsymbol{e}_\tau^{\ H} \boldsymbol{R}_\ell \, \boldsymbol{e}_\tau \tag{19}$$

where $\ell = 0, \ldots, N_A$ denotes the anchor index, the superscript $\left(\cdot\right)^H$ denotes the transpose complex conjugate, and $\boldsymbol{R}_\ell$ is the sample covariance matrix computed by averaging over properly aligned frequency samples,

$$\boldsymbol{R}_\ell = \frac{1}{N_f} \sum_{k=1}^{N_f} \boldsymbol{Y}_k^{(\ell)} \left(\boldsymbol{Y}_k^{(\ell)}\right)^H \tag{20}$$

where $\boldsymbol{Y}_k^{(\ell)}$ denotes the received signals from the ℓ-th anchor in the frequency-domain, and N_f is the number of observed frames. In threshold-based TOA estimators, the TOA estimate is the first peak in the periodogram P_ℓ that exceeds a certain threshold. Selection of the threshold is always a critical factor that may drop the performance drastically if it is not chosen properly. The proposed solution, instead of setting a threshold and keeping the first peak above it, selects all prominent peaks and the associated potential TOAs are used to find the unknown target position. All the potential TOAs are gathered in the central processing unit and the combination of TOAs that minimize the trilateration cost function are the ones selected. In other words, the position estimate will be the one that minimizes the following cost function:

$$\hat{\boldsymbol{p}} = \arg\min_{p} \left[\min_k \sum_{\ell=1}^{N_A} \left(\tau_\ell \left[k_\ell\right] c - || \, \boldsymbol{p} - \boldsymbol{p}_\ell \, ||^2 \right) \right] \tag{21}$$

where $\boldsymbol{k} = \left[k_1 \ldots k_\ell \ldots k_{N_A}\right]$, $k_\ell \in \left\{1, \ldots, m_\ell\right\}$, m_ℓ is the number of peaks detected in the ℓ-th anchor node and $\tau_\ell[k_\ell]$ is the potential k_ℓ-th TOA detected in the ℓ-th anchor node.

However, not all selected peaks are equally likely to be the true TOA. For example in PDP shown in Figure 6, the last peak to be the true TOA would mean that all the previous peaks were noise or pulse sidelobes of that peak and such thing is highly unlikely. Somehow, we must add information related to the energy received from each peak and the relative position of the peaks in the PDP. Therefore, it is convenient to weight each peak by the probability of that peak being the true TOA. By weighting the peaks we improve the overall performance of the position estimation technique because the number of peaks gets reduced (last peaks will have probability 0 so we remove them) so it will be easier to minimize the cost function, and moreover we reduce the probability of choosing paths that do not represent the position of the target.

Figure 6. PDP realization

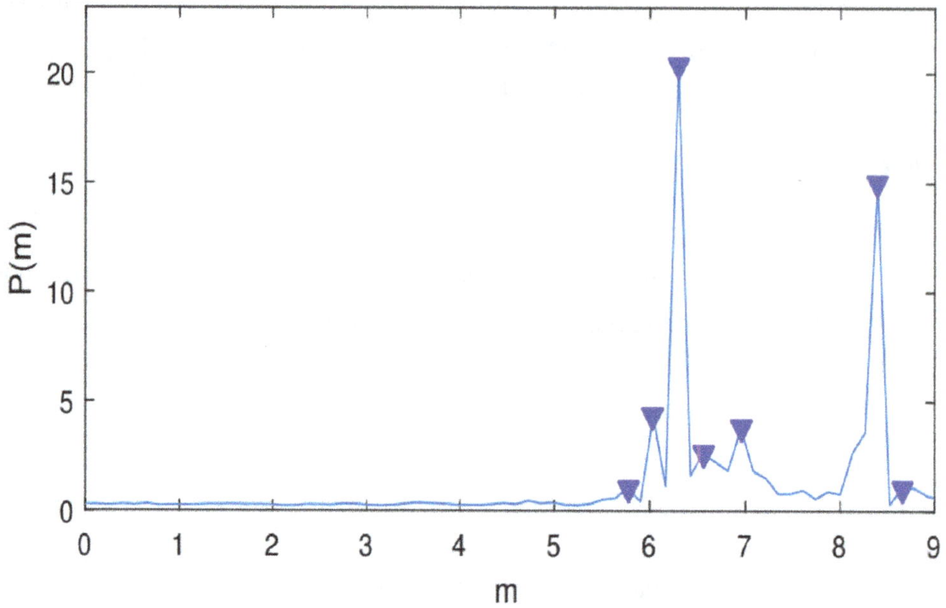

However, as it happened with threshold selection (in the pure TOA PDP approach), finding these probabilities is not an easy task. As a practical workaround, we proposed and heuristic weighting criterion based on the distribution of the energy of the TOA,

$$\theta_{peak_i} = \prod_{k=1}^{i-1}\left(1-\theta_{peak_k}\right)\Pr\left(E_{TOA} < E_{peak_i}\right) \tag{22}$$

where E_{TOA} is the random variable that represents the energy of the TOA (its distribution can be found empirically) and E_{peak_i} is the energy of the *i-th* detected peak. The computed weights are introduced into the cost function (1.15) in a way that peaks associated with higher probability yield a lower value, that is why we subtract them to the original cost function. In other words the position estimate is given by,

$$\hat{\boldsymbol{p}} = \arg\min_{p}\left[\min_{k}\sum_{\ell=1}^{N_A}\left(\tau_\ell\left[k_\ell\right]c - \|\boldsymbol{p}-\boldsymbol{p}_\ell\|^2\right) + \lambda\left(-\sum_{\ell=1}^{N_A}\theta_\ell\left[k_\ell\right]\right)\right] \tag{23}$$

where $\theta_\ell\left[k_\ell\right]$ is the weight of the k_ℓ peak of the ℓ-th beacon and the parameter λ is a trade-off between consistency and probability of the delays.

When it comes to solving the cost function in (1.17) we must deal with a function that depends on discrete variables (the variable k). Addressing this discrete optimization problem by brute force, which is enumerating all the possible combinations, is not feasible as the number of combinations grows exponentially with the number of anchor nodes. Therefore, we propose the use of *Branch and Bound*

(Clausen 1999) algorithm to try to reduce the search space. *Branch and Bound* is a tree-search based algorithm where each node of the tree is a sub-problem, then after solving a sub-problem we can set a bound about how good the final solution (leaves of the tree) could potentially be. If that bound is better than an already found solution we have to expand this node and explore its children. Otherwise, we can prune that node, thus reducing the search space.

From this point onwards, we will refer to the anchor nodes as beacons and the nodes of a tree as nodes.

In our particular problem, the elements of the tree would be:

- Nodes: A particular trilateration using fewer beacons than the available. The depth of the tree is equal to the number of beacons.
- Lower bound: The cost of that particular trilateration.
- Children: The children of a given parent node are the trilaterations using the same parameters of the parent node but adding the TOA information of the new beacon (the number of children is equal to the number of potential TOAs estimated by the new beacon).
- Leaves: A possible solution of the problem. There are as many leaves as number of combinations.

The way to explore the tree is a key factor to find the solution quickly. The proposed technique is the best-first search strategy which explores the tree by expanding the most promising node.

Algorithm 1 is a pseudo-code implementation of the proposed algorithm, where the *Priority queue* is a queue where each element has an associated priority (in our case the lower the bound the higher the priority), then an element with high priority is served before an element with low priority, and *TreeNode* is an structure that stores the beacons used, the potential TOA used for each beacon, the value of the bound and the estimated position if only these beacons were available.

CS-Based Two Steps Multi-TOA Estimator

The CS-based two steps Multi-TOA estimator aims to modify first stage of the latter scheme by exploiting the underlying sparsity present in the temporal domain of UWB channels, a large portion of multipath reflections are very attenuated and the dominant reflections are a few compared to the overall dimensionality of the Nyquist sampled received signal. A CS reconstruction technique is employed to perform the multiple TOA estimation, in other words, a set of candidate or prominent delays for the TOA is extracted at each receiving node. At the second step, after all the multiple sets of candidate TOAs from every anchor are gathered at a central processing unit, the described trilateration method based on a Branch and Bound algorithm is used to find the target's position.

Consider again a subset of observation vector in the frequency domain at the $l-th$ anchor node:

$$\breve{\boldsymbol{Y}}_k^{(\ell)} = \Phi \boldsymbol{Y}_k^{(\ell)} \tag{24}$$

where Φ is the sensing matrix introduced in previous sections which, in this case, it randomly selects k samples of $\breve{\boldsymbol{Y}}_k^{(\ell)}(k < N)$ assuming a uniform distribution. This matrix Φ is given by randomly selecting k rows of the identity matrix $\boldsymbol{I}_N$.

To motivate the discussion of the CS linear model and the choice of the sparsifying dictionary, the following linear arrangement of the correlation matrix is proposed:

Algorithm 1. Computation of position

```
Define: Priority queue PQ, struct TreeNode
Select a beacon arbitrarily, call it b1
for all TOAs in b1 do .                         ▶ First Tree layer
        Create a TreeNode, call it node
        Set node.bound = 0
        Set rest of node parameters
        Put node into PQ
end for
Solution = null
J = ∞
while PQ is not empty do
        Remove most priority node of PQ, call it topNode
        if topNode.bound < J then
                Explore children of topNode
                Put children into PQ
                for all children do
                        if child is leaf & child.bound < J then
                                update J
                                Solution = child
                        end if
                end for
        end if
end while
```

$$\breve{\boldsymbol{R}}^{(\ell)} = \sum_{i=0}^{N_\tau-1} \sigma_{\tau_i} \breve{\boldsymbol{e}}_{\tau_i} \breve{\boldsymbol{e}}_{\tau_i}^{H} + \breve{\boldsymbol{R}}_{\varepsilon}^{(\ell)} \tag{25}$$

where $\breve{\boldsymbol{R}}^{(\ell)} \in \mathbb{C}^{k \times k}$ is the compressed version of the sample covariance matrix. $\breve{\boldsymbol{R}}_{\varepsilon}^{(\ell)}$ represents imperfections of the model and σ_{τ_i} is the energy indicator of delay τ_i. These values σ_{τ_i} are concatenated in a column vector of dimensions $N_\tau \times 1$ to form $\sigma = \left[\sigma_{\tau_1} \ldots \sigma_{\tau_{N_\tau}}\right]$. Basically, the proper physical interpretation of σ is the distribution of the received signal power over the different propagation delays. Finally, vector denotes the compressive steering vector corresponding to the delay τ_i.

The model in (27) can be conveniently rewritten into a sparse notation as follows,

$$\breve{\boldsymbol{r}}^{(\ell)} = \boldsymbol{A}\sigma_l + \breve{\boldsymbol{r}}_{\varepsilon}^{(\ell)} \tag{26}$$

where $\breve{\boldsymbol{r}}^{(\ell)}$ is a $k^2 \times 1$ vector formed by the concatenation of the columns of $\breve{\boldsymbol{R}}$. From now on, to clarify notation, the concatenation of columns will be denoted with the operator $vec(\cdot)$. Therefore $\breve{\boldsymbol{r}}^{(\ell)}$

$= \mathrm{vec}\left(\breve{\boldsymbol{R}}^{(\ell)}\right)$. The columns of $\boldsymbol{A}$ contain the vectorized candidate covariance matrices corresponding to the different tentative delays that conform the temporal scanning grid, and is defined as,

$$\boldsymbol{A} = \left[\breve{\boldsymbol{r}}\left(\tau_0\right) \quad \breve{\boldsymbol{r}}\left(\tau_1\right)\cdots\breve{\boldsymbol{r}}\left(\tau_{N_\tau-1}\right)\right] \tag{29}$$

where $\breve{\boldsymbol{r}}\left(\tau_i\right) = \mathrm{vec}\left(\breve{\boldsymbol{e}}_{\tau_i}\breve{\boldsymbol{e}}_{\tau_i}{}^{H}\right)$. The variable $\breve{\boldsymbol{r}}_{\varepsilon}$ encompasses undesired contributions and imperfections of the model. Note that the elements different from zero of vector

$$\boldsymbol{\sigma} = \left[\sigma\left(\tau_0\right) \quad \sigma\left(\tau_1\right)\cdots\sigma\left(\tau_{N_\tau-1}\right)\right]^T \tag{30}$$

correspond to the grid point where we find PDP values different from zero, therefore they represent possible TOA candidates. The sparse vector $\boldsymbol{\sigma}$ can be recovered from the compressive measurements by solving the following optimization problem,

$$\min_{\sigma_l}\left\|\boldsymbol{\sigma}_l\right\|_{l_1} \text{ subject to } \quad \breve{\boldsymbol{r}}^{(\ell)} \approx \boldsymbol{A}\boldsymbol{\sigma}_l \tag{31}$$

There are many CS reconstruction algorithms that are able to compute (31). Orthogonal Matching Pursuit algorithm (OMP) is considered due to its simplicity and greediness. It has an iterative nature, meaning that the algorithm needs to be halted according to some rule or stopping criteria. Many of them are based on the knowledge of the level of noise present in the measurements, which very often is unavailable information. For this particular application, the number of iterations will be fixed. This parameter, from now on named K, will also represent the level of sparsity present in the recovered signal, having exactly K non-zero entries, and will play a fundamental role on the performance of the proposed algorithm. Rather than concerning on an accurate recovery of $\boldsymbol{\sigma}$, the goal of OMP, in this paper, is to extract a set of significant delay paths $\left\{\tau_0,\ldots,\tau_{K-1}\right\}$ in an attempt to incorporate the TOA delay information. Algorithm 2 describes the OMP method used to estimate the multiple-TOAs.

In line 5, $\boldsymbol{A}_{\Gamma^i}$ denotes the matrix composed by the columns defined in the set Γ^i. The term $\boldsymbol{A}_{\Gamma^i}^{\dagger}$ represents the pseudo inverse of $\boldsymbol{A}_{\Gamma^i}$. Finally, λ denotes the time resolution of the dictionary, i.e. the ability to resolve paths, and is related to the sampling period. The number of columns of the dictionary is related to the time resolution in the following manner: $N_\tau = \frac{T_f}{\lambda}$. By a simple examination of Algorithm 2, it can be seen that the number of identified paths by the CS procedure corresponds to the number of iterations K as well.

Numerical Results

Next, we present some numerical experiments to evaluate the performance of the proposed method and its comparison to a non-CS benchmark TOA algorithm. We focus on the IEEE 802.15.4a UWB channel

Algorithm 2. OMP based Multi-TOA estimator

Input: Measurements $\breve{\boldsymbol{r}}^{(\ell)}$ at the ℓ-th anchor, Dictionary $\boldsymbol{A}$, Number of searched paths K

Output: $\left\{\tau_0,\ldots,\tau_{K-1}\right\}$ after K iterations

Initialization $\mathrm{y} = \breve{\boldsymbol{r}}^{(\ell)}, \mathrm{e}^0 = \breve{\boldsymbol{r}}^{(\ell)}, \mathrm{I}^0 = \varnothing$

while $i \leq K$ **do**

$l = \underset{l=1,\ldots,N_T}{\operatorname{argmax}} \left|\boldsymbol{a}_l, \boldsymbol{e}^i\right|$ {finding the column in dictionary **A** with maximum correlation with residual.}

$\mathrm{I}^i = \mathrm{I}^{i-1} \cup l$

$\boldsymbol{e}^{i+1} = \boldsymbol{e}^i - \boldsymbol{A}_{\mathrm{I}^i} \boldsymbol{A}_{\mathrm{I}^i}{}^{\dagger} \boldsymbol{y}$.

i=i+1

end while

$\left\{\tau_0,\ldots,\tau_{K-1}\right\} = \mathrm{I}^{\mathrm{K}}\lambda$

models, a standard developed with actual real measurements as described in (Molisch et al. 2004). In particular, we focus in the CM3 LOS Office model which is an indoor office environment characterized to exhibit dense multipath. The simulation setup is configured as follows: A single target is randomly allocated in a $6x6m^2$ room with four beacons placed at the corners and is continuously transmitting. The transmitted pulse is a baseband Gaussian monocycle of duration T_p. The received signal is low pass filtered to avoid aliasing and then sampled at 2 GHz at each beacon. This set-up is performed using a total of 500 distinct realizations of the IEEE 802.15.4a CM3 channel model and with different target's positions.

The time of frame T_f denotes the time between consecutive pulses. The total observation window or integration time at the receiver is determined by the coherence time of the fading channel and it limits the number of frames that can be used for averaging. Finally, the time resolution of the dictionary is set to be multiple of the sampling period, defined as $T_s = \frac{1}{F_s}$. Therefore by decreasing the resolution, the estimator accuracy can be improved at the expense of a gain in coherence of the dictionary, which can reduce the performance of the CS algorithm.

To strictly focus on the performance behavior due to compression and remove the effect of insufficient data records, the size of the compressed observations was forced to be the same for any compression rate. Thus, for a high compression rate, the estimator took samples for a larger period of time. In Table 1, this effect is illustrated:

The metrics of performance will be the empirical CDF of the TOA positioning error and the Root Mean Square Error (RMSE) of the estimator, defined as $RMSE\left[m\right] = \sum_{j=1}^{J} \hat{\boldsymbol{p}}_j - \boldsymbol{p}_j{}^2$, where J denotes the number of considered realizations. The main simulation parameters are provided in Table 2:

For the benchmark algorithm, it is considered a threshold-based TOA estimation technique introduced in (Navarro & Najar 2011), where the peak corresponding to the TOA value in the PDP is chosen

Table 1. Relationship between ρ and N_f

ρ	1	0.75	0.5	0.25
κ	128	96	64	32
N_f	128	171	256	512

Table 2. Simulation Parameters

Sampling Frequency F_s	2 GHz
Pulse duration T_p	1 ns
Number of realizations J	500
Signal To Noise Ratio	4 dB
Dictionary resolution λ	T_s
Anchors N_A	4

as the first peak value that is greater than the detection threshold. For evaluation purposes, an off-line calibration of the threshold per noise level for this particular channel model and thus we are comparing our method to a TOA algorithm with an optimum threshold in the MSE sense.

The position error of Multi-TOA and TOA-based two-step approach when no compression is considered is depicted in Figure 7 and Figure 8in terms of the CDF for Office LOS (CM3) and Office NLOS (CM4) scenarios, respectively. The parameter λ in (25) for the Multi-TOA approach is obtained empirically. For comparison purposes, both Figure 7 and Figure 8 show the CDF for the two-steps when the threshold is deviated a 30% and a 40% from the optimum value. These curves show the sensibility of the two-step to the threshold value. From both Figure 7 and Figure 8, it can be concluded that the proposed Multi-TOA gives better performance than the TOA-based two-step approach. Comparing the results depicted in Figure 7 and Figure 8, it can also be concluded that the proposed algorithm is robust to the NLOS propagation impairments.

Regarding the performance of the CS-based Multi-TOA algorithm, one of the main objectives is to reduce the amount of processed data at the front-end system without greatly compromising the accuracy of the estimator.

The number of searched paths, K, in the CS-based TOA algorithm is unknown and needs to be determined through an optimization process. Moreover the choice of K also affects the computational complexity of the strategy as we are overloading the next stage. We will analyze the contribution of this parameter, named K, on the positioning error.

If the number of paths searched, K, in the algorithm is high then there is a potential problem of picking the wrong atom (column) in the CS dictionary due to the noise, however, if the K, is too small then

Figure 7. Empirical CDF of the position error using 4 beacons for Multi-TOA and 2-Steps TOA. Evaluated in Office LOS (CM3)

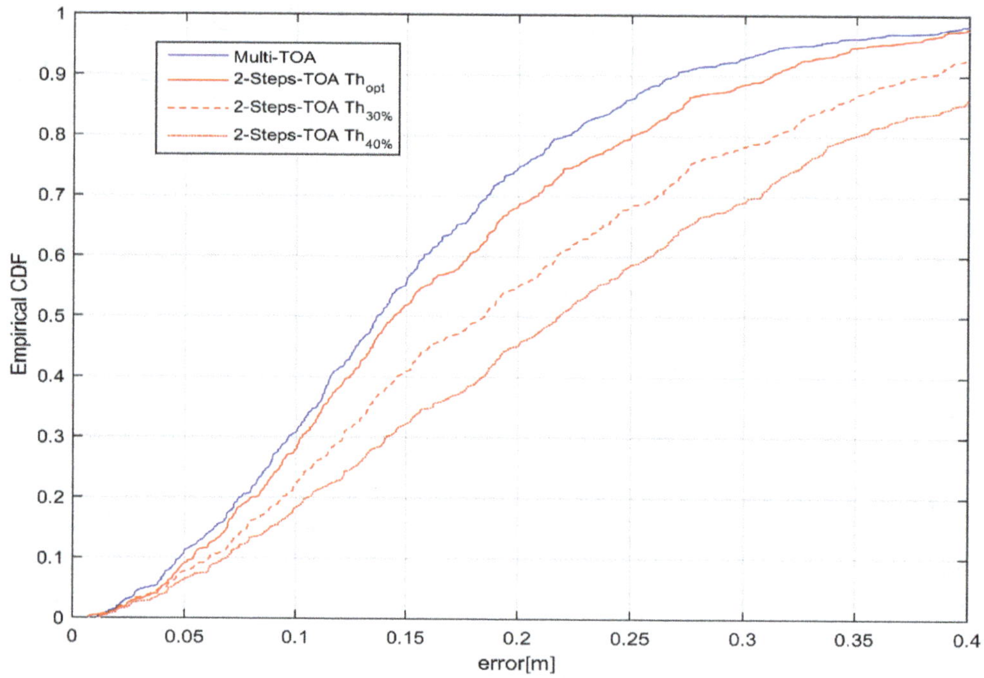

Figure 8. Empirical CDF of the position error using 4 beacons for Multi-TOA and 2-Steps TOA. Evaluated in Office NLOS (CM4)

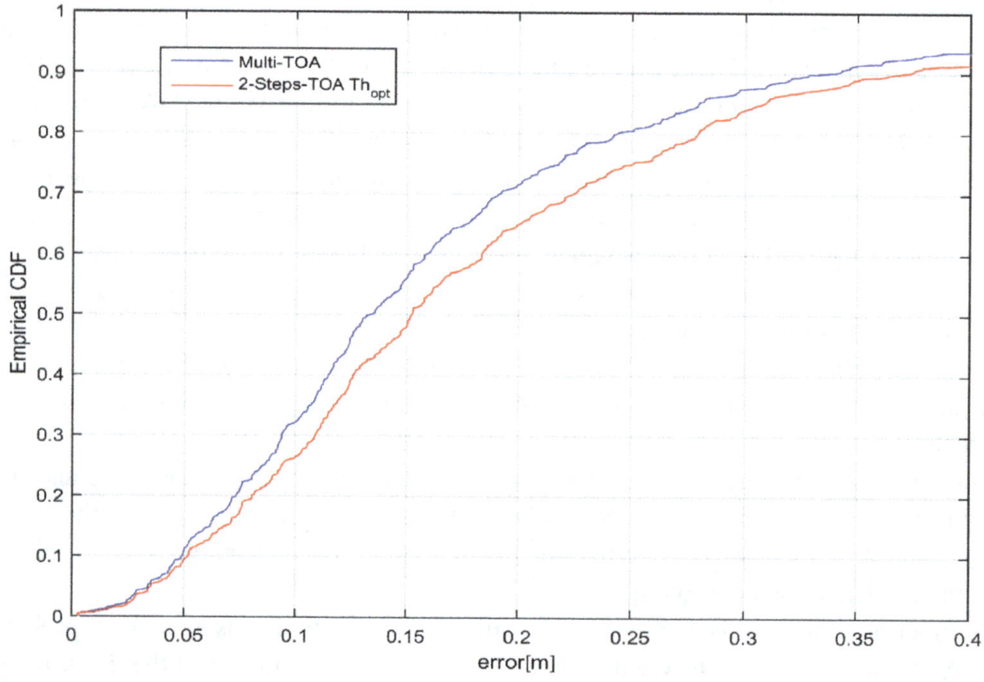

we may miss the true TOA, due to the possibility of earlier paths being weaker than the first path. Therefore, an optimal value is found at $K = 7$ in Figure 9 in the RMSE sense.

The effect of the compression rate on the CS-based Multi-TOA is evaluated and compared with the benchmark TOA and standard Multi-TOA algorithms for a couple of different levels of sparsity to observe the tradeoff between both CS variables. The compression rate is defined as $= \frac{\kappa}{N}$. Then, the original sampling frequency that was determined by the pulse duration, can be rewritten as $F_s' = \rho F_s$ with $0 < \rho \leq 1$.

In typical CS applications, the tradeoff between level of sparsity and compression rate is clearly a problem. The number of measurements needed for the CS algorithm to succeed scales linearly with the level of sparsity. In our case, the working point of the sparsity ratio $\delta = \frac{K}{N}$ is pretty low, therefore this tradeoff does not have an impact on the performance. Nonetheless, the inherent degradation of the estimation due to effect of high compression is clearly present on the plots. In both, Figure 10 and Figure 11, for $\rho = 0.25$ and below, the CS-based Multi-TOA suffers a significant loss of performance as it is unable to reconstruct the correct TOA paths.

DPE FOR IR-UWB SYSTEMS

In general, positioning in IR-UWB systems is solved implementing two-step approach as described in the previous section. In IR-UWB the distance measurements are obtained from TOA measurements, which

Figure 9. RMSE variation with K for a given compression rate, ρ=0.5

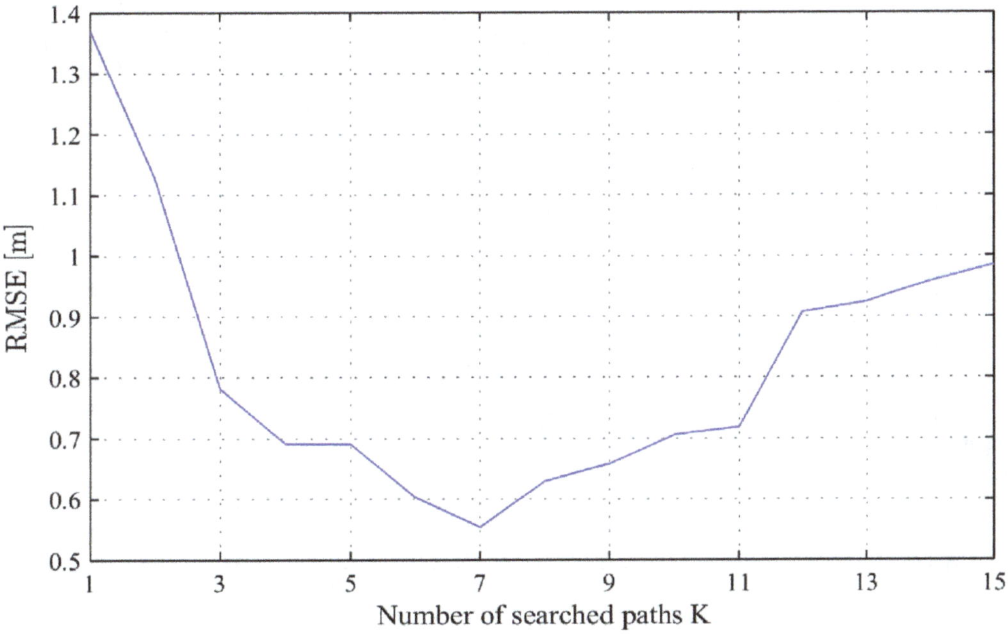

Figure 10. Empirical CDF of error for a given number of searched paths K=7

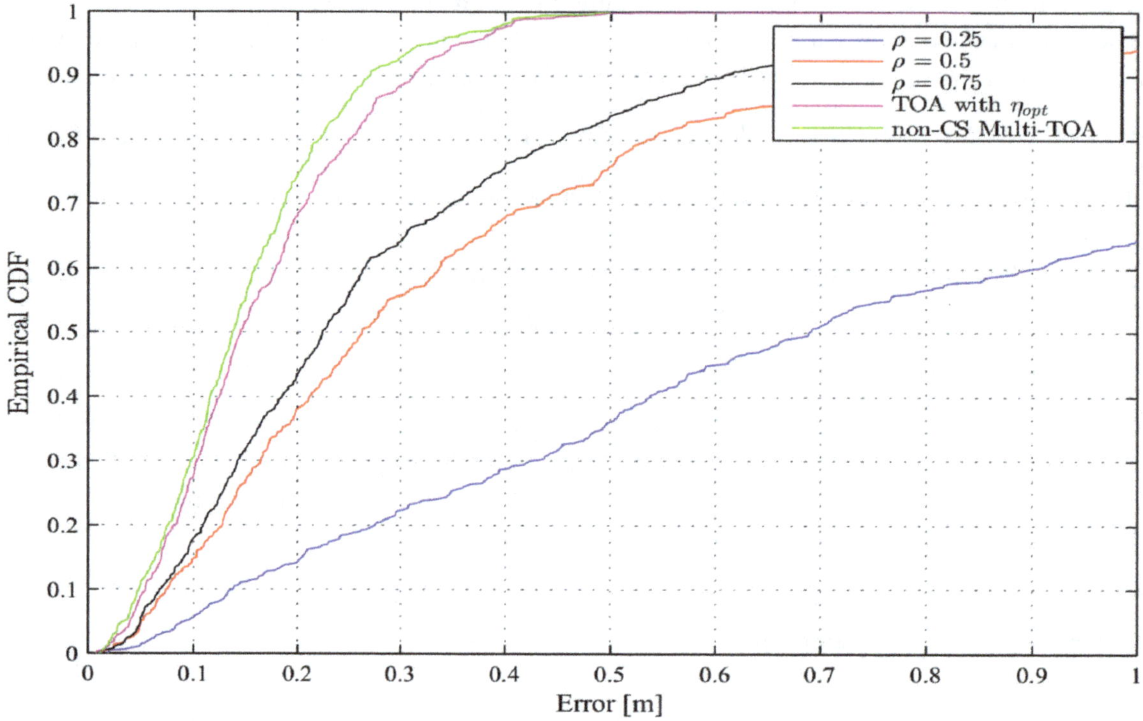

Figure 11. Empirical CDF of error for a given number of searched paths K=10

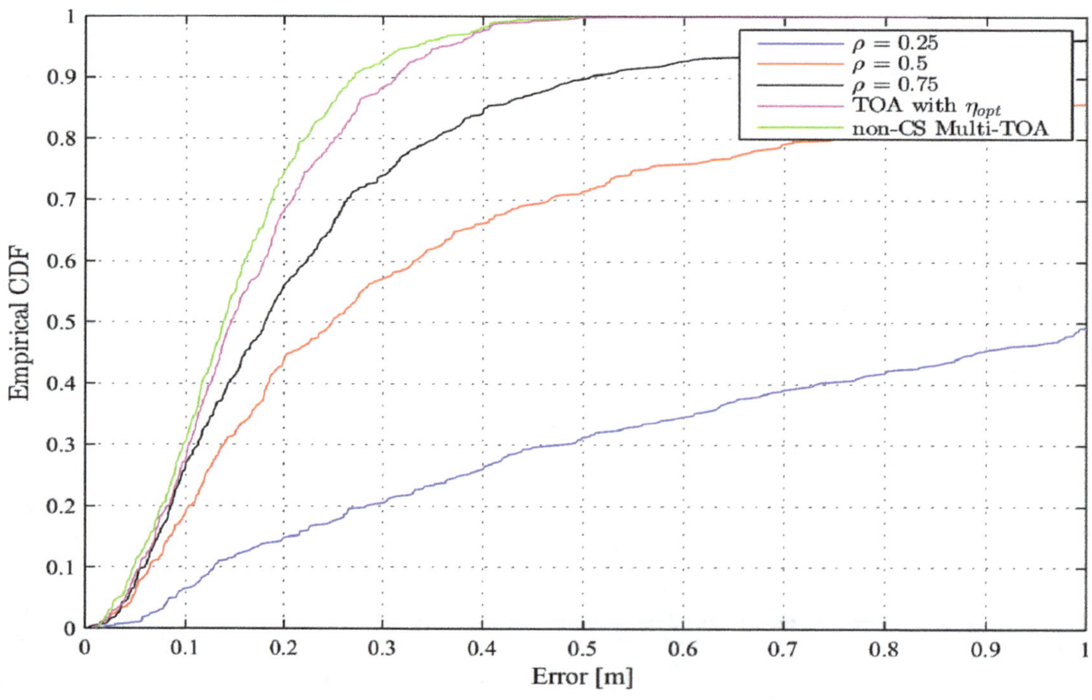

is proven to be the optimum measurement choice in the sense that it achieves the lower CRB (Gezici et al. 2005). The measure distance is given by,

$$\tilde{\rho}_l = \frac{\tilde{\tau}_{0,l}}{c} \tag{32}$$

with c being the speed-of-light and $\tilde{\tau}_{0,l}$ the TOA measurement. Therefore, the variance of the distance estimates σ_l^2 depends on the variance of the TOA measurements, which is characterized by the CRLB (Gezici et al. 2005).

The positioning problem can then be formulated as a Least Squares (LS) problem, where the goal is to minimize the squared error. More specifically, the estimated position can be expressed as indicated in (9).

However, recent works claim that this approach to positioning can be potentially improved, especially in challenging scenarios, if one treats the problem as a whole. That is, to convert the two-step approach into a single-step procedure where digitized signals are combined to solve directly for the position coordinates. This positioning approach is known as DPE, which main features and basic principle has been introduced in previous sections.

DPE has been mainly proposed for narrowband emitters positioning in (Weiss 2004) and for passive geolocation (Weiss 2011) applying MLE. In UWB scenarios, the dense multipath channel poses additional challenges to positioning algorithms, also for DPE approach. (Eric et al. 2008) propose applying the MUSIC algorithm for solving self-position using a DPE approach. The solution though requires knowledge of the channel attenuations for LOS paths, which poses difficulties in practice. (Navarro et al. 2013) introduced a DPE approach based on the generalization of the PDP for TOA estimation, which has proven of high accuracy for conventional two-stage positioning. The motivation for this particular DPE scheme is two-fold: on one hand it can reuse the frequency domain receiver architecture allowing assessment of similar receiver capabilities but different positioning implementations; on the other hand, it does not require calibration as in the two-step approach where proper threshold setting is required.

The problem formulation is applicable to either positioning or localization applications. In a localization set-up, the L anchor nodes obtain ranging measurements to determine the position of a target node, instead of being the target node the one which performs ranging measurements from the anchors (positioning), and determines his own position.

To formulate the DPE problem, we take the frequency domain samples $\mathbf{Y}_k$ defined in (19) and extend it to the multiple anchor case. To this aim, we define the observation frequency sample vector $\mathbf{Y}$, as the concatenation of all received signals from the N_A anchors. More specifically,

$$\mathbf{Y} = \left[\mathbf{Y}^{(1)} \quad \mathbf{Y}^{(2)} \quad \ldots \quad \mathbf{Y}^{(N_A)}\right]^T \tag{33}$$

Dropping the symbol and pulse indexes to ease notation, the signal model can be written in terms of the LOS contribution (first term)

$$\mathbf{Y} = \overline{\mathbf{S}} \cdot \mathbf{e}_{\mathbf{p}} + \tilde{\mathbf{V}} \tag{34}$$

where the steering vectors are now explicitly expressed in terms of the spatial coordinates and the final unknown position $\mathbf{p}$, instead of the distance measurements. That is,

$$\mathbf{e}_{\mathrm{p}} = \left[\mathbf{e}_{f(\mathbf{p}_1)} \quad \mathbf{e}_{f(\mathbf{p}_2)} \quad \cdots \quad \mathbf{e}_{f\left(\mathbf{p}_{N_A}\right)} \right]^T \tag{35}$$

with

$$\mathbf{e}_{f(\mathbf{p}_\ell)} = \left[1 \quad e^{-j\frac{\omega_0}{c}\|\mathbf{p}-\mathbf{p}_\ell\|} \quad \ldots \quad e^{-j\frac{\omega_0}{c}(N-1)\|\mathbf{p}-\mathbf{p}_\ell\|} \right]^T \tag{36}$$

$\overline{\mathbf{S}}$ is the block diagonal matrix of the train of pulses scaled by the LOS fading coefficient,

$$\overline{\mathbf{S}} = \begin{bmatrix} h_{0,1}\mathbf{S} & \cdots & \mathbf{0} \\ \mathbf{0} & & \vdots \\ \vdots & \ddots & \mathbf{0} \\ \mathbf{0} & \cdots & h_{0,N_A}\mathbf{S} \end{bmatrix} \tag{37}$$

Extending principle of the PDP presented in (21) we obtained a DPE estimator that aims maximizing the following cost function

$$\hat{\mathbf{p}} = \arg\max_{\mathbf{p}\in\mathbb{R}^2} \mathbf{e}_{\mathrm{p}}^{\;H} \boldsymbol{R} \mathbf{e}_{\mathrm{p}} \tag{38}$$

Where now the correlation matrix $\mathbf{R} = \mathbf{Y}\mathbf{Y}^H$ is of higher dimension. We shall remark that the proposed solution is not the optimum DPE solution but a sub-optimal scheme that aims to extend the reduced complexity super-resolution ranging algorithm in (21) to the higher dimensional problem associated to the Euclidean space. The cost function $\Phi\left(\mathbf{p}\right) \triangleq \mathbf{e}_{\mathrm{p}}^{\;H} \boldsymbol{R} \mathbf{e}_{\mathrm{p}}$ is referred to as Power-Position Profile (PPP), given that it provides a characterization of the signal energy distribution with respect to the spatial position.

The optimization of (38) is a non-convex function which can be solved by grid search over the two-dimensional space with $\hat{\mathbf{p}} = \left(\hat{x}, \hat{y}\right)$. More sophisticated methods can be employed to improve search efficiency.

Numerical Results

To evaluate the performance, we next provide some illustrative numerical results. We consider several scenarios where anchors experiences different channel conditions (LOS/NLOS) in order to evaluate accuracy and robustness against NLOS. There is also variation in the number of available nodes. Figure

12 shows some arbitrary node distribution of anchors (labelled by numbers from 1 to 9) and target node, for different number of active reference nodes (inactive nodes are depicted in grey color).

The specific node positions are specified in Table 3. The assessment of DPE is done under realistic channel models provided in the standard IEEE 802.15.4a. (Molisch et al. 2004). For each anchor a channel is generated according to environment conditions: some with uniform set-up, e.g. all nodes in LOS and others with mixed models including some nodes in LOS and some in NLOS. The selected scenarios aim also to illustrate how increasing the number of anchor nodes and improving the geometry, affects accuracy and robustness.

As an illustrative example of the benefits of the DPE versus the standard two-step TOA positioning, figure 13 shows the empirical cumulative density function of the position error for a scenario with five and nine reference nodes, respectively. The numerical results are obtained considering the propagation channels models developed within the framework of the IEEE 802.15.4a, which are characterized to exhibit dense multipath components, especially in indoor scenarios (Molisch et al. 2004) (Molisch, 2005). In particular figure 13 depicts the position error in an Industrial LOS set-up, which corresponds

Figure 12. Scenario configuration: 4, 5 or 9 anchors, respectively. The target remains in the same position for all scenarios.

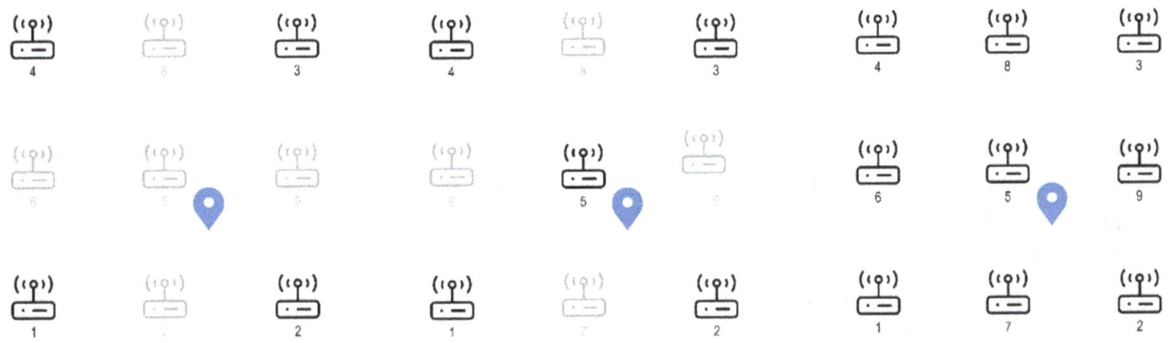

Table 3. Anchors and target position in meters

Anchors	$\mathbf{p}_1 = (0,0)$ $\mathbf{p}_2 = (0,6)$ $\mathbf{p}_3 = (6,6)$ $\mathbf{p}_4 = (6,0)$	$\mathbf{p}_5 = (3,3)$ $\mathbf{p}_6 = (3,0)$ $\mathbf{p}_7 = (0,3).$ $\mathbf{p}_8 = (3,6).$ $\mathbf{p}_9 = (6,3)$
Target	$\mathbf{p} = (4,2)$	

Figure 13. Empirical CDF of the position error for DPE and Two-steps TOA-based positioning as a function of the number of anchor nodes. Evaluated in Industrial LOS (CM7).

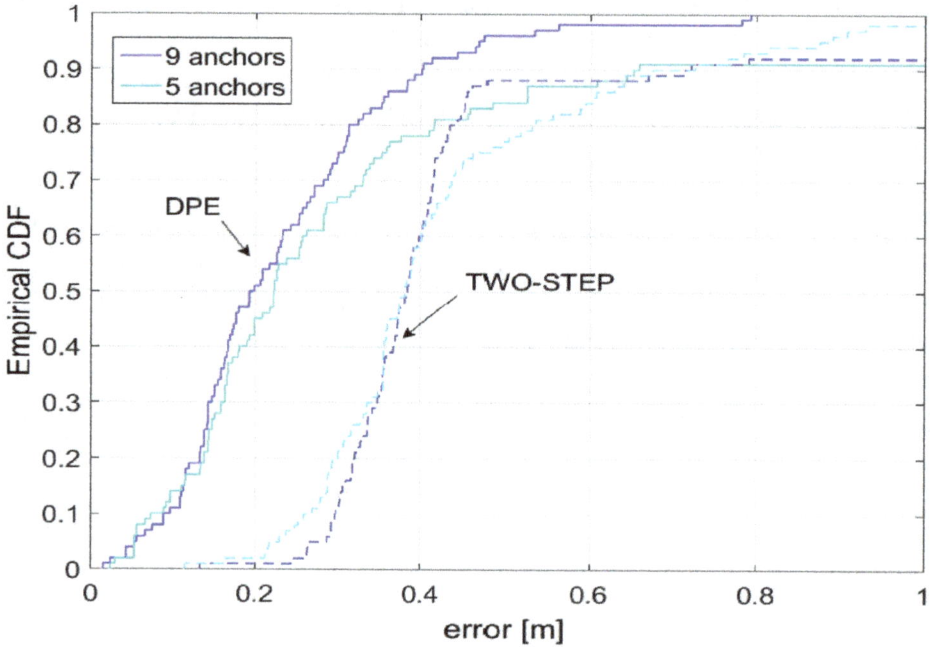

to channel mode CM7. Under this scenario, where LOS component is dominant, the DPE solution clearly outperforms the two-stage positioning. This gain is observed for both configurations, low and high number of anchor nodes.

By representing a single realization of the power-position profile, shown in Figure 14 for the same channel mode CM7, one observes as the number of nodes increases, the reduction of side-lobes and secondary local maxima. This reduction is also significant for channel model CM3 (Office LOS) with stronger multipath components than in CM7, as depicted in Figure 15. Thus increasing the number of anchor nodes is proved effective to mitigate multipath effect in DPE solution. Although the scheme is sensitive to strong multipath condition it shows high robustness to signal strength. The performance accuracy remains stable in a wide range of SNR, down to low vales. In the example, nodes operate at an average SNR of 0 dB.

The assessment of DPE for IR-UWB under realistic channel models shows potential improvements with respect to two-step standard positioning. Recalling the advantages, we conclude that the DPE solution does not require the estimation of channel coefficients. Nor it requires calibration. This is opposed to the two-step approach which requires proper threshold selection to obtain accurate performance, which in practice, is difficult since threshold varies with propagation conditions. However, there are several issues that required further investigation. On the one hand DPE under strong multipath components suffers performance degradation since it creates local maxima in the cost function that may lead to position errors. Therefore, exploring other multipath mitigation techniques besides increasing the number of nodes or improving the Geometric Dilution of Precision (GDOP) is an issue of further investigation. Approaches proposed for the two-step positioning solution are not directly applicable.

Figure 14. Power-position profile realization evaluated for all nodes in Industrial LOS (CM7) for DPE with 4 and 9 anchors, respectively

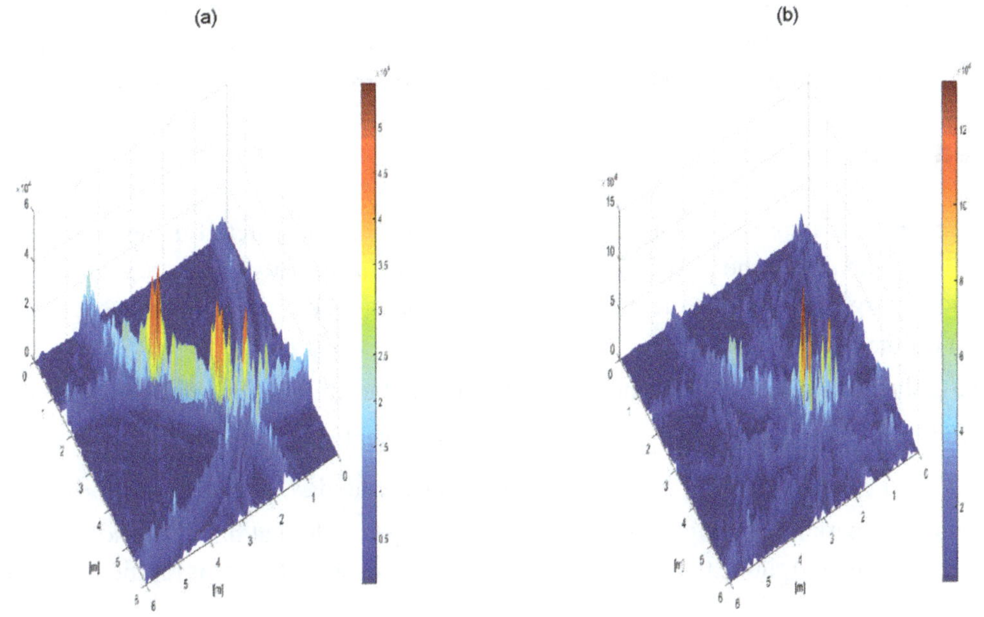

Figure 15.Power-position profile realization evaluated for all nodes in Office LOS (CM3) for DPE with 4 and 9 anchors, respectively

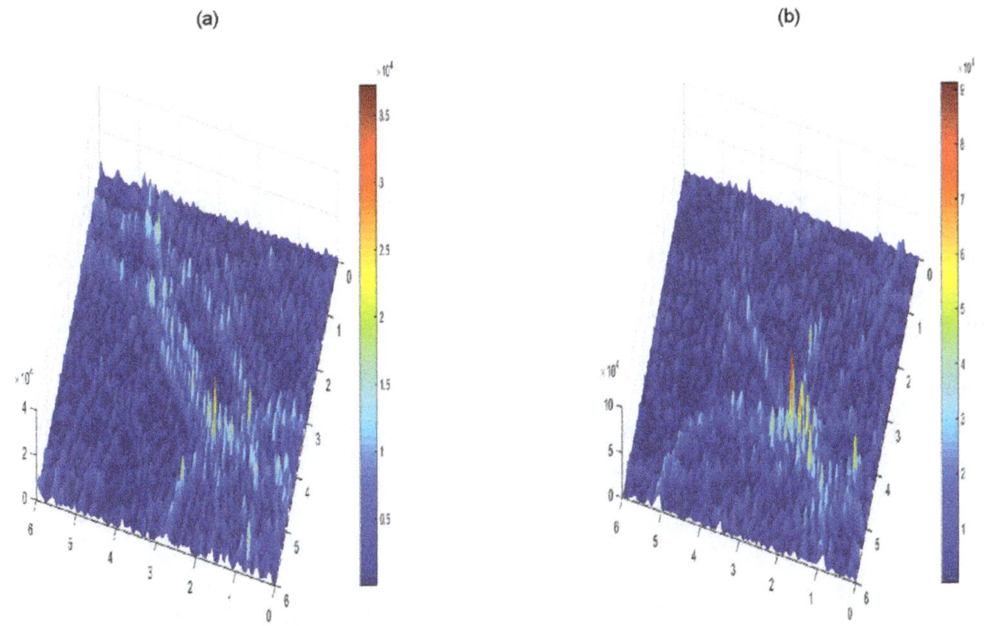

On the other hand, DPE relies on having the full vector of the sampled-signal observations. In practical systems, such requirement can imply a significant communication burden to the system. However, one way to reduce complexity it to apply CS techniques to reduce the problem dimension and to improve complexity performance by replacing brute-force search by compressed-sensing algorithms that exploit the inherent sparsity of the problem. Details following this approach is introduced in the following section.

CS-Based DPE

CS theory states that a sparsely representable signal can be reconstructed using very few number of measurements compared to the signal dimension. Based on the sparsity of the target area, i.e., the number of unknown targets is small in the discrete spatial domain, let us consider the subset of the complete observation defined in (26), $\breve{\mathbf{Y}}_k^{(\ell)}$.

Assuming the target area is partitioned into a finite number of positions to be evaluated, $N_x \times N_y$, in the X-axis and the Y-axis, the scenes can be represented by the indicator function $\sigma\left(k,l\right)$, $k = 0,\ldots,N_x - 1$, $l = 0,\ldots,N_y - 1$. Let $\boldsymbol{\sigma}'$ be the concatenated $N_x N_y \times 1$ scene indicator vector corresponding to the spatial sampling grid. For the positioning problem at hand, the scene vector $\boldsymbol{\sigma}'$, can be can be seen as the output of the PPP defined in (38) for each of the geographical locations that are evaluated.

The sparsity of the power-position profile introduced in (38) motivates the investigation of CS-based reconstruction of the scene from compressive measurements. To linearly relate the compressive measurements with the energy in the power-position profile we propose the following model,

$$\breve{\mathbf{R}} = \sum_{i=0}^{N_x N_y - 1} \sigma_{p_i} \breve{\mathbf{e}}_{\mathbf{p}_i} \breve{\mathbf{e}}_{\mathbf{p}_i}^H + \breve{\mathbf{R}}_\varepsilon \tag{39}$$

where $\breve{\mathbf{R}} \in \mathbb{C}^{\kappa N_A \times \kappa N_A}$ is the compressed sample covariance matrix computed as $\breve{\mathbf{R}} = \frac{1}{N_f}\sum_{k=1}^{N_f} \breve{\mathbf{Y}}_k \breve{\mathbf{Y}}_k^H$, with $\breve{\mathbf{Y}}_k = \left[\breve{\mathbf{Y}}_k^{(1)} \cdots \breve{\mathbf{Y}}_k^{(\ell)} \cdots \breve{\mathbf{Y}}_k^{(N_A)}\right]^T$ denoting the concatenation of compressive signals from all anchors, $\breve{\mathbf{R}}_\varepsilon$ represents imperfections of the model and σ_{p_i} is the energy indicator function at location $\boldsymbol{p}_i$. Vector $\breve{\mathbf{e}}_{\mathbf{p}_i} = \left[\boldsymbol{\Phi}\mathbf{e}_{f_1(\boldsymbol{p}_i)} \cdots \boldsymbol{\Phi}\mathbf{e}_{f_{N_A}(\boldsymbol{p}_i)}\right]^T$ denotes the compressive steering vector corresponding to the grid location $\boldsymbol{p}_i$.

The model in (39) can be conveniently rewritten into a sparse notation as follows,

$$\breve{\mathbf{r}} = \mathbf{A}'\boldsymbol{\sigma}' + \breve{\mathbf{r}}_\varepsilon \tag{40}$$

where $\breve{\mathbf{r}}$ is a $\left(\kappa N_A\right)^2 \times 1$ vector formed by the concatenation of the columns of $\breve{\mathbf{R}}$. From now on, to clarify notation, the concatenation of columns will be denoted with the operator $\mathrm{vec}\left(\bullet\right)$. Therefore,

$\breve{\mathbf{r}} = \mathrm{vec}\left(\breve{\mathbf{R}}\right)$. The columns of $\mathbf{A}'$ contain the vectorized candidate covariance matrices corresponding to the different tentative target locations $\boldsymbol{p}_i$ that conform the spatial scanning grid, and is defined as,

$$\mathbf{A}' = \left[\breve{\mathbf{r}}\left(\boldsymbol{p}_0\right) \quad \breve{\mathbf{r}}\left(\boldsymbol{p}_1\right) \cdots \quad \breve{\mathbf{r}}\left(\boldsymbol{p}_{N_x N_y \times 1}\right)\right] \tag{41}$$

where $\breve{\mathbf{r}}\left(\boldsymbol{p}_i\right) = \mathrm{vec}\left(\breve{\mathbf{e}}_{\mathbf{p}_i} \breve{\mathbf{e}}_{\mathbf{p}_i}^H\right)$. The variable $\breve{\mathbf{r}}_\varepsilon$ encompasses undesired contributions and imperfections of the model. Note that the elements different from zero of vector

$$\boldsymbol{\sigma}' = \left[\sigma\left(\boldsymbol{p}_0\right) \quad \sigma\left(\boldsymbol{p}_1\right) \cdots \quad \sigma\left(\boldsymbol{p}_{N_x N_y \times 1}\right)\right]^T \tag{42}$$

correspond to the grid point where we find PDP values different from zero.

The sparse vector $\boldsymbol{\sigma}'$ can be recovered from the compressive measurements by solving the following optimization problem,

$$\min_{\boldsymbol{\sigma}'} \left\|\boldsymbol{\sigma}'\right\|_{l_1} \text{ subject to } \quad \breve{\mathbf{r}} \approx \mathbf{A}'\boldsymbol{\sigma}' \tag{43}$$

As mentioned in previous sections, there are plenty of algorithms available in the literature to solve the optimization problem in (43). In this paper, we choose Orthogonal Matching Pursuit (OMP) (Tropp & Gilbert 2007) to solve (43), which is an iterative algorithm known to provide a fast and easy to implement solution.

Numerical Results

The proposed CS-based DPE has been analyzed by means of Monte-Carlo MATLAB-based simulations with the aim of evaluating its potential gains over conventional DPE and conventional two-step approach in IR-UWB. For numerical evaluation we considered the channel models developed within the framework of the IEEE 802.15.4a (Molisch et al. 2004). In particular we evaluated the localization performance for the CM3 Office LOS and CM7 Industrial LOS channel models, whose propagation channels are characterized to exhibit dense multipath components, especially the CM3. Without loss of generality, we present preliminary results for a simplified scenario with a single target. The transmitted pulse was a Gaussian monocycle pulse with duration $T_p = 1\,\mathrm{ns}$.The observation window T_f was equal to 56ns. The non-compressive sampling rate is given by 2 GHz, which translated into an observation window of N = 128 samples. The sampling rates of $\breve{\mathbf{Y}}_k^{(\ell)}$ and $\mathbf{Y}_k^{(\ell)}$ are related through the compression rate $\rho = \frac{\kappa}{N}$.

To strictly focus on the performance behavior due to compression and remove the effect of insufficient data records, the size of the compressed observations was forced to be the same for any compression rate. Thus, for a high compression rate, the estimator took samples for a larger period of time. The simulation parameters are summarized in Table 4, where N_f denotes the number of observed data

Table 4. Simulation Parameters

ρ	**1**	**0.50**	**0.25**	**0.13**
κ	128	64	32	16
N_f	128	256	1024	2048

frames. The position algorithm was evaluated in a 2D setting with the target randomly placed within a square room of 6 x 6 m2. The area of interest was divided into a 200 x 200 grid and the resolution was 0.03m.

First, to illustrate the random sampling idea we tested with a single realization to motivate the discussion, where the target node was placed at $\boldsymbol{p} = \left[5.27 \quad 4.03\right]^T$ and $N_A = 4$ anchor nodes placed at the corners of the room. We considered the CM3 Office LOS propagation environment and an average SNR at each anchor node of 4 dB. The SNR at each anchor is defined as,

$$SNR = \frac{1}{\sigma_w^2} \cdot \frac{1}{\kappa N_f} \sum_{k=1}^{N_f} \left(\breve{\mathbf{Y}}_k^{(\ell)}\right)^H \breve{\mathbf{Y}}_k^{(\ell)} \tag{44}$$

The simulated room geometry is depicted in Figure 16(a). Figure 16(c) shows the result obtained with the conventional two-step procedure, using all $\mathrm{N} \cdot \mathrm{N}_\mathrm{A}$ observations ($\rho = 1$). Although the estimated node position (indicated by a black cross in the figure), can be seen clearly around the correct node position (indicated by a blue point) in Figure 16(c), it must be noted that the positioning accuracy of trilateration greatly depends on the accuracy of each range estimate given by the TOA estimation at each node. One of the challenges in TOA estimation is detecting the peak corresponding to the direct propagation path, which corresponds to the first peak in the PDP. In order to find the first peak, it is important to determine the detection threshold. The first peak was chosen as the first of those peaks that were greater than the detection threshold. The approach followed in the paper for the evaluation of the threshold-based TOA algorithm was to perform an off-line calibration of the threshold per noise level such that it provides minimum MSE. This optimum threshold was then used in the simulations to perform the Monte Carlo trials. Figure 16(b) shows the power range profile obtained at each of the anchor nodes, where the corresponding threshold has been depicted in red. Notice that this is the best threshold selection, and thus we are comparing our approach to a clairvoyant two-steps method that is able to select the optimum threshold in the MSE sense. It is expected that using any of the threshold selection methods in the literature, the performance should decrease. Considering the complete data volume, i.e., $\rho = 1$, conventional DPE provides the cost function depicted in Figure 16(d). Figure 16(d) exhibits few local maximums of similar amplitude in close vicinity of the true node position, thereby making the detection of the global maximum cumbersome. For the proposed CS-based DPE method with 25% of randomly selected data ($\rho = 0.25$) generates the target space vector shown in Figure 16(f) after convenient rearrangement, where a single peak associated with the true target position (indicated by a gray circle centered at the true target position in the figure) can be easily identified. For comparison purposes, the

Figure 16. Example: (a) Target space, (b) Power delay profiles for the TOA estimation step (ρ = 1), (c) Two-Step Approach (ρ = 1), (d) Conventional DPE (ρ = 1), (e) Least-square solution for CS-based DPE with ρ = 0.25 and (f) Proposed CS-based DPE for ρ = 0.25.

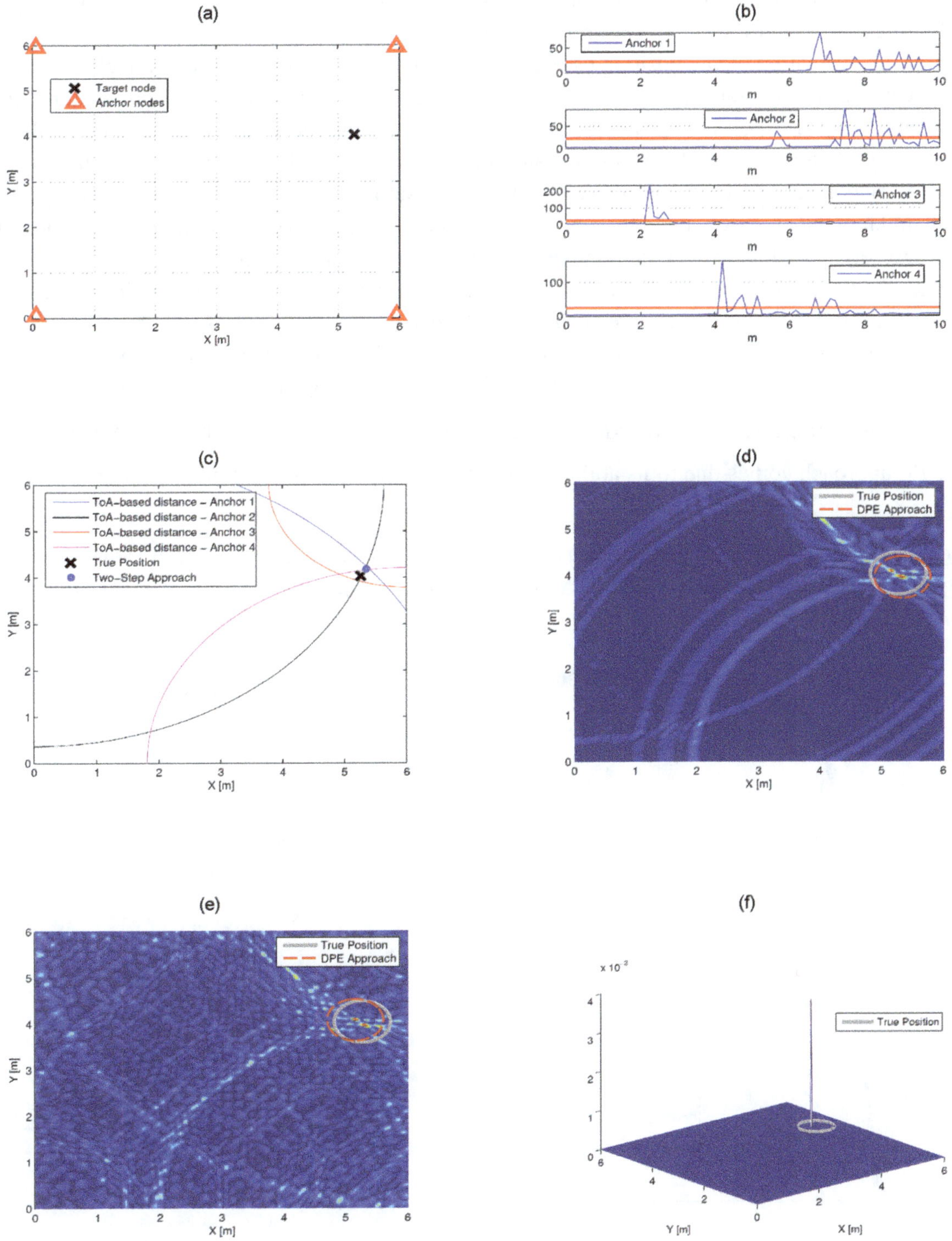

result obtained with conventional DPE using the same reduced data set is shown in Figure 16(e). It can be observed that conventional DPE generates many artifacts in close vicinity of the global maximum (indicated by a gray circle centered at the true target position in the figure) corresponding to the true target location. The less cluttered representation obtained with CS-based DPE underscores the importance of forcing sparse solution via l1-normminimization.

For the evaluation of the position estimation accuracy, we considered 200 channel realizations for each scenario. To illustrate the advantage of merging the two-step approach into a single-step estimation problem, Figure 17 shows the position error in terms of the cumulative density function (CDF) of conventional DPE and two-step TOA-based positioning for under CM7 Industrial LOS scenario, an average SNR at each anchor node of 4dB and without compression ($\rho = 1$). It is worth mentioning, however, that similar results were obtained for other SNRs (from 0dB to 10dB). From Figure 17, we observe that the CDF corresponding to the DPE approach is always above that achieved by the conventional two-step approach, in accordance to (Weiss 2004) (Amar & Weiss 2008) (Gusi, Closas, Mallat & Vandendorpe 2014). Generally speaking, the more anchor nodes participate in the positioning, the more accurate the resulting location estimate. Figure 17(a) and Figure 17(b) depict the positioning error obtained considering 4 and 6 anchor nodes, respectively. In the 6 anchor scenario, we use the same anchor distribution depicted in Figure 16(a) plus two extra anchors in $p_5 = \begin{bmatrix} 3 & 0 \end{bmatrix}^T$ and $p_6 = \begin{bmatrix} 3 & 6 \end{bmatrix}^T$. Since the strength of DPE approach comes due to its ability to jointly process signals at different anchor nodes, it is not

Figure 17. Empirical CDF of the position error for conventional DPE and two-step TOA-based positioning ($\rho = 1$) evaluated in IEEE 802.15.4a Industrial LOS (CM7): (a) 4 anchor nodes, (b) 6 anchor nodes.

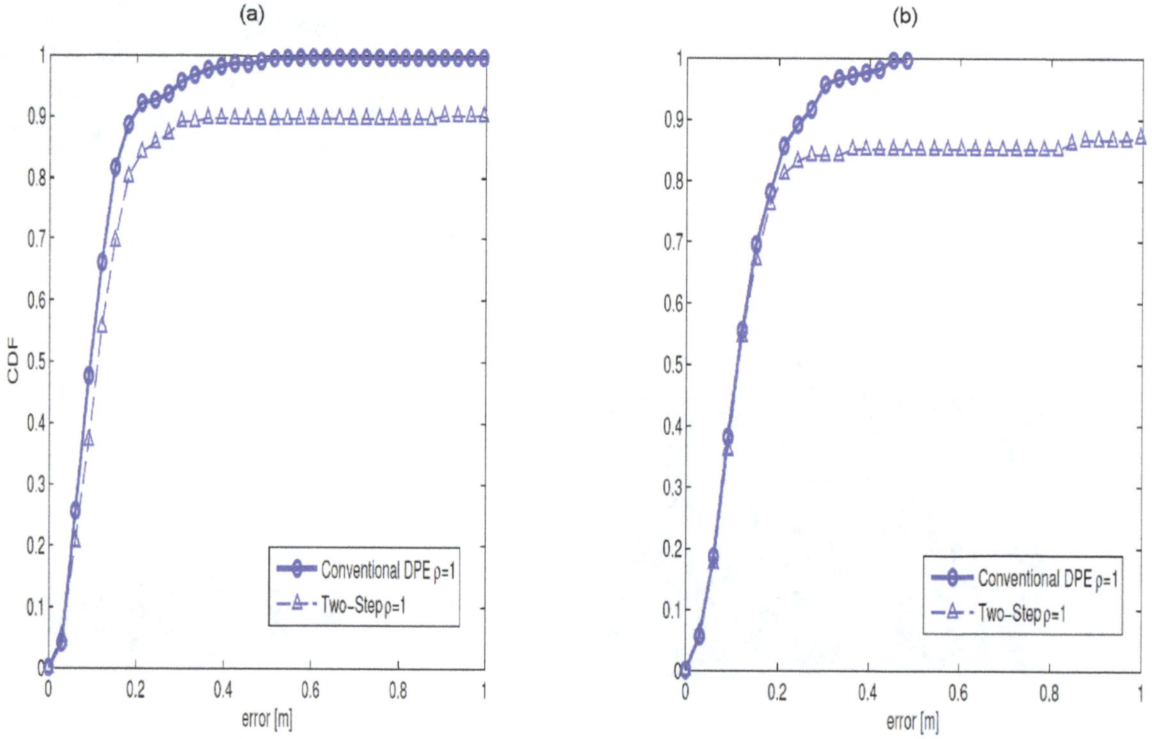

surprising that increasing the number of anchor nodes translates into a slightly better performance. However, it can be observed that the curves corresponding to the two-step approach degrade with the number of anchor nodes. We attribute this behavior to the difficulties in choosing a proper threshold detector for each anchor node. One advantage of DPE versus two-step positioning is that DPE does not require any calibration, in the sense that two-step scheme requires the proper setting of the threshold value, which in practice varies with the propagation conditions.

To test the effect of reducing acquired samples, Figure 18(a) and Figure 18(b) show the CDF of the proposed CS-based DPE under CM7 Industrial LOS and CM3 Office LOS channel, respectively, an average SNR at each anchor node of 4dB and for different compression rates. On the one hand, from Figure 18, it can be concluded that, in a large percentage (about 80% under CM3 and 100%under CM7), the proposed CS-based DPE with half of the original samples provides the same localization capabilities as conventional DPE, i.e., the position of the global maximum is not affected by the reduction of data. However, on the other hand, as expected, an excessive reduction in the number of measurements translates into a degradation of the localization performance. Finding the value of κ that ensures perfect recovery of $\boldsymbol{\sigma}$ is a problem of longstanding interest in the CS community (Candes & Wakin 2008), (Donoho 2006). Existing results in this regard showed that the minimum κ depends on the number of non-zero components of the sparse vector $\boldsymbol{\sigma}$, which in our case is a function dependent on the multipath components present in the propagation channel under consideration. In future work, we will study the relationship between the multipath and the minimum number of measurements.

Figure 18. Empirical CDF of the position error for conventional DPE and CS-DPE for different compression rates: (a) Idustrial LOS (CM7), (b) Office LOS (CM3).

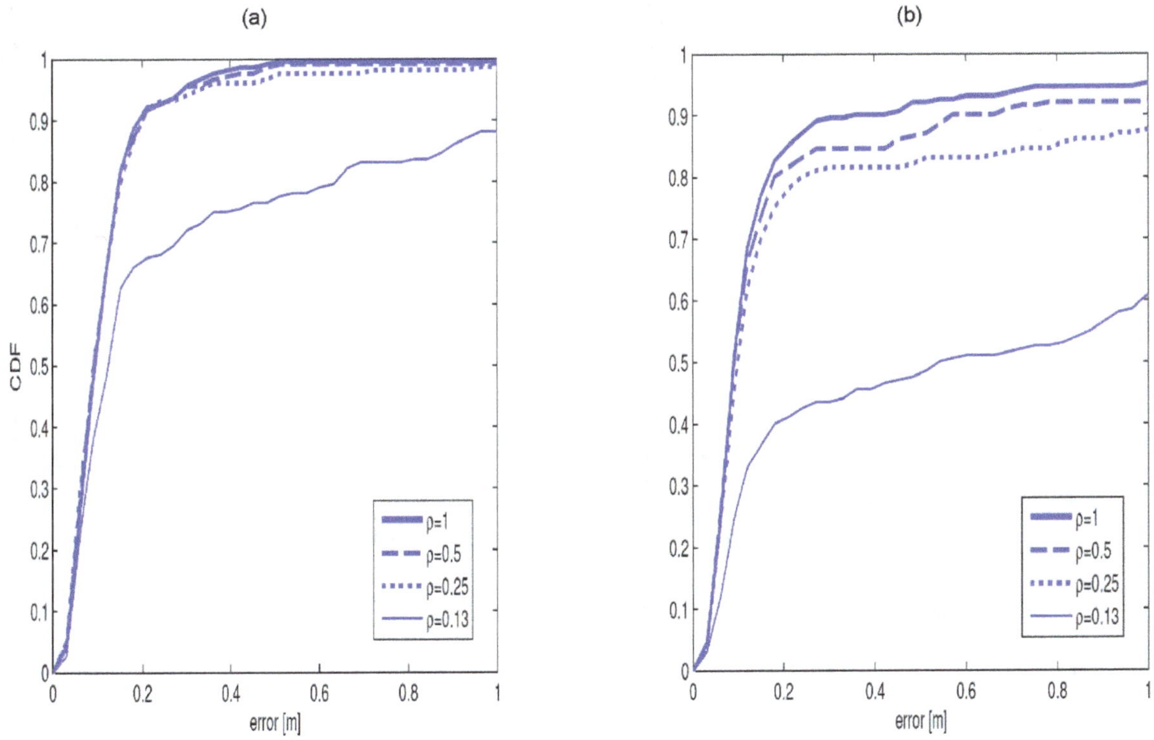

Generally, Industrial LOS (CM7) is a friendlier environment compared to Office LOS (CM3), where the much denser multipath poses additional challenges for the positioning problem. The latter is clearly evident when Figure 18(a) is compared to Figure 18(b). It is in CM3 where both the conventional DPE and the proposed CS-based DPE experience larger difficulties to accurately estimate the node position due to the strong and dense multipath components.

To illustrate the performance difference of these two propagation scenarios, Figure 19 shows the PDP obtained with conventional DPE ($\rho = 1$) and the CS-based reconstructed target space with $\rho = 0.5$ using 4 anchor nodes with an average SNR at each anchor node of 0dB. This single realization is used to motivate the discussion. Denser multipath propagation causes more interference between the first path and other paths in the resulting target space of the conventional DPE. This statement is confirmed by the comparison of Figure 19(a) and Figure 19(b), where conventional DPE is evaluated in CM7 and CM3, respectively. The PDP obtained with conventional DPE in CM7 have less clutter residuals compared to the corresponding CM3 PDP, leading to an enhanced localization performance. The ambiguity caused by the CM3 strong multipath components also affects the localization performance of CS-based DPE, as evident from the comparison of Figure 19(c) and Figure 19(d), where CS-based DPE is evalu-

Figure 19. PDP realization with 4 anchors and SNR=0dB for: (a) Conventional DPE with ρ = 1evaluated in CM7 Industrial LOS, (b) Conventional DPE with ρ = 1evaluated in CM3 Office LOS, (c) CS-based DPE with ρ = 0.5 evaluated in CM7 Industrial LOS and (d) CS-based DPE with ρ = 0.5 evaluated in CM3 Office LOS.

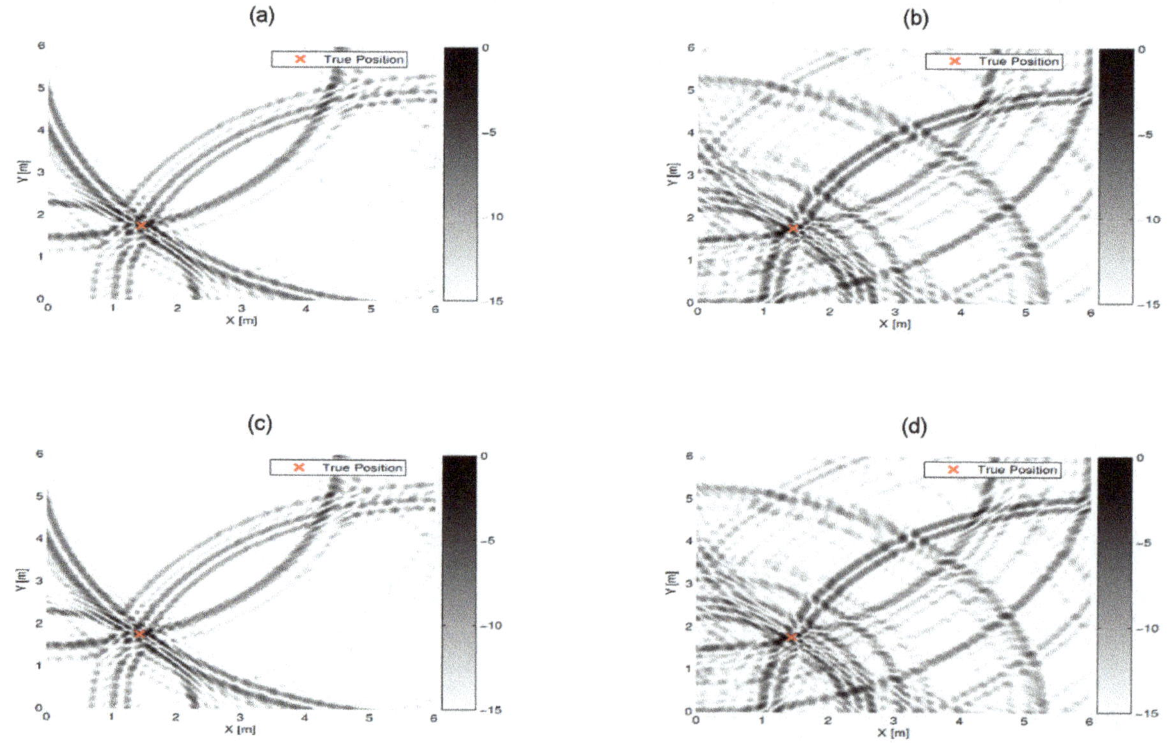

ated in CM7 and CM3, respectively. The estimated position obtained with CS-based DPE is located further away from the true target position when considering IEEE 802.15.4a CM3 due to the dense multipath.

Next, the performance in low SNR scenarios is evaluated by means of the probability of detection (P_d) versus SNR curves in order to illustrate both conventional DPE and the proposed CS-based DPE robustness against noise. Theoretical derivation of the probability of false alarmversus the probability of detection, requires the statistical distribution of the localization algorithm when the target node is present (Hypothesis H1) and with absence of target node (HypothesisH0). To find the empirical distribution for H1 and H0, we recorded 200 Monte Carlo simulations, each in the presence of a target node (H1), and 200 records without the target node (H0), keeping the maximum value obtained by the resulting PDP in conventional DPE and the value different from zero of σ' in CS-based DPE. Figure 20shows the normalized distributions ofH0 and H1 for a particular example of CS-based DPE under CM3 Office LOS channel, with 4anchor nodes, an average SNR of -20dB and $\rho = 0.25$. The approximated probability densityfunction computed using a kernel smoothing method is also plotted in Figure 20(a). These kernel-based nonparametric density estimates were used for computation of the probability figures. Figure 20(b) shows the P_d versus SNR results of the conventional DPE compared to the proposedCS-based DPE for $\rho = 0.25$ and for a fixed probability of false alarm $P_{fa} = 0.001$. It is clear that, even for low compression rates, the proposed sparsity-based DPE approach is able to reliably localize the target node in very low SNR transmission. It can be observed from Figure 20(b) how the curves progressively deteriorate as ρ and SNR decrease.

Figure 20. (a) Theoretical and empirical distributions of H1 and H0 for CS-based DPE with ρ = 0.52, (b) Probability of detection versus SNR for P_{fa} = 0.001.

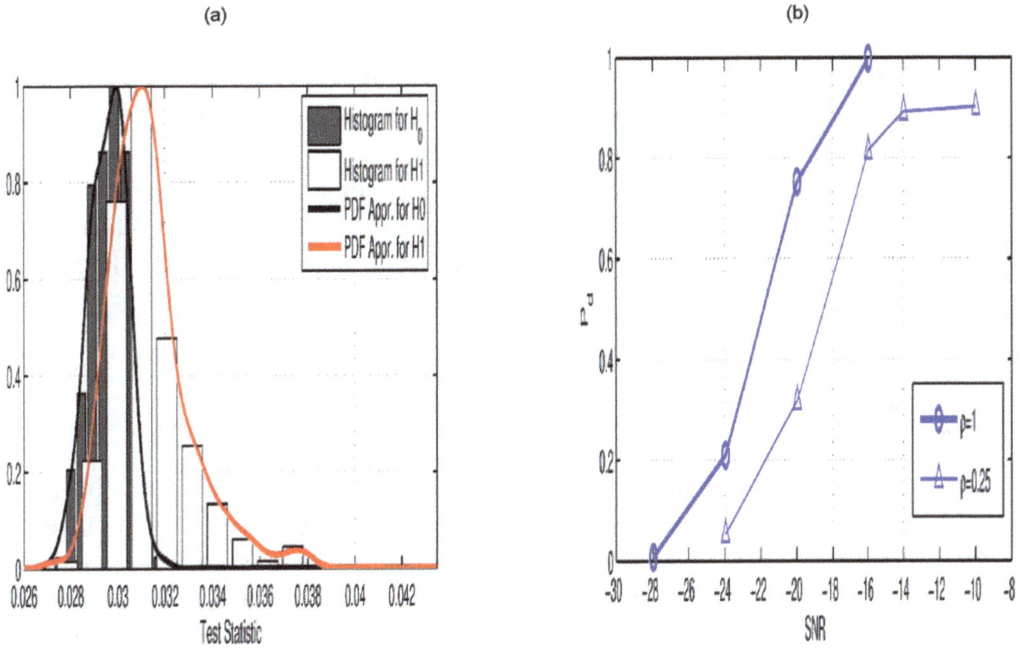

FUTURE RESEARCH DIRECTIONS

Although the techniques presented in this chapter have tackled some of the main problems of IR-UWB localization, such as the high sampling rate and the threshold selection to identify the first arriving path, there are yet many opportunities for extending and improving the proposed techniques. This section discusses some of the topics that we consider particularly interesting for extension.

- One of the challenging topics in any source localization problem is the estimation of the unknown position in the presence of multipath reflections. Therefore, an interesting path for research is the multipath effect in both the CS and non-CS based DPE strategies. DPE-based techniques under strong multipath components suffer performance degradation since the multipath created several local maxima in the PPP that may lead to position errors. How to mitigate the multipath effect or how to contructively combine it to improve the localization accuracy is one of the issues left for future research.
- In addition and implicitly considered on the multipath effect, is the Non-Line-of-Sight (NLOS) conditions, which refers to the case when the direct LOS between the target node and the anchor is blocked. In this case, only reflections of the UWB pulse from scatterers reach the receiving node and, thus, the delay of the first arriving pulse does not represent the true TOA. This, and how to overcome this situation, is an interesting research topic to be investigated in future actions.
- Moreover, sparsity based localization is an emerging topic in the signal processing community and comparison of the proposed sparsity based localization algorithms with recently proposed techniques should be carried out in order to judiciously position the proposed techniques within the literature.
- Although CS and sparsity-based processing is expected to ease the implementation of IR-UWB techniques in general by relaxing the sampling bottleneck, still the proposed techniques have some open issues related to practical implementation on real systems. An important challenge faced by the CS-based techniques proposed in this chapter is the large computational complexity that results from matrices $\mathbf{A}$ and $\mathbf{A}'$, whose columns contain correlation matrices of high dimension (samples x anchor nodes). As a consequence, the optimization problem in (31) and (44) are computationally intensive.
- Last but not least, there has been an emerging trend in combining measurements of different nature to compute the estimated location of nodes. These are known as hybrid techniques and the most prominent one is the technique that combined TOA measurements with RSS measurements, which are easily available in mobile terminals connected to a base station.

CONCLUSION

In this chapter, we tackle the problem of localization in IR-UWB under realistic channel models for two-step multi-TOA-based procedure and DPE approach. For the two-step multi-TOA-based method, we exploit the temporal sparsity of the UWB received signal in order to apply CS reconstruction on the temporal domain and at each anchor separately. We showed with numerical simulations that the most significant received paths are detected and efficiently used in the next step for the multi-trilateration, thus avoiding the threshold based detection.

As a step forward, direct-positioning approaches are reviewed and compressive sensing solutions introduced to alleviate DPE's computational cost, particularly in what regards to the amount of information to be sent to a central fusion unit. Simulation results are provided based on realistic channel models, which show the benefits of DPE with respect to two-step procedures in terms of accuracy and prove the efficiency of CS in reducing the data to be processed without compromising the location estimation accuracy.

REFERENCES

Alereon Inc. (2017). *Alereon Demonstrates Military Wireless Personal Area Network*. Retrieved March 16, 2017, from: http://www.alereon.com/?page_id=2992

Amar, A., & Weiss, A. J. (2008), New asymptotic results on two fundamental approaches to mobile terminal location. *Proceedings of International Symposium on Communications, Control and Signal Processing (ISCCSP)*, 1320–1323. doi:10.1109/ISCCSP.2008.4537430

Boyd, S., & Vandenberghe, L. (2004). Convex Optimization. Cambridge, UK: Cambridge University Press. doi:10.1017/CBO9780511804441

Candes, E. J., Romberg, J., & Tao, T. (2006). Robust Uncertainty Principles: Exact Signal Reconstruction From Highly Incomplete Frequency Information. *IEEE Transactions on Information Theory*, *52*(2), 489–509. doi:10.1109/TIT.2005.862083

Candes, E. J., & Tao, T. (2006). Near Optimal Signal Recovery from Random Projections: Universal Encoding Strategies. *IEEE Transactions on Information Theory*, *52*(12), 5406–5425. doi:10.1109/TIT.2006.885507

Candes, E. J., & Wakin, M. B. (2008). An Introduction to Compressed Sampling. *IEEE Signal Processing Magazine*, *25*(2), 21–30. doi:10.1109/MSP.2007.914731

Cevher, V., Duarte, M., & Baraniuk, R. (2008). Distributed Target Localization Via Spatial Sparsity. *Proceedings of European Signal Processing Conference (EUSIPCO).*

Chen, S. S., Donoho, D. L., & Saunders, M. A. (1999). Atomic Decomposition by Basis Pursuit. *SIAM Journal on Scientific Computing*, *20*(1), 33–61. doi:10.1137/S1064827596304010

Clausen, J. (1999). Branch and bound algorithms-principles and examples. Department of Computer Science, University of Copenhagen.

Closas, P., Fernandez-Prades, C., & Fernandez-Rubio, J. A. (2007). Maximum Likelihood Estimation of Position in GNSS. *IEEE Signal Processing Letters*, *14*(5), 359–362. doi:10.1109/LSP.2006.888360

Closas, P., Fernandez-Prades, C., & Fernandez-Rubio, J. A. (2009). Cramer-Rao Bound Analysis of Positioning Approaches in GNSS Receivers. *IEEE Transactions on Signal Processing*, *57*(10), 3775–3786. doi:10.1109/TSP.2009.2025083

Closas, P., Fernandez-Prades, C., Fernandez-Rubio, J. A., Wis, M., Veccione, G., Zanier, F., & Crisci, M. et al. (2015). Evaluation of GNSS Direct Position Estimation in Realistic Multipath Channels. *Proceedings of the ION GNSS International Technical Meeting.*

Dampf, J., Pany, T., Bär, W., Winkel, J., Stöber, C., Fürlinger, K., Closas, P. & Garcia-Molina, J.A. (2015). More than we ever dreamed possible: Processor technology for GNSS software receivers in the year 2015. *Inside GNSS Magazine, 10*(4), 62–72.

Dardari, D., Chong, C. C., & Win, M. (2008). Threshold-based time-of-arrival estimators in UWB dense multipath channels. *IEEE Transactions on Communications, 56*(8), 1366–1378. doi:10.1109/TCOMM.2008.050551

Dardari, D., Closas, P., & Djuric, P. (2015). Indoor tracking: Theory, methods, and technologies. *IEEE Transactions on Vehicular Technology, 64*(4), 1263–1278. doi:10.1109/TVT.2015.2403868

Dardari, D., Conti, A., Ferner, U., Giorgetti, A., & Win, M. (2009). Ranging with Ultrawide Bandwidth Signals in Multipath Enviroments. *Proceedings of the IEEE, 97*(2), 404–537. doi:10.1109/JPROC.2008.2008846

Decawave. (2017). *Indoor Positioning Systems*. Retrieved March 16, 2017, from http://www.decawave.com/

Donoho, D. (2006). Compressed Sensing. *IEEE Transactions on Information Theory, 52*(4), 1289–1306. doi:10.1109/TIT.2006.871582

Eric, M., Dukic, M., & Vucic, D. (2008), Direct Position Estimation of UWB Transmitters in Multipath Conditions. *Proceedings of IEEE International Conference on Ultra-Wide Band (ICUWB)*, 241-244.

Gezici, S., Zhi Tian, , Giannakis, G. B., Kobayashi, H., Molisch, A. F., Poor, H. V., & Sahinoglu, Z. (2005). Localization via ultra-wideband radios: A look at positioning aspects for future sensor networks. *IEEE Signal Processing Magazine, 22*(4), 70–84. doi:10.1109/MSP.2005.1458289

Gholami, M. R., Reza, M., Rydstrom, M., & Strom, E. G. (2010). Positioning of Node using Plane Projection onto Convex Sets. *Proceedings of Wireless Communications and Networking Conference (WCNC)*. doi:10.1109/WCNC.2010.5506717

Gholami, M. R., Ström, E. G., Sottile, F., Dardari, D., Conti, A., Gezici, S., & Spirito, M. A. (2010). Static positioning using UWB range measurements. *Proceedings of Future Network and Mobile Summit.*

Gusi-Amigo, A., Closas, P., Mallat, A., & Vandendorpe, L. (2014), Cramer-Rao Bound Analysis of UWB based Localization Approaches. *Proceedings of IEEE International Conference on Ultra-Wideband (ICUWB)*. doi:10.1109/ICUWB.2014.6958943

Kay, S. M. (1993). *Fundamentals of Statistical Signal Processing: Estimation Theory*. Upper Saddle River, NJ: Prentice Hall.

Ke, W., & Wu, L. (2012). Sparsity-Based Multi-Target Direct Positioning Algorithm Based on Joint-Sparse Recovery. *Progress in Electromagnetics Research, 27*, 99–114. doi:10.2528/PIERC11110704

Lagunas, E., Navarro, M., Closas, P., & Najar, M. (2014), Spatial Sparsity Based Direct Positioning for IR-UWB in IEEE 802.15.4a Channels. *Proceedings of IEEE International Conference on Ultra-Wide Band (ICUWB)*. doi:10.1109/ICUWB.2014.6958999

Lee, J. Y., & Scholtz, R. A. (2002). Ranging in a dense multipath environment using an UWB radio link. *IEEE Journal on Selected Areas in Communications*, *20*(9), 1677–1683. doi:10.1109/JSAC.2002.805060

Li, X., & Pahlavan, K. (2004). Super-resolution TOA estimation with diversity for indoor geolocation. *IEEE Transactions on Wireless Communications*, *3*(1), 224–234. doi:10.1109/TWC.2003.819035

Lingwen, Z., & Zhenhui, T. (2010). A Novel UWB Signal Sampling Method for Localization based on Compressive Sensing. *China Communications*, *7*(1), 65–72.

Mallat, S. G., & Zhang, Z. (1993). Matching Pursuits with Time-Frequency Dictionaries. *IEEE Transactions on Signal Processing*, *41*(12), 3397–3415. doi:10.1109/78.258082

Molisch, A., Balakrishnan, K., Chong, C., Emami, S., Fort, A., Karedal, J., ... Siwiak, K. (2004). *P802.15-04-0662-00-004a, IEEE 802.15.4a channel model - Final Report*. IEEE.

Navarro, M., Closas, P., & Najar, M. (2013), Assessment of Direct Positioning for IR-UWB in IEEE 802.15.4a Channels. *Proceedings of IEEE International Conference Ultra-WideBand (ICUWB)*, 55-60. doi:10.1109/ICUWB.2013.6663822

Navarro, M., & Najar, M. (2011). Frequency Domain joint TOA and DOA estimation in IR-UWB. *IEEE Transactions on Wireless Communications*, *10*(10), 3174–3184. doi:10.1109/TWC.2011.072511.090933

Pallas, M. A., & Jourdain, G. (1991). Active high resolution time delay estimation for large BT signals. *IEEE Transactions on Signal Processing*, *39*(4), 781–788. doi:10.1109/78.80899

Ping, W., Huailin, R., & Fuhua, F. A Method of Indoor Multi-Path IR-UWB Localization Based on Bayesian Compressed Sensing, *Proceedings of IEEE International Conference on Signal Processing (ICSP)*, 56–59.

Pourhomayoun, M., Fowler, M., & Wu, N. (2012). Spatial Sparsity Based Emitter Localization. *Proceedings of Conference on Information Sciences and Systems (CISS)*.

Riad, S. M. (1986). The deconvolution problem: An overview. *Proceedings of the IEEE*, *74*(1), 82–85. doi:10.1109/PROC.1986.13407

Rydstrom, M., Strom, E. G., & Svensson, A. (2006), Robust Sensor Network Positioning Based on Projections onto Circular and Hyperbolic Convex Sets (POCS). *Proceedings of IEEE Workshop on Signal Processing Advances in Wireless Communications*, 1-5. doi:10.1109/SPAWC.2006.346399

Tropp, J. A. (2004). Greed is Good: Algorithmic Results for Sparse Approximation. *IEEE Transactions on Information Theory*, *50*(10), 2231–2242. doi:10.1109/TIT.2004.834793

Tropp, J. A., & Gilbert, A. C. (2007). Signal Recovery From Random Measurements Via Orthogonal Matching Pursuit. *IEEE Transactions on Information Theory*, *53*(12), 4655–4666. doi:10.1109/TIT.2007.909108

Ubisense. (2017). *Location Intelligence solutions*. Retrieved March 16, 2017, from http://www.ubisense.net/en/

Vaughan, R. G., & Scott, N. L. (1999). Super-resolution of pulsed multipath channels for delay spread characterization. *IEEE Transactions on Communications*, *47*(3), 343–347. doi:10.1109/26.752811

Weiss, A. J. (2004). Direct position determination of narrowband radio frequency transmitters. *IEEE Signal Processing Letters*, *11*(5), 513–516. doi:10.1109/LSP.2004.826501

Weiss, A. J. (2011). Direct Geolocation of Wideband Emitters Based on Delay and Doppler. *IEEE Transactions on Signal Processing*, *59*(6), 2513–2521. doi:10.1109/TSP.2011.2128311

Weiss, A. J., & Amar, A. (2005). Direct Position Determination of Multiple Radio Signals. *EURASIP Journal on Advances in Signal Processing*, (1): 653549.

Yang, D., Li, H., & Peterson, G. (2013, January). Compressive Sensing Based Sub-mm Accuracy UWB Positioning Systems: A Space-Time Approach. *Digital Signal Processing*, *23*(1), 340–354. doi:10.1016/j.dsp.2012.07.012

Yang, L., & Giannakis, G. B. (2005). Timing ultra-wideband signals with dirty templates. *IEEE Transactions on Communications*, *53*(11), 1952–1963. doi:10.1109/TCOMM.2005.858663

KEY TERMS AND DEFINITIONS

Analog-to-Digital Converter: Electronic device that converts an analog signal into a digital signal.

Cramer-Rao Lower Bound: CRLB is a lower bound on the variance of any unbiased estimator. The CRLB tells us the best we can ever expect to be able to do.

Fading: Fading is variation of the attenuation of a signal with different variables. These variables include time, geographical position, and radio frequency.

Impulse-Radio Ultra-Wideband: Radio technology that uses a very low energy level for short-range, high-bandwidth communications over a large portion of the radio spectrum.

Maximum Likelihood Estimator: In statistics, the ML estimator is a method of estimating the parameters of a statistical model given observations, by finding the parameter values that maximize the likelihood of making the observations given the parameters.

Nyquist Rate: Refers to the sampling period that is equivalent to twice the bandwidth of a bandlimited function or a bandlimited channel.

Power Delay Profile: Signal representation based on the intensity of a signal as a function of time delay.

Sparsity: Term used in signal processing to describe signals with high percentage of samples equal or approximately zero.

Trilateration: Centralized beacon-based localization algorithm that uses parameter estimates from three different sensor nodes to compute the 2D position of the target node.

Chapter 8
Integrated Positioning With UWB and INS

Kai Wen
Wuhan University, China

Kegen Yu
Wuhan University, China

ABSTRACT

The chapter studies the positioning techniques based on ultra wideband (UWB) and low cost inertial measurement unit (IMU) with a focus on the fundamental theories of integrated positioning based on UWB and IMU. Details are provided for multilateral positioning theory of UWB, UWB calibration method, IMU error analysis, inertial navigation algorithm, and Kalman filter (KF) theory. Particularly, to mitigate the influence of non-line-of-sight (NLOS) propagation on positioning accuracy, a NLOS mitigation method based on fuzzy theory is presented. Meanwhile, the detailed data fusion processes of loosely coupled and tightly coupled systems are introduced and performance evaluation is conducted using experimental data.

INTRODUCTION

Indoor location techniques can roughly divided into three categories, namely self-contained location (SCL), fast networking location (FNL) and all-source location (ASL). SCL technique refers to that position (and velocity and attitude in some scenarios) can be obtained by a single device independently without additional infrastructures. The typical representation of SCL is inertial navigation system (INS). SCL technique is appealing due to simple operation and wide availability, but at the cost of a high requirement of the performance of devices, that leads to either high cost implementation or low precision location performance. FNL technique depends on infrastructures to implement positioning or navigation, such as Bluetooth, Wireless Fidelity (WiFi), ZigBee and Ultra-Wideband (UWB), which universally achieves low positioning accuracy. Although UWB can achieve sub-meter precision based on low cost sensors, the reliability and accuracy will be affected in complex propagation environment. ASL technique realizes positioning and navigation by making use of complementary characteristics of different technologies

DOI: 10.4018/978-1-5225-3528-7.ch008

to achieve a high accuracy at an acceptable implementation cost. ASL has been drawing significant attention in the field of indoor location, especially for affordable high accuracy commercial applications.

UWB refers to any radio signal with an absolute bandwidth larger than 500 MHz or a relative bandwidth larger than 20% (Federal Communications Commission, 2002), which is one of the most promising technologies for indoor location due to the properties of centimeter ranging accuracy in line-of-sight (LOS) scenarios, strong multipath resistance and to some extent penetrability through wall, and hence is usually used for accurate indoor fast networking location[1]. However, UWB indoor location systems have two main deficiencies. The first is that it is easy to suffer from non-line-of-sight (NLOS) propagation in complex environments, leading to degraded positioning accuracy. The other deficiency is that it is unable to provide precise information about orientation and velocity, although roughly velocity may be calculated using estimated position coordinates. ASL technique shows potential to solve the problems by combining UWB and other complementary technologies, which is currently an active research area of UWB-based indoor location. The goal of this chapter is to investigate issues in the integration of UWB and INS as well as to discuss a new NLOS mitigation method for mobile tracking. Firstly fundamentals of UWB and INS systems are described, including the basic positioning principle of each system, and the system error analysis and mitigation. Then the chapter introduces a few machine learning based NLOS mitigation methods, and presents a new NLOS mitigation method based on fuzzy theory. In addition, two schemes for UWB and INS integration are introduced and performance evaluation is performed with experimental data collected in very recent experiments. Finally, the chapter is concluded and future research directions are recommended.

SYSTEM FUNDAMENTALS

Ultra-Wideband Positioning System

Positioning

An UWB positioning system typically consists of a network of some fixed units with known coordinates and mobile units whose positions are to be determined as shown in Figure 1. Both fixed units and mobile units can transmit and receive UWB signal. According to the requirement, either navigation (positioning) or tracking (monitoring) can be enabled. Navigation refers to obtaining the location information of a mobile unit by itself, while tracking refers to that the position of mobile unit is determined by network or system in a monitoring center. No matter whether navigation or tracking is involved, the fundamental principle of position determination is basically the same.

In an UWB system, the use of very short pulse typically in the order of 1ns results in a high resolution in time and consequently in range, which is a remarkable advantage in the time related measurement. Therefore UWB based position calculation usually exploits time of arrival (TOA), time difference of arrival (TDOA), and two-way time of fly (TW-TOF) techniques. Angle of arrival (AOA) technique is usually unsuited for UWB systems, while received signal strength technique is much inferior to time related techniques. Taking example for the TW-TOF technique, the true distance d between two communication units can be expressed as

Figure 1. An UWB setup consists of four fixed units and one mobile unit

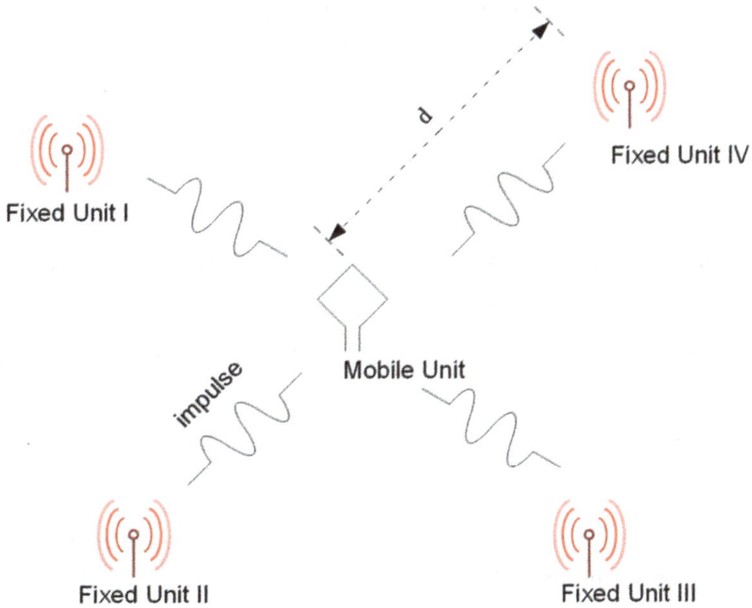

$$d = \frac{c\hat{\tau} + \delta}{2} \tag{1.1}$$

where $\hat{\tau}$ is the estimated signal round-trip time from one unit to the other, c represents the speed of light, and δ is the ranging error. Assuming the measured distance is $\hat{d}$, (1.1) can be rewritten as

$$\hat{d} = \frac{1}{2}c\hat{\tau} = d - \frac{1}{2}\delta \tag{1.2}$$

The position of mobile unit can be determined by multilateration using a position estimation technique with multiple simultaneous distance measurements (at least three for 2D, four for 3D). A range of estimation techniques can be used for position determination, such as direct calculation method, least squares (LS), double maximum likelihood, and kalman filter (KF). The details of various methods are not described here, but they can be found in many references such as (Yu & Oppermann, 2002; Yu, Sharp & Guo, 2009).

Calibration

Calibration is to find out the deterministic offset in the measurement system of interest. For a high precision measurement instrument, calibration has usually been completed and the deterministic offset has been compensated by manufacturers. In some cases, manufacturers also may provide a calibration function module for each instrument (e.g. PlusON® products built by Time Domain Inc.) so that users can simply compensate the deterministic offset in a device using a test before an application. However, in the case of high accuracy positioning or positioning with low cost devices, users still need to conduct calibration for each device. For UWB location system, calibration is mainly for the clock offset.

For position estimation based on time measurement, the measurement equation can be modeled as

$$t_{r,i} = t_{t,i} + \frac{\left\| \mathbf{p}_r - \mathbf{p}_{t,i} \right\|_2}{c} + \Delta t_r + \Delta t_{t,i} + \varsigma_i + \eta_i \tag{1.3}$$

where $t_{r,i}$ is the signal arrival time recorded by a receiver, $t_{t,i}$ is the signal transmission time, $\mathbf{p}_r$ and $\mathbf{p}_{t,i}$ are the position vectors of the receiver and i th transmitter respectively, c is the propagation velocity of pulse signal, Δt_r and $\Delta t_{t,i}$ are the clock offsets of the receiver and i th transmitter respectively, ς_i is the time delay induced by NLOS, and η_i is the measurement noise which is considered as the independently and identically distributed (i.d.d.) Gaussian noise.

Suppose that measurements are carried out without the effect of NLOS ($\varsigma = 0$). Meanwhile, $t_{t,i}$, $\mathbf{p}_r$ and $\mathbf{p}_{t,i}$ are known and the signal arrival time $t_{r,i}$ is measured. Thus, only parameters Δt_r and $\Delta t_{t,i}$ are unknown in (1.3). By differencing between measurement equation for the i th transmitter and that for the j th transmitter, the clock offset of receiver is eliminated and the differential measurement equation can be written as

$$t_{r,ij} = t_{t,i} - t_{t,j} + \frac{\left\| \mathbf{p}_r - \mathbf{p}_{t,i} \right\|_2}{c} - \frac{\left\| \mathbf{p}_r - \mathbf{p}_{t,j} \right\|_2}{c} + \Delta t_{t,i} - \Delta t_{t,j} + \eta_{ij} \tag{1.4}$$

where $t_{r,ij} = t_{r,i} - t_{r,j}$ and $\eta_{ij} = \eta_i - \eta_j$. Suppose that there are k transmitters in an UWB setup. Then, there are $k(k-1)/2$ independent measurement equations (e.g., $i = 1, 2, \cdots, k-1$, $j = i+1, i+2, \cdots, k$ in (1.4)). Rewrite (1.4) as

$$u_{ij} = \Delta t_{t,i} - \Delta t_{t,j} + \eta_{ij} \tag{1.5}$$

with

$$u_{ij} = t_{r,ij} - t_{t,i} + t_{t,j} - \frac{\left\| \mathbf{p}_r - \mathbf{p}_{t,i} \right\|_2}{c} + \frac{\left\| \mathbf{p}_r - \mathbf{p}_{t,j} \right\|_2}{c} \tag{1.6}$$

The simultaneous measurement equations can be written in a compact form as

$$\mathbf{u} = \mathbf{A}\Delta\mathbf{t} + \boldsymbol{\eta} \tag{1.7}$$

where $\mathbf{u} = [u_{12}\ u_{13} \cdots u_{1k}\ u_{23}\ u_{24} \cdots u_{2k} \cdots u_{(k-1)k}]^T$, $\boldsymbol{\eta} = [\eta_{12}\ \eta_{13} \cdots \eta_{1k}\ \eta_{23}\ \eta_{24} \cdots \eta_{2k} \cdots \eta_{(k-1)k}]^T$, $\mathbf{A}$ is the coefficient matrix with $k(k-1)/2$ rows and k columns and $\Delta\mathbf{t} = [\Delta t_{t,1}\ \Delta t_{t,2} \cdots \Delta t_{t,k}]^T$. Then the simple but effective weighted LS estimator can be used to obtain a solution to the offset parameter vector as

$$\Delta \mathbf{t} = \left(\mathbf{A}^T \mathbf{W} \mathbf{A}\right)^{-1} \mathbf{A}^T \mathbf{W} \mathbf{u} \tag{1.8}$$

where $\mathbf{W}$ is the weighting matrix. In the case where η_{ij} are mutually independent Gaussian random variables, $\mathbf{W}$ may be set to be a diagonal matrix as

$$\mathbf{W} = diag\left\{\sigma_{12}^2, \sigma_{13}^2, \cdots, \sigma_{1k}^2, \sigma_{23}^2, \sigma_{24}^2, \cdots, \sigma_{2k}^2, \cdots, \sigma_{(k-1)k}^2\right\} \tag{1.9}$$

where σ_{ij}^2 is the variance of noise η_{ij}.

Inertial Navigation System

Basic Concepts

At the end of the Second World War the prototype of INS was designed for rocket flight control. The technology is a self-contained navigation technique which uses measurements provided by an inertial measurement unit (IMU) consisting of inertial sensors (accelerometer and gyroscope) to determine the position, orientation and velocity of an object at any time relative to those at the starting time (Woodman, 2007). Initially an IMU is quite complicated whose inertial sensors are mounted on a platform isolating from any external rotational motion in the help of gimbals. The design is beneficial to achieve a simple transform from acceleration into velocity and position. However the mechanical complexity as well as the size and weight of an IMU are increased, resulting in a high cost. At the end of 1960s, the strap-down inertial navigation system (SINS) was introduced, which does not need the stable platform whose function in INS is replaced by mathematical operations. Although the computational complexity of SINS is significantly greater, it is not a significant issue due to the advance in computing hardware and software. The reduction in mechanical complexity, size, weight and cost has been an outstanding advantage of SINS and has widened the range of possible applications of IMU.

IMU can be broadly classified into three categories: strategic-grade, navigation-grade and consumption-grade, mainly based on the performance of the gyroscope used. For consumption-grade IMU the gyro drift rate is in the scope of $1 \sim 10^\circ/h$ or more and accelerometer bias ranges between 0.1 and 1.0 mg or more (Greenspan, 1995). Recently popular consumption-grade IMU is manufactured by the micro-electro mechanical system (MEMS) technique. Hence MEMS is generally used to refer to the consumption-grade IMU in the field of navigation. For convenience, MEMS refers to MEMS-based IMU from now in this chapter. More information on MEMS can be found in (Titterton & Weston, 2005).

Errors in MEMS

MEMS errors comprise deterministic errors, such as errors in alignment, scale factor and accelerometer linearity, and stochastic errors. In practice a few errors can be neglected because of their trivial influences on the positioning performance. Usually scale factor error, constant bias, bias instability, and angular/velocity random walk have to been considered in the case of low cost MEMS. Here attention is only paid on the later three, which are commonly discussed in the MEMS-related literature.

Constant Bias

Constant bias is the output value of an inertial sensor under the state without undergoing any external force. In the case of a uniaxial accelerometer, if the constant bias is δ, the influence on position can be obtained by double integral

$$\hat{p}\left(t_k\right) = \int_0^{t_k}\int_0^{t_k}\left[\alpha\left(t\right)+\delta\right]dt = \int_0^{t_k}\int_0^{t_k}\alpha\left(t\right)dt + \frac{1}{2}\delta t_k^2 \tag{1.10}$$

where $a(t)$ is the actual output of the accelerometer, corrupted with white noise, $\hat{p}(t_k)$ is the device position calculated with $a(t)$ at time t_k under the presence of constant bias δ (here we assume that the initial device position is the same as the origin of coordinate system). It is clearly seen that the influence of constant bias on position estimation grows quadratically with time. In the similar way, it can be shown that the constant bias of a uniaxial gyroscope would cause an error in orientation, which grows linearly with time.

If there are no other errors except for zero-mean white noise, constant bias is equivalent to the average of a long output sequence of the gyroscope or accelerometer under the state without undergoing any external force. Therefore, in practice, the constant bias can be roughly obtained by averaging a large number of measurements when the involved sensor is isolated from any external force in advance, and then the bias is subtracted from the in-situ measurements during navigation or tracking. The method is feasible for gyroscopes but impracticable for accelerometers due to the presence of gravitational acceleration whose effect can be confused with constant bias. To remove the gravitational effect, the accurate orientation information is required at the current instant. Note that the constant bias may change once the device is rebooted and the variation magnitude can be described by the bias repeatability.

Bias Instability

Bias instability indicates that the “constant bias” of an inertial sensor may change over time in fixed conditions. In some literature bias instability is also called flicker noise because the bias variance over time is caused by the flicker noise in electronic components (Woodman, 2007). Usually bias instability is described by a random walk model. Consider the bias instability of a uniaxial accelerometer and assume that the noise causing the constant bias wander is white noise. Then, the bias at time t_k is

$$b_k = b_{k-1} + \int_{t_{k-1}}^{t_k}\gamma dt \tag{1.11}$$

where b_{k-1} and b_k are the constant biases of accelerometer at the time t_{k-1} and t_k respectively, and γ is the white noise variable whose mean value is zero and standard deviation is σ. According to the rectangular integral rule and supposing that there are N noise samples over the time span of integral and that the noise sampling interval is ε, (1.11) can be expressed as

$$b_k = b_{k-1} + \sum_{i=0}^{N-1} \varepsilon\gamma_i \tag{1.12}$$

Then, mean and variance of b_k can be calculated as

$$E(b_k) = E(b_{k-1} + \sum_{i=0}^{N-1} \varepsilon\gamma_i) = b_{k-1} + \varepsilon\sum_{i=0}^{N-1} E(\gamma_i) = b_{k-1} \tag{1.13}$$

$$\mathrm{var}(b_k) = \mathrm{var}(b_{k-1} + \sum_{i=0}^{N-1} \varepsilon\gamma_i) = \varepsilon^2\sum_{i=0}^{N-1} \mathrm{var}(\gamma_i) = N\varepsilon^2\sigma^2 = \sigma^2\varepsilon t \tag{1.14}$$

where t is the time span from time $t_{t_{k-1}}$ to time t_k. That is, the constant bias at time t_k is a random variable with expected value b_{k-1} and standard deviation $\sigma\sqrt{\varepsilon t}$. Theoretically we can know the mean and standard deviation of constant bias at any time once bias initial value and noise are specified. However, the constant bias does not really behave as a strict random walk model, so that it is unreasonable to constrain the constant bias within a specific range when the actual value may vary freely. Hence, the random walk model is only an approximation for the true variation process. By the way, the classification of IMU levels usually uses the bias instability of the gyroscope involved.

Angle/Velocity Random Walk

Angle and velocity random walk errors come from integral of white noise sequences contained in the output of gyroscope and accelerometer respectively, which own the property of random wander. Taking example for gyroscope, angle random walk error $\delta\omega_k$ at time t_k can be expressed as

$$\delta\omega_k = \int_{t_{k-1}}^{t_k} \lambda dt \tag{1.15}$$

where λ is white noise process with zero mean and standard deviation σ . Supposing that there are N noise samples over the time span of integral and that the noise sampling interval is ε, we can obtain $E(\delta\omega_k) = 0$, $\mathrm{var}(\delta\omega_k) = \sigma^2\varepsilon t$ based on (1.13) and (1.14), where $t = t_k - t_{k-1}$. It is seen that the mean of angle random walk error is zero and the standard deviation is proportional to $\sqrt{t}$, which indicate the influence of noise in the gyroscope output on angle. For velocity random walk error, the similar conclusions can be obtained.

Since angle and velocity random walk error describe the magnitude of white noise in the output of gyroscope and accelerometer, they are commonly used as the index for device performance. It is common for manufactures to specify noise using the standard deviation of angle/velocity random walk error when t equals one hour, i.e.

$$\sigma_{angle} = \sigma\sqrt{\varepsilon} \tag{1.16}$$

For example, $\sigma_{angle} = 12$ (units $^\circ/\sqrt{h}$) means that the standard deviation of angle output error would be $12\sqrt{t}$ degree after t hour. Occasionally, the noise level of gyroscope is also described in terms of power spectral density (PSD) or noise density (ND). The relationships between different indexes are (Woodman, 2007)

$$\sigma_{angle} = \frac{1}{60}\sqrt{PSD} = \frac{1}{60}ND \tag{1.17}$$

where the units of PSD and ND are $(^\circ/h)^2 \,/\, Hz$ and $^\circ/h \,/\, \sqrt{Hz}$ respectively.

Allan Deviation Analysis Technique

Deterministic errors are determined and compensated by calibration operation. An accurate calibration usually takes a long time and requires special instruments. Meanwhile, some deterministic errors for MEMS are unrepeatable when the device is rebooted. Therefore, it is unnecessary to determine the parameters of deterministic errors by an accurate calibration operation in advance, but the process is online carried out based on mathematical equations. A large number of online calibration methods can be found in MEMS-related literatures.

On the other hand, stochastic errors are dealt with stochastic models. For a linear time-invariant system, by having knowledge of the outputs and assuming white noise inputs, it is possible to characterize the unknown model. To build the proper error model, statistics of the stochastic error have to be analyzed. Currently, for MEMS, the common error analysis method is Allan deviation analysis technique, which is briefly introduced as follows.

Assuming a sample sequence $\{x_1, x_2, \cdots, x_N\}$, consisting of N consecutive measurements with sampling interval δ, is divided into k $(k < N \,/\, 2)$ groups. Here it is assumed *N* is a multiple of *k*. Each group has the same number of samples, $m = N \,/\, k$, and the same time length, $\tau = m\delta$. The average of the i th group, $\overline{x}_i$, is simply given by

$$\overline{x}_i = \frac{1}{m}\sum_{j=(i-1)\cdot m+1}^{i\cdot m} x_j,\ \ i = 1, 2, \cdots, k \tag{1.18}$$

Then, Allan variance for the mean sequence $\{\overline{x}_1, \overline{x}_2, \cdots, \overline{x}_k\}$ can be calculated as

$$\sigma^2(\tau) = \frac{1}{2(k-1)}\sum_{i=1}^{k-1}(\overline{x}_{i+1} - \overline{x}_i)^2 \tag{1.19}$$

By changing the number of group, a large number of Allan variances can be calculated. Then all Allan variances are drawn as an Allan variance plot. Individual errors can be identified by finding various slopes in the Allan variance plot. The different slopes in the plot represent the errors in different parts of the sample sequence. The method is readily operated in time domain and is developed based on spectral analysis. The relationship between Allan variance and PSD of the intrinsic random process is (El-Sheimy, Hou & Niu, 2008)

$$\sigma^2(\tau) = 4\int_0^\infty s_x(f)\frac{\sin^4(\pi f\tau)}{(\pi f\tau)^2}df \tag{1.20}$$

where $s_x(f)$ represents the PSD of random process x. For INS error analysis, the x is usually considered as a sum of five mutually independent colored noise processes in the output of an inertial sensor. If PSD of each noise process is linear, the PSD of x is given by

$$s_x(f) = \sum h_a f^a \tag{1.21}$$

where a is the slope of the PSD of one noise process and h_a is the noise coefficient. In practice, the value of a is an integer ranging from -2 to +2. Substituting (1.21) into (1.20), the Allan variance can be written as a sum of variances of individual errors in time domain

$$\sigma^2(\tau) = \sigma_Q^2(\tau) + \sigma_{ARW/VRW}^2(\tau) + \sigma_{RRW/RW}^2(\tau) + \sigma_{BI}^2(\tau) + \sigma_{TR}^2(\tau) \tag{1.22}$$

where $\sigma^2(\tau)$ is the Allan variance of inertial sensor random error, σ_Q^2 is the variance of quantization noise, $\sigma_{ARW/VRW}^2$ is the variance of angle random walk error or the velocity random walk error, $\sigma_{RRW/RW}^2$ is the variance of rate random walk error or random walk error, σ_{BI}^2 is variance of the bias instability error, and σ_{TR}^2 is the variance of gyroscope trend or accelerometer trend. Allan variance analysis technique is easy to understand and operate, but the method only divides all errors into many typical categories, without considering the errors caused by other factors, such as linearity and temperature stability, leading to a limited ability in error analysis.

INS Algorithm in Random 3D Cartesion Coordinate Frame

According to Newton's law of motion, the accelerometer contained in IMU will generate accelerations once an external force imposes on the inertial sensor. By integrating the accelerations, the changes of velocity and position relative to the initial time can be obtained. With the known initial velocity and position, it is trivial to obtain the velocity and position at any time. In fact, the computation is more complicated than that because the inertial sensor measures the parameters in a coordinate frame which is different from the coordinate frame for navigation solution. Therefore, orientation information provided by gyroscope is required to implement the transformation between different coordinate frames.

Meanwhile, the effect of gravity and earth rotation on measurements of IMU has to be removed. The coordinate frames involved in indoor location application basically include:

1. Sensor coordinate frame (s) is usually a right-handed rectangular coordinate system that is body-fixed to the sensors device and is used to output rate-of-turn and acceleration.
2. Body coordinate frame (b) is a rectangular coordinate system with origin and orientation fixed relative to the vehicle body. The output of inertial sensors needs to be mapped to the body coordinate system to describe the state change of body. Usually the body coordinate system should be aligned with the sensor coordinate system in the installation of sensors to simplify the projection operation.
3. Inertial coordinate frame (i) is a coordinate system which is absolutely stationary or moves at a constant speed. The measurements of gyroscope and accelerometer are relative to the inertial coordinate system. In the case of navigation near the surface of earth, we often set the origin of inertial coordinate system at the center of earth, the x-axis point to the vernal equinox, the z-axis point to the earth's rotation axis and y-axis complete the right-hand system.
4. Navigation coordinate frame (n) is used to express the state of body in some navigation applications. Geographic coordinate system is generally used as a navigation coordinate system in navigation application near the surface of earth, such as for aircraft or ship. In indoor location application, a random 3D Cartesion coordinate frame (l) can be used as the navigation coordinate system.
5. Geographic coordinate frame (g) is a coordinate system where the x-axis is along with the prime vertical orientation toward east, the y-axis is along with the meridian orientation toward north, and the z-axis is along with the normal direction toward up.

Besides the transform between different coordinate frames, a few errors, such as sculling error, rotation error, and non-communicative error, are troublesome. Fortunately, for indoor location applications, some factors can be treated as a constant or ignored because their influence is trivial when the INS is used in a small region. Define a local coordinate frame, which is aligned with the body coordinate system at the initial time. The INS algorithm is deduced as follows. Firstly, the sensed angular rates are integrated to compute the direction cosine matrix (DCM) $\mathbf{C}_b^l(t)$ which indicates the transformation from b-frame to l-frame at current time. The $\mathbf{C}_b^l(t)$ can be calculated as following differential equation

$$\dot{\mathbf{C}}_b^l\left(t\right) = \mathbf{C}_b^l\left(t\right) \cdot \mathbf{w}_{lb}^b\left(t\right) \times \tag{1.23}$$

where "$\cdot$" at the top of DCM $\mathbf{C}_b^l(t)$ denotes the time derivative operation, the superscript "l" and "b" indicate l-frame and b-frame respectively, $\mathbf{w}_{lb}^b(t) \times$ is the skew-symmetric matrixs of $\mathbf{w}_{lb}^b(t)$, and $\mathbf{w}_{lb}^b(t)$ is the rotation vector of b-frame relative to l-frame. It is complex to directly solve the differential equation, but a recursive form can be used to determine the transformation matrix based on the matrix chain-multiplication

$$\mathbf{C}_{bk}^{lk} = \mathbf{C}_{l(k-1)}^{lk} \mathbf{C}_{b(k-1)}^{l(k-1)} \mathbf{C}_{bk}^{b(k-1)} \tag{1.24}$$

where the subscript "k" represents time t_k, $\mathbf{C}_{bk}^{b(k-1)}$ is rotation matrix of b-frame from time t_k to t_{k-1} in the i-frame, $\mathbf{C}_{l(k-1)}^{lk}$ is rotation matrix of l-frame from time t_{k-1} to t_k in the i-frame, and $\mathbf{C}_{bk}^{lk}$ and $\mathbf{C}_{b(k-1)}^{l(k-1)}$ are the rotation matrix from b-frame to l-frame in the i-frame at time t_k and t_{k-1} respectively. $\mathbf{C}_{l(k-1)}^{lk}$ and $\mathbf{C}_{bk}^{b(k-1)}$ can be calculated using the rotation angle $\Delta\theta_l$ and $\Delta\theta_b$

$$\Delta\theta_l = \int_{t_{k-1}}^{t_k} \mathbf{w}_{ie}^l(t)dt, \quad \Delta\theta_b = \int_{t_{k-1}}^{t_k} \mathbf{w}^b(t)dt \tag{1.25}$$

where $\mathbf{w}_{ie}^l(t)$ is the projection of earth rotation angular rate in l-frame and $\mathbf{w}^b(t)$ is the output of gyroscope. The value of $\mathbf{w}_{ie}^l(t)$ is relative with the geographic longitude $L(t)$. For applications in a small region, $L(t)$ is treated as a constant or the effect of earth rotation is ignored, especially for low cost MEMS. Meanwhile, assume that outputs of gyroscope between time t_{k-1} and t_k are relative to the b-frame at time t_{k-1}. For high accuracy applications, the influence of b-frame variation over time, known as non-communicative error, must be considered.

The computed transformation matrix $\mathbf{C}_b^l(t)$ is then used to rotate the measurements of accelerometer from the b-frame to l-frame. Note that the output of accelerometer is affected by earth gravity and the effect must be compensated. The total accelerations can be described as

$$\dot{\mathbf{v}}^l(t) = \mathbf{C}_b^l(t)\mathbf{f}^b(t) - 2\mathbf{w}_{ie}^l(t)\mathbf{v}^l(t) + \mathbf{g}^l(t) \tag{1.26}$$

where $\mathbf{f}^b(t)$ denotes the sensed acceleration measurement, $\mathbf{v}^l(t)$ represents the mobile velocity in l-frame, and $\mathrm{g}^l(t)$ is the mapping of gravity acceleration in l-frame. Since mobile velocity is small in indoor scenarios, the effect of $2\mathbf{w}_{ie}^l(t)\mathbf{v}^l(t)$ usually is rather small so that it can be ignored. Solving the simple linear first-order equation produces the mobile velocity estimate.

Then, the position $\mathbf{p}^l(t)$ in l-frame can be readily determined by

$$\dot{\mathbf{p}}^l(t) = \mathbf{v}^l(t) \tag{1.27}$$

Figure 2 shows the detailed flowchart for determining the attitude, velocity and position of the mobile in a local coordinate system using measurements by MEMS.

Since a number of errors are included in the output of IMU and the compensation from calibration or stochastic error model is incomplete, the results from INS algorithm will diverge soon because of the accumulation of residual error. Currently there are three main methods to limit the divergence of navigation results: (1) combine IMU with other sensors, such as UWB, global navigation satellite system (GNSS), magnetometer, and vision; (2) limit the error and noise level using IMU array; (3) correct periodically the error of IMU by external information, such as zero velocity update (ZUPT) and coordinate update (CUPT). IMU array method is usually to integrate multiple MEMS-based IMUs in a single printed circuit board so that they sense the same motion characteristics. The original outputs of individual IMU are combined to reduce the stochastic errors and noise level in the ultimate output.

Figure 2. Detailed solution process in l-frame for MEMS. The function box with dashed line is omitted in indoor location applications, and here it is given only for showing the full process of INS algorithm.

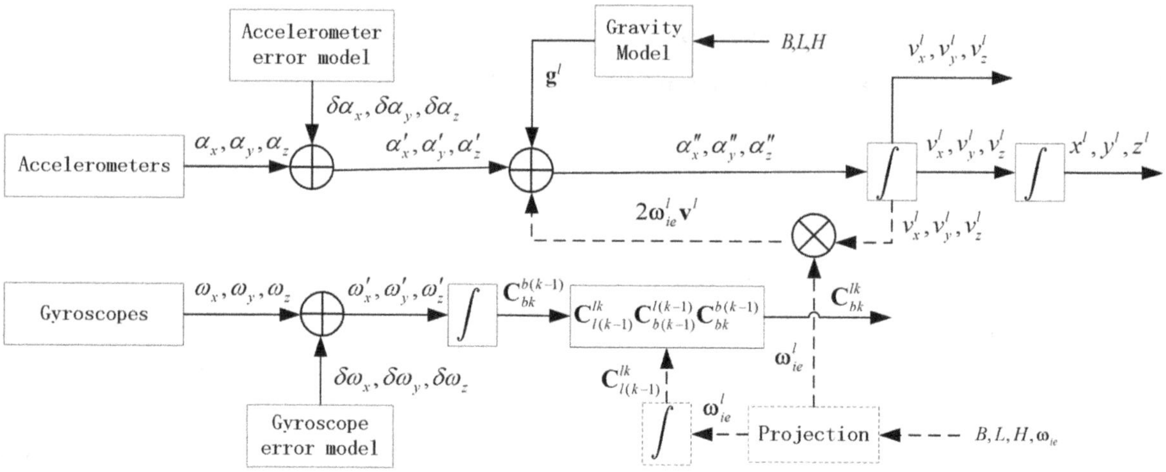

Besides, multiple MEMS-based IMUs can also be mounted on separate boards. In such case, the measurements of different IMUs cannot directly be combined together. However the position estimates from the different IMUs can be related to each other once the position relationship among boards are known, and thus the stochastic errors can still be reduced (Skog, Nilsson & Handel 2014). A significant issue associated with the method is how to combine the measurements from different IMUs. Inappropriate combination may prevent the potential performance gain to be attained (Bancroft & Lachapelle, 2011; Martin, Groves, Newman & Faragher, 2013). Both ZUPT and CUPT use external known information to limit the accumulation of residual error and correct the accumulated error. ZUPT is widely used for pedestrian dead reckoning (PDR). The basic idea of ZUPT for PDR is to find out the short period of static state by detecting the process of one's steps, and the zero-velocity is used for error correction. The use of frequent zero velocity update eliminates the problem of error accumulation over time and thus significantly extends the ability of an IMU to maintain accuracy within the desired bound for a longer period. The static state detection is generally accomplished by analyzing the signature of the outputs of accelerometer or gyroscope. However, since the method typically results in a 1%-2% positional error, biomechanical ground reaction sensor is be used for step detection to improve the positioning accuracy (Bebek, Suster, Rajgopal, Fu, Huang, Cavusoglu, Young, Mehregany, Van Den Bogert & Mastrangelo, 2010; Guo, Bebek, Cavusoglu, Mastrangelo &Young, 2015).

NLOS IDENTIFICATION AND MITIGATION

Non-Line-of-Sight Propagation

NLOS propagation is one of the most difficult barriers for UWB accurate positioning. NLOS propagation usually refers to that the direct path signal transmits through one or more obstacles before arriving at the receiver. However in the literature, there is often no clear distinction between such a NLOS propagation and multipath propagation. That is, all propagations under the absence of LOS path, when transmitted signal arrives at the receiver through penetrated, reflected, diffracted, or scattered paths as shown in Figure 3, are called NLOS propagation (Khodjaev, Park & Malik, 2010).

In the case of NLOS propagation, the signal energy is decayed and the arrival time is delayed, resulting in errors for time measurement-based and signal strength-based ranging. For ranging based on time measurement, arrival time usually is estimated by correlation technique or match filter technique. Due to multipath effect, the detected correlation peak will shift, resulting in an error in the signal arrival time estimate (Li, He & Xu, 2016). UWB technology owns a high time resolution, due to the use of a very short pulse of less than a few nanoseconds, greatly enhancing the ability to distinguish the direct path and delayed paths. However, extra propagation time caused by penetrating a medium which is not a free space may be a more serious problem in signal arrival time estimation. For instance, in the case of penetrating a wall, the extra distance (ranging error) can be approximately calculated by

$$\Delta s = \frac{\delta(\sqrt{\xi} - 1)}{\cos\theta} \tag{1.28}$$

where δ and ξ are the thickness and relative dielectric constant of the wall respectively, and θ represents the incident angle.

Overview of Basic NLOS Mitigation Methods

NLOS mitigation is a popular research subject in the field of UWB location technology, which may be considered as two separate operations (identification and mitigation). The purpose of NLOS identification is usually to aid NLOS mitigation, and thus, for convenience, in this chapter we refer to the two operations as NLOS mitigation together. Traditional NLOS mitigation techniques can be classified into three categories, namely range measurement based methods, signal channel characteristic analysis based methods, and position based methods.

Figure 3. LOS and NLOS propagation cases in ideal conditions

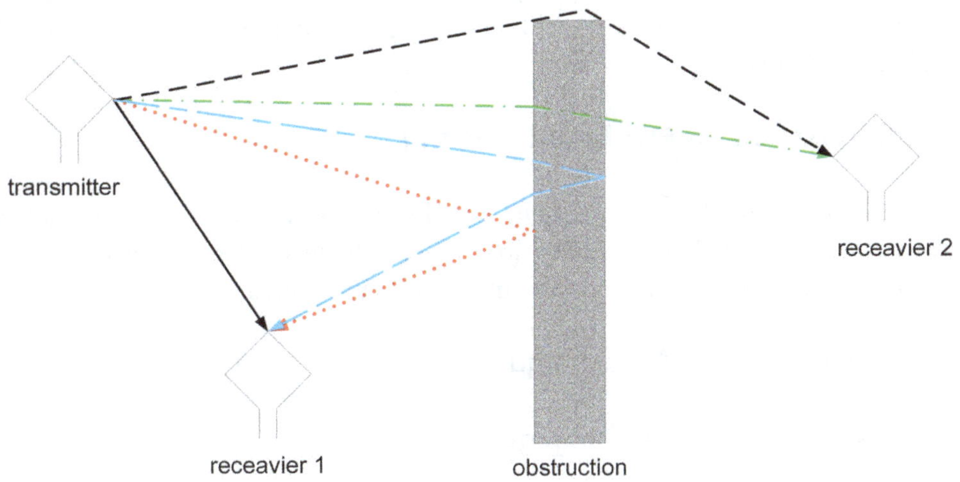

Range based methods mainly use the probability density function (PDF) of range measurements to distinguish LOS and NLOS, while in the absence of the priori PDF, the identification is only based on the variance of range measurements (Borras, Hatrack & Mandayam, 1998). The methods are simple but are limited by the requirement of priori statistical knowledge and the presence of time latency. Meanwhile, the PDF of range measurements will change with the surroundings, so the methods are only suitable for static applications. Although some improved schemes of the methods can be used for dynamic applications, extra assumptions are required, and the accuracy and robustness of LOS identification is low (Borras, Hatrack & Mandayam, 1998; Venkatraman & Caffery, 2002; Yu & Dutkiewicz, 2013). Channel based methods use the channel impulse response (CIR) to distinguish LOS and NLOS, including two popular kinds of methods. The first kind of methods use the PDF or cumulative distribution function (CDF) of the received power envelope, which has Ricean distribution under LOS condition and has Rayleigh distribution under NLOS condition (Lakhzouri, Lohan, Hamila & Renfors, 2003). The other kind of methods use the characteristics of received signal, such as amplitude, signal-noise-ratio (SNR), mean excess delay, root-mean-square (RMS) delay spread, kurtosis and so on, for NLOS identification. Some channel based methods only utilize a single metric to determine NLOS (Shimizu, Masui, Ishii & Sakawa, 2000; Yu et al., 2009; Xiao, Wen, Markham, Trigoni, Blunsom, & Frolik, 2015), while others combine multiple metrics to determine NLOS (Guvenc, Chong, Watanabe & Inamura, 2008; Silva & Hancke, 2016). The schemes based on a single metric are usually to compare the metric of received signal to a priori threshold value, while the multi-metrics based schemes usually use the joint likelihood ratio test function to identify NLOS. Similar to range based method, a priori knowledge (threshold value or jointly PDF) is required, so the methods are unsuited for a wide range of applications as well. While both the range based and channel based methods identify NLOS propagation before estimating position, position-based methods identify NLOS in the process of position determination and may use the calculated position coordinates to determine NLOS. When redundant ranges are available, by comparing the position estimates produced by different subsets of ranges, it is likely to identify the range measurement under NLOS condition (Casas, Marco & Guerrero, 2006). Or by exploiting extra environment information, such as geometries, map and the continuity of a path, NLOS also can be identified using the calculated position (Jo, Lee, Ha & Kang, 2006; Gustafson, Elwell & Soltz, 2006; Lakhzouri et al., 2003). The position based methods seem to be more flexible and robust because empirical knowledge is not needed, and the performance mainly depends on algorithms and auxiliary information used. However, in some cases where there are no redundant ranges or multiple ranges belong to NLOS, the methods would be ineffective.

NLOS Mitigation Based on Machine Learning

In recent years, due to the stimulation of artificial intelligence study, machine learning technology has been extensively investigated and applied in various fields. The technology has also been exploited in NLOS mitigation and four relevant methods are introduced in the section.

Support Vector Machine (SVM) Technique

A binary linear classifier is a function of the form

$$l(\mathbf{x}) = sign\left[y(\mathbf{x})\right] \in \{-1,+1\} \tag{1.29}$$

with

$$y(\mathbf{x}) = \mathbf{w}^T \varphi(\mathbf{x}) + w_0 \tag{1.30}$$

where $\mathbf{x}$ is the input (sample data vector, e.g. channel characteristics) on which classification is conducted depending, $\varphi(\cdot)$ is a predetermined function mapping vectors from the input space to a very high-dimension feature space, $sign(\cdot)$ is the sign function, $\mathbf{w} = [w_1\ w_2\ \cdots\ w_n]^T$ and w_0 are unknown parameters of the classifier, $y(\mathbf{x})$ is the projection of input in classification space. Once $\mathbf{w}$ and w_0 are obtained, (1.29) is used to carry out classification. The solution method of the classifier is to find the hyper plane $y(\mathbf{x}) = 0$ which determines a difrection where the distance between the projections of the training vectors of two different classes is maximal. The distance is given by

$$\rho(\mathbf{w}, w_0) = \min_{\{\mathbf{x}:y(\mathbf{x})=1\}} \frac{\mathbf{x}\cdot\mathbf{w}}{\|\mathbf{w}\|} - \max_{\{\mathbf{x}:y(\mathbf{x})=-1\}} \frac{\mathbf{x}\cdot\mathbf{w}}{\|\mathbf{w}\|} \tag{1.31}$$

Assuming that a set of training data $\{\mathbf{x}_k, l(\mathbf{x}_k)\}_{k=1}^N$, where $\mathbf{x}_k \in R^n$ and $l(\mathbf{x}_k) \in \{-1,+1\}$ are the input and output respectively, the hyper plane can be obtained by solving the following optimization problem

$$\begin{aligned} &\arg\min_{\mathbf{w},w_0} \frac{1}{2}\|\mathbf{w}\|^2 \\ &s.t.\quad l(\mathbf{x}_k)\, y(\mathbf{x}_k) \geq 1,\ k = 1, 2, \cdots, N \end{aligned} \tag{1.32}$$

Typically, we cannot guarantee all data are separable. So, to allow the SVM to tolerate misclassification, (1.32) is modified as

$$\begin{aligned} &\arg\min_{\mathbf{w},w_0,\xi} \frac{1}{2}\|\mathbf{w}\|^2 + \frac{1}{2}\lambda\sum_{k=1}^{N}\xi_k, \\ &s.t.\quad l(\mathbf{x}_k)\, y(\mathbf{x}_k) \geq 1 - \xi_k\ \ and\ \ \xi_k \geq 0,\ k = 1, 2, \cdots, N \end{aligned} \tag{1.33}$$

where ξ_k are the slack variable that allows the SVM to tolerate misclassification and λ controls the trade-off between reducing training error and reducing model complexity. To solve the optimization problem with additional inequality constraints, we must find and solve its Lagrangian dual of the original problem. However it is tricky to solve the optimization problem because the Lagrangian dual is a quadratic program. In order to simplify the solution, the inequality in (1.33) can be replaced by equality in the form of (1.34) and the improved approach is termed Least Squares SVM (LS-SVM) (Marano, Gifford, Wymeersch & Win, 2010).

$$\begin{aligned} &\arg\min_{\mathbf{w},w_0,\xi} \frac{1}{2}\|\mathbf{w}\|^2 + \frac{1}{2}\lambda\sum_{k=1}^{N}\xi_k^2, \\ &s.t.\quad l(\mathbf{x}_k)y(\mathbf{x}_k) = 1-\xi_k,\ \ k=1,2,\cdots,N \end{aligned} \tag{1.34}$$

For NLOS mitigation, sample data consisting of channel characteristics and propagation conditions (+1 represents the NLOS and -1 represents the LOS), are used to train the classifier. When the classifier is obtained, we can determine whether the range is conducted under NLOS condition based on the channel characteristics. Similarly, we can use sample data consisting of channel characteristics and ranging errors to train a regressor by

$$\delta r(\mathbf{x}) = \mathbf{w}'^{T}\psi(\mathbf{x}) + w_0' \tag{1.35}$$

where $\delta r(\mathbf{x})$ is the ranging error, $\psi(\cdot)$ is a predetermined function, and $\mathbf{w}'^{T}$ and w_0' are the unknown parameters of the regressor. The ranging error can be compensated with the predicted ranging error by the regressor under NLOS condition. The main drawback of this method is that the number of support vectors (SVs) typically increases linearly with the size of training data.

Relevance Vector Machine (RVM) Technique

RVM is invented by Michacl E.Tipping in 2000. The technique and the SVM are similar; both of them own the same function form and transform a nonlinear problem in a low dimensional space to a linear problem in a high dimensional space. But SVM builds its learning machine based on the structural risk minimization principle, while RVM builds its learning machine within the Bayesian framework. Compared with SVM, RVM has a number of advantages, including that the outputs of classifier and regressor are both a numerical value and a probability, the selection of kernel function for classifier and regressor is unlimited by the Mercer condition, and it is more suitable for online use because the number of relevance vectors (RVs) slowly increases with the size of the training data.

Similar to SVM, the binary linear classifier of RVM can be expressed in the same form as (1.29) and (1.30). Define

$$t(\mathbf{x}) = y(\mathbf{x}) + \eta \tag{1.36}$$

where η is Gaussian noise, $\eta \sim N(0,\sigma^2)$. Here we combine the $\mathbf{w}$ and w_0 in (1.30) to form an expanded vector $\mathbf{w}'' = [w_0\ w_1\ \cdots\ w_n]^T$.

For a set of given training sample data $\{\mathbf{x}_k, l(\mathbf{x}_k)\}_{k=1}^{N}$ where $\mathbf{x}_k$ represents the n-dimensional channel characteristic vector and $l(\mathbf{x}_k)$ is the propagation condition (+1 represents the NLOS and -1 represents the LOS), $t(\mathbf{x}_k)$ is subject to $t(\mathbf{x}_k) \sim N(y(\mathbf{x}_k), \sigma^2)$. By assuming that sequence $\{t(\mathbf{x}_k)\}_{k=1}^{N}$ are i.d.d, the likelihood function of training sample data can be expressed easily. However if we directly determine $\mathbf{w}''$ using the maximum likelihood (ML) method, it can be found that most of the elements in $\mathbf{w}''$ are nonzero so that the phenomenon known as "overfitting" occurs. Therefore in RVM technique $\mathbf{w}''$ is determined by the maximum posterior probability (MAP) algorithm.

Assume that vector $\mathbf{w}''$ has a multivariate Gaussian distribution

$$\mathbf{w}'' \sim N_{n+1}(\mathbf{0}, \mathbf{\Lambda}) \tag{1.37}$$

with

$$\Lambda = diag\left\{\lambda_0, \lambda_1, \cdots, \lambda_n\right\} \tag{1.38}$$

where $N_{n+1}(\mathbf{0}, \mathbf{\Lambda})$ represents a $n+1$ dimensional real Gaussian distribution with mean vector $\boldsymbol{\mu} = \mathbf{0}$ and covariance matrix $\mathbf{\Lambda}$.

For a given input $\mathbf{x}_i$, the posterior probability of NLOS can be written as logistic sigmoid function on $y(\mathbf{x}_i)$ as (Nguyen, Jeong, Shin & Win, 2015)

$$P\left(l\left(\mathbf{x}_i\right) = 1 \middle| \mathbf{w}''\right) = h\left[y\left(\mathbf{x}_i\right)\right] = h\left[\mathbf{w}^T \boldsymbol{\varphi}\left(\mathbf{x}_i\right) + w_0\right] \tag{1.39}$$

with

$$h(t) = \frac{1}{1 + e^{-t}} \tag{1.40}$$

Then the likelihood function for all training data can be written as

$$P\left(l\left(\mathbf{x}_1\right), l\left(\mathbf{x}_2\right), \cdots, l\left(\mathbf{x}_N\right) \middle| \mathbf{w}''\right) = \prod_{i=1}^{N} \left\{ h\left(\mathbf{w}^T \boldsymbol{\varphi}\left(\mathbf{x}_i\right) + w_0\right)^{(1+l(\mathbf{x}_i))/2} \left[1 - h\left(\mathbf{w}^T \boldsymbol{\varphi}\left(\mathbf{x}_i\right) + w_0\right)\right]^{(1-l(\mathbf{x}_i))/2} \right\} \tag{1.41}$$

Based on the Bayesian formula,

$$P\left(\mathbf{w}'' \middle| l\left(\mathbf{x}_1\right), l\left(\mathbf{x}_2\right), \cdots, l\left(\mathbf{x}_N\right)\right) \propto P\left(l\left(\mathbf{x}_1\right), l\left(\mathbf{x}_2\right), \cdots, l\left(\mathbf{x}_N\right) \middle| \mathbf{w}''\right) P\left(\mathbf{w}''\right),$$

the coefficient vector $\mathbf{w}''$ can be approximated as the mode of the posterior distribution and given by

$$\arg\max_{\mathbf{w}''} \left\{ \sum_{i=1}^{N} \left\{ \frac{1 + l\left(\mathbf{x}_i\right)}{2} \ln h\left[y\left(\mathbf{x}_i\right)\right] + \frac{1 - l\left(\mathbf{x}_i\right)}{2} \ln\left\{1 - h\left[y\left(\mathbf{x}_i\right)\right]\right\} \right\} - \frac{1}{2} \mathbf{w}''^T \mathbf{\Lambda}^{-1} \mathbf{w}'' \right\} \tag{1.42}$$

Then iteratively reweighted least squares (IRLS) algorithm is usually adopted to solve the optimization problem in (1.42). Similarly, a regressor for NLOS error prediction can be obtained using RVM technique. The RVM technique avoids some deficiencies of SVM, but the technique owns a higher computational complexity and takes more time to train the classifier or regressor so that it may not be suited for scenarios where low computational complexity is required.

Principal Component Analysis (PCA) Technique

PCA is a widely used technique which is invented by Karl Pearson in 1901. The basic idea of PCA technique is to obtain the eigenvectors and eigenvalues of a set of data based on Eigen-decomposition of covariance matrix. The method is suitable for complex data analysis. In the NLOS mitigation research, PCA technique is also introduced into range determination for UWB ranging (Savic, Larsson, Ferrer-Coll & Stenumgaard, 2015).

Assume that the complex impulse response of the channel is given by

$$h(t) = \sum_{i=1}^{N} a_i \delta(t - \tau_i) \tag{1.43}$$

where $a_i \in \mathbb{C}$, N is the number of detected paths, and τ_i is the delay of ith path. The power delay profile (PDP) of the channel is defined as

$$p(t) = \begin{cases} \left|h(t)\right|^2, & \left|h(t)\right|^2 > P_{TH} \\ 0, & others \end{cases} \tag{1.44}$$

where P_{TH} is a threshold level for excluding the effect of thermal noise. A number of channel parameters can be extracted from the PDP, such as time of arrival (TOA), received signal strength (RSS), and RMS delay spread. Based on one or more of these parameters, range can be determined. Usually the estimated TOA τ_1 is used to estimate the range simply by

$$\hat{d} = c\tau_1 = d + \delta \tag{1.45}$$

where c is the signal transmission speed, $\hat{d}$ is the estimate of true distance d, and δ is the estimation error which can be significant under NLOS propagation. In the case of range determination using multi-parameters, PCA technique can be used for parameter choice. Firstly, all channel parameters $\alpha_i (i = 1, 2, \cdots, k)$ abstracted from (1.44) are normalized to be zero-mean

$$\tilde{\alpha}_i = \frac{\alpha_i - \mu_i}{\sigma_i} \tag{1.46}$$

where α_i is ith received channel parameter, μ_i and σ_i are the mean and the standard deviation of α_i respectively. Define vector $\mathbf{a} = [\tilde{\alpha}_1\ \tilde{\alpha}_2\ \cdots\ \tilde{\alpha}_k]^T$, and transform the k-dimensional vector to an n-dimensional vector (n≫k) in a feature space using a non-linear transformation function $\phi(\cdot)$.

Assuming that the mean of $\phi(\mathbf{a})$ is zero, the covariance matrix of $\phi(\mathbf{a})$ is given by $\mathbf{C} = E[\phi(\mathbf{a})\phi^{\mathrm{T}}(\mathbf{a})]$. The principal components of $\mathbf{C}$ can be determined using the following eigenvalue expansion

$$\mathbf{K}\mathbf{v}_i = \lambda_i n \mathbf{v}_i \tag{1.47}$$

where $\mathbf{K}$ is an $n \times n$ kernel matrix with $(\mathbf{K})_{ij} = k(\mathbf{a}_i, \mathbf{a}_j)$, $\mathbf{a}_i$ and $\mathbf{a}_j$ come from sample data. The eigenvalues and eigenvectors of $\mathbf{K}$ are given by $\lambda_i n$ and $\mathbf{v}_i$ respectively, and $\mathbf{v}_i = [v_{i1}\ v_{i2}\ \cdots\ v_{in}]^T$. Then the projection of a test point $\mathbf{a}$ onto ith eigenvector is given by

$$y_i(\mathbf{a}) = \sum_{j=1}^{n} v_{ij} k(\mathbf{a}, \mathbf{a}_j) \tag{1.48}$$

where $k(\mathbf{a}, \mathbf{a}_j)$ is the kernel function. The standard PCA owns a linear kernel $k(\mathbf{a}, \mathbf{a}_j) = \mathbf{a}^T \mathbf{a}_j$. However, PCA projects the data in a linear sub-space, so it is not useful if the data lies on a nonlinear manifold. Therefore a nonlinear version of PCA, known as kernel PCA (kPCA), is widely used. The common kernel functions for kPCA include polynomial kernel and Gaussian radial basis function. A p-order polynomial kernel is given by

$$k(\mathbf{a}, \mathbf{a}_j) = (\mathbf{a}^T \mathbf{a}_j + 1)^p \tag{1.49}$$

where p is determined empirically. Assume that the principle component $y_i(\mathbf{a})$ is related to the true range as

$$y_i(\mathbf{a}) = y_i = b_{0,i} + b_{1,i} d + \eta_i \tag{1.50}$$

where d is the true range, $b_{0,i}$ and $b_{1,i}$ are model parameters which can be obtained by curve fitting using training samples, and η_i is considered as a zero-mean Gaussian noise whose variance σ_i^2 can also be determined using the training samples. In the case of M principal components corresponding to the M largest eigenvalues are chosen, the likelihood function can be obtained as

$$p\left(y_1, y_2, \cdots, y_M \middle| d\right) = \prod_{i=1}^{M} p\left(y_i \middle| d\right) = \prod_{i=1}^{M} \frac{1}{\sigma_d \sqrt{2\pi}} \exp\left(-\frac{\left(d - \mu_d\right)^2}{2\sigma_d^2}\right) \tag{1.51}$$

with

$$\mu_d = \frac{1}{b_{1,i}}\left(y_i - b_{0,i}\right), \sigma_d = \frac{\sigma_i}{b_{1,i}} \tag{1.52}$$

It can be derived that the minimum mean-square-error estimate of the distance is given by

$$\hat{d}_{kPCA} = \sigma^2_{kPCA} \sum_{i=1}^{M} \left(b_{1,i} \frac{y_i - b_{0,i}}{\sigma_i^2} \right) \tag{1.53}$$

with

$$\sigma^2_{kPCA} = \left[\sum_{i=1}^{M} \left(\frac{b^2_{1,i}}{\sigma_i^2} \right) \right]^{-1} \tag{1.54}$$

The method does not involve NLOS identification and it can artfully mitigate the NLOS effect on ranging. However the complexity of range computation is increased as well as the method have to be implemented together with hardware design.

K-Nearest Neighbor (K-NN) Technique

K-Nearest Neighbor (K-NN) technique is one of the simplest nonparametric learning methods and is widely used in classification and regression problems. Assuming that n channel characteristics are used for NLOS identification, an input vector $\mathbf{x} = [x_1\ x_2\ \cdots\ x_n]^T$ can be formed. If there are N sample data sets, $\{\mathbf{x}_k, \delta r_k\}_{k=1}^N$, where $\mathbf{x}_k \in R^n$ is the channel characteristic vectors and δr_k is the corresponding ranging error. For each received $\mathbf{x}' = [x_1'\ x_2'\ \cdots\ x_n']^T$, the difference between $\mathbf{x}'$ and $\mathbf{x}_k$ can be expressed by the Euclidean distance

$$d_k = \left\| \mathbf{x}' - \mathbf{x}_k \right\| \tag{1.55}$$

All Euclidean distances are combined to form a new set of $\{d_1, d_2, \cdots, d_N\}$. By sorting the set in ascending order and then choosing the first K minimum Euclidean distances from the sorted set, the corresponding K ranging errors are expressed as $\{\delta r_1', \delta r_2', \cdots, \delta r_K'\}$. Then the range measurement can be corrected as

$$\tilde{r} = \hat{r} + \sum_{i=1}^{K} w_i \cdot \delta r_i' \tag{1.56}$$

where $\hat{r}$ is the measured range, $\tilde{r}$ is the corrected range, and w_i is the weight which emphasize the degree of importance of each correction.

K-NN technique is extremely simple to implement because training process is not needed and only sample data is used. However, the method would result in low accuracy in the presence of sparse or inhomogeneous sample data.

NLOS Mitigation Based on Fuzzy Theory

The obvious disadvantages of SVM/RVM-based methods are that the training process is time-consuming and the computational complexity is high, which hamper their utilization in low cost and low complexity scenarios. PCA technique directly mitigates the influence of NLOS in the process of distance determination within the device. Therefore, the technique has to be implemented together with hardware design, which would be impracticable for many applications. K-NN technique is extremely simple to implement because training process is not needed and only sample data is used, and thus has great potentiality in lost cost applications. However, the method would result in low accuracy in the presence of sparse or inhomogeneous sample data. To achieve an accurate NLOS mitigation, a large number of samples have to be collected. Meanwhile, sample data need to be updated frequently as the environment changes. An improved method based on K-NN technique by introducing fuzzy theory is proposed in this section.

Overview

Indoors, although ranging errors are different under different NLOS propagation conditions, such as propagating around a corner, penetrating through a door or a wall, obstructed by pedestrians and so on, the fundamental cause leading to ranging errors is two-fold, power decay and time delay. It is natural to use some typical NLOS circumstances to represent all the NLOS propagation conditions. In this case, we only need to collect sample data under different types of propagation conditions, whereas the K-NN method collects samples over uniformly distributed locations. Meanwhile, in most cases, the environment change is usually accompanied by changes in spatial layout but not in types of propagation conditions. Therefore, the proposed method can achieve NLOS mitigation using sparse samples as well as less update operations. The method is also practical because NLOS propagation may not be introduced by a single obstacle, but several obstacles simultaneously. For instance, signal may penetrate through a concrete wall as well as a wood board, or be disturbed by pedestrians and furniture simultaneously. Hence it is necessary to consider both the case of single obstacle and that of multiple obstacles.

To determine an accurate compensating error under a specific propagation condition using sample data coming from representative propagation conditions, fuzzy theory is introduced. The method considers the range correction as a combination of ranging errors under different NLOS propagation conditions or the ranging error coming from a single propagation condition by fuzzy judgment. Therefore, compared with the K-NN method, the proposed method is more flexible in determining range correction, while range correction always comes from K samples for K-NN method.

Algorithm Description

Propagation conditions are broadly divided into m categories $C = \{c_0, c_1, c_2, \cdots, c_{m-1}\}$, where c_0 is the propagation condition under LOS, and $c_i (i > 0)$ represents the ith kind of NLOS propagation condition. Assume that the sample data consist of m subsets, $D = \{D_0, D_1, D_2, \cdots, D_{m-1}\}$, where $D_i = \{\mathbf{x}_k^i, \delta(\mathbf{x}_k^i)\}_{k=1}^{N_i}$ represents the set of samples under NLOS propagation condition c_i. $\mathbf{x}_k^i$ is a vector consisting of signal characteristics of a range measurement, $\delta(\mathbf{x}_k^i)$ is the corresponding ranging error and N_i is the number of measurements (each measurement produces a characteristic vector $\mathbf{x}$ and a ranging error $\delta(\mathbf{x})$). $\mathbf{x}_k^i$ can be written as

$$\mathbf{x}_k^i = \left[x_{k,1}^i \; x_{k,2}^i \; \cdots \; x_{k,n}^i\right]^T \tag{1.57}$$

where $x_{k,j}^i$ is the j th signal characteristic of the k th characteristic sample set under NLOS propagation condition c_i.

Assume that the influences of different signal characteristics on NLOS identification are seen as a vector

$$\mathbf{w} = \left[w_1 \; w_2 \; \cdots \; w_n\right]^T, \; \sum_{i=1}^{n} w_i = 1 \tag{1.58}$$

where w_i is the influence degree of signal characteristic x_i, whose value may be set empirically.

For a given range measurement $\hat{r}$ with a set of signal characteristics $\mathbf{x}' = [x_1' \; x_2' \; \cdots \; x_n']^T$, the probability that $\hat{r}$ is measured under propagation condition c_i is calculated as

$$\mu\left(c_i\right) = \mathbf{w}^T\mathbf{u}\left(c_i\right) \tag{1.59}$$

with

$$\mathbf{u}\left(c_i\right) = \left[\mu\left(c_i, x_1'\right) \; \mu\left(c_i, x_2'\right) \cdots \mu\left(c_i, x_n'\right)\right]^T \tag{1.60}$$

where $\mu(c_i)$ is the membership which indicates that the possibility that $\mathbf{x}'$ belongs to propagation condition c_i determined together by all signal characteristics. Similarly, $\mu(c_i, x_j')$ represents the membership which indicates that the probability that $\mathbf{x}'$ belongs to propagation condition c_i determined only by a single characteristic x_j'. $\mu(c_i, x_j')$ is calculated as the following membership function

$$\mu\left(c_i, x_j'\right) = 1 - \left|\frac{x_j' - x_j^i}{\max\left\{x_j', x_j^i\right\}}\right| \tag{1.61}$$

Under propagation condition c_i, the number of x_j^i which may be used in (1.61) is N^i. The selection which sample is used in (1.61) is determined by the rule

$$\min\left\{d_k^i\right\}_{k=1}^{N_i} \tag{1.62}$$

with

$$d_k^i = \left\| \mathbf{x}' - \mathbf{x}_k^i \right\| \tag{1.63}$$

Therefore a NLOS identification result is a fuzzy set of C, which can be expressed as

$$\underset{\sim}{B} = \left\{ \frac{\mu\left(c_0\right)}{\Sigma}, \frac{\mu\left(c_1\right)}{\Sigma}, \ldots, \frac{\mu\left(c_{m-1}\right)}{\Sigma} \right\} \tag{1.64}$$

with

$$\Sigma = \mu\left(c_0\right) + \mu\left(c_1\right) + \cdots + \mu\left(c_{m-1}\right) \tag{1.65}$$

As mentioned above, it is flexible for the fuzzy theory based NLOS mitigation technique to determine the compensating error. Propagation condition is considered as c_i if the following condition is satisfied

$$\mu\left(c_i\right) > \beta \tag{1.66}$$

where $\beta > 0$ is a threshold which is set empirically. In this case, the measured range is corrected as

$$\tilde{r} = \hat{r} + \delta(\mathbf{x}_j^i) \tag{1.67}$$

Alternatively, the propagation is considered as a combination of multiple conditions and the measured range is corrected as

$$\tilde{r} = \hat{r} + \sum_{i=0}^{m-1} \frac{\mu\left(c_i\right)}{\Sigma} \delta\left(\mathbf{x}_j^i\right) \tag{1.68}$$

In (1.67) and (1.68), $\delta(\mathbf{x}_j^i)$ is the corresponding ranging error of the selected sample $\mathbf{x}_j^i$ under propagation c_i by rule (1.62). Note that $\delta(\mathbf{x}_j^0) = 0$.

Meanwhile, due to the limited number of chosen representative propagation conditions, the proposed method is incapable for handling some special scenarios, therefore another threshold κ is set empirically to exclude these situations

$$\mu\left(c_i\right) < \kappa, \forall\, i \in \left\{0, 1, 2, \cdots, \text{m-1}\right\} \tag{1.69}$$

Once the above inequality is satisfied, we have to abandon the distance measurement. Note that the condition (1.69) enables the proposed method to discern the abnormal measurements as well.

Performance Evaluation

One experiment was conducted over an area of 18m×65m to evaluate the performance of the fuzzy theory based NLOS mitigation technique described above. The location scenario is located on the second floor of a teaching building within Wuhan University. This is a complex indoor environment, containing wall, corridor, pedestrian, wood door, classroom desk and so on. In the experiment, representative propagation conditions are divided into six categories, $C = \{c_0, c_1, c_2, c_3, c_4, c_5\}$, namely indoor LOS, propagating around a corner, penetrating through a wood door, penetrating through a wall, propagating in a corridor, and obstructed by pedestrians. The experiment consists of two stages. In the first stage, sample data under six propagation conditions were collected. In the second stage, a ranging test and a dynamic tracking test were conducted. Experimental scene is shown in Figure 4 where the circles denote positions of fixed units and the squares stand for positions of the mobile unit in the ranging test, while the diamonds denote positions of fixed units and the dashed line stands for a about 10m×5.5m trajectory of a mobile unit in the dynamic tracking test. For both the dynamic tracking test and ranging test, PulsON® 410 RCM (ranging and communication module) devices built by Time Domain were used for logging the measured ranges based on TW-TOF and the channel scan data (discrete signal). In the dynamic tracking test, an UWB transceiver was installed on a moving cart which was pulled along the trajectory manually, as shown in the following section.

Figure 5 shows the received channel scan data and the corresponding CIRs under six different propagation conditions. Both signal amplitudes and patterns of recorded waveforms are quite different among different propagation conditions. The signal amplitudes are obviously reduced under propagating around a corner and penetrating through a wood door or a wall, compared with those under LOS propagation. In the case of propagating around a corner or when the signal is obstructed by pedestrians, the recorded waveforms change more significantly. When penetrating a wall the noise level is considerably higher. The CIRs show significant differences between different cases as well, as shown on the right panel of Figure 5. Figure 6 shows three received waveforms as well as three CIRs when penetrating through a wall with three different propagation distances. It turns out that the received waveforms and CIRs are very similar when penetrating through the same type of obstacles.

Figure 4. Distribution of tested locations

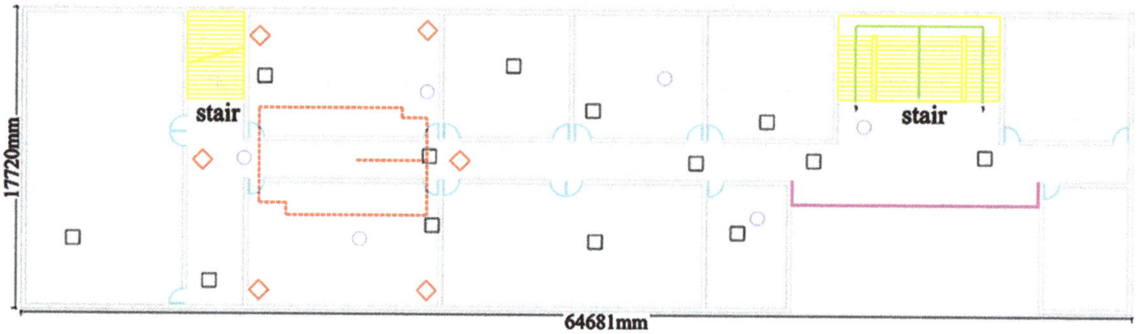

Figure 5. Channel scan data and CIRs under different propagation conditions

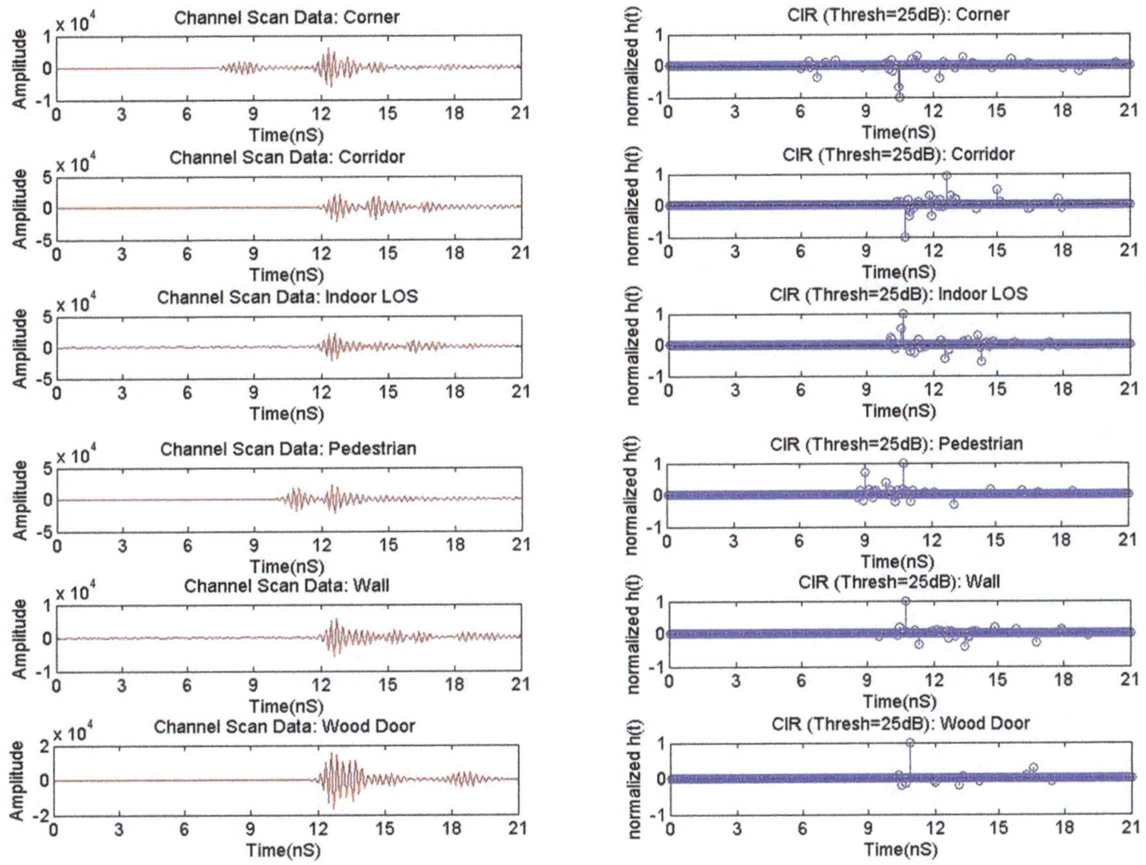

Currently, machine learning based NLOS mitigation methods commonly extract feature parameters from CIR. However it is impracticable for most of the commercial applications because the devices involved usually do not have the function to obtain CIR. On the other hand, the channel scan data is easy to be obtain by most of the off-the-shelf devices. Therefore, it is proposed to directly use signal characteristics from channel scan data in the proposed NLOS mitigation technique. Figure 7 shows the Euclidean distances calculated as (1.63) based on the same features abstracted from CIR and channel scan data under six different range measurements when penetrating a wall. Similarly, Figure 8 shows the calculated Euclidean distances under six different propagation conditions when the range is seven meters. The Euclidean distance is inversely proportional to the membership indicating the degree that the measured range belongs to a propagation condition. Therefore Figure 7 and Figure 8 imply that the identification results are consistent when calculating signal characteristics with CIRs and channel scan data.

In this experiment, signal characteristics used include total energy, maximum amplitude, rise time (RT), RMS delay spread, and kurtosis, represented by vector $\mathbf{x}_k^i = [x_{k,1}^i \; x_{k,2}^i \; x_{k,3}^i \; x_{k,4}^i \; x_{k,5}^i]^T$. The influence of each factor, w_i in (1.58), is considered as the same and is set to be 0.2. The threshold values β in (1.66) and κ (1.69) are set to be 0.8 and 0.3 respectively. The number of range measurements N_i in each representative propagation condition ranges from four to seven. Figure 9 and Figure 10 show the

Figure 6. Channel scan data and CIRs under different propagation distances

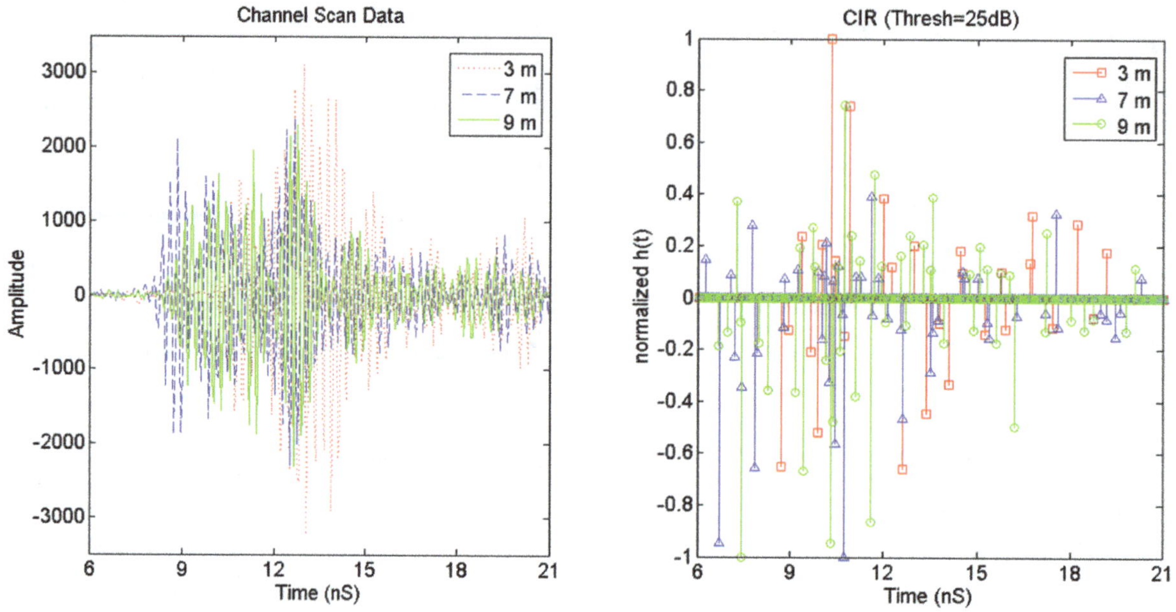

Figure 7. Calculated Euclidean distances based on CIR and channel scan data under six different range measurements when penetrating a wall

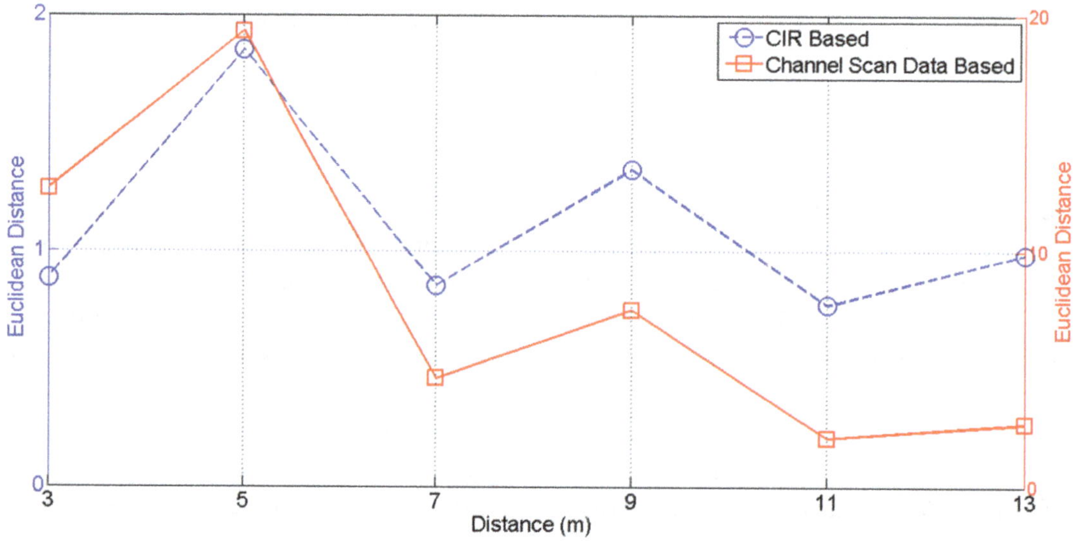

empirical cumulative distribution functions (CDFs) and distribution of ranging errors without and with NLOS mitigation. We can see that the ranging errors are reduced significantly using the proposed method. About 80% of the ranging errors of original range measurements are below 0.5 meters, while about 98% of the ranging errors with NLOS identification and mitigation are below 0.5 meters. Note that the number of corrected ranging errors is smaller than that of original ranging errors in Figure 10.

Figure 8. Calculated Euclidean distances based on CIR and channel scan data under six different propagation conditions when the range is seven meters

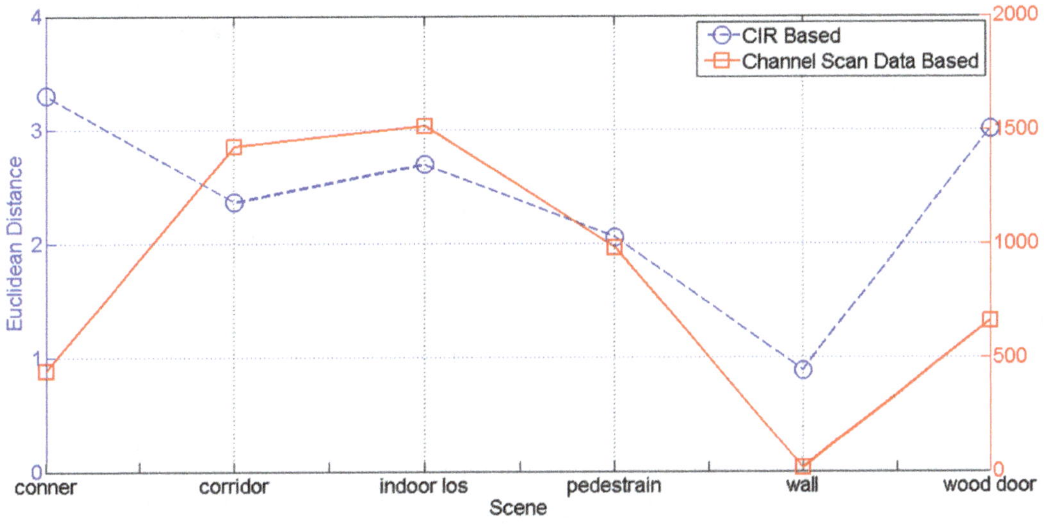

Figure 9. The empirical CDFs of ranging errors without and with NLOS mitigation

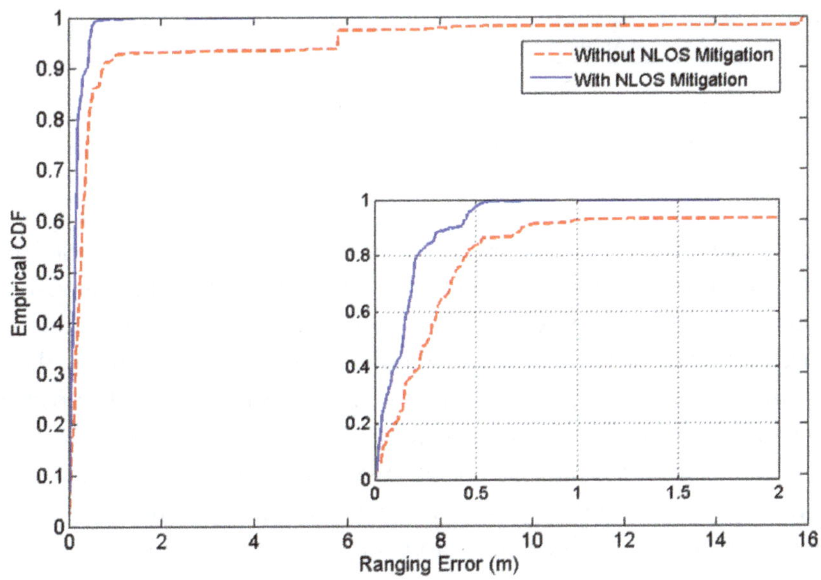

That is because measurements are abandoned if the memberships under all representative propagation conditions are less than 0.3. Meanwhile, some outliers in the original range measurements are observed in Figure 10, whose ranging errors reach a few meters even up to 16 meters. But after NLOS mitigation, nearly all outliers are abandoned. Table 1 shows the mean and the standard deviation of ranging errors and root mean square error (RMSE) of measured ranges without and with NLOS mitigation. Results show that the RMSE can be reduced from 2.58 to 0.25 meters, implying that the difference between corrected range and true range is significantly smaller.

Figure 10. Ranging error distribution without and with NLOS mitigation

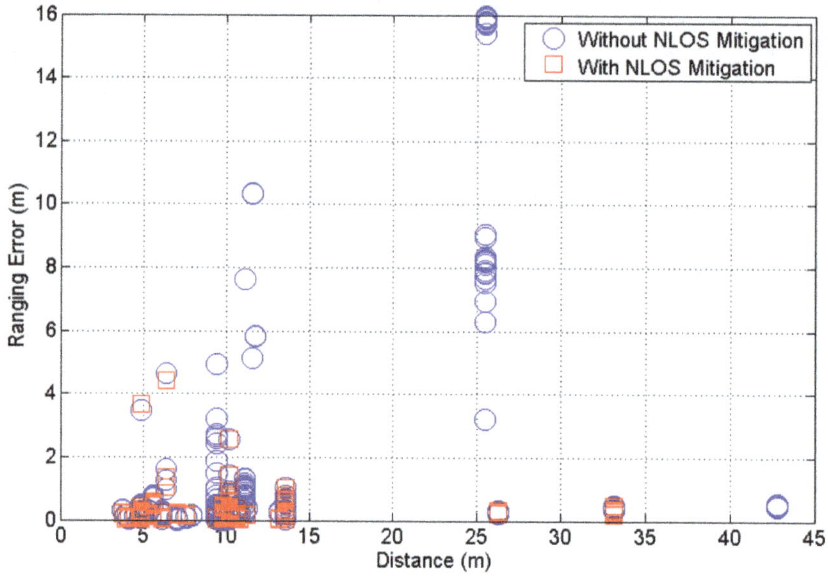

Table 1. The statistics of measured ranges without and with NLOS mitigation

	Mean (m)	Standard deviation (m)	RMSE (m)
without mitigation	0.85	2.44	2.58
with mitigation	0.16	0.20	0.25

In the dynamic tracking test, a measured range is improved by the proposed NLOS mitigation technique before it is used for position determination. Figure 11 shows the 2D positions estimated by UWB multilateration along a trajectory. The bold solid line represents the estimated trajectory when the measured ranges are corrected by the proposed technique, while the solid line is the estimated trajectory without NLOS mitigation. The results show that with NLOS mitigation the positioning accuracy is significantly improved and nearly no abnormal position estimates are observed. Note that there are 850 position points on the trajectory, but only 771 position estimates are produced after NLOS mitigation. That is because if the number of valid range measurements after NLOS mitigation is less than three, a valid position estimate cannot be obtained. Therefore we should avoid using the technique in scenarios where position estimate is required for every sampling instant.

TWO KALMAN FILTER BASED DATA FUSION APPROACHES

Other effective method, to mitigate the influence of NLOS on UWB positioning accuracy, is to introduce other positioning techniques. INS is a technique independent of environmental conditions, so that the positioning performance is not affected by NLOS or multipath. Meanwhile, INS technique does not rely

Figure 11. Trajectories calculated by UWB multilateration

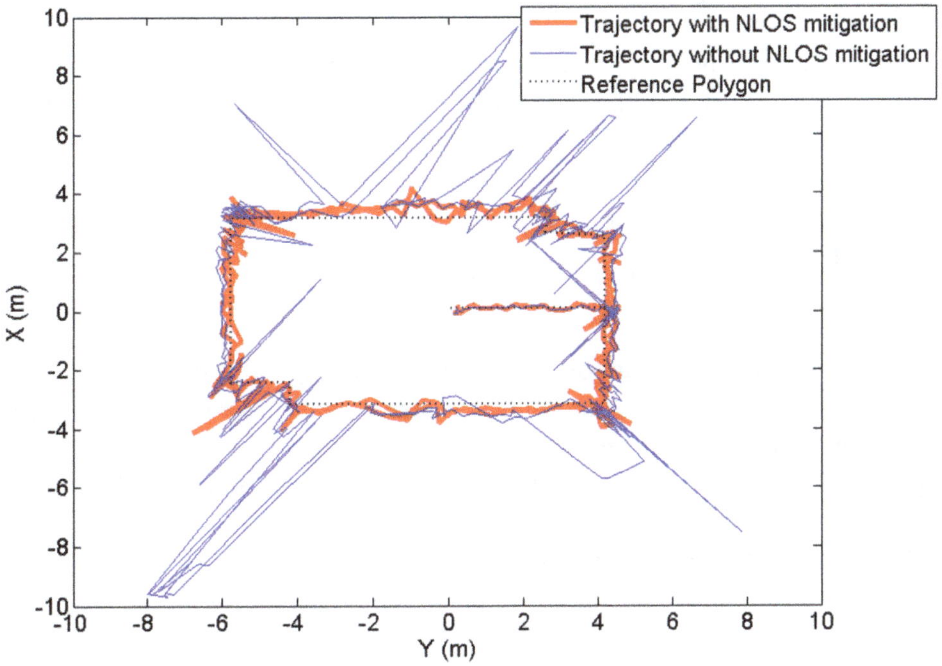

on any pre-installed infrastructure and is convenient to use at any time and at any place. However due to error accumulation, the position error increases with time and divergence occurs quickly. Positioning by integrating UWB and INS can make full use of advantages of both UWB and INS, improving both positioning accuracy and usage in complex environments.

Kalman Filtering Estimator

Kalman Filter

Kalman filter (KF) is an estimator originally designed for solving the linear-quadratic problem, i.e., estimating the instantaneous "state" of a linear dynamic system perturbed by white noise (Grewal &Andrews 2008). The technique is developed on the basis of Wiener-Kolmogorov filter theory, transforming the filtering problem from frequency domain to time domain by applying the notion of state variables to the Wiener-Kolmogorov filtering problem. As a result, a simple and user-friendly filtering method is achieved. One difference between KF and Wiener-Kolmogorov filter is that the former not only can be used for estimation problem with continuous measurements but also for discrete time problem in either stationary or non-stationary conditions.

In KF theory, a system is expressed by a model consisting of state equations, which describe the state transformation from current time to next time with system input and noise disturbance, and measurement equations which describe the transformation from state variables to measurement variables. Usually the state equations are built based on the Newton's laws of mechanics and gravitation, implying that the state is expressed in a continuous form. A typical linear state equation is written as following differential form

$$\dot{\mathbf{x}}(t) = \mathbf{F}(t)\mathbf{x}(t) + \mathbf{C}(t)\mathbf{u}(t) + \mathbf{w}(t) \tag{1.70}$$

where matrix $\mathbf{F}(t)$ is the dynamic coefficient matrix, $\mathbf{C}(t)$ is the input coupling matrix (ICM), $\mathbf{u}(t)$ is the deterministic input vector, $\mathbf{x}(t)$ is the state vector, and $\mathbf{w}(t)$ is the white noise process subject to

$$\begin{aligned} &E\left(\mathbf{w}(t)\right) = \mathbf{0}, \\ &E\left(\mathbf{w}(t_1)\mathbf{w}^T(t_2)\right) = \mathbf{Q}(t_1)\delta(t_2 - t_1) = \begin{cases} \mathbf{Q}(t_1) & t_2 = t_1 \\ \mathbf{0} & t_2 \neq t_1 \end{cases} \end{aligned} \tag{1.71}$$

where $\mathbf{Q}(t)$ is the covariance matrix of noise $\mathbf{w}(t)$. Although originally state equations are expressed in continuous time, KF has to be transformed into a discrete time form for implementing on a computer. It may be convenient to build the state equation in continuous time first and then convert it to discrete time form in practice. The conversion can be achieved as (Grewal &Andrews 2008)

$$\mathbf{x}(t_k) = \mathbf{\Phi}(t_k, t_{k-1})\mathbf{x}(t_{k-1}) + \mathbf{\Gamma}(t_k)\mathbf{u}(t_k) + \mathbf{w}(t_k) \tag{1.72}$$

with

$$\begin{cases} \mathbf{\Phi}(t_k, t_{k-1}) = \exp(\int_{t_{k-1}}^{t_k} \mathbf{F}(s)ds) \\ \mathbf{\Gamma}(t_k) = \int_{t_{k-1}}^{t_k} \mathbf{\Phi}(t_k, s)\mathbf{C}(s)ds \\ \mathbf{Q}(t_k) = \mathbf{\Phi}(t_k, t_{k-1})\left[\int_{t_{k-1}}^{t_k} \mathbf{\Phi}^{-1}(s, t_{k-1})\mathbf{Q}(s)\mathbf{\Phi}^{-T}(s, t_{k-1})ds\right]\mathbf{\Phi}^T(t_k, t_{k-1}) \end{cases} \tag{1.73}$$

where $\mathbf{\Phi}(t_k, t_{k-1})$ is the state transition matrix (STM) from time t_{k-1} to t_k, $\mathbf{\Gamma}(t_k)$ is the discrete-time ICM, $\mathbf{w}(t_k)$ is the white noise vector, and $\mathbf{Q}(t_k)$ is the covariance matrix of $\mathbf{w}(t_k)$. In (1.70), $\mathbf{F}(t)$ and $\mathbf{C}(t)$ are either time variant or time invariant. When both $\mathbf{F}(t)$ and $\mathbf{C}(t)$ are time invariant, the system is called time invariant system, and (1.72) is simplified as

$$\mathbf{x}_k = \mathbf{\Phi}\mathbf{x}_{k-1} + \mathbf{\Gamma}\mathbf{u}_k + \mathbf{w}_k \tag{1.74}$$

Meanwhile, the relationship between system state parameters and measurements is

$$\mathbf{z}_k = \mathbf{H}\mathbf{x}_k + \mathbf{v}_k \tag{1.75}$$

where $\mathbf{z}_k$ is measurement vector at time t_k, $\mathbf{H}$ is the measurement sensitivity matrix (MSM) which is assumed time invariant here, and $\mathbf{v}_k$ is a white noise vector subject to

$$E\left(\mathbf{v}_k\right) = \mathbf{0},$$
$$E\left(\mathbf{v}_k \mathbf{v}_j^T\right) = \mathbf{R}_k \delta_{kj} = \begin{cases} \mathbf{R}_k & k = j \\ \mathbf{0} & k \neq j \end{cases} \tag{1.76}$$

where $\mathbf{R}_k$ is the covariance matrix of $\mathbf{v}_k$. Based on the state equation (1.74) and the measurement equation (1.75) as well as the assumed noise property in (1.71) and (1.76), KF can be deduced using the criterion of minimizing the mean-square error of estimates as

$$\begin{cases} \hat{\mathbf{x}}_{k|k-1} = \mathbf{\Phi} \hat{\mathbf{x}}_{k-1|k-1} + \mathbf{\Gamma} \mathbf{u}_k \\ \mathbf{P}_{k|k-1} = \mathbf{\Phi} \mathbf{P}_{k-1|k-1} \mathbf{\Phi}^T + \mathbf{Q}_k \\ \mathbf{K}_k = \mathbf{P}_{k|k-1} \mathbf{H}^T \left(\mathbf{H} \mathbf{P}_{k|k-1} \mathbf{H}^T + \mathbf{R}_k\right)^{-1} \\ \hat{\mathbf{x}}_{k|k} = \hat{\mathbf{x}}_{k|k-1} + \mathbf{K}_k \left(\mathbf{z}_k - \mathbf{H} \hat{\mathbf{x}}_{k|k-1}\right) \\ \mathbf{P}_{k|k} = \left(\mathbf{I} - \mathbf{K}_k \mathbf{H}\right) \mathbf{P}_{k|k-1} \end{cases} \tag{1.77}$$

where $\hat{\mathbf{x}}_{k|k-1}$ is the predicted state estimate based the $\hat{\mathbf{x}}_{k-1|k-1}$ and input $\mathbf{u}_k$ at time t_k, $\hat{\mathbf{x}}_{k|k}$ is the state estimate at time t_k, $\mathbf{P}_{k|k-1}$ is the covariance matrix of the error in the predicted estimate $\hat{\mathbf{x}}_{k|k-1}$, $\mathbf{P}_{k|k}$ is the covariance matrix of the error in the estimate $\hat{\mathbf{x}}_{k|k}$ and $\mathbf{K}_k$ is the gain matrix. According to the equations, once the initial value of $\mathbf{x}_{0|0}$ and $\mathbf{P}_{0|0}$ are given as well as the known noise statistical information of $\mathbf{Q}_k$ and $\mathbf{R}_k$, KF can deduce the state estimate $\mathbf{x}_{k|k}$ at any time t_k when a measurement $\mathbf{z}_k$ is available.

Extended Kalman Filter

KF is built on the fact that the system state is described by a linear equation and the relationship between system state vector and measurement vector is linear. However in a large number of potential applications, the linear condition is hard to satisfy. It is interesting that although KF is invented in the linear form originally, its first application is for a nonlinear system[2]. Applying KF for nonlinear problems involves linear approximation by assuming knowledge of an approximate solution and by describing the deviations from the reference by linear equations (Sorenson, 1970). Considering a simple discrete time invariant nonlinear system model with system input

$$\begin{aligned} \mathbf{x}_k &= \mathbf{f}\left(\mathbf{x}_{k-1}, \mathbf{u}_k\right) + \mathbf{w}_k \\ \mathbf{z}_k &= \mathbf{h}\left(\mathbf{x}_k\right) + \mathbf{v}_k \end{aligned} \tag{1.78}$$

where the definitions of variables $\mathbf{x}_k$, $\mathbf{z}_k$, $\mathbf{u}_k$, $\mathbf{w}_k$ and $\mathbf{v}_k$ in (1.78) are the same as those in (1.74) and (1.75). Ignoring the influence of noise, we can obtain the approximate state vector $\tilde{\mathbf{x}}_k = \mathbf{f}(\tilde{\mathbf{x}}_{k-1}, \mathbf{u}_k)$ and observation vector $\tilde{\mathbf{z}}_k = \mathbf{h}(\tilde{\mathbf{x}}_k)$. The equations are expanded at the approximate values $\tilde{\mathbf{x}}_{k-1}$ and $\tilde{\mathbf{x}}_k$ using Taylor series

$$\begin{cases} \mathbf{x}_k \approx \mathbf{f}(\tilde{\mathbf{x}}_{k-1}, \mathbf{u}_k) + \sum_{n=1}^{\infty} \frac{\mathbf{f}^n\left(\tilde{\mathbf{x}}_{k-1}, \mathbf{u}_k\right)}{n!}\left(\mathbf{x}_{k-1} - \tilde{\mathbf{x}}_{k-1}\right) + \mathbf{w}_k \\ \mathbf{z}_k \approx \mathbf{h}\left(\tilde{\mathbf{x}}_k\right) + \sum_{n=1}^{\infty} \frac{\mathbf{h}^n\left(\tilde{\mathbf{x}}_k\right)}{n!}\left(\mathbf{x}_k - \tilde{\mathbf{x}}_k\right) + \mathbf{v}_k \end{cases} \tag{1.79}$$

with

$$\mathbf{f}^n\left(\tilde{\mathbf{x}}_{k-1}, \mathbf{u}_k\right) = \left.\frac{\partial \mathbf{f}^n}{\partial^n \mathbf{x}}\right|_{\tilde{\mathbf{x}}_{k-1}, \mathbf{u}_k}, \quad \mathbf{h}^n\left(\tilde{\mathbf{x}}_k\right) = \left.\frac{\partial \mathbf{h}^n}{\partial^n \mathbf{x}}\right|_{\tilde{\mathbf{x}}_k} \tag{1.80}$$

Then the linearized state equation and measurement equation can be approximately written as

$$\begin{cases} \Delta\mathbf{x}_k = \sum_{n=1}^{\infty} \frac{\mathbf{f}^n\left(\tilde{\mathbf{x}}_{k-1}, \mathbf{u}_k\right)}{n!} \Delta\mathbf{x}_{k-1} + \mathbf{w}_k \\ \Delta\mathbf{z}_k = \sum_{n=1}^{\infty} \frac{\mathbf{h}^n\left(\tilde{\mathbf{x}}_k\right)}{n!} \Delta\mathbf{x}_k + \mathbf{v}_k \end{cases} \tag{1.81}$$

with

$$\Delta\mathbf{x}_k = \mathbf{x}_k - \tilde{\mathbf{x}}_k, \ \Delta\mathbf{z}_k = \mathbf{z}_k - \tilde{\mathbf{z}}_k \tag{1.82}$$

In practice, one retains the first order items and abandons higher order items. Although the linear approximation opens a door to use KF for nonlinear problems, the computational complexity is increased because approximate values need to be calculated in each stage. Meanwhile, the linear approximation will lead to an undesirable divergence problem, which may produce an unusable result. To reduce the linear approximation error which may lead to divergence and the computational complexity, extended Kalman filter (EKF) can be used. The principle of EKF is to linearize the nonlinear state and observation equations by employing the best estimates of state vector as the reference values used at each stage as follows

$$\begin{cases} \mathbf{x}_k \approx \mathbf{f}\left(\hat{\mathbf{x}}_{k-1|k-1}, \mathbf{u}_k\right) + \left.\frac{\partial \mathbf{f}}{\partial \mathbf{x}}\right|_{\hat{\mathbf{x}}_{k-1|k-1}, \mathbf{u}_k}\left(\mathbf{x}_{k-1} - \hat{\mathbf{x}}_{k-1|k-1}\right) + \mathbf{w}_k \\ \mathbf{z}_k \approx \mathbf{h}\left(\hat{\mathbf{x}}_{k|k-1}\right) + \left.\frac{\partial \mathbf{h}}{\partial \mathbf{x}}\right|_{\hat{\mathbf{x}}_{k|k-1}}\left(\mathbf{x}_k - \hat{\mathbf{x}}_{k|k-1}\right) + \mathbf{v}_k \end{cases} \tag{1.83}$$

where the definitions of variables $\hat{\mathbf{x}}_{k-1|k-1}$ and $\hat{\mathbf{x}}_{k|k-1}$ are the same as those in (1.77). Define

$$\begin{cases} \mathbf{\Phi}_k = \left.\frac{\partial \mathbf{f}}{\partial \mathbf{x}}\right|_{\hat{\mathbf{x}}_{k-1|k-1}, \mathbf{u}_k}, \ \boldsymbol{\phi}_k = \mathbf{f}\left(\hat{\mathbf{x}}_{k-1|k-1}, \mathbf{u}_k\right) - \left.\frac{\partial \mathbf{f}}{\partial \mathbf{x}}\right|_{\hat{\mathbf{x}}_{k-1|k-1}, \mathbf{u}_k} \cdot \hat{\mathbf{x}}_{k-1|k-1} \\ \mathbf{H}_k = \left.\frac{\partial \mathbf{h}}{\partial \mathbf{x}}\right|_{\hat{\mathbf{x}}_{k|k-1}}, \quad \mathbf{y}_k = \mathbf{h}\left(\hat{\mathbf{x}}_{k|k-1}\right) - \left.\frac{\partial \mathbf{h}}{\partial \mathbf{x}}\right|_{\hat{\mathbf{x}}_{k|k-1}} \cdot \hat{\mathbf{x}}_{k|k-1} \end{cases} \tag{1.84}$$

Then (1.83) can be approximately written as

$$\begin{cases} \mathbf{x}_k = \mathbf{\Phi}_k \mathbf{x}_{k-1} + \boldsymbol{\phi}_k + \mathbf{w}_k \\ \mathbf{z}_k - \mathbf{y}_k = \mathbf{H}_k \mathbf{x}_k + \mathbf{v}_k \end{cases} \tag{1.85}$$

Similar to (1.77), the EKF can be implemented by

$$\begin{cases} \hat{\mathbf{x}}_{k|k-1} = \mathbf{f}\left(\hat{\mathbf{x}}_{k-1|k-1}, \mathbf{u}_k\right) \\ \mathbf{P}_{k|k-1} = \mathbf{\Phi}_k \mathbf{P}_{k-1|k-1} \mathbf{\Phi}_k^T + \mathbf{Q}_k \\ \mathbf{K}_k = \mathbf{P}_{k|k-1} \mathbf{H}_k^T \left(\mathbf{H}_k \mathbf{P}_{k|k-1} \mathbf{H}_k^T + \mathbf{R}_k\right)^{-1} \\ \hat{\mathbf{x}}_{k|k} = \hat{\mathbf{x}}_{k|k-1} + \mathbf{K}_k\left(\mathbf{z}_k - \tilde{\mathbf{y}}_k - \mathbf{H}_k \mathbf{x}_k\right) = \hat{\mathbf{x}}_{k|k-1} + \mathbf{K}_k \tilde{\mathbf{y}}_k \\ \mathbf{P}_{k|k} = \left(\mathbf{I} - \mathbf{K}_k \mathbf{H}_k\right) \mathbf{P}_{k|k-1} \end{cases} \tag{1.86}$$

with

$$\tilde{\mathbf{y}}_k = \mathbf{z}_k - \mathbf{h}\left(\hat{\mathbf{x}}_{k|k-1}\right) \tag{1.87}$$

where the definitions of $\mathbf{P}_{k|k-1}$, $\mathbf{P}_{k|k}$, $\mathbf{K}_k$, $\mathbf{Q}_k$ and $\mathbf{R}_k$ are the same as those in (1.77). The advantage of EKF is that there is no need for approximation computation for each stage. However, the method is only effective when the estimate errors $\mathbf{x}_{k-1} - \hat{\mathbf{x}}_{k-1|k-1}$ and $\mathbf{x}_k - \hat{\mathbf{x}}_{k|k-1}$ are small. Since the matrices $\mathbf{P}_{k-1|k-1}$ and $\mathbf{P}_{k|k-1}$ are still linear approximations of the true error covariance matrices, convergence cannot always be guaranteed. Therefore in practical applications EKF does not ensure the elimination of divergence problem and some alternative filtering methods are developed, such as unscented Kalman filter and particle filter.

Loosely-Coupled Method

Loosely-coupled approach refers to a solution where measurements from one or several of the individual sensors or systems are pre-processed before they are used to obtain the final result. This approach is easy to understand and use, however, when an individual system does not have a solution, it may not be able to perform the data fusion. A loosely-coupled data fusion scheme based on KF is described as follows.

Assume that the state variables are three-dimensional position, velocity and attitude, which are combined to form a state vector $\mathbf{x} = [x\ y\ z\ v_x\ v_y\ v_z\ \alpha\ \beta\ \gamma]^T$, the measurements of MEMS are 3-axes acceleration $\mathbf{a} = [a_x\ a_y\ a_z]^T$ and 3-axes angular rate $\boldsymbol{\omega} = [\omega_x\ \omega_y\ \omega_z]^T$, and the measurement of UWB is range value. Firstly, the state equation can be strictly described according to the Newton's second law of motion as

$$\dot{\mathbf{x}}(t) = \mathbf{F}\mathbf{x}(t) + \mathbf{C}(t)\mathbf{u}(t) + \mathbf{G}(t)\mathbf{w}(t) \tag{1.88}$$

with

$$\mathbf{F} = \begin{pmatrix} \mathbf{0}_{3\times3} & \mathbf{I}_{3\times3} & \mathbf{0}_{3\times3} \\ \mathbf{0}_{3\times3} & \mathbf{0}_{3\times3} & \mathbf{0}_{3\times3} \\ \mathbf{0}_{3\times3} & \mathbf{0}_{3\times3} & \mathbf{0}_{3\times3} \end{pmatrix}, \quad \mathbf{C}(t) = \mathbf{G}(t) = \begin{pmatrix} \mathbf{0}_{3\times3} & \mathbf{0}_{3\times3} & \mathbf{0}_{3\times3} \\ \mathbf{0}_{3\times3} & \mathbf{C}_b^l(t) & \mathbf{0}_{3\times3} \\ \mathbf{0}_{3\times3} & \mathbf{0}_{3\times3} & \mathbf{C}_b^l(t) \end{pmatrix} \tag{1.89}$$

where $\mathbf{x}(t)$ is the current state of system, $\mathbf{F}$ is the constant coefficient matrix, $\mathbf{C}(t)$ is the current ICM, $\mathbf{G}(t)$ is the process noise coupling matrix, $\mathbf{C}_b^l(t)$ is the current rotation matrix from b-frame to l-frame, $\mathbf{w}(t)$ is the current system noise process, $\mathbf{I}_{3\times3}$ is the 3×3 identity matrix, and $\mathbf{0}_{3\times3}$ represents zero matrix of dimensions 3×3. According to (1.73), the discrete state equation can be written as

$$\mathbf{x}_k = \boldsymbol{\Phi}\mathbf{x}_{k-1} + \boldsymbol{\Gamma}_k\mathbf{u}_k + \mathbf{G}_k\mathbf{w}_k \tag{1.90}$$

with

$$\begin{cases} \boldsymbol{\Phi} = \mathbf{I}_{9\times9} + \Delta\tau\cdot\mathbf{F}, \\ \boldsymbol{\Gamma}_k \approx \left(\mathbf{I}_{9\times9}\cdot\Delta\tau + \dfrac{1}{2}\mathbf{F}\cdot\Delta\tau^2\right)\mathbf{C}_{k-1} \\ \mathbf{G}_k = \mathbf{C}_{k-1} \\ \mathbf{C}_{k-1} = \begin{pmatrix} \mathbf{0}_{3\times3} & \mathbf{0}_{3\times3} & \mathbf{0}_{3\times3} \\ \mathbf{0}_{3\times3} & \mathbf{C}_b^l(k-1) & \mathbf{0}_{3\times3} \\ \mathbf{0}_{3\times3} & \mathbf{0}_{3\times3} & \mathbf{C}_b^l(k-1) \end{pmatrix} \end{cases} \tag{1.91}$$

where $\boldsymbol{\Phi}$ is the STM, $\boldsymbol{\Gamma}_k$ and $\mathbf{G}_k$ are the discrete-time ICM and the noise coupling matrix at time t_k respectively, $\mathbf{w}_k$ is the system noise vector at time t_k which is assumed to be Gaussian white noise, $\mathbf{I}_{9\times9}$ is the 9×9 identity matrix, $\mathbf{C}_b^l(k-1)$ is the rotation matrix from b-frame to l-frame at time t_{k-1} and $\Delta\tau$ is the sampling interval. The statistics of $\mathbf{w}_k$ are provided by offline testing or are determined empirically. Note that the deterministic input vector $\mathbf{u}(t)$ needs to be corrected using the predetermined error models or dynamic error models as

$$\mathbf{u}(t) = \begin{pmatrix} \mathbf{0}_{3\times1} \\ \mathbf{a}\left(t\right) - \delta\mathbf{a}\left(t\right) \\ \boldsymbol{\omega}\left(t\right) - \delta\boldsymbol{\omega}\left(t\right) \end{pmatrix} \quad (1.92)$$

where $\delta\mathbf{a}(t)$ and $\delta\boldsymbol{\omega}(t)$ are the corrections of $\mathbf{a}(t)$ and $\boldsymbol{\omega}(t)$ from error models respectively, and $\mathbf{0}_{3\times1}$ is a 3×1 column vector of zeros. The parameters of dynamic error models can be estimated together with the state parameters as described in the following tightly-coupled section.

The measurement equation is given by

$$\mathbf{z}_k = \mathbf{H}\mathbf{x}_k + \mathbf{v}_k \quad (1.93)$$

with

$$\mathbf{H} = \left[\mathbf{I}_{3\times3} \quad \mathbf{0}_{3\times3} \quad \mathbf{0}_{3\times3}\right] \quad (1.94)$$

where $\mathbf{z}_k$ is the vector of UWB range measurements and $\mathbf{v}_k$ is measurement noise vector which is assumed to be Gaussian white noise and the statistics of $\mathbf{v}_k$ are obtained by measurement samples.

KF is then used to obtain the state estimate when measurements are available. It is worth noting that the selection of state variables can be different for a practical problem. The use of more state variables will result in higher computational complexity. Meanwhile, some state parameters may not be corrected under the limited types of measurements, such as the attitude in $\hat{\mathbf{x}}_{k|k-1}$. The phenomenon is easy to see if one analyzes $\mathbf{K}_k(\mathbf{z}_k - \mathbf{H}\hat{\mathbf{x}}_{k|k-1})$ or the system model in which the UWB range measurements are only related to position and position can be related to velocity but not to attitude. The overall process of loosely-coupled data fusion for UWB and MEMS is described in Figure 12. The solution results in Figure 12 include: position estimated by UWB multilateration; position, velocity and attitude estimated by INS algorithm; position, velocity and attitude estimated by KF.

Tightly-Coupled Method

A tightly-coupled approach refers to directly determine the final result using the original measurements provided by all individual sensors. Compared with loosely-coupled method, the tightly-coupled method owns more advantages. On one hand, solving the positioning result of individual system is avoided. On the other hand, tightly-coupled method is more robust in complex environments. This is because, for UWB location applications, when the number of available ranges in a complex environment is less than three for 2D positioning or four for 3D positioning, no UWB-based position estimate is produced, which does not enable data fusion for loosely-coupled method. The INS-based position estimate will soon be

Figure 12. The data fusion process of loosely-coupled method

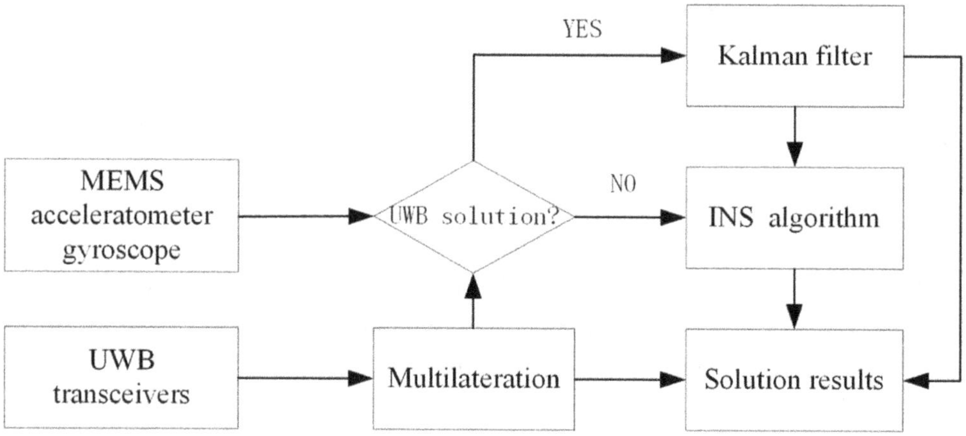

divergent without aid from UWB. However the tightly-coupled method can enable data fusion with a smaller number of range measurements than the minimum number for UWB multilateration. As a result, tightly-coupled method is more suitable for applications in complex environments.

As defined in loosely-coupled method, the state parameters are $\mathbf{x} = [x\ y\ z\ v_x\ v_y\ v_z\ \alpha\ \beta\ \gamma]^T$, the outputs of MEMS are 3-axes acceleration $\mathbf{a} = [a_x\ a_y\ a_z]^T$ and 3-axes angular rate $\boldsymbol{\omega} = [\omega_x\ \omega_y\ \omega_z]^T$. Similarly, considering the STM is time invariant, discretized state equation and measurement equation can be built as

$$\mathbf{x}_k = \boldsymbol{\Phi}\mathbf{x}_{k-1} + \boldsymbol{\Gamma}_k\mathbf{u}_k + \mathbf{G}_k\mathbf{w}_k \tag{1.95}$$

$$\mathbf{z}_k = \mathbf{h}(\mathbf{x}_k) + \mathbf{v}_k \tag{1.96}$$

where $\mathbf{x}_k$, $\boldsymbol{\Phi}$, $\boldsymbol{\Gamma}_k$, $\mathbf{u}_k$, $\mathbf{G}_k$ and $\mathbf{w}_k$ are defined the same as those in (1.90); $\mathbf{z}_k$ and $\mathbf{v}_k$ are defined the same as those in (1.93). To use the EKF, the nonlinear measurement equation is linearized as

$$\mathbf{z}_k - \mathbf{y}_k = \mathbf{H}_k\mathbf{x}_k + \mathbf{v}_k \tag{1.97}$$

with

$$\mathbf{H}_k = \left.\frac{\partial \mathbf{h}}{\partial \mathbf{x}}\right|_{\hat{\mathbf{x}}_{k|k-1}},\ \mathbf{y}_k = \mathbf{h}\left(\hat{\mathbf{x}}_{k|k-1}\right) - \left.\frac{\partial \mathbf{h}}{\partial \mathbf{x}}\right|_{\hat{\mathbf{x}}_{k|k-1}} \cdot \hat{\mathbf{x}}_{k|k-1} \tag{1.98}$$

The errors of accelerators and gyroscopes are modeled as random walks

$$\begin{aligned}\delta\mathbf{a}_{k+1} &= \delta\mathbf{a}_k + \eta_k \\ \delta\omega_{k+1} &= \delta\omega_k + \zeta_k\end{aligned} \tag{1.99}$$

where $\delta\mathbf{a}_{k+1}$ and $\delta\mathbf{a}_k$ are the biases in the output of accelerator at time t_{k+1} and t_k respectively, $\delta\omega_{k+1}$ and $\delta\omega_k$ are the biases in the output of gyroscope at time t_{k+1} and t_k respectively, η_k is the noise of accelerator, and ζ_k is the noise of gyroscope. η_k and ζ_k are assumed to be i.i.d Gaussian noise processes.

Combining (1.93) and (1.99) produces the expanded state equation as

$$\mathbf{x}'_k = \Phi'\mathbf{x}'_{k-1} + \Gamma'_k\mathbf{u}'_k + \mathbf{G}'_k\omega'_k \tag{1.100}$$

with

$$\begin{cases} \mathbf{x}'_k = \begin{pmatrix} \mathbf{x}_k \\ \delta \end{pmatrix}, \ \Phi' = \begin{pmatrix} \Phi & \mathbf{M} \\ \mathbf{0}_{6\times 9} & \mathbf{I}_{6\times 6} \end{pmatrix}, \ \Gamma'_k = \begin{pmatrix} \Gamma_k & \mathbf{0}_{9\times 6} \\ \mathbf{0}_{6\times 9} & \mathbf{I}_{6\times 6} \end{pmatrix}, \ \mathbf{u}'_k = \begin{pmatrix} \mathbf{u}_k \\ \mathbf{0}_{6\times 1} \end{pmatrix} \\ \omega'_k = \begin{pmatrix} \omega_k \\ \mathbf{e}_k \end{pmatrix}, \ \mathbf{G}'_k = \begin{pmatrix} \mathbf{G}_k & \mathbf{0}_{9\times 6} \\ \mathbf{0}_{6\times 9} & \mathbf{I}_{6\times 6} \end{pmatrix}, \ \mathbf{M} = \begin{pmatrix} \frac{1}{2}t^2\mathbf{I}_{3\times 3} & \mathbf{0}_{3\times 3} \\ t\mathbf{I}_{3\times 3} & \mathbf{0}_{3\times 3} \\ \mathbf{0}_{3\times 3} & t\mathbf{I}_{3\times 3} \end{pmatrix} \end{cases} \tag{1.101}$$

where $\delta = [\delta\mathbf{a}_k \ \delta\mathbf{w}_k]^T$, $\mathbf{e}_k = [\eta_k \ \zeta_k]^T$, $\mathbf{0}_{m\times n}$ represents the $m \times n$ zero matrix and $\mathbf{I}_{m\times n}$ represents the $m \times n$ identify matrix.

The overall process of data fusion for UWB and MEMS using tightly-coupled method is illustrated in Figure 13. The solution results in Figure 13 include: position, velocity and attitude estimated by INS algorithm; position, velocity and attitude estimated by EKF.

Performance Evaluation

Loosely-Coupled Method

An experiment was conducted in the underground garage of a library within Wuhan University. Four UWB ranging and communication devices are fixed at locations as shown in the left panel of Figure 14. The fifth UWB device and the MEMS are fixed on a small mobile cart (as shown in the right panel of Figure 14) which is pushed manually. A laptop is also placed on the cart for UWB and IMU operation and data recording. The used UWB positioning system have been described in the section "NLOS Mitigation Based on Fuzzy Theory"; the used MEMS "MTi-G-700 GPS/INS" with a typical in-run bias stability of $12° / h$ is manufactured by Xsens, which logs three-axis acceleration and angular rate at a rate of 100Hz. In the experiment, the motion state of mobile cart is divided into three stages. Firstly, moves the cart clockwise along the circumference of a reference rectangle, and after returning to the

Figure 13. The data fusion process of tightly-coupled method

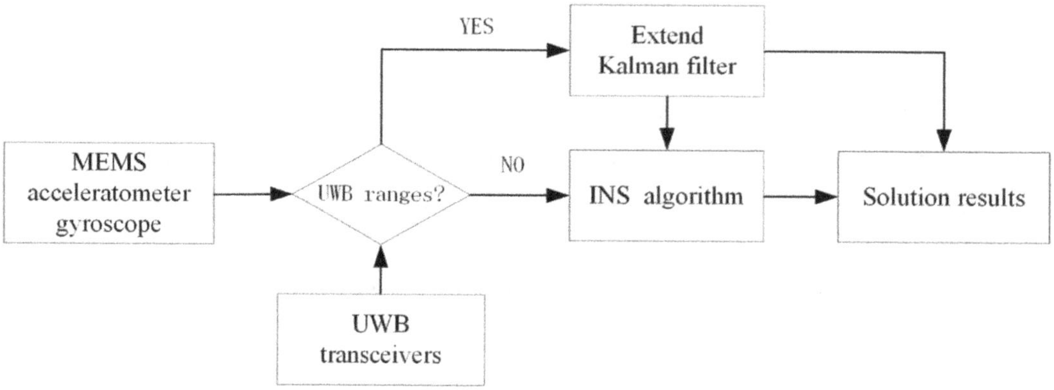

starting point moves again along the two short edges and diagonals of the reference rectangle in a figure-eight pattern. The moving procedure is repeated three times. Then, an about five minutes break is taken. Finally, the mobile cart moves again three times in the same way but in the opposite direction. Signal blockage by the body is avoided by keeping a pose lower than the height of cart. The reference rectangle and moving trajectories calculated by UWB multilateration and loosely-coupled method are shown in Figure 15. The calculated trajectories in Figure 15 are smoothed using the moving average method at time window length of one second. Note that all data involved in graphical results in this section are smoothed in the same way, while all statistics are calculated using the original results.

To measure the positional error (i.e. the distance between the true and estimated positions), the true position of the mobile unit on the reference trajectory is needed. However it is difficult to find a device which can record a more accurate dynamic tracking trajectory. In fact, the UWB is often seen as the most accurate available low-cost positioning system indoors. Although a strategic-grade IMU can be used to achieve the goal, it is extremely expensive. Therefore a method based on the measurement of cross-track error is used for positional error estimation (Sharp, Yu & Sathyan, 2012). The method can obtain roughly cumulative distribution function (CDF) of 2D positional errors using the probability density function (PDF) of cross-track errors. Meanwhile, statistics of 2D positional errors, such as,

Figure 14. Experimental scene and setup

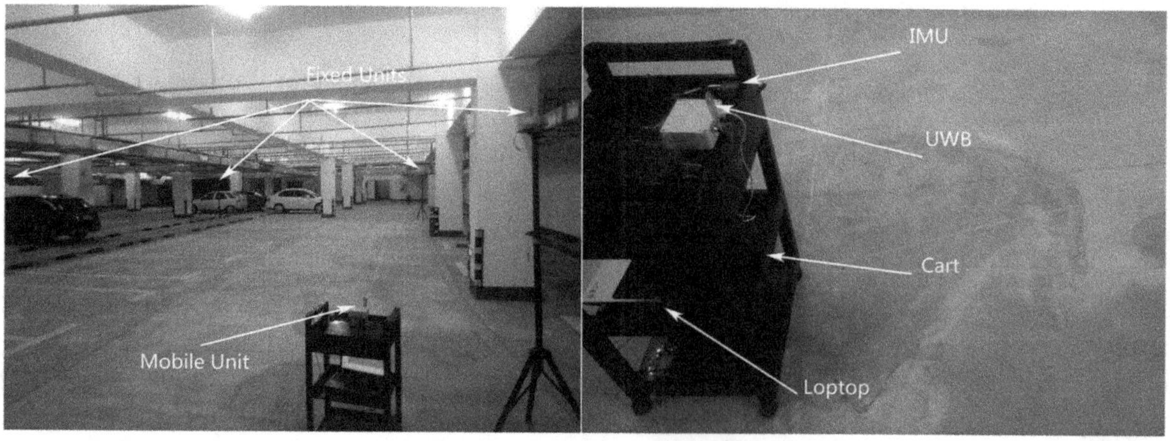

Figure 15. 2D trajectories calculated by UWB multilateration and loosely-coupled method

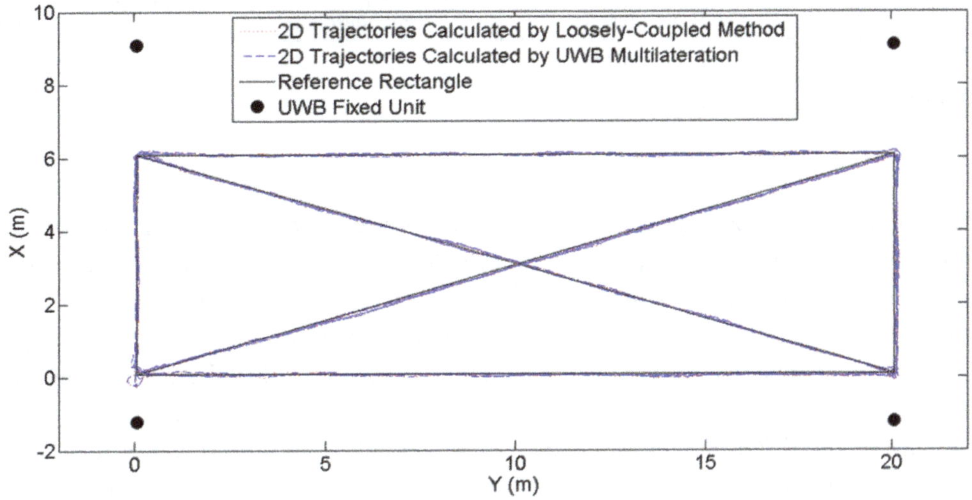

mean, standard deviation and RMSE, can be estimated based on cross-track errors as well. Figure 16 shows the CDF of estimated positional errors. The results show that the positional error distribution of loosely-coupled method has a similar pattern to that of UWB multilateration. For positional errors below 0.1 meter, the CDF for UWB multilateration is about 92%, while that is about 85% for loosely-coupled method. Meanwhile, statistics of estimated positional errors are shown in Table 2, which indicate that the loosely-coupled method performs slightly inferior as well.

Figure 16. CDFs of estimated positional errors
(PDF of measured cross-track errors is fitted by Gaussian model)

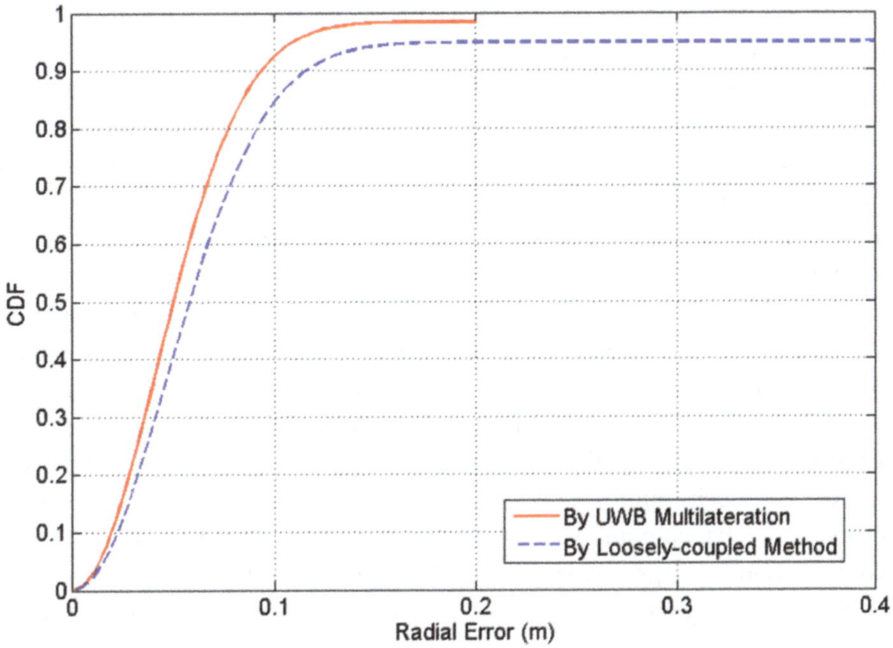

Table 2. Mean and standard deviation of estimated positional errors and RMSE of positions

	Mean (m)	Standard Deviation (m)	RMSE (m)
UWB multilateration	0.066	0.041	0.078
Loosely-coupled method	0.074	0.048	0.088

Velocities calculated by UWB multilateration and loosely-coupled method are shown in Figure 17. 2D velocities by two methods are compared in Figure 18, in which the dotted line is the mean velocity of the whole trajectory. In this experiment, the cart was pushed at a speed as steady as possible and thus the mean value can roughly reflect the true velocity. However, around the corner, sharp changes in velocity may occur. It seems that both the calculated velocities are slightly larger than the true velocity, revealed clearly by the phenomena that the calculated velocities both are non-zero during the static state.

Assuming $\Delta\mathbf{x}$ is the difference between calculated velocities and the mean of actual velocity and $\Delta\mathbf{y}$ is the difference between the mean of actual velocity and true velocities, therefore $\Delta\mathbf{x}+\Delta\mathbf{y}$ is the difference between calculated velocities and true velocities. Meanwhile, assuming $\Delta\mathbf{x}$ is independent of $\Delta\mathbf{y}$, the mean and standard deviation of $\Delta\mathbf{x}+\Delta\mathbf{y}$ are

$$\begin{cases} u_{\Delta x+\Delta y} = u_{\Delta x} + u_{\Delta y} \\ \sigma_{\Delta x+\Delta y} = \sqrt{\sigma_{\Delta x}^2 + \sigma_{\Delta y}^2} \end{cases} \tag{1.102}$$

where $u_{\Delta x}$ and $u_{\Delta y}$ are the means of $\Delta\mathbf{x}$ and $\Delta\mathbf{y}$ respectively, and $\sigma_{\Delta x}$ and $\sigma_{\Delta y}$ are the standard deviations of $\Delta\mathbf{x}$ and $\Delta\mathbf{y}$ respectively. According to (1.102), it is clearly that the mean and standard deviation of $\Delta\mathbf{x}$ can reflect the mean and standard deviation of $\Delta\mathbf{x}+\Delta\mathbf{y}$, although $\sigma_{\Delta x}$ is less than $\sigma_{\Delta x+\Delta y}$. Therefore, we can see from Table 3 that the UWB-based velocity estimate is slightly better than that based on loosely-coupled method. However, UWB-based velocity is calculated by two adjacent position samples resulting in that the update frequency of velocity is limited by UWB sampling rate, while loosely-coupled method can be used for some situations with a high update velocity.

Figure 19 compares the attitude estimate performance between loosely-coupled method and INS algorithm. It is mentioned early (see section "Loosely-coupled Method ") that the calculated attitude by the coupled method (whether loosely-coupled or tightly-coupled method) is not corrected by data fusion algorithm and is independent of the UWB range measurements. The slight difference between attitude by loosely-coupled method and that by INS algorithm is only caused by different solution processes.

As shown in Figure 16 and Table 2, the positioning accuracy by UWB multilateration is slightly superior to that by loosely-coupled method. It seems to be a paradoxical conclusion with an expected improvement in positioning accuracy mentioned in the beginning of this section. Actually, only in some special cases, where the positioning accuracy achieved by UWB multilateration is lower than that by INS algorithm in a very short time (the duration must be enough short to guarantee a valid restriction for INS divergence), the expected positioning accuracy improvement can be achieved. Meanwhile, to achieve the objective, an adaptive KF technique is required to tune the noise parameters of UWB measurements. However, due to the general low precision of the MEMS used, the loosely-coupled method is usually used for other goal of achieving similar but not higher positioning accuracy when UWB encounters an

Figure 17. (a). Velocity estimate by UWB multilateration (b). Velocity estimate by loosely-coupled method

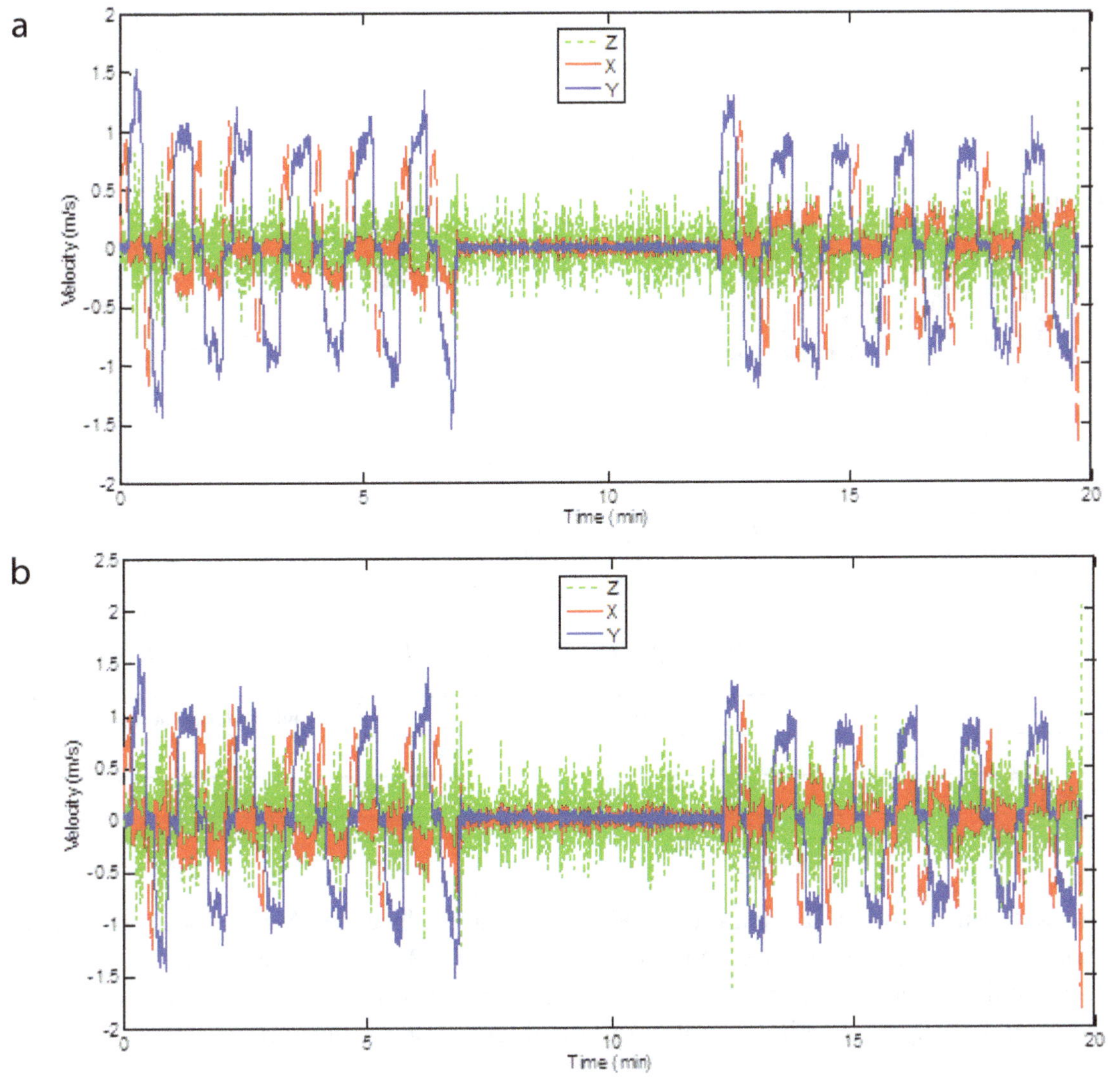

outage for a short time. Therefore, in general, we directly use the UWB solution for positioning, and fusion solution is only used in the case where UWB multilateration is failed. Figure 20 shows the calculated 2D trajectories when UWB encounters an outage for five different time durations. The result show that when outage period is 0.5s, 1s, 3s, and 5s, the calculated trajectory deviates from that by UWB multilateration by up to around 0.1m, 0.3m, 1m, and 1.6m, respectively. This is in accordance with the expectation that the deviation quickly increases with increased UWB outage period.

Assume the case that UWB only can obtain a valid position estimate after the duration of 0.5s, during which UWB fails to provide a position estimate and the position is calculated by INS algorithm. Figure 21 shows the comparison between CDF of positional error estimates in this case and that by UWB multilateration at a normal interval of 0.2s. Statistics of estimated positional errors are given in Table 4. The

Figure 18. Comparison between 2D velocity estimates by different methods

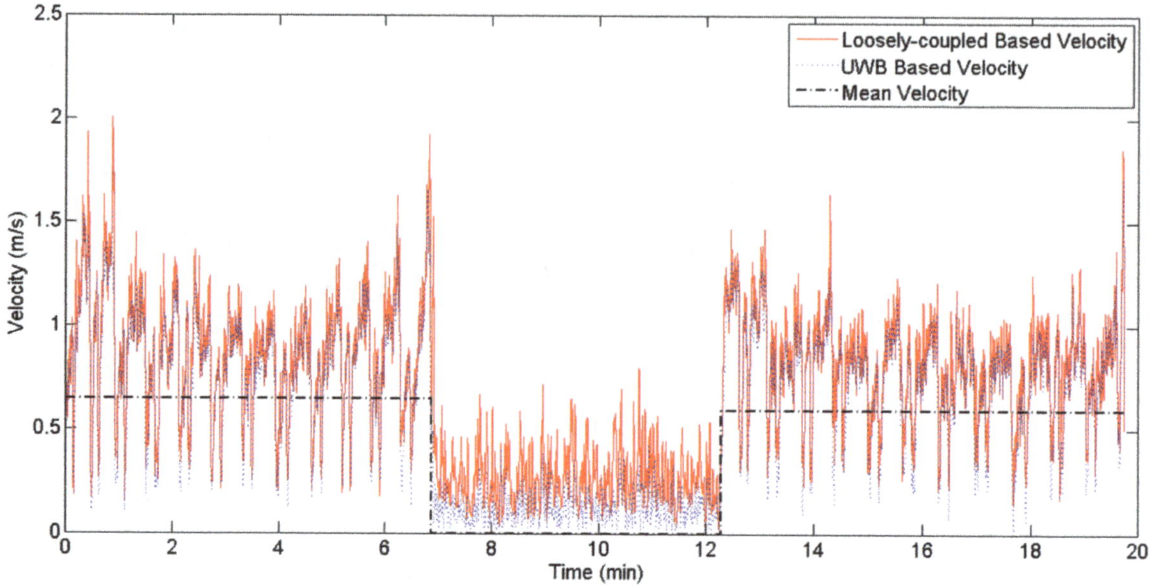

Table 3. Mean and standard deviation of differences between calculated velocities and mean velocity

	Mean (m/s)	standard deviation (m/s)
UWB based	0.595	0.400
Loosely-coupled based	0.688	0.420

results show that the positioning accuracy is obviously degraded since the RMSE of positions is increased from 0.08 to 0.16 meters. However even in the presence of UWB outage the estimated positional errors nearly are within 0.5 meters and about 80% of errors are under 0.1 meter, which indicate the positioning accuracy would satisfy the requirement of most of high precision applications.

Tightly-Coupled Method

The same data used in the evaluation of loosely-coupled method is used to evaluate the tightly-coupled method. Figure 22 shows calculated 2D trajectories using the tightly-coupled method when four different numbers of ranges are used separately. In the case where only one or two UWB devices provide valid distance measurements, the method does not produce valid position estimate, although the positional error is not greater than some value. The CDFs of estimated positional errors when using four or three ranges are shown in Figure 23. It can be roughly seen that the method performs well when either three or four UWB devices are able to provide valid distance measurements, although the positioning accuracy with four devices is slightly better. As expected, the CDF of estimated positional errors below a certain value with four ranges is slightly larger than that with three ranges. The statistics of estimated positional errors are given in Table 5, which indicate that the positioning performance of tightly-coupled method with three and four ranges are similar.

Figure 19. (a). Attitude estimate by INS algorithm (b). Attitude estimate by loosely-coupled method

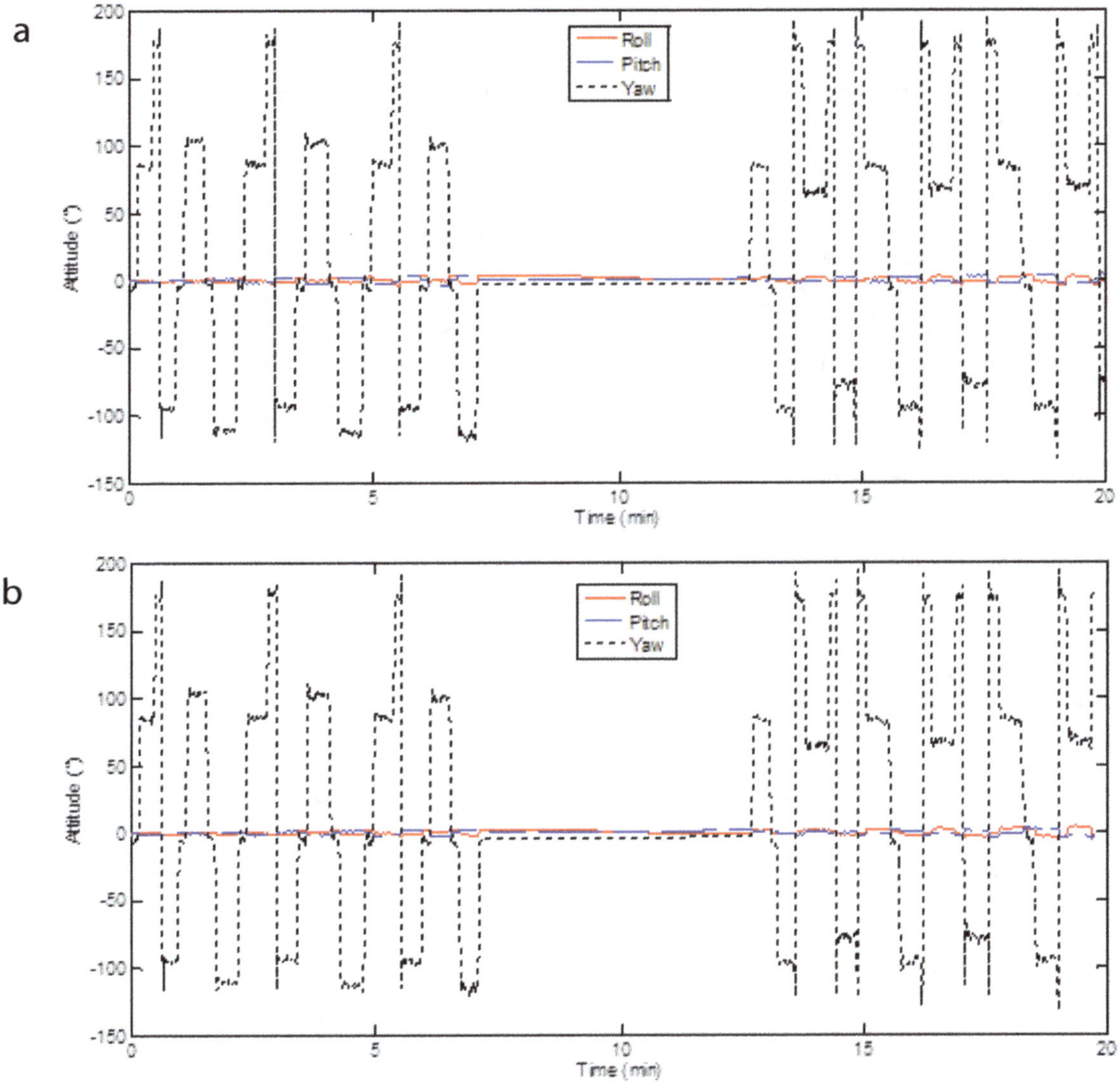

To compare positioning performance in a fair way, the same output error models of accelerometer and gyroscope are used for both loosely-coupled and tightly-coupled methods. According to Figure 23 and Table 5, it is seen that the positioning accuracy of tightly-coupled method whether four or three UWB devices provide valid distance measurements are both slightly better than that of loosely-coupled method, but are both inferior to that of UWB multilateration. Similarly to loosely-coupled method, the tightly-coupled method is generally used for some cases where the number of valid UWB range measurements is less than the minimum required or for cases where attitude information or high-frequency position information is needed rather than for improving the positioning accuracy.

Other parameters calculated by tightly-coupled method under four valid distance measurements (i.e. velocity, attitude and bias) are shown in Figure 24. The velocity and attitude estimates can roughly reflect

Figure 20. 2D trajectories calculated by UWB multilateration and loosely-coupled method when UWB has an outage for different time durations

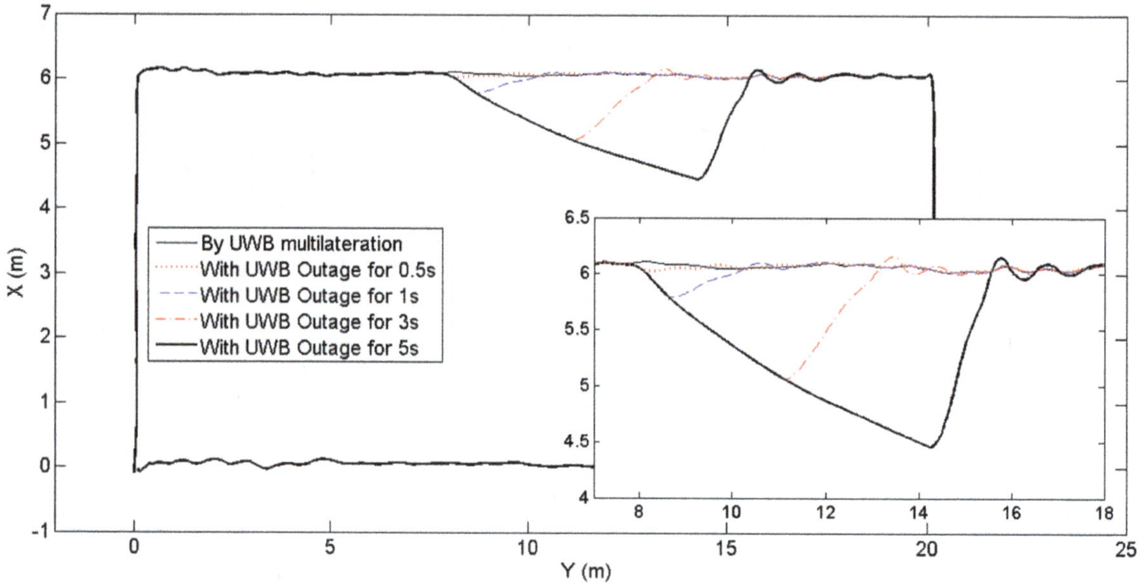

Figure 21. CDFs of estimated positional errors
(PDF of measured cross-track errors is fitted by Gaussian model)

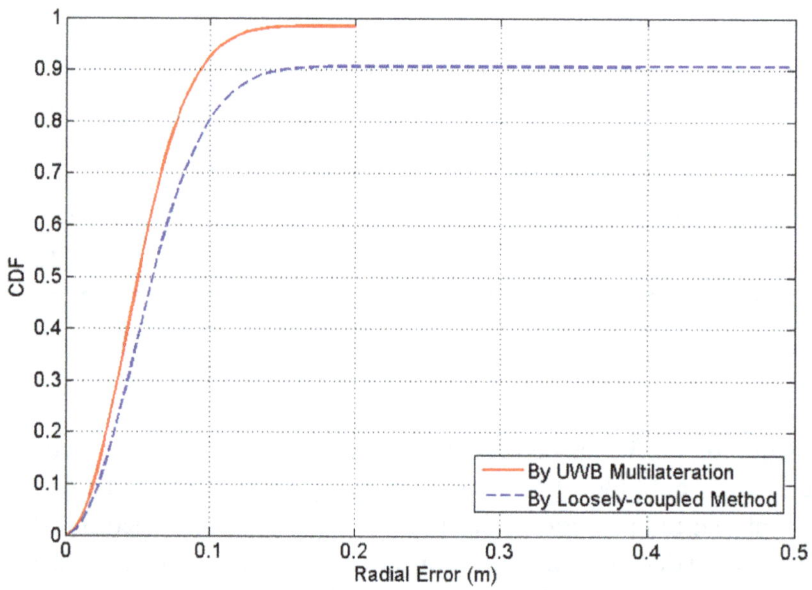

Table 4. Mean and standard deviation of estimated positional errors and RMSE of positions

	Mean (m)	standard deviation (m)	RMSE (m)
By UWB multilateration	0.066	0.041	0.078
By loosely-coupled method	0.121	0.108	0.162

Figure 22(a). 2D trajectories calculated when the number of ranges is 4 and 3 (b). 2D trajectories calculated when the number of ranges is 2 and 1

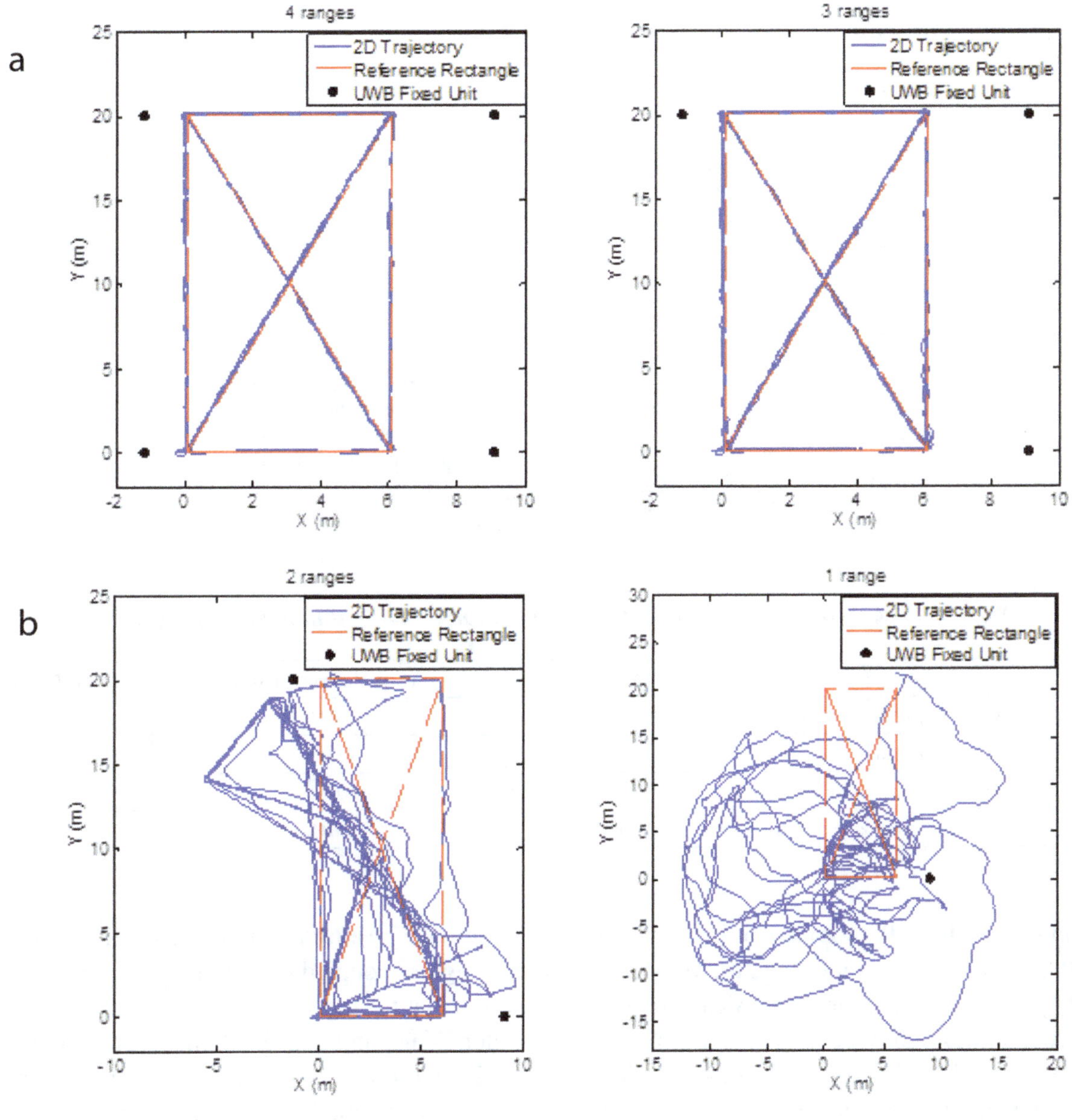

the actual situation by an intuitive judgment based on the physical scene. It can be seen that the acceleration bias in x-axis grows greatly with time, up to 0.5 m / s^2 within 20 minutes. The z-axis bias is completely negligible and the y-axis component changes slowly around -0.1 m / s^2. Each of the three bias components of the gyroscope is a constant which is smaller than 0.2 degrees over the duration of 20 minutes. The method is invalid to estimate the variation of angular rate bias because the UWB range measurements do not provide correction information for angular rate bias as for attitude estimate.

Figure 23. CDFs of estimated positional errors
(PDF of measured cross-track errors is fitted by Gaussian model)

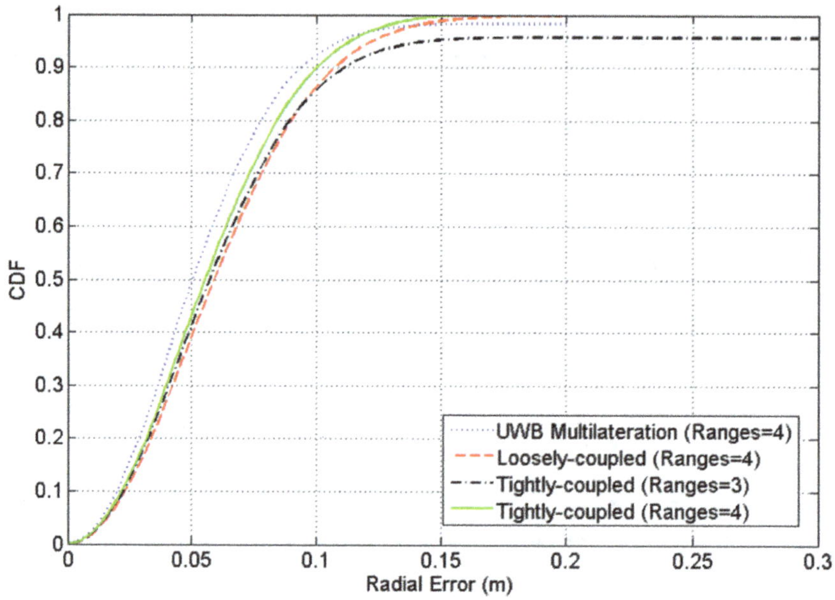

Table 5. Mean and standard deviation of estimated positional errors and RMSE of positions

	Mean (m)	standard deviation (m)	RMSE (m)
Tightly-coupled (4 ranges)	0.074	0.043	0.085
Tightly-coupled (3 ranges)	0.072	0.048	0.087
Loosely-coupled (4 ranges)	0.077	0.048	0.091
UWB multilateration (4 ranges)	0.066	0.041	0.078

Figure 25 shows the comparison between 2D velocity estimates by loosely-coupled and tightly-coupled methods. During the static state, the velocity estimate by tightly-coupled method is closer to zero than that by loosely-coupled method. Meanwhile, Table 6 shows that the mean and standard deviation of velocity errors by tightly-coupled method are slightly smaller. Which indicate that tightly-coupled method is better than loosely-coupled method in velocity estimation although the difference between velocities by two methods is minor. Note that the attitude information by two methods are calculated in the same way, hence they are exactly the same.

CONCLUSION

This chapter investigated positioning in complex indoor environments with a focus on the integration of UWB and MEMS. The fundamental principles of UWB and MEMS positioning were studied. Loosely-coupled and tightly-coupled schemes were developed based on KF and EKF, respectively. Also, a NLOS mitigation technique was introduced for UWB positioning in NLOS environments. Two experiments were

Figure 24. (a). Calculated velocity information (b). Calculated attitude information (c). Calculated bias information

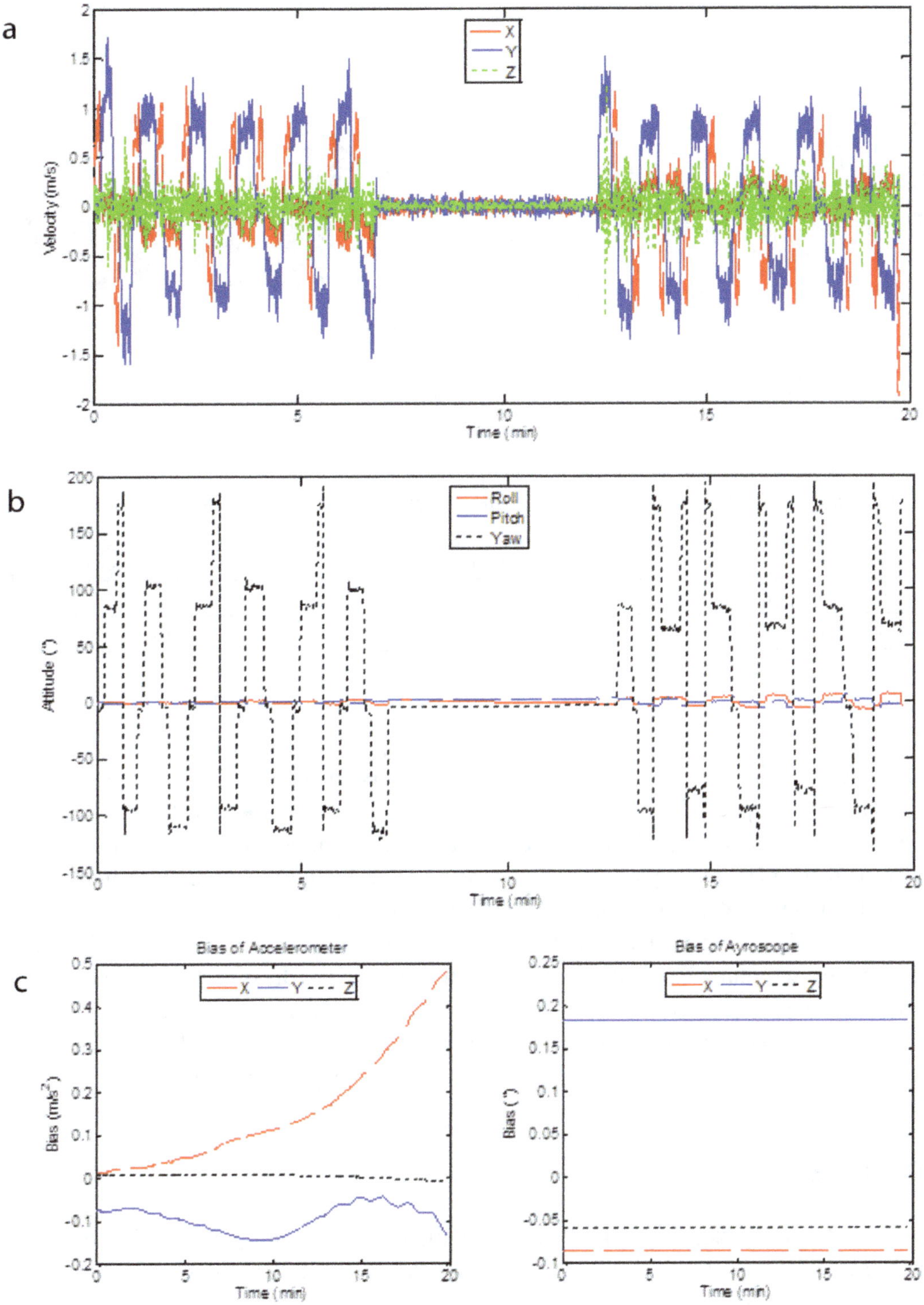

Figure 25. Comparison between 2D velocity estimates by different methods

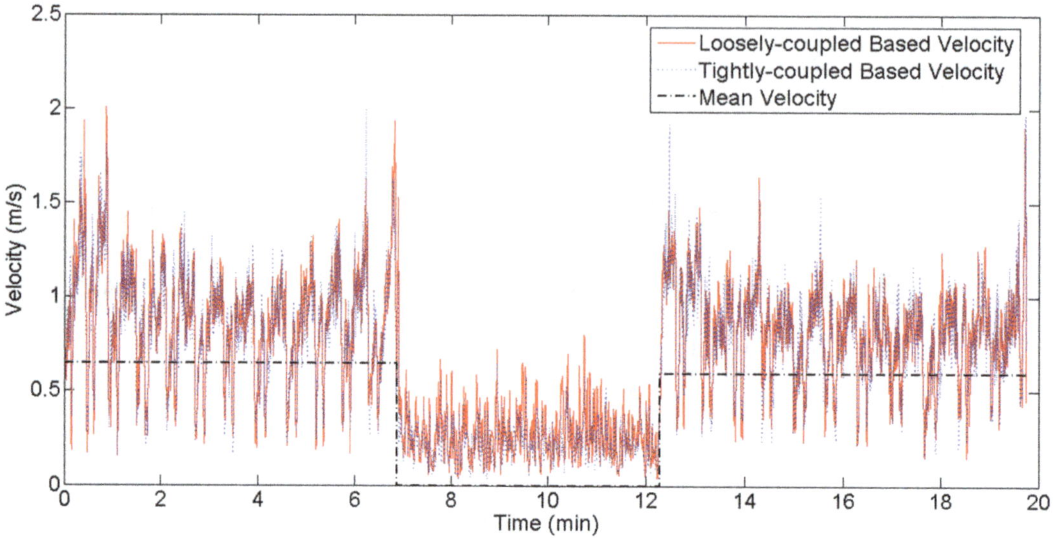

Table 6. Mean and standard deviation of differences between calculated velocities and mean velocity

	Mean (m/s)	standard deviation (m/s)
Loosely-coupled based	0.679	0.420
Tightly-coupled based	0.657	0.405

conducted in a teaching building and an underground garage for performance evaluation of the NLOS mitigation technique and the two coupled methods, respectively. The results from the first experiment show that using the proposed NLOS mitigation technique, RMSE of measured ranges can be reduced from 2.58 to 0.25 meter. The results from the second experiment demonstrate that the RMSE of estimated positional errors by loosely-coupled is 9.1cm, and that by tightly-coupled method are 8.5cm and 8.7cm when the number of valid range measurements is four and three, respectively. The combination of the proposed NLOS mitigation technique and the tightly-coupled method is currently under investigation. There are still a number of other issues which need to be addressed. For instance, the proposed NOLS mitigation technique needs to be enhanced to improve accuracy and usability. New techniques are required to tackle the divergence problem of EKF caused by inaccurate UWB range measurements. These problems are worth further investigation, which are the ongoing research issues of the authors.

REFERENCES

Bancroft, J. B., & Lachapelle, G. (2011). Data fusion algorithms for multiple inertial measurement units. *Sensors (Basel)*, *11*(7), 6771–6798. doi:10.3390/s110706771 PMID:22163985

Bebek, O., Suster, M. A., Rajgopal, S., Fu, M. J., Huang, X., Cavusoglu, M. C., & Mastrangelo, C. H. et al. (2010). Personal navigation via high-resolution gait-corrected inertial measurement units. *IEEE Transactions on Instrumentation and Measurement*, *59*(11), 3018–3027. doi:10.1109/TIM.2010.2046595

Borrás, J., Hatrack, P., & Mandayam, N. B. (1998). Decision theoretic framework for NLOS identification. In *Proceedings of Vehicular Technology Conference* (vol. 2, pp.1583-1587). Piscataway, NJ: IEEE.

Casas, R., Marco, A., Guerrero, J. J., & Falcó, J. (2006). Robust estimator for non-line-of-sight error mitigation in indoor localization. *EURASIP Journal on Applied Signal Processing*, *2006*(1), 1–8. doi:10.1155/ASP/2006/43429

El-Sheimy, N., Hou, H., & Niu, X. (2007). Analysis and modeling of inertial sensors using Allan variance. *IEEE Transactions on Instrumentation and Measurement*, *57*(1), 140–149. doi:10.1109/TIM.2007.908635

Federal Communications Commission. (2002). *Revision of part 15 of the commission's rules regarding ultra-wideband transmission systems, first report and order* (ET Docket 98-153). Adopted Feb.14, 2002, Released Apr. 22, 2002. Author.

Marano, S., Gifford, W. M.,Wymeersch, H., & Win, M. Z., (2010). NLOS identification and mitigation for localization based on UWB experimental data. *IEEE Journal on Selected Areas in Communications*, *28*(7), 1026–1035. doi:10.1109/JSAC.2010.100907

Greenspan, R. L. (1995). Inertial navigation technology from 1970–1995. *Navigation*, *42*(1), 165–185. doi:10.1002/j.2161-4296.1995.tb02334.x

Grewal, M. S., & Andrews, A. P. (Eds.). (2008). *Kalman filtering: theory and practice using MATLAB* (3rd ed.). Piscataway, NJ: Wiley Blackwell. doi:10.1002/9780470377819

Guo, Q., Bebek, O., Cavusoglu, M. C., Mastrangelo, C. H., & Young, D. J. (2015). A personal navigation system using MEMS-based high-density ground reaction sensor array and inertial measurement unit. In *Proceedings of Transducers - 2015, International Conference on Solid-State Sensors, Actuators and Microsystems* (pp.1077-1080). Piscataway, NJ: IEEE. doi:10.1109/TRANSDUCERS.2015.7181113

Gustafson, D. E., Elwell, J. M., & Soltz, J. A. (2006). Innovative indoor geolocation using RF multipath diversity. In *Proceedings of Position, Location, and Navigation Symposium, 2006 IEEE/ION* (vol.2006, pp.904-912). Piscataway, NJ: IEEE. doi:10.1109/PLANS.2006.1650690

Guvenc, I., Chong, C. C., Watanabe, F., & Inamura, H. (2008). NLOS identification and weighted least-squares localization for UWB systems using multipath channel statistics. *EURASIP Journal on Advances in Signal Processing*, (1): 36.

Jo, Y. H., Lee, J. Y., Ha, D. H., & Kang, S. H. (2006). Accuracy enhancement for UWB indoor positioning using ray tracing. In *Proceedings of Position, Location, and Navigation Symposium, 2006 IEEE/ION* (vol.2006, pp.565-568). Piscataway, NJ: IEEE.

Khodjaev, J., Park, Y., & Malik, A. S. (2010). Survey of NLOS identification and error mitigation problems in UWB-based positioning algorithms for dense environments. *Annals of Telecommunications -. Annales des Télécommunications*, *65*(5-6), 301–311. doi:10.1007/s12243-009-0124-z

Lakhzouri, A., Lohan, E. S., Hamila, R., & Renfors, M. (2003). Extended Kalman filter channel estimation for line-of-sight detection in WCDMA mobile positioning. *EURASIP Journal on Advances in Signal Processing*, *2003*(13), 1268–1278. doi:10.1155/S1110865703306018

Li, X., He, J., Xu, L., & Wang, Q. (2016). The effect of multipath and NLOS on TOA ranging error and energy based on UWB. In *Proceedings of IEEE International Conference on Consumer Electronics-Taiwan* (pp.1-2). Piscataway, NJ: IEEE. doi:10.1109/ICCE-TW.2016.7520939

Martin, H., Groves, P., Newman, M., & Faragher, R. (2013). A new approach to better low-cost MEMS IMU performance using sensor arrays. *Proceedings of ION GNSS*, 2125-2142.

Nguyen, T. V., Jeong, Y., Shin, H., & Win, M. Z. (2015). Machine learning for wideband localization. *IEEE Journal on Selected Areas in Communications*, *33*(7), 1357–1380. doi:10.1109/JSAC.2015.2430191

Savic, V., Larsson, E. G., Ferrer-Coll, J., & Stenumgaard, P. (2015). Kernel methods for accurate UWB-based ranging with reduced complexity. *IEEE Transactions on Wireless Communications*, *15*(3), 1783–1793. doi:10.1109/TWC.2015.2496584

Sharp, I., Yu, K., & Sathyan, T. (2012). Positional accuracy measurement and error modeling for mobile tracking. *IEEE Transactions on Mobile Computing*, *11*(6), 1021–1032. doi:10.1109/TMC.2011.119

Shimizu, H., Masui, H., Ishii, M., & Sakawa, K. (2000). LOS and NLOS path-loss and delay characteristics at 3.35 GHz in a residential environment. In *Proceedings of Antennas and Propagation Society International Symposium* (vol. 2, pp. 1142-1145). Piscataway, NJ: IEEE. doi:10.1109/APS.2000.875424

Silva, B., & Hancke, G. P. (2016). IR-UWB-based non-line-of-sight identification in harsh environments: Principles and challenges. *IEEE Transactions on Industrial Informatics*, *12*(3), 1188–1195. doi:10.1109/TII.2016.2554522

Skog, I., Nilsson, J. O., & Handel, P. (2014). An open-source multi inertial measurement unit (MIMU) platform. In *Proceedings of International Symposium on Inertial Sensors and Systems* (Vol.20, pp.1-4). Piscataway, NJ: IEEE. doi:10.1109/ISISS.2014.6782523

Sorenson, H. W. (1970). Least-squares estimation: From Gauss to Kalman. *Spectrum IEEE*, *7*(7), 63–68. doi:10.1109/MSPEC.1970.5213471

Titterton, D., & Weston, J. (2005). Strapdown inertial navigation technology. *Aerospace & Electronic Systems Magazine IEEE*, *20*(7), 33–34. doi:10.1109/MAES.2005.1499250

Venkatraman, S., & Caffery, J. (2002). Statistical approach to non-line-of-sight BS identification. In *Proceedings of International Symposium on Wireless Personal Multimedia Communications* (vol.1, pp.296-300). Piscataway, NJ: IEEE. doi:10.1109/WPMC.2002.1088180

Woodman, O. J. (2007). *An introduction to inertial navigation*. Technical report, University of Cambridge Computer Laboratory, Cambridge CB3 OFD, United Kingdom, UCAM-CL-TR-696, ISSN 1476-2986.

Xiao, Z., Wen, H., Markham, A., Trigoni, N., Blunsom, P., & Frolik, J. (2015). Non-line-of-sight identification and mitigation using received signal strength. *IEEE Transactions on Wireless Communications*, *14*(3), 1689–1702. doi:10.1109/TWC.2014.2372341

Yu, K., & Dutkiewicz, E. (2013). NLOS identification and mitigation for mobile tracking. *IEEE Transactions on Aerospace and Electronic Systems, 49*(3), 1438–1452. doi:10.1109/TAES.2013.6557997

Yu, K., & Oppermann, I. (2004). Positioning. In I. Oppermann, M. Hämäläinen, & J. Iinatti (Eds.), *UWB theory and applications* (pp. 175–196). Hoboken, NJ: John Wiley & Sons, Ltd.

Yu, K., Sharp, I., & Guo, Y. J. (Eds.). (2009). *Ground-based wireless positioning*. Piscataway, NJ: Wiley-IEEE Press. doi:10.1002/9780470747988

ENDNOTES

1 Note that in this chapter the mentioned UWB is limited in UWB technology based on impulse radio.

2 Kalman filter was early used in the trajectory estimation and control problem for the Apollo project at the Ames Research Center of NASA.

Chapter 9
Navigation Based on Sensors in Smartphones

Guenther Retscher
TU Wien, Austria

Allison Kealy
RMIT University, Australia

ABSTRACT

With the increasing ubiquity of smartphones and tablets, users are now routinely carrying a variety of sensors with them wherever they go. These devices are enabling technologies for ubiquitous computing, facilitating continuous updates of a user's context. They have built-in MEMS-based accelerometers for ubiquitous activity monitoring and there is a growing interest in how to use these together with gyroscopes and magnetometers to build dead reckoning (DR) systems for location tracking. Navigation in complex environments is needed mainly by consumer users, private vehicles, and pedestrians. Therefore, the navigation system has to be small, easy to use, and have reasonably low levels of power consumption and price. The technologies and techniques discussed here include the fusion of inertial navigation (IN) and other sensors, positioning based on signals from wireless networks (such as Wi-Fi), image-based methods, cooperative positioning systems, and map matching (MM). The state-of-the-art of MEMS-based location sensors and their integration into modern navigation systems are also presented.

INTRODUCTION

Modern smartphones are equipped with a large number of sensors, including Global Navigation Satellite Systems (GNSS), Wireless Fidelity (Wi-Fi; a.k.a. WLAN), Bluetooth, accelerometers, magnetometers, cameras, microphones, and health monitors. These sensors are natural candidates for sensing the context in which a user is situated. Automatic access to such context information can be exploited towards localization. The increasing ubiquity of location-aware devices, fueled by the consumer smartphone market has pushed the need for robust GNSS-like positioning capabilities in GNSS difficult environments. In particular, Micro-Electro-Mechanical Systems (MEMS) technology has revolutionised user localization and MEMS inertial sensors and systems have become indispensable to the future of navigation. Navigation

DOI: 10.4018/978-1-5225-3528-7.ch009

itself has become much broader than just providing simple solutions to LBS (Location Based Services) queries such as "Where am I?" or "How to get from a start point A to a destination B?" It has moved into new areas such as games, geolocation, mobile mapping, virtual reality, tracking, health monitoring and context awareness. These functionalities are being further enhanced by ongoing progress in MEMS technology. Their small size, low power and weight, and low cost have led to increased use, new applications, increased mobility, increased integration (hence better performance) and extended operation. When integrated with GNSS and other sensors, the integrated system enhances the localization performance in GNSS denied environments where the satellite signal is either totally blocked or attenuated. The combination of the two systems exploits their complementary characteristics. These integrated navigation technologies and methods have become indispensable in many applications like car navigation, human motion modelling, first-responder personal navigation, UAV (Unmanned Aerial Vehicle) navigation, and portable navigation. In this chapter the emphasis is on applicability and performance for navigation of a mobile client. It provides details of the performance capabilities of MEMS-based sensors in combination with wireless positioning technologies for localization of a mobile client and gives indicative performance levels as well as inherent limitations. It is emphasised that no single sensor or technique is currently able to meet the stringent positioning requirements for the increasing number of safety and liability critical mass market applications. Integration and sensor fusion of MEMS and wireless technologies is a necessity to guarantee an acceptable level of performance. Thus, this chapter reviews the range of MEMS sensors and wireless positioning techniques available in current mobile devices, such as smartphones and tablets.

The chapter is structured as follows: Firstly, the background of MEMS sensor technology is briefly discussed followed by a detailed description of the operational principle of MEMS-based location sensors. Then, individual sensors based localization techniques and systems are described, including dead reckoning (DR), inertial navigation (IN), map matching and Kalman filtering for continuous user localization. Also, activity detection and altitude determination using barometric pressure sensors are studied. Next, the combination with other techniques, such as wireless options (Wi-Fi and Bluetooth) and vision-aided sensors (digital cameras), are discussed. Sensor fusion techniques and cooperative positioning are briefly studied. Finally, a brief description of typical applications, i.e., pedestrian and vehicle navigation is presented, and a summary concludes this chapter.

MEMS-BASED LOCATION SENSORS

MEMS are miniaturized mechanical and electro-mechanical devices and structures that are made using micro-fabrication techniques. They integrate electrical and mechanical components with feature sizes in the micrometer-scale, which can be fabricated using integrated circuit batch-processing technologies (Gad-el-Hak, 2001). MEMS have allowed the development of several microdevices such as accelerometers (Li et al., 2011a), gyroscopes (Che et al., 2010), micromirrors (Li et al., 2011b), and pressure sensors (Mian and Law, 2010). The physical dimensions of MEMS devices can range from below 1 µm to several mm. The types of MEMS devices vary from relatively simple structures having no moving elements, to complex electromechanical systems with multiple moving elements under the control of integrated microelectronics. MEMS assembly involves joining two or more components to form a MEMS device. A basic requirement of automated micro-assembly is that the system must be able to transport micro-scale parts and components and to manipulate them so that precise spatial relations (with micro-scale tolerances) can be established. For example, in die alignment and parts insertion or in packaging processes such as die bonding, device sealing, etc., the operating tolerances are very small and therefore, assembly and positioning tasks require very high precision and repeatability. In addition, assembly techniques for the macro domain do not directly translate to the micro-domain because of the dominance of effects such as Van der Waals forces, surface tension, and electrostatic forces (Hsieh, 2013). Apart from size reduction, MEMS technology offers many benefits such as batch production and cost reduction, lightweight, low-power consumption, ruggedization, design flexibility, high sensitivity and high resolution within limits (Herrera-May et al., 2009). The reduction in size of the sensing elements, however, creates challenges for attaining good performance. In this section a selection of sensors relevant for positioning are discussed in detail.

Current smartphones have several embedded sensors that can be used for localization. In Android devices, for instance, the sensors are accessed and managed by the SensorManager class of API programs. If a sensor is initialized the SensorManager delivers data to the system (Vogel, 2016). An accelerometer and magnetic field sensor are standard sensors in current generation mobile devices. Android has provided a virtual orientation sensor up to version 2.2. The disadvantage of this orientation sensor was that the determination of the azimuth (heading of the user) was computational intensive and not very accurate if the roll value of the smartphone was not zero. Nowadays, gyroscopes are routinely found in mobile devices and they can be meaningfully combined with the magnetometer. The obtained sensor orientation values refer to the measured data in the respective axes of the device coordinate frame as illustrated in Figure 1 on the left. As illustrated, the Y-axis can be oriented to point in the forward direction, the Z-axis upwards and the X-axes form a right-handed coordinate system with the other two axes. The orientation sensor axes are dependent on the smartphone's location in space which is represented in a global coordinate system (see Figure 1 on the right). Then the orientation of the device in the global coordinate system is represented by a rotation matrix that is a function of the attitude parameters, i.e.,

Figure 1. Smartphone coordinate frame (left) and global coordinate system (right)

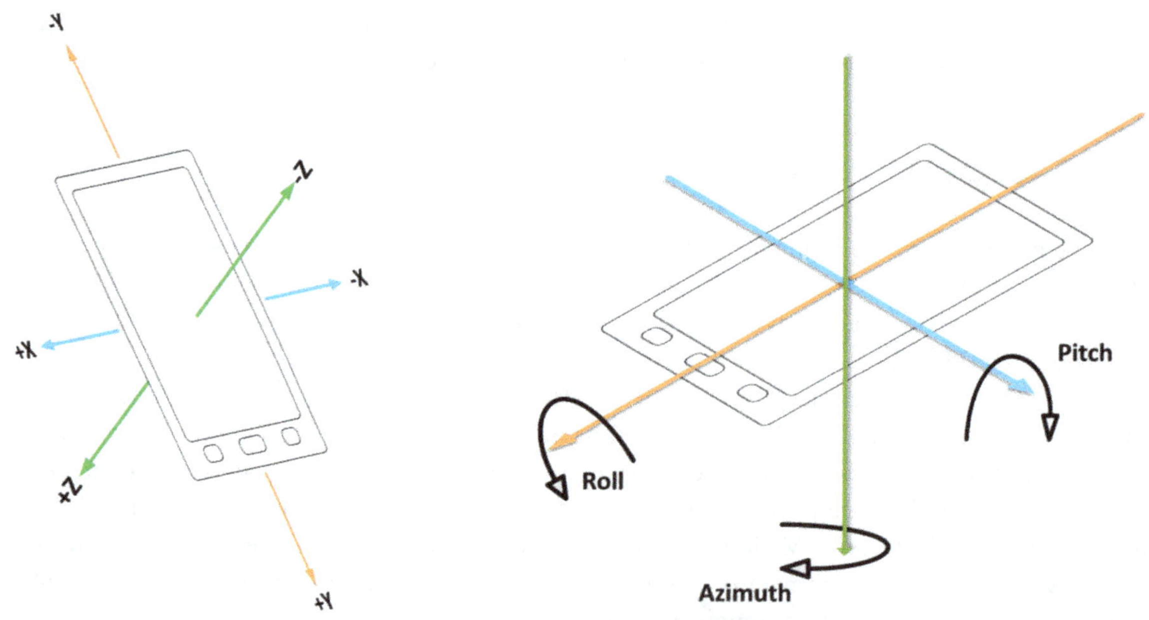

azimuth (heading), pitch and roll. These three angles can be measured dynamically by MEMS orientation sensors. For a better understanding of the operational principles of the smartphone sensors, their characteristics and functioning are described more in detail in the following.

Accelerometers

MEMS accelerometers are either the pendulous/displacement mass type or the resonator type (Yazdi et al., 1998). The first types use closed-loop capacitive sensing and electrostatic forcing whereas the second types are based on the resonance principle of operation. Both of them can detect acceleration in two primary ways: (1) the displacement of a hinged or flexure-supported proof mass under acceleration results in a change in a capacitive or piezoelectric readout or (2) the change in frequency of a vibrating element is caused by a change in the element's tension induced by a change of loading from a seismic proof mass. The force rebalance pendulous or lateral displacement accelerometers are a monolithic silicon structure consisting of a torsional pendulum with capacitive readout and an electrostatic torque. The pendulum is supported by a pair of flexure pivots and the readout and torquing electrodes are built into the device beneath the tilt plate. The output of the angle sensor is integrated and then used to drive the torque to maintain the tilt plate in a fixed nulled position. The torque required to maintain this balance is proportional to the input acceleration. The second type of MEMS accelerometers are the resonant accelerometers or Vibrating Beam Accelerometers (VBAs). They are an open-loop device in that the proof mass is not rebalanced to its center position when force is applied (Barbour, 2010; Barbour and Schmidt, 2001). The working principle in this case is that the acceleration moves the seismic proof mass out of position and changes the gaps of the interdigitated differential comb capacitor arrangement. The resulting capacitance changes are then converted into an electrical signal (Leinenbach et al., 2010). For accuracy, it relies on the scale-factor stability inherent in the material properties. The pendulous types can meet a wide performance range from 1 mg and for tactical systems down to 25 μg. Resonant accelerometers or VBAs can reach higher performance down to 1 μg (Barbour, 2010; Barbour and Schmidt, 2001).

Gyroscopes

For MEMS INS attaining suitable gyro performance is more difficult to achieve than accelerometer performance. Numerous MEMS gyroscopes have been developed or are under development (Shkel, 2001). Fundamentally, MEMS gyros fall into four major areas, i.e., (1) the vibrating beams, (2) the vibrating plates, (3) the ring resonators, and (4) the dithered accelerometers. Gyroscopes are usually built as hybrid solutions with sensor and electronics as two separate chips (Leinenbach et al., 2010). The operational principle for all vibratory gyroscopes is based on the utilization of the Coriolis force. If a mass is vibrated sinusoidally in a plane, and that plane is rotated at some angular rate Ω, then the Coriolis force causes the mass to vibrate sinusoidally perpendicular to the frame with amplitude proportional to the angular rate Ω. Measurement of the Coriolis-induced motion provides knowledge of the angular rate Ω. This angular rate measurement is the underlying principle of all quartz and silicon micromachined gyros (Barbour, 2010). These gyroscopes are usually designed as an electronically driven resonator which are often fabricated out of a single piece of quartz or silicon. They operate in accordance with the dynamic theory that when an angular rate Ω is applied to a translating body, a Coriolis force is generated. When this angular rate is applied to the axis of a resonating tuning fork, its tines experience a Coriolis force, which then produces torsional forces about the sensor's axis. These forces, which are proportional to the applied angular rate, cause displacements that can be measured capacitively in a silicon instrument or piezoelectrically in a quartz instrument. The output is then demodulated, amplified and digitized to form the device output. Their extremely small size, combined with the strength of silicon, makes them ideal for very high acceleration applications (Barbour and Schmidt, 2001). For purely surface-micromechanical gyroscopes, given their small sizes and capacitances, monolithic integration is a potential option to be considered not mainly for cost but for performance arguments (Leinenbach et al., 2010).

Combined Solutions

The difficulty in producing high performing, small gyroscopes has created further interest in all-accelerometer systems (also known as gyro-free). Two approaches are typically used. In the first approach, the Coriolis effect is exploited and typically, three opposing pairs of monolithic MEMS accelerometers are dithered on a vibrating structure (or rotated). This approach allows the detection of the angular rate Ω. In the second approach, the accelerometers are placed in fixed locations and used to measure angular acceleration. This approach is known as direct approach (Zorn, 2002). Both approaches have in common that the accelerometers also measure linear acceleration, which enables to provide a full navigation solution. In the direct approach, however, the need to make one more integration step makes it more vulnerable to bias variations and noise, so the output errors grow by an order of magnitude faster over time than when using a conventional Inertial Measurement Unit (IMU). However, these devices only provide navigation grade performance, and are most useful in GNSS aided applications (Barbour, 2010). The concept of a tactical grade all-accelerometer IMU requires accelerometers with accuracies on the order of nano-g's or better, and with large separation distances (Zorn, 2002). Barbour (2010) stated that the use of all-accelerometer navigation for GNSS-unavailable environments will not be viable until the far future, if ever. In this case an augmentation with other absolute positioning techniques, such as Wi-Fi, Bluetooth, etc., is required. Then accurate and reliable dead reckoning (DR) will be possible. Further sensor size reduction is possible through the combination of two in-plane (X- and Y-axis) and one out-of-plane (Z-axis) sensors on one chip. These multi-axes gyroscopes and accelerometer chips result in IMUs as small as 0.2 cm^3 in size (Barbour, 2010).

Magnetic Field Sensors

The local magnetic field is the measured quantity of magnetic microsensors based on MEMS. These sensors exploit the Lorentz force for measuring magnetic fields and can use different sensing types, such as (1) capacitive, (2) optical, or (3) piezoresistive. Most sensors with piezoresistive detection have had an easy signal processing and a straightforward fabrication process. However, temperature fluctuations have affected their performance. Optical readout systems have allowed microsensors with a reduction in the electronic circuitry and immunity to electromagnetic interference. MEMS sensors with capacitive sensing have presented little dependence on the temperature, but have needed vacuum packaging and complex electronic circuitry (Herrera-May, 2011). By the change of electric resistance of the material on which pressure or strain is applied a force on the sensor is generated. Thereby the measured magnetic field or actually the magnetic flux density (in the unit Tesla [T]) is in general a measure for the exercised force on the sensor. The earth magnetic field has a flux density of around 30 to 60 μT. Electrical devices in the surrounding environment generate itself a magnetic field that can absolutely reach the magnitude of the earth's magnetic field and therefore can significantly influence the sensor. Hence, magnetic field sensor measurements are often unreliable in buildings due to the presence of many electronic devices, wires, etc. Therefore a combination with gyroscope measurements is usually performed. The disadvantage of gyroscope measurements, on the other hand, is that they can only provide a short time stability and show large drift rates which accumulate quickly.

Barometric Pressure Sensors

Altimeters inside phones and navigation systems demand small-size, low-cost and high-accuracy MEMS barometric pressure sensors (Müller et al., 2010). The key element is a diaphragm containing piezoresistors which can be formed by ion implantation or in-diffusion. The operational principle is that applied pressure deflects the diaphragm and thereby changes the resistance of the piezoresistors. By arranging the piezoresistors in a Wheatstone bridge, an output signal voltage can be generated (Bao, 2004). The measurement sensitivity of the pressure sensor is determined by the strain at the bottom plane of the diaphragm whereby larger strain leads to higher sensitivity (Sharma and Inoue, 2010). Such type of altimeters can enable altitude determination of the mobile user, e.g. to determine the correct floor in a multi-storey building.

MEMS Sensor Potential

For many navigation applications, improved accuracy and/or performance is not necessarily the most important issue, but meeting performance at reduced cost and size is. In particular, a small navigation sensor size allows the introduction of guidance, navigation, and control into applications previously considered out of reach. In this context, the small size, extreme ruggedness, and potential for very low-cost and weight means of MEMS gyroscopes and accelerometers have been, and will be, able to utilize inertial guidance systems; a situation that was unthinkable before MEMS. The reduction in size of the sensing elements, however, creates challenges for attaining good performance. In general, the performance of MEMS IMUs continues to be limited by gyro performance (Weinberg and Kourepenis, 2006), which is typically at around 10 to 30 deg/h, rather than by accelerometer performance, which has demonstrated tens of micro-g or better. Thus, MEMS has struggled to reach high accuracy tactical grade quality (Barbour, 2010).

Comparison of Positioning Sensor Technologies

Table 1 compares the most commonly used location sensors and systems in mobile devices classified depending on their positioning capability (absolute or relative), information type, navigation information and typical achievable positioning accuracy. Specifications and achievable positioning accuracies are based on information and values taken from Faragher and Harle (2014), Hightower and Boriello (2001), Liu et al. (2007), Mautz (2012), Retscher (2007), Retscher (2016) and Stojanović and Stojanović (2014). The geometric location information type refers to the category where the user's location is expressed in form of coordinates that identify the location on a map. On the other hand, symbolic location expresses a location in a natural language way, such as in the engineering university building, in the office, on the third floor of a building, etc. Descriptive information using cameras are, for instance, grey scales, and geometric information direction or ranges from the camera to the object. For absolute location determination, a shared reference grid for all located objects is used, while with relative localization it is meant

Table 1. Specifications of the most commonly used location sensors and techniques in smartphones

Type	Sensor	Positioning Capability	Information Type	Navigation Information	Typical Accuracy	Characteristics
MEMS	accelerometer	relative	geometric	a_{tan}, a_{rad}, a_z	< 0.03 m/s^2	subject to drift calibration should be made when absolute positions are available
	gyroscope	relative	geometric	azimuth (heading) α	0.5° – 3°	short term accuracy stability not subject to external disturbances but to drifts calibration should be made when absolute positions are available
	barometric pressure sensor	relative	geometric	ΔZ	1 – 3 m	requires calibration by a given initial height to provide heights with respect to a global reference frame
	temperature sensor	–	–	T	0.2° – 0.5° C	requirement for barometric height conversion
Others	magnetometer	relative	geometric	Azimuth (heading) α	0.5° – 3°	measurement of the azimuth in respective to the earth's magnetic field long-term accuracy stability subject to magnetic disturbances sensitive to tilt
	GNSS	absolute	geometric	X, Y, Z	m-level	absolute position in global coordinate frame high accuracy of position and velocity estimation medium attitude accuracy
	Wi-Fi card	absolute	geometric	X, Y	3 – 5 m	location fingerprinting absolute position in local frame signal attenuation due to distance penetration through walls multipath affected interference from other users in 2.4 GHz frequency band possible
	Bluetooth	absolute	geometric/ symbolic	X, Y	few m 3 – 5 m	for proximity measurement using Bluetooth Low Energy iBeacons
	camera	absolute	descriptive/ geometric/ symbolic	X, Y, Z	m-level	scene analysis image database required overlap and conjugate points required

that the positions of the user are determined relatively to a start point, for instance, using DR with the MEMS location sensors. As can be seen from the specifications in the Table 1 typical positioning accuracies and the characteristics vary quite significantly and all of the sensors and positioning methods provide a specific performance. Due to their additional complementary characteristics a meaningful combination in form of a hybrid solution yields obviously to the best performance for localization of a mobile smartphone user.

MEMS BASED LOCALISATION TECHNIQUES

The trajectory of a mobile user is obtained from position, velocity, acceleration and orientation measurements as a function of time. The measurement system can consist of a GNSS receiver or other wireless positioning techniques, accelerometers, gyroscope, magnetometer, etc. It is essential that for the calculation of the trajectory a hybrid combination of these different sensors is performed whereby the sensors must guarantee a number of conditions, such as a high availability, a high measuring and sampling frequency, a quick measurement and response time, the possibility for highly automatic measurement and data acquisition, the ability for sensor synchronization, a good stability of the scale and offset, to name the most important ones. These requirements can be fulfilled by autonomous sensors which form a self-confined system and do not rely on external parameters. GNSS receivers, for instance, cannot guarantee to fulfill this requirement completely as they require tracking of at least four satellites to obtain a 3D position fix.

The sampling frequency is defined by the sampling theorem which says that the sampling frequency must be at least double than the highest existing frequency in the signal. With irregular movements the sampling frequency is dependent on the smallest time interval between the movement changes of a moving user. To determine the optimum measuring frequency it is necessary to have knowledge about the kind of movement to be determined. As kinematic observations do not provide redundancy a higher sampling frequency has to be chosen whereby a balanced signal-to-noise ratio has to be taken as a basis. The response and measurement time should lie a multiple time under the defined measuring frequency. Time synchronization of the sensors is also an essential aspect for sensor fusion. Due to the lack of redundancy the stability properties of the sensors cannot be controlled internally and must therefore be controlled by sensors with complementary properties. The sensor drifts are more or less time depending and may be reduced through calibration and an analytical description of the drift behavior. Estimation theory, in general, and Kalman filtering (see e.g. Gelb, 1974; Retscher and Mok, 2004; Welch and Bishop, 2006) in particular, provide a theoretical framework for combining information from various sensors. By properly combining the information from an Inertial Navigation System (INS) and other absolute positioning systems, the errors in position and velocity are stable. Furthermore, INS can be used as a short-term fallback while GNSS signals are unavailable, for example in an urban canyon. In the following, the operational principle of DR, IN and Map matching (MM) is briefly reviewed followed by a discussion of activity detection of a pedestrian user as well as altitude determination in indoor environments.

Dead Reckoning (DR)

The accelerometers integrated in the mobile device can be used to estimate the distance travelled from the accelerations made while in motion and gyroscopes in combination with magnetometers to obtain the current heading of the user. Starting from a known position, e.g. determined by GNSS or other wireless absolute positioning technique (see Table 1), the current position of the user can then be dead reckoned using observations of the inertial sensors (Mok et al., 2011). Thus, it is possible to predict positions of the user by projecting an ordered course and speed from a known present position. In other words, the inertial sensors can be used to make a relative position estimate by means of DR, where the position is estimated based on a known position and the displacements made since then. DR techniques differ from other localisation techniques because the position is always calculated relative to the previously calculated position and no correlation with the real position can be made.

DR can provide the best available information on position, however, is subject to significant cumulative errors, i.e., either compounding, multiplicatively or exponentially, due to many factors as both velocity and direction must be accurately known at all instants for position to be determined accurately. New positions are calculated solely from previous values, any errors and uncertainties of the process are cumulative, so the error and uncertainty in the position grow with time. Thus, an update with an absolute positioning technique is required from time to time part way through the journey. Wireless localization and vision-aided systems are suitable options in complex environments (see the next section).

Inertial Navigation (IN)

The basic principle of an Inertial Navigation (IN) is based on DR in 3D space where from the measurement of the distance travelled and the corresponding azimuth a position change between consecutive measurement epochs is derived. Using sequential calculation the resulting discretised trajectory is obtained. For the determination of the 3D distance differences and orientation angles between following positions two types of sensors are employed, i.e., accelerometers and gyroscopes, which are combined to a specific system group. While doing this the observations of the accelerometers and gyroscope of the IN system (INS) are converted into new observations regarding position and orientation of the system. Then six degrees of freedom, i.e., three translations and three rotations, have to be estimated to be able to determine a movement in space. From this it follows that an INS in its elementary shape consists of three orthogonal arranged motion sensors (accelerometers) for determination of the acceleration vector and three orthogonal arranged rotation sensors (rate-gyroscopes) for attitude determination measuring angular velocity and linear acceleration respectively. Hence, with an INS the position, orientation, and velocity (direction and speed of movement) of a moving user can be continuously calculated via DR. An INS can detect a change in its geographic position (a move east or north, for example), a change in its velocity (speed and direction of movement), and a change in its orientation (rotation about an axis). It does this by measuring the linear acceleration and angular velocity applied to the system.

IN suffer from integration drift, i.e., small errors in the measurement of acceleration and angular velocity are integrated into progressively larger errors in velocity, which are compounded into still greater errors in position. Since the new position is calculated from the previous calculated position and the measured acceleration and angular velocity, these errors accumulate roughly proportionally to the time since the initial position was input. An error in the acceleration measurements causes an error in distance which grows squarely with time due to double integration. Thus, the position must be periodically corrected by input from some other type of navigation system. Accordingly, INS are usually used to supplement other navigation systems, providing a higher degree of accuracy than is possible with the use of any single system.

Map Matching (MM)

The process of localizing a user's positions from sensor(s) to its actual road position on a map it is called map matching (MM). MM can also be defined as a process that uses spatial information such as the road network to determine the spatial reference of the user's location (Quddus et al., 2003), or also as the technique that combines electronic map with location information to obtain the real world location of a user. These algorithms can be used not only to provide the vehicle's position in the correct segment but also can improve the positioning accuracy if good spatial road network data are available (Ochieng et al., 2004).

The common map matching techniques traditionally use spatial information in an attempt to improve navigation accuracy. A feature-based MM is thereby most descriptive and originates from the field of pattern recognition. The search area is limited by the current location determined by the absolute positioning sensors. The current vehicle position can then be derived from this matching process. Other

MM procedures derive the position from curvature pattern of the road elements and mutual allocation by cross correlation. In a simplified way the distance travelled is identified in the digital road map and MM algorithm assigns the current vehicle location to the road in the map using position corrections. For that purpose a search area is defined using the distance travelled and its estimated accuracy to match all possible trajectories to the road map. The MM algorithm yields then the most likely trajectory under consideration of the accuracy and plausibility of the estimated parameters. Thus, MM enables an improvement of the vehicle trajectory obtained from the positioning sensors. Apart from using simple searching techniques more complex MM algorithms use fuzzy logic, Kalman filtering and belief theory. Quddus et al. (2007) has categorized them into four groups, i.e., (1) geometric, (2) topological, (3) probabilistic and (4) advanced MM algorithms. The geometrics are those algorithms that make use of geometric information of the spatial road network data considering only the shapes of the links, regardless how they were connected. On the other hand, topological MM algorithms look into the relationship of the entities (points, lines and polygons), as adjancy, connectivity and containment. Probabilistics consider the error of the positioning solution, acquired from the sensors (absolute positioning technique and/or INS and/or DR). Elliptical or rectangular confidence regions are built using the error variances of positions and the identification of the roads is made by selecting the segments within this confidence regions. More advanced MM algorithms make use other refined mathematical approaches such as Kalman filtering, particle filter or fuzzy logic models.

Although the main purpose of MM approaches is to accurately display vehicle location on a GIS-based digital map, accurate road map information integrated with other sensors such as cameras, GNSS, etc., can improve positioning accuracy. For some Intelligent Transportation System (ITS) applications, however, these algorithms are still not yet capable of supporting the navigation requirements. Quddus et al. (2007) pointed out that the main limitations are (1) initial identification of vehicle positions, (2) matching positioning fixes in complex road lay-out (such as Y-junctions and fly-overs), and (3) performance evaluation in dense urban areas. To fully develop a reliable and robust navigation system for complex environments, improvements in the algorithms and integration of sensors are still needed to allow safe and continuous navigation.

Activity and Step Detection

Low-cost inertial and motion sensors embedded in smartphones have provided a new platform for dynamic activity pattern inference (Saeedi and El-Sheimy, 2015). Human activity recognition aims to recognize the motion of a person from a series of observations from the user's body and environment. Randell and Muller (2000), for instance, used a single biaxial accelerometer for classifying six activities (i.e., walking, running, sitting, walking upstairs, walking downstairs, and standing). Lee and Mase (2002) developed an activity and location recognition system using a combination of a bi-axial accelerometer, compass, and gyroscope. In many studies, however, sensors are usually worn on the human body for activity detection and up to now only a limited number of studies use a smartphone to collect data for activity recognition (Saeedi and El-Sheimy, 2015).

Technically the cycle of walking is across a stride (two steps), although a sensor sited along the body's mid-axis can exhibit a one-step period if the user's gait is symmetric. Typical stride frequencies are around 1 to 2 Hz, a range that few activities other than walking exhibit. Peak detection (Kim et al., 2004; Randell et al., 2003) or zero-crossing counting (Beauregard, 2006; Goyal et al., 2011) on low-pass filtered accelerometer signals can be used to find specific steps. In general, vertical acceleration provides better performance (Marschollek et al., 2008), but depends on being able to isolate the orthogonal accelerations in the global frame. This is difficult to do on a smartphone that is not firmly attached to the body. Instead, the signal magnitude is often substituted (Brajdic and Harle, 2013). In the following, the use of smartphone inertial sensors for step detection is discussed briefly.

The smartphone accelerometers recognize the acceleration of the device in three axes as shown in Figure 1. Different motion sequences can thereby be ascertained. If a smartphone is held horizontally in the hand during a forward motion then an acceleration in Y-axis is induced. Figure 2 shows an example of the recorded acceleration of a walking person in Z-axis. As can be seen the acceleration values show significant maxima and minima. From these peaks it is possible to count the steps of the walking user. An easy approach to count the steps is to recognize one step from the excess of a certain threshold. Then a step is detected when this value first gets lower than a defined threshold and subsequently upon a second threshold. Zero-crossings can also be employed for step detection. Apart from these two approaches another type of method takes advantage of the sinusoidal-type pattern of acceleration signals to detect step events. In these cases, frequency domain analysis, such as fast Fourier transform (FFT), is applied to analyze acceleration signal series (Liu, 2012). FFT results usually show a strong frequency peak in the range of 0.5 to 2 Hz when walking. This corresponds to a walking frequency of 0.5 to 2 steps per second (Mezentsev, 2005).

Figure 2. Typical recording of accelerometer sensor data of a walking user in Z-axis

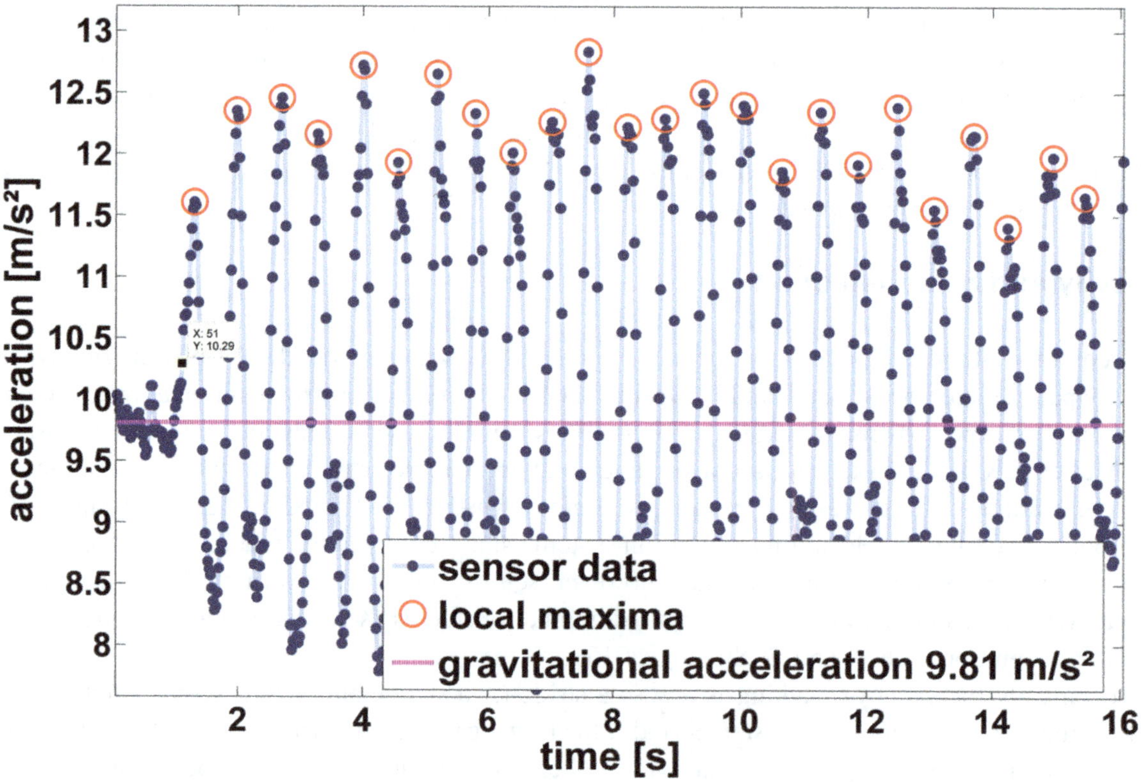

Correction for the gravity effect on the X-, Y- and Z-axes of the smartphone's local coordinate system is the key to correct the determination of accelerometer-derived distance travelled. The MEMS-based three-axis accelerometer allows the device to detect the force applied along the three axes in order to accomplish specific functions based on predefined configurations. The mobile device can be oriented in the form that one of the axes is aligned in the direction of the movement or heading (e.g. the Y-axis), and the positive X-axis is pointing rightward and the positive Z-axis is upward (compare Figure 1). When the Y-axis is horizontal, the gravity effect will be fully reflected on the Z-axis. When the Y-axis is inclined at an angle α, the gravity effect $\vec{g}$. wl be the projection of the gravity vector $\boldsymbol{g}$ onto the inclined Y-axis with the magnitude $|g| \cdot \cos \alpha$ (Mok et al., 2011). However, a user will not carry a phone in hands all the time but will most likely put it in the trouser pocket. Therefore, most existing step detection algorithms cannot be used directly, but adjustments have to be made to take into account this position of the accelerometers on the body. The step detection algorithm applied by Bylemans et al. (2009) to the accelerometer data uses the values of the most vertical axis of one of the accelerometers to detect a pattern in. Because a phone can be put in a person's trouser pocket in any direction, with any side up or down, the accelerations are observed to determine which axis is the most vertical one of both accelerometers and their three axes. The acceleration of the axis that is pointing directly to the earth has a value of 1 g due to gravity. So if the smartphone is lying flat on a table, with the display side up, then the Z-axes of both accelerometers would theoretically have a value of 1000 mg. If the phone is put crooked (not along one of the axes) in someone's pocket, the values will be lower than 1000 mg. So to detect which accelerometer is the most vertical axis, the absolute average of the last 30 samples, or 1.2 seconds, of all three axes of the two accelerometers of which the absolute value is closest to 1 g, is the most vertical axis and the accelerometer to use.

To obtain the distance travelled from step detection also the step length needs to be estimated. Liu (2012) distinguishes four models, i.e., constant models, a cluster of linear models, non-linear multivariable models or non-parametric analytical techniques. The constant model is least accurate as an error of 5 cm in step length – which is common between steps even for the same person – represents already a relative error of 15% compared to a 35 cm step length. In the second linear model using clusters uses a correlation between step length and temporal variables, such as step interval and frequency. Consequently, the resulting step length estimation only follows a time variant process leading to a limited improvement of accuracy for the estimation of the distance travelled as terrain slop, body height of the person and trajectory curvature are not considered. Non-linear models try to consider these influencing factors in a way that they consider step length as a nonlinear function of these variables. In addition to these parametric models non-parametric analytical techniques use, for example, fuzzy logic, wavelet analysis or artificial neural networks (ANNs) (Moafipoor, 2009).

Altitude Determination

Of increasing significance is determination of the altitude of the user, for instance, if a user should be located on the correct floor in a multi-storey building. Such a task can be accomplished with the barometric pressure sensor (altimeter). The inertial sensors can usually only provide reliable 2D localization capabilities and they can now be augmented with this sensor. Furthermore, if only three GNSS satellites are visible providing only a 2D positioning solution the use of an altimeter can lead to localization capabilities in 3D. Altitude determination with a barometric pressure sensor can be performed relatively from a given start height (e.g. obtained from GNSS outside the building or from a known height point in the indoor environment) (Retscher and Thienelt, 2004). As the user walks inside the building and up the stairs or elevator to other floors the altitude is determined relatively to this given start value. From the difference and changes of the air pressure in the sensor altitude differences can be calculated using a simple relationship between the pressure changes and height differences. This process involves pressure observations at the two stations which geographic location is approximately known. For the conversion of the air pressure in a height difference also the mean value of the temperature at both stations is required (see Retscher, 2007). MEMS infrared thermometers are also increasingly found in smartphones and they can provide mean temperature differences between the stations required for the conversion of the air pressure changes into altitude differences. Changes in pressure can then be converted to changes in height. Figure 3 shows an example were one smartphone users was using an elevator from the ground to the sixth floor in a multi-storey office building. The user stopped in every floor for a short time of less than a minute. As can be seen the smartphones altimeter is capable to determine the correct floor and the relative height between them.

COMBINATION OF MEMS AND OTHER SENSOR SYSTEMS

For the majority of smartphone navigation systems, it is the combination of MEMS sensors with other sensor systems, such as wireless localization techniques or even vision-aided technologies that provides

Figure 3. Altimeter readings of a smartphone in a multi-storey office building

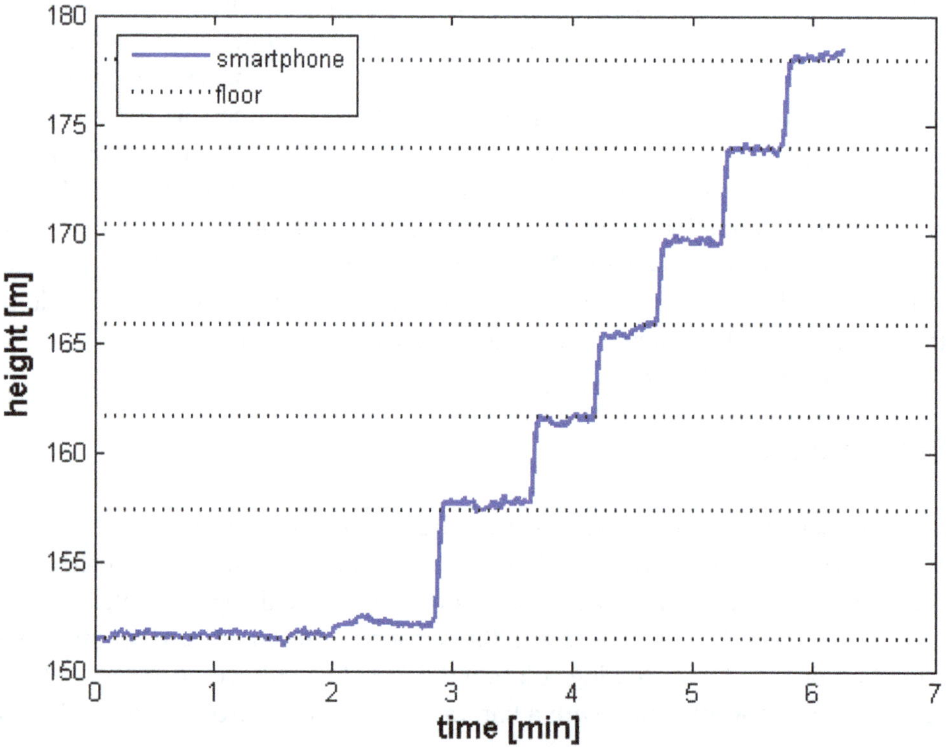

the optimal solution. In this section, the integration with wireless options and vision-based systems is briefly discussed.

Integration With Wireless Options

Apart from GNSS, which is mainly not available in complex environments, wireless options, such as Wi-Fi and Bluetooth (see e.g. Chen at al., 2012; Faragher and Harle, 2014; Kotanen et al., 2003), can be alternatives (see Table 1). On the one hand, MEMS sensors can be used to provide relative positioning information for continuous localization, and on the other hand, wireless localization techniques provide an absolute localization capability in either a local or global coordinate frame. Thereby relative position measurement methods via DR determine location by integrating a sequence of measurements over time, which means that the current location estimate depends on previous estimates. Since location estimates are obtained by the integration of sensor measurements, this leads to the unbounded accumulation of location errors over time. Location estimates determined from absolute position measurements are independent of any previous estimates because location is obtained from a single or a set of measurements without integrating measurements over time. The combination of these two technologies therefore takes advantages of the strengths of both with the end result of a more robust position solution. In the following, the use of Wi-Fi and Bluetooth is briefly reviewed as these technologies are commonly employed and they are therefore representative. A further discussion of all possible location sensors would be out of the scope of this book chapter.

Various wireless standards have been established. Among them, the standards for Wi-Fi, IEEE 802.11b and wireless PAN, IEEE 802.15.1 (Bluetooth), are used more widely for measurement and localization. The bandwidth of 802.11b is 11 Mbits and it operates at 2.4 GHz frequency (Wang et al., 2006). It is originally a technology for short-range wireless data communication and is typically deployed as an ad-hoc network in a hot-spot fashion in the areas where a wireless Internet access is needed. IEEE 802.11 is currently the most utilized Wi-Fi technology. Wireless networks are built by attaching a device called access point (AP) to the edge of a wired network. The transmission power is the major factor that has direct influence on the effective range. Hence, the received signal strengths (RSSs) and MAC (Media Access Control) addresses of the APs are location-dependent information that can be adopted for positioning purpose. An observable associated with a MAC address of an AP consist of the following information: (1) the unique MAC address of the RF transmitter, (2) the location of the RF transmitter, and (3) the effective range of the signal, or the size of the signal coverage area of the RF transmitter (Chen et al., 2012). For localization of a mobile device either cell-based solutions or (tri)lateration and location fingerprinting are commonly employed. A comprehensive description of these localization techniques is given in Retscher (2016). The advantage of the utilization of Wi-Fi is that no designated infrastructure has to be built up. It is a prominent so-called 'signal-of-opportunity' and therefore widely used.

Bluetooth is considered a cable replacement for mobile devices and it is mainly designed to maximize the ad hoc networking functionality (Wang et al., 2006). Compared to Wi-Fi, the gross bit rate is lower (1 Mbps), and the range is shorter (typically around 10 m). On the other hand, Bluetooth is a 'lighter' standard, highly ubiquitous (embedded in most phones, etc.) and supports several other networking services in addition to IP. For positioning either tags or Bluetooth Low Energy (BLE) iBeacons are common. Bluetooth tags are small size transceivers. As any other Bluetooth device, each tag has a unique ID.

This ID can be used for localization of the tag (Prasithsangaree et al., 2002). iBeacon is a Bluetooth low energy protocol developed by Apple and compatible hardware transmitters, typically so-called beacons, broadcast their identifier to nearby portable electronic devices. The technology enables mobile devices to perform actions when in close proximity to an iBeacon whereby a universally unique identifier picked up by a compatible App or operating system is transmitted. The identifier and several bytes sent with it can be used to determine the device's physical location. Localization is based on proximity sensing and cell-based solutions.

Positioning Methods Using Wireless Signals

Location sensing can be classified into cell-based positioning (also sometimes referred to as proximity technique in the literature), (tri)lateration and angulation, hyperbolic lateration, scene analysis, location fingerprinting, dead reckoning (DR) and hybrid solutions. Retscher (2016) provides a comprehensive overview of these techniques applied to indoor localization. In the following only the principles of the most relevant methods for the localization of smartphones are briefly discussed.

The simplest and most straightforward methods are cell-based approaches (so-called cell-ID (identification) or cell-of-origin (COO)) which are based on the cell identity and the location associated to it (Trevisiani and Vitaletti, 2004; Chen, 2012). The location of a smartphone, for instance, is described in relation to the vicinity to the location of known object(s), such as cell towers in a cellular phone network, Wi-Fi APs or Bluetooth tags, where the location of the known object or the symbolic cell-ID defines the location of the device. Positioning accuracies thereby depend on the size of the associated cell (Retscher et al., 2012).

Lateration uses range measurements to known locations where either one- or two-way travel times are converted into ranges between the unknown and known position. The former is also referred to as time of arrival (TOA) and the later as round trip time of flight (RTOF) (Liu et al., 2007). The location of the user is then obtained with the intersection of at least three spherical surfaces where the centers are the known locations and the radii the measured ranges. Apart from direct measurement of travel times ranges can also be derived from RSS measurements. This can be the case in Wi-Fi positioning. The ranges are then derived based on the nature of the RSS. Theoretically, the RSS decreases with the transmitted energy propagating into space. Path loss models can be employed to establish the relationship between the RSS and propagating distances (Retscher et al., 2012). A differential approach proposed by the first author of this chapter is referred to as Differential Wi-Fi in analogy to the well-known Differential GPS operational principle. Instead of theoretical path loss models this approach utilizes continuous RSS scans carried out during the localization determination of the mobile user to improve the positioning accuracies and reliability of the solution (Retscher and Tatschl, 2016; Retscher et al., 2017). The major attempt is to minimize the spatial and temporal variations and propagation effects of the Wi-Fi signals by the use of additional reference stations (RSs) during the measurements. Range corrections are deduced if RSs are deployed at certain AP locations or in a network in the area of interest. A low-cost implementation of the RSs is realized using a number of Raspberry Pi units, which serve simultaneously as both APs and RSs – emitting and scanning Wi-Fi signals. DWi-Fi outperforms common lateration methods in complex environments, such as in buildings where a damping of the Wi-Fi signals is caused by different objects and walls.

RSS measurements can also be employed directly for positioning using so-called location fingerprinting. The principle of operation is that first features (so-called fingerprints) of a scene or signal are collected in a training phase on a number of reference points (RPs) and then the user's location is estimated by matching current measurements with the closest predetermined location fingerprints in the positioning phase (Mok and Retscher, 2007). RSS-based fingerprinting was first employed for Wi-Fi positioning with the system RADAR in 2000 (Bahl and Padmanabhan, 2000). The main problems in fingerprinting, however, are that the construction of a fine radio map representing RSS distributions in the area of interest leads to high workload and heterogeneous mobile devices measure RSS differently. A novel approach developed for an integration of Wi-Fi positioning and IN leads to a significant reduction of workload by a factor of 4 compared to standard Wi-Fi fingerprinting while still keeping the same level of achievable positioning accuracies. Instead of RPs distributed in a regular grid in the area of interest significant waypoints are selected on whom the training measurements are carried out. These waypoints are referred to as intelligent checkpoints (iCPs) and they are the selected RPs which have to be passed along the way while navigating from a start point to the destination. The iCPs are twofold intelligent because of the fact that they depend on their meaningful selection and because of their logical sequence in their correct order. Thus, always the following iCP is known due to a vector graph allocation in the fingerprinting database (DB) and only a small limited number of iCPs needs to be tested when matching the current RSS scans in the positioning phase. These iCP detection can be employed for absolute positioning to update the inertial smartphone sensors and to reduce their drift rates (Retscher and Hofer, 2017). To use an iCP to correct a position, the passing time has to be determined as precise as possible. The implemented iCP/INS algorithm uses specific information about the structure of a building in the case of indoor navigation. For iCPs along a corridor, for instance, conditions can be defined to make sure that iCPs could only be passed in the movement direction. This is to prevent them from being misrecognized, in case they are passed. In that, the obtained values of the orientation sensor or the step length are calibrated. Then the step length is automatically corrected if someone climbs the stairs in a building, for instance.

Vision-Aided Navigation System Integration

Smartphones have also built-in cameras that can be used for localization. In general, a camera is independent from other sensors and the noise in images is not cumulating over time as well as also free from an infrastructure installation. Visual positioning means the usage of information obtained from images to resolve the user position or ease the resolving of the user position by integrating the visual information with position measurements obtained with other methods. The visual information may also be used to make the position solution more accurate and more widely available. Either an absolute position of the user by utilizing reference images in a database or a relative position and location change by examining consecutive images is obtained. Regardless whether the visual-based positioning is based on either one of the two approaches, the first step of the pre-processing are similar. This includes to retrieve the required camera parameters (Ruotsalainen and Kuusniemi, 2012).

When the camera is carried by a pedestrian, the motion of the camera relates to the motion of the pedestrian, for instance, calculated from consecutive images. Integrating the motion information obtained from the images derives a navigation system with increased availability and accuracy (Ruotsalainen et al., 2011). In the following, only a few particularly selected approaches are briefly discussed as those are most commonly employed when using smartphones. The visual device is mainly used as complementary sensor that collaborates with other on-board sensors (Li et al., 2013). The theory behind this is that either features in the images or vanishing points of straight lines are used.

Specific tags or significant patters, for instance, can be detected to determine the user's location. QR codes are an example which can be distributed within a building and detected in the images by scene analysis. Furthermore, the matching of perspective images of the environment captured by the smartphone camera carried by a person, to prerecorded images or videos which have been collected to build up 3D models stored in an image/video database can be performed (Retscher, 2016). The databases contain images of recognizable features in the surroundings attached with position information (see e.g. Aoki et al., 1999; Robertson and Cipolla, 2004; Steinhoff et al., 2007; Zhang and Kosecka, 2006). When a match between images in the database and the ones taken by a smartphone camera is found, the absolute position may be obtained. The information contained in images can then be used to recognize when the camera is pointing at a known object. Visual features in a close surrounding environment may also be detected and relative positioning information obtained by detecting the motion of the features in consecutive images (Ruotsalainen, 2013). This procedure is also referred to as vision-aiding. Motion of the features enables computation of the heading change and translation of the camera between consecutive images. This motion information may be further integrated with e.g. GNSS or other sensor measurements to obtain complementary information for positioning. As mentioned above, images provide accurate motion information, because they do not suffer from accumulating measurement errors like inertial sensors, or signal obstruction like GNSS. The database based procedure is though laborious due to the a priori preparations and is restricted to the predefined region (Ruotsalainen et al., 2011).

Hide et al. (2009), for instance, used the so-called Image Bag-of-Words algorithm to build a prior map of the area to be navigated. Images are captured at known positions during an initial 'mapping' survey and added to the Bag-of-Words database. During the navigation process, this database can then be queried wherever self-localization is needed to search for images that appear to be of the same place, hence providing occasional position updates when a location with an earlier position estimate is observed. These schemes are robust to changes in scale, illumination, camera position, and small changes in the scene, and databases of hundreds of thousands of images can be queried in real-time with error rates of just a few percent, depending on the frequency of update required and the nature of the environment (Cummins and Newman, 2008). This information, together with an estimate of the uncertainty in the difference from the mapped camera location, is used to update a Kalman filter for trajectory determination and correct the position and velocity estimate. Such a use of position updates for pedestrian navigation has many advantages such as being low cost and passive, not requiring a dedicated infrastructure, and being feasible for use over wide areas (Hide et al., 2009).

Indoor surroundings, like offices and public buildings, however, are usually poor with features. The lighting is also often constricted. An approach to overcome these drawbacks is to use straight lines, like borders of floors and walls, to calculate vanishing points. The algorithms for vanishing point calculations, however, are mostly computationally heavy. Ruotsalainen et al. (2011), for example, introduced an indoor pedestrian navigation system using a rapid algorithm utilizing visual-aided heading of the motion for smartphones. In this algorithm, the change in heading is calculated using vanishing points with a frequency of 1 Hz which fulfills the real-time requirements set for navigation for smartphones. The system also integrates the visual-aiding information with measurements obtained from other sensors using an Extended Kalman filter.

SENSOR FUSION TECHNIQUES AND COOPERATIVE POSITIONING

Sensor Fusion Techniques

The fusion of the observations made by the sensors embedded in smartphone is key to achieving robust positioning performance. The most widely used fusion algorithm is the Kalman filter. It is a recursive algorithm that uses a series of prediction and measurement update steps to obtain an optimal, in a minimum variance sense, estimate of the state vector (Gelb, 1974). As the theory of Kalman filtering may be found in many textbooks it is not further discussed in this chapter; but a brief summary of the operational principle of two selected sensor fusion techniques which are widely employed nowadays is provided in the following.

Particle filters, or also known as Monte Carlo localization (MCL), have become widely used, for example, in robot localization in indoor environments. Their principle of operation is that starting with a set of particles in the prediction phase each set is applied through a motion model by sampling of the particles. This approach results in a new set particles at the following epoch which approximate a random sample from the predictive density. In the next step, the update phase is applied where measurements are taken into account to weigh all of the sampling sets. Then the particles are computed by re-sampling from the weighted set. These two phases are repeated recursively for the subsequent steps (Dellaert et al., 1999). The MCL has some advantages over other algorithms, such as Kalman filtering. MCL is able to represent multi-modal distributions which is useful for self-localization and it is relatively easy to implement (Hasnur Rabiain et al., 2013).

Another algorithm, adopted from the domain of wireless sensor networks (WSNs) is the SPAWN algorithm. The SPAWN algorithm makes use of factor graphs (FG) and sum product algorithm (SPA) where the FG is a method of graphically represent a factorization of a Bayesian network while the SPA is a message passing algorithm for performing inference on the FG. Consider a WSN consisting a set of nodes and a set of anchors. Each node elaborates its information from the previous step from its last position estimation. Then it receives messages from visible anchors and neighboring nodes. Using the new information, it updates its positional estimation and shares it with its neighbors. The messages shared among the nodes represent the probability density function. This makes the approach a truly distributed algorithm that is highly suitable for an increasing number of cooperative positioning applications (Wymeersch et al., 2009).

Cooperative Positioning (CP)

An increasing number of personal mobility scenarios require the localization of a group of end users or individuals within a group or neighborhood. Often, this is the case in adverse operating conditions and complex environments such as inside of large buildings, train stations, etc., where only limited positioning information is available. A solution to this problem is referred to as Cooperative (or Collaborative) Positioning (CP). In CP, end user localization is performed using its own measurement data plus any additional information coming from neighboring users in the form of inter-nodal ranges or other means. As the multiple users are operating in close proximity, it is possible for these users to exchange some or all of their measurements and navigation state estimates in order to help themselves and the other cooperating users (Retscher et al., 2017).

In most CP implementations it is helpful if the users have the ability to measure the range between themselves and other cooperating users directly (Nilsson et al., 2013), such as an implementation of the 802.15.4a ranging protocols via a 2.4 GHz or UWB (Ultra-Wide Band) radio. In a dynamic network environment sub-networks of users (i.e., nodes) navigating jointly, can be created ad-hoc whereby some nodes may be part of different sub-networks. Still, information exchange among the sub-networks must be assured. Conceptually, the sub-networks can consist of nodes of equal hierarchy or may contain a master node (also called anchor node) that will normally have a better set of sensors and will be collecting measurements from all client nodes to perform the CP solution (Grejner-Brzezinska et al., 2009). The key components of a CP network system are: (1) the inter-nodal ranging sub-system, (2) the optimisation of the dynamic network configuration, (3) the time synchronisation, (4) the optimum distributed sensor aperture size for a given number of nodes, (5) the communication sub-system, (6) the selection of master or anchor nodes, and (7) the network topology. For further information about the technical details for the use in Positioning, Navigation and Timing (PNT) applications the reader is referred to the paper of Kealy et al. (2015). It must be noted that these requirements must be met in any case for applications ranging from Intelligent Transportation Systems (ITS), vehicle collision avoidance systems, personal/pedestrian navigation and Location-based services (LBS).

In the context of smartphones it may be possible to implement CP by using in-built communications capabilities to exchange data with nearby users. The data provided could be processed in a decentralized fashion whereby each user (smartphone) utilizes for example, the range information between users - either independently, or in a centralized or quasi centralized mode whereby the data from groups of cooperating users is processed at a single point and then distributed back to the members. The former decentralized approach suffers from the potential for the formation of error loops which require careful de emphasis of the received data in some situations, while the latter centralized approach requires more data band-width at lower latencies and more processing power at the centralized node. Regardless of the approach selected or the technologies used to achieve it, the inclusion of CP should result in the triple benefit of improving absolute positioning accuracy, improving relative positioning accuracy, while also providing improved situational awareness of the other vehicles operating in close proximity to each other, making the task of navigating in complex environments both more accurate and safer for all cooperating users.

APPLICATIONS

Two major applications of MEMS-based location sensors in mobile devices are pedestrian and vehicle navigation. The principle of operation and applicability of MEMS as well as the current challenges are briefly described in this section.

Pedestrian Navigation

For pedestrian navigation the MEMS accelerometers can be used to estimate the distance travelled from the accelerations made while walking and magnetometers and gyroscopes to obtain the current heading of the user. Following the DR concept for relative positioning this navigation process is referred to as pedestrian dead reckoning (PDR). When working with accelerations, two approaches can be applied to measure the linear displacement, namely integration of the accelerations or step detection combined with step size estimate. In the first case, the distance travelled can be theoretically calculated by integrating the accelerations once for velocity, twice for distance. Due to the double integration, however, any error in the signal will propagate rapidly, so the drift on the received signals from the accelerometer makes it impossible to use integration for walks of more than a few seconds. A technique that could be used to overcome this is the so-called Zero Velocity Update (ZUPT), where the velocity is reset to zero between every consecutive step when the foot is stationary for a small amount of time. This way any error produced during one step has no influence on the steps that follow. ZUPT can only be used when the accelerometer is placed on the foot, where advantage can be taken of the stationary period between footsteps. In step detection the distance travelled is obtained from step counts by processing the fluctuating vertical accelerations, which cross zero twice with every step. When the number of steps and the step size are acquired, the distance can be calculated by multiplication (Bylemans et al., 2009). Step detection procedures described above provide the distance travelled of a pedestrian more reliable.

Vehicle Navigation in Intelligent Transport Systems (ITS)

As discussed above, DR is subject to significant cumulative errors due to many factors. If a vehicle odometer is used in classical built-in vehicle navigation systems the distance travelled is derived from the measurements of wheel sensors in the two not powered wheels of the vehicle. With wheels sensors the direction change from the distance difference of the curve-inwards and outwards wheel can also be obtained. This calculation of the direction change can additionally be supported using measurements of a magnetometer and/or sensors which measure the steering wheel's impact angle. As displacement is measured by the number of rotations of a wheel with an odometer, any discrepancy between the actual and assumed diameter, due perhaps to the degree of inflation and wear, will be a source of error.

Smartphones are nowadays increasingly used for vehicle navigation. It should not be disregarded that in vehicle navigation, higher sensor dynamics occur which results in higher demands for the localization sensor system. Very demanding positioning requirements for cooperative ITS (CITS) have emerged in response to the major transportation challenges for society today. At the core of all CITS is the positioning and communications technologies used to connect people, vehicles, infrastructure, etc. As CITS evolve and become increasingly safety-liability critical, it is essential that the required performance levels for positioning (accuracy, integrity and availability) are assured. Major goals are to specify and characterize the requirements set for the system by the environment, users and regulations for the urban traffic. In addition to the conventional sensors, such as inertial sensors and odometers, the usability of emerging technologies for augmenting or replacing GNSS and improved routing, has to be achieved. The technologies will include at least image-based navigation, cooperative driving systems and crowdsourcing traffic information for improved routing. In addition, different integration algorithms for providing the navigation solution are applied.

FUTURE RESEARCH DIRECTIONS

At present, navigation is mainly based on the use of GNSS, providing good performance in open outdoor environments. However, navigation solutions with sufficient accuracy and integrity are needed in complex environments, such as urban canyons or indoors, where GNSS is significantly degraded or unavailable. Navigation in complex environments is needed across an increasing range of safety and liability critical consumer users, private vehicles and pedestrians. Therefore, the navigation system has to be small, effortless to use as well as have reasonable low levels of power consumption and price. For overcoming the in this chapter mentioned navigation challenges, research has been very active for decades for finding a suitable set of other methods for augmenting or replacing the use of GNSS in positioning. Especially pedestrian and vehicle navigation have high demands and require therefore further research in these directions.

It is evident here, that no single sensor or technique is able to meet the positioning requirements for the increasing number of safety and liability critical mass-market applications. Integration is one approach to improving performance level, however, to make a significant step change in high performance positioning in GNSS difficult environments, higher performance level are required from MEMS and wireless technologies. The scope of such integration is not only to improve the accuracy and the precision of the solution, but also to provide a means to increase the robustness and the reliability of the positioning procedure to threats typical of the urban environment and indoors, such as signal obscuring, multipath and radio-frequency interference. Wireless networks positioning based on utilization of wireless networks has become popular in recent years especially in dense urban and indoor environments. These terrestrial positioning technologies are being consolidated and several works for the hybridization with GNSS core positioning is being studied, following to a certain extent the integration process of GNSS and INS. However, research is still needed to define proper error models, tailored integration algorithms and other topics to fully exploit the diverse source of information.

Different sources that may be used for integrated positioning are wireless options, and growing importance is being achieved by cameras that are increasingly used for positioning for a wide field of applications at all levels of accuracy. The success of optical methods originates from miniaturization and advancement in the technology of detectors (e.g. CCD sensors). In parallel, there has been an increase in data transmission rates and computational capabilities of processing equipment as well as profound development of algorithms in image processing. On intelligent vehicles, for instance, cameras are of high interest to be used independently or, preferably, hybridized with GNSS, since they are equipping more and more frequently the new vehicles and therefore can provide valuable information at a very low additional cost.

A further challenging topic is cooperative positioning, which is of specific value in the urban environment thinking to connected vehicles, vehicular networks and pedestrians. Peer-to-peer and cooperative positioning are one of the major evolutions for mass-market positioning, bringing together capabilities of satellite navigation and communication systems. It is well-known that smartphones already provide user position leveraging both GNSS and information collected through the communication network (e.g. Assisted-GNSS using additional data from mobile communication networks). However, exploiting the exchange of information among close users can attain further benefits which can be of specific benefit, to any scenario where full visibility of the satellite constellation cannot be granted (Dovis et al. 2014).

Furthermore, regarding MEMS sensors future commercial markets will need multifunctional sensors on a single chip for measuring several parameters such as magnetic field, pressure, acceleration, and temperature (Herrera-May et al., 2011). Tactical grade performance of the MEMS INS is one of the main goals to be achieved.

CONCLUSION

MEMS sensors are now essential components of modern smartphones and tablets. Wireless sensor technology allows MEMS sensors to be integrated with signal-conditioning and radio units to form 'motes' with extremely low cost, small size and low power requirement. New miniaturized sensors and actuators based on MEMS are available on the market or in the development stage. Today's smartphone sensors can include MEMS-based accelerometers, microphones, gyroscopes, temperature and humidity sensors, light sensors, proximity and touch sensors, image sensors, magnetometers, barometric pressure sensors and capacitive fingerprint sensors. They have been integrated to wireless sensor nodes (Wang et al., 2006). Smart devices have built-in motion (or inertial) sensors, such as accelerometers and/or gyroscopes or magnetometers. However, these sensors were not intended for navigation purposes at the beginning, for instance, the accelerometers are used primarily for applications such switching the display from landscape to portrait as well as gaming. These embedded sensors, however, are natural candidates for localization of a mobile client. MEMS accelerometers and gyroscopes, for instance, can be employed for localization in combination with absolute positioning technologies, such as GNSS or other wireless technologies. It is demonstrated in this chapter that these sensors have a high potential for localization of mobile users in combination with absolute positioning methods, such as wireless options (e.g. Wi-Fi) or vision-aided approaches using the phones digital camera.

ACKNOWLEDGMENT

The basics and fundamental principle of operation of MEMS-based sensors and their use for mobile smartphone users were presented and published for the first time at the ION Pacific PNT 2017 Conference in Honolulu, Hawaii, USA (Kealy and Retscher, 2017). The authors thank Mr. Hannes Hofer for providing Figures 1 and 2.

REFERENCES

Aoki, H., Schiele, B., & Pentland, A. (1999), Realtime personal positioning system for wearable computers. *Proceedings of 3rd IEEE International Symposium on Wearable Computers ISWC 1999*, 37-44. doi:10.1109/ISWC.1999.806642

Bao, M.-H. (2004). *Micro mechanical transducers* (2nd ed.). Amsterdam: Elsevier.

Barbour, N. M. (2010). *Inertial navigation sensors*. Report, Charles Stark Draper Laboratory (P-4994) Cambridge, MA 02139 USA. Retrieved January 2017, from: http://oai.dtic.mil/oai/oai?verb=getRecord&metadataPrefix=html&identifier=ADA581016

Barbour, N. M., & Schmidt, G. (2001). Inertial sensor technology trends. *IEEE Sensors Journal*, *1*(4), 332–339. doi:10.1109/7361.983473

Beauregard, S. (2006). A helmet-mounted pedestrian dead reckoning system. *Proceedings of International Forum on Applied Wearable Computing IFAWC*, *6*, 1–11.

Brajdic, A., & Harle, R. (2013). Walk detection and step counting on unconstrained smartphones. *Proceedings of ACM International Joint Conference on Pervasive and Ubiquitous Computing UbiComp'13*, 225-234. doi:10.1145/2493432.2493449

Bylemans, I., Weyn, M., & Klepal, M. (2009). Mobile phone-based displacement estimation for opportunistic localisation systems. *Proceedings of 3rd International Conference on Mobile Ubiquitous Computing, Systems, Services and Technologies UBICOMM'09*, 113-118. doi:10.1109/UBICOMM.2009.23

Che, L., Xiong, B., Li, Y., & Wang, Y. (2010). A novel electrostatic-driven tuning fork micromachined gyroscope with a war structure operating at atmospheric pressure. *Journal of Micromechanics and Microengineering*, *20*(1), 15–25. doi:10.1088/0960-1317/20/1/015025

Chen, R., Pei, L., Liu, J., & Leppäkoski, H. (2012). WLAN and Bluetooth positioning in smart phones. In R. Chen (Ed.), *Ubiquitous positioning and mobile location-based services in smart phones* (pp. 44–68). Hershey, PA: IGI Global. doi:10.4018/978-1-4666-1827-5.ch003

Cummins, M., & Newman, P. (2008). Accelerated appearance-only SLAM. In *Proceedings of IEEE International Conference on Robotics and Automation* (pp. 1828-1833). IEEE Xplore.

Dellaert, F., Fox, D., Burgard, W., & Thrun, S. (1999). Monte Carlo localization for mobile robots. In *Proceedings of IEEE International Conference Robotics and Automation 1999* (vol. 2, pp. 1322-1328). IEEE.

Dovis, F., Chiasserini, C. F., Musumeci, L., & Borgiattino, C. (2014). Context-aware peer-to-peer and cooperative positioning. *Proceedings of International Conference on Localization and GNSS 2014 ICL-GNSS 2014*, 1-6.

Faragher, R., & Harle, R. (2014). An analysis of the accuracy of Bluetooth Low Energy for indoor positioning applications. In *Proceedings of 27th International Technical Meeting of The Satellite Division of the Institute of Navigation ION GNSS+ 2014* (pp. 201-210). Institute of Navigation.

Gad-el-Hak, M. (2001). Introduction. In M. Gad-el-Hak (Ed.), *The MEMS handbook*. CRC Press. doi:10.1201/9781420050905.sec1

Gelb, A. (1974). *Applied . Cambridge, Massachusetts, USA, and London*. UK: The M.I.T. Press.

Goyal, P., Ribeiro, V., Saran, H., & Kumar, A. (2011). Strap-down pedestrian dead-reckoning system. In *Proceedings of International Conference on Indoor Positioning and Indoor Navigation IPIN'11* (pp. 1-7). IEEE Xplore.

Grejner-Brzezinska, D. A., Toth, C. K., Li, L., Park, J., Wang, X., Sun, H., & Zheng, Y. F. et al. (2009). Positioning in GPS-challenged environments: Dynamic sensor network with distributed GPS aperture and internodal ranging signals. In *Proceedings of ION GNSS Conference* (pp. 111-123). Institute of Navigation.

Hasnur Rabiain, A., Kealy, A., Retscher, G., Alam, N., Dempster, A., Grejner-Brzezinska, D. A., & Gikas, V. et al. (2013). Cooperative positioning algorithms and techniques for land mobile applications. *Proceedings of 8th International Symposium on Mobile Mapping Technology*.

Herrera-May, A. L., Aguilera-Cortés, L. A., García-Ramírez, P. J., & Manjarrez, E. (2009). Resonant magnetic field sensors based on MEMS technology. *Sensors (Basel)*, *9*(10), 7785–7813. doi:10.3390/s91007785 PMID:22408480

Herrera-May, A. L., Aguilera-Cortés, L. A., García-Ramírez, P. J., Mota-Carrillo, N. B., Padrón-Hernández, W. Y., & Figueras, E. (2011). Development of resonant magnetic field microsensors: Challenges and future applications. In O. Minin (Ed.), *Microsensors* (pp. 65–84). Rijeka, Croatia: InTech.

Hide, C., Botterill, T., & Andreotti, M. (2009). An integrated IMU, GNSS and image recognition sensor for pedestrian navigation. In *Proceedings of ION GNSS Conference*. Institute of Navigation.

Hightower, J., & Boriello, G. (2001). *A survey and taxonomy of location systems for ubiquitous computing. Technical Report*. University of Washington, Department of Computer Science and Engineering.

Hsieh, S.-J. (2013). Challenges and opportunities in MEMS development, assembly and applications. *Assembly Automation*, *33*(2), 2. doi:10.1108/aa.2013.03333baa.001

Kealy, A., & Retscher, G. (2017). MEMS and wireless options in cellular phones for user localization. *Proceedings of ION Pacific PNT*.

Kealy, A., Retscher, G., Toth, C. K., Hasnur-Rabiain, A., Gikas, V., Grejner-Brzezinska, D. A., & Moore, T. et al. (2015). Collaborative navigation as a solution for PNT applications in GNSS challenged environments – Report on field trials of a joint FIG/IAG working group. *Journal of Applied Geodesy*, *9*(4), 244–263. doi:10.1515/jag-2015-0014

Kim, J. W., Jang, H. J., Hwang, D.-H., & Park, C. (2004). A step, stride and heading determination for the pedestrian navigation system. *Journal of Global Positioning Systems*, *3*(1-2), 273–279. doi:10.5081/jgps.3.1.273

Kotanen, A., Hännikäinen, M., Leppäkoski, H., & Hämälainen, T. D. (2003). Positioning with IEEE 802.11b wireless LAN. *Proceedings of 14th International Symposium on Personal, Indoor and Mobile Radio Communications PIMRC*, 2218-2222. doi:10.1109/PIMRC.2003.1259110

Lee, S., & Mase, K. (2002). Activity and location recognition using wearable sensors. *IEEE Pervasive Computing*, *1*(3), 24–32. doi:10.1109/MPRV.2002.1037719

Leinenbach, C., Kattelus, H., & Knechtel, R. (2010). Surface micromachining. In V. Lindroos, M. Tilli, A. Lehto, & T. Motooka (Eds.), *Handbook of silicon based MEMS materials and technologies* (pp. 473–487). William Andrew Applied Science Publisher. doi:10.1016/B978-0-8155-1594-4.00028-0

Li, L., Stankovic, V., Stankovic, L., Li, L., Cheng, S., & Uttamchandani, D. (2011a). Single pixel optical imaging using a scanning MEMS mirror. *Journal of Micromechanics and Microengineering*, *21*(2), 025022. doi:10.1088/0960-1317/21/2/025022

Li, X., Wang, J., & Li, T. (2013). Seamless positioning and navigation by using geo-referenced images and multi-sensor data. *Sensors (Basel)*, *13*(7), 9047–9069. doi:10.3390/s130709047 PMID:23857267

Li, Y., Zheng, Q., Hu, Y., & Xu, Y. (2011b). Micromachined piezoresistive accelerometer based on a asymmetrically gapped cantilever. *Journal of Microelectromechanical Systems*, *20*(1), 83–94. doi:10.1109/JMEMS.2010.2100024

Liu, H., Darabi, H., Banerjee, P., & Liu, J. (2007). Survey of wireless indoor positioning techniques and systems. *IEEE Transactions on Systems, Man and Cybernetics. Part C, Applications and Reviews*, *37*(6), 1067–1080. doi:10.1109/TSMCC.2007.905750

Liu, J. (2012). Hybrid positioning with smart phones. In R. Chen (Ed.), *Ubiquitous positioning and mobile location-based services in smart phones* (pp. 159–194). Hershey, PA: IGI Global. doi:10.4018/978-1-4666-1827-5.ch007

Marschollek, M., Goevercin, M., Wolf, K.-H., Song, B., Gietzelt, M., Haux, R., & Steinhagen-Thiessen, E. (2008). A performance comparison of accelerometry-based step detection algorithms on a large, non-laboratory sample of healthy and mobility-impaired persons. *Proceedings of 30th Annual International Conference of the IEEE Engineering in Medicine and Biology Society EMBS '08*, 1319-1322. doi:10.1109/IEMBS.2008.4649407

Mautz, R. (2012). Indoor positioning technologies. Swiss Geodetic Commission, Geodetic-Geophysical Reports of Switzerland, 86, 134 pgs.

Mezentsev, O. A. (2005). *Sensor aiding of HSGPS pedestrian navigation* (PhD Dissertation). Geomatics Engineering, University of Calgary, Canada.

Mian, A., & Law, J. (2010). Geometric optimization of a van der pauw structure based MEMS pressure sensor. *Microsystem Technologies*, *16*(11), 1921–1929. doi:10.1007/s00542-010-1124-7

Moafipoor, S. (2009). *Intelligent personal navigator supported by knowledge-based systems for estimating dead reckoning navigation parameters* (PhD Dissertation). The Ohio State University, USA.

Mok, E., & Retscher, G. (2007). Location determination using WiFi – fingerprinting versus WiFi – trilateration. *Journal of Location Based Services*, *1*(2), 145–159. doi:10.1080/17489720701781905

Mok, E., Retscher, G., Wang, D., & Xia, L. (2011). Use of smartphones for tracking and trip recording. *Proceedings of 8th International Symposium on Location-Based Services*, 137-152.

Müller, G., Friedberger, A., & Knese, K. (2010). Porous silicon based MEMS. In V. Lindroos, M. Tilli, A. Lehto, & T. Motooka (Eds.), *Handbook of silicon based MEMS materials and technologies* (pp. 409–431). William Andrew Applied Science Publisher. doi:10.1016/B978-0-8155-1594-4.00025-5

Nilsson, J. O., Zachariah, D., Skog, I., & Händel, P. (2013). Cooperative localization by dual foot-mounted inertial sensors and inter-agent ranging. *EURASIP Journal on Advances in Signal Processing*, *2013*(1), 164. doi:10.1186/1687-6180-2013-164

Ochieng, W. Y., Quddus, M. A., & Noland, R. B. (2004). Map-matching in complex urban road networks. *Brazilian Journal of Cartography*, *55*(2), 1–14.

Prasithsangaree, P., Krishnamurthi, P., & Chrysanthis, P. K. (2002). On indoor position with wireless LANs. *Proceedings of 13th IEEE International Symposium on Personal, Indoor, and Mobile Radio Communications PIMRC 2002*, 2, 720-724. doi:10.1109/PIMRC.2002.1047316

Quddus, M. A., Ochieng, W. Y., & Noland, R. B. (2007). Current map-matching algorithms for transport applications: State-of-the-art and future research directions. *Transportation Research Part C, Emerging Technologies*, *15*(5), 312–328. doi:10.1016/j.trc.2007.05.002

Quddus, M. A., Ochieng, W. Y., Zhao, L., & Noland, R. B. (2003). A general map matching algorithm for transport telematics applications. *GPS Solutions*, *7*(3), 157–167. doi:10.1007/s10291-003-0069-z

Randell, C., Djiallis, C., & Muller, H. (2003). Personal position measurement using dead reckoning. In: *Proceedings of International Symposium on Wearable Computers ISWC '03* (pp. 166-173). IEEE Computer Society.

Randell, C., & Muller, H. (2000). Context awareness by analysing accelerometer data. *Proceedings of International Symposium on Wearable Computers*, 175-176.

Retscher, G. (2007). Augmentation of indoor positioning systems with a barometric pressure sensor for direct altitude determination in a multi-storey building. *Journal of Cartography and Geographic Information Science CaGIS*, *34*(4), 305–310. doi:10.1559/152304007782382909

Retscher, G. (2016). Indoor navigation. In E.W. Grafarend (Ed.), Encyclopedia of geodesy. Springer International Publishing Switzerland. doi:10.1007/978-3-319-02370-0_9-1

Retscher, G., & Hofer, H. (2017). Wi-Fi location fingerprinting using an intelligent checkpoint sequence. *Journal of Applied Geodesy*, *11*(3). doi:10.1515/jag-2016-0030

Retscher, G., Hofer, H., Kealy, A., Gikas, V., & Obex, F. (2017). Cooperative localization in indoor environments using constrained differential Wi-Fi and UWB measurements. In *Proceedings of ION GNSS++ Conference*. Institute of Navigation.

Retscher, G., & Mok, E. (2004). Sensor fusion and integration using an adapted Kalman filter approach for modern navigation systems. *Survey Review*, *37*(292), 439–447. doi:10.1179/sre.2004.37.292.439

Retscher, G., & Tatschl, T. (2016). Indoor positioning using Wi-Fi lateration – Comparison of two common range conversion models with two novel differential approaches. In *Proceedings of 2016 4th International Conference on Ubiquitous Positioning, Indoor Navigation and Location Based Services UPINLBS*. IEEE Xplore. doi:10.1109/UPINLBS.2016.7809967

Retscher, G., & Thienelt, M. (2004). NAVIO – A navigation and guidance service for pedestrians. *Journal of Global Positioning Systems CPGPS*, *3*(1-2), 208–217. doi:10.5081/jgps.3.1.208

Retscher, G., Zhu, M., & Zhang, K. (2012). RFID positioning. In R. Chen (Ed.), *Ubiquitous positioning and mobile location-based services in smart phones* (pp. 69–95). Hershey, PA: IGI Global. doi:10.4018/978-1-4666-1827-5.ch004

Robertson, D., & Cipolla, R. (2004). An image-based system for urban navigation. *Proceedings of the British Machine Vision Conference*, 260-272.

Ruotsalainen, L. (2013). *Vision-aided pedestrian navigation for challenging GNSS environments* (PhD Dissertation). Publications of the Finnish Geodetic Institute.

Ruotsalainen, L., & Kuusniemi, H. (2012). Visual positioning in a smartphone. In R. Chen (Ed.), *Ubiquitous positioning and mobile location-based services in smart phones* (pp. 130–158). Hershey, PA: IGI Global. doi:10.4018/978-1-4666-1827-5.ch006

Ruotsalainen, L., Kuusniemi, H., & Chen, R. (2011). Visual-aided two-dimensional pedestrian indoor navigation with a smartphone. *Journal of Global Positioning Systems, 10*(1), 11–18. doi:10.5081/jgps.10.1.11

Saeedi, S., & El-Sheimy, N. (2015). Activity recognition using fusion of low-cost sensors on a smartphone for mobile navigation application. *Micromachines, 6*(8), 1100–1134. doi:10.3390/mi6081100

Sharma, P., & Inoue, A. (2010). Metallic glass. In V. Lindroos, M. Tilli, A. Lehto, & T. Motooka (Eds.), *Handbook of silicon based MEMS materials and technologies* (pp. 447–472). William Andrew Applied Science Publisher. doi:10.1016/B978-0-8155-1594-4.00027-9

Shkel, A. (2001). *Micromachined gyroscopes: Challenges, design solutions, and opportunities. Smart Structures and Materials* (Vol. 4334). SPIE.

Steinhoff, U., Omerčević, D., Perko, R., Schiele, B., & Leonardis, A. (2007). How computer vision can help in outdoor positioning. In *Proceedings of the European conference on Ambient intelligence 2007* (pp. 124-141). Springer-Verlag Berlin Heidelberg. doi:10.1007/978-3-540-76652-0_8

Stojanović, D., & Stojanović, N. (2014). Indoor localization and tracking: Methods, technologies and research challenges. *Facta Universitatis, Series. Automatic Control and Robotics, 13*(1), 57–72.

Trevisiani, E., & Vitaleti, A. (2004). Cell-ID location technique, limits and benefits: an experimental study. *Proceedings of 6th IEEE Workshop on Mobile Computing Systems and Applications WMCSA 2004*, 51-60. doi:10.1109/MCSA.2004.9

Vogel, L. (2016). *Getting started with Android development – Tutorial.* Retrieved March 2017, from: http://www.vogella.com/tutorials/Android/article.html

Wang, N., Zhang, N., & Wang, M. (2006). Wireless sensors in agriculture and food industry—Recent development and future perspective. *Computers and Electronics in Agriculture, 50*(1), 1–14. doi:10.1016/j.compag.2005.09.003

Weinberg, M., & Kourepenis, A. (2006). Error sources in in-plane silicon tuning fork gyroscopes. *Journal of Micromechanical Systems, 15*(3), 479–491. doi:10.1109/JMEMS.2006.876779

Welch, G., & Bishop, G. (2006). *An introduction to the Kalman filter.* Report, UNC-Chapel Hill TR 95-041.

Wymeersch, H., Lien, J., & Win, M. (2009). Cooperative localization in wireless networks. *Proceedings of the IEEE, 97*(2), 427–450. doi:10.1109/JPROC.2008.2008853

Yazdi, N., Ayazi, F., & Najafi, K. (1998). Micromachined inertial sensors. *Proceedings of the IEEE, 86*(8), 1640–1659. doi:10.1109/5.704269

Zhang, W., & Kosecka, J. (2006). Image based localization in urban environments. In *3rd International Symposium on 3D Data Processing, Visualization and Transmission* (pp. 33-40). IEEE Computer Society.

Zorn, A. (2002). GPS-aided all-accelerometer inertial navigation. In *Proceedings of International Technical Meeting of Satellite Division of the Institute of Navigation.* Institute of Navigation.

KEY TERMS AND DEFINITIONS

Accelerometer: A device that measures physical acceleration experienced by an object whereby the measured acceleration is relative to a free-fall, or inertial, observer who is momentarily at rest relative to the object being measured.

Barometer: Any instrument that measures atmospheric pressure. An altimeter can determine the altitude from measurements of the barometric pressure differences.

Bluetooth: A wireless technology standard for exchanging data over short distances (using short-wavelength UHF radio-waves in the ISM band) from fixed and mobile devices, and building personal area networks (PANs). Using tags and Bluetooth Low Energy beacons Bluetooth is useable for positioning.

Dead Reckoning (DR): Process of calculating one's current position by using a previously determined position and advancing that position based upon known or estimated speeds over elapsed time and course.

Gyroscope: A spinning wheel or disc in which the axis of rotation is free to assume any orientation by itself.

Inertial Navigation System (INS): A navigation system that keeps track of its position by accurately measuring acceleration with accelerometers and rotation with gyroscopes. Unlike other navigation systems, an INS does not depend on external (radio) measurements.

Kalman Filter: A recursive estimation algorithm that uses a series of measurements observed over time, containing statistical noise and other inaccuracies. The filter produces estimates of unknown variables that tend to be more precise than those based on a single measurement alone, by using Bayesian inference and estimating a joint probability distribution over the variables for each timeframe.

Lateration: Is the process of determining absolute and relative locations by measurement of ranges. In 2D geometry, when it is known that a point lies on two curves such as two circles, then the circle centers and the two radii provide sufficient information to narrow the possible locations to two. If three known locations are involved it is referred to as trilateration.

Location Fingerprinting: A positioning process of finding a location from which fingerprints stored in a database have the best match to the snap-shot of the currently observed measurements. Using wireless access points or other base stations, the intensity of the RSS can be measured leading to a position solution via fingerprinting.

Magnetometer: An instrument for measuring the intensity of a magnetic field, especially the earth's magnetic field. A compass is a simple example of a magnetometer, one that measures the direction of an ambient magnetic field.

Map Matching (MM): The process of localizing vehicle's positions from sensor(s) in its actual road position on a map.

MEMS: Micro-Electro-Mechanical Systems, or MEMS, is a technology that in its most general form can be defined as miniaturized mechanical and electro-mechanical elements (i.e., devices and structures) that are made using the techniques of microfabrication. The critical physical dimensions of MEMS devices can vary from well below one micron on the lower end of the dimensional spectrum, all the way to several millimeters.

Received Signal Strength (RSS): Is an indication of power level from a transmitter received by an antenna.

Sensor Fusion: It is combining of sensory data or data derived from disparate sources such that the resulting information has less uncertainty than would be possible when these sources were used individually. The data sources for a fusion process are not specified to originate from identical sensors. The term uncertainty reduction in this case can mean more accurate, more complete, or more dependable.

Visual-Aiding: The usage of position information obtained from images to ease the resolving of the user position or to make the positioning solution more accurate or more widely available.

Wireless Fidelity (Wi-Fi): A technology for wireless local area networking with devices based on the IEEE 802.11 standards to extend or substitute for a wired local area network, such as Ethernet. It is originally a technology for short-range wireless data communication and is typically deployed as an ad-hoc network in a hot-spot fashion in the areas where a wireless Internet access is needed. Using RSS measurements in can also be employed for localization.

Chapter 10
Multi-Sensor Integrated Navigation in Urban and Indoor Environments:
GNSS, Inertial, and Range Sensors Integration

Mohamed Mohamed Atia
Carleton University, Canada

ABSTRACT

The art of multi-sensor processing, or "sensor-fusion," is the ability to optimally infer state information from multiple noisy streams of data. One major application area where sensor fusion is commonly used is navigation technology. While global navigation satellite systems (GNSS) can provide centimeter-level location accuracy worldwide, they suffer from signal availability problems in dense urban environment and they hardly work indoors. While several alternative backups have been proposed, so far, no single sensor or technology can provide the desirable precise localization in such environments under reasonable costs and affordable infrastructures. Therefore, to navigate through these complex areas, combining sensors is beneficial. Common sensors used to augment/replace GNSS in complex environments include inertial measurement unit (IMU), range sensors, and vision sensors. This chapter discusses the design and implementation of tightly coupled sensor fusion of GNSS, IMU, and light detection and ranging (LiDAR) measurements to navigate in complex urban and indoor environments.

INTRODUCTION

Currently, the dominant navigation technology is the Global Navigation Satellite Systems (GNSSs) (Misra & Enge, 2011) (Farrell, 2008) (Ahmed, 2002). GNSS-based localization has been significantly improved in the recent decade by having more satellite constellations and by developing augmentation networks (Misra & Enge, 2011) (e.g. Wide Area Augmentation Systems "WAAS", Space Based Augmentation Systems "SBAS"). In addition to the U.S. system (the global positioning system "GPS"), several other

DOI: 10.4018/978-1-5225-3528-7.ch010

satellite-based systems have been deployed or redeployed such as the Russian system (Global Navigation Satellite System "GLONASS"), the Chinese system (BeiDou Navigation Satellite System "BDS"), and the European satellite navigation system "Galileo". Having several satellite constellations increases the availability and coverage of positioning service. For example, the GLONASS has better coverage in the northern latitudes which supplements the limited coverage of GPS satellites in these high latitude areas. Furthermore, increased number of satellites enhances the satellites-user geometry which consequently enhance location accuracy. While GNSSs provide accurate absolute positioning under open-sky, it suffers from signal attenuation and multi-path in dense urban areas. In indoor areas, signals transmitted by GNSS satellites are often not receivable anymore. Therefore, navigation in dense urban and indoor environments is still a challenge that cannot be addressed by a single navigation technology under reasonable costs and affordable infrastructures.

To fill this gap, several alternatives to GNSS have been proposed. One of the most common alternative to GNSS is Inertial navigation systems (INS) (Farrell, 2008) (Savage, 2007) (El-Sheimy, Hou, & Niu, 2007) (Titterton & Weston, 2004). INS is one of the earliest navigation technologies that can be used to design and develop self-contained independent navigation systems that can work in complex environments. The general concept of the integration (i.e. fusion) of GNSS and INS is the online recursive calibration of INS sensors for errors during strong GNSS coverage which is utilized later to enhance INS standalone performance when GNSS is lost or attenuated. However, inertial sensors (especially low-cost ones) have complex stochastic error characteristics that changes dynamically. Therefore, even a well-calibrated INS cannot sustain reliable accuracy for long periods without external measurements due to unavoidable sensors errors accumulation as the chapter will demonstrate later. To develop robust navigation systems that can sustain for long periods without GNSS, more alternatives are needed.

In mapping technology, range and vision sensors (Groves, 2013) (Geiger, Ziegler, & Stiller, 2011) are used to develop a map of the surroundings. Current mapping technology depends on an accurate positioning system to project range/vision sensors on a reference frame and, hence, construct a map of the environment. Without a reliable positioning system, an accurate map cannot be developed. To overcome this issue, the Simultaneous Localization and Mapping (SLAM) (Harmat, Trentini, & Sharf, 2015) (Huang & Dissanayake, 2007) (Luo & Lai, 2014) technology is used. To implement accurate SLAM, commonly post-processing of collected data is needed to perform a global optimization (including loop closures and bundle adjustments) to generate a precise map and an accurate record of the travelled trajectory. While SLAM is a beautiful approach that solves both localization and mapping at the same formulation, it may suffer from scalability and real-time challenges in large urban and indoor areas due to the accumulation of large amount of map features. Real-time SLAM-based localization may be feasible using fast SLAM algorithms such as "fastSLAM" (Montemerlo, Thrun, Koller, & Wegbreit, 2002) or if a map of the area pre-exists (Lynen, et al., 2015). However, in highly dynamic environments, maps may need frequent updates. To address this challenge, the integration of range/vision sensors in the online positioning process (i.e. filtering) has been proposed (Perera, Wijesoma, & Adams, 2010) (Luo & Lai, 2014). The idea is to use mapping sensors (either range or vision) as "motion" sensors (not only as environment perception/mapping sensors) and to fuse their motion information with other common sensors (e.g. GPS/INS). Therefore, this chapter will focus on the fusion of relative motion information extracted from LiDAR sensors with GNSS/INS assembly to enhance positioning in dense urban and indoor environments.

The main objective of the chapter is to discuss the design and implementation of tightly-coupled multi-sensors integrated navigation systems that go beyond GNSS/INS integration. As a case study, the chapter describes how relative motion estimation from mapping sensors (LiDAR in this work) can be fused with tightly-coupled GNSS/INS assembly to enhance the navigation accuracy and sustainability in complex dense urban and indoor environments without the need for a pre-defined map of the area. The chapter will study each individual sensor to analyze the merits and the limitations of each. Afterwards it will then show how sensor fusion algorithms can be used to enhance individual sensors performance and to improve the overall positioning and navigation accuracy in complex dense urban and indoor areas.

BACKGROUND

To describe Position, Velocity and Attitude (PVA) information, they need to be referenced to a system of coordinate frames. The common coordinate frames used in this chapter are Body Frame (*B*), Local Level Navigation Frame (*L*), Earth Centered Earth Fixed (*ECEF*). All frames are described in Figure 1, Figure 2, and Table 1.

Kinematic Modeling of 3D Motion of Rigid Bodies

3D motion of rigid bodies is fully characterized by translational and rotational rates. An inertial sensor measures 3D motion using accelerometers and gyroscopes sensors. To measure full 3D motion, 6 degrees of freedom (DOF) inertial sensor (3 accelerometer axes and 3 gyroscopes axes) are needed. Translational rate can be inferred from acceleration (measured in m/s^2) and rotational rate is simply the angular velocity measured in rad/s or degrees/s. The performance and accuracy of inertial sensors depends on the material, the physics and fabrication process of the sensor. The physics and types of inertial sensors are beyond the scopes of this chapter. Commonly, what navigation researchers and engineers need to understand to use inertial sensors properly is the measurements, grade and types of errors these sensors exhibit. An Inertial

Figure 1. Body Frame (B)

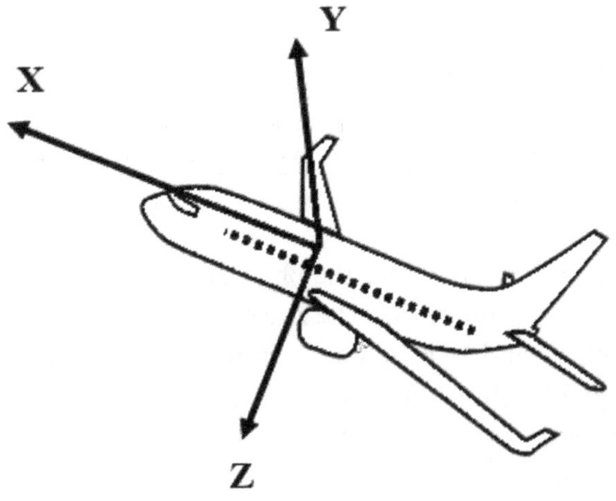

Figure 2. Local Level Navigation Frame (L) and Earth Centered Earth Fixed (ECEF) Frame

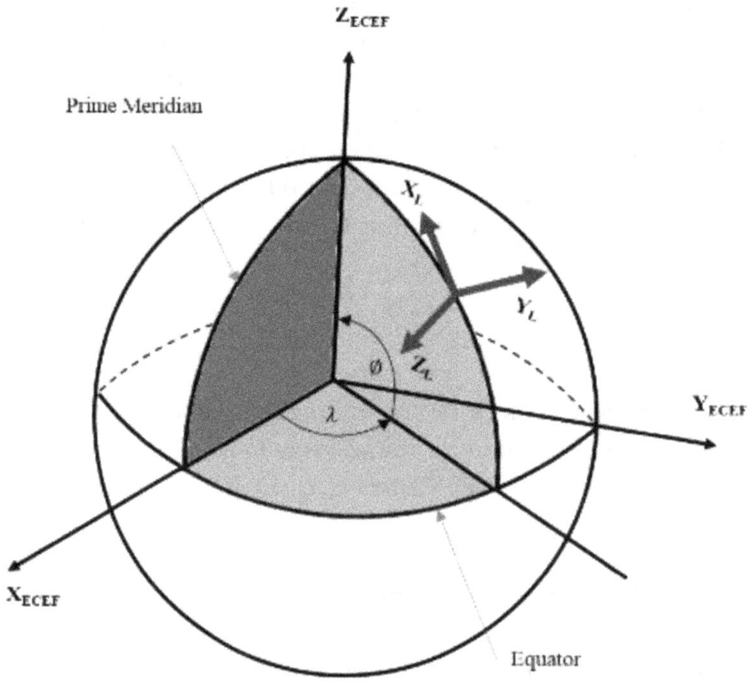

Table 1. Reference Coordinate Frames

Frame	Definition
B: body frame	Origin: Body center of mass X: Longitudinal (forward) direction Y: Transversal (lateral) direction Z: Down (vertical) direction
L: local-level navigation frame	Always leveled with respect to Earth ellipsoid surface Origin: Vehicle location X: True north direction Y: East direction Z: Down vertical direction
E: Earth- Centered Earth-Fixed (ECEF)	Rotates with Earth Origin: Center of the Earth Z: extends through the North Pole. X: passes through the intersection of the equatorial plane and the prime meridian. Y: completes the right-hand coordinate system in the equatorial plane.

measurement unit (IMU) is a device that wraps the physical sensor and includes several preprocessing algorithms to produce accurate measurements. The reason is that the physical characteristics of inertial sensors materials change with several factors such as humidity and temperature. Therefore, one of the most common preprocessing algorithm inside an IMU is temperature calibration that compensate the raw measurements for temperature variation. That is why some IMU may provide temperature readings in addition to inertial measurements.

Inertial Sensor Measurements

This section will show typical measurements and analyze common errors of a MEMS-based IMU. MEMS stands for micro-electro-mechanical system which is the name of the technology used to manufacture such sensors. The advantage of MEMS IMU is the small size, light weight, and low cost. It can be found in smartphones, tablets, gaming, and virtual reality devices. However, the downside of MEMS IMU is the accuracy as will be shown shortly. Figure 3 shows a MEMS IMU board that includes the MPU9250 IMU from Invensense and a microcontroller (SAMD21: 32-bit ARM Cortex-M0). The microcontroller can read measurements from the IMU.

MPU9250 IMU includes a motion processor that correct the measurements for temperature and other errors such as vibration effects. However, as we will see shortly, measurements output is still contaminated by different types of errors. Figure 4 shows IMU measurements in static condition (the IMU was placed on a flat table and hold static for 5 hours). The measurements were taken while x-axis point to vertical upward. An ideally vertically aligned upward accelerometer should measure exactly the gravity $\left(g \cong 9.8m / s^2\right)$ while leveled horizontal axis should measure zero. All gyroscope measurements should measure exactly zero (technically speaking it will measure the Earth rotation rate

Figure 3. Razor 9DOF IMU from SparkFun

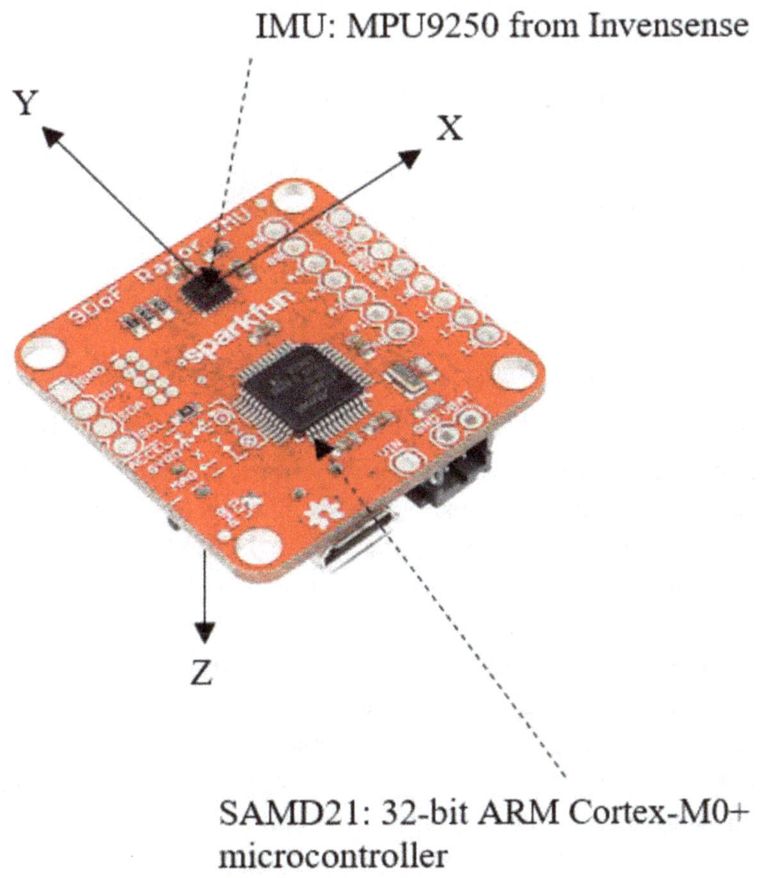

Figure 4. Gyroscope and accelerometer measurements of a stationary IMU for 5 hours

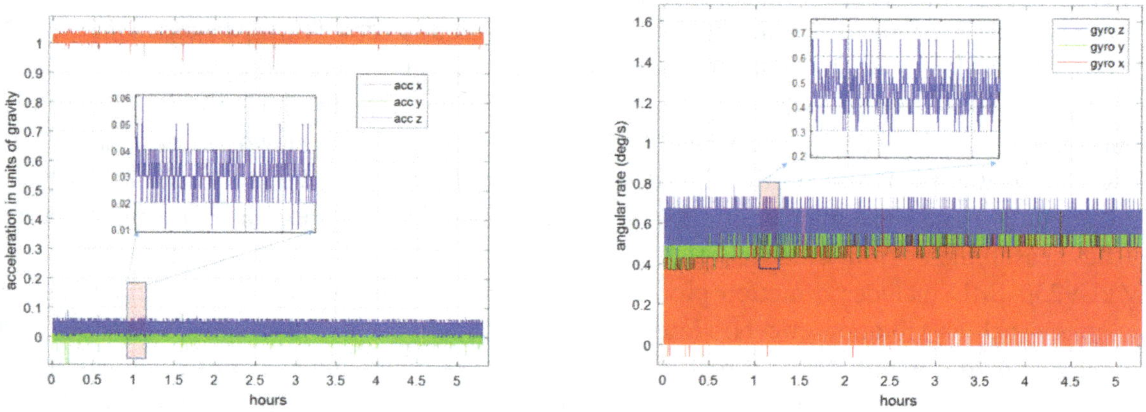

"$w^e \cong \frac{2\pi}{24\,hours} = 7.2722 * 10^{-4}\, rad\,/\,s$" projected on the 3 axes of the IMU frame). However, measurements displayed in Figure 4 shows noisy measurements. The nature of the motion processing equations (will be discussed shortly) leads to the integration of such noisy measurements which accumulates unbounded drifts of the estimated PVA. In the coming section, the effect of such noises on the estimated 3D PVA will be shown.

Inertial Sensor Errors and Measurement Model

Common types of IMU errors include deterministic bias, scale factor, random noise, and axis non-orthogonality. A typical measurement model of IMU is given by:

$$\hat{m}(t) = N(1 + s(t))m(t) + b(t) + n(t) \tag{1}$$

where $\hat{m}(t)$ is the noisy measurement, $m(t)$ is the true measurement, N is the axis-non-orthogonality error matrix, $s(t)$ is the scale factor, $b(t)$ is the bias, and $n(t)$ is Gaussian random noise. Biases and scale factors are deterministic error components that vary from run to run and commonly modelled in sensor fusion algorithms to be estimated by external measurements (e.g. GPS or LiDAR). The random part is commonly reported by manufacturers under the "noise spectral density" or "random walk" names (Naser El-Sheimy, 2008). For example, a gyroscope random walk is called angular random walk (ARW) (Farrell, 2008) and it is measured in $^{o}/\sqrt{s}$ which is an indication of expected angular error if sensor output is integrated over time. The expected angular error is proportional to the square root of integration time (IEEE, 1997). A common method to characterize error components of IMUs is to processes a long stationary history of IMU data using Allan Variance method (Naser El-Sheimy, 2008) (IEEE, 1997). Allan variance is a statistical method that was originally developed by David W. Allan (Allan, 1966) to study frequency fluctuations of an atomic oscillator. A 5-hour stationary IMU data recorded at 100Hz and processed by Allan Variance algorithm for the gyroscope measurements are shown in Figure 5.

Figure 5. Allan Variance results for gyroscope measurements from the MPU9250 IMU

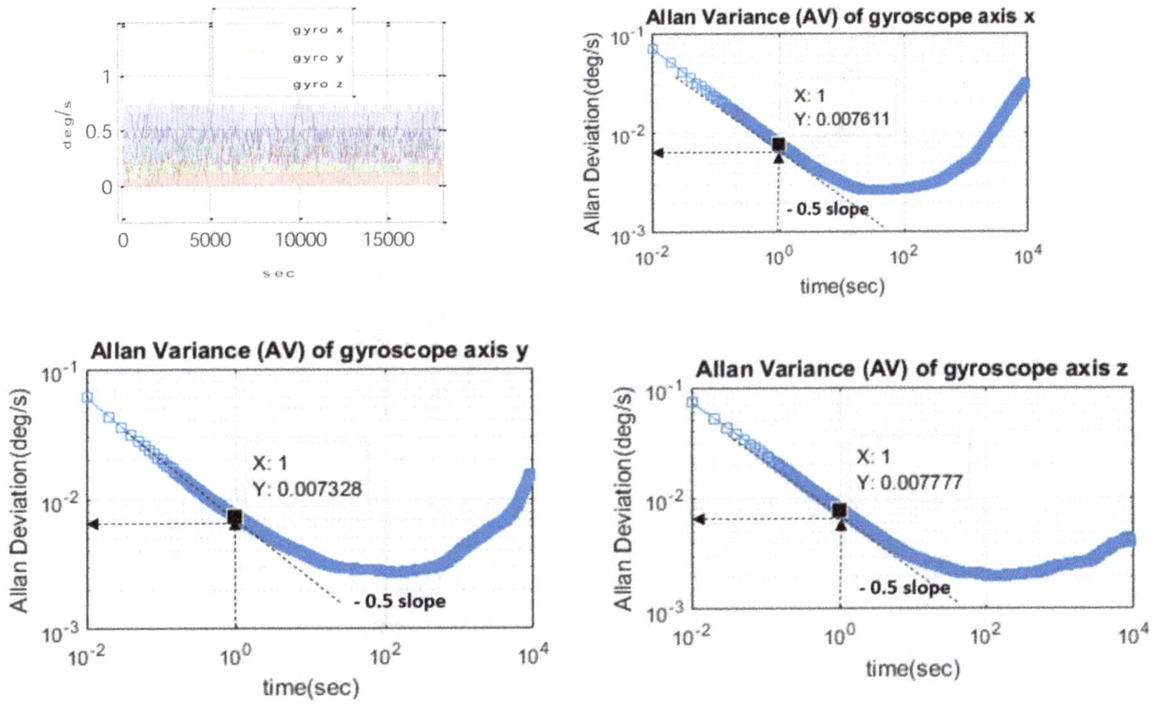

According to the datasheet of the MPU9250, gyroscope ARW is 0.001 $°/\sqrt{s}$ at 100Hz sampling rate. To verify this value, ARW can be estimated from Allan Variance curve by taking the intersection of the Allan variance curve at -0.5 slope with the time value of 1s which shows that the gyroscopes AWR is approximately 0.007 $°/\sqrt{s}$.

3D IMU-Based Kinematic Model

The kinematic model (also known as system model, transition model, and mechanization equations) is a model that processes inertial measurements as input and calculate the 3D PVA of the object as output. The most common kinematic model in modern navigation systems is Strapdown (Savage, 2007) where the IMU is rigidly attached to the body. The Strapdown kinematic model consists of orientation model and velocity model. Velocity is further integrated into position.

3D Orientation Estimation

Estimation of 3D orientation of a rigid body frame (*B*) with respect to a reference local level navigation frame (*L*) is known as Attitude and heading reference system (AHRS) (Farrell, 2008). Orientation can be described in several forms such as direction cosine matrix $C_B^L \in \mathbb{R}^{3x3}$, quaternion vector $q_B^L \in \mathbb{R}^4$, rotation vector $\varnothing_B^L \in \mathbb{R}^3$, or Euler angles (roll: *r*, pitch: *p*, and azimuth: *A*) (Savage, 2007). An illustration graph that shows Euler angles representation is shown in Figure 6. The orientation kinematic

Figure 6. Illustration of Orientation in terms of Euler Angles

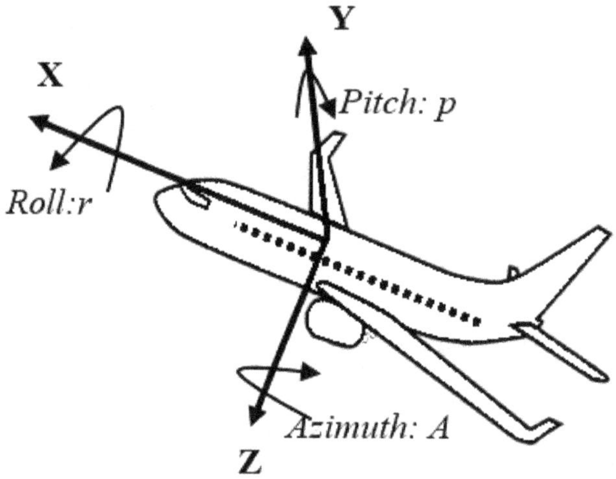

model calculates 3D orientation given ω_{LB}^{B}: the angular rate of frame *B* with respect to frame *L* as represented in frame *B*). Orientation kinematic models are based on the following fundamental differential equation (Savage, 2007):

$$\dot{C}_B^L = C_B^L\left(\omega_{LB}^{B}x\right) \tag{2}$$

Where $\left(\boldsymbol{\omega}_{LB}^{B}x\right)$ is the skew matrix of $\boldsymbol{\omega}_{LB}^{B} = \left[\omega_{LBX}^{B} \quad \omega_{LBY}^{B} \quad \omega_{LBY}^{B}\right]^T$ defined by:

$$\left(\omega_{LB}^{B}x\right) = \begin{bmatrix} 0 & -\omega_{LB_Z}^{B} & \omega_{LB_Y}^{B} \\ \omega_{LB_Z}^{B} & 0 & -\omega_{LB_X}^{B} \\ -\omega_{LB_Y}^{B} & \omega_{LB_X}^{B} & 0 \end{bmatrix} \tag{3}$$

While popular Euler angles are easily readable for users, direction cosine matrix is convenient in transforming vectors between frames. If we define $cp = \cos\left(p\right), sp = \sin\left(p\right), cr = \cos\left(r\right)$, $sr = \sin\left(r\right), cA = \cos\left(A\right)$, $sA = \sin\left(A\right)$, direction cosine matrix can be obtained from Euler angles as follows (Savage, 2007):

$$C_B^L = \begin{bmatrix} cpcA & -crsA + srspcA & srsA + crspcA \\ cpsA & crcA + srspsA & -srcA + crcpsA \\ -sp & srcp & crcp \end{bmatrix} \tag{4}$$

Although direction cosine matrix rate equation looks simple, it suffers from singularity due to non-linear trigonometric functions. Therefore, quaternion domain is preferred as a stable and singularity-free way to represent rotations. q_B^L has four components (*a*,*b*,*c*, and *d*). $a = cos\left(\frac{\varnothing_B^L}{2}\right)$ represents the scalar part of q_B^L and $[b\ c\ d]^T$ is the vector part given by $sin\left(\frac{\varnothing_B^L}{2}\right)\left[l \quad m \quad n\right]^T$ where $\left[l \quad m \quad n\right]^T$ is the rotation vector $\varnothing_B^L$. q_B^L satisfies the following constraints:

$$a^2 + b^2 + c^2 + d^2 = 1 \tag{5}$$

In addition to stable singularity-free rotation representation, additional attractive feature of quaternion is that the orientation differential equation is linear as follows:

$$\dot{q}_B^L = \frac{1}{2} q_B^L W_{LB}^B \tag{6}$$

where W_{LB}^B is defined as:

$$W_{LB}^B = \left[0 \quad \omega_{LB}^B\right]^T \tag{7}$$

Multiplication in (6) is "quaternion multiplication" defined by:

$$\dot{q}_B^L = \frac{1}{2}\begin{bmatrix} a & -b & -c & -d \\ b & a & -d & c \\ c & d & a & -b \\ d & -c & b & a \end{bmatrix} W_{LB}^B \tag{8}$$

The direction cosine matrix can be obtained from quaternion vector as follows:

$$\begin{aligned}
C_B^L(1,1) &= \left(a^2 + b^2 - c^2 - d^2\right) \\
C_B^L(1,2) &= 2(bc - ad) \\
C_B^L(1,3) &= 2(bd + ac) \\
C_B^L(2,1) &= 2(bc + ad) \\
C_B^L(2,2) &= \left(a^2 - b^2 + c^2 - d^2\right) \\
C_B^L(2,3) &= 2(cd - ab) \\
C_B^L(3,1) &= 2(bd - ac) \\
C_B^L(3,2) &= 2(cd + ab) \\
C_B^L(3,3) &= \left(a^2 - b^2 - c^2 + d^2\right)
\end{aligned} \tag{9}$$

Therefore, most AHRS uses the quaternion-based kinematic model to calculate 3D orientation. The kinematic model in (6) can be rearranged as follows:

$$\dot{q}_B^L = \frac{1}{2}\left(W_{LB}^B x\right)\begin{bmatrix} a & b & c & d \end{bmatrix}^T \tag{10}$$

Where

$$\left(W_{LB}^B x\right) = \begin{bmatrix} 0 & -\omega_{LB_X}^B & -\omega_{LB_Y}^B & -\omega_{LB_Z}^B \\ \omega_{LB_X}^B & 0 & \omega_{LB_Z}^B & -\omega_{LB_Y}^B \\ \omega_{LB_Y}^B & -\omega_{LB_Z}^B & 0 & -\omega_{LB_X}^B \\ \omega_{LB_Z}^B & \omega_{LB_Y}^B & -\omega_{LB_X}^B & 0 \end{bmatrix} \tag{11}$$

Typically, raw gyroscopic measurements need to be compensated for errors (e.g. biases in this implementation). However, for a platform on or close to Earth surface, raw gyroscopic measurements need to be compensated also for earth-rotation rate (ω_{IE}^L) and for any motion of the local level navigation frame (i.e. transport rate "ω_{EL}^L"; the angular rate that will keep the local navigation frame horizontal while moving over the ellipsoid surface of Earth). Therefore, we define the following angular rate measurement vector as follows:

$$\hat{\omega}_{LB}^B = \omega_{LB}^B - b_g - C_L^B \omega_{IL}^I \tag{12}$$

where b_g is the gyroscope biases and

$$\omega_{IL}^I = \omega_{IE}^L + \omega_{EL}^L \tag{13}$$

The earth rotation rate can be calculated as follows:

$$\omega_{IE}^I = \begin{bmatrix} w^e \cos\left(\varphi\right) & 0 & -w^e \sin\left(\lambda\right) \end{bmatrix} \tag{14}$$

Where w^e is Earth rotation and φ, λ are latitude and longitude coordinate of the platform. The transport rate is defined by:

$$\omega_{EL}^L = \begin{bmatrix} \frac{v_e}{\left(R_n + h\right)} & \frac{-v_n}{\left(R_m + h\right)} & \frac{v_e}{\left(R_n + h\right)}\tan\left(\varphi\right) \end{bmatrix} \tag{15}$$

Where h is the height (or altitude) of the vehicle, R_n and R_m are the normal and meridian radii of curvature of Earth ellipsoid at current latitude. Velocity in L frame is defined by:

$$v^L = \begin{bmatrix} v_n & v_e & v_d \end{bmatrix} \tag{16}$$

Due to random sensor errors, we define gyroscope measurements as follows:

$$\hat{\omega}_{LB}^B = \omega_{LB}^B - b_g - \omega_{IL}^B + G_g w \tag{17}$$

where G_g is a noise shaping parameter and w is a zero mean unity variance Gaussian noise vector. By including bias and random noise terms, the orientation kinematic model will be in the form:

$$\dot{q}_B^L = \frac{1}{2}\left(\hat{W}_{LB}^B x\right) q_B^L \tag{18}$$

Where

$$\widehat{W}_{LB}^B = \begin{bmatrix} 0 & \omega_{LB}^B - b_g - \omega_{IL}^B + G_g w \end{bmatrix}^T \tag{19}$$

By solving the orientation kinematic differential equation, the quaternion q_B^L is calculated which can be converted into transformation matrix or Euler angles as described earlier.

3D Velocity and Position Estimation

Velocity is estimated by simply projecting the accelerometer measurements $a_{SF}^B = \begin{bmatrix} f_x & f_y & f_z \end{bmatrix}^T$ from body *B* frame to local level navigation frame *L* and integrate the results once after compensating for gravity force. However, because *L* is a rotating frame, a_{SF}^B must be compensated also for Coriolis effect. Coriolis effect is the acceleration the accelerometer will measure due to the rotation of *L* frame. Therefore, the velocity equation is calculated as follows (Savage, 2007)

$$\dot{v}^L = C_B^L a_{SF}^B + g^L - \left(\omega_{EL}^L + 2\omega_{IE}^L\right) \mathrm{x}\, v^L \tag{20}$$

Where g^L is the gravity vector in *L* frame. Geodetic position can be estimated by:

$$\dot{\varphi} = \frac{v_n}{\left(R_m + h\right)} \tag{21}$$

$$\dot{\lambda} = \frac{v_e}{\left(R_n + h\right)\cos\left(\varphi\right)} \tag{22}$$

$$\dot{h} = -v_d \tag{23}$$

IMU Noise Effects

Ideally, processing the measurements from a stationary IMU should generate zero PVA. However, due to noise, PVA drifts without bounds. In common filtering scenarios, non-orthogonality, scale factor, and bias errors can be estimated. The random noise part is then left up without compensation. Figure 7 shows the results of processing approximately one minute of IMU data at 100Hz setting all axis measurements to zero except gyro-z as seen in the figure. In one minute, we have 0.3° maximum azimuth error which corresponds to ARW of $0.3^o / \sqrt{60s} \cong 0.0387^o / \sqrt{s}$. This shows that the ARW reported in datasheets or estimated by Allan Variance is sometimes optimistic. The effect of gyro-z error on position and velocity errors is also shown in Figure 7.

Figure 7. Errors due to gyro-z random noise

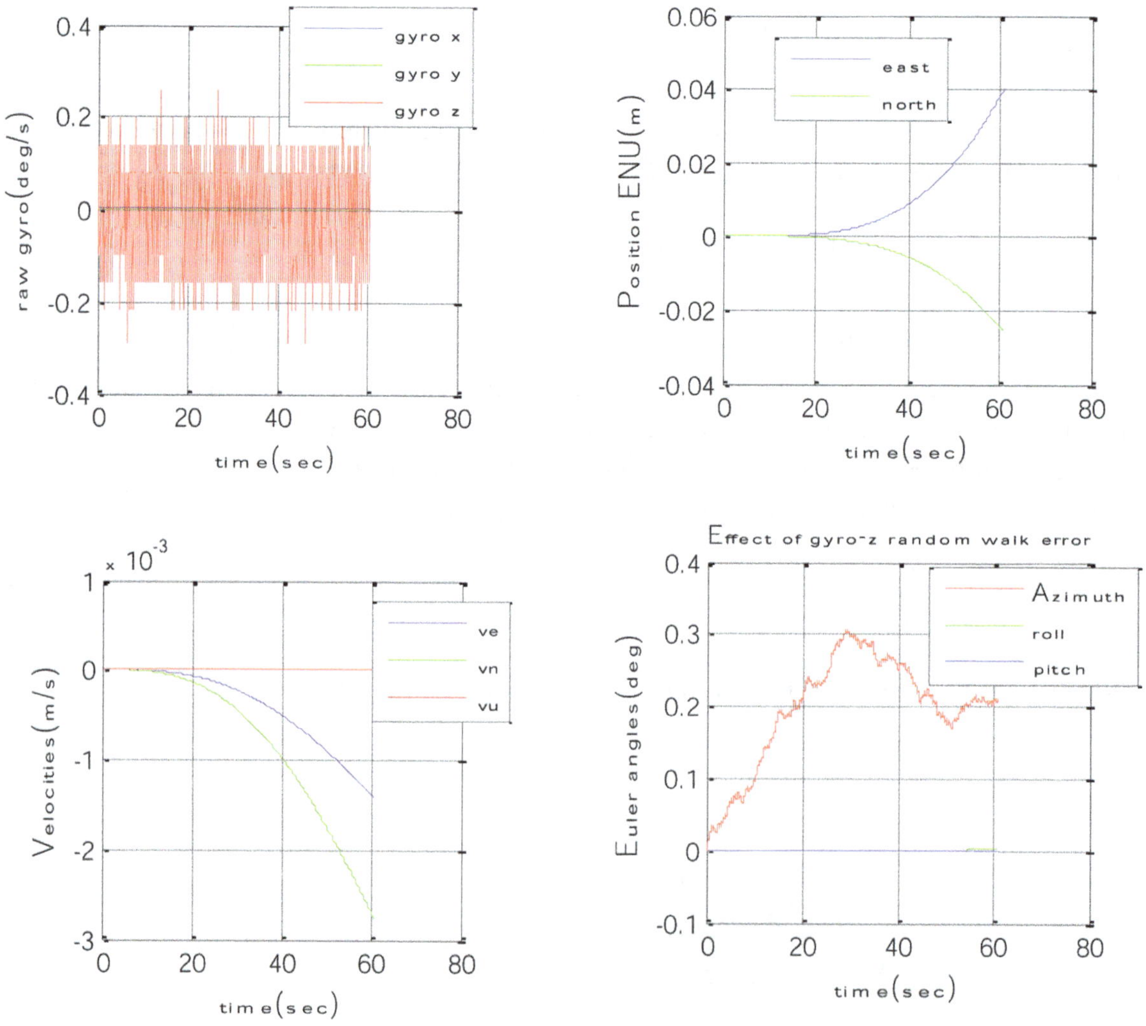

The same experiment was performed but with only gyro-x is non-zero as seen in Figure 8. The error in this case is extremely larger. The reason is that a gyro-x error will lead to tilt angles errors (roll angle in this case). An erroneous roll will project part of the gravity ($sin(r)g)sin(r)g$). on y-axis which is a significant acceleration error that leads to huge error in velocity and, consequently, position errors as seen in Figure 8. Figure 9 shows the effect of z-axis accelerometer noise (called velocity random walk: VRW).

Figure 10 shows the effect of Earth rotation rate if not compensated for in the kinematic model equations. Ignoring the Earth rotation rate effect leads to erroneous rotation around north and vertical axis which is projected on the body frame causing the system to generate erroneous angles (heading and roll). As mentioned above, a small roll error will lead to significant acceleration error which consequently cause velocity and position errors as seen in Figure 10.

Figure 8. Errors due to small tilt (pitch/roll) errors generated by gyro random noise

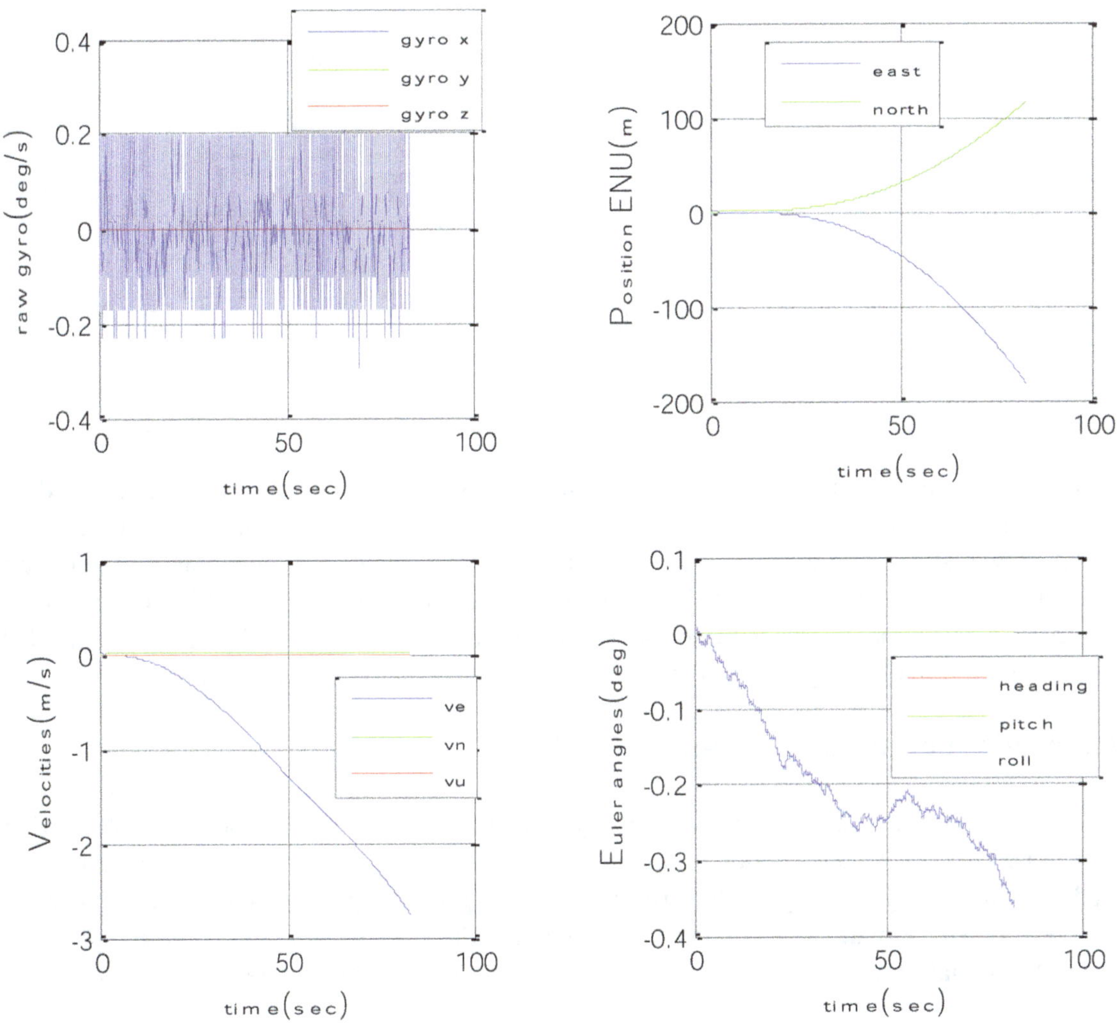

Figure 9. Errors due to accelerometer random noise

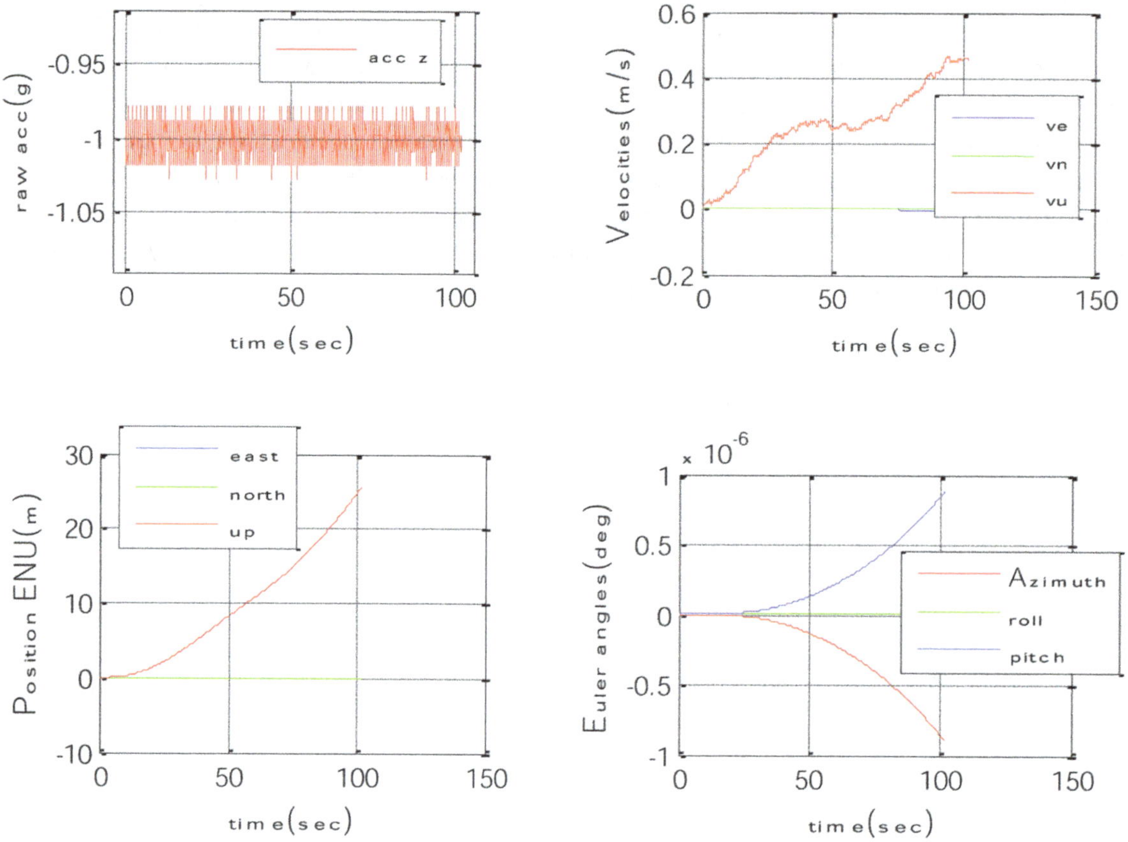

Commonly, the most significant error source after random walks is the bias. Uncompensated biases will accumulate more significant errors in short periods. To show the effect of biases, Figure 11 and Figure 12 show a simulation test where artificial IMU measurements were generated to generate the shown reference track in Figure 11. The simulation was performed adding a small bias to gyroscope and accelerometer measurements. Due to these various errors which accumulate with integration, IMU alone cannot be used to provide long-term stable accurate navigation in complex environments.

Global Navigation Satellite System (GNSS)

The GNSS-based positioning systems (El-Rabbany, 2006) (Grewal, Weill, & Andrews, 2007) (Misra & Enge, 2011) consists of a constellation of space vehicles (SVs "satellites") that transmit time-stamped encoded radio frequency (RF) signals toward Earth. These time-stamped RF signals can be received by RF receivers which estimate the receiving time. Knowing the transmission and receiving times of the signal and the signal speed, the distances between SVs and the receiver can be measured simply by:

$$r = c\left(\tau_{RX} - \tau_{TX}\right) \tag{24}$$

Figure 10. Errors due to neglecting Earth rotation rate

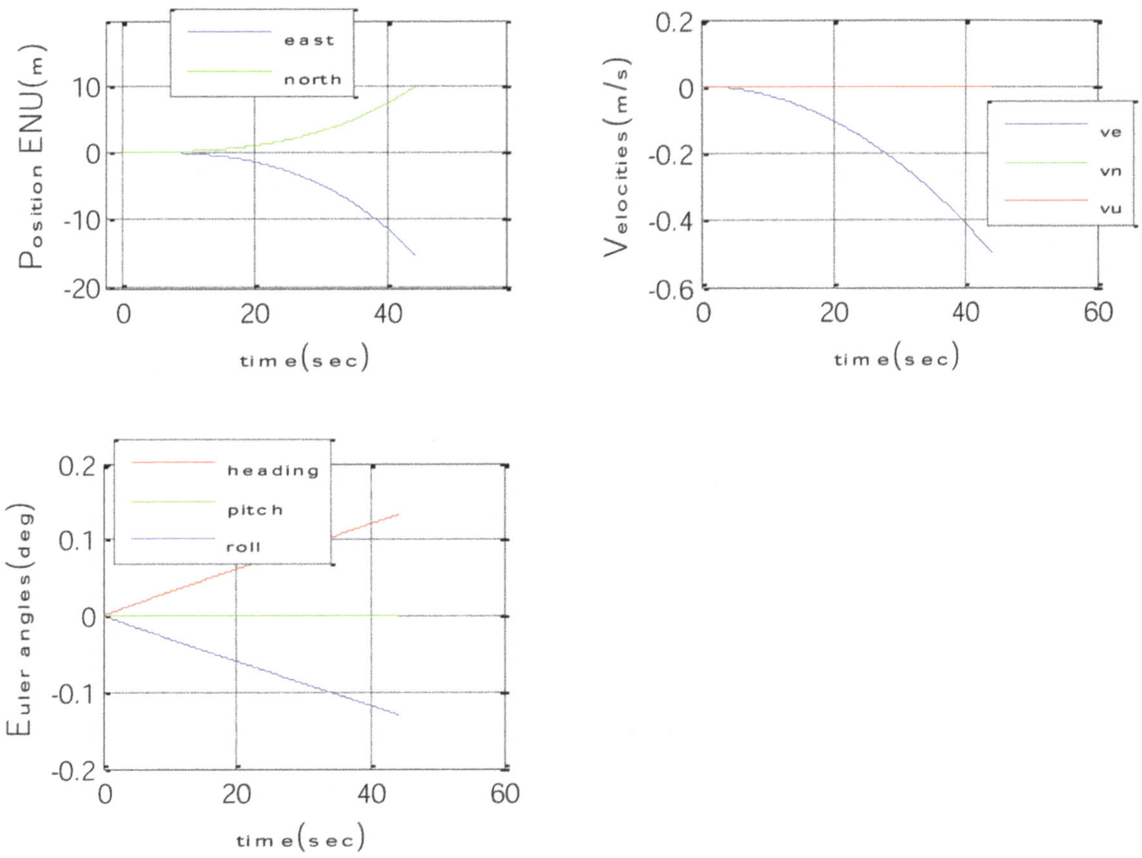

Where r is the true distance between the satellite and the receiver, c is the speed of light, τ_{RX} is signal receiving time and τ_{TX} is signal transmission time. The distance to three SVs are sufficient to determine the 3D location of the receiver with respect to the SVs locations. If SVs locations are known with respect to a reference frame (e.g. ECEF), user location can be determined with respect to ECEF and hence to any other coordinate or reference frame. This process is called trilateration. In practice, a fourth SV is needed to estimate receiver clock bias as will be explained shortly.

For velocity calculation, fortunately, the RF signal Doppler is proportional to the relative velocity between the SV and the user receiver. The Doppler is a physical phenomenon that make the receiver receive an RF signal with a frequency shifted from the original transmission frequency as follows (Misra & Enge, 2011):

$$D = \frac{\dot{\rho}}{\lambda} \tag{25}$$

Where $\dot{\rho}$ is rate of change of distance between the satellite and the receiver. λ is the wave length of the RF carrier signal transmitted by the satellite. Like 3D position calculation, knowing the Doppler

Figure 11. Three minutes simulation results with a 0.2 deg/s gyro-z-bias

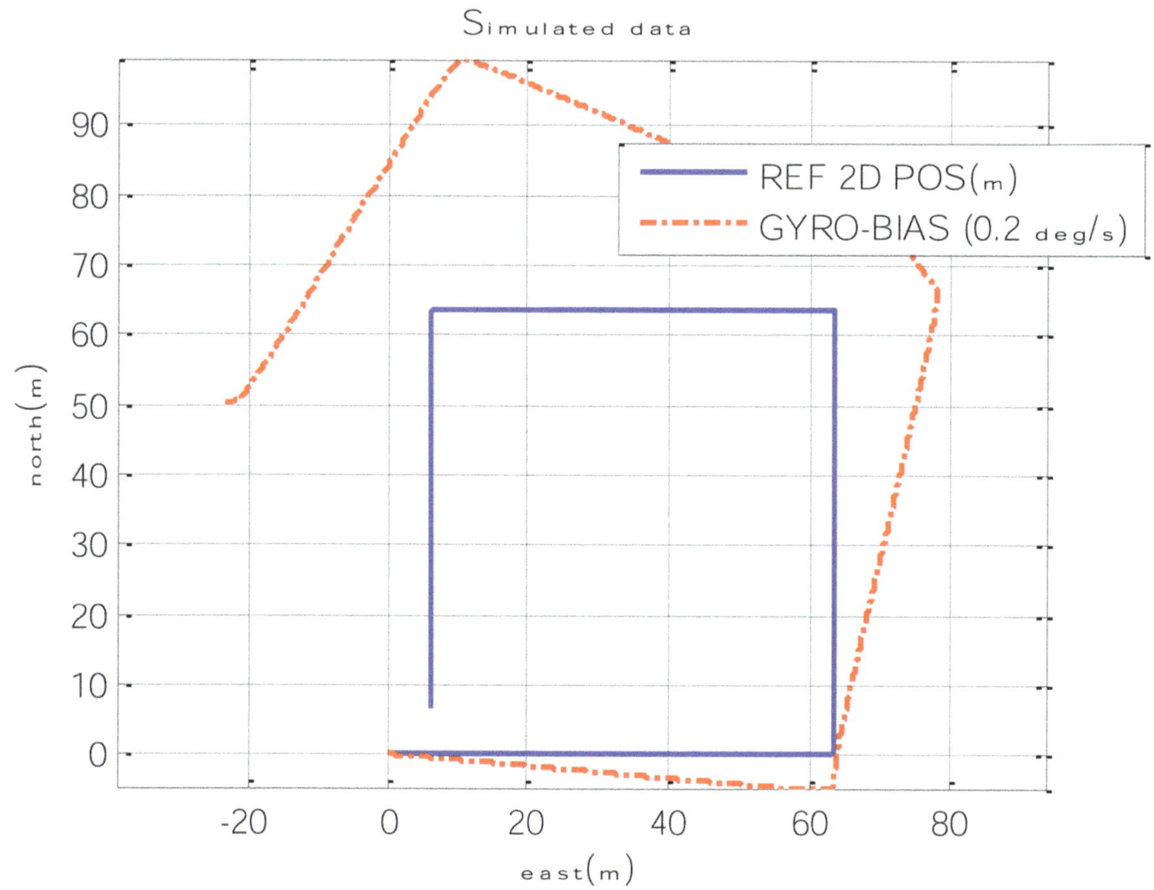

from at least three SVs along with their velocities, the user receiver velocity can be calculated. However, a fourth SV is needed to estimate the receiver clock drift. The detailed position and velocity calculation will be given shortly. To process RF from multiple SVs, SVs must be identified properly by the receivers. In the U.S system (Global Positioning System "GPS"), user receiver identifies SVs by the binary code (commonly called pseudo-random number "PRN"). PRN is code sequence that is unique to each SV. Although all SVs transmit on the same frequency, individual SVs signals are decoupled by multiplying the incoming signal with the specific SV PRN code sequence (Akos, Borre, & Bertelsen, 2007). This scheme is called code division multiplexing access (CDMA). In the Russian system (GLONASS), each SV transmits on a separate specific frequency which makes the SV identification more straight forward. This scheme is called frequency division multiplexing access (FDMA).

GNSS Errors

While distance and Doppler equations are extremely simple, several major challenges make the estimation of user position and velocity complex. Some of these major challenges are briefly discussed in the following sub-sections.

Figure 12. Three minutes simulation results with a 0.01 m/s² acc-z-bias

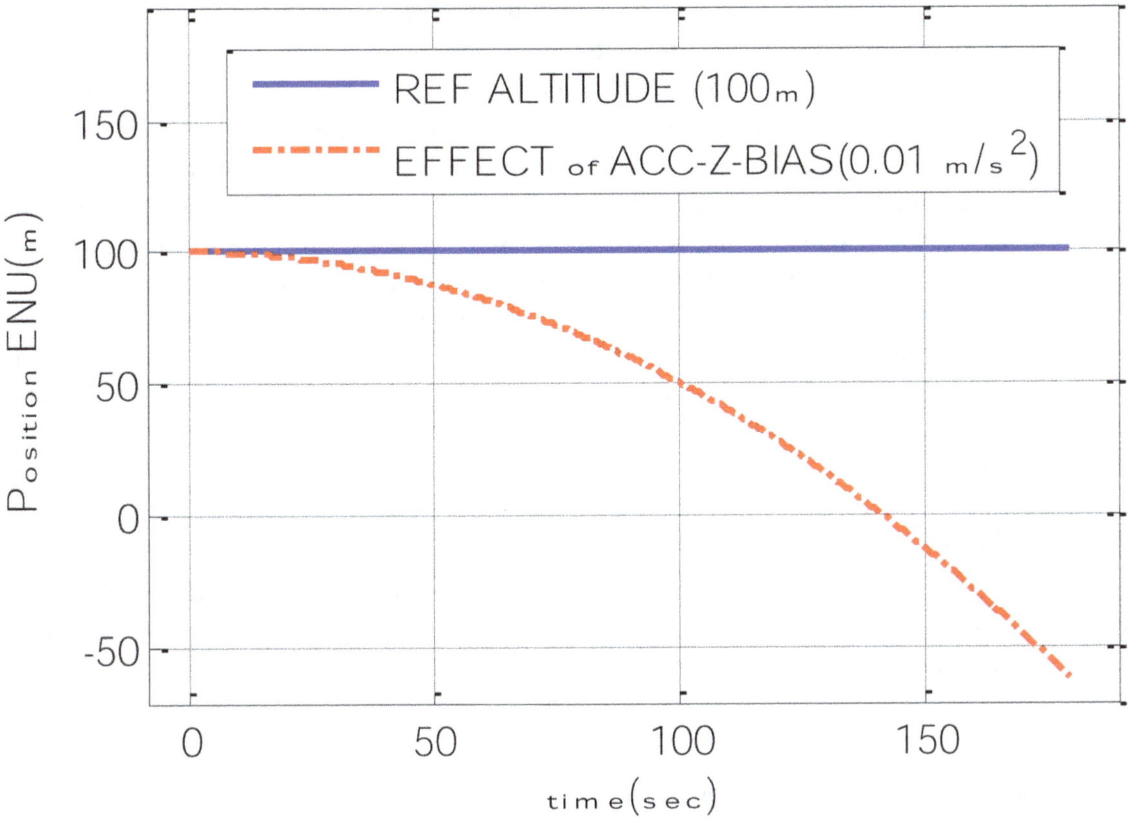

Figure 13. GNSS and Trilateration Concept

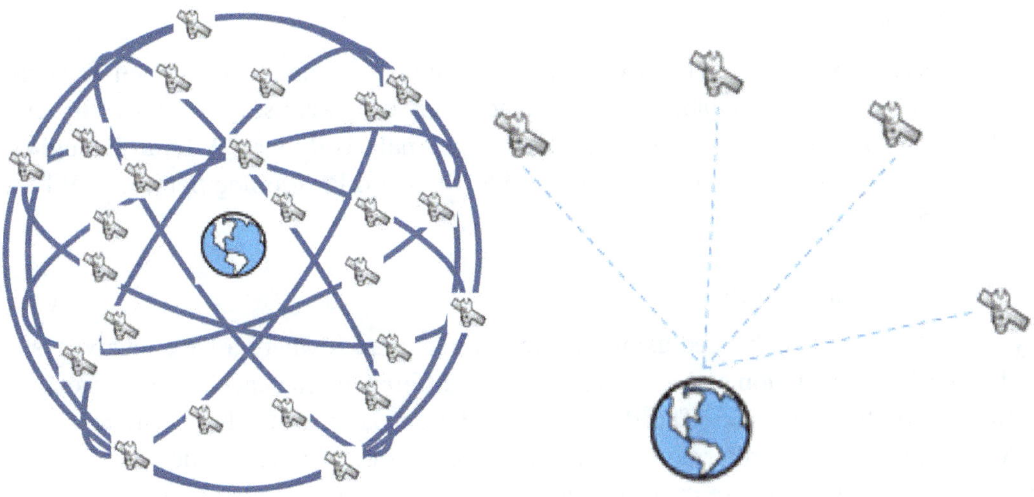

Timing Errors

For precise distance estimate, τ_{TX} and τ_{RX} must be estimated extremely accurately (a 1 µs time error corresponds to a 300m distance error when multiplied by speed of light "$c \cong 3 * 10^8 m / sc \cong 3 * 10^8 m / s$"). To achieve nanosecond time accuracy, SVs utilize expensive atomic clocks. For the receivers, the cost must be maintained reasonably low to enable scalable and commercial applications. Therefore, low-cost receivers have less expensive clocks based on conventional electronic crystal oscillators. Receiver clock errors (e.g. biases and drifts) can then be estimated by observing more than 3 SVs. Therefore, to estimate full 3D position, velocity, and receiver clock errors, at least 4 SVs must be visible. The more SVs visible by the receiver, the more accurate the estimation of the 3D location, velocity and receiver clock bias/drifts. Another timing-related challenge is the synchronization to a global common timing system. To keep all SVs synchronized to a master clock, all SVs clocks are monitored by a network of ground stations that constantly correct SVs clock biases and drifts (Groves, 2013) (Misra & Enge, 2011). The corrections are sent along with the navigation message so the receiver can compensate for these drifts in the calculation of distances.

Atmospheric Errors

Another challenge is the atmospheric effects that apply delays on the transmitted RF signals. These atmospheric delays must be compensated in distance calculation since they are not corresponding to the true distance between the SVs and user receivers.

Other Errors

Random noise due to electronic circuit errors will also affect the estimation of τ_{RX} and D causing fluctuation of the estimated distance and Doppler. Due to all these effects, the estimated distances and Doppler are not called "range" and "range-rates". Instead, they are called pseudo-range and pseudo-range rate.

Standard GNSS Accuracy

For standard GPS technology, a position accuracy of 3-5 meters is common under open-sky where direct LOSs to more than 4 SVs are available. This is the accuracy of typical single-point-positioning (SPS) operation mode. According to the FAA GPS Performance Analysis Report (William J. Hughes Technical Center, July 31, 2014), horizontal accuracy of GPS (Standard Positioning Service "SPS") is within 3.351m with a 95% confidence level.

GNSS Accuracy Enhancement

Accuracy can be extremely enhanced using differential techniques where a base receiver is used as a reference known-location station and measurements are differenced to cancel out common errors such as tropospheric and SV clock errors. This differential technique is called DGNSS (Groves, 2013) (Misra & Enge, 2011) (e.g. DGPS). Furthermore, accuracy can be extremely enhanced down to few centimeters when high-resolution carrier phase measurements are used instead of code sequence. A carrier-phase measurement has approximately 19cm resolution (this is the GPS L1 carrier wavelength). However, in obstructed environments where direct line of sight is not available, signal multipath causes random errors that are difficult to detect and mitigate. For these reasons, the measurement model of a typical GNSS pseudo-range measurement is given by:

$$\rho = r + c\Delta t_{r} - c\Delta t_{sv} + cI_{sv} + cT_{sv} + \varepsilon \tag{26}$$

Where ρ is the measured range, r is the true range, Δt_{r} is the receiver clock bias, Δt_{sv} is the SV clock bias, I_{sv} is the Ionospheric delay, T_{sv} is the Tropospheric delay, and ε is a random noise. Similarly, the pseudo-range rate measurement model is given by:

$$\dot{\rho} = \dot{r} + c\dot{\Delta} t_{r} - c\dot{\Delta} t_{sv} + c\dot{I}_{sv} + c\dot{T}_{sv} + v \tag{27}$$

Where $\dot{\rho}$ is the measured pseudo-range rate, $\dot{r}$ is the true pseudo-range rate, $\Delta \dot{t}_{r}$ is the receiver clock drift, $\Delta \dot{t}_{sv}$ is the SV clock drift, $\dot{I}_{sv}$ and $\dot{T}_{sv}$ are atmospheric drifts, and v is a random noise.

Standard Positioning Service (SPS)

Having a set of pseudo-range and pseudo-range rate measurements, receiver position and velocity can be estimated using a nonlinear least-square algorithm (Groves, 2013). This scheme is called single point or standard positioning service (SPS) which is the standard method used in low-cost receivers. A nonlinear least-square algorithm finds the optimal solution for a problem in the following form:

$$y = f\left(x\right) \tag{28}$$

Where y are observation (in our case, a set of pseudo-range and pseudo-range rate measurements) and x is the user location, velocity, receiver clock bias, and receiver clock drift. Therefore, x is a vector given by:

$$x = \begin{bmatrix} X^{ECEF} & Y^{ECEF} & Z^{ECEF} & \dot{X}^{ECEF} & \dot{Y}^{ECEF} & \dot{Z}^{ECEF} & b & d \end{bmatrix}^{T} \tag{29}$$

Where b is receiver clock bias (in meters), and d is receiver clock drift (in meters/s). The reader will notice that SV clock bias and atmospheric delays (I_{sv} and T_{sv}) are not modeled in (29). In practice, receivers estimate I_{sv} T_{sv} in a separate process after decoding the navigation message and extracting the SVs parameters (Misra & Enge, 2011). SV clock error parameters are also embedded in the navigation message. To solve the equation system of (28), an initial guess x_0 is given and the system is linearized around this given initial guess so we have:

$$\delta y = \frac{\partial f\left(x\right)}{\partial x}\delta x \tag{30}$$

The initial guess x_0 is used to predict pseudo-range and pseudo-range rate measurements (e.g. to obtain y_0) which is compared to observed measurements y to obtain an observation error(δy). Because equation (30) is linear, we can use linear least-square algorithm to estimate δx_0. Once δx_0 is estimated, a new guess is updated by:

$$x_{k+1} = x_k + \delta x_k \tag{31}$$

This process continues in iterations until the δy is sufficiently small (which means our estimate x_{k+1} produces measurements close to the observed measurements). The key equation in this algorithm is the nonlinear function $f\left(x\right)$ which can be given by:

$$\rho\left(X^{ECEF}, Y^{ECEF}, Z^{ECEF}\right) = \sqrt{\left(X^{ECEF} - X_{SV}^{ECEF}\right)^2 + \left(Y^{ECEF} - Y_{SV}^{ECEF}\right)^2 + \left(Z^{ECEF} - Z_{SV}^{ECEF}\right)^2} + b \tag{32}$$

$$\dot{\rho}\left(\dot{X}^{ECEF}, \dot{Y}^{ECEF}, \dot{Z}^{ECEF}\right) = 1^{SV}.\left[\left(\dot{X}^{ECEF} - \dot{X}_{SV}^{ECEF}\right) \quad \left(\dot{Y}^{ECEF} - \dot{Y}_{SV}^{ECEF}\right) \quad \left(\dot{Z}^{ECEF} - \dot{Z}_{SV}^{ECEF}\right)\right]^T + d \tag{33}$$

Where

$$1^{SV} = \frac{\left[\left(X^{ECEF} - X_{SV}^{ECEF}\right) \quad \left(Y^{ECEF} - Y_{SV}^{ECEF}\right) \quad \left(Z^{ECEF} - Z_{SV}^{ECEF}\right)\right]^T}{\sqrt{\left(X^{ECEF} - X_{SV}^{ECEF}\right)^2 + \left(Y^{ECEF} - Y_{SV}^{ECEF}\right)^2 + \left(Z^{ECEF} - Z_{SV}^{ECEF}\right)^2}} \tag{34}$$

$1^{SV}1^{SV}$ is the LOS vector from user position to SV position described in ECEF frame
If *M* SVs are visible, the overall linearized model can be given by:

$$\delta \mathbf{y} = \begin{bmatrix} \delta\rho^{SV_1} \\ \vdots \\ \delta\rho^{SV_M} \\ \delta\dot{\rho}^{SV_1} \\ \vdots \\ \delta\dot{\rho}^{SV_M} \end{bmatrix} = \begin{bmatrix} (1^{SV_1})^T & 1 & 0_{3\times1} & 0 \\ \vdots & \vdots & \vdots & \vdots \\ (1^{SV_M})^T & 1 & 0_{3\times1} & 0 \\ 0_{3\times1} & 0 & (1^{SV_1})^T & 1 \\ \vdots & \vdots & \vdots & \vdots \\ 0_{3\times1} & 0 & (1^{SV_M})^T & 1 \end{bmatrix} \left[\delta\mathbf{x}\right] = H\delta\mathbf{x} \tag{35}$$

The optimal least square solution of equation (33) is given by:

$$\delta\mathbf{x} = \left(H^T H\right)^{-1} H^T \delta\mathbf{y} \tag{36}$$

Accuracy of GNSS solution can be affected by the geometry of the visible SVs. The more diverse the SVs (spread out in the sky), the more accurate the results. The shaded areas in Figure 14 is called the delusion of precision (DOP). As can be seen from Figure 14, DOP depends on the geometry of the SVs as seen by the user receiver. Therefore, the geometry is also a factor that affects the accuracy of the GNSS. It can be easily seen from equation (34) that the geometry is expressed in the design matrix H which shows how errors in pseudo-ranges $\delta y \delta y$ are projected into errors in the estimated user position $\delta x \delta x$.

GPS Stationary Test

To get a feeling of standard SPS GPS accuracy, a stationary test was performed in an open-sky area in a well-surveyed position where number of visible GPS SVs ranged from 4 to 9. The test results are shown in Figure 15, Figure 16, and Figure 17. The test shows positioning accuracy, number of satellites, and position DOP. The horizontal position root mean square error (RMSE) in this test is 1.8019m. However, other advanced GNSS technology such as DGPS can achieve centimeter-level accuracy.

While GNSS and INS are the most common positioning technologies, they cannot be used to map or detect the surroundings. Therefore, other sensors are used. The most common sensors used for mapping are range and vision sensors. While they are mainly used for mapping, range, vision, and other sensors such as RFID or WiFi are key to overcome GNSS limitations when it comes to dense urban or deep indoor areas where GNSS do not work at all and INS drifts cannot be calibrated. In the next sections, the concept of using range sensors for tracking object's position and orientation is explained.

Figure 14. The effect of geometry on position accuracy

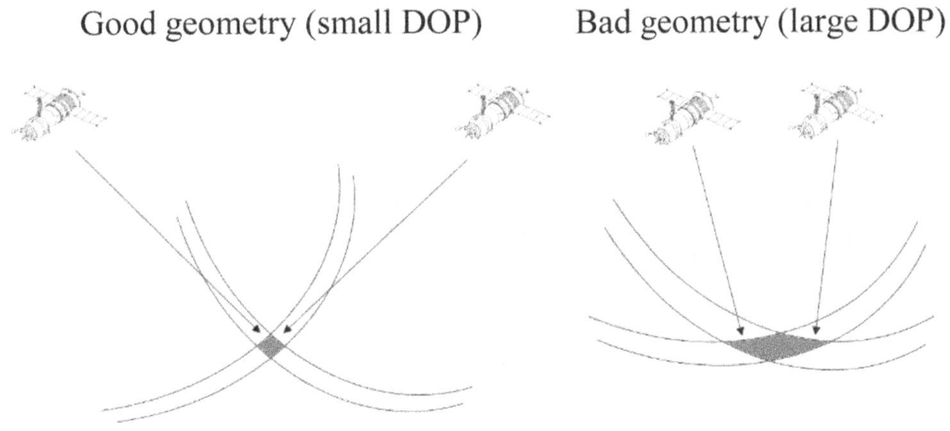

Figure 15. GPS Stationary Test: 3D Positioning Errors (absolute error in meters)

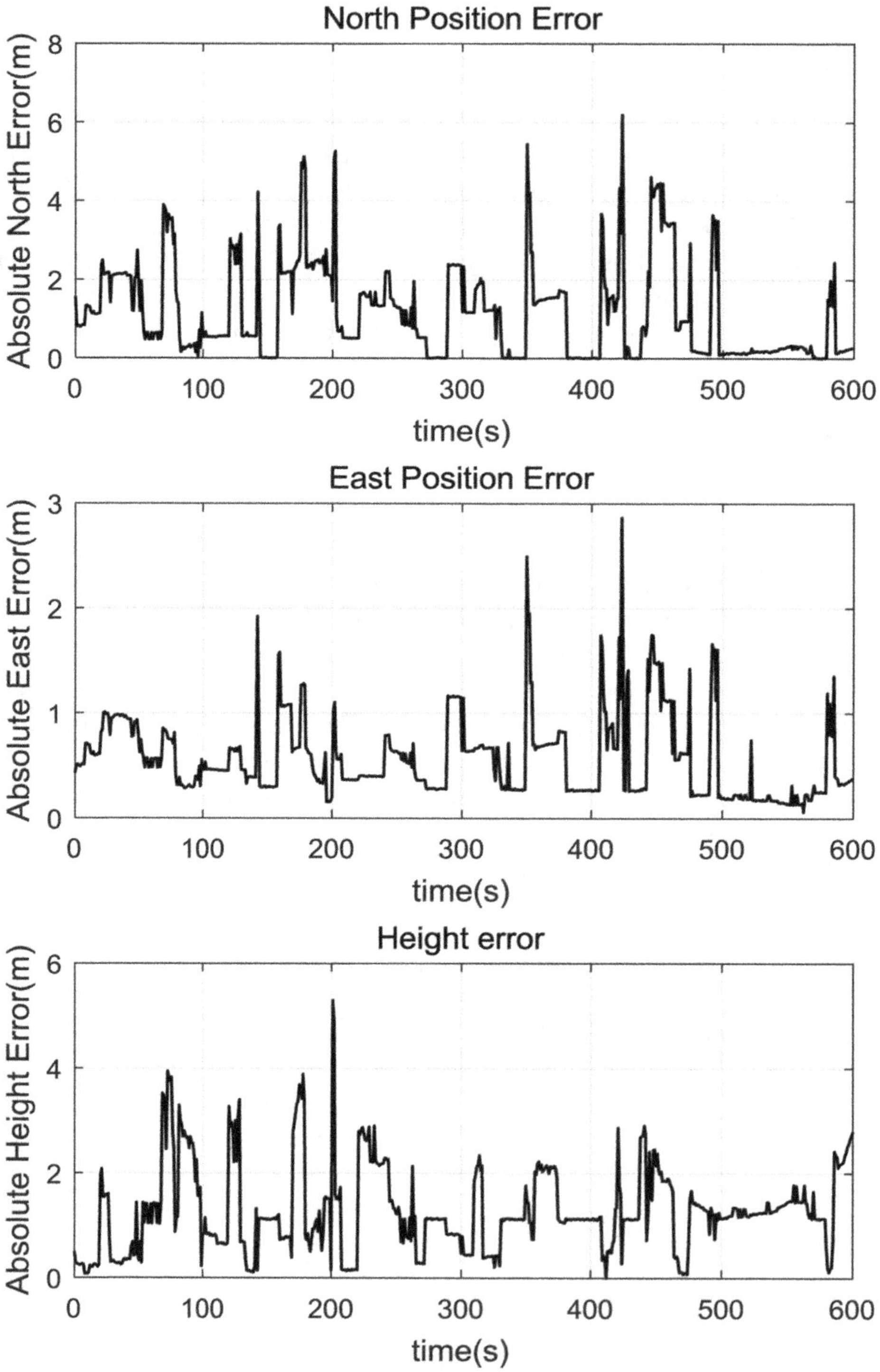

Figure 16. GPS Stationary Test: Number of SVs

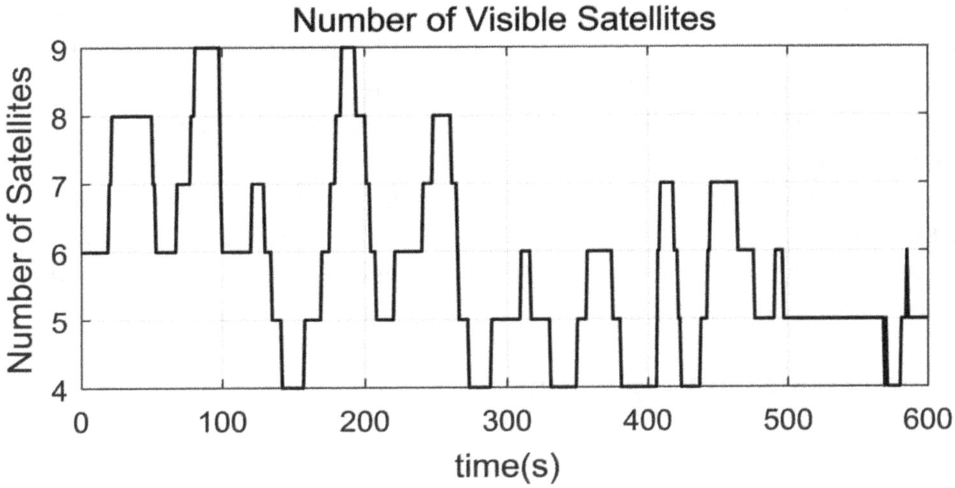

Figure 17. GPS Stationary Test: Position DOP

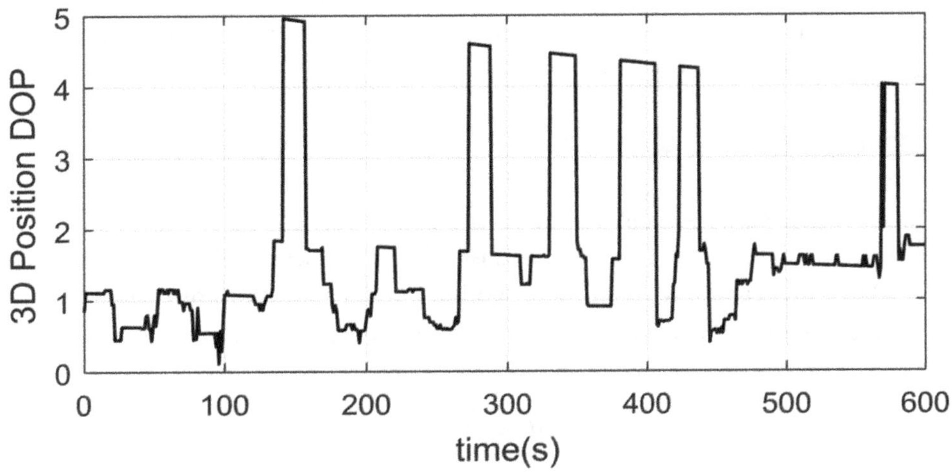

SOLUTIONS AND RECOMMENDATIONS

To augment GNSS/INS in dense and urban indoor areas, a continuous source of external aiding measurements that can assist the IMU is necessary. As described in the beginning of the chapter, this work will investigate the fusion of LiDAR with GNSS/INS sensors to improve overall performance in dense urban and indoor environments.

Range Sensors

Common range sensors such as LiDAR and RADARs (Groves, 2013) (Censi, 2008) can measure the distance to the surroundings by analyzing the reflection of a transmitted signal on the environment objects and surfaces. The accuracy of range sensors is characterized by the precision of the measurement. Figure 18 shows how 2D range measurement can be characterized by two parameters δd: distance precision and $\delta\alpha$: angular precision. In 3D, precision will be described by a conic shape parameters. Low-cost range sensors such as RADARs have low angular and range precision (high δd and large $\delta\alpha$). Therefore, they are commonly used for obstacle detection, object recognition, and collision avoidance but they cannot be used for self-precise positioning.

The most commonly used range sensor for localization is LiDAR. The laser scanner used in this chapter is a LMS111 from SICK company (SICK, 2008). It can achieve precise mm-range precision with accurate 0.5 angular resolution to cover 270° field of view. Figure 19 shows the LMS111 mounted on experimental unmanned ground vehicle (UGV).

Range-Based Localization Techniques

Map Correlation-Based Localization

One way to estimate position and orientation from LiDAR is to correlate sensor readings with a map as Figure 20 shows. While this method is straightforward, it needs a fresh up-to-date map which may not be available in all scenarios. Furthermore, in some areas, such as corridors, correlation with the map may lead to ambiguous positioning. However, with the help of other sensors such as IMU and odometer (speed measurements), this ambiguity can be resolved.

Features Tracking-Based Localization

Another method to track range sensor motion is the estimation of the relative motion (rotation and translation) by tracking scanned features between consecutive scans. Figure 21 illustrates how two consecutive LiDAR scans can be used to track motion. At given initial pose (Figure 21: a right), the sensor measures a 2D map of the surroundings that can be represented in the sensor body-frame as 2D

Figure 18. Range Measurement Precision Parameters

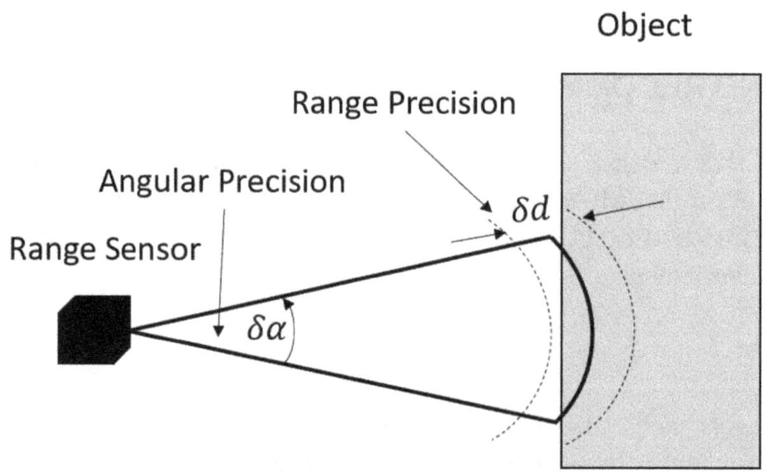

Figure 19. SICK LMS111 Scanner mounted on Husky A200 Unmanned Ground Vehicle

data points (Figure 21: a left). When the sensor moves, a second scan can be obtained (Figure 21: b). The two scans are overplayed in Figure 21: c. As seen in Figure 21: c left graph, some data points from first scan will disappear from the field of view in the second scan. Similarly, some other new points will appear in the new scan. However, some common data points will remain in field of view and they will stay visible in both scans. These common data-points are shown in Figure 21: c right graph. The small arrows show how the coordinates of these common data point have moved from the first scan to the second scan. If the transformation that transformed the coordinates of these common data points between the two scans can be estimated, the platform motion can be estimated. Several methods have been studied to infer motion from range scans. A common method is called iterative Closest Point (ICP) method which is described in the next subsection.

Iterative Closest Point (ICP) Point-Scan Matching Method

ICP-method estimates relative motion from direct point-scan matching between consecutive range scans. ICP method was first introduced in (Besl & Mckay, 1992) and thereafter numbers of variants have been proposed (Rusinkiewicz & Levoy., 2001). A data point P_i can be described by $\begin{bmatrix} X_i & Y_i \end{bmatrix}$ in scan *i* and by $\begin{bmatrix} X_{i+1} & Y_{i+1} \end{bmatrix}$ in scan *i+1*. Transformation between the coordinates in scans *i* and *i+1* can be described by a transformation matrix characterized by rotation angle (ΔA) and translational motion (T^B: $\begin{bmatrix} \Delta_x & \Delta_y \end{bmatrix}$). In order to infer ΔA and T^B, correct correspondences is needed. Correspondence is simply a link that

Figure 20. Localizing a 2D LiDAR scanner by correlating scan measurements with an indoor map

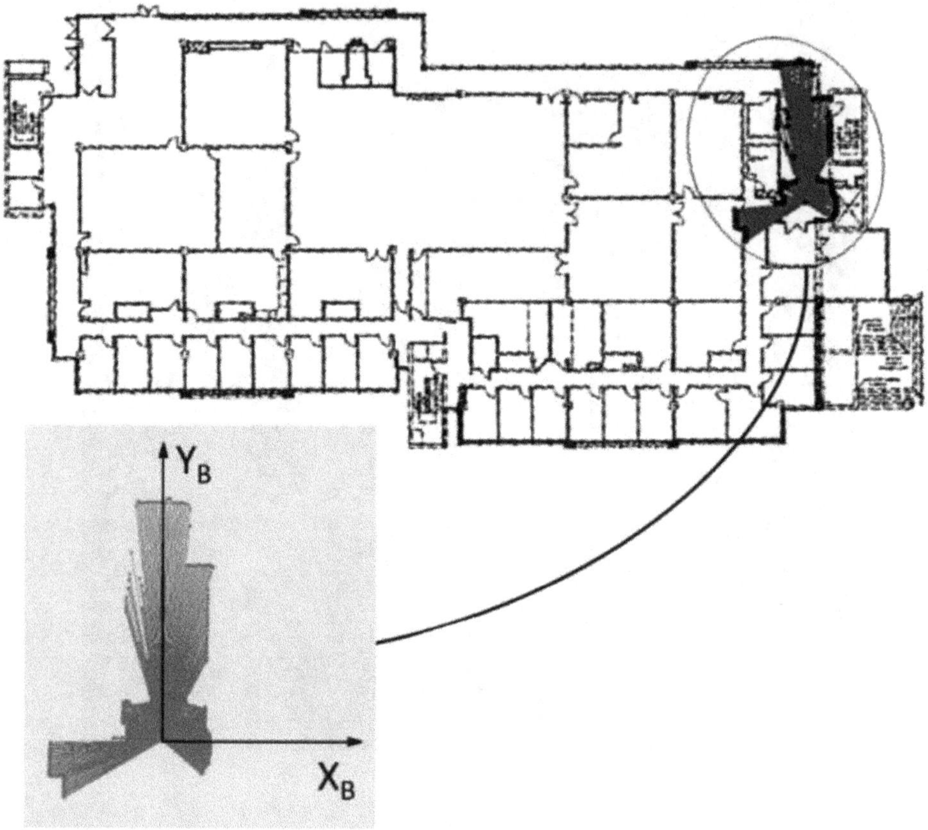

determines for each data point in scan *i* what the "corresponding" data point in the new scan (scan *i+1*). If several data points with correct correspondences are available, they can be projected from scan *i* to scan *i+1* using the following matrix transformation:

$$\begin{bmatrix} X_i \\ Y_i \end{bmatrix} = \begin{bmatrix} \cos(\Delta A) & \sin(\Delta A) \\ -\sin(\Delta A) & \cos(\Delta A) \end{bmatrix} \begin{bmatrix} X_{i+1} \\ Y_{i+1} \end{bmatrix} + \begin{bmatrix} \Delta_x \\ \Delta_y \end{bmatrix} \tag{37}$$

where ΔA is the relative azimuth (or heading) change and T^B is the translational motion described as a vector in the body frame in the old scan. Equation (37) can be written as:

$$P_i = RP_{i+1} + T^B \tag{38}$$

Estimating relative motion between scans can then be modelled as an optimization problem in rotation (R) and translation (T^B) as follows:

Figure 21. Motion Tracking from Consecutive LiDAR Scans
(a) Scan 1, (b) Scan 2, (c) Correspondences

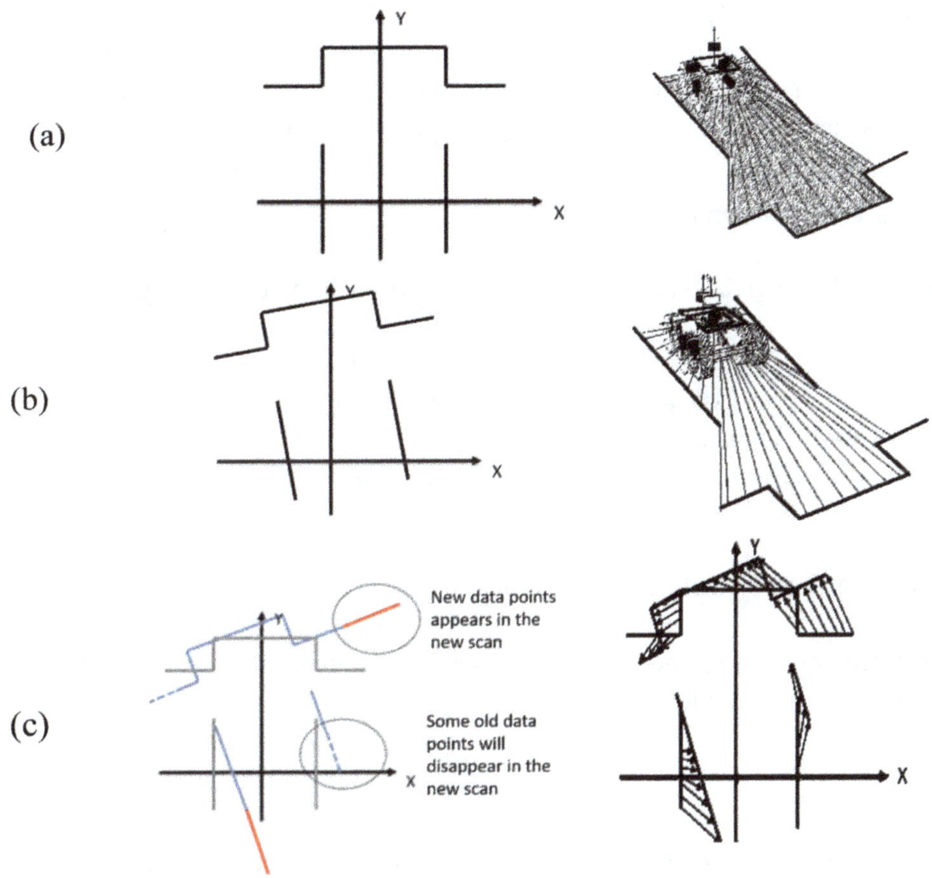

$$J\left(T,R\right)=\sum_{i\in C}\left\|P_i-\left(RP_{i+1}+T^B\right)\right\|^2 \tag{39}$$

Finding Correspondences Between Two LiDAR Consecutive Scans

If an initial relative motion guess can be obtained from inertial sensor or odometry, a k-d tree technique can be applied to find correspondence (closest points pairs) in old scan for each point in the new scan. Considering the existence of outliers and noise, not all the closest point pairs are used to compute the transformation. Instead, a threshold is commonly applied to filter out some of the matched points. There are several methods to establish data points correspondence such as the method described in (Besl & Mckay, 1992) and (Censi, 2008). Once correspondences are established, the cost function $J\left(T,R\right)$ in (37) can be optimized using a least-squares algorithm as described in (Censi, 2008).

Range-Measurements Errors

By concatenating relative motions, a 2D trajectory of the platform can be estimated. However, due to measurements noise and correspondence errors, the estimated 2D trajectory is not always accurate. As seen in Figure 22 where reflection of LiDAR beams through open doors or on glass surfaces cause outlier data points that are difficult to track through consecutive scans and consequently leads to erroneous relative motion estimation when used as measurement in optimizing $J\left(T, R\right)$ in (37). To overcome these limitations, sensor fusion is used.

Range-Aided GNSS/INS Systems for Urban and Indoor Navigation

This section will show step-by-step how to design a sensor fusion system to track ground vehicles in complex environments such as dense urban and indoor areas using LiDAR-aided GNSS/INS platform. A brief summary about Kalman filter and how it is applied to perform sensor fusion is given first. Then, the application of the described Kalman filter in Range-aided GNSS/INS navigation problem will be described.

Integrated Navigation Problem Formulation

The navigation problem can be modelled as a dynamic system of states vector $x(t)$ where states are evolving based on a system model:

$$\dot{x}(t) = f(x(t), u(t)) + w(\mathrm{t}) \tag{40}$$

And external observations are modeled by a measurement model as follows:

Figure 22. LiDAR scan outliers

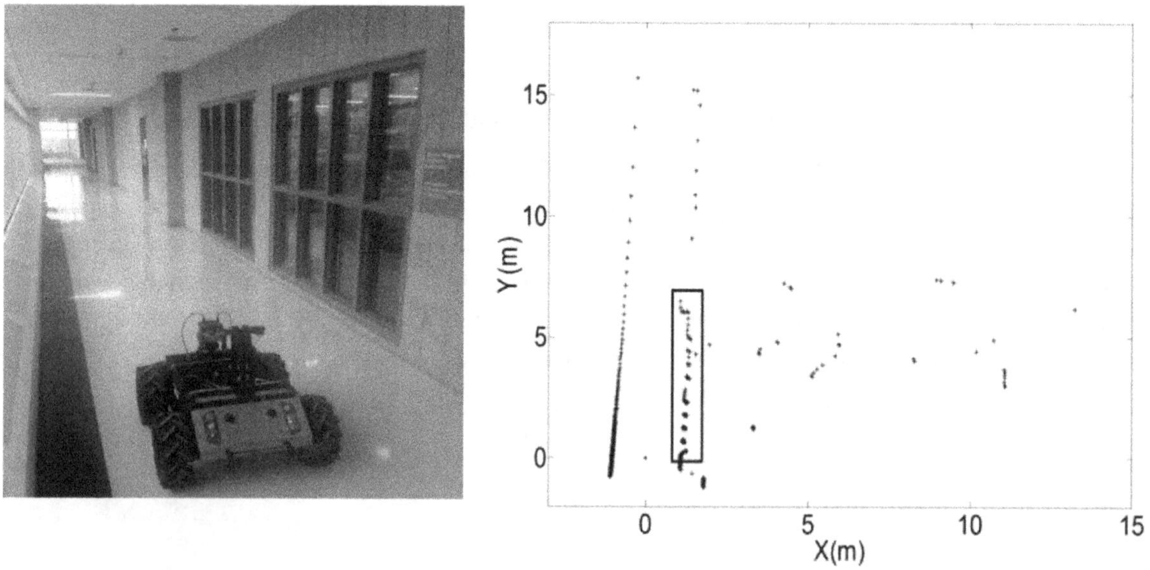

$$y(t) = h(x(t)) + v(\mathrm{t}) \tag{41}$$

where $f(.)$ is a nonlinear dynamic model, $w(\mathrm{t})$ is a stochastic system noise vector, $u(\mathrm{t})$ is a control signal vector, $y(t)$ is external measurements vector, $h(.)$ is a nonlinear measurement model and $v(\mathrm{t})$ is a stochastic measurement noise vector. Kalman filter can be applied only to linear systems. Therefore, system and measurement models are linearized around a nominal trajectory using first order Taylor series expansion as follows:

$$\delta\dot{x}(t) = F(t)\delta x(t) + Gw(t) \tag{42}$$

$$\delta y(t) = H(t)\delta x + \mathrm{v}(t) \tag{43}$$

Where

$$F(t) = \frac{\partial f(x(t), u(t))}{\partial x^T} \tag{44}$$

$$G(t) = \frac{\partial f(x(t), u(t))}{\partial w^T} \tag{45}$$

$$H(t) = \frac{\partial h(x(t))}{\partial x^T} \tag{46}$$

For the system of (40-41), Kalman Filter (Farrell, 2008) (Savage, 2007) provides an optimal estimation of δx assuming $w(t)$ and $\mathrm{v}(t)$ are zero-mean Gaussian noise vectors with covariance matrices defined by:

$$Q(t) = \left\langle w(t)w(t)^T \right\rangle \tag{47}$$

$$R(t) = \left\langle \mathrm{v}(t)\mathrm{v}(t)^T \right\rangle \tag{48}$$

where $\langle . \rangle$ is the expectation operator and δx is error vector with zero-mean and a covariance matrix *P* defined by:

$$P(t) = \left\langle \delta x(t) \delta x(t)^T \right\rangle \tag{49}$$

In discrete form, using zero-hold discretization where derivative is approximated by:

$$\delta \dot{x}_k \approx \frac{\delta x_{k+1} - \delta x_k}{T} \tag{50}$$

where T is the sampling period, equations (40-41) can be written in discrete form as follows:

$$\delta x_k = (I + F_k T)\delta x_{k-1} + w_{k-1} \tag{51}$$

$$\delta y_k = H_k \delta x_k + \mathrm{v}_k \tag{52}$$

The optimal estimation of the error vector, δx_k, given measurements, y_k, is calculated using two steps: prediction and measurement update. In prediction, the states and their error covariance matrix are predicted as follows:

$$x_{k+1} = f(x_k, u_k) \tag{53}$$

$$P_{k+1} = (I + F_{k+1}T)P_k(I + F_{k+1}T)^T + G_{k+1}Q_{k+1}G_{k+1}{}^T T^2 \tag{54}$$

In the update step (when some external measurements y_{k+1} become available, the states are corrected as follows:

$$K_{k+1} = P_{k+1}\mathrm{H}_{k+1}^T(\mathrm{H}_{k+1}P_{k+1}\mathrm{H}_{k+1}^T + R_{k+1})^{-1} \tag{55}$$

$$x_{k+1} = x_{k+1} + K_{k+1}\left[y_{k+1} - h(x_{k+1})\right] \tag{56}$$

$$P_{k+1} = (I - K_{k+1}H_{k+1})P_{k+1} \tag{57}$$

The prediction-update steps are performed recursively to estimate the trajectory. More details about Kalman filter derivation and approximations can be found in (Farrell, 2008) (Groves, 2013) and (Savage, 2007). In the next subsections, the steps to formulate the range-aided GNSS/IMU navigation problem into a Kalman filter problem is discussed.

System Measurements

System measurements are the observables represented by $y(t)$ in the measurement model in equation (41). These are the observables that Kalman filter will use to infer the full navigation state including sensors errors. In the presented system, two basic measurements are used, GNSS and LiDAR. In open-sky, GNSS are fused. When the vehicle approaches the indoor area, LiDAR measurements are applied. The measurements from GNSS are simply the pseudo-range and pseudo-range rate molded in (26-27). GNSS measurement model is described in (35). The LiDAR measurements are the relative translation (T^B) and heading change (ΔA) estimated from LiDAR scans. To estimate T^B and ΔA of the moving vehicle from LiDAR scans, all LiDAR measurements must be projected to a horizontally leveled frame (we will call it I^m). This is important if the vehicle moves in 3D with relatively large values of roll and pitch angles. The leveling transformation from body frame B to I^m can be performed by rotating around y-axis by roll angle and around x-axis by pitch angle leading to the following final rotation matrix from B to I^m:

$$C_B^{I^m} = R(p)R(r) = \begin{bmatrix} \cos r & 0 & \sin r \\ \sin p \sin r & \cos p & -\sin p \cos r \\ -\cos p \sin r & \sin p & \cos p \cos r \end{bmatrix} \tag{58}$$

For any LiDAR data point $\left(x^B, y^B, z^B\right)$, it can be projected to I^m using $C_B^{I^m}$. Because a 2D LiDAR is used, z^B is be set to zero. Once two consecutive scans of LiDAR data points are available in the I^m frame, data-points correspondences in the two scans can be established (Censi, 2008) and ICP algorithm can be used to estimate T^B and ΔA as described in equations (37-39).

System Models

System model predicts system state represented by $x(t)$ in (40). System states include position, velocity, orientation, IMU biases, GNSS receiver clock bias, and relative motion variables T^B and ΔA. System models for position, velocity, and orientation are described in (18-23). In the following subsections, the system models for T^B, ΔA, IMU biases, and GNSS receiver clock bias are described.

System Model for Relative Motion

Relative motion is predicted from IMU. The IMU-based relative motion between two LiDAR consecutive scans is then used in the ICP algorithm as an initial guess to solve the optimization problem described by equation (39). This initial condition is simply initial values for T^B and ΔA to facilitate finding correct data-points correspondences and to help ICP-algorithm to converge quickly minimizing the cost function. The IMU-based relative motion change between two LiDAR consecutive scans can be estimated by processing the outputs of the vertical gyroscope after transforming it into the I^m frame where LiDAR scans are processed. The angular velocity in the I^m frame is given by:

$$\omega_{LB}^{I^m} = \begin{bmatrix} \omega_x^{I^m} & \omega_y^{I^m} & \omega_z^{I^m} \end{bmatrix}^T = C_b^{I^m} \omega_{LB}^B \tag{59}$$

$$\omega_{LB}^{I^m} = \begin{bmatrix} \omega_x^{I^m} & \omega_y^{I^m} & \omega_z^{I^m} \end{bmatrix}^T = C_b^{I^m} \omega_{LB}^{B} \quad (60)$$

The vertical rotation rate in I^m frame is then given by:

$$\omega_z^{I^m} = -\omega_{LB_x}^{B} \cos p \sin r + \omega_{LB_y}^{B} \sin p + \omega_{LB_z}^{B} \cos p \cos r \quad (61)$$

Therefore, the vehicle azimuth changes during the time interval T in I^m frame is:

$$\Delta A = \omega_z^{I^m} T \quad (62)$$

And the position changes along x-axis and y-axis are of in I^m frame are given by:

$$\Delta x = v \cos p \sin \Delta A T \quad (63)$$

$$\Delta y = v \cos p \cos \Delta A T \quad (64)$$

where $v = \sqrt{v_e^2 + v_n^2}$ is the horizontal vehicle speed.

System Model for IMU errors

To complete the IMU-related system model equations, gyroscope and accelerometer biases need to be modeled. The common model for how gyroscope and accelerometer biases change with time are commonly modeled as Gauss-markov process (Groves, 2013) which can be described in the following general form:

$$\delta \dot{b}_\omega(t) = -\beta_\omega \delta b_\omega(t) + \sqrt{2\sigma_\omega^{\,2} \beta} w(t) \quad (65)$$

$$\delta \dot{b}_a(t) = -\beta_a \delta b_a(t) + \sqrt{2\sigma_a^{\,2} \beta} w(t) \quad (66)$$

where σ_ω, σ_a are the standard deviation, β_ω, β_a are the reciprocal of the time constant of the autocorrelation function (i.e. Gauss-markov process parameters) (Groves, 2013) (Farrell, 2008) of the gyroscope and accelerometer biases respectively and $w(t)$ is a unity standard deviation zero-mean Gaussian noise.

System Model for GNSS Receiver Clock Bias

The GPS receiver clock bias δb_r and drift δd_r are simply modeled as a random walk process (Farrell, 2008) given by:

$$\delta \dot{b}_r = \delta d_r \tag{67}$$

$$\delta \dot{d}_r = \sigma_d w(t) \tag{68}$$

where $\sigma_d w(t)$ is zero-mean Gaussian noise.

Error States, Transition and Noise Shaping Matrices

The overall system model is defined by equations (18-23) and (62-68). The combined system error states are given by:

$$\delta x = \begin{bmatrix} \delta p^L & \delta v^L & \delta q_{LB}^B & \delta b_\omega & \delta b_a & \delta \Delta x & \delta \Delta y & \delta \Delta A & \delta b_r & \delta d_r \end{bmatrix}^T \tag{69}$$

Where δp^L is error in position, δv^L is error in velocity, δq_{LB}^B is error in orientation quaternion, δb_ω is error in gyroscope biases, δb_a is error in accelerometer biases, $\delta \Delta x$, $\delta \Delta y$ are error in translational relative motion in body-frame, $\delta \Delta A$ is error in heading change, and δb_r , δd_r are receiver clock bias and drifts respectively. By applying Taylor series expansion to overall system model described in (18-23) and (62-68), linearization with respect to state variable x will lead to the transition matrix F defined by (44) of the overall integrated system given by:

$$F = \begin{bmatrix} F_{IMU} & 0 \\ 0 & F_{GPS} \end{bmatrix} \tag{70}$$

where

$$F_{IMU} = \begin{bmatrix} F_{pp} & F_{pv} & 0_{3\times4} & 0_{3\times3} & 0_{3\times3} & 0_{3\times1} & 0_{3\times1} & 0_{3\times1} \\ 0_{3\times3} & 0_{3\times3} & F_{vq} & 0_{3\times3} & -R_{b(q)}^l & 0_{3\times1} & 0_{3\times1} & 0_{3\times1} \\ 0_{4\times3} & 0_{4\times3} & \frac{1}{2}F_{qq} & F_{qb\omega} & 0_{4\times3} & 0_{4\times1} & 0_{4\times1} & 0_{4\times1} \\ 0_{3\times3} & 0_{3\times3} & 0_{3\times4} & F_{b\omega} & 0_{3\times3} & 0_{3\times1} & 0_{3\times1} & 0_{3\times1} \\ 0_{3\times3} & 0_{3\times3} & 0_{3\times4} & 0_{3\times3} & F_{bf} & 0_{3\times1} & 0_{3\times1} & 0_{3\times1} \\ 0_{1\times3} & F_{\Delta xv} & 0_{1\times4} & 0_{1\times3} & 0_{1\times3} & 0 & 0 & F_{F_{\Delta x \Delta A}} \\ 0_{1\times3} & F_{\Delta yv} & 0_{1\times4} & 0_{1\times3} & 0_{1\times3} & 0 & 0 & F_{F_{\Delta y \Delta A}} \\ 0_{1\times3} & 0_{1\times3} & 0_{1\times4} & F_{\Delta A b\omega} & 0_{1\times3} & 0 & 0 & 0 \end{bmatrix} \tag{71}$$

and

$$F_{GPS} = \begin{bmatrix} 0 & 1 \\ 0 & 0 \end{bmatrix} \quad (72)$$

Note that the terms that are divided by the square of the earth curvature radius are ignored. Therefore, the elements in F matrix are given as follows: F_{pp} is 3 x 3 matrix where all elements are zero except $F_{pp}(2,1)$ which is given by:

$$F_{pp}(2,1) = \frac{v_e \tan\varphi}{\left(R_N + h\right)\cos\varphi} \quad (73)$$

F_{pv} is 3 x 3 matrix given by:

$$F_{pv} = \begin{bmatrix} 0 & \frac{1}{R_m + h} & 0 \\ \frac{1}{\left(R_n + h\right)\cos\varphi} & 0 & 0 \\ 0 & 0 & 1 \end{bmatrix} \quad (74)$$

Other entries are defined as follows:

$$F_{vq} = \begin{bmatrix} F_{vq1} & F_{vq2} & F_{vq3} & F_{vq4} \\ -F_{vq4} & -F_{vq3} & F_{vq2} & F_{vq1} \\ F_{vq3} & -F_{vq4} & -F_{vq1} & F_{vq2} \end{bmatrix} \quad (75)$$

Where

$$F_{vq1} = 2\left(a(f_x - b_{fx}) - d(f_y - b_{fy}) + c(f_z - b_{fz})\right) \quad (76)$$

$$F_{vq2} = 2\left(b(f_x - b_{fx}) + c(f_y - b_{fy}) + d(f_z - b_{fz})\right) \quad (77)$$

$$F_{vq3} = 2\left(-c(f_x - b_{fx}) + b(f_y - b_{fy}) + a(f_z - b_{fz})\right) \quad (78)$$

$$F_{vq4} = 2\left(-d(f_x - b_{fx}) - a(f_y - b_{fy}) + b(f_z - b_{fz})\right) \tag{79}$$

and

$$F_{qq} = \begin{bmatrix} 0 & -(\omega_x - b_{\omega x}) & -(\omega_y - b_{\omega y}) & -(\omega_z - b_{\omega z}) \\ \omega_x - b_{\omega x} & 0 & \omega_z - b_{\omega z} & -(\omega_y - b_{\omega y}) \\ \omega_y - b_{\omega y} & -(\omega_z - b_{\omega z}) & 0 & \omega_x - b_{\omega x} \\ \omega_z - b_{\omega z} & \omega_y - b_{\omega y} & -(\omega_x - b_{\omega x}) & 0 \end{bmatrix} \tag{80}$$

$$F_{qb\omega} = -\frac{1}{2}\begin{bmatrix} -b & -c & -d \\ a & -d & c \\ d & a & -b \\ -c & b & a \end{bmatrix} \tag{81}$$

$$F_{b\omega} = diag\left(-\beta_{\omega x}, -\beta_{\omega y}, -\beta_{\omega z}\right) \tag{82}$$

$$F_{bf} = diag\left(-\beta_{fx}, -\beta_{fy}, -\beta_{fz}\right) \tag{83}$$

Where *diag(v)* constructs a matrix of diagonal elements of vector *v*, $\beta_{\omega x}, \beta_{\omega y}, \beta_{\omega z}$ and $\beta_{fx}, \beta_{fy}, \beta_{fz}$ are the reciprocal of the time constant of the Gauss-markov process of the gyroscope and accelerometer biases respectively. The remaining elements of the transition matrix are defined as follows:

$$F_{\Delta xv} = \frac{\cos p \sin \Delta A}{v}\begin{bmatrix} v_e & v_n & v_u \end{bmatrix} \tag{84}$$

$$F_{\Delta x \Delta A} = v \cos p \cos \Delta A \tag{85}$$

$$F_{\Delta yv} = \frac{\cos p \cos \Delta A}{v}\begin{bmatrix} v_e & v_n & v_u \end{bmatrix} \tag{86}$$

$$F_{\Delta y \Delta A} = -v \cos p \sin \Delta A \tag{87}$$

$$F_{\Delta Ab\omega} = \begin{bmatrix} \cos p \sin r & -\sin p & -\cos p \cos r \end{bmatrix} \tag{88}$$

Linearizing the system of equation (18-23) and (62-68) with respect to noise vector $w(t)$ will result into the noise shaping matrix G defined by (45). All elements of G are zeros except the following elements:

$$G(4:6,4:6) = R_b^l(q) \tag{89}$$

$$G(7:10,1:3) = -F_{qb\omega} \tag{90}$$

$$G(11:13,7:9) = diag(\sqrt{2\sigma_{\omega x}^2 \beta_{\omega x}}, \sqrt{2\sigma_{\omega y}^2 \beta_{\omega y}}, \sqrt{2\sigma_{\omega z}^2 \beta_{\omega z}}) \tag{91}$$

$$G(14:16,10:12) = diag(\sqrt{2\sigma_{fx}^2 \beta_{fx}}, \sqrt{2\sigma_{fy}^2 \beta_{fy}}, \sqrt{2\sigma_{fz}^2 \beta_{fz}}) \tag{92}$$

$$G(19,1:3) = -F_{\Delta Ab\omega} \tag{93}$$

$$G(21,21) = \sigma_d \tag{94}$$

where σ_d is the standard deviation of random walk process of the receiver clock bias.

Measurement Model

Measurement models relates the difference between system prediction of observables and external measurements which is represented by $\delta y(t)$ in (43).

GPS Measurement Model

In outdoor environments where GPS signals might be received, pseudo-range and pseudo-range rates from GPS are fused with the IMU at the rate of 1 Hz. The measurement model for GPS/IMU tightly coupled scheme is:

$$\delta y_{GPS} = \begin{bmatrix} \delta\rho^{SV_1} \\ \vdots \\ \delta\rho^{SV_M} \\ \delta\dot{\rho}^{SV_1} \\ \vdots \\ \delta\dot{\rho}^{SV_M} \end{bmatrix} = \begin{bmatrix} \rho_{IMU}^{SV_1} - \rho_{GPS}^{SV_1} \\ \rho_{IMU}^{SV_2} - \rho_{GPS}^{SV_2} \\ \vdots \\ \rho_{IMU}^{SV_M} - \rho_{GPS}^{SV_M} \\ \dot{\rho}_{IMU}^{SV_1} - \dot{\rho}_{GPS}^{SV_1} \\ \dot{\rho}_{IMU}^{SV_2} - \dot{\rho}_{GPS}^{SV_2} \\ \vdots \\ \dot{\rho}_{IMU}^{SV_M} - \dot{\rho}_{GPS}^{SV_M} \end{bmatrix} = \begin{bmatrix} H_{M\times3}^{\rho} & 0_{M\times3} & 0_{M\times10} & 1 & 0 \\ 0_{M\times3} & H_{M\times3}^{\rho} & 0_{M\times10} & 0 & 1 \end{bmatrix} \delta x \tag{95}$$

where

$$H_{M\times3}^{\rho} = \begin{bmatrix} (1^{SV_1})^T \\ \cdot \\ \cdot \\ \cdot \\ (1^{SV_M})^T \end{bmatrix} \begin{bmatrix} -(R_n + h)\sin\varphi\cos\lambda & -(R_n + h)\cos\varphi\sin\lambda & \cos\varphi\cos\lambda \\ -(R_n + h)\sin\varphi\sin\lambda & (R_n + h)\cos\varphi\cos\lambda & \cos\varphi\sin\lambda \\ (R_n + h)\cos\varphi & 0 & \sin\varphi \end{bmatrix} \tag{96}$$

and

$$\dot{\rho}_{IMU}^{SV_i} = \sqrt{\left(X_{IMU}^{ECEF} - X_{SV_i}^{ECEF}\right)^2 + \left(Y_{IMU}^{ECEF} - Y_{SV_i}^{ECEF}\right)^2 + \left(Z_{IMU}^{ECEF} - Y_{SV_i}^{ECEF}\right)^2} \tag{97}$$

$$\dot{\rho}_{IMU}^{SV_i} = 1^{SV} \cdot \left[\left(\dot{X}_{IMU}^{ECEF} - \dot{X}_{SV_i}^{ECEF}\right)\left(\dot{Y}_{IMU}^{ECEF} - \dot{X}_{SV_i}^{ECEF}\right)\left(\dot{Z}_{IMU}^{ECEF} - \dot{X}_{SV_i}^{ECEF}\right)\right]^T \tag{98}$$

LiDAR Measurements

The measurement model for LiDAR measurements is given by:

$$\delta y_{LiDAR} = \begin{bmatrix} \Delta x_{lidar} - \Delta x_{ins} \\ \Delta y_{lidar} - \Delta y_{ins} \\ \Delta A_{lidar} - \Delta A_{ins} \end{bmatrix} = I_{3\times3} \begin{bmatrix} \delta\Delta x \\ \delta\Delta y \\ \delta\Delta A \end{bmatrix} \tag{99}$$

A note about redundancy of δq_{LB}^{B}:

It is important to mention the issue of the redundancy constrains of quaternion error vector due to the quaternion constrains ($a^2 + b^2 + c^2 + d^2 = 1$). Therefore, the error covariance matrix of the Kalman filter will have a singularity. To resolve this issue, error states can be reduced by removing the scalar component of the quaternion (the 'a' component) and replacing it by the following equivalent relationship obtained from quaternion constraints:

$$a = \sqrt{1 - \left(b^2 + c^2 + d^2\right)} \tag{100}$$

When the Kalman filter estimates the errors in *b, c,* and *d,* the error in *a* can be estimated using the following formula which is derived from (53) by applying Taylor series linearization:

$$\delta a \simeq \frac{1}{a}\begin{bmatrix} b & c & d \end{bmatrix}\begin{bmatrix} \delta b \\ \delta c \\ \delta d \end{bmatrix} \tag{101}$$

Experiments and Results

The system was tested on the UGV shown in Figure 23, ("Husky A200") from Clearpath Robotics Inc Waterloo-based. This UGV is equipped with CHR-UM6 MEMS IMU (CHRobotics, n.d.), LMS 111 SICK laser scanner, Ublox EKV-7 GPS receiver and WiFi-interface. The UGV was wirelessly teleoperated and data were collected and post processed. The measurement covariance matrices R_{k+1} for both GPS pseudo-range/pseudo-range rate measurements and LiDAR relative motion measurements, and all biases/drifts stochastic noise parameters used in the *G* matrix have been empirically tuned to achieve the best performance. The experimental area included open-sky, dense urban, and indoor sections in the department of Electrical and Computer Engineering in the Royal Military College of Canada. The dense urban area includes a ramp as shown in Figure 24.

GPS positioning results are shown in Figure 25. GPS accuracy deteriorates when the UGV approaches the building. When GPS accuracy deteriorates, the fusion algorithm will discard GPS measurements and accept LiDAR measurements. If LiDAR measurements are not applied, the solution will use only IMU inside the building which leads to drifts as shown in Figure 26. If LiDAR relative motion estimation is applied in the fusion algorithm, the solution is improved as can be visually seen in Figure 27. Biases estimation results are shown in Figure 28 and Figure 29.

Figure 23. Experimental UGV

Figure 24. Experimental Environment in the Royal Military College of Canada

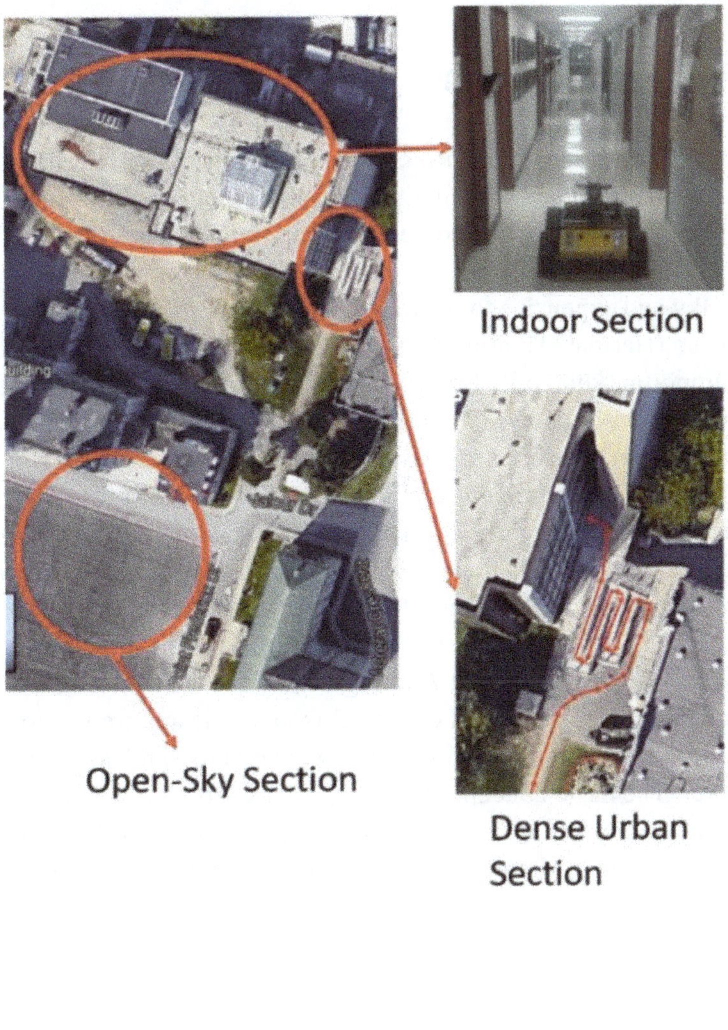

Figure 25. GPS Results (left). Reference Solution (Right)

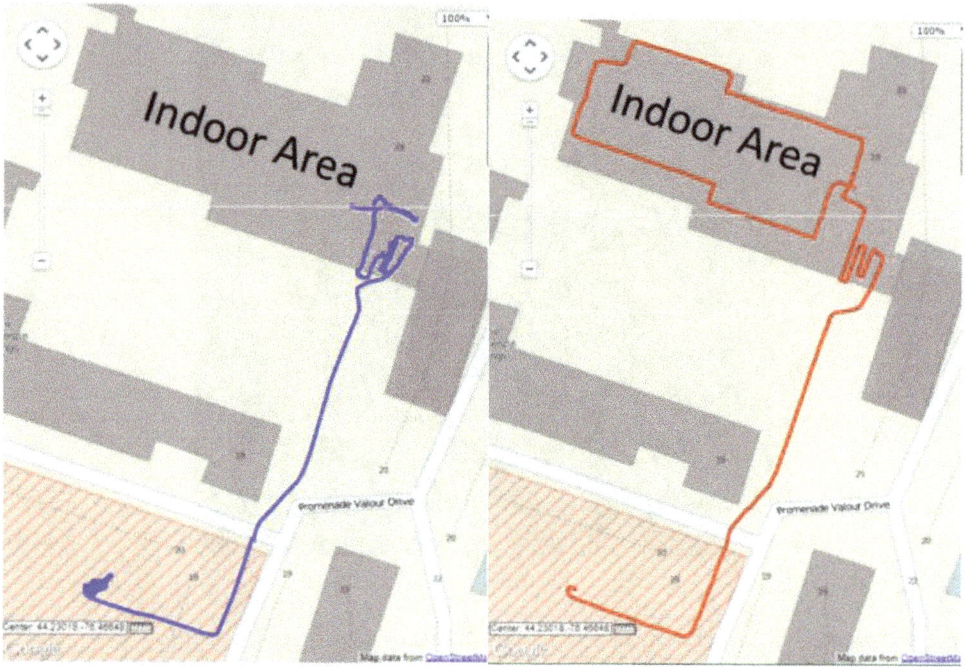

FUTURE RESEARCH DIRECTIONS

Future trends in sensor fusion and navigation in complex environments include the augmentation or entire replacement of traditional EKF by soft-computing intelligent techniques such as fuzzy logic, artificial intelligence, and deep learning methods. Another common trend is the utilization of more sensors. For example, in the case presented in this chapter, while LiDAR sensors are accurate and they provide long-range, they are still bulky and expensive to be used in productions for common transportation vehicles. Therefore, vision sensors (Harmat, Trentini, & Sharf, 2015) (Horgan, Flannery, & Toal, 2009) are becoming increasingly important due to its low-cost, smaller size, lower-power consumption, and richer information (most LiDAR sensors provide 2D scans while vision sensors scan the entire environment in 3D). The vision sensors can be used in the same way the LiDAR sensors are being used. The difference is in the structure of raw measurements. While LiDAR provides "range", vision sensors provides "pixels". If visual maps are available, correlation with such maps can be used as an absolute localization technique. Otherwise, the point-scan matching explained in LiDAR can be extended to visual data to provide visual features extraction and tracking. Because the vision data is a projection of the 3D scene on a planar 2D pixel domain, the processing is more complex than the case with LiDAR. However, the concept is still the same. Consider the point *P* in Figure 30. If it can be seen from two different positions (C1 and C2), the relative motion from C1 to C2 (rotation R and translation T) can be estimated using the art of photogrammetry. Photogrammetry is a wide rich research area and its details are beyond the scope of this chapter. The basic concept utilized in relative motion estimation from visual features tracked between consecutive frames is the Epipolar constraints (Hartley & Zisserman, 2003) (Geiger, Ziegler, & Stiller, 2011) (Harmat, Trentini, & Sharf, 2015) is briefly discussed. Epipolar constraints equations are given by:

Figure 26. GPS/IMU Solution

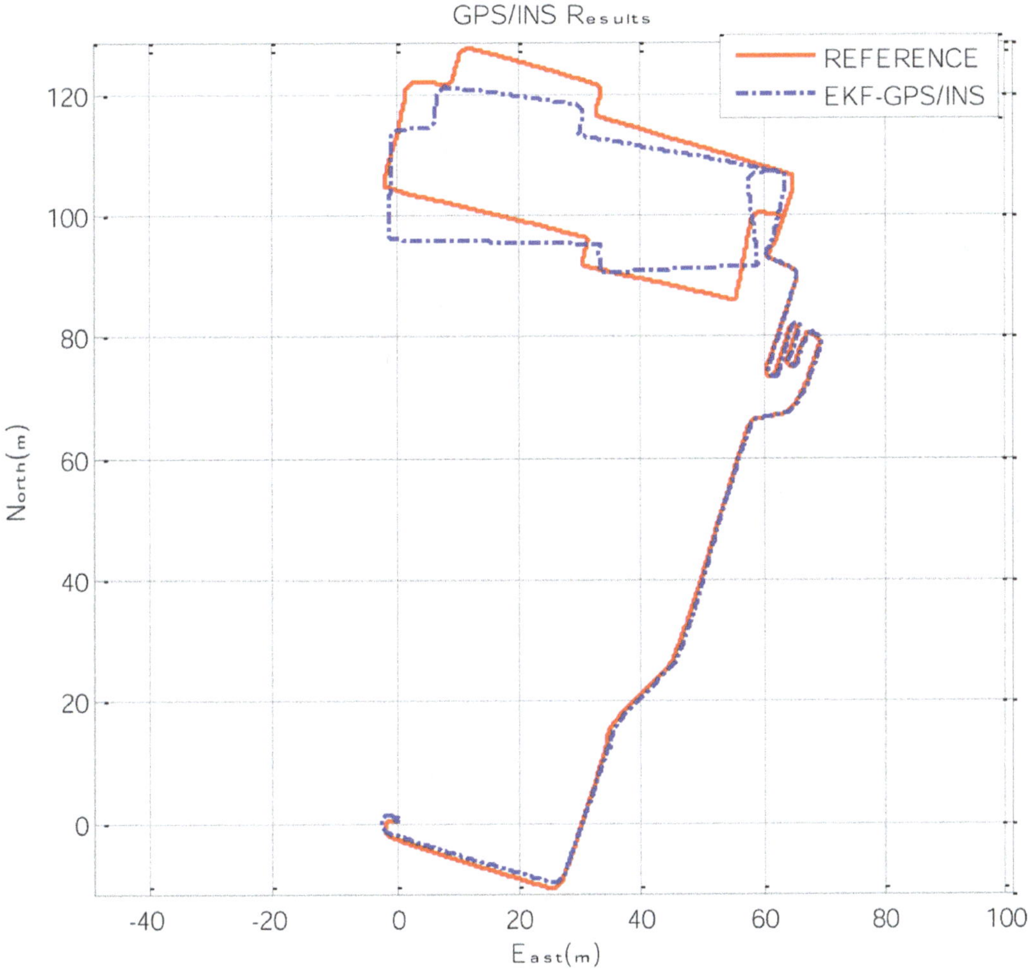

$$x_2 \otimes \left(Rx_1 + T \right) = 0 \tag{102}$$

$$x_2 \otimes Ex_1 = 0 \tag{103}$$

$$E = \begin{bmatrix} R_{1,1} & R_{1,2} & R_{1,3} & T_x \\ R_{2,1} & R_{2,2} & R_{2,3} & T_y \\ R_{3,1} & R_{3,2} & R_{3,3} & T_z \\ 0 & 0 & 0 & 1 \end{bmatrix} \tag{104}$$

Figure 27. GPS/IMU/LiDAR Solution

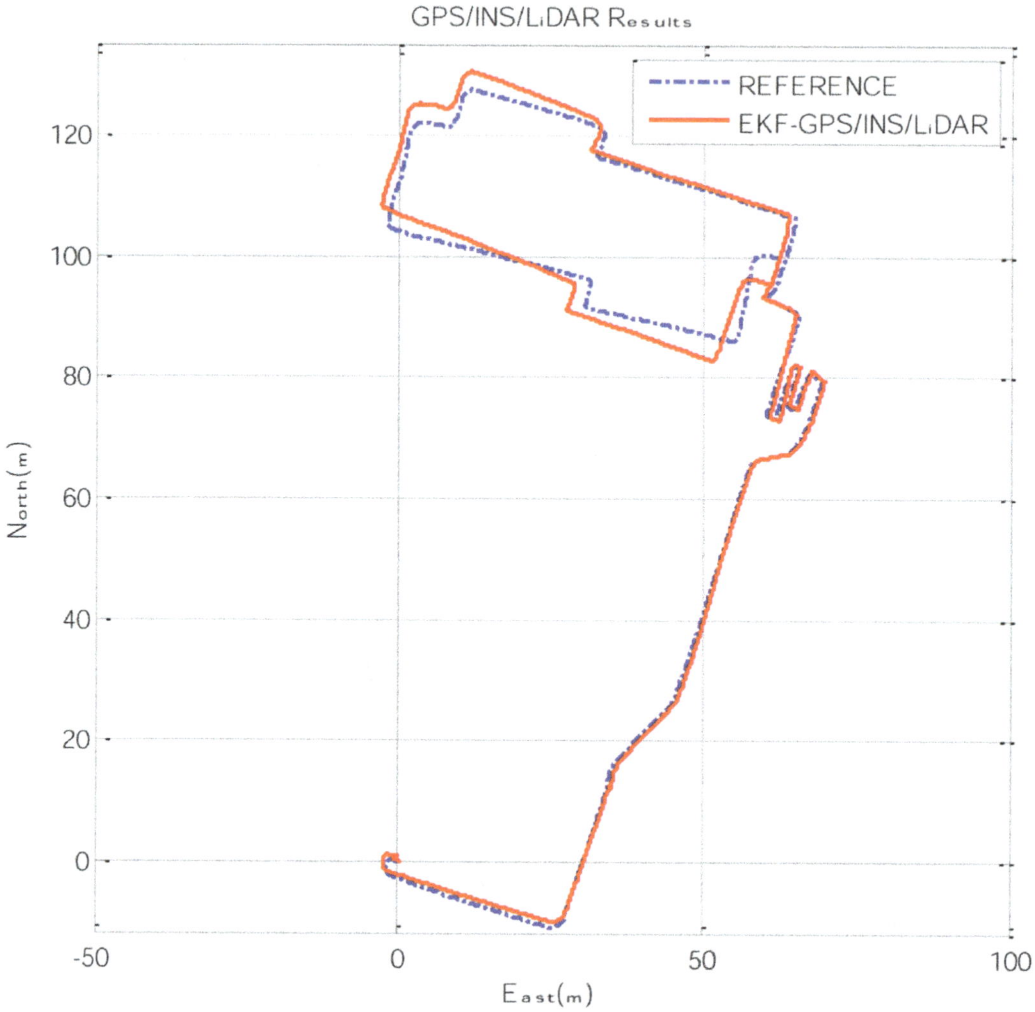

Where x_1 is the 3D world coordinate of P as seen in the first camera location and x_2 is the coordinate of P as seen from the second camera location. Because all what we have is just a 2D image frame with pixel locations of features, we need a method to transform from 2D pixel location to 3D world coordinate. The conversion is governed by the camera calibration equation described by:

$$p = \underbrace{\begin{bmatrix} u \\ v \\ d \end{bmatrix}}_{pixel} = \underbrace{\begin{bmatrix} f_x & 0 & c_x \\ 0 & f_y & c_y \\ 0 & 0 & 1 \end{bmatrix}}_{Calibration} \underbrace{\begin{bmatrix} X \\ Y \\ Z \end{bmatrix}}_{World} \tag{105}$$

where u,v is pixel location, 'd' is the depth factor, X,Y,Z is the world coordinate, $f_x f_y$ is the focal lengths of the camera, $c_x c_y$ is the center point of the camera. Parameters $f_x f_y$ and $c_x c_y$ can be estimated by common

Figure 28. Accelerometer biases

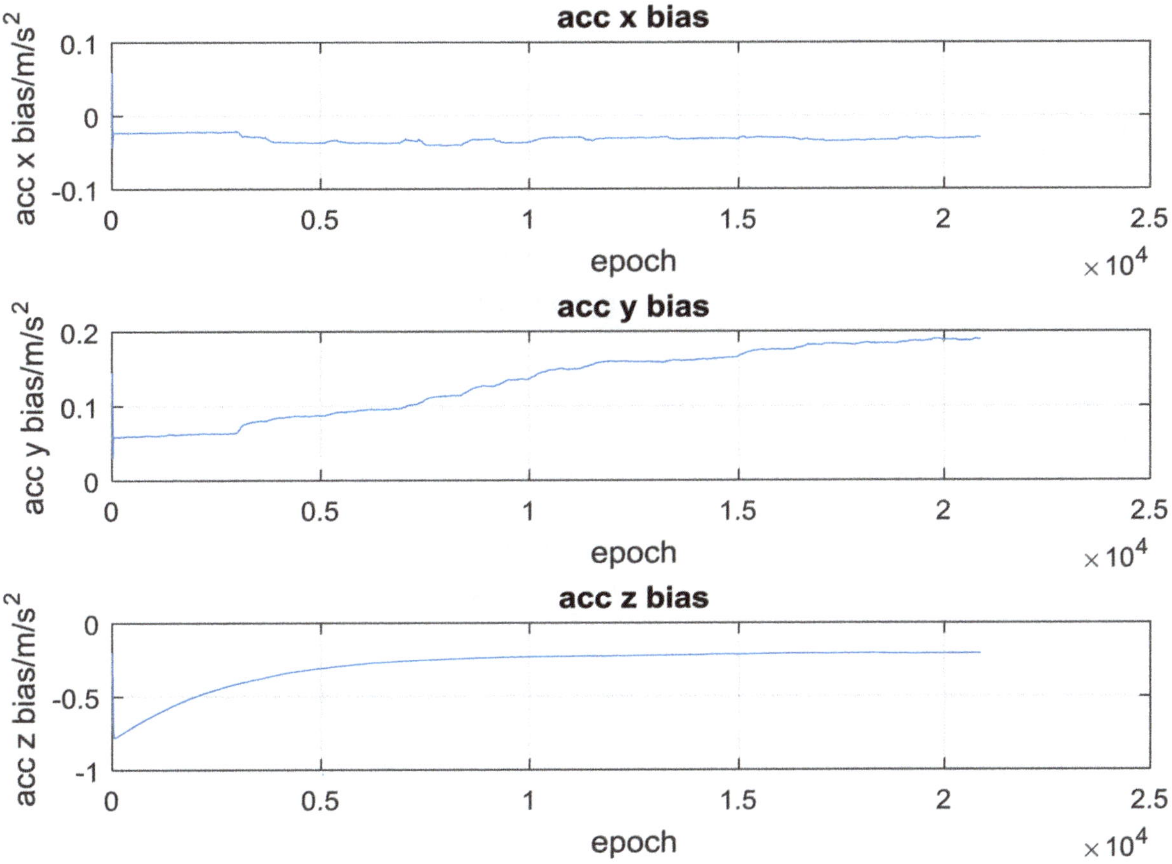

camera calibration methods which are now standard library in many scientific computation tools such as Matlab. However, the depth *"d"* is still a main challenge to estimate in visual tracking systems since several 3D world features can be projected on the same pixel location. This is called "scale" ambiguity which is a challenge in vision-aided positioning. One way to resolve this scale ambiguity is to use other sensors such as IMU or LiDAR to provide a guess about the relative motion and hence a scale factor can be estimated and used later-on on future processing. Another method is to use multiple cameras as proposed in (Harmat, Trentini, & Sharf, 2015). For more details about motion tracking from visual information, readers are referred to (Hartley & Zisserman, 2003), (Geiger, Ziegler, & Stiller, 2011) and (Harmat, Trentini, & Sharf, 2015).

CONCLUSION

This chapter demonstrated the characteristics, advantages, integration, and limitations of two main navigation technologies; GNSS and INS. While GNSS/INS technology enhance overall performance in dense urban areas, it is not sufficiently sustainable indoors. Therefore, the chapter described how mapping sensors (e.g. LiDAR) can be used to enhance the dead-reckoning accuracy of INS when GNSS is attenu-

Figure 29. Gyroscope biases

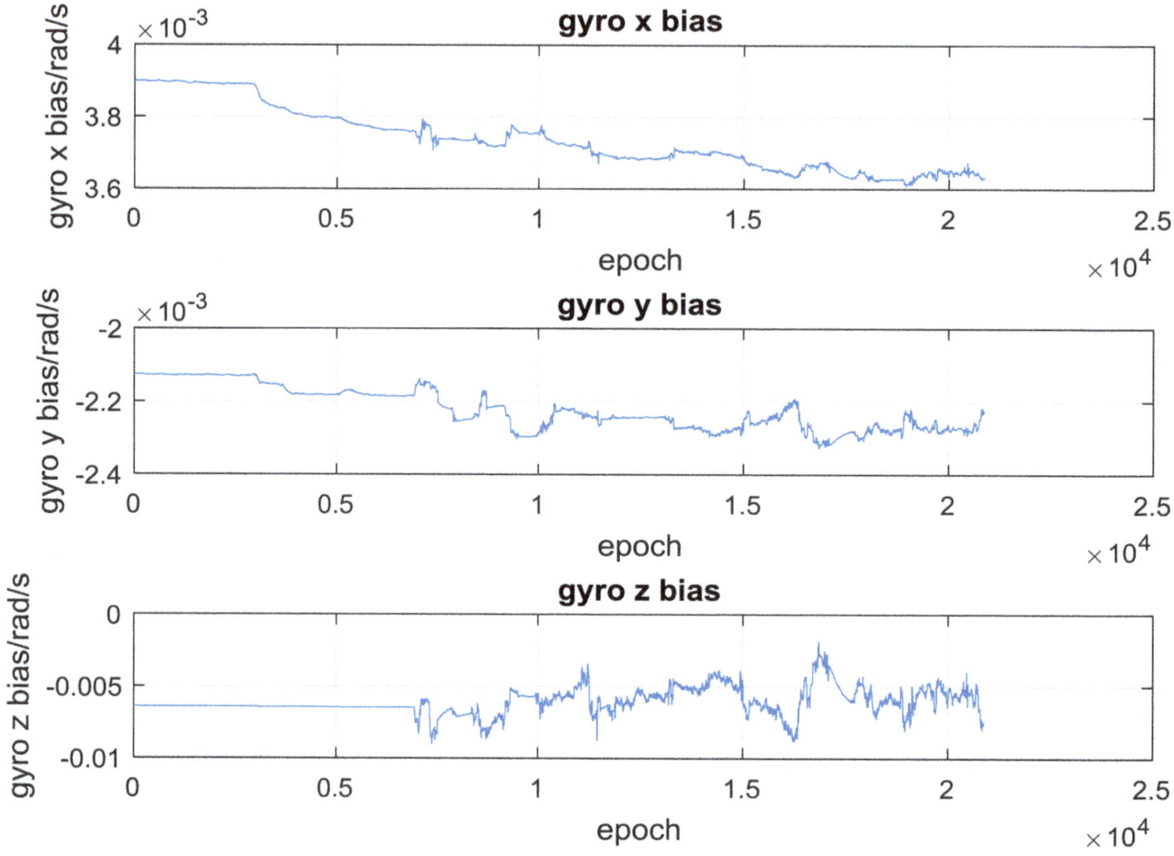

Figure 30. Visual Tracking Concept

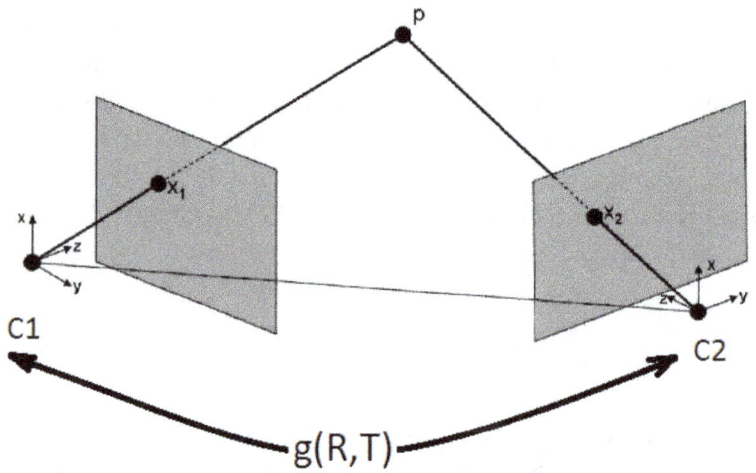

ated or completely lost. EKF design and implementation steps were provided along with experimental verification in a typical scenario that includes open-sky, dense urban, and complex indoor environment. Experimental results showed significant enhancement without the need for a pre-defined map of the area.

ACKNOWLEDGMENT

The author is grateful to the Robotics and Control Laboratory in the Royal Military College of Canada, for facilitating the collection of the Husky A200 UGV data used in this work.

REFERENCES

Ahmed, E.-R. (2002). *Introduction to GPS: The Global Positioning System.* Artech House.

Akos, D. M., Borre, K., & Bertelsen, N. (2007). *A Software-Defined GPS and Galileo Receiver: A Single-Frequency Approach.* Springer.

Allan, D. W. (1966). Statistics of atomic frequency standards. *IEEE Proceedings*, 221-230.

Besl, P. J., & Mckay, N. D. (1992). A method for registration of 3-D shapes. *IEEE Transactions on Pattern Analysis and Machine Intelligence*, *14*(2), 239–256. doi:10.1109/34.121791

Censi, A. (2008). An ICP variant using a point-to-line metric. *IEEE International Conference on Robotics and Automation*. doi:10.1109/ROBOT.2008.4543181

CHRobotics. (n.d.). Retrieved 4 18, 2014, from https://www.chrobotics.com/products/inertial-and-orientation-sensors

El-Rabbany, A. (2006). *Introduction to GPS: The Global Positioning System.* Artec House Inc.

El-Sheimy, N., Hou, A., & Niu, X. (2007). Analysis and Modeling of Inertial Sensors Using Allan Variance. *IEEE Transactions on Instrumentation and Measurement*, *57*(1), 140–149. doi:10.1109/TIM.2007.908635

Farrell, J. A. (2008). *Aided Navigation, GPS with High Rate Sensors.* New York: McGraw Hill.

Geiger, A., Ziegler, J., & Stiller, C. (2011). Stereoscan: Dense 3D reconstruction in real-time. *IEEE Intelligent Vehicles Symposium (IV).*

Grewal, M., Weill, S., & Andrews, A. (2007). *Global Positioning Systems, Inertial Navigation, and Integration.* Wiley & Sons Inc. doi:10.1002/0470099720

Groves, P. D. (2013). *Principles of GNSS, Inertial, and Multisensor Integrated Navigation Systems.* Artech House.

Harmat, A., Trentini, M., & Sharf, I. (2015). Multi-Camera Tracking and Mapping for Unmanned Aerial Vehicles in Unstructured Environments. *Journal of Intelligent & Robotic Systems*, *78*(2), 291–317. doi:10.1007/s10846-014-0085-y

Hartley, R., & Zisserman, A. (2003). *Multiple View Geometry in Computer Vision*. Cambridge University Press.

Horgan, J., Flannery, F., & Toal, D. (2009). Towards real time vision based UUV navigation using GPU technology. *Oceans*, 2009.

Huang, S., & Dissanayake, G. (2007). Convergence and Consistency Analysis for Extended Kalman Filter Based SLAM. *IEEE Transactions on Robotics*, *23*(5), 1036–1049. doi:10.1109/TRO.2007.903811

IEEE. (1997). *IEEE Standard Specification Format Guide and Test Procedure for Single-Axis Interferometric Fiber Optic Gyros*. IEEE Standard Specification Format.

Luo, R., & Lai, C. C. (2014). Multisensor Fusion-Based Concurrent Environment Mapping and Moving Object Detection for Intelligent Service Robotics. *IEEE Transactions on Industrial Electronics*, *61*(8), 4043–4051. doi:10.1109/TIE.2013.2288199

Lynen, S., Sattler, T., Bosse, M., Hesch, J., Pollefeys, M., & Siegwart, R. (2015). Get Out of My Lab: Large-scale, Real-Time Visual-Inertial Localization. In L. E. Kavraki, D. Hsu, & J. Buchli (Eds.), *Robotics: Science and Systems XI* (p. 37). Rome, Italy: RSS. doi:10.15607/RSS.2015.XI.037

Misra, P., & Enge, P. (2011). *Global Positioning System*. Signals, Measurements, and Performance.

Montemerlo, M., Thrun, S., Koller, D., & Wegbreit, B. (2002). FastSLAM: A factored solution to the simultaneous localization and mapping problem. *Proceedings of the National Conference on Artificial Intelligence*, 593-598.

Naser El-Sheimy, H. H. (2008). Analysis and Modeling of Inertial Sensors Using Allan Variance. *IEEE Transactions on Instrumentation and Measurement*, *57*(1), 140–149. doi:10.1109/TIM.2007.908635

Perera, L. D., Wijesoma, W. S., & Adams, M. D. (2010). SLAM With Joint Sensor Bias Estimation: Closed Form Solutions on Observability, Error Bounds and Convergence Rates. *IEEE Transactions on Control Systems Technology*, *18*(3), 732–740. doi:10.1109/TCST.2009.2026165

Rusinkiewicz, S., & Levoy, M. (2001). Efficient Variants of the ICP Algorithm. *Third International Conference on 3D Digital Imaging and Modeling (3DIM)*. doi:10.1109/IM.2001.924423

Savage, P. G. (2007). *Strapdown Analytics*. Strapdown Associates Inc.

SICK. (2008). *Operating instructions: LMS100/111/120 laser measurement systems*. SICK AG.

Titterton, D. H., & Weston, J. L. (2004). Strapdown Inertial Navigation Technology (2nd ed.). The Institution of Electrical Engineers.

William J. Hughes Technical Center. (2014). *Global Positioning System (GPS), Standard Positioning Service (SPS), Performance Analysis Report #86*. William J. Hughes Technical Center, NSTB/WAAS T&E Team.

Chapter 11
Multisensor Integration and Data Fusion for Positioning

Yong Li
Navextech Technology Ltd, Australia

Wei Jiang
Beijing Jiaotong University, China

Ling Yang
Tongji University, China

Chris Rizos
University of New South Wales, Australia

ABSTRACT

This chapter presents an alternative framework to the traditional centralised Kalman filtering (CKF) approach for implementing the optimal state estimation algorithm in support of multisensor integration. The data fusion algorithm is implemented via a series of transformations of vectors in the so-called information space (iSpace). This chapter describes how the conventional decentralised Kalman filtering (DKF) algorithm can be derived using a unified approach. A new global optimal fusion (GOF) algorithm is derived using the iSpace approach. Two case studies are presented to illustrate applications of the multisensor algorithms for GNSS/Locata/INS and GNSS/WiFi/INS integration.

INTRODUCTION

To implement a practical multisensor navigation system, the challenges include, but are not limited to, the time-synchronisation of sensor outputs, communications in the sensor network, and computational load allocation for local and/or central processors. Nevertheless, the core technology for multisensory navigation integration is the algorithm, which combines the data from multiple sensors so as to obtain the optimal estimate of the states of the host platform. Some fundamental questions related to the optimal state estimation in multisensor systems include:

DOI: 10.4018/978-1-5225-3528-7.ch011

- What information (e.g. the data from the sensors and apriori knowledge) can be used in the estimation?
- How to compare the accuracies of different algorithms? Which is the better criterion for multisensor data fusion, the centralised Kalman filter (CKF) or the two-level optimal criterion?
- Is it necessary to develop a unified theoretical framework to deal with the multisensor data fusion problem?
- Does the framework provide a means for understanding multisensor data fusion as well as conventional optimal estimations?

The conventional approach for optimal estimation of a multisensor system is the centralised Kalman filtering approach, which, as the name implies, processes all sensor measurements at a central processor. CKF has some inherent defects, such as heavy computational load and poor fault-tolerant capability. To deal with the above shortcomings, decentralised Kalman filtering (DKF) methods have been developed for applications such as distributed control systems and integrated navigation. A typical DKF configuration is a spider-net style sensor network, in which a processor at the centre of the network is used to manage all sensor nodes. Each node consists of the sensor and its accompanying local processor. The local estimates are sent to the central processor where they are fused to obtain the globally optimal estimate (Hashemipour *et al* 1988).

Data fusion techniques can also be applied for target tracking applications, and are classified either as low-level "scan" fusion or high-level "track-to-track" fusion, which are achieved by combining observations from different sensors and combining estimates from sensor sites, respectively. Similar to the DKF in multisensor navigation applications, the advantages of easy fault detection and low computational load make the track-to-track fusion architecture a popular choice. The key issue for track-to-track fusion is strong correlation between the tracks because they have a common prediction error resulting from a common process model (Bar-Shalom 1981). Systems that ignore such correlation would be expected to have poor performance. Many methods have been developed to address the correlation problem (Bar-Shalom 1981; Aeberhard *et al* 2012). However such methods still cannot achieve better performance than the low-level "scan" fusion method.

To address the fundamental questions listed above, a new framework was recently developed for multisensor integration, which provides a unified approach to derive the multisensor fusion algorithms using the information space approach (Li 2014). The optimal fusion is implemented by a series of transformations between the information spaces (*i*Spaces). The transformations map the source information vectors from the measurement *i*Spaces to the estimate *i*Space to generate the fused information vector. The *i*Space approach provides a means by which the accuracies of different algorithms can be compared on the same theoretical basis. In fact there are multiple sources of information available for estimation in a multisensor system. An algorithm can use either some or all of the information resources to derive the solution, and therefore algorithms using different source information will have different estimation accuracies although they all are optimal in terms of their own *i*Spaces. From this point of view the CKF is no longer necessarily superior to other algorithms. It is just one of the optimal fusion algorithms in the *i*Space framework instead of providing the global optimality criterion of the multisensor data fusion.

During the past few decades many positioning and navigation systems and technologies have been developed to address a variety of user requirements under different application environments. The Global Navigation Satellite System (GNSS) technology is able to provide accurate position, velocity and time data, on a 24/7 basis, without any impact of mission length or time since update. However it is most effective in open outdoor environments where there is no (or at least very little) signal blockage, interference or attenuation. In contrast, the inertial navigation system (INS) technology comprises a self-contained single hardware package with sensors that have relatively high short-term stability, and are immune to jamming or signal outages because they do not make use of any external transmitted signals. Because of the high degree of complementarity, a combination of GNSS and INS technologies could not only mitigate the weakness of GNSS, but also bound the INS error that otherwise would grow with time when the INS operates without GNSS solutions. This powerful synergy between GNSS and INS makes the integration of the two technologies widespread and indispensable.

Although an integrated GNSS/INS provides augmented solutions its performance is still limited by the quality of GNSS measurements, and the geometric strength of the satellite constellation used in the solutions. When GNSS signals are blocked or there is interference, disrupting availability beyond a certain period of time, the INS sensor error will grow with time and impact on positioning accuracy, reliability and availability. In the case of longer GNSS outages, as in the case of indoor applications, additional sensors are needed as a replacement of GNSS to aid the INS so that continuous navigation can be maintained. The WiFi-based indoor positioning technology and terrestrial-based positioning technology such as Locata are introduced in this chapter to complement GNSS/INS, especially in environments where GNSS is unavailable.

The additional terrestrial ranging signals can significantly enhance the performance of the total system in a number of ways, including reducing the dilution of precision (DOP) to improve the accuracy, availability and reliability of the navigation solutions. By being included within a conventional GNSS/INS integrated system as an independent, but complementary, component to address issues such as GNSS signal blockage or interference, such terrestrial systems assure the continuity of navigation solutions. Moreover, in some severe GNSS signal attenuation cases, as in indoor positioning, where GNSS cannot be used for precise positioning, such terrestrial technology could be used to ensure continuous navigation as a "GNSS-alternative" technology, as long as the issue of multipath in indoor environments can also be addressed.

Since 2002, Locata Corporation has been developing a terrestrial-based positioning technology known as "Locata". Locata uses a "local constellation of transmitters" to provide continuous time-synchronised ranging signals. In a multisensor navigation strategy, with the addition of more sensors, more information becomes available. In a GNSS/Locata/INS configuration each sub-system could provide position, velocity and acceleration information to users. The navigation system needs to fuse the solutions from all the sensors.

This chapter introduces the *i*Space framework and unifies the multisensor data fusion algorithms through the information space approach. Two case studies are presented to illustrate how the algorithms are applied to the challenge of multisensor integration: GNSS/Locata/INS, and GNSS/WiFi/INS. The chapter consists of 5 sections. Section 1 "Introduction", presents the background to multisensor data fusion and the motivation underpinning the *i*Space approach. Section 2 "Information space theory and data fusion algorithms", describes the information space framework for multisensor data fusion, providing a unified approach to derive the optimal estimation algorithms such as CKF, the Federal Kalman filter (FKF) and the global optimal fusion (GOF) algorithm. Section 3 "GNSS/Locata/INS integration", introduces the Locata technology, together with a description of a robust georeferencing system based on the loosely-coupled integrated GNSS, Locata, and INS sub-systems, able to satisfy centimetre-level accuracy requirements for both open and GNSS signal occluded environments. Three integrated algorithms – CKF, FKF, and GOF – are applied to implement such a triple-integrated system. Section 4

"GNSS/WiFi/INS integration" describes this triple-integrated system as applicable for indoor pedestrian navigation. To realise seamless navigation in outdoor and indoor environments, the GNSS/WiFi/INS integrated configuration can effectively improve the WiFi stand-alone positioning performance without increasing the computational burden. In outdoor environments, the GNSS/INS sub-systems can effectively provide the navigation solution, whereas for indoor environments the positioning and attitude solutions from the INS will be aided with the WiFi positioning solutions. To further mitigate the INS ramp errors, step detection is implemented during the indoor navigation phase. The CKF is used to implement the GNSS/WiFi/INS multisensor configuration. Section 5 concludes with the "Summary" of the contents of this chapter.

INFORMATION SPACE THEORY AND DATA FUSION ALGORITHMS

In a multisensor navigation system, the GNSS/INS sub-systems can be considered the fundamental component because both GNSS and INS have the capability of providing the full navigation set of parameters: position, velocity and attitude. The other sub-system (such as Locata) can provide supplementary outputs to enable the integrated system to operate effectively in any environment or scenario (Li and Rizos 2010; Jiang *et al* 2013). Most state-of-the-art GNSS/INS systems are designed to estimate the INS solution errors using the GNSS measurement data through either loosely- or tightly-coupled integration architectures.

In addition to the multiple information sources provided by the sensor outputs, there are also other information sources, from knowledge about the movement – such as the INS error propagation, physical laws, kinematic equations, and constraints of the movement (such as velocity constraints for land vehicles, people, etc), and others.

System Statement

A multisensor system can be defined as a system which has more than one sensor providing redundant measurements for estimating the system states. The local system associated with each sensor can be modelled as:

$$\mathbf{x}_i(t_{k+1}^i) = \mathbf{f}_i[\mathbf{x}_i(t_k^i), t_k^i] + \mathbf{g}_i[\mathbf{u}_i(t_k^i), t_k^i] + \mathbf{w}_i(t_k^i) \tag{1}$$

$$\mathbf{z}_i(t_k^i) = \mathbf{h}_i[\mathbf{x}_i(t_k^i), t_k^i] + \mathbf{v}_i(t_k^i) \tag{2}$$

where there are n local systems ($i = 1,2,\cdots,n$), $\mathbf{x}_i(k) \in R^{n_i}$ is the local state of the system; $\mathbf{u}_i(k) \in R^{r_i}$ is a known control vector; $\mathbf{z}_i$ is the measurement vector of the #i sensor; and $\mathbf{w}_i(k) \in R^{p_i}$ and $\mathbf{v}_i(k) \in R^{m_i}$ are the noise terms (all the noises can be assumed Gaussian white); $\mathbf{f}_i(\cdot)$, $\mathbf{g}_i(\cdot)$, and $\mathbf{h}_i(\cdot)$ are the state transition function, control function, and the measurement function, respectively, associated with #i local system.

The time of each local system represented by Equations (1) and (2) is denoted by a different superscript. Each local system has its own time frame, for example GNSS time is aligned to UTC time; whereas the INS may have its own internal clock and the output is therefore time-marked against this internal clock. Once time-synchronisation among the local systems is achieved, all local systems can be then referred to the same time frame:

$$\mathbf{x}_i(k+1) = \mathbf{f}_i[\mathbf{x}_i(k), k] + \mathbf{g}_i[\mathbf{u}_i(k), k] + \mathbf{w}_i(k) \tag{3}$$

$$\mathbf{z}_i(k) = \mathbf{h}_i[\mathbf{x}_i(k), k] + \mathbf{v}_i(k) \tag{4}$$

In additional to the local systems there may be a global system equation that describes the behaviour of the global state of the system. The global state vector can be the common states of the system such as the position, velocity and attitude of the platform. However, the local state vectors could consist of the common states plus the sensor biases. The global system equation can be the physical law which governs the behaviour of the system, which in general can be written as:

$$\mathbf{x}(k+1) = \mathbf{f}[\mathbf{x}(k), k] + \mathbf{w}(k) \tag{5}$$

where $\mathbf{x}(k) \in R^n$ is the global state of the system; $\mathbf{f}(\cdot)$ is the global state transition function; and $\mathbf{w}(k) \in R^p$ is the global process noise.

The matrix $\mathbf{S}_i$ maps the global state vector to the local state vector:

$$\mathbf{S}_i : \mathbf{x} \rightarrow \mathbf{x}_i, \text{ or } \mathbf{x}_i = \mathbf{S}_i\mathbf{x} \tag{6}$$

In most navigation applications the non-linearity of the systems is not strong and can be approximated by first-order linearisation, such as in the case of the GNSS range measurements and the INS error equation used in the tight integration of GNSS and INS (Wendal *et al* 2006). The system models can be written in the linear forms by ignoring the control term:

$$\mathbf{x}(k+1) = \mathbf{\Phi}(k)\mathbf{x}(k) + \mathbf{w}(k) \tag{7}$$

$$\mathbf{z}_i(k) = \mathbf{H}_i(k)\mathbf{x}_i(k) + \mathbf{v}_i(k) \tag{8}$$

where Φ_i is the local state transition matrix; $\mathbf{\Phi}$ is the global state transition matrix; and H_i is the local measurement matrix. If the non-linearity cannot be accounted for by first-order approximation, the Sigma-point Kalman filtering (SPKF) or particle filtering approach can be used (Wendal *et al* 2006; Ruan *et al* 2008).

Definitions and Theorems of *i*Space Framework

For a multisensor system there are multiple information sources which can be used in the estimation problem, and include:

- The measurements from sensors, denoted as $\mathbf{z}_i (i = 1, 2, \cdots, n)$.
- The local state predictions from local filters, denoted as $\vec{\mathbf{x}}_i (i = 1, 2, \cdots, n)$.
- The local state estimates from local filters, denoted as $\hat{\mathbf{x}}_i (i = 1, 2, \cdots, n)$.
- The global state prediction, denoted as $\vec{\mathbf{x}}$.
- The global state estimate, denoted as $\hat{\mathbf{x}}$.

Definition 1: Assuming $\hat{\mathbf{x}}_i$ and $\hat{\mathbf{x}}_j$ are two random vectors, their covariance matrices are $\hat{\mathbf{P}}_i$ and $\hat{\mathbf{P}}_j$ respectively, and their correlation matrix is $\hat{\mathbf{P}}_{ij}$ if (Li 1997):

- $\hat{\mathbf{P}}_{ij} = \mathbf{0}$, $\hat{\mathbf{x}}_i$ and $\hat{\mathbf{x}}_j$ are *orthogonal to* each other;
- $\hat{\mathbf{P}}_{ij} = \hat{\mathbf{P}}_i \quad or \quad \hat{\mathbf{P}}_{ij} = \hat{\mathbf{P}}_j$, $\hat{\mathbf{x}}_i$ and $\hat{\mathbf{x}}_j$ are *parallel to* each other; and
- $\hat{\mathbf{P}}_i \neq \mathbf{0} \quad and \quad \hat{\mathbf{P}}_{ij} \neq \hat{\mathbf{P}}_i \quad and \quad \hat{\mathbf{P}}_{ij} \neq \hat{\mathbf{P}}_j$, $\hat{\mathbf{x}}_i$ and $\hat{\mathbf{x}}_j$ are *oblique to* each other.

Two vectors are independent if they are orthogonal or oblique.

Theorem 1 shows how the optimal estimate can be obtained from a set of information sources. The mathematical proof of this theorem can be found in Appendix A of Li (2014).

Theorem 1 (optimal fusion theorem): Θ is a random vector space (RVS), which is expanded by a set of unbiased orthogonal vectors $\hat{\mathbf{x}}_i$(i=1,2,⋯,n) with covariance matrix $\hat{\mathbf{P}}_i$:

$$\Theta = \{\hat{\mathbf{x}} \big| \hat{\mathbf{x}} = \sum_{i=1}^{n} \mathbf{C}_i \hat{\mathbf{x}}_i\} \tag{9}$$

where C_i is weighted matrix of $\hat{\mathbf{x}}_i$. If the matrix $\sum_{i=1}^{n} (\mathbf{S}_i^T \hat{\mathbf{P}}_i^{-1} \mathbf{S}_i)$ is non-singular, the vector $\hat{\mathbf{x}}$ is the optimal estimate of x (Li 1997; 2013):

$$\hat{\mathbf{x}} = \hat{\mathbf{P}}(\sum_{i=1}^{n} \mathbf{S}_i^T \hat{\mathbf{P}}_i^{-1} \hat{\mathbf{x}}_i) \tag{10}$$

with the covariance matrix:

$$\hat{\mathbf{P}} = (\sum_{i=1}^{n} \mathbf{S}_i^T \hat{\mathbf{P}}_i^{-1} \mathbf{S}_i)^{-1} \tag{11}$$

where $\mathbf{S}_i$ is the mapping matrix between x and $\hat{\mathbf{x}}_i$ i.e., $\mathbf{x} = \mathbf{S}_i \mathbf{x}_i$.

From Equation (11) it can be seen that the more information that is used, the larger the trace of $\hat{\mathbf{P}}^{-1}$, or the smaller trace of $\hat{\mathbf{P}}$, therefore the higher accuracy of the estimate. Hence the following conclusion can be drawn:

Conclusion 1: The more information resources that are used, the higher the accuracy of the estimate.

Theorem 1 is fundamental to the state estimation problem for a multisensor system. Note that the optimal estimate Equation (10) is derived from a set of orthogonal bases. Hence finding a set of orthogonal bases is the key to applying this theorem.

Although there are multiple information sources that can be used in data fusion, some of them cannot be the bases of the space simultaneously because the parallel vectors cannot provide additional information to the estimation, as indicated by the theorem below.

Theorem 2 (parallel theorem): Two vectors $\hat{\mathbf{x}}_i$ and $\hat{\mathbf{x}}_j$ are parallel to each other. Both of them are estimates of state vector x. Θ_1 is the space expanded by either $\hat{\mathbf{x}}_i$ or $\hat{\mathbf{x}}_j$. Θ is the space expanded by both $\hat{\mathbf{x}}_i$ and $\hat{\mathbf{x}}_j$. The accuracy of the optimal estimate in Θ_1 is at the same level of accuracy as that in Θ.

The proof of Theorem 2 is given in Appendix B of Li (2014). Theorem 2 shows that among parallel vectors only one of them can be the base of the space. In other words, arbitrary two bases of a space must be orthogonal or oblique to each other.

Theorem 3: The state predictions and the state estimates are parallel to each other.

The proof of Theorem 3 is given in Appendix C of Li (2014). Theorem 3 shows that the state predictions and the state estimates cannot be selected as bases of the same space simultaneously, applying Theorem 2.

Theorem 1 shows how to obtain the optimal estimate when the bases are orthogonal to each other. However, when two bases are oblique to each other, Theorem 4 indicates how to orthogonalise them.

Theorem 4 (orthogonalisation theorem): Two unbiased vectors $\hat{\mathbf{x}}_1$ and $\hat{\mathbf{x}}_2$ are oblique to each other.

Two corresponding unbiased orthogonal vectors $\breve{\mathbf{x}}_1$ and $\breve{\mathbf{x}}_2$ can be always constructed as follows:

$$\breve{\mathbf{x}}_1 = \hat{\mathbf{x}}_1 \tag{12}$$

$$\breve{\mathbf{x}}_2 = \mathbf{K}_1\breve{\mathbf{x}}_1 + \mathbf{K}_2\hat{\mathbf{x}}_2 \tag{13}$$

$$\mathbf{K}_1 = -\mathbf{K}_2\hat{\mathbf{P}}_{21}\hat{\mathbf{P}}_1^{-1} \tag{14}$$

$$\mathbf{K}_2 = [\mathbf{I} - \hat{\mathbf{P}}_{21}\hat{\mathbf{P}}_1^{-1}\mathbf{S}_{21}]^{-1} \tag{15}$$

where $\mathbf{S}_{21}$ is the mapping matrix between $\mathbf{x}_1$ and $\mathbf{x}_2$, i.e. $\mathbf{x}_2 = \mathbf{S}_{21}\mathbf{x}_1$. The proof of Theorem 4 is given in Appendix D of Li (2014).

Transformation of *i*Vectors

According to Theorem 1, the optimal estimation of the multisensor system Equations (10) and (11) can be rewritten in the following information form (Li 2014):

$$\hat{\mathbf{P}}^{-1}\hat{\mathbf{x}} = \sum_{i=1}^{n} \mathbf{S}_i^T \hat{\mathbf{P}}_i^{-1} \hat{\mathbf{x}}_i \tag{16}$$

$$\hat{\mathbf{P}}^{-1} = \sum_{i=1}^{n} \mathbf{S}_i^T \hat{\mathbf{P}}_i^{-1} \mathbf{S}_i \tag{17}$$

The inverse of the covariance matrix is the Fisher's information matrix, in terms of a Gaussian unbiased estimate, in the sense of minimum mean-squared error (Frieden 2004; Mutambara 1998). Denote $\Im$ as the information space which is the collection of the random vectors of $\mathbf{x}$. Similarly, $\Im_i$ is the information space which is the collection of the random vectors of $\mathbf{x}_i$. The information matrix (*i*Matrix) can be denoted as:

$$D(\hat{\mathbf{x}}) = \hat{\mathbf{P}}^{-1} \text{and } D(\hat{\mathbf{x}}_i) = \hat{\mathbf{P}}_i^{-1} \tag{18}$$

The information matrix $D(\hat{\mathbf{x}}_i)$ can be transformed from the space of $\Im_i$ to the space of $\Im$:

$$D(\hat{\mathbf{x}}_i | \Im) = \mathbf{S}_i^T \hat{\mathbf{P}}_i^{-1} \mathbf{S}^T \tag{19}$$

This is hereafter referred to as the 'projection matrix' $D(\hat{\mathbf{x}}_i | \Im)$, which means "the projection of the information matrix of $\hat{\mathbf{x}}_i$ in the space of $\Im$". Matrix $\mathbf{S}_i$ is the same as in Equation (5).

The random vector is left-multiplied by its associated information matrix to form the information vector (*i*Vector):

$$\left|\hat{\mathbf{x}}\right\rangle = D(\hat{\mathbf{x}})\hat{\mathbf{x}} \text{ and} \left|\hat{\mathbf{x}}_i\right\rangle = D(\hat{\mathbf{x}}_i)\hat{\mathbf{x}}_i \tag{20}$$

The *i*Vector can be transformed from the space of $\Im_i$ to the space of $\Im$:

$$\left|\hat{\mathbf{x}}_i, \Im\right\rangle = \mathbf{S}_i^T \left|\hat{\mathbf{x}}_i\right\rangle \tag{21}$$

This is hereafter referred to as the 'projection vector' $\left|\hat{\mathbf{x}}_i, \Im\right\rangle$, which means "the projection of the information vector of $\hat{\mathbf{x}}_i$ in the space of $\Im$".

By applying the definitions above, Equations (16) and (17) can be rewritten:

$$\left|\hat{\mathbf{x}}\right\rangle = \sum_{i=1}^{n} \left|\hat{\mathbf{x}}_i, \Im\right\rangle \tag{22}$$

$$D(\hat{\mathbf{x}}) = \sum_{i=1}^{n} D(\hat{\mathbf{x}}_i \left| \Im \right.) \tag{23}$$

It can be seen that the optimal fusion is equivalent to the sum of the projection vectors, and the associated estimate covariance is obtained by the sum of the projection matrices. Figure 1 illustrates the projection of the information vector of $\left|\hat{\mathbf{x}}_i\right\rangle$ from *i*Space $\Im_i$ to $\Im$.

In a multisensor system there are several sources of information (Table 1): measurements for the estimation, measurements from sensors, the local estimates and predictions from the local filters, the global estimate and prediction from the global filter.

Figure 1. Projection of iVector $\left|\hat{\mathbf{x}}_i\right\rangle$ from iSpace I_i to I

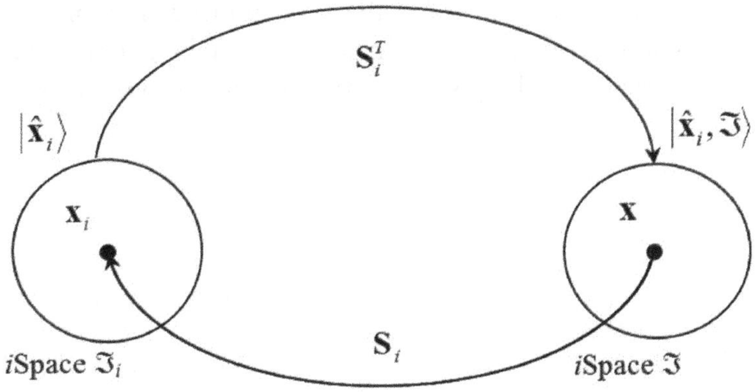

Table 1. Sources of information for multisensor estimation.

Source information	Notation	*i*Space	*i*Vector	*i*Matrix	Source
Local estimate		$\Im_i$	$\left\|\hat{\mathbf{x}}_i\right\rangle$	$D(\hat{\mathbf{x}}_i)$	Local filter
Global estimate	$\hat{\mathbf{x}}$	$\Im$	$\left\|\hat{\mathbf{x}}\right\rangle$	$D(\hat{\mathbf{x}})$	Global filter
Local prediction	$\vec{\mathbf{x}}_i$	$\Im_i$	$\left\|\vec{\mathbf{x}}_i\right\rangle$	$D(\vec{\mathbf{x}}_i)$	Local filter
Global prediction	$\vec{\mathbf{x}}$	$\Im$	$\left\|\vec{\mathbf{x}}\right\rangle$	$D(\vec{\mathbf{x}})$	Global filter
Measurements		$\Re_i$	$\left\|\mathbf{z}_i\right\rangle$	$D(\hat{\mathbf{z}}_i) = \mathbf{R}_i^{-1}$	Sensor

The measurement vector $\mathbf{z}_i$ has a covariance of $\mathbf{R}_i$. Among these vectors, the measurement vectors are orthogonal to each other if the noises of the sensors are uncorrelated with respect to each other. The prediction vectors and the measurement vectors are orthogonal to each other if the system noises and the measurement noises are uncorrelated with respect to each other.

In fact it can be seen that Equations (22) and (23) are the Gauss-Markov estimates when n=1 (with only one measurement vector $|\mathbf{z}\rangle$), and they are the Kalman filter estimates when n=2 (with one measurement vector $|\mathbf{z}\rangle$ and the prediction vector $|\hat{\mathbf{x}}\rangle$).

Optimal Multisensor Fusion Algorithms

Using the *i*Space approach all the optimal data fusion algorithms presented in the literature can be derived, such as the CKF, the parallel Kalman filter (Hashemipour *et al* 1988; Roy *et al* 1991), the federated Kalman filter (FKF), and others.

Centralised Kalman Filter

The CKF has a typical configuration as shown in Figure 2, estimating the global state vector by constructing a global measurement vector when all of the measurements from the sensors have arrived.

The CKF uses the measurement vectors and the global prediction vector to obtain the global estimate vector. From Equations (22) and (23), the CKF's information form is written as:

$$|\hat{\mathbf{x}}\rangle = |\vec{\mathbf{x}}\rangle + \sum_{i=1}^{n} |\mathbf{z}_i, \Im\rangle \tag{24}$$

$$D(\hat{\mathbf{x}}) = D(\vec{\mathbf{x}}) + \sum_{i=1}^{n} D(\mathbf{z}_i |\Im) \tag{25}$$

Figure 2. Structure of CKF algorithm

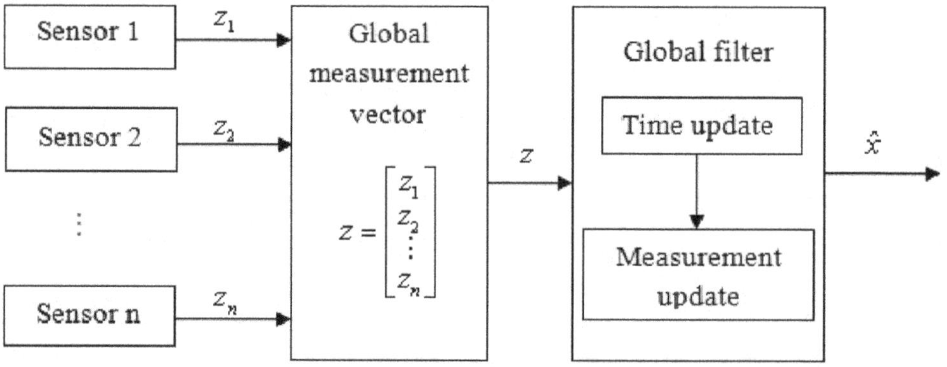

Note that the prediction vector has the same information space as the estimate vector. Therefore the projections of the prediction vector and its matrix are equal to *i*Vector $\left|\vec{\mathrm{x}}\right\rangle$ and *i*Matrix $D(\vec{\mathrm{x}})$ respectively.

Federated Kalman Filter

The decentralised integration architecture was developed to overcome the limitations of the centralised filtering approach. The most commonly used decentralised filtering algorithm is the FKF (Calson and Berarducci 1994; Calson 1996). It is usually regarded as a two-stage data processing algorithm, composed of a set of local filters and one master filter, as illustrated in Figure 3. The local filters work in parallel and independent of each other, and their solutions are periodically fused by the master filter to yield a global solution. This decentralisation characteristic of the FKF ensures a low computation load and a good fault detection and isolation ability. However the accuracy decreases compared with the CKF. In order to achieve the same level of accuracy as the CKF, the information sharing approach is applied in the FKF. The basic concept of the information distribution approach is to divide the global state prediction and its covariance across the local filters, implement the parallel local processing with the sensor measurements, and then fuse the updated local information in the master filter to generate the global estimation.

The FKF is suitable for the case when the local filters and the fusion centre share the same system model (Carlson 1988; 1996), which results in the mapping matrix from local state vector to the global state vector being an identity matrix, i.e. $\mathbf{S}_i = \mathbf{I}$.

The FKF's fusion algorithm needs the local estimates so that Equations (22) and (23) can be used directly. In order to ensure the same accuracy as that of the CKF, the local state prediction and its covariance matrix must be calculated from the global state prediction and its covariance:

Figure 3. Structure of the FKF algorithm

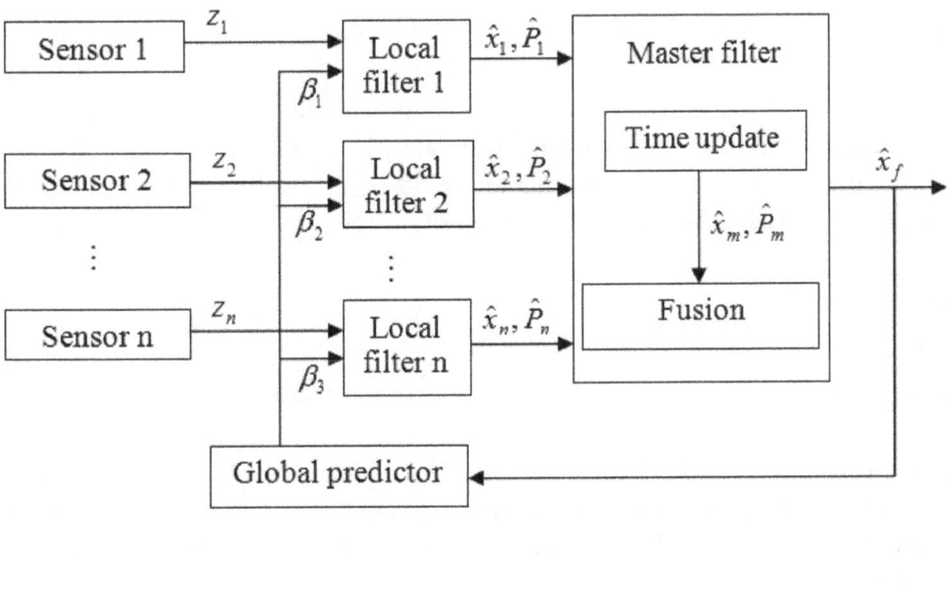

$$\vec{\mathbf{x}}_i = \vec{\mathbf{x}} \tag{26}$$

$$\vec{\mathbf{P}}_i = \frac{1}{\beta_i}\vec{\mathbf{P}} \tag{27}$$

$$\sum_{i=1}^{N}\beta_i = 1 \tag{28}$$

where β_i is the information-sharing factor associated with the #i local system. This is the so-called information-sharing principle (Carlson 1988; 1996). In fact one can substitute the local Kalman filter's fusion Equations (27) and (28) into Equations (22) and (23):

$$\left|\hat{\mathbf{x}}\right\rangle = \sum_{i=1}^{n}(\left|\vec{\mathbf{x}}_i, \Im\right\rangle + \left|\mathbf{z}_i, \Im\right\rangle) \tag{29}$$

$$D(\hat{\mathbf{x}}) = \sum_{i=1}^{n}[D(\vec{\mathbf{x}}_i \left|\Im\right.) + D(\mathbf{z}_i \left|\Im\right.)] \tag{30}$$

The CKF's fusion algorithm Equations (24) and (25) can then be obtained by substituting Equations (26)-(28) into Equations (29) and (30). The configuration of the FKF is illustrated in Figure 3. In general the FKF requires that the local state predictions and their covariances be assigned according to Equations (26)-(29) at each filtering step. The local state estimates are then fused using Equations (22) and (23).

Global Optimal Fusion

According to Theorem 3, the state estimate and the state prediction are parallel to each other, and thus only one of them can be used as the independent information for the estimation. The CKF uses the global prediction and the measurements for the estimation. In other words, the local state predictions are not used. For a multisensor system the maximum amount of independent information is 2n+1:

- Measurements $\mathbf{z}_i$ $(i = 1, 2, \cdots, n)$
- Local predictions $\vec{\mathbf{x}}_i$ $(i = 1, 2, \cdots, n)$
- Global prediction $\vec{\mathbf{x}}$

If the measurement noises and the local system noises are uncorrelated with respect to each other, the measurements and the local predictions are orthogonal to each other. Furthermore, the measurements and the local predictions comprise the local Kalman filtering fusion, thus they can be replaced by the local state estimates.

The global state prediction and the local state estimates are oblique rather than orthogonal to each other because of the correlation between Kalman filtering state estimates of the two preceding epochs. In fact, the current global state prediction is obtained from the previous state estimate:

$$\vec{\mathbf{x}}(k) = \Phi\hat{\mathbf{x}}(k-1) = \Phi[\mathbf{C}(k-1)\hat{\mathbf{x}}(k-1) + \sum_{i=1}^{n}\mathbf{D}_i(k-1)\hat{\mathbf{x}}_i(k-1)] \tag{31}$$

Obviously the present global state prediction $\vec{\mathbf{x}}(k)$ is correlated with the previous local state estimate $\hat{\mathbf{x}}_i(k-1)$. Therefore, $\vec{\mathbf{x}}(k)$ and $\hat{\mathbf{x}}_i(k)$ are correlated because $\hat{\mathbf{x}}_i(k-1)$ and $\hat{\mathbf{x}}_i(k)$ are correlated. As a result, the optimal fusion of Equations (30) and (31) cannot be applied directly before $\vec{\mathbf{x}}$ and $\hat{\mathbf{x}}_i$ are orthogonalised. By applying Theorem 4, the present global state prediction $\vec{\mathbf{x}}$ can be orthogonalised. The orthogonalised global state prediction $\overline{\mathbf{x}}$ and the local state estimates can be optimally fused using Theorem 1:

$$\hat{\mathbf{x}} = \hat{\mathbf{P}}(\overline{\mathbf{P}}^{-1}\overline{\mathbf{x}} + \sum_{i=1}^{n}\mathbf{S}_i^T\hat{\mathbf{P}}_i^{-1}\hat{\mathbf{x}}_i) \tag{32}$$

$$\hat{\mathbf{P}} = (\overline{\mathbf{P}}^{-1} + \sum_{i=1}^{n}\mathbf{S}_i^T\hat{\mathbf{P}}_i^{-1}\mathbf{S}_i)^{-1} \tag{33}$$

$$\overline{\mathbf{x}} = \mathbf{K}\vec{\mathbf{x}} + \sum_{i=1}^{n}\mathbf{K}_i\hat{\mathbf{x}}_i \tag{34}$$

$$\overline{\mathbf{P}} = \mathbf{K}\vec{\mathbf{P}}\mathbf{K}^T + \sum_{i=1}^{n}\mathbf{K}_i\hat{\mathbf{P}}_i\mathbf{K}_i^T + \sum_{i=1}^{n}\mathbf{K}\overline{\mathbf{P}}_i\mathbf{K}_i^T + \sum_{i=1}^{n}\mathbf{K}_i\overline{\mathbf{P}}_i^T\mathbf{K}^T \tag{35}$$

$$\mathbf{K} = [\mathbf{I} - \Phi\hat{\mathbf{P}}(-1)\sum_{i=1}^{n}(\mathbf{S}_i^T\Phi_i^T\vec{\mathbf{P}}_i^{-1}\mathbf{S}_i)]^{-1} \tag{36}$$

$$\mathbf{K}_i = -\mathbf{K}\Phi\hat{\mathbf{P}}(-1)\mathbf{S}_i^T\Phi_i^T\vec{\mathbf{P}}_i^{-1} \tag{37}$$

$$\overline{\mathbf{P}}_i = \Phi\hat{\mathbf{P}}(-1)\mathbf{S}_i^T\Phi_i^T\vec{\mathbf{P}}_i^{-1}\hat{\mathbf{P}}_i \tag{38}$$

where $\hat{\mathbf{P}}(-1)$ is the covariance of the global state estimate of the previous epoch. Equations (32)-(38) are the globally optimal fusion (GOF) algorithm, which optimally utilises the largest set of information bases of the multisensor system.

The CKF is optimal in the space which is expanded by the local measurements and the global prediction. However, the GOF is optimal in the space which is expanded by the measurements, the local state predictions, and the global state prediction. The CKF space is therefore a sub-set of the GOF space. The GOF uses more information sources than the CKF. According to Conclusion 1, the GOF must have a higher accuracy than the CKF.

GNSS, INS, AND LOCATA INTEGRATION

Overview of the Locata Technology

The use of terrestrial ranging technology can be traced back earlier than GPS. Before the U.S. launched the first GPS satellites it tested the system concept and initial GPS user equipment with ground-based transmitters which broadcast GPS-like ranging signals, referred to as "pseudolites" (Harrington and Dolloff 1976). From the early 1980's this technology began to be considered as a complement for GPS operations (Beser and Parkinson 1982; Klein and Parkinson 1984). "Locata" has been under development as a terrestrial ranging system since 2002. The fundamentals of the Locata technology, and some results of earlier testing, have been comprehensively reported in the literature (e.g. Barnes *et al* 2003; Choudhury *et al* 2009a; Li and Rizos 2010; Montillet *et al* 2009). Such studies have verified that this technology can be used for a wide range of positioning applications, including precise indoor positioning (Barnes *et al* 2003; Rizos *et al* 2010), slow structural displacement monitoring (Choudhury *et al* 2009a), and kinematic navigation (Bertsch *et al* 2009; Li and Rizos 2010).

Locata is a ground-based positioning system which transmits ranging signals at frequencies in the 2.4GHz Industrial, Scientific, and Medical (ISM) radio band. Such ranging signals, from transmitters known as "LocataLites", can be tracked by a user receiver. A Locata network – or "LocataNet" – consists of a number of time-synchronised LocataLite transceivers located within or around a defined service area (Barnes *et al* 2004, Rizos *et al* 2010). The "user segment" includes any number of fixed or moving Locata user receivers (or "rovers") operating within the service area, deriving position and time using signals transmitted by the LocataNet.

The LocataLite device consists of an integrated Locata transmitter and receiver, is currently based on Field Programmable Gate Array (FPGA) technology, as shown in Figure 4.

It has been reported that the operating range of a LocataNet can be up to 50 kilometres, when transmitting with adequate signal power (Locata Corporation 2014). Locata has some features that distinguish it from both GNSS technology and conventional pseudolites. These include:

- **Time-Synchronisation:** All LocataLites in a LocataNet are synchronised to a master station in the network, either in a direct link or in a cascaded mode, using the TimeLoc time synchronisation process. TimeLoc maintains phase difference and compensates the clock difference among all transmitting signals in a LocataNet. By doing this, the time difference can be kept to below a few nanoseconds and the frequency stability is less than one part per billion (ppb) (Gauthier *et al* 2013). This implies that there is effectively no transmitter clock error.

Figure 4. LocataLite: receiver (left) and transmitter (right)

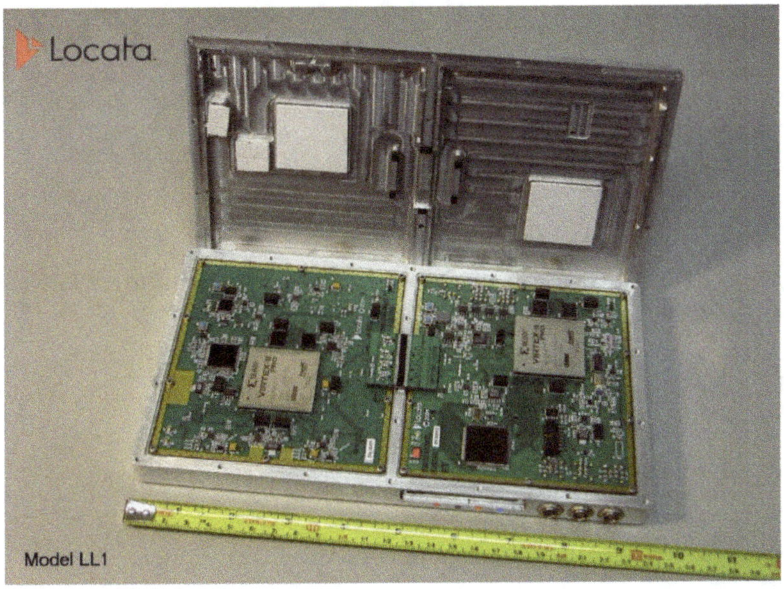

- **Licence-Free and Frequency Diversity:** LocataLites broadcast signals on two frequencies in the licence-free 2.4GHz ISM band. The two S-band frequencies (S1 – 2.41428GHz and S6 – 2.46543GHz) provide frequency diversity to aid multipath mitigation. For each LocataLite transceiver, two spatially separated antennas are used, hence a cluster of four distinct signals are transmitted by each LocataLite.
- **Time Reference:** The LocataNet can synchronise to any time source, such as GPS (or other GNSS), providing a 1 pulse-per-second (PPS) time reference. Alternatively the LocataNet can operate completely autonomously by using its own time reference generated by the master LocataLite.
- **Stronger Signals:** Typical Locata receiver signal strengths range from -60dBm to -105dBm, which means that Locata signals can penetrate into buildings and dense forests, and provide a positioning service in GNSS-denied areas. Vulnerability to signal jamming is also considerably reduced.
- **TD/CDMA:** The Locata signal ultilises a Time Division Multiple Access (TDMA) scheme for transmissions to supplement the Code Division Multiple Access (CDMA) modulation scheme. The LocataNet pseudo-random spreading codes, derived from the GPS C/A-codes, run at ten times the rate of the GPS C/A-code chip rate but with only a 1/10 transmit duty cycle. The added signal orthogonality introduced by assigning different time slots to different transmitters ensures sufficient signal discrimination to overcome the so-called "near-far" problem (Cobb 1997).

Locata Measurement Model

Similar to GNSS, the range measurements of Locata are of two types: pseudorange and carrier phase. The carrier phase measurements are more precise than the pseudorange measurements. The basic Locata observation equation between receiver and LocataLite channel i can be written as:

$$p^i = \rho^i + c \cdot dT + d^i_{trop} + v^i_p \tag{39}$$

$$\lambda \cdot \varphi^i = \rho^i + c \cdot dT + \lambda \cdot N^i + d^i_{trop} + +v^i_\varphi \tag{40}$$

where

p^i and φ^i are the pseudorange measurement and carrier phase measurement between Locata receiver and LocataLite i, respectively;

λ is the wavelength of the Locata carrier phase, depending on the transmitting frequency;

c is the velocity of light in vacuum;

ρ^i is the true or geometric range between Locata receiver to LocataLite i;

dT is the receiver clock error;

d^i_{trop} is the troposphere delay;

N^i is the Locata carrier phase ambiguity; and

v^i_p and v^i_φ are residual errors due to multipath, scattering and any other source on pseudorange and carrier phase, respectively.

Note that there is no transmitter clock error in the observation equations because of the tight time synchronisation of the LocataLites. Nor is there an ionospheric delay term. The tropospheric delays in d^i_{trop} must be calculated by an appropriate tropospheric model (Choudhury *et al* 2009b). The Locata ambiguities are typically estimated as floating-point values which can be resolved using either a known-point-initialisation (KPI) (Mervart *et al* 2008) or on-the-fly (OTF) resolution technique (Amt 2006; Bertsch *et al* 2009; Jiang *et al* 2013).

Locata technology also provides Doppler frequency measurements. The Doppler frequency is measured in both the S1 and S6 frequency bands. When scaled by the corresponding wavelength, the range-rate measurement, which reflects the relative velocity between user and LocataLite, can be obtained. The measurement equation is:

$$\dot{p}^i = \dot{\rho}^i + c \cdot d\dot{T} + \dot{d}^i_{trop} + v^i_{\dot{p}} \tag{41}$$

where

$\dot{p}^i$ is the range-rate measurement derived from Doppler measurements;

$\dot{\rho}^i$ is the relative velocity between the Locata receiver and the LocataLite i;

$d\dot{T}$ is the receiver clock drift;

c is the velocity of electromagnetic radiation in vacuum;

$\dot{d}^i_{trop}$ is the troposphere drift error; and

$v^i_{\dot{p}}$ is the combined error residual.

The generalised linear observation equation for Locata positioning represented by Equation (41) can be simplified as a Gauss-Markov model. The optimal estimation of unknown states can be obtained by least-squares (LS) estimation applying only the measurement model, or by a Kalman filter (KF) using both dynamic and measurement models.

GNSS, Locata, and INS Integrated System

In general, with limited GNSS availability, the navigation accuracy can only be maintained at the metre-level for no more than a few minutes even if a reasonably high quality INS is used. Many fusion algorithms have been investigated to mitigate this problem, however during longer GNSS outages navigation outputs may not meet the requirements for many applications (Han and Wang 2011; Jaradat and Abdel-Hafez 2014). Additional sensors, integrated with GNSS and INS, could assure the accuracy and continuity of the navigation solutions (Grejner-Brzezinska *et al* 2011). The terrestrial-based "Locata" technology uses a "local constellation" of ranging signals to address local user positioning requirements.

In order to satisfy the requirements of high accuracy and high reliability, GNSS and Locata measurements are combined with those of the INS using a loosely-coupled approach. A loosely-coupled integration system uses solved-for positions and velocities as the measurement inputs to the integration algorithm, that then uses them to estimate the INS errors. The integrated navigation solution is the INS navigation solution, corrected with the Kalman filter estimates of its errors. The two main advantages of loosely-coupled integration are its relative simplicity and redundancy. The architecture is simple in that it can be used with any INS and any GNSS/Locata user equipment, making it well suited for retrofit applications. Moreover, the standalone GNSS and Locata solutions are usually available, in addition to the integrated solution, which enables basic parallel solution integrity monitoring (Groves 2008). The GOF is adopted here to integrate GNSS, Locata and INS, as shown in Figure 5. In the proposed GNSS/Locata/INS multisensor configuration the radio navigation systems have their own drawbacks. For example, the Locata signals are range-limited, and the GNSS signals can be easily jammed. Neither Locata nor GNSS alone can fully satisfy the fault-tolerant requirements of the system, thus a well-designed integration filter is needed.

Figure 5. System configuration of GOF-based GNSS/Locata/INS integration

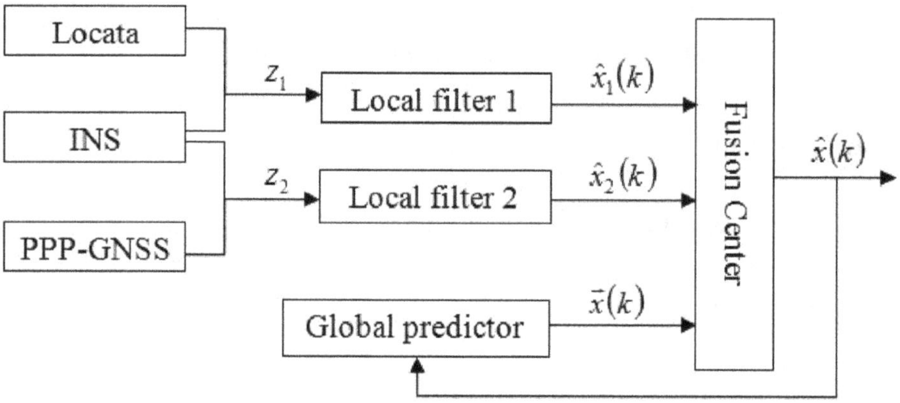

Models for GNSS/Locata/INS Integration

To implement the integration system, the INS error model should be defined. Several INS error models have been described in the literature (e.g., Arshal 1987; Goshen-Meskin and Bar-Itzhack 1992; Lee *et al* 1998), however this chapter utilises the psi-angle error model (Goshen-Meskin and Bar-Itzhack 1992):

$$\begin{cases} \delta\dot{r}^n = \delta v^n - \omega_{en}^n \times \delta r^n \\ \delta\dot{v}^n = -\left(\omega_{ie}^n + \omega_{in}^n\right) \times \delta v^n + \psi \times f^n + \nabla^n + \delta g^n \\ \dot{\psi} = \omega_{in}^n \times \psi + \varepsilon^n \end{cases} \tag{42}$$

where the subscripts “i”, “e”, and “n” represent the earth-centred inertial frame (i-frame), the earth-centred earth-fixed frame (e-frame) and the n-frame, respectively; ω_{en}^n is the angular rate vector of the n-frame with respect to the e-frame, resolved in the n-frame; ω_{ie}^n is the angular rate vector of the e-frame with respect to the i-frame, resolved in the n-frame; ω_{in}^n is the angular-rate vector, depends on the quality and geometry of the navigation frame with respect to the inertial frame, resolved in the navigation frame; δr^n, δv^n and ψ are the position, velocity and attitude error vectors, respectively; ∇^n is the accelerometers’ error vector; δg^n is the error of the computed gravity vector; and ε^n is the gyrocopes’ drift error.

Since the Locata and GNSS systems use different antennas, the lever arm correction needs to be applied. Here the positions of the Locata antenna and GNSS antenna are both referred to the inertial measurement unit (IMU) reference point, with the lever arm estimated as unknown parameters at each epoch. Therefore the estimated state vector consists of 21 navigation error states, which could be written in three sub-vectors, $x_{\dot{E}rv}$, $x_{\mu\nabla}$ and x represent the sub-vectors of motion component, gyro and accelerometer errors and level arm components, respectively:

$$x(t) = \begin{bmatrix} x_{\psi rv} & x_{\varepsilon\nabla} & x_{\eta} \end{bmatrix} \tag{43}$$

specifically

$$x_{\psi rv} = \begin{bmatrix} \psi_N & \psi_E & \psi_D & \delta r_N & \delta r_E & \delta r_D & \delta v_N & \delta v_E & \delta v_D \end{bmatrix}$$

$$x_{\varepsilon\nabla} = \begin{bmatrix} \varepsilon_x & \varepsilon_y & \varepsilon_z & \nabla_x & \nabla_y & \nabla_z \end{bmatrix}$$

$$x_{\eta} = \begin{bmatrix} \delta\eta_{Lx} & \delta\eta_{Ly} & \delta\eta_{Lz} & \delta\eta_{Gx} & \delta\eta_{Gy} & \delta\eta_{Gz} \end{bmatrix}$$

where the subscripts "N", "E", "D" denote the north, east, and down axes in the n-frame, respectively; the subscripts "x", "y", "z" represent the direction of the roll, pitch, and yaw axes in the body frame, respectively; and the last six symbols $\delta\eta_*$ represent the unknown Locata and GNSS lever arm components.

The system dynamic model is (Groves 2008):

$$\begin{bmatrix} \dot{x}_{\psi rv} \\ \dot{x}_{\varepsilon\nabla} \\ \dot{x}_{\eta} \end{bmatrix} = \begin{bmatrix} F_{11} & F_{12} & 0_{9\times6} \\ 0_{6\times9} & 0_{6\times6} & 0_{6\times6} \\ 0_{6\times9} & 0_{6\times6} & 0_{6\times6} \end{bmatrix} \begin{bmatrix} x_{\psi rv} \\ x_{\varepsilon\nabla} \\ x_{\eta} \end{bmatrix} + \begin{bmatrix} 0_{9\times1} \\ \acute{E}_{\varepsilon\nabla} \\ 0_{6\times1} \end{bmatrix} \tag{44}$$

where $\omega_{\varepsilon\nabla}$ is the noise vector of the accelerometer and gyro errors, which are modelled as a random biases; and the sub-matrices F_{11} and F_{12} of the system matrix F are calculated based on the INS error model:

$$F_{11} = \begin{bmatrix} F_{\psi\psi} & F_{\psi r} & F_{\psi v} \\ 0 & F_{rr} & F_{rv} \\ F_{v\psi} & F_{vr} & F_{vv} \end{bmatrix}, \quad F_{12} = \begin{bmatrix} -C_b^n & 0_{3\times3} \\ 0_{3\times3} & 0_{3\times3} \\ 0_{3\times3} & C_b^n \end{bmatrix} \tag{45}$$

where

$$F_{11} = \begin{bmatrix} F_{\psi\psi} & F_{\psi r} & F_{\psi v} \\ 0 & F_{rr} & F_{rv} \\ F_{v\psi} & F_{vr} & F_{vv} \end{bmatrix}, \quad F_{12} = \begin{bmatrix} -C_b^n & 0_{3\times3} \\ 0_{3\times3} & 0_{3\times3} \\ 0_{3\times3} & C_b^n \end{bmatrix}$$

$$F_{\psi\psi} = \begin{bmatrix} 0 & -\left(\omega_{ie}\sin L + \dfrac{v_E \tan L}{R_n + h}\right) & \dfrac{v_N}{R_m + h} \\ \omega_{ie}\sin L + \dfrac{v_E}{R_n + h} & 0 & \omega_{ie}\cos L + \dfrac{v_E}{R_n + h} \\ -\dfrac{v_N}{R_m + h} & -\left(\omega_{ie}\cos L + \dfrac{v_E}{R_n + h}\right) & 0 \end{bmatrix},$$

$$F_{\psi r} = \begin{bmatrix} -\omega_{ie}\sin L & 0 & -\dfrac{v_E}{\left(R_n + h\right)^2} \\ 0 & 0 & \dfrac{v_N}{\left(R_m + h\right)^2} \\ -\left(\omega_{ie}\cos L + \dfrac{v_E}{R_n + h}\sec^2 L\right) & 0 & \dfrac{v_E \tan L}{\left(R_n + h\right)^2} \end{bmatrix}, \quad F_{\psi v} = \begin{bmatrix} 0 & \dfrac{1}{R_n + h} & 0 \\ -\dfrac{1}{R_m + h} & 0 & 0 \\ 0 & -\dfrac{\tan L}{R_n + h} & 0 \end{bmatrix},$$

$$F_{rr} = \begin{bmatrix} 0 & 0 & \dfrac{-v_N}{\left(R_m + h\right)^2} \\ \dfrac{v_E \tan L \sec L}{R_n + h} & 0 & -\dfrac{v_E \sec L}{\left(R_n + h\right)^2} \\ 0 & 0 & 0 \end{bmatrix}, \quad F_{rv} = \begin{bmatrix} \dfrac{1}{R_m + h} & 0 & 0 \\ 0 & \dfrac{1}{\left(R_n + h\right)\cos L} & 0 \\ 0 & 0 & -1 \end{bmatrix},$$

$$F_{v\psi} = \begin{bmatrix} 0 & -f_z & f_y \\ f_z & 0 & -f_x \\ -f_y & f_x & 0 \end{bmatrix},$$

$$F_{vr} = \begin{bmatrix} -v_E\left(2\omega_{ie}\cos L + \dfrac{v_E \sec^2 L}{R_n + h}\right) & 0 & \dfrac{{v_E}^2 \tan L}{\left(R_n + h\right)^2} - \dfrac{v_N v_D}{\left(R_m + h\right)^2} \\ 2\omega_{ie}\left(v_N \cos L - v_D \sin L\right) + \dfrac{v_N v_E \sec^2 L}{R_m + h} & 0 & -v_E \dfrac{v_N \tan L + v_D}{\left(R_m + h\right)^2} \\ 2\omega_{ie} v_E \sin L & 0 & \dfrac{{v_N}^2}{\left(R_m + h\right)^2} + \dfrac{{v_E}^2}{\left(R_n + h\right)^2} \end{bmatrix},$$

$$F_{vv} = \begin{bmatrix} \frac{v_D}{R_m + h} & -2\left(\omega_{ie} \sin L + \frac{v_E \tan L}{R_n + h}\right) & \frac{v_N}{R_m + h} \\ 2\omega_{ie} \sin L + \frac{v_E \tan L}{R_n + h} & \frac{v_N \tan L + v_D}{R_m + h} & 2\omega_{ie} \cos L + \frac{v_E}{R_m + h} \\ 0 & -2\left(\omega_{ie} \cos L + \frac{v_E}{R_n + h}\right) & 0 \end{bmatrix},$$

in which f_x, f_y and f_z are the accelerometer-sensed specific force in the three directions in the body frame.

In the proposed GNSS/Locata/INS integration the local filters and global filter share the same system model. Thus for the local filters the state vectors are also represented by Equation (45). In case of the measurement updates, both Locata and GNSS can provide the carrier phase-based position and velocity solutions to correct the INS errors. Due to the lever arm uncertainty, the positions of both Locata antenna and GNSS antenna need to be corrected to the IMU reference point (Bar-Itzhack 1977):

$$r_*^n = r_{INS}^n + C_b^n \eta_*^b \tag{46}$$

$$v_*^n = v_{INS}^n + \left(\omega_{ie}^n \times + \omega_{en}^n \times\right) C_b^n \eta_*^b - C_b^n \left(\eta_*^b \times\right) \omega_{ib}^b \tag{47}$$

where the subscript * denotes the Locata and GNSS antennas. Using the above model the computed position and velocity at the Locata antenna are:

$$\hat{r}_*^n = \hat{r}_{INS}^n + \hat{C}_b^n \eta_{*0}^b - \hat{C}_b^n \left(\eta_{*0}^b \times\right) \psi + \hat{C}_b^n \delta\eta_*^b \tag{48}$$

$$\hat{v}_*^n = \hat{v}_{INS}^n + \hat{C}_b^n \left(\omega_{ib}^b \times\right) \eta_{0*}^b + \hat{C}_b^n \left(\omega_{ib}^b \times\right) \delta\eta_*^b - \left[\left(\hat{C}_b^n \left(\omega_{ib}^b \times\right) \eta_{0*}^b\right) \times\right] \tag{49}$$

where $\eta_{0^*}^b$ is the initial estimate of the lever arm of Locata and GNSS.

The linearised measurement models of the Locata local filter and GNSS local filter can now be written as (Li etal 2012):

$$\begin{bmatrix} \delta r_L^n - \delta r_{INS}^n - C_b^n \eta_{L0}^b \\ \delta v_L^n - \delta v_{INS}^n - C_b^n \left(\omega_{ib}^b \times\right) \eta_{L0}^b \end{bmatrix} = \begin{bmatrix} \hat{C}_b^n \left(\eta_{L0}^b \times\right) & I_{3\times3} & 0_{3\times3} & 0_{3\times6} & \hat{C}_b^n & 0_{3\times3} \\ -\left[\left(\hat{C}_b^n \left(\omega_{ib}^b \times\right) \eta_{L0}^b\right) \times\right] & 0_{3\times3} & I_{3\times3} & 0_{3\times6} & \hat{C}_b^n \left(\acute{E}_{ib}^b \times\right) & 0_{3\times3} \end{bmatrix} x(t) + v(t) \tag{50}$$

$$\begin{bmatrix} \delta r_G^n - \delta r_{INS}^n - C_b^n \eta_{G0}^b \\ \delta v_G^n - \delta v_{INS}^n - C_b^n \left(\omega_{ib}^b \times\right) \eta_{G0}^b \end{bmatrix} = \begin{bmatrix} \hat{C}_b^n \left(\eta_{G0}^b \times\right) & I_{3\times3} & 0_{3\times3} & 0_{3\times6} & 0_{3\times3} & \hat{C}_b^n \\ -\left[\left(\hat{C}_b^n \left(\omega_{ib}^b \times\right) \eta_{G0}^b\right)\times\right] & 0_{3\times3} & I_{3\times3} & 0_{3\times6} & 0_{3\times3} & \hat{C}_b^n \left(\omega_{ib}^b \times\right) \end{bmatrix} x(t) + v(t) \tag{51}$$

The covariance of the measurement noise in the local filter is defined by the general accuracy of the local Locata and GNSS solutions:

$$R_L = diag\left((\sigma_{Lp})^2 \quad (\sigma_{Lp})^2 \quad (\sigma_{Lp})^2 \quad (\sigma_{Lv})^2 \quad (\sigma_{Lv})^2 \quad (\sigma_{Lv})^2 \right)$$

$$R_G = diag\left((\sigma_{Gp})^2 \quad (\sigma_{Gp})^2 \quad (\sigma_{Gp})^2 \quad (\sigma_{Gv})^2 \quad (\sigma_{Gv})^2 \quad (\sigma_{Gv})^2 \right)$$

where the first three are the covariance of position in units of m^2, and the last three are the covariance of the velocity in units of $(m/s)^2$.

Test of GNSS/Locata/INS Integrated System

The test was conducted on October 2012 on Sydney Harbour, on the fringes of Sydney's CBD where the multipath effect is quite severe. The LocataNet comprised eight LocataLites, six of them installed along the shore, and the remaining two were located further away, with one on the Harbour Bridge (LL7) and the other at Kirribilli (LL8). The LocataNet was time-synchronised with the Master LocataLite (LL3), itself configured for synchronisation with GPS time. The rover trajectory and the location of the LocataLites are shown in Figure 6.

The rover experiment setup was installed on a survey vessel provided by the Sydney Ports Authority. A Locata user unit and a Leica GNSS dual-frequency receiver were installed, with antenna locations as shown in Figure 7. The distance between Locata and GNSS antenna centres was measured beforehand to be 0.24m. A survey-grade Applanix POS-MV INS was also fixed on the vessel, with performance parameters as listed in Table 2.

The GNSS integer ambiguity-fixed differential carrier phase positioning solutions computed by the Leica Geo Office (LGO) software (Leica Geosystems AG 2010) served as the ground truth. The velocity reference data were computed using the MV-POSView software, with a stated horizontal velocity accuracy of better than 0.03m/s (Applanix Corporation 2009). In order to conduct performance evaluation, all the raw data from the Locata rover, GNSS receiver and INS were collected and post-processed. Locata and GNSS data output rates were set to 10Hz, and inertial measurement rate was set to 200Hz. Both Locata and the INS were synchronised with GPS time.

The trajectory of the rover consisted of continuous circular tracks lasting 15 minutes for each circuit. During this period both Locata and GPS signals were tracked – 7-8 GPS satellites and 4-8 LocataLites (with a total of 11-32 Locata signals). Excluding one satellite and two Locata signals as the reference in the GNSS/Locata system, a total 15-37 single-differenced observables were obtained. The DOP values

Figure 6. LocataNet configuration and rover trajectory for 2012 Sydney Harbour test

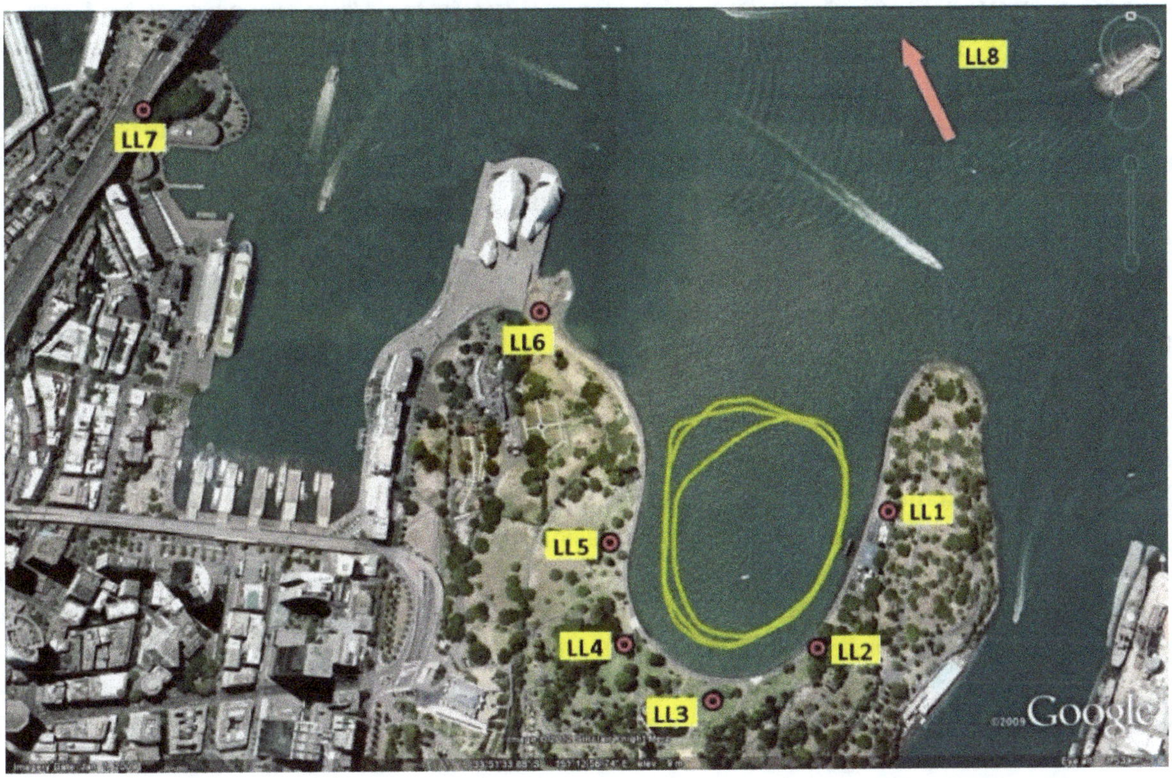

Figure 7. Test vessel and rover experiment setup for 2012 Sydney Harbour test

Table 2. Specifications of the INS used in the 2012 Sydney Harbour test (Applanix Corporation 2009).

	Gyroscopes	Accelerometers
Bias	1°/h ~ 5°/h	50 ~ 500 μg
Scale factor	-	0.01% ~ 0.02%
Random noise	0.07°/sqrt(h)	-

confirmed that the system geometry was "good" during the test, and the average Vertical DOP (VDOP) was 0.66, the DOPs in the north (NDOP) and in the east (EDOP) were 0.31 and 0.36, respectively.

In order to evaluate the performance of the CKF, FKF and GOF algorithms, position and velocity solutions calculated by the three data fusion algorithms were analysed, and are summarised in Figures 8-10. As the LocataNet was almost planar, the vertical component observability by Locata is not as good as for the horizontal components. The tide height data, measured and recorded by the Fort Denison Tide Gauge, was used to constrain the height in the processing, and the experiment analysis below focused on the horizontal direction components. From the position plots in the three plots it can be seen that all three algorithms can achieve accurate and stable horizontal positioning solutions.

By comparing the three plots one can see that the performance of the CKF and the FKF are similar, while the solution using the GOF is more accurate and stable. The Root Mean Square (RMS) and Distance Root Mean Square (DRMS) values for the horizontal direction components are listed in Table 3 – the accuracy of all three solutions are better than 0.08m. The CKF and the FKF have similar accuracy with a DRMS of about 0.08m. However the GOF achieves a slightly higher accuracy with the DRMS of 0.07m, which is an improvement of 12.5% in comparison with the CKF and the FKF.

For the velocity solution there are similar conclusions to those of the position solution. The velocity from the GOF has a higher accuracy than the CKF and the FKF derived velocities. The FKF solution is slightly worse than that of the CKF. The GOF improves the horizontal velocity DRMS by 33.3% in comparison with the CKF and the FKF derived velocities.

In order to investigate further the performance of the data fusion algorithms, the corresponding aposteriori estimate error covariance $\hat{P}(k)$ of every epoch was also computed. Figures 11-13 show the

Figure 8. Position and velocity error of the CKF system

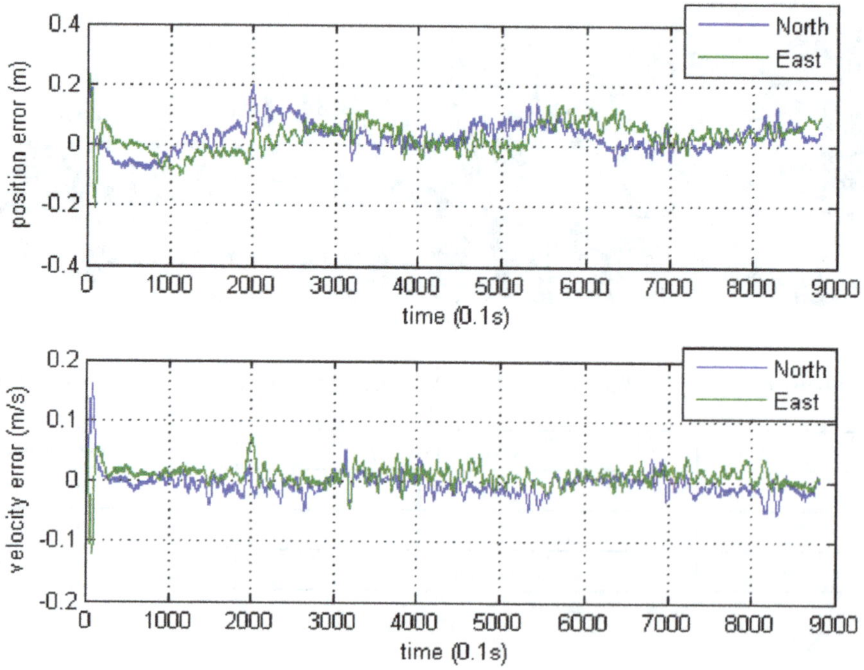

Figure 9. Position and velocity error of the FKF system

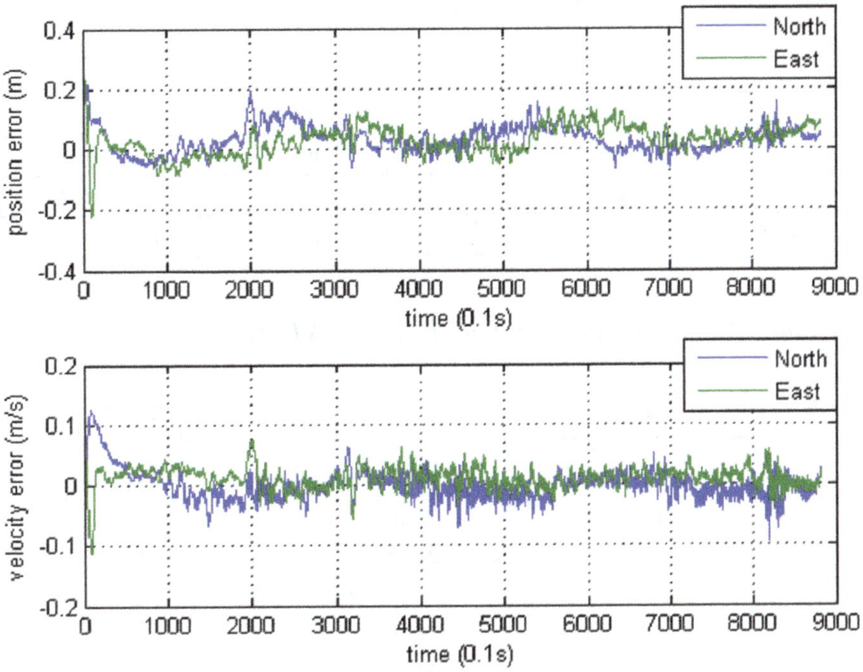

Figure 10. Position and velocity error of the GOF system

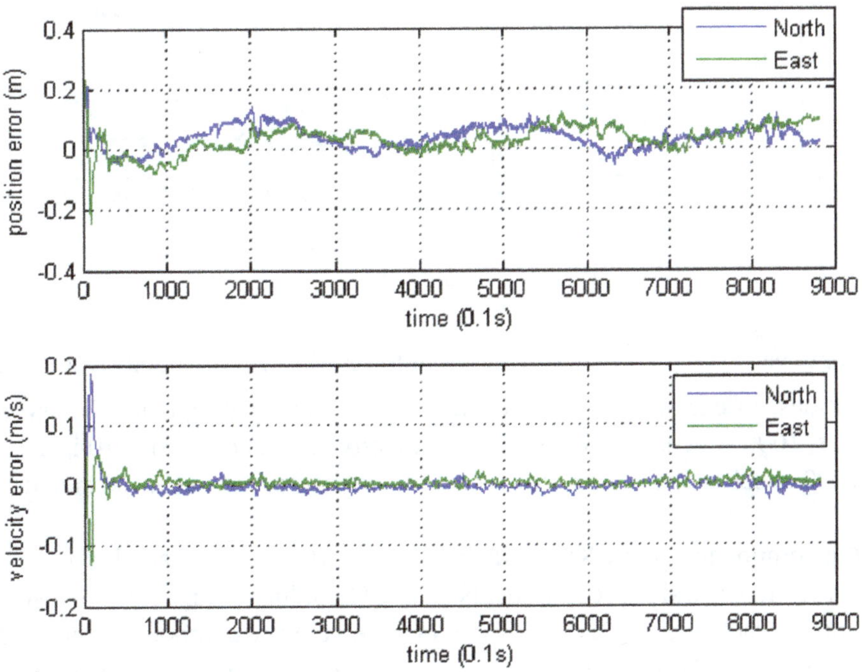

Table 3. Position and velocity analysis of the three filtering options.

			CKF	FKF	GOF
Position (m)	RMS	North	0.06	0.06	0.05
		East	0.06	0.06	0.05
	DRMS		0.08	0.08	0.07
Velocity (m/s)	RMS	North	0.02	0.03	0.02
		East	0.02	0.02	0.01
	DRMS		0.03	0.03	0.02

Figure 11. Position and velocity precision evaluation of the CKF system

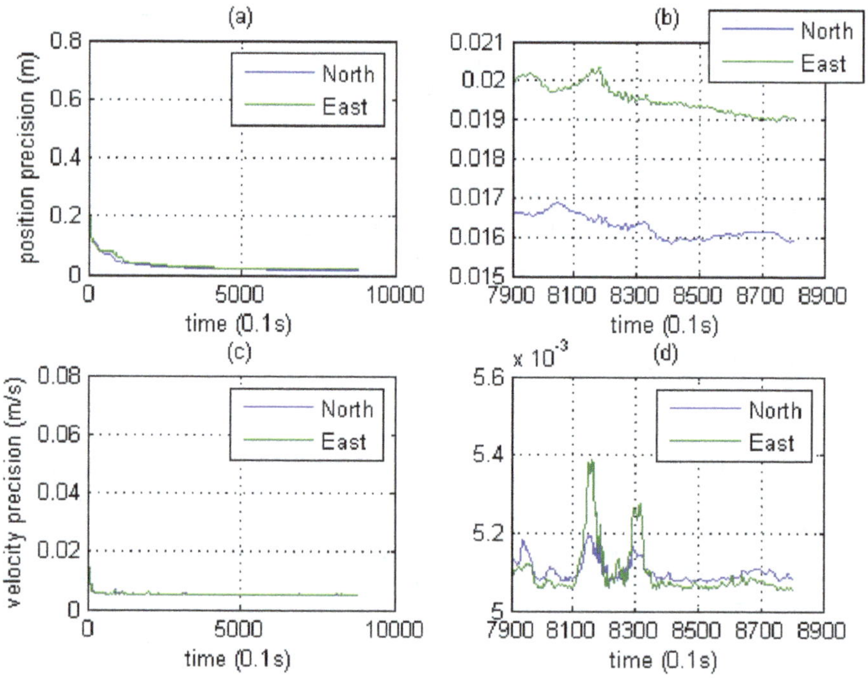

square root of error covariance $\sqrt{\hat{P}\left(k\right)}$ (w.r.t sqrtP) of the position and velocity, for the CKF, the FKF, and the GOF solutions. In each plot, (a) is the sqrtP of position at all epochs, (b) shows the sqrtP of position over the last 1000 epochs, (c) is the sqrtP of velocity at all epochs, and (d) is the sqrtP of velocity over the last 1000 epochs. The blue colour and green colour represent the north and east directions, respectively.

For the position components, the CKF's and the FKF's sqrtP are close and both finally converge to less than 0.02m for the north and east components. The GOF solution converges to a similar value, but better than those produced by the CKF or the FKF. For the velocity component, the GOF solution has an obvious improvement, i.e. from 5×10^{-3} m/s to 4×10^{-3} m/s for both the north and east components.

Figure 12. Position and velocity precision evaluation of the FKF system

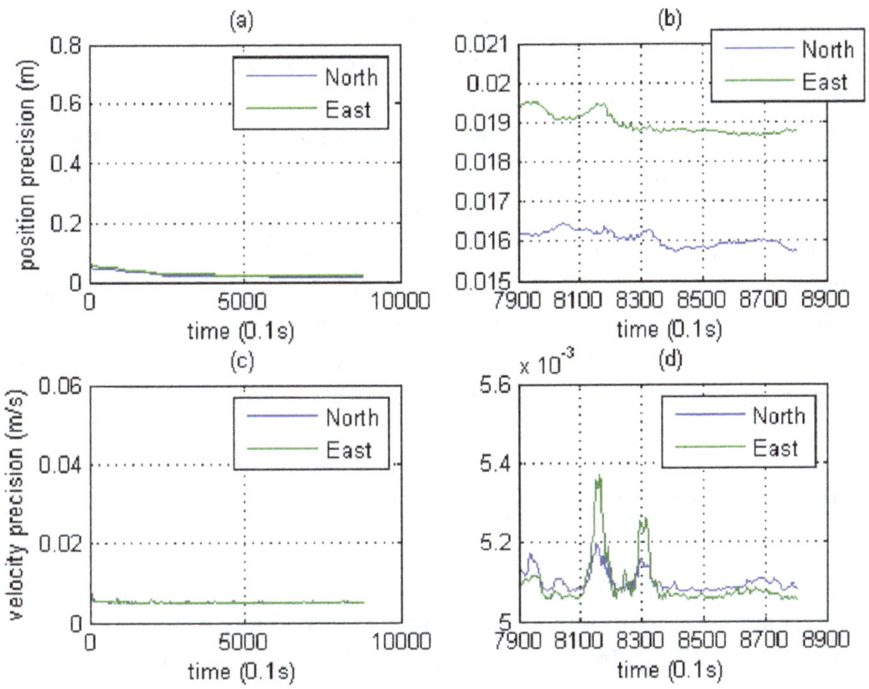

Figure 13. Position and velocity precision evaluation of the GOF system

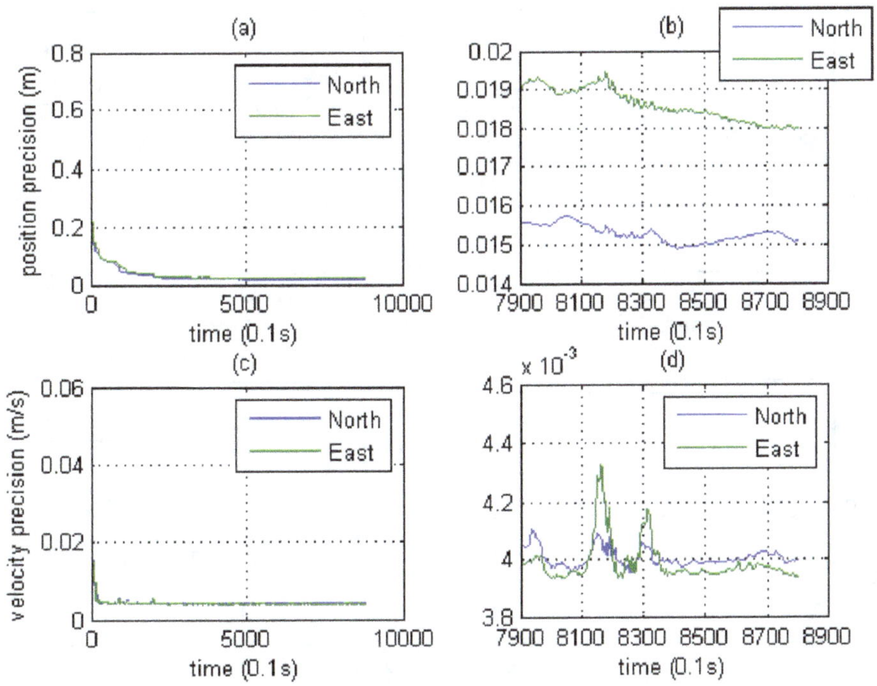

It can therefore be concluded that the GOF estimate of position and velocity error covariance is improved in comparison with the conventional CKF and FKF algorithms.

To evaluate the fault-tolerance capability of the GOF-based triple-integration system, GNSS outlier failures were simulated. The performance of the GOF-based system under a GNSS outlier failure scenario is shown in Figure 14. In particular, at 250s, 260s, 270s, 280s, 290s, and 300s, the 1m and 0.5m/s additive errors were introduced to the GNSS position and velocity outputs, respectively. As the additive error were introduced to the GNSS position and velocity output, which would be further introduced as an unexpected noise into the filter input. The previous defined measurement noise matrix R therefore could not correctly describe the error-affected measurement. The plots (a) and (c) in Figure 14 are the horizontal position and velocity results for all epochs, (b) and (d) are the plots zoomed-in for the period from 240s to 310s (to show the six outliers more clearly). It can be seen that the proposed system has stable performance and is not affected by such failures. This demonstrates that the GOF-based triple-integration system is able to provide a slightly better outlier-tolerant navigation solution.

GNSS, INS, AND WIFI INTEGRATION

To realise seamless navigation in outdoor and indoor environments, an integrated GNSS/INS/WiFi system can improve the WiFi standalone positioning performance without increasing the computational burden. In outdoor environments the GNSS/INS sub-system can effectively provide the navigation solution.

Figure 14. Fault-tolerance evaluation of GOF-based triple integration system

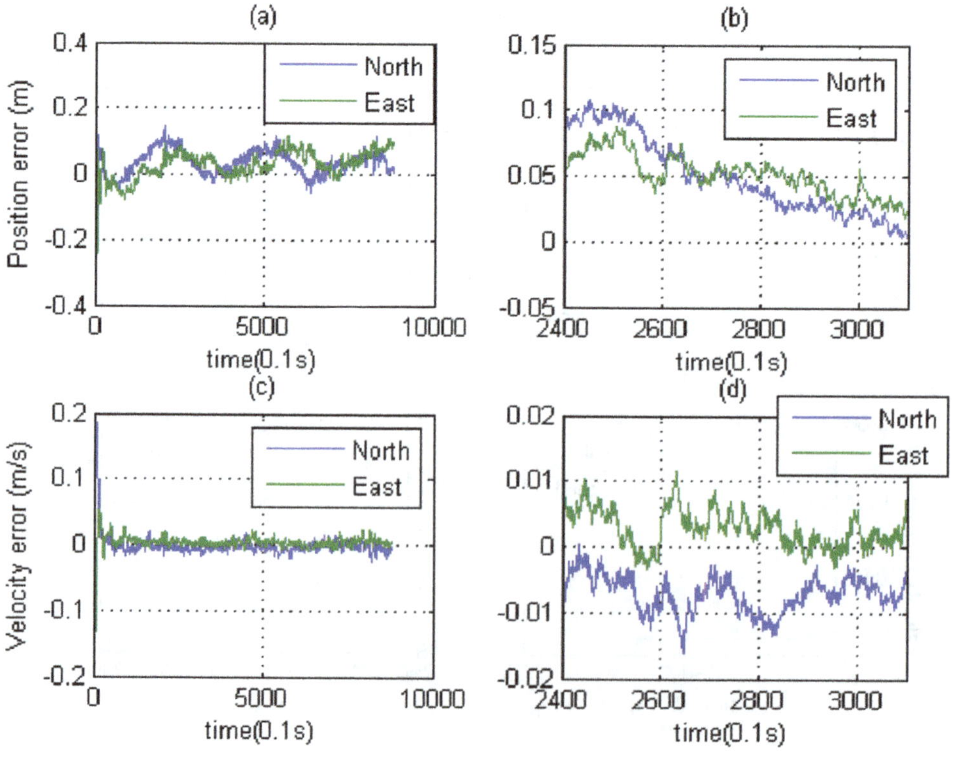

In indoor environments the positioning and attitude solutions from the INS will be aided by the WiFi positioning. To further mitigate the INS ramp errors, step detection is also implemented during indoor navigation. The CKF is used to implement the GNSS/INS/WiFi integration.

GNSS/WiFi/INS Integration Algorithm

Step detection, and accumulation of the counted steps, is a widely-used method for pedestrian navigation. By using this method the sensor noises, offset and tilt variation can be avoided, but most of the useful information from the accelerometer and gyroscope measurements are wasted. In this section the information sources are integrated in an optimal manner. To restrain the large sensor noises and biases some constraints based on pedestrian properties are applied.

With measurements and constraints from external sources the GNSS/WiFi/INS integration architecture is indicated in Figure 15 with the following steps:

Step 1: INS mechanisation update at each epoch, and update the Kalman filter prediction states and variance matrix.

Step 2: Pedestrian walking status detection - If the pedestrian is static, get velocity pseudo-measurements by ZUPTs constraint; or if the user is going straight ahead, get yaw pseudo-measurements from heading constraint.

Step 3: Step detection – If one step is finished, get position measurements by step integration.

Figure 15. Integration flowchart for GNSS/WiFi/INS integration procedure.

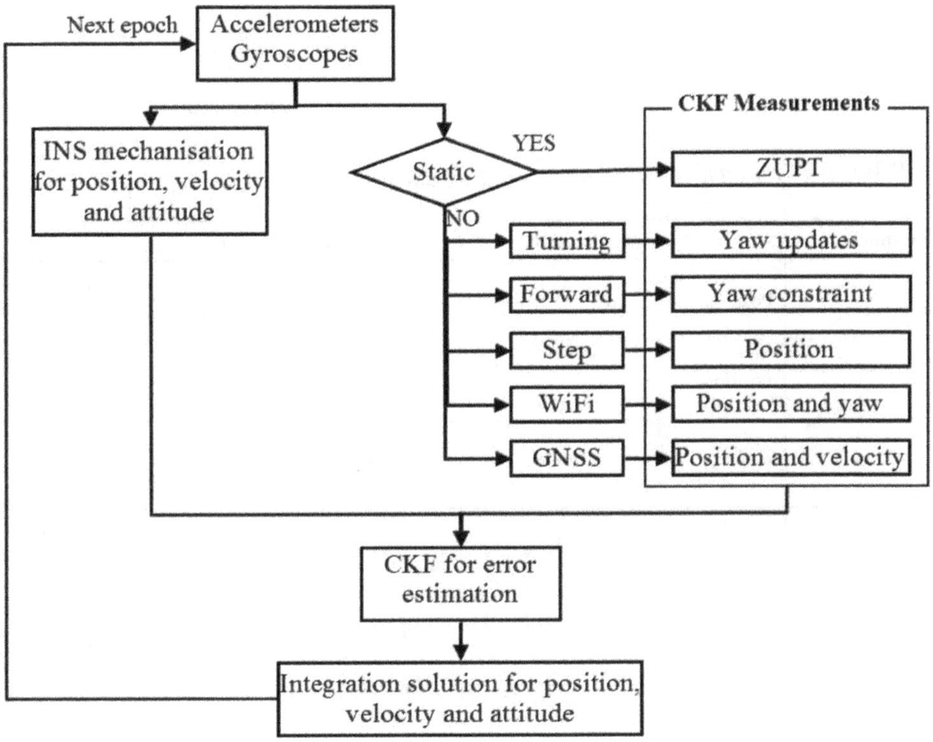

Step 4: If GNSS solution is available at this epoch, get position measurements.
Step 5: If WiFi solution is available at this epoch, get position and yaw measurements – the yaw measurements are based on the direction of the straight corridors.
Step 6: Combining all measurements from Step 2 to Step 5, generate the CKF solution with the state prediction function from Step 1.
Step 7: State correction and go to next epoch.

GNSS, WiFi and INS measurements and error characteristics are summarised in Table 4.

To implement the integration systems, the INS psi-angle error model is used (Eq. (42)). The estimated state vector consists of 15 navigation error states, which could be written in two sub-vectors, $x_{\psi rv}$ and $x_{\varepsilon\nabla}$ represent the sub-vectors of motion component (attitude, position and velocity errors, 9×1 in total), gyro and accelerometer errors (6×1 in total), as described in Eq. (43). The system dynamic model is given in Eq.(44) without the level arm components.

INS Pedestrian Dead-Reckoning Algorithms

Pedestrian dead-reckoning (PDR) exploits the kinematics of human movement, and can be defined as the process of estimating the present position by projecting travelled distance and azimuth from a known starting point. The common PDR mechanisation is to detect and count steps, estimate each step's length, and propagate a new position using the absolute heading. This approach slightly differs from classical inertial navigation techniques where position propagation is calculated by integrating measured acceleration twice, but it is well adapted to cope with the complex characteristics of human movement.

With the initial coordinates (x0, y0), and initial heading angle ψ_0, the horizontal coordinates are propagated as:

Table 4. GNSS/WiFi/INS measurements used in an integrated system.

Information sources	Measurements	Design matrix	noise
GNSS	Position and velocity	$H=\left[I_{6\times6},0_{6\times9}\right]$	$v\sim N\left(0,R_{6\times6}\right)$
WiFi	Position	$H=\left[I_{3\times3},0_{3\times12}\right]$	$v\sim N\left(0,R_{3\times3}\right)$
Step detection	Position	$H=\left[I_{3\times3}\quad 0_{3\times12}\right]$	$v\sim N\left(0,R_{3\times3}\right)$
Turning detection	Yaw	$H=\left[0_{1\times8}\quad 1\quad 0_{1\times6}\right]$	$v\sim N\left(0,R_{1\times1}\right)$
INS-ZUPT	velocity	$H=\left[0_{3\times3}\quad I_{3\times3}\quad 0_{3\times9}\right]$	$v\sim N\left(0,R_{3\times3}\right)$

$$\begin{cases} x_k = x_{k-1} + s_k * \cos\psi_k \\ y_k = x_{k-1} + s_k * \sin\psi_k \end{cases} \quad (52)$$

where s_k is the step length with index k. The heading angle update depends on the pedestrian's turning status, which is either constant or derived from the angular-rate sensed by the gyroscopes.

When a pedestrian walks up and down stairs, the step integration is required on both horizontal and vertical directions:

$$\begin{cases} x_k = x_{k-1} + s * \cos\psi_k \\ y_k = x_{k-1} + s * \sin\psi_k \\ z_k = z_{k-1} + h \end{cases} \quad (53)$$

where (x_k, y_k, z_k) are the coordinates in the local navigation frame (east-north-up), and s and h are, respectively, the horizontal and vertical length of each stair step.

Step Detection

The walked distance is obtained by counting the pedestrian steps. The estimation of the distance is divided into two parts: step detection and the actual step length estimation. Analysing the characteristic acceleration signals, steps can be detected from the movements of the walker. The occurrence of a step is based on the observation of a repeating and detectable pattern in the signal. The pattern is searched using either the vertical acceleration component (Jirawimut *et al* 2003; Goyal *et al* 2011), or the norm of the 3-D acceleration (Gusenbauer *et al* 2010; Park *et al* 2012). In this chapter the norm of the three acceleration components is used:

$$a_i = \sqrt{a_{x_i}^2 + a_{y_i}^2 + a_{z_i}^2} \quad (54)$$

where i is the index of the measurement sample; $a_{x_i}, a_{y_i}, a_{z_i}$ are the acceleration components at three orthogonal direction; and a_i is the norm value.

Step detection here is based on the peak point detection, i.e. the instances when the value of the acceleration norm increases to the top value or falls to the bottom value. A typical acceleration norm pattern generated by walking movement is shown in Figure 16, where the step starts are also indicated. The signal goes through a wave-like cycle during one step, however, due to signal noise, the wave does not always corresponds with a step. To mitigate the influence of signal noise, the step detection algorithm includes two constraints:

- The threshold for minimum time interval of the detected step (Perttula et al 2014).
- The threshold for minimum signal amplitude of the detected step (Jahn et al 2010).

Figure 16. Step detection based on peak. Vertical red lines show the instants of step starts.

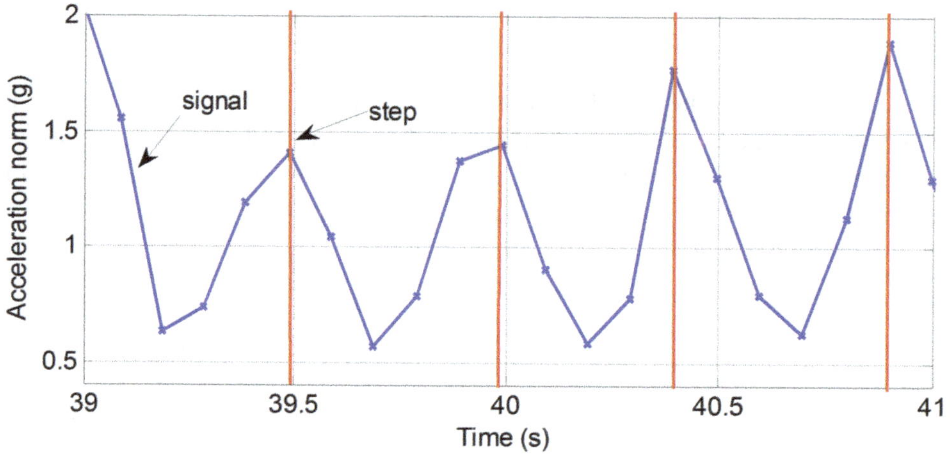

When both the thresholds are satisfied, an actual step is assumed to have been detected. The algorithm considers reaching peak points as possible new step starts and checks the constraints, to accept or reject the start. The end of the current step is found in two situations:

- The start of the next step is accepted.
- The start of the next step does not follow the current step in the predefined time frame.

The latter occurs when the user stops walking or changes their walk status. The algorithm outputs the start and end times of the step either when the start of the next step is detected, or when the step end is concluded from the detection timeout.

Step Length Estimation

The step length is not a constant value, and is influenced by a number of factors, such as walking speed, variance of the accelerometer readings in a pace, as well as the ground inclination (Ladetto & Merminod 2002; Cho & Park 2006). Research has shown a tight coupling between the step frequency and the step length (Sun *et al* 2009; Leppäkoski *et al* 2013). If a pedestrian moves faster, both the step frequency and the step length increase, and vice versa. Levi and Judd (1999) initially argued that step length is influenced linearly by step frequency. Therefore a linear curve-fitting algorithm can be used to estimate the step length:

$$\Delta s_k = \frac{c_1}{\Delta t_k} + c_0 \tag{55}$$

where *k* is the index of the detected step; Δt_k is the time interval of the step; and c_0 and c_1 are constant parameters, respectively.

To determine the parameters, the pedestrian was asked to walk a fixed distance of 40m over 30 separate trials, with different speed. For each test, the time and number of the steps were recorded. The average step length and step frequency can be computed and then used to estimate parameter c_0 and c_1. However, the parameters used here may vary for different groups of pedestrians, hence a calibration procedure should be implemented for each person.

The step length also depends on the specific walk status. When the user is climbing stairs, the user's step length is constrained by the length of the stair steps rather than the natural gait of the pedestrian, and the vertical coordinate is also updated by the height of the stair steps.

Heading Determination

The heading angle changes gradually in a turn and remains essentially unchanged when the pedestrian walks in a straight line. However, a low-cost INS can only maintain heading accuracy for a very short period of time even when the pedestrian walks in a straight line. Although use of a magnetometer may help to maintain heading stability, the magnetic measurement is easily disturbed by external interference sources (Borestein *et al* 2009; Jiménez *et al* 2010).

Assuming the initial heading angle is known, the heading will be only updated in a turn. Due to the obvious signal noises it is difficult to distinguish different states on the basis of raw signals alone. Here a moving window is used to resample the signal, so that the random noise can be largely mitigated. The signal properties during the different walking states can be expressed by:

$$\begin{cases} E\left(\omega_i\right) = 0, D\left(\omega_i\right) = \sigma_\omega^2, \quad i = 1, \cdots, m, m+t+1, \cdots n \\ E\left(\omega_i\right) = \omega_0, D\left(\omega_i\right) = \sigma_\omega^2, i = m+1, \cdots, m+t \end{cases} \tag{56}$$

where ω_i is the angular-rate measurement along the vertical axis at the *i*th epoch; "*E*" and "*D*" are the expectation and variance function, respectively; ω_0 is a value which is not equal to zero; and σ_ω^2 is a positive value. Equation (56) indicates that the pedestrian walks straight ahead for the first *m* epochs, followed by a turn from m to *m+t* epochs.

The filtered signal is denoted as:

$$\omega_i' = \frac{\sum_{i=i-k+1}^{i} \omega_i}{k} \tag{57}$$

where *k* is the length of the filtering window; and ω_i' is the filtered value of ω_i. Combining with Equation (56) it can be deduced that:

$$\begin{cases} E\left(\omega_i^{'}\right) = 0, \quad i = 1, \cdots, m \\ E\left(\omega_i^{'}\right) = \dfrac{i-m}{k}\omega_0, \quad i = m+1, \cdots, m+\left(k \le t\right) \\ E\left(\omega_i^{'}\right) = \omega_0, \quad i = m+k+1, m+t\left(k \le t\right) \\ E\left(\omega_i^{'}\right) = \dfrac{k-\left(i-m-t\right)}{k}\omega_0, i = m+t+1, \cdots, m+t+k-1\left(k \le t\right) \\ E\left(\omega_i^{'}\right) = 0, i = m+t+k, \cdots, n\left(k \le t\right) \end{cases} \tag{58}$$

where $D\left(\omega_i^{'}\right) = \frac{\sigma_\omega^2}{k}, i = 1, \cdots, n$. Equation(58) indicates that the expectation of the filtered angular-rate measurement is not equal to zero during the first k epochs after turning. Thus the influence of this period should be subtracted from the detected period.

When a turn is detected, the heading change estimation also relies on step detection. The integration limits for the angular-rate during one step are used to estimate the heading change:

$$\Delta\varphi_k = \frac{1}{f_s}\sum_{i=k_1}^{k_n}\omega_i \tag{59}$$

where k is the index of the detected step; k_1 and k_n are the time indices of the start and end epoch of the step, respectively; and f_s is the data sample rate. The integration of gyro outputs restarts at the beginning of each step.

Once the walk distance and heading update in the step are determined, the pedestrian position is computed by starting from initial coordinates and initial heading angle ψ_0. The heading and coordinates are propagated by:

$$\begin{pmatrix} \psi_k \\ x_k \\ y_k \end{pmatrix} = \begin{pmatrix} \psi_{k-1} + \Delta\varphi_k \\ x_{k-1} + \Delta s_k \cos\psi_k \\ y_{k-1} + \Delta s_k \sin\psi_k \end{pmatrix} \tag{60}$$

Experimental Test of GNSS/WiFi/INS Positioning

An experiment was conducted on the campus of UNSW, Sydney, Australia. Data were collected in both outdoor and indoor environments. The nAX5 -- a MEMS-INS/GPS device from navextech (http://www.navextech.com/) – was used to collect inertial measurements and GPS data. In the test the nAX5 was carried on the person's back and the WiFi signals were collected using a laptop carried by the person. The nAX5 was configured to output 10Hz IMU updates.

WiFi Positioning

WiFi (WLAN based on 802.11 standards) is widely used in public areas such as airports, office buildings, hospitals, hotels, shopping centres, and so forth. Considering the fact that the WiFi chips can now be found in almost all consumer devices (smartphones, tablets, computers, and others), it is relatively easy to implement WiFi positioning using existing WLAN access point infrastructure. Several location methods have been proposed in the past decades, which can be divided into two basic categories: time-of-arrival (TOA)/time-difference-of-arrival (TDOA) methods (Guvenc & Chong 2015), and received signal strength (RSS) methods. The trilateration approach (Xiao *et al* 2015) and the fingerprinting approach (Cheng *et al* 2014) are two classes of RSS-based positioning techniques.

The precision of the fingerprinting database largely depends on the divergences of the signal strength at different access points (APs). To analyse the RSS variability of the WiFi signals at each Reference Point (RP), an experiment was carried out with four WiFi APs servicing an office area. Data at 54 points on a grid point of 0.6m were collected. The RSS values are shown in Figure 17. The smaller the distance between a RP and an AP, the stronger the signal strength. In addition, the RSS interval bounded by the standard deviation (STD) values shows that the signals are much more stable at RPs which are far away from the AP . This means the signal variability among adjacent RPs are much more significant, contributing to better positioning accuracy.

The tests were conducted along the office corridor and the environment was relative stable. In the indoor environments, the signals are easily blocked by human bodies, walls or furniture (Akyildiz et al 2009), so it is difficult to assure the positioning accuracy and system stability by this method. To improve WLAN positioning performance some enhanced models considering Rayleigh fading effects (Akyildiz et al 2009), shadow fading effects (Salo et al 2007) or multipath effects caused by walls (Chen & Kobayashi 2002) have been investigated.

Two experiments were conducted in order to assess the navigation performance of a WiFi positioning system under different indoor environments. In these experiments the distance between two RPs was on

Figure 17. RSS at each RP from WiFi APs

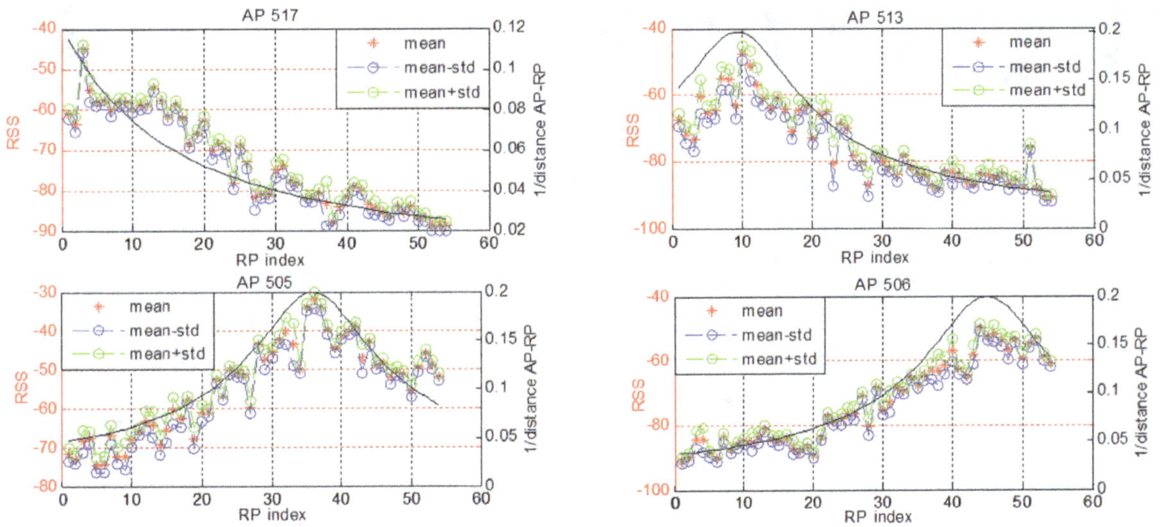

average 1m. The RSS was sampled at each RP for 1 minute. The first experiment was conducted on the lower ground level of the Electrical Engineering Building at UNSW. The building layout is shown in Figure 18 (left). The second experiment was conducted on level 2 of the Eastgardens Shopping Centre, Sydney, Australia. The building layout is shown in Figure 18 (right). Experiments 1 and 2 were conducted in environments typical of offices and shopping centres. In an office building rooms are often located on both sides of central corridors. A feature of shopping centres is that shops are located around one side of circular corridors. Figure 19 shows the positioning results of these two experiments, where the blue marks are results obtained using traditional fingerprinting methods, and red marks are the results from the proposed method.

It can be seen from Figure 19 (left) that the WiFi positioning results with and without topological information are similar in Experiment 1. In contrast, Figure 19 shows an obvious improvement when using topological information in Experiment 2, where the red points have a much smoother trajectory than the blue points. The variation in performance of these two experiments is likely due to the different characteristics of the building corridors. In Experiment 1 there are only two parallel corridors which are more than 30m apart and connected by a perpendicular corridor. It is nearly impossible to miss-locate one point to erroneous corridor which is not connected with the current corridor. However, the building layout in Experiment 2 is more complex. There are seven corridors, and some of them are parallel but close to each other. As a result the possibility of locating a point in a wrong corridor is high. By considering the topological relationships among corridors, the probability of miss-matching can be reduced.

Figure 18. Building layout in Experiment 1 (left) and 2 (right)

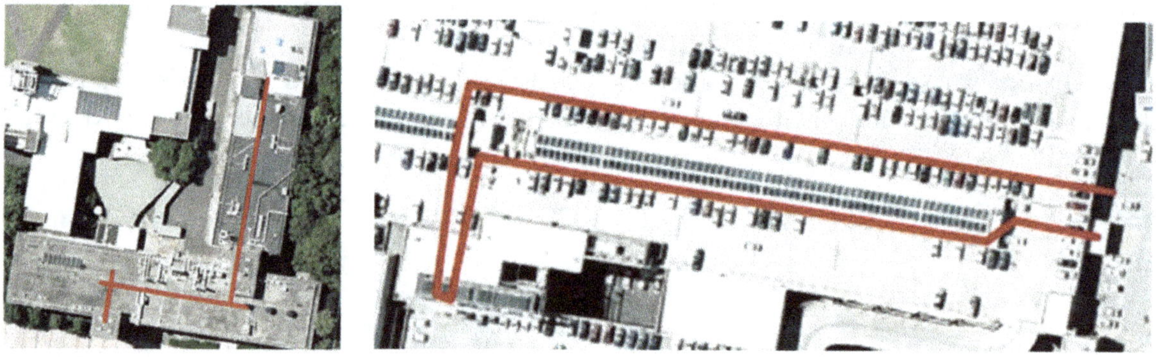

Figure 19. WiFi-only positioning results of Experiment 1 (left) and 2 (right)

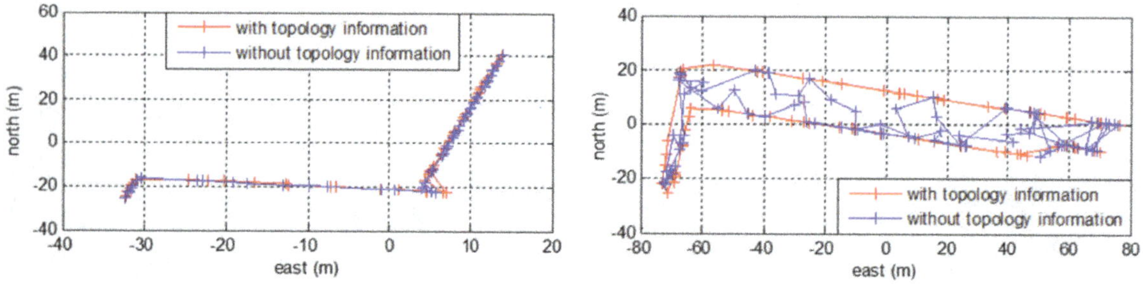

GNSS/WiFi/INS Integrated System

The integration flowchart are given in Figure 15, and the KF state function and measurement function are given in Table 4 and Eq.(44), respectively.

The position and heading solutions derived by different integrated systems from Experiment 1 are shown in Figure 20, where the red line is the GPS/INS solution without walk status constraints, the blue and green lines are results from the integrated GPS/WiFi/INS system, without and with map-based information, with both constrained by walk status detection.

The red line in Figure 20 (left) shows that the INS-only solution diverges quickly in an indoor environment. The blue line indicates that upon adding position corrections from WiFi the divergence can be controlled, such that the trajectory is much more consistent with the geometry of the corridors. The green line indicates the best performance in terms of position accuracy and stability, when considering constraints from walk status detection and context feature in the integrated system.

Figure 20 (right) shows the attitude solutions without and with topological constraints. The red lines are the references solution from the building orientation information. It can be seen that the green points are much more consistent, while the yaw angle shown by the blue points are neither accurate nor stable even when the person is walking in a straight line along a corridor. This further indicates that the accuracy and stability of attitude estimation is improved by considering building orientation information in the fingerprinting database.

SUMMARY

Through the framework of the information space, multisensor data fusion can be expressed simply as the sum of a series of transformations of the information vectors. All the information sources – the state predictions, the state estimates, and the measurements – are treated equally in the information space.

Figure 20. Solutions from Experiment 1
(red:GPS/INS; blue/green: GPS/WiFi/INS without/with topology constraints).

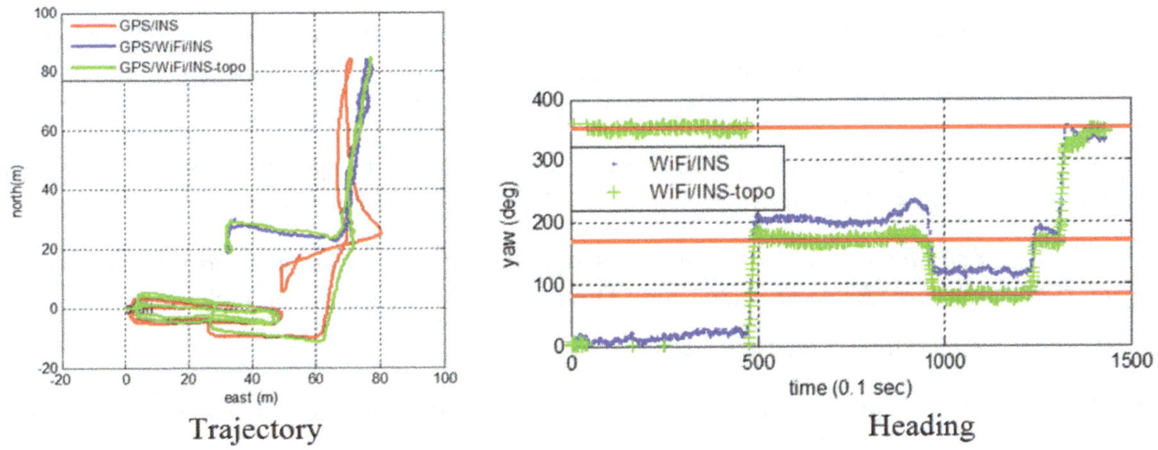

The transformation of an information vector can be easily implemented by the transpose of the mapping matrix between the state vectors.

The framework has demonstrated its efficiency in deriving decentralised Kalman filters such as the federated Kalman filter (FKF). The global optimal fusion (GOF) algorithm is derived from this framework, which maximally utilises the information sources that the multisensor system can provide. For multisensor systems there are multiple information sources which can be used. Utilising different sources as the bases of the estimation space leads to different algorithms. Compared to the centralised Kalman Filter (CKF) and the FKF, the GOF uses all of the information sources that a multisensor system provides including the local predictions.

The design is based on the loosely-coupled integrated GNSS, terrestrial ranging technology, and INS sub-systems. To satisfy both accuracy and reliability requirements, three integration algorithms – CKF, FKF, and GOF – are introduced and are implemented into a triple-integrated GNSS/Locata/INS system. The three data fusion algorithms have been evaluated using data from a field test conducted on Sydney Harbour. An analysis of the results has demonstrated that the integrated system is able to provide centimetre-level accuracy. A comparison of the CKF, FKF and GOF solutions verifies the theoretical analysis result, i.e. that the FKF and CKF solutions have the same accuracy, and the GOF solution has a slightly higher accuracy. Moreover, to evaluate the reliability of the GOF algorithm, a series of outliers were simulated. The results confirm the small outlier fault-tolerant capability of the GOF-based triple-integration method.

A seamless pedestrian navigation system based on INS, GNSS and WiFi sub-systems is described. The information sources such as the walking status detection, step detection, and the turns, are integrated in a CKF. With the step and heading angle constraints the solution accuracy and stability can be maintained when GNSS and WiFi signals are blocked. By integrating WiFi with INS, the long-term navigation stability for indoor environments is further improved. Experimental results have verified that by adding building/corridor topological information into the fingerprinting database the searching and matching reliability is improved. By using the more accurate position measurements, as well as the orientation constraints provided by WiFi, the performance of the integrated navigation system is also improved.

REFERENCES

Aeberhard, M., Schlichtharle, S., Kaempchen, N., & Bertram, T. (2012). Track-to-track fusion with asynchronous sensors using information matrix fusion for surround environment perception. *IEEE Transactions on Intelligent Transportation Systems*, *13*(4), 1717–1726. doi:10.1109/TITS.2012.2202229

Akyildiz, I. F., Sun, Z., & Vuran, M. C. (2009). Signal propagation techniques for wireless underground communication networks. *Physical Communication*, *2*(3), 167–183. doi:10.1016/j.phycom.2009.03.004

Amt, J. R. H. (2006) *Methods for aiding height determination in pseudolite-based reference systems using batch Least-Squares Estimation* (Master's Thesis). Air Force Institute of Technology.

Applanix Corporation. (2009). *POS MV V4 User guide*. Author.

Arshal, G. (1987). Error equations of inertial navigation. *Journal of Guidance, Control, and Dynamics*, *10*(4), 351–358. doi:10.2514/3.20225

Bar-Itzhack, I. Y. (1977). Navigation computation in terrestrial strapdown inertial navigation systems. *IEEE Transactions on Aerospace and Electronic Systems, 13*(6), 679–689. doi:10.1109/TAES.1977.308509

Bar-Shalom, Y. (1981). On the track-to-track correlation problem. *IEEE Transactions on Automatic Control, 26*(2), 571–572. doi:10.1109/TAC.1981.1102635

Barnes, J., Rizos, C., & Kanli, M. (2004) *Indoor industrial machine guidance using Locata: A pilot study at BlueScope Steel. 60th Annual Meeting of the U.S. Institute of Navigation*, Dayton, OH.

Barnes, J., Rizos, C., Wang, J., Small, D., Voight, G., & Gambale, N. (2003) LocataNet: The positioning technology of the future? *6th International Symposium on Satellite Navigation Technology Including* Mobile Positioning and Location Services, Melbourne, Australia.

Bertsch, J., Choudhury, M., Rizos, C., & Kahle, H. G. (2009) *On-the-fly ambiguity resolution for Locata.* International Symposium on GPS/GNSS, Gold Coast, Australia.

Beser, J., & Parkinson, B. W. (1982). The application of NAVSTAR differential GPS in the civilian community. *Navigation Journal of U.S. Institute of Navigation, 29*(2), 107–135. doi:10.1002/j.2161-4296.1982.tb00795.x

Borestein, J., Ojeda, L., & Kwanmuang, S. (2009). *Heuristic reduction of gyro drift in IMU-based personnel tracking systems.* SPIE Defense, Security and Sensing Conference, Orlando, FL.

Carlson, N. A. (1996) *Federated filter for computer-efficient, near-optimal GPS integration. IEEE Position Location and Navigation Symposium*, Atlanta, GA. doi:10.1109/PLANS.1996.509093

Carlson, N. A., & Berarducci, M. P. (1994). Federated Kalman filter simulation results. *Navigation, Journal of the U.S. Institute of Navigation, 41*(3), 297–321. doi:10.1002/j.2161-4296.1994.tb01882.x

Chen, Y., & Kobayashi, H. (2002) Signal strength based indoor geolocation. *1st IEEE Int. Conference on Communications*, 1, 436–439.

Cheng, J. T., Yang, L., Li, Y., & Zhang, W. (2014). Seamless Outdoor/Indoor Navigation with WIFI/GPS Aided Low Cost Inertial Navigation System. *Elsevier Editorial System(tm) for Physical Communication. SI: Indoor Navigation & Tracking., 13*, 31–43.

Cho, S. Y., & Park, C. G. (2006). MEMS based pedestrian navigation system. *Journal of Navigation, 59*(1), 135–153. doi:10.1017/S0373463305003486

Choudhury, M., Harvey, B. R., & Rizos, C. (2009b) *Tropospheric correction for Locata when known point ambiguity resolution technique is used in static survey – is it required?* International Symposium on GPS/GNSS, Gold Coast, Australia.

Choudhury, M., Rizos, C., & Harvey, B. R. (2009a) *A survey of techniques and algorithms in deformation monitoring applications and the use of the Locata technology for such applications.* 22nd International Technical Meeting of the Satellite Division of the U.S. Institute of Navigation, Savannah, GA.

Cobb, H. S. (1997). *GPS pseudolites: Theory, design and applications* (Ph.D thesis). Stanford University, Stanford, CA.

Frieden, B. R. (2004). Science from Fisher information – a unification. Cambridge University Press.

Gauthier, J. P., Glennon, E. P., Rizos, C., & Dempster, A. G. (2013). *Time transfer performance of Locata – Initial results. 45th Annual Precise Time and Time Interval System and Application Meeting*, Bellevue, WA.

Goshen-Meskin, D., & Bar-Itzhack, I. Y. (1992). Unified approach to inertial navigation system error modelling. *Journal of Guidance, Control, and Dynamics*, *15*(3), 648–653. doi:10.2514/3.20887

Grejner-Brzezinska, D. A., Toth, C. K., Sun, H., Wang, X., & Rizos, C. (2011). A robust solution to high-accuracy geolocation: Quadruple integration of GPS, IMU, pseudolite, and terrestrial laser scanning. *IEEE Transactions on Instrumentation and Measurement*, *60*(11), 3694–3708. doi:10.1109/TIM.2010.2050981

Groves, P. D. (2008). *Principles of GNSS, Inertial and Multisensory Integrated Navigation Systems*. Artech House.

Gusenbauer, D., Isert, C., & Krösche, J. (2010) Self-contained indoor positioning on off-the-shelf mobile devices. *Proc. ICIPIN*, 1–9. doi:10.1109/IPIN.2010.5646681

Guvenc, I., & Chong, C. C. (2009). A survey on TOA based wireless localization and NLOS mitigation techniques. *IEEE Communications Surveys and Tutorials*, *11*(3), 107–124. doi:10.1109/SURV.2009.090308

Han, S., & Wang, J. (2011). Quantization and colored noises error modeling for inertial sensors for GPS/INS integration. *IEEE Sensors Journal*, *11*(6), 1493–1503. doi:10.1109/JSEN.2010.2093878

Harrington, R. L., & Dolloff, J. T. (1976) *The inverted range: GPS user test facility. IEEE Position Location and Navigation Symposium*, San Diego, CA.

Hashemipour, H. R., Roy, S., & Laub, A. J. (1988). Decentralized structures for parallel Kalman filtering. *IEEE Transactions on Automatic Control*, *AC-33*(1), 88–94. doi:10.1109/9.364

Jahn, J., Batzer, U., Seitz, J., Patino-Studencka, L., & Boronat, J. (2010). Comparison and evaluation of acceleration based step length estimators for handheld devices. *Proc. ICIPIN*, 1–6. doi:10.1109/IPIN.2010.5646888

Jaradat, M. A. K., & Abdel-Hafez, M. F. (2014). Enhanced, delay dependent, intelligent fusion for INS/GPS navigation system. *IEEE Sensors Journal*, *14*(5), 1545–1554. doi:10.1109/JSEN.2014.2298896

Jiang, W., Li, Y., & Rizos, C. (2013). On-the-fly Locata/inertial navigation system integration for precise maritime application. *Measurement Science & Technology*, *24*(10), 105104. doi:10.1088/0957-0233/24/10/105104

Jiménez, A. R., Seco, F., Prieto, J. C., & Guevara, J. (2010) *Indoor Pedestrian Navigation using an INS/EKF framework for Yaw Drift Reduction and a Foot-mounted IMU. 7th IEEE Workshop on Positioning Navigation and Communication*, Dresden, Germany. doi:10.1109/WPNC.2010.5649300

Jirawimut, R., Ptasinski, P., Garaj, V., Cecelja, F., & Balachandran, W. (2003). A method for dead reckoning parameter correction in pedestrian navigation system. *IEEE Transactions on Instrumentation and Measurement*, *52*(1), 209–215. doi:10.1109/TIM.2002.807986

Klein, D., & Parkinson, B. W. (1984). The use of pseudo-satellites for improving GPS performance. *Navigation, Journal of the U.S. Institute of Navigation, 31*(4), 303–315. doi:10.1002/j.2161-4296.1984.tb00881.x

Ladetto, Q., & Merminod, B. (2002) *Digital magnetic compass and gyroscope integration for pedestrian navigation. 9th Saint Petersburg International Conference on Integrated Navigation Systems*, Saint Petersburg, Russia.

Lee, H. K., Lee, J. G., Rho, Y. K., & Park, C. G. (1998). Modelling quaternion errors in SDINS: Computer frame approach. *IEEE Transactions on Aerospace and Electronic Systems, 34*(1), 289–300. doi:10.1109/7.640286

Leica Geosystems AG. (2010). *Leica Geo Office Datasheet*. Author.

Leppäkoski, H., Collin, J., & Takala, J. (2013). Pedestrian navigation based on inertial sensors, indoor map, and WLAN signals. *Journal of Signal Processing Systems for Signal, Image, and Video Technology, 71*(3), 287–296. doi:10.1007/s11265-012-0711-5

Levi, R., & Judd, T. (1999). *Dead reckoning navigational system using accelerometer to measure foot impacts*. US Patent 5583776.

Li, Y. (1997). *Decentralized filtering theory and its application to the spacecraft navigation using GPS* (PhD dissertation). Northwestern Polytechnical University, China.

Li, Y. (2014). Optimal multisensor integrated navigation through information space approach. *Physical Communication. SI: Indoor Navigation and Tracking, 13*, 44–53.

Li, Y., Li, Y., Rizos, C., & Xu, X. S. (2012) Observability Analysis of SINS/GPS during In-Motion Alignment Using Singular Value Decomposition. Advanced Materials Research, 433 – 440, 5918-5923.

Li, Y., & Rizos, C. (2010), *Seamless navigation through a Locata-enhanced GPS and INS integrated system*. International Symposium on GPS/GNSS, Taipei, Taiwan.

Locata Corporation. (2014). *LocataNet positioning signal interface control document*. Available at http://www.locata.com/wp-content/uploads/2014/07/Locata-ICD-100E.pdf

Mervart, L., Lukes, Z., Rocken, C., & Iwabuchi, T. (2008) *Precise point positioning with ambiguity resolution in real-time*. 21st International Technical Meeting of the Satellite Division of the U.S. Institute of Navigation, Savannah, GA.

Montillet, J., Roberts, G. W., Hancock, C., Meng, X., Ogundipe, O., & Barnes, J. (2009). Deploying a Locata network to enable precise positioning in urban canyons. *Journal of Geodesy, 83*(2), 91–103. doi:10.1007/s00190-008-0236-7

Mutambara A.G.O. (1998). *Decentralized estimation and control for multisensor systems*. CRC.

Park, J., Kim, Y., & Lee, J. (2012). Waist mounted pedestrian dead-reckoning system. *Proc. ICURAI*, 335–336.

Perttula, A., Leppakoski, H., Kirkko-Jaakkola, M., Davidson, P., Collin, J., & Takala, J. (2014). Distributed Indoor Positioning System with Inertial Measurements and Map Matching. *IEEE Transactions on Instrumentation and Measurement*, *63*(11), 2682–2695. doi:10.1109/TIM.2014.2313951

Rizos, C., Roberts, G. W., Barnes, J., & Gambale, N. (2010) Locata: A new high accuracy indoor positioning system. *1st International Conference on Indoor Positioning and Indoor Navigation*, 441 – 447.

Roy, S., Hashemi, R. H., & Laub, A. J. (1991). Square root parallel Kalman filtering using reduced-order local filters. *IEEE Transactions on Aerospace and Electronic Systems*, *AES-27*(2), 276–289. doi:10.1109/7.78303

Ruan, Y., Willett, P., Marrs, A., Palmieri, F., & Marano, S. (2008). Parctical fusion of quantized measurements via particle filtering. *IEEE Transactions on Aerospace and Electronic Systems*, *AES-44*(1), 15–28. doi:10.1109/TAES.2008.4516986

Salo, J., Vuokko, L., El-Sallabi, H. M., & Vainikainen, P. (2007). An additive model as a physical basis for shadow fading. *IEEE Transactions on Vehicular Technology*, *56*(1), 13–26. doi:10.1109/TVT.2006.883797

Sun, Z., Mao, X., Tian, W., & Zhang, X. (2009). Activity classification and Dead Reckoning for Pedestrian Navigation with Wearable Sensors. *Measurement Science & Technology*, *20*(1), 015203. doi:10.1088/0957-0233/20/1/015203

Wendel, J., Metzger, J., Moenikes, R., Maier, A., & Trommer, G. F. (2006). A performance comparison of tightly coupled GPS/INS navigation systems based on extended and sigma point Kalman filters, *NAVIGATION. Journal of the Institute of Navigation*, *53*(1), 21–31. doi:10.1002/j.2161-4296.2006.tb00368.x

Xiao, Z., Wen, H., Markham, A., Trigoni, N., Blunsom, P., & Frolik, J. (2015). Non-Line-of-Sight Identification and Mitigation Using Received Signal Strength. *IEEE Transactions on Wireless Communications*, *14*(3), 1689–1702. doi:10.1109/TWC.2014.2372341

Chapter 12
Weak GNSS Signal Acquisition for Indoor Environments

Seung-Hyun Kong
KAIST, South Korea

ABSTRACT

High sensitivity and fast acquisition are two important goals that must be considered in the development of signal processing techniques for a GNSS acquisition function to meet the demands for LBS in GNSS-challenged environments, such as indoor and urban canyon. This chapter introduces the fundamentals of GNSS acquisition functions, GNSS acquisition techniques for new GNSS signals, and GNSS acquisition techniques achieving high sensitivity and fast acquisition. Therefore, this chapter contains useful information for engineers who study the fundamentals and principles of GNSS acquisition and the state-of-the-art GNSS signal acquisition techniques for weak signals.

INTRODUCTION

In 1996, the U.S. Federal Communication Commission (FCC) issued the emergency 911 phase II requirements that mandated enhanced positioning accuracy of mobile phones and immediate delivery of the positioning results to the public safety answering point (PSAP) (Hatfield, 2002). Since then, the global positioning system (GPS) has been the enabling technology for E911 service and many mobile location based services (LBS).

However, the majority of mobile users are in urban and indoor areas, where GNSS (global navigation satellite systems including GPS) signal propagation is often obstructed by trees, glass, and walls, and as a result, received GNSS signals can be severely attenuated and it becomes very difficult to detect the signals for positioning (Kaplan, 2005; Van Diggelen, 2009) and LBS. The received signal strength (RSS) of GPS signals, expressed in carrier to noise ratio (C/N_0), can be lower than 5dB-Hz when the receiver is located at deep inside indoor areas, whereas, the same signal may have C/N_0 of around 45dB-Hz in outdoor line-of-sight (LOS) environments (Kaplan, 2005; Van Diggelen, 2009). Consequently, the GNSS signal detection in dense urban and indoor environments requires a significant sensitivity improvement, which can be achievable with a very long coherent correlation interval (a.k.a. dwell time)

DOI: 10.4018/978-1-5225-3528-7.ch012

at a great cost, for example, very high computational cost and receiver complexity (Kaplan, 2005; Van Diggelen, 2009; Parkinson & Spilker, 1996). As a result, low computational fast GNSS acquisition is a challenge for GNSS receivers in GNSS-challenged environments. In this viewpoint, higher sensitivity and faster acquisition are two conflicting goals for a GNSS acquisition function, and both of these goals must be considered in the development of GNSS signal processing techniques to meet the demands for LBS in GNSS-challenged environments. This chapter introduces GNSS acquisition techniques from this viewpoint.

This chapter introduces the fundamentals of GNSS acquisition, modifications to the GNSS acquisition techniques for new GNSS signals, and low computational fast GNSS acquisition techniques with high sensitivity. In detail, this chapter consists of 5 sections is as follows. Section 2 introduces the conventional single channel GNSS acquisition techniques and the fundamentals of GNSS signal acquisition. In section 3 and section 4, low computational (fast acquisition) GNSS signal processing techniques for high sensitivity standalone GNSS receivers and multi-satellite acquisition techniques are studied, respectively. And section 5 and section 6 discuss additional GNSS acquisition techniques for the new GNSS signals with (data-free) pilot channels and assisted GNSS (A-GNSS) technique, respectively.

FUNDAMENTALS OF GNSS SIGNAL ACQUISITION

GNSS (signal) acquisition function searches and detects incoming GNSS signals to initialize the tracking function that actually produces fine GNSS signal measurements for positioning. In general, the GNSS signals are spread by periodic long spreading codes, and the GNSS satellite motion relative to the receiver generates a Doppler frequency shift in the incoming signal as large as 5kHz to 10kHz depending on the receiver's motion. Since the precise time and location of a GNSS receiver is not initially known, the GNSS acquisition function of the receiver cannot estimate the code phase and Doppler frequency of the incoming signals. Therefore, the task of the GNSS acquisition function is to identify the prompt code phase and Doppler frequency of the incoming GNSS signals, as quickly as possible, within a resolution which is fine enough to initialize the tracking function successfully. Since the code phase and Doppler frequency of an incoming GNSS signal can have any value within the 2-dimensional (2D) hypothesis plane (aka, the search space) which includes all possible combinations of code phase and Doppler frequency hypothesis, the acquisition function needs to test each hypothesis within the 2D plane using a matched filter (e.g., correlator) (Kaplan, 2005; Van Diggelen, 2009; Parkinson & Spilker, 1996) until the detection of the incoming GNSS signal.

In practice, the number of hypotheses to test is very large in GNSS and can be too large (see Section 2.1), and the acquisition process can be both time consuming and computationally expensive. The number of code phase hypotheses N_{Hc} is two times the spreading code length L_c, because the code phase search step size Δ_τ is set to a half code chip in most GNSS receivers. However, the number of Doppler frequencies N_{Hf} grows due to both the user dynamics relative to the satellite motion and the Doppler frequency search step size Δ_f which is inversely proportional to the coherent correlation interval T_{co} (Kaplan, 2005; Van Diggelen, 2009; Parkinson & Spilker, 1996). In general, a longer T_{co} is required to detect a weaker GNSS signal and the number of Doppler frequency hypotheses increases consequently. As a result, both the acquisition time for weaker GNSS signals and the acquisition complexity increase exponentially due to the longer correlation interval required to test each hypothesis and the larger number of (Doppler frequency) hypotheses in the search space. Note that the increase in acquisition complexity may not be a critical problem for GNSS receivers in open-sky outdoor environments, where signals are received at a moderate or strong level.

To lessen the acquisition time and acquisition complexity, a GNSS receiver needs to select the appropriate integration technique, search strategy, and detection scheme based on the GNSS signal modulation scheme, received signal strength (RSS), available hardware resources and computational capacity of the receiver, required mean acquisition time (MAT, i.e., expected average time for a signal acquisition), etc. In the following, we introduce the conventional integration techniques, conventional search strategies, and detection schemes that involve the detection variables and detection thresholds.

GNSS Signal Model and Integration Techniques

The matched filter used for hypothesis testing in GNSS receivers is the auto-correlation function (ACF) (Kaplan, 2005; Van Diggelen, 2009; Parkinson & Spilker, 1996) that performs correlations between the incoming satellite signal $y(t)$ that is down-converted to an IF frequency f_I and a receiver replica signal $x(t)$ for a combination of code phase hypothesis κ and Doppler frequency hypothesis λ. In BPSK modulated GNSS signals such as GPS L1-C/A (coarse acquisition), $y(t)$ and $x(t)$ can be expressed as

$$y(t) = AD(t-\tau)P(t-\tau)e^{j(2\pi(f_I+f_D)t+\theta)} + v(t), \tag{1}$$

$$x(t;\kappa,\lambda) = P(t-\kappa\Delta_\tau)e^{j2\pi(f_I+\lambda\Delta_f)t} \tag{2}$$

where A is the signal amplitude, $D(t)$ and $P(t)$ represent the bi-phase modulated data at a data rate $R_b \left(=\frac{1}{T_b}\right)$ and spreading code (of length L_c) at a code rate of $R_c \left(=\frac{1}{T_c}\right)$ with a period T_P, respectively, θ is an unknown carrier phase, $v(t)$ represents a complex AWGN (additive white Gaussian noise) process with two-sided power spectral density $\frac{N_0}{2}$, and Δ_τ and Δ_f are the code phase and Doppler frequency search step sizes, respectively. In this section, we use the GPS L1 C/A signal as an example of one of the most representative GNSS signals, where R_b=50bps, R_c=1.023MHz, L_c=1023, T_P=1ms, T_b = 20ms, and $T_c \cong 1\mu$ s. To identify the prompt code phase τ and the prompt Doppler frequency f_D of the incoming signal, the acquisition function performs correlations for each combination of (κ, λ), until a correlation output is found to be larger than the detection threshold γ. Since $T_P \ll T_b$ in GPS L1 C/A, $D(t)$ can be assumed constant for a coherent dwell time $T_{co} = T_p$ to simplify the GNSS signal acquisition into a search process over the 2D hypothesis plane (or search space). The correlation output between the samples of $y(t)$ and $x(t)$ over an interval T_{co} can be expressed as

$$R[\eta,\kappa,\lambda]=\sum_{n=0}^{N_{CO}-1} y[n+\eta]x[n;\kappa,\lambda], \tag{3}$$

where N_{co} is the number of samples received during the interval T_{co} for a sampling frequency f_s, η represents the index of the incoming signal segment of N_{co} samples being correlated, and $R\left[\eta,\kappa,\lambda\right]=R^I\left[\eta,\kappa,\lambda\right]+jR^Q\left[\eta,\kappa,\lambda\right]$ in Figure 1. In general, the acquisition function employs an (post-correlation) integration function $G\left(\cdot\right)$ that integrates consecutive correlation outputs to further increase acquisition sensitivity. The integration techniques for $G\left(\cdot\right)$ are introduced below.

Coherent Integration Technique

When the coherent correlation interval T_{co} is only 1ms ($=T_1$), $D\left(t\right)$ can be assumed constant, since T_b is 20ms ($\gg T_1$). Therefore, the consecutive coherent correlation outputs can be coherently added for N_i ($1\le N_i<20$) times to build a coherently integrated detection variable $Z=Z_{co}$ as

$$Z_{CO}[\kappa,\lambda]=\left|\sum_{\eta=0}^{N_i-1} R[\eta,\kappa,\lambda]\right|^2. \tag{4}$$

Note that N_i is called the accumulation length, and that Z_{co} is equivalent to the detection variable for a coherent correlation with a long interval N_iT_{co}. In general, this technique amplifies the signal peak (i.e., Z_{co} for the correct hypothesis: $Z_{co}\mid H_1$) by N_i times. However, when $N_i\gg 1$, there can be bit transition(s) during the interval N_iT_{co}, which often causes a significant degradation of $Z_{co}\mid H_1$.

Figure 1. A diagram for the correlation and integration function to test an individual hypothesis at [κ , λ].

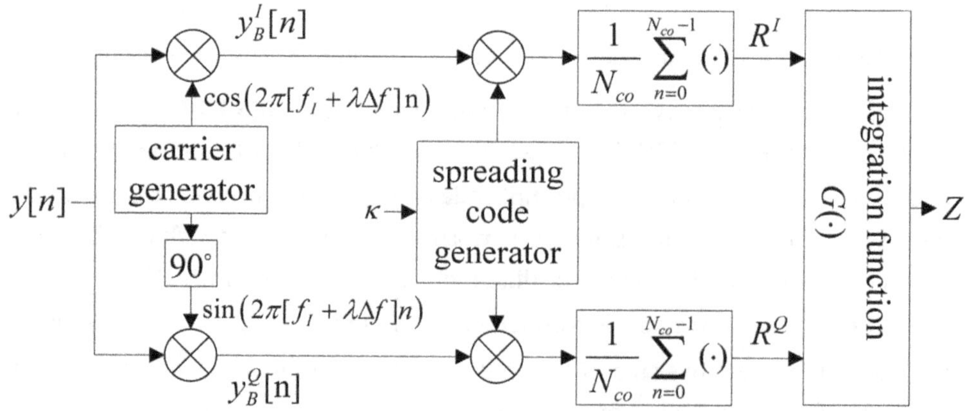

Non-Coherent Integration Technique

To avoid the bit transition problem in the long coherent integration, an acquisition function may choose a T_{co} ($\geq T_1$) that is much smaller than T_b and non-coherently combine the N_i consecutive coherent correlation outputs by taking the sum of squares of each coherent correlation output to build a detection variable Z_{nc} as

$$Z_{nc}[\kappa,\lambda] = \sum_{\eta=0}^{N_i-1} \left| R[\eta,\kappa,\lambda] \right|^2, \tag{5}$$

which is immune to both the bit transition problem and the phase shift in the consecutive correlation outputs. However, the resulting increase in the signal peak after the N_i times accumulation is much less than N_i times, and it suffers from squaring loss (Van Diggelen, 2009) for weak GNSS signals.

Differential Integration Technique

Another possible scheme to cope with the bit transition problem is to multiply the prompt (at time instant η) coherent correlation output with the conjugate of the previous one (at time instant η -1) and coherently accumulate the results to build a differential coherently integrated detection variable (Ávila Rodriguez, Pany, & Eissfeller, 2004)

$$Z_{di}^1[\kappa,\lambda] = \left| \sum_{\eta=1}^{N_i-1} R[\eta,\kappa,\lambda] \times R^*[\eta-1,\kappa,\lambda] \right|^2, \tag{6}$$

where $(\cdot)^1$ represents one span. This technique exploits the fact that the signal components in the consecutive correlation outputs (e.g., $R[\eta_1,\tau,\lambda]$ and $R[\eta_2,\tau,\lambda]$) that are separated by the unit correlation interval (i.e., η_2 - η_1 = 1 span) has the same (or similar) phase, while the noise components are independent. However, there can be a monotonous phase shift between the consecutive correlation outputs, which can result in an SNR loss in Z_{di}^1. To cope with the phase shift, modified general differential combination (MGDC) technique (Ta, Qaisar, Dempster, & Dovis, 2012) uses

$$Z_{MGDC}[\kappa,\lambda] = \sum_{m=1}^{N_i-1} \left| Z_{di}^m[\kappa,\lambda] \right| = \sum_{m=1}^{N_i-1} \left| \sum_{\eta=m}^{N_i-1} R[\eta,\kappa,\lambda] \times R^*[\eta-m,\kappa,\lambda] \right|, \tag{7}$$

that non-coherently accumulates the differential coherently integrated detection variables of all possible spans (from 1 to N_i -1 span). The MGDC shows a higher robustness against the phase shift than the general difference combination (GDC) technique (Corazza & Pedone, 2007) which coherently accumulates the differential coherently integrated detection variables of all possible spans. It is found that the MGDC technique is useful for combining partial correlation outputs of GNSS signals that have a very long spreading code length, such as L2C, without degrading the acquisition performance due to the bit transitions (Dovis & Ta, 2012). The SNR improvement with Z_{MGDC} is higher than Z_{nc} but much lower than Z_{co} because of the noise amplification caused by the multiplication of noise components in different correlation outputs.

Note that the detection variables in the above integration techniques are different and have different statistical distributions (in the presence and absence of the signal), and, as a result, each of the above integration techniques have different probabilities of detection (P_d), miss-detection (P_m), and false alarm (P_f) in the individual hypothesis test. Note also that the detection variables in (4), (5), (6), and (7) have magnitude-squared function but can have a different non-linear function (Van Diggelen, 2009).

Search Strategy for Lower Acquisition Complexity

In general, the code phase search step size Δ_τ is set to a half chip ($\mathrm{T}_c / 2$), so that the total numberof code phase hypotheses is

$$N_{Hc} = 2L_c \tag{8}$$

and since the Doppler frequency is searched from $-f_{\max}^D$ to $f_{\max}^D$ about f_I with a step size Δ_f, the total number of Doppler frequency hypotheses is

$$N_{Hf} = 2\left\lceil f_{\max}^D / \Delta_f \right\rceil + 1, \tag{9}$$

where $\lceil \cdot \rceil$ is the ceiling operator. Therefore, for a common choice of $\Delta_f = 1/\left(2T_{co}\right)$ and $f_{\max}^D = 5\text{kHz}$, (while $\Delta_f = 2/\left(3T_{co}\right)$ and $f_{\max}^D = 10\text{ kHz}$ can be used for dynamic receivers), the maximum number of hypothesis to test becomes

$$N_H = 2L_c \times \left(2\left\lceil f_{\max}^D T_{co}\right\rceil + 1\right) \cong 4L_c\left\lceil 2f_{\max}^D T_{co}\right\rceil. \tag{10}$$

In the serial search, for an example, a GPS L1 C/A signal acquisition function uses a single correlator with $N_i = 1$ and (usually) $T_{co} = 1\text{ms}$ to test $N_H \left(\cong 41000\right)$ hypotheses serially. Because the true (correct) hypothesis H_1 can be any one of the N_H hypotheses, the mean and worst case acquisition times are $\frac{1}{2}N_H T_{co} = 20.5$ s and $N_H T_{co} = 41$ s, respectively. The mean acquisition time (MAT, μ_T) can grow exponentially for acquisition functions using the serial search strategy when, for example, N_i times, longer coherent correlation interval is used; N_i times longer N_{co} to test each hypothesis and N_i times more Doppler frequency hypotheses due to the smaller Δ_f result in an increased MAT by N_i^2

times. Therefore, it is necessary to have an effective search strategy to reduce MAT for the given hardware resources and computational capacity in the GNSS receiver. In following subsections, some searching strategies are discussed.

Complete Search Strategy

In the complete search strategy, a GNSS receiver tests all of the hypotheses in the search space and identifies the true hypothesis H_1. Testing all hypotheses in parallel may require a massive amount of correlators, so the complete search strategy may have the highest acquisition complexity. Conventionally, the maximum integration output in the search result, $\max\{Z\}$, is used as a detection variable to compare with the detection threshold. This is similar to the hybrid search strategy that is presented below. However, there are different detection variables and detection schemes also used for the complete search strategy and for the hybrid search strategy, as discussed below.

Hybrid Search Strategy

In the hybrid search strategy, a smaller number of hypotheses than N_H are tested in parallel. As an example, a hybrid search strategy simultaneously testing $M\ (=N_{Hc})$ code phase hypotheses for a Doppler frequency hypothesis using M parallel correlators is depicted in Figure 2, where 'D' represents a

Figure 2. Time-domain hybrid search with MTC.

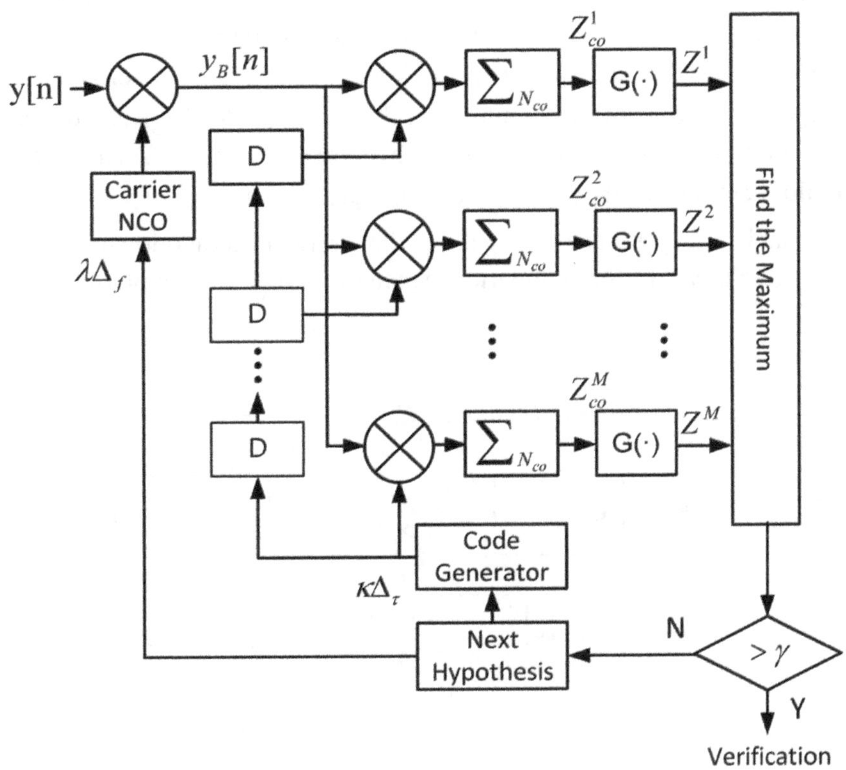

half chip delay. The time-domain hybrid search in Figure 2 can achieve about M times smaller MAT than $\mu_{T,S}$ when the incoming signals are strong enough. The time-domain hybrid search in Figure 2 can also be realized in the frequency-domain as shown in Figure 3.

The incoming signal samples $y\left[n\right]$ and receiver replica samples $x\left[n\right]$ are collected for one code period (T_p) to build y and x vectors of size N_{co}, respectively, and it is found that

$$rev[x] \circledast y = IDFT\left\{DFT\left\{x\right\}^{*} \cdot DFT\left\{y\right\}\right\} = IDFT\left\{X^{*} \cdot Y\right\}, \tag{11}$$

where $\circledast$ is the circular convolution operator, $\text{rev}\left[x\right] = \left\{x\left[n\right] \mid n = N_{co}, N_{co} - 1, \ldots, 1\right\}$, DFT$\{\cdot\}$ and IDFT$\{\cdot\}$ are the discrete Fourier transform (DTF) and inverse DFT (IDTF) operations, respectively, $\left(\cdot\right)^{*}$ represents the complex conjugate, $X = \text{DFT}\left\{x\right\}$, and $Y = \text{DFT}\left\{y\right\}$. Each IDFT$\left\{X^{*} \cdot Y\right\}$ result in (11) is used for the signal detection, and similarly to the complete search strategy there are a couple of detection variables and detection schemes used for the hybrid search strategy in practice.

The frequency-domain hybrid search can be implemented in a DSP (digital signal processor) chip, so it becomes useful to reduce the computational cost and MAT for weaker GNSS signal acquisition. In the literature, the mean acquisition computation (MAC, μ_C, i.e., expected average computational cost for a signal acquisition) is introduced as a measure of the mean computational cost for the acquisition function (Kim & Kong, 2014a). In practice, X^{*} is pre-generated and stored in a memory of a receiver and FFT (fast Fourier transform) and IFFT (inverse FFT) are used instead of DFT and IDFT, respectively. This effectively reduces the computational cost of acquisition. Figure 3 shows the FFT-based hybrid search technique (FFT-based technique in short), and due to the circular convolution in the FFT-based correlation (Van Diggelen, 2009), x is padded with N_{co} (or more) zeros to make the length of X equal to an integer power of 2 and y should be of the same size to X. Also, the circular shifting of Y in Figure 3 is to shift the Doppler frequency hypothesis to the next one. As a result, the overall number of complex multiplications for the FFT-based technique has an order of $\text{O}\left(N_{co}\log_2 N_{co}\right)$.

When the test variable is larger than the threshold, the verification function tests the detected hypothesis multiple times again to verify the acquisition. Since the detected code phase and Doppler frequen-

Figure 3. Frequency-domain hybrid search with MTC.

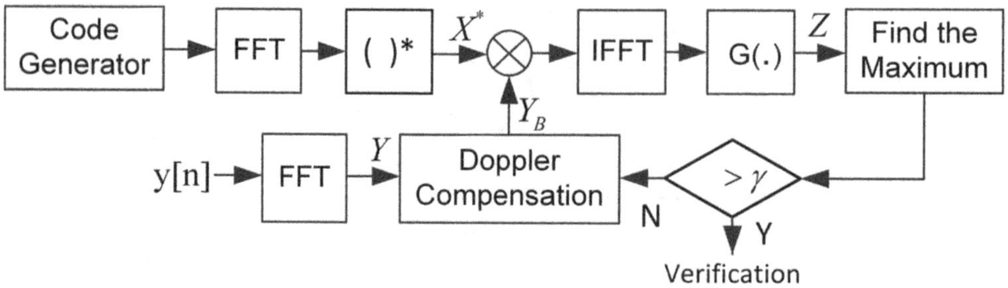

cy by the acquisition function have errors as large as $\frac{\Delta_\tau}{2}$ and $\frac{\Delta_f}{2}$, respectively, the pull-in search process is used to find a finer Doppler frequency estimate, and then the signal tracking function estimates the precise code phase and Doppler frequency of the incoming signal.

Detection Schemes and Detection Threshold

In GNSS, a detection scheme involves a detection variable and a detection rule for a reliable signal detection from the search result. In practice, the selection of the detection variable depends on the search strategy as discussed in the following.

Threshold Crossing (TC)

In the serial search, the integration function $G(\cdot)$ output Z in Figure 1 resulting from the current hypothesis test, is used as a detection variable, and the detection is declared when the detection variable is found to be larger than the detection threshold γ. This detection rule is called the threshold crossing (TC) (Viterbi, 1995) and is used only for the serial search. The overall system performance of the acquisition function using TC can be expressed in system probabilities, such as the detection P_D, miss-detection P_M, and false alarm P_F probabilities, are found in (Borio, 2008). Note that the system probabilities are different from those of the individual hypothesis test, since the system probabilities (P_D, P_M, and P_F) represent the performance of the overall acquisition function while the probabilities P_d, P_m, and P_f are related to the detection performance for each individual hypothesis test.

Maximum Threshold Crossing (MTC) and Maximum-to-Second-Maximum-Ratio (MTSMR)

For acquisition functions employing the hybrid or complete search strategies, the maximum of the integration function $\mathrm{G}(\cdot)$ output (in other words, the maximum of the search result) is often used as the detection variable and compared to the detection threshold, which is the maximum threshold crossing (MTC) scheme (Corazza G. E., 1996) shown in Figures 2 and 3. Note that, in the hybrid search strategy, the maximum of the integration function output is only a local maximum. In some cases, MTSMR (maximum to second maximum ratio) (Geiger, Vogel, & Soudan, 2012) of the search result can be used as the detection variable, where the second maximum should be at least one chip apart from the maximum. In (Kim & Kong, 2014b), it is found that the system probabilities, P_D, P_M, and P_F, of the acquisition function using MTC (Borio, 2008) are very similar to those using MTSMR.

Detection Threshold

The detection threshold γ has a critical effect on the acquisition performance, since it affects both the probabilities (i.e., P_d, P_m, and P_f) of the individual hypothesis test and the system probabilities of the acquisition function. The conventional rule for determining the detection threshold γ is to find the γ that limits P_f under a small constant level (e.g., 0.01 or less). This is known as the constant false alarm rate (CFAR) (Parkinson & Spilker, 1996). This rule is effective, since the false alarm costs a large penalty in terms of acquisition time, and the detection threshold γ for a low CFAR leads to the optimal acquisition performance (i.e., minimum MAT μ_T or MAC μ_C) for GNSS signals with high SNR. Note that the MAT μ_T and MAC μ_C are functions of γ, P_d, P_m, P_f, penalty time, N_{Hc}, N_{Hf}, N_i, and T_{co}, and that the minimum $\mu_{T,\min}$ and minimum $\mu_{C,\min}$ can be found by taking the derivative of μ_T and μ_C with respect to γ equal to zero, respectively. However, in (Kim & Kong, 2014b), it is found that the algebraically determined optimal detection threshold γ achieves a much smaller MAT μ_T and MAC μ_C than those obtained by the CFAR-based detection threshold for weak GNSS signals. Note also that to determine γ for all of the detection variables except MTSMR, knowledge of the average integration function output at incorrect hypotheses, i.e., $\mathrm{E}\left[Z \mid H_0\right]$, is required.

Factors Affecting the GNSS Acquisition Performance

In addition to the detection variable, search strategy, and detection scheme, the acquisition performance depends on a number of other factors. These include receiver dependent factors and receiver independent factors. For example, front-end filtering, sampling rate f_s, oscillator stability, and quantization levels belong to the receiver dependent factors, and multipath and NLOS, interference signals, user dynamics, and code Doppler, ionospheric and tropospheric delays, and the C/N$_0$ of the incoming signal (Kaplan, 2005; Van Diggelen, 2009; Parkinson & Spilker, 1996) are receiver independent factors. Among the receiver dependent factors, the sampling rate and oscillator stability are critical design factors for achieving low computational cost and fast GNSS signal acquisition. Since $N_{co} = T_{co}f_s$, a higher sampling rate increases N_{co}, which increases the SNR (i.e., the post-correlation SNR) at the integration function output, but the MAC μ_C increases at the same time.

On the other hand, when T_{co} becomes too long for low cost quartz crystal oscillator used in many

GNSS receivers (Ward, 1996), the frequency drift during T_{co} degrades the post-correlation output SNR. In addition to the above factors, the parameters of the signal structure, such as data bit width, modulation scheme, secondary code, and relative power allocation between the pilot and data channels, have a strong influence on the acquisition complexity, too.

Challenges for Weak GNSS Signal Acquisition Function

Figure 4 shows the normalized complete search result for $T_{co} = 1\text{ms}$ and $N_i = 1$, where the correct (true) hypothesis H_1 is at $\kappa = 1226$ and $\lambda = 11$. As shown, there are distinctive false peaks in the search result (i.e., large $Z \mid H_0$'s) and when the signal peak (i.e., the integration output at the correct hypothesis $Z \mid H_1$) is not relatively large due to the low C/N$_0$ of the incoming signal and when N_iN_{co} is not large enough, the false peaks can cause false alarms. In fact, the false peaks are from the non-zero output of the autocorrelation between the spreading codes of the different delay, and from the noise added in the received signal. In practice, the size of the false peaks can be suppressed by using larger N_{co} and N_i. However, increasing N_{co} is restricted by the data bit transition, and increasing N_i is less effective and costs much larger MAT than increasing N_{co}. As an example, when $N_t = N_iN_{co}$ signal samples are used, the coherent integration technique produces a signal peak power proportional to N_t^2, while the non-coherent integration technique has a signal peak power proportional to $N_iN_{co}^2$. Therefore,

Figure 4. A complete search result over the 2D plane.

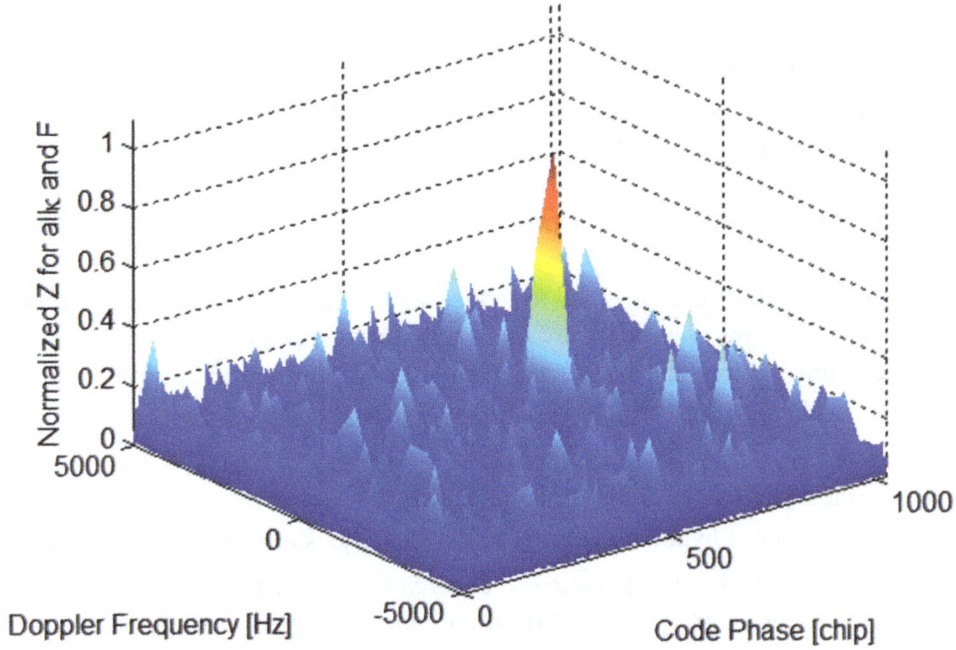

the coherent integration technique produces N_i times higher signal peak power. However, the coherent integration technique requires a longer acquisition time and higher computational complexity due to the increased number of Doppler frequency hypotheses by N_i times. For the FFT-based technique in Figure 3, the total number of complex multiplications for the coherent and non-coherent integration techniques are $N_t(4\log_2 N_t + 7)$ and $N_t(4\log_2 N_{co} + 7)$, respectively, for a Doppler frequency hypothesis, but the coherent integration technique has N_i times more Doppler frequency hypotheses. Consequently, improving the GNSS acquisition sensitivity for weak signals conflicts with another primary goal of GNSS receivers - low acquisition complexity. Therefore, a trade-off between the sensitivity and acquisition complexity should be necessary for mass-market receivers. However, achieving the two goals at the same time is still a challenge in the development of high sensitivity GNSS receivers.

LOW COMPUTATION TECHNIQUES ACHIEVING FAST ACQUISITION AND HIGH SENSITIVITY

While higher acquisition sensitivity can be achieved by increasing N_i and N_{co} and by channel combining in $G(\cdot)$, low computational fast acquisition techniques are of great interest when $N_t(= N_i N_{co} = f_s T_t)$ should be very large. Among various low computational fast acquisition techniques introduced in the literature, such as averaging techniques, hypothesis compression techniques, fast computation algorithms (for FFT-based techniques), and multi-satellite acquisition techniques, we investigate hypothesis compression techniques, fast computation algorithms, and multi-satellite acquisition techniques in this section. The averaging techniques are to reduce the size of the incoming signal samples and may require a special search strategy for Doppler frequency hypothesis. Hypothesis compression techniques (Kim & Kong, 2014a; Li, Cui, Lu, & Feng, 2008) are to compress the 2D hypothesis plane into a smaller size, which can be realized with a compressed receiver replica signal that consists of multiple receiver replica signals of different hypotheses. The fast computation algorithms include data segmentation, a new signal processing technique for the FFT-based technique, and a new search strategy.

Hypothesis Compression Techniques

While the sample compression techniques are to reduce the size of signal samples to process, hypothesis compression techniques are to reduce the size of the search space by compressing a group of hypotheses into one compressed hypothesis. In fact, the hypothesis compression can be realized by combining receiver replica signals with multiple hypotheses. Because a compressed receiver replica $x_c\left[n\right]$ is made up of multiple hypotheses, when the compressed hypothesis is detected (in the coarse acquisition), there should be the second process (i.e., fine acquisition) to identify the correct hypothesis contained in $x_c\left[n\right]$. There are two types of hypothesis compression techniques: the folding technique (Li, Cui, Lu, & Feng, 2008) compresses hypotheses separated in the search space, and the two-dimensional compressed correlator (TDCC) technique (Kim & Kong, 2014a) compresses neighboring hypotheses in the search space into a hypothesis.

Folding Technique

The folding technique (Li, Cui, Lu, & Feng, 2008) shown in Figure 5 is useful in the acquisition of GNSS signals with long spreading code. The receiver replica signal $x\left[n\right]$ of a long spreading code of period length L_c is divided into $N_F(\gg 1)$ equal length segments that are combined to construct a short compressed code of length L_c / N_F. Note that each code of the compressed sequence $x_{c,F}\left[n\right]$ is a sum of N_F code chips at every $N_{F,x}$ chips apart (e.g., $x_{c,F}\left[n\right] = x\left[n\right] + x\left[n + N_{F,x}\right] + \ldots + x\left[n + \left(L_{lc} - N_{F,x}\right)\right]$). The compressed code is used for the receiver replica in the FFT-based technique, so that the receiver can test $N_{c,F}$ code phase hypotheses simultaneously and may reduce MAT μ_T and MAC μ_C by nearly $N_{c,F}$ times. However, since there is only one correct code phase in the group of $N_{c,F}$ code phases found

Figure 5. (Dual)-Folding technique using hybrid search and MTC.

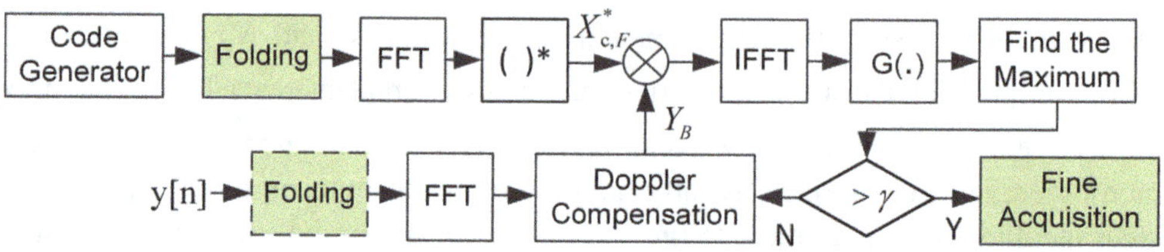

in the (coarse) acquisition, the receiver should perform another search over the $N_{c,F}$ individual hypotheses and identifies the true hypothesis.

In the dual-folding technique (Li, Cui, Lu, & Feng, 2008), both the incoming signal and the replica signal are folded to reduce the MAT and MAC more, however, the technique suffers from a larger SNR loss as the compression size increases. Note that the green blocks in Figure 5 represent the additional functions and modifications to the FFT-based technique in Figure 3, which is used in other figures throughout the chapter.

TDCC Technique

The two-dimensional compressed correlator (TDCC) compresses a group of C_c (>1) neighboring code phase hypotheses and C_f (>1) neighboring Doppler frequencies hypotheses into one compressed hypothesis for a coarse acquisition. By this method the search space is reduced by $C_c C_f$ times smaller than the original size. Similar to the folding technique, the compression by TDCC causes an increase of noise power by $C_c C_f$ times but allows the accumulation of the signal energy contained in the neighboring hypotheses around the correct hypothesis in the acquisition.

As an example, the neighboring code phase hypothesis and the neighboring Doppler frequency hypothesis at Δ_τ and Δ_f apart from the correct hypothesis, respectively, contains about a half of the correlation peak when $\Delta_\tau = \mathrm{T}_c / 2$ and $\Delta_f = 1/(2T_{co})$. It is also found (Kim & Kong, 2014b) that the phase of the signal peak in the correlation output increases linearly by $\pi / 2$ [rad] for every 500 Hz. The TDCC technique utilizes this property to coherently combine the correlation outputs from $[C_c \times C_f]$ neighboring hypotheses, which can be realized by using a compressed receiver replica signal that is constructed by coherently combining C_c code replica signals and phase compensated C_f Doppler carrier signals. The compressions in the code phase domain and Doppler frequency domain are illustrated with the FFT-based technique employing TDCC in Figure 6, where the frequency domain processing

Figure 6. FFT-based TDCC using hybrid search and MTC.

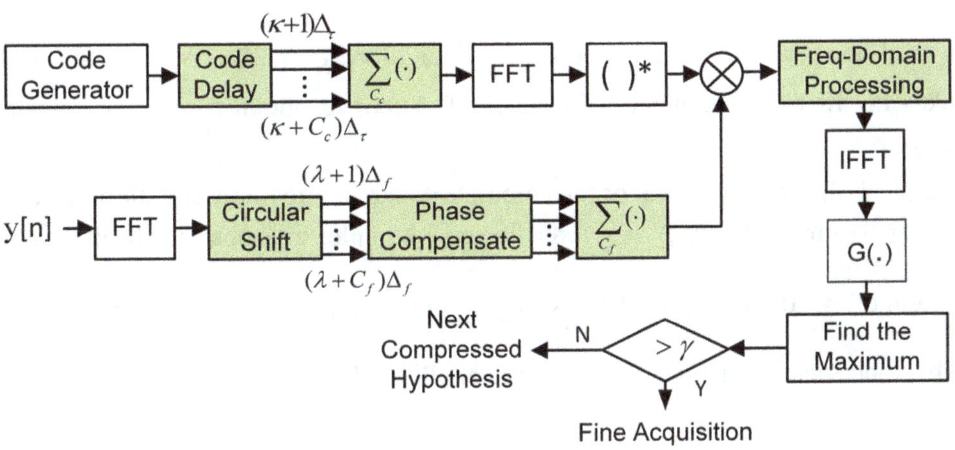

(Freq-Domain Processing block) is an additional signal processing algorithm to further reduce computational complexity. In the fine acquisition step, the TDCC technique employs $C_c C_f$ parallel correlators to perform hypothesis testing for each hypothesis contributed to the detected compressed hypothesis, and the hypothesis with the maximum of the $C_c C_f$ correlation outputs is identified as the correct hypothesis. The TDCC technique can achieve about C_f times lower MAT μ_T and MAC μ_C for strong incoming signals. However, in general, there is an increasing SNR loss with respect to the size of $C_c C_f$.

Fast Computation Algorithms for FFT-Based Technique

Recent fast computation algorithms makes use of the repetition of the GNSS spreading codes to reduce the computational cost in the FFT-based technique by employing additional signal processing techniques.

FFT-Based Partial Correlation Technique

When the spreading code period is too long, it becomes computationally expensive to use the FFT-based technique. In this case, the FFT-based partial correlation that obtains partial correlation outputs for signal segments can be used to reduce the computational cost. In (Yang, 2005; Psiaki, 2004), the segmented FFT computation algorithm is devised for the FFT-based long coherent correlation of the GPS L2-CL signals by the following four steps (Yang, 2005). In the following, it is assumed that the incoming signal $y(t)$ and receiver replica spreading code $x(t)$ are sampled and collected over a period $T_P\left(=T_t\right)$ and then divided into N_i segments, where each segment ($k=1,2,\ldots,N_i$) has N_{co} samples over T_{co}, i.e., $T_t = N_i T_{co}$. The first two steps are FFT-based correlation between $y[n]$ and $x[n]$ with zero-padding. Let y_k and x_k be the k-th segment of incoming signal samples and spreading code samples, respectively.

Step 1: In the first step, y_k is zero-padded and FFT'd (i.e., FFT applied) to obtain Y_k for all k.

Step 2: In the second step, pairs of consecutive x_k (i.e., $[x_1, x_2]$, $[x_2, x_3]$, …, $[x_{N_i-1}, x_{N_i}]$, $[x_{N_i}, x_1]$) are FFT'd to obtain X_{rl} for $r,l \in \{1,2,\ldots,N_i\}$.

Step 3: The third step performs the IFFT of $Y_k\left(X_{rl}\right)^*$ for all $r,l \in \{1,2,\ldots,N_i\}$ and $k \in \{1,2,\ldots,2N_i\}$, and takes the first N_{co} elements of each IFFT output to obtain the partial correlation output $R_{k,r} = y_k \circledast x_r$.

Step 4: Since the l-th partial segment of the complete coherent circular correlation output between the $N_i N_{co}$ signal samples y and $N_i N_{co}$ spreading code samples x is obtained from the point-by-point summation of N_i IFFT outputs, i.e., $\sum_{k=1}^{N_i} R_{l+k-1,k}$, the complete coherent integration result can be found by concatenating the partial segments for all $l=1,2,\ldots,N_i$ in the fourth step.

Note that the third and fourth steps can be repeated for circular shifted Y_k, since the circular shift of Y_k by one-sample is a Doppler frequency shift of $\Delta_f = 1/\left(2T_{co}\right)$ in the incoming signal. However, the circular shift in the unit of Δ_f is only a coarse search. To perform the fine search in (Psiaki, 2004), a carrier signal of frequency δf is multiplied to the incoming signal and then the technique performs the complete four step algorithm with the slight frequency compensated signal. In (Psiaki, 2004), multiple δf values are selected uniformly in $\left[-1/\left(4T_{co}\right), 1/\left(4T_{co}\right)\right]$ for each coarse search step. In practice, there can be N_i fine search hypotheses and $22 f_{\max}^{D} T_{co}$ ($\cong 400$ for GPS L2-CL) coarse search hypotheses. As a result, the overall computational complexity has an order of $O\left(f_{\max}^{D} T_{co} N_i N_t \log_2 N_{co}\right)$.

SDHT-Based Fast Acquisition Technique

The SDHT (Synthesized Doppler-Frequency Hypothesis Testing) technique in Figure 7 exploits the idea that a long coherent integration result over multiple code periods can be obtained from multiple coherent integration outputs over a code period. In the SDHT technique (Kong, 2015), when the coherent correlation length is short enough (e.g., N_1 for 1ms), the correlation output (of all code phases) for a Doppler frequency hypothesis (e.g., $\left(\lambda + \frac{1}{2}\right)\Delta f$) can be directly estimated from the correlation output for a neighboring Doppler frequency hypothesis (e.g., $\lambda \Delta f$ or $\left(\lambda + 1\right)\Delta f$) using a simple phase compensation. The SDHT technique can be useful, when a receiver tries to double the coherent integration length (i.e., $N_i = 2$) for higher sensitivity.

For example, when a receiver obtains N_1 signal samples and performs a coherent correlation with $T_{co} = T_1$ (=1msec) for the Doppler frequencies at every 500Hz from -5kHz to 5kHz (for example, -5000Hz, -4500Hz, -4000Hz, … 5000Hz) but no signal peak is found, the receiver may need to double the coherent dwell time to $2T_1$. In such a case, a receiver employing the SDHT technique utilizes the next N_1 samples to perform an additional coherent search with $T_{co} = T_1$ for the same tested Doppler

Figure 7. SDHT-based fast acquisition technique. The green blocks are required algorithms to the FFT-based technique in Figure 3

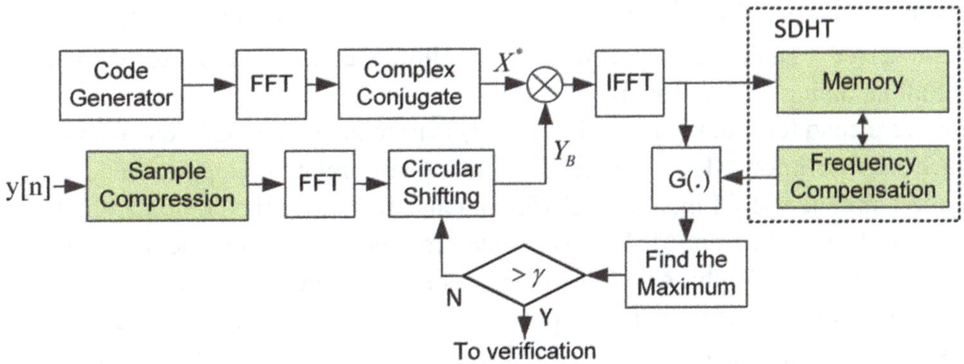

frequencies and coherently adds the results to the previous correlation results to find the correlation output for $N_i = 2$ at the same Doppler frequencies. When no signal is found, the receiver employs the SDHT technique to obtain coherent correlation outputs for Doppler frequencies (for example, -4750Hz, -4250Hz, …, 4750Hz) in between the above tested Doppler frequencies. As a result, the receiver finds a coherent correlation output for an overall coherent interval $2T_1$ at Doppler frequency hypotheses at -5000Hz, -4750Hz, -4500Hz, …, 4750Hz, and 5000Hz. This process can be repeated for N_i =4,8,…, until the receiver acquires the signal.

It should be noted that since the SDHT technique does not assume a data-free (wiped-off) incoming signal and finds consecutive correlation outputs $R\left[\eta,:,:\right]$ for $\eta = 1,2,\ldots$, it may be possible to apply a data bit search procedure at the end of the SDHT technique.

Multi-Satellite Acquisition Technique

In contrast to the LOS GNSS signals in outdoor environments, satellite signals in the GNSS-challenged environments are arriving at the receiver with low C/N_0 levels. In the multi-satellite (MS) acquisition technique (Progri, Bromberg, & Michalson, 2004; Kong, 2014), a receiver searches multiple satellite signals to acquire the signals simultaneously.

In (Progri, Bromberg, & Michalson, 2004), an iterative MS maximum likelihood (MSML) acquisition technique is introduced for a standalone GPS receiver, where the technique tests each code phase and Doppler frequency hypothesis with a fine resolution and combines the ACF outputs of all GPS satellites in view coherently to maximize the SNR of the combined ACF outputs. The iterative MSML technique shows a significant SNR improvement in detecting weak GPS signals in the presence of strong GPS signals, however, the iterative algebraic algorithm is computationally exhaustive. In (Kong, 2014), a low computational (and) coherent MSML acquisition technique for A-GPS receivers in asynchronous networks is introduced. Since the coherent MSML estimation problem is to find the solution Ψ to

$$\arg\max_{\Psi} \exp\left(\frac{-\left[y(t)-\sum_{s=1}^{S} x_s(t)\right]^2}{N_{0,all}}\right) = \arg\max_{\Psi}\left(2\int_0^{T_{co}} y(t)\sum_{s=1}^{S} x_s(t)dt - \int_0^{T_{co}}\left[\sum_{s=1}^{S} x_s(t)\right]^2 dt\right), \tag{12}$$

where $(.)_s$ represents the s-th GNSS satellite among S visible satellites for $1 \le s \le S$, $N_{0,all}$ is the noise power in the received signal, and Ψ includes the amplitude, code phase, Doppler frequency, and carrier phase of all incoming GNSS signals.

Note that the second term in the right side of (12) is to reduce the interference between the satellite signals, which requires knowledge of the relative code phase, relative Doppler frequency, relative carrier phase, and amplitude of all incoming GNSS signals. To reduce the computational complexity for the problem in (12), the coherent MSML technique employs three component algorithms; a reduced search algorithm for the code phase and Doppler frequency within the cell coverage area, a low computational algebraic algorithm to find the complex amplitude (i.e., amplitude and carrier phase) of each GNSS signals from a linear equation of $2S$ variables, and an efficient and effective approximation of the interference term in (12). The reduced search algorithm narrows the 2D hypothesis search space down

Figure 8. Coherent MSML acquisition technique

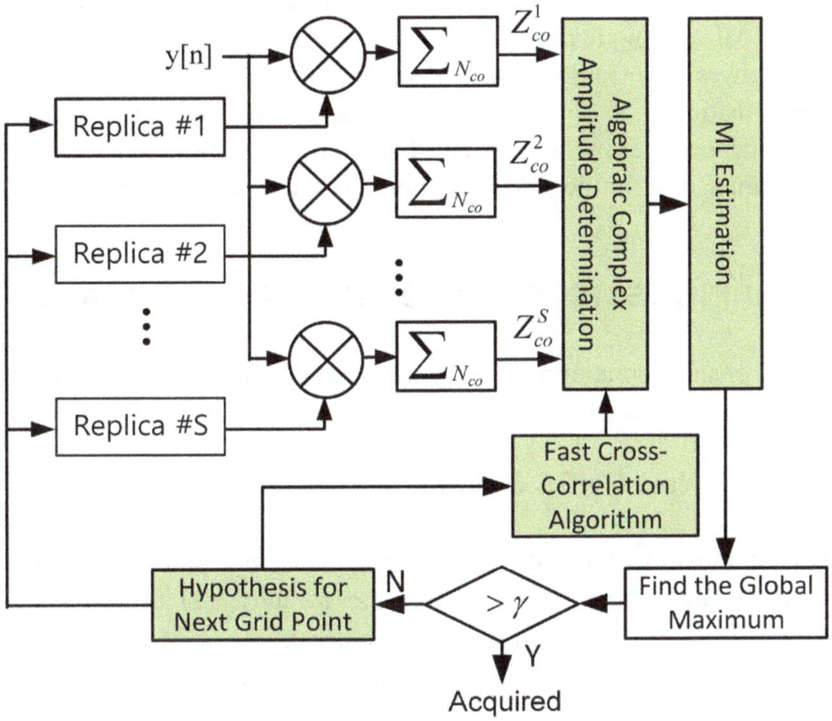

to a small number of hypotheses that correspond to the grid points of the (base station's) cell coverage area, and, therefore, the code phase and Doppler frequency search is actually performed in a small 2D (or 3D) position domain.

Usage and Comparison of the Acquisition Techniques

In the application perspective, the sample compression techniques are only for the time-domain signal processing, while the hypothesis compression techniques are for both the time-domain and frequency-domain signal processing. And there is a difference between the folding technique and the TDCC technique in utilizing the techniques: the folding technique is suitable when the code length of the target GNSS signal is much larger than the coherent integration (i.e., $T_{co} \ll T_p$), while the TDCC technique is useful when the coherent correlation interval is the same to the code length (i.e., $T_{co} = T_p$).

On the other hand, the fast computation algorithms are only for the FFT-based technique and useful when T_{co} has to be very large and the improvement of SNR in the post-integration output needs to be maximized. This is different from the hypothesis compression techniques, where SNR degradation in the post-integration output is unavoidable. However, the fast computation algorithms have different applications. The FFT-based partial correlation technique is useful for GNSS signals with a very long code period T_p, while the SDHT-based fast acquisition technique is useful when there are multiple code periods of the target GNSS signal within the intended coherent correlation interval T_{co}.

Unlike the other techniques, the MSML technique can acquire a number of GNSS signals simultaneously, but the computational cost can be much larger than other techniques introduced in this chapter. This is because the MSML technique performs correlations for every GNSS signals and compensates the cross-correlation between the GNSS signals. Therefore, the MSML technique may be useful when there is additional information, such as the range of code phases and Doppler frequencies of the incoming GNSS signals, available to the GNSS receiver. This can be possible in the assisted GNSS technique, however, the computational cost for the MSML technique may be still very high.

CHANNEL COMBINING TECHNIQUES FOR NEW GNSS SIGNALS

This section introduces channel combining (or in other words, joint channel acquisition) techniques for new GNSS signals with pilot channels.

Channel Structure of New GNSS Signals

Figure 9 shows a diagram of an acquisition function with channel combining for QPSK (Quadrature Phase Shift Keying) modulated GNSS signals. The channel combining algorithms can be integrated into the integration function $G(\cdot)$ in Figure 1 as shown in Figure 9.

In general, an incoming GNSS signal with a pilot channel can be expressed as

Figure 9. A diagram for the correlation, combining, and integration function to test an individual hypothesis at [κ , λ]. The dotted lines are for the pilot channel, the solid lines are for the data channel, $(\cdot)_D$, $(\cdot)_P$, $(\cdot)^I$, and $(\cdot)^Q$ represent the data channel, pilot channel, real and imaginary components, respectively.

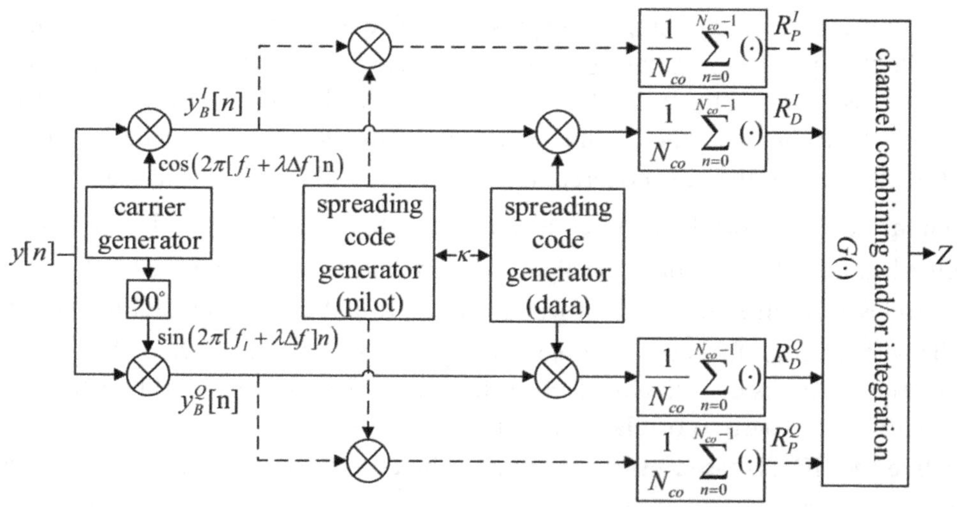

$$y(t) = \begin{cases} A_D D(t-\tau) S_D(t-\tau) SC_D(t-\tau) \cos(2\pi(f_I + f_D)t + \theta) & \textit{for QPSK modulation} \; (13a) \\ + A_P P_P(t-\tau) S_P(t-\tau) SC_P(t-\tau) \sin(2\pi(f_I + f_D)t + \theta) + v(t), & \\ & \\ \left[A_D D(t-\tau) P_D(t-\tau) SC_D(t-\tau) + A_P P_P(t-\tau) SC_P(t-\tau)\right] & \textit{for in-phase signals} \; (13b) \\ \times \cos(2\pi(f_I + f_D)t + \theta) + v(t), & \end{cases}$$

where $(\cdot)_D$ and $(\cdot)_P$ are for the data and pilot channels, respectively, $A_{(\cdot)}$ represents the amplitude, $P(t)$ and $S(t)$ are the binary phase modulated primary and secondary spreading codes, respectively, and $SC(t)$ is for the subcarrier signal. In addition, the length and the chip rate of $P(t)$ are denoted as L_c and $R_c\left(=1/T_c\right)$, respectively. Note that (13a) is for GPS L5, Galileo E5a and E5b, where pilot and data channels are modulated by QPSK (quadrature phase shift keying), and (13b) is for GPS L1C, GPS L2C, Galileo E1 OS (open service), and BeiDou B1, where signals are separated by different spreading codes or time division multiplexing. The channel parameters of some of the GNSS signals belong to (1), (13a), and (13b) are summarized in Table 1 (Foucras, Ngayap, Bacard, & Ekambi, 2016), where the unspecified units of R_b, R_s (encoded symbol rate), L_c, R_c are bps, sps, chips, MHz, respectively, and CBOC($\pm$) $=\sqrt{10/11}\text{sinBOC}(1,1) \pm \sqrt{1/11}\text{sinBOC}(1,1)$, and sinBOC(m,1)= sign[sin($2\pi m R_c t$)].

Many of the new GNSS signals have data-free (pilot) and data channels which are separated as expressed in (13a) using a 90° phase difference, orthogonal codes (13b), or time division multiplexing (TDM) (13b). For example, GPS L5 and Galileo E5a and E5b adopt QPSK modulation, GPS L1C, Galileo E1 OS, and BeiDou B1 utilize orthogonal codes, and GPS L2C has the TDM scheme for the channel separation.

When the incoming GNSS signal has a composite channel structure, the acquisition function may utilize the integration function $G(\cdot)$ to combine the signal power in the composite channel (i.e., channel combining) to increase the acquisition sensitivity. However, the channel combining increases the acquisition complexity and the benefit from the channel combining may be small when most of the signal power is in the pilot channel. In addition, when most of the signal power is carried in the low frequency component, e.g., sinBOC(1,1), and the rest is in the high frequency component, e.g., sinBOC(6,1), combining the two frequency components may not be beneficial to improve the sensitivity, considering the increased acquisition complexity. In the following, we introduce channel combining techniques for new GNSS signals in (13a) and (13b).

Channel Combining Techniques for QPSK-Modulated GNSS Signals

Some of the new GNSS signals, such as GPS L5 and Galileo E5a and E5b have the QPSK (quadrature phase shift keying) modulation for pilot and data channels of equal power. Therefore, it can be useful to combine the signal powers in the two channels using the integration function $G(\cdot)$ for joint acquisition. There are a few channel combining techniques for pilot and data channels separated by 90° phase difference; non-coherent, semi-coherent, and differentially coherent combining techniques (Borio, O'Driscoll, & Lachapelle, 2010).

Table 1. Channel Parameters of some GNSS signals belong to (1), (13a), and (13b)

	A_P / A_D	$D(t)$	$P_P(t)$	$P_D(t)$	$S_P(t)$	$S_D(t)$	$SC_P(t)$	$SC_D(t)$	
GPS L1 C/A	-	R_b =50	-	L_c =1023 R_c =1.023	-	-	-	-	(1)
GPS L2C	CM and CL codes are chip-by-chip multiplexed	R_s =50	L_c =767,250 R_c =0.5115	L_c =10230 R_c =0.5115	-	-	-	-	(13b)
GPS L5	1	R_s =50	L_c =10230 R_c =10.23	L_c =10230 R_c =10.23	L_c =20 R_c =0.001	L_c =10 R_c =0.001	-	-	(13a)
Galileo E1 OS	1	R_s =250	L_c =4092 R_c =1.023	L_c =4092 R_c =1.023	L_c =25 R_c =250Hz	-	CBOC(+)	CBOC(-)	(13b)
Galileo E5a	1	R_s =50	L_c =10230 R_c =10.23	L_c =10230 R_c =10.23	L_c =100 R_c =0.001	L_c =20 R_c =0.001	-	-	(13a)

Non-Coherent Channel Combining

Non-coherent channel combining simply adds the magnitude (or magnitude squared) of the separate correlation outputs of the pilot and data channels (Borio, O'Driscoll, & Lachapelle, 2010), so that the detection variable becomes

$$Z_{ncc}[\kappa,\lambda] = \sum_{\eta=0}^{N_i-1} \left\{ \left| R_D[\eta,\kappa,\lambda] \right|^2 + \left| R_P[\eta,\kappa,\lambda] \right|^2 \right\}, \tag{14}$$

Similar to non-coherent integration, non-coherent channel combining technique makes a relatively small SNR increase comparing to the coherent combining technique, and suffers from the squaring loss (Van Diggelen, 2009).

Semi-Coherent Channel Combining

Since the relative phase between the data channel and the pilot channel is either +90° or -90° due to the unknown data and the secondary code modulating the channels, an incoming GNSS signal can have one of the two combinations: $R_D + jR_P$ and $R_D - jR_P$, where $R_D (= R_D^I + jR_D^Q)$ and $R_P (= R_P^I + jR_P^Q)$ are the correlation outputs of data and pilot channels, respectively, and $(\cdot)^I$ and $(\cdot)^Q$ represent the in-phase and quadrature-phase components, respectively (see Figure 9). Therefore, the detection variable of the semi-coherent channel combining can be expressed as

$$Z_{scc}[\kappa,\lambda] = \sum_{\eta=0}^{N_i-1} \max\left\{\left|R_+[\eta,\kappa,\lambda]\right|^2, \left|R_-[\eta,\kappa,\lambda]\right|^2\right\}, \tag{15}$$

where $R_\pm\left[\eta,\kappa,\lambda\right] = R_D\left[\eta,\kappa,\lambda\right] \pm jR_P\left[\eta,\kappa,\lambda\right]$. Therefore, in the semi-coherent combining, the integration function $G(\cdot)$ selects the combination producing larger magnitude (or magnitude squared) between the two possible combinations. In (Borio, O'Driscoll, & Lachapelle, 2010), semi-coherently combined and non-coherently integrated detection variable produced by $G(\cdot)$ is analyzed.

Differentially Coherent Channel Combining

Denoting the unknown phase of the data and pilot channels as φ and $\varphi \pm 90°$, respectively, multiplying the data channel correlation output R_D with the complex conjugate of the pilot channel correlation output R_P^* results in imaginary components (i.e., the resulting phase becomes $\mp 90°$). And the magnitude of the result can be accumulated to build the detection variable of the differentially coherent channel combining technique as

$$Z_{dcc}[\kappa,\lambda] = \sum_{\eta=0}^{N_i-1} \left|\mathrm{Im}\{R_D[\eta,\kappa,\lambda]R_P^*[\eta,\kappa,\lambda]\right|, \tag{16}$$

Therefore, in (Borio, O'Driscoll, & Lachapelle, 2010), the detection variable with differentially coherent combining is obtained from the magnitude of the imaginary components of the result, $\left|Im\left\{R_D R_P^*\right\}\right|$, where $Im\{\cdot\}$ is a function taking the imaginary part.

Channel Combining Technique for In-Phase GNSS Signals

GPS L1C, Galileo E1 OS (open service), and the candidate BeiDou B1 signals have pilot and data channels that are separated by spreading codes, but GPS L1C and Galileo E1 OS and most of the candidate BeiDou B1 signals are transmitted in a single in-phase channel. In fact, the candidate BeiDou B1 pilot channel has a small sinBOC(6,1) modulated high frequency signal component in the quadrature phase.

However, in (Foucras, Ngayap, Bacard, & Ekambi, 2016), the advantage of combining the two channels to improve the acquisition sensitivity is found to be small when the signal power is not evenly distributed between the two channels. In addition, combining the two channels increases the receiver complexity considerably. For example, the GPS L1C and BeiDou B1 have 75% of their signal power in the pilot channel, and remaining 25% in the data channel. The acquisition performance, in terms of detection probability in the individual hypothesis test, P_d, when only the pilot channel is used is slightly lower than that when the both channels are used. In addition, since the pilot channels of GPS L1C and BeiDou B1 have about 88% of the total signal power in the sinBOC(1,1) and the remaining 12% in sinBOC(6,1), a receiver may discard the higher frequency component (i.e., sinBOC(6,1)) and process the signal component in sinBOC(1,1) at the cost of a small sensitivity loss; 1.1dB for GPS L1C and 0.56dB for BeiDou B1 signals.

GPS L2C signal has a time division multiplexing (TDM) of data modulated CM (civil moderate) code and CL (civil long) code of T_p=20ms and T_p=1.5s, respectively. Since the two channels are separated in time and the CL code acquisition requires a high acquisition complexity, some GPS L2C acquisition techniques leverage CM code acquisition to acquire the CL code (Kim & Kong, 2015); once the CM code is acquired, there are only 75 CL code phase hypotheses left. In some studies (Yang, 2005; Psiaki, 2004), partial correlation algorithms are utilized to reduce computational cost in the FFT-based technique for L2C signals (introduced in Section 4), and, in (Ta, Qaisar, Dempster, & Dovis, 2012), MGDC is used for a partial acquisition of L2C signals.

In case of the Galileo E1 OS, E5a, and E5b and GPS L5 signals, the pilot and data channels have equal amount of signal power, which makes it effective to combine both channels for higher acquisition sensitivity (Ta, Dovis, Margaria, & Presti, 2010; Margaria, Dovis, & Mulassano, 2008; Bastide, Julien, Macabiau, & Roturier, 2002; Hegarty, 2006; Pini, Viviani, & Lo Presti, 2010). In the case of the Galileo E1 OS signal, however, both channels have sinBOC(1,1) and sinBOC(6,1) components carrying about 91% and 9% of the total signal power, respectively. Therefore, it may be useful and efficient for a receiver acquiring Galileo E1 OS signal to utilize both channels with the sinBOC(6,1) component discarded in order to reduce the acquisition complexity.

ASSISTED ACQUISITION TECHNIQUE

To enhance acquisition sensitivity and to reduce the TTFF (time-to-first-fix: time to make a first position fix in a cold start) of GNSS receivers integrated into the mobile phones and connected to the cellular networks, the GNSS receivers may exploit GNSS satellite signal observations made by the network. As shown in Figure 10, in the Assisted GNSS (A-GNSS) systems (Monnerat, Couty, Vincent, Huez, & Chatre, 2004), a cellular network, equipped with an assistance server and a reference GNSS station, obtains and delivers GNSS signal observations to assist with two major tasks of the GNSS receivers connected through mobile phones: acquisition and positioning.

The acquisition assistance sent from the assistance server supplies the satellite ID, expected Doppler frequency and Doppler rate, Doppler uncertainty, code phase, integer code phase (the number of code periods from the last bit boundary), code phase search window, GPS bit number, azimuth, and elevation (Harper, 2010) to the connected GNSS receiver (i.e., the A-GNSS receiver). These are the essential information elements (IE) needed to indicate (with uncertainties) the Doppler frequency and the prompt

Figure 10. Overall view of the Assisted GNSS system.

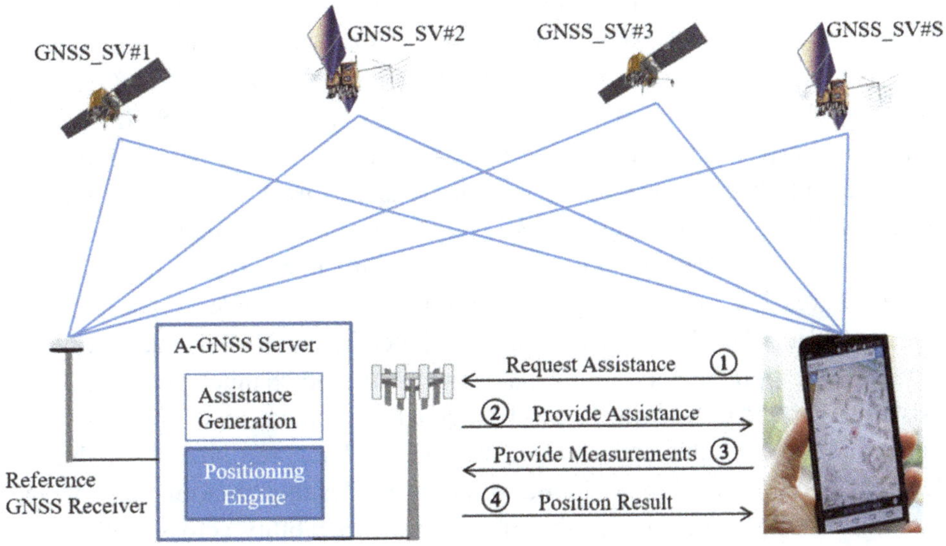

code phase of the incoming GNSS signal expected to be observed at the A-GNSS receiver. These information makes it possible for the acquisition function in the A-GNSS receiver to narrow the whole search space down to a small (and local) space that contains a much smaller number of hypotheses than N_H (the total number), which improves the acquisition performance in terms of speed and complexity. Other assistance data supplied to the A-GNSS receiver includes the reference time, reference location (usually, the base station location), differential corrections, navigation model (i.e., satellite ephemeris and clock corrections), ionospheric model, UTC (coordinated universal time) model, and real-time integrity (Harper, 2010).

In practice, there are two operational modes of the A-GNSS technique, the Mobile Station (MS)-assisted and the MS-based. In the MS-assisted mode, an A-GNSS receiver acquires and measures incoming GNSS signals and the measurements are sent to the assistance server for positioning. Whereas, in the MS-based mode, the A-GNSS receiver not only acquires and measures incoming GNSS signals but also calculates its position (TR45, 1999, October). The acquisition assistance is only delivered in the MS-assisted mode. In the MS-based mode, a mobile phone can quickly download the other assistance data through the cellular traffic channel in advance, so that the A-GNSS receiver can save multiple seconds of the time required to download the slow data from GNSS satellites directly. Then the A-GNSS receiver exploits the downloaded data to define the range of code phase and that of Doppler frequency, within which the incoming GNSS signals can be detected (Van Diggelen, 2009).

While the principle of A-GNSS has been standardized accordingly to the generations of cellular networks, the benefits of A-GNSS actually reply on the clock and frequency accuracy of the downlink pilot channel. In the cellular networks synchronized to GPS, the downlink pilot signal, for example, the pilot channel of the 2G CDMA (2nd Generation Code Division Multiple Access) network and cell-specific reference signal (CRS) of LTE (long term evolution) network in TDD (time division duplexing) mode, can be used as a fine frequency reference to stabilize the mobile phone oscillator and as an accurate time reference. In this case, an A-GNSS receiver defining a smaller search space based on the assistance only needs to consider the Doppler frequency uncertainties due to the user motion in addition to the Doppler uncertainty sent by the assistance server. Note that the timing uncertainty due to the downlink signal propagation from the base station to the mobile phones at the cell edge is taken into account when the code phase uncertainty is generated by the assistance server. On the other hand, in asynchronous cellular networks (to GPS) such as the WCDMA (wideband-CDMA) and LTE systems in FDD (frequency division duplexing) mode (3GPP TS 45.010 v8.0.0, 2008; 3GPP TS 25.101 v8.0.0, 2008), the Doppler frequency uncertainty in the search space needs to be large enough to include the oscillator frequency error caused by the error in the downlink carrier frequency and the frequency tracking function in mobile phones.

The recent A-GNSS standard developed by 3GPP (3rd Generation Partnership Project) supports new features such as multi-GNSS, differential corrections, real-time integrity, satellite-based augmentation systems, ephemeris extension, and carrier phase measurements (Fernandez-Prades, Closas, Vila'Valls, Naharro, & Sanjose, 2012). The A-GNSS standard developed by the 3GPP is the control plane A-GNSS standard that defines the signaling and protocols of the controlled channels for the cellular network to deliver the assistance data and positioning results between the assistance server and mobile phones, which makes the connection more reliable and robust against network congestion and suitable for emergency use. In parallel to the 3GPP, OMA (Open Mobile Alliance) has developed a user plane A-GNSS standard, called SUPL (secure user plane location), that insures interoperability across devices, operators, and countries. For example, an A-GNSS receiver can have a secure TCP/IP connection to the SUPL enabled assistance server and bypass traditional cellular network elements (Van Diggelen, 2009).

HIGH SENSITIVITY GNSS RECEIVER TECHNIQUES

To achieve a successful deep indoor positioning, a GNSS receiver needs to be able to detect incoming GNSS signals through concrete walls. This requires a very high sensitivity, to acquire extremely weak signals, which can be much less than 10dB-Hz. To this end, the coherent dwell time T_{co} should be as large as possible. The conventional GNSS acquisition function using $T_{co} = T_1$ and $f_s = 2R_c$ to detect outdoor LOS signals of C/N_0=40dB-Hz or higher has a processing gain (PG) of 43dB, whereas by comparison an indoor GNSS receiver trying to detect GNSS signals of C/N_0=10dB-Hz or less should have more than 73dB PG. In practice, an indoor GNSS receiver must have a PG of 80dB or more for the required sensitivity, so, T_{co} should be a second or more and $f_s \gg 2R_c$. In principle, there are at least two possible approaches to achieve a very high sensitivity to detect weak GNSS signals. One utilizes sensor-fusion to support the acquisition function, and the other exploits recent sensitivity improving signals processing techniques.

Multisensor-Based Indoor Positioning

DINGPOS is a multisensor-based indoor positioning platform (Vecchione, et al., 2010;Pany, et al., 2010) utilizing an Extended Kalman Filter (EKF) to integrate GNSS (GPS L1 and L5 and Galileo E1 and E5a), a micro-electro-mechanical-system (MEMS) subsystem comprised of an inertial measurement unit (IMU), a magnetometer, a barometer, and WiFi for a proximity location. The DINGPOS platform uses a high-performance Oven Controlled Crystal Oscillator (OCXO) with 25ppb temperature stability, and receives assistance information (from an external assistance server) that includes the list of visible satellites, their Doppler frequencies and ephemeris, and the position of the server. Utilizing the assistance and OCXO, the DINGPOS platform may only need to search the Doppler frequency of each satellite within the range of -500Hz to 500Hz with a 25Hz search step size. The acquisition flow of the DINGPOS platform is shown in Figure 11. For an input sample data, the platform performs the Double-FFT (Vecchione, et al., 2010) and stores the correlation output in a 3D matrix whose dimensions represent the code phase, Doppler frequency, and bit transition hypothesis. The Double-FFT method allows a joint FFT-based correlation and a fine frequency search. Multiple 3D matrices are non-coherently accumulated to improve the acquisition sensitivity. In the next step, the noise floor in correlation outputs for different Doppler frequencies is normalized before the signal peak detection to avoid false alarms at incorrect Doppler frequencies. When the maximum peak above the detection threshold (i.e., MTC) is found, the DINGPOS platform examines the found peak to check whether the peak is due to the ISI of strong satellites or not. As shown in Figure 11, when the signal peak is not detected or is due to the ISI, the DINGPOS platform starts the process again with a new set of input samples. Otherwise, the DINGPOS platform performs a refined acquisition, in which it obtains five correlation points around the ACF peak to estimate a more accurate peak point and its code phase.

In some indoor tests (Vecchione, et al., 2010), the DINGPOS platform exhibited position errors of about 7m using satellite signals of C/N_0=28dB-Hz, which takes less than 1 second with a GPU-based computing system. In another indoor test (Pany, et al., 2010), the OCXO and MEMS subsystem are calibrated in outdoor environments in advance, and during the indoor tests, the MEMS subsystem is used to estimate user dynamics to generate the μ-trajectory (over 1~2 seconds) of user motion. The μ-trajectory is then used to generate a receiver replica signal with very accurate code phase and carrier phase changes along the user motion, which enables a fine coherent correlation so that the DINGPOS platform is able to acquire GNSS signals of C/N_0 around 10dB-Hz.

Figure 11. DINGPOS signal processing flow.

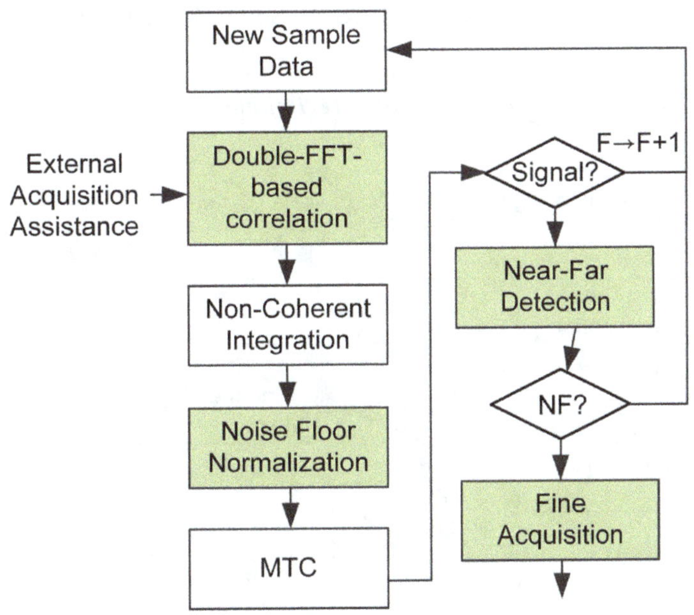

SDHT in Assisted GNSS

A SDHT-based (Kong, 2015) deep indoor A-GPS technique is developed and tested at the KAIST (Korea Advanced Institute of Science and Technology, Rep. of Korea) campus. Figure 12 depicts the signal processing flow of the technique. An external (A-GPS) server equipped with a reference GPS station on the rooftop of the tallest building on the campus broadcasts the list of visible satellites, the Doppler frequency, code phase, ephemeris, and pseudo-range rate of each listed satellite, and the position of the reference GPS station. A C/A signal centered at f_I =4.092MHz is sampled at f_s =16.368MHz using a SiGe GN3S Sampler V2 (mode 1) to produce successive T_{1000} (=1 second) long input samples $y[n]$. In the beginning, a post-processing computing platform (with i7-2600 CPU and 4 Cores) obtains the $y[n]$ from the sampler, down-converts a target GPS signal to the near baseband using the Doppler frequency provided by the server, and divides the input into T_{1000} / T_1 (=1000) segments (each segment provides a PG of about 72dB). Then, the computing platform applies sample averaging to the segments so that each output segment has 4 compressed samples per a chip, and the T_{1000} / T_1 output segments are re-arranged to form a compressed matrix M_i of size $[(T_{1000} / T_1) \times 4092]$. Each row of the matrix is applied to the coherent FFT-based correlation, and each column of the resulting output matrix M_o is cross-correlated with a data bit sequence (provided by the server) represented at every T_1 seconds. When no MTC is found larger than the detection threshold, the SDHT technique is applied to each row of M_o to find the coherent correlation results for the Doppler frequency hypotheses at every 0.5Hz within [-100, +100]Hz. When the target GPS signal is detected, the computing platform performs several T_{1000} long parallel time-domain coherent correlations with the input samples $y[n]$ around the detected hypothesis for a fine acquisition. The computing platform finds the weighted least-square fit, which has the same rate of code phase change recently observed by the server, to the recent pseudo-range (PR) measurements obtained by the computing platform in the last 10 seconds in order to generate a smoothed PR estimate

Figure 12. SDHT-based A-GPS Indoor Positioning Technique

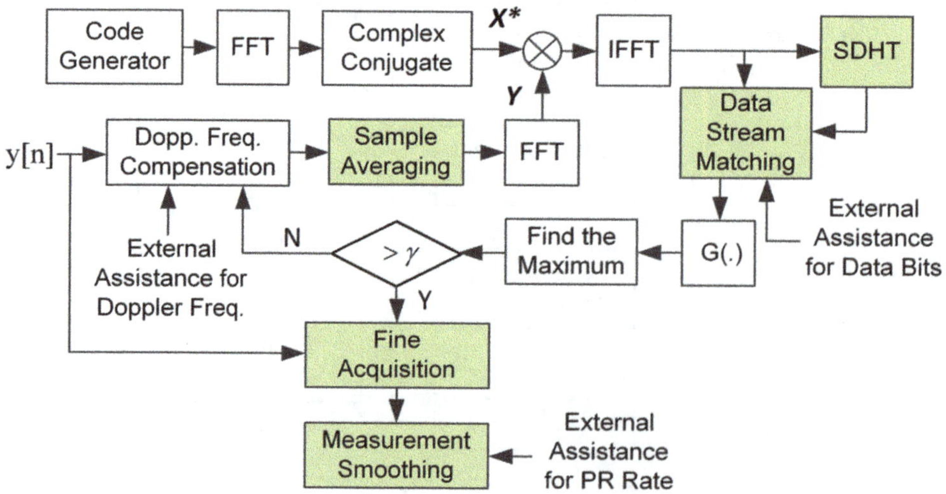

Figure 13. (a) Indoor test locations. (b) Position fixes before smoothing. (c) Position fixes after smoothing.

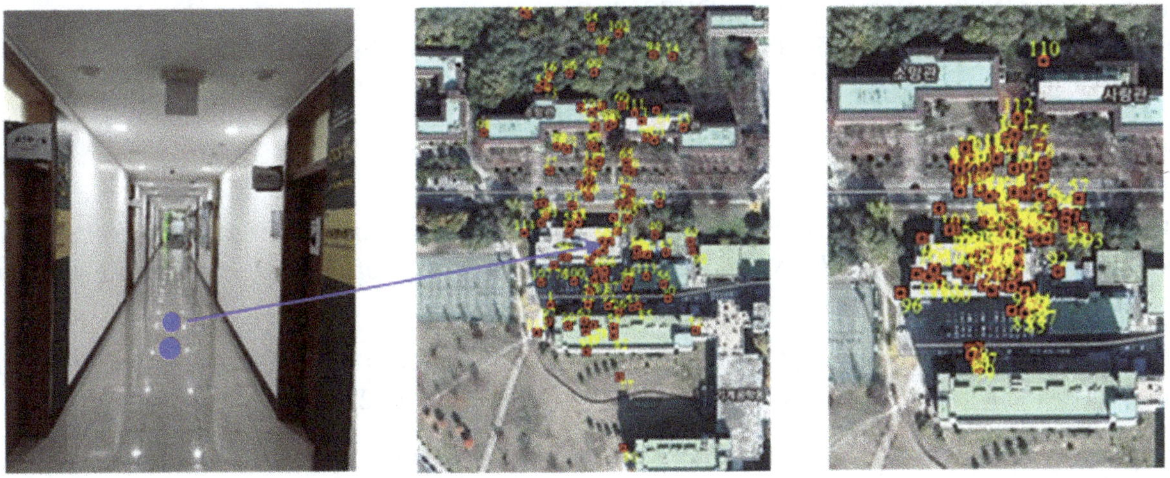

at every T_{1000}. This is effective, since the distance between the server and the receiver (i.e., the computing platform) is small enough (less than 400m).

Real indoor environment tests in a static scenario are performed inside a 3-story concrete building (N7-5) on the KAIST campus. The geo-coordinates of the building is 36°22'23.64"N, 127°21'27.70"E. The building floor has $16.8\text{m} \times 45.4$ m in size, and the building has concrete walls except for the south glass wall. A 40m long straight narrow inner corridor running in the West-East direction (Figure 13(a)) down the center of the building is surrounded by thick (23cm) concrete walls. The blue circles in Figure 13(a) indicate the two (2m apart) test locations on the east side of the 3rd floor corridor. In addition to the concrete wall obstruction, there is 7-story tall concrete building to the south, a hill to the north, a 5-story red brick building on the north hill, and a 3-story building to the east completely blocks the low elevation GPS signals, so that there is no chance of receiving good GPS signals at the test locations. The time to first fix (TTFF) of the computing platform is 22.1 seconds, and the post-correlation SNR is from 5dB to 15dB. The mean successful positioning error in 122 out of 250 positioning trials is 62.1m before the PR smoothing, and is 18.8m after the PR smoothing, as shown in Figure 13(b) and Figure 13(c), respectively. In the figures, the small red dots (with yellow number marks) indicate the position fixes.

CONCLUSION AND FUTURE DIRECTIONS

To meet the growing demand for GNSS positioning in GNSS-challenged environments, high acquisition sensitivity and fast acquisition have become the two most desired features for GNSS receivers. In the first part of this chapter, we have introduced the fundamentals of the GNSS acquisition function and some of the new acquisition techniques for new GNSS signals using pilot and data channels in QPSK or in the same phase. In the second part, we have investigated fast GNSS acquisition techniques, which achieve both low computational cost and high sensitivity, for standalone GNSS and assisted GNSS receivers. As discussed in relation to the state-of-the-art acquisition techniques for GNSS-based indoor positioning, signal processing techniques for higher sensitivity and faster acquisition are still under development with the goal of delivering reliable positioning performance. However, GNSS acquisition technologies are also evolving rapidly, and new GNSS signals will become available and effective in the near future. Accordingly, GNSS positioning in the GNSS-challenged environments is expected to become more reliable and accurate, making GNSS more useful for indoor LBS in the near future.

REFERENCES

Ávila Rodriguez, J. A., Pany, T., & Eissfeller, B. (2004). A theoretical analysis of acquisition algorithms for indoor positioning. *Proceedings of the 2nd ESA Workshop on Satellite Navigation User Equipment Technologies NAVITEC.*

Bastide, F., Julien, O., Macabiau, C., & Roturier, B. (2002). Analysis of L5/E5 Acquisition, Tracking and Data Demodulation Thresholds. *Proceedings of ION GPS*, 2196-2207.

Borio, D., Camoriano, L., & Lo Presti, L. (2008, July). Impact of GPS acquisition strategy on decision probabilities. *IEEE Transactions on Aerospace and Electronic Systems*, *44*(3), 996–1011. doi:10.1109/TAES.2008.4655359

Borio, D., O'Driscoll, C., & Lachapelle, G. (2010, January). Composite GNSS signal acquisition over multiple code periods. *IEEE Transactions on Aerospace and Electronic Systems*, *46*(1), 193–206. doi:10.1109/TAES.2010.5417156

Corazza, G. E. (1996, September). On the MAX/TC criterion for code acquisition and its application to DS-SSMA systems. *IEEE Transactions on Communications*, *44*(9), 1173–1182. doi:10.1109/26.536923

Corazza, G. E., & Pedone, R. (2007, November). Generalized and average likelihood ratio testing for post detection integration. *IEEE Transactions on Communications*, *55*(11), 2159–2171. doi:10.1109/TCOMM.2007.908531

Dovis, F., & Ta, T. H. (2012). High sensitivity techniques for GNSS signal acquisition. In Global Navigation Satellite Systems: Signal, Theory and Applications. InTech. doi:10.5772/29453

Fernandez-Prades, C., Closas, P., Vila'Valls, J., Naharro, M. F., & Sanjose, A. T. (2012). Assisted GNSS in LTE-Advanced Networks and its Application to Vector Tracking Loops. *Proceedings of ION ITM*, 1462-1476.

Foucras, M., Ngayap, U., Bacard, F., & Ekambi, B. (2016). Acquisition performance comparison of new generation of GNSS BOC-modulated signals. *Proceedings of ION GNSS*, 107-119.

Geiger, B., Vogel, C., & Soudan, M. (2012, April). Comparison between ratio detection and threshold comparison for GNSS acquisition. *IEEE Transactions on Aerospace and Electronic Systems*, *48*(2), 1772–1779. doi:10.1109/TAES.2012.6178098

3. GPP TS 25.101 v8.0.0. (2008, January). Radio transmission and reception (FDD)(Release 8).

3. GPP TS 45.010 v8.0.0. (2008, May). Radio subsystem synchronization (Release 8).

Harper, N. (2010). *Server-Side GPS and Assisted-GPS in Java.* Artech House.

Hatfield, D. N. (2002). *A report on technical and operational issues impacting the provision of wireless enhanced 911 services*. Federal Communications Commission.

Hegarty, C. J. (2006). Optimal and Near-optimal Detectors for Acquisition of the GPS L5 Signal. *Proceedings of the 2006 National Technical Meeting of The Institute of Navigation*, 717-725.

Kaplan, E. D. (2005). *Understanding GPS: Principles and Applications* (2nd ed.). Artech House.

Kim, B., & Kong, S.-H. (2014, May). Design of FFT-based TDCC for GNSS Acquisition. *IEEE Transactions on Wireless Communications*, *13*(5), 2798–2808. doi:10.1109/TWC.2014.040714.131884

Kim, B., & Kong, S.-H. (2014, May). Determination of Detection Parameters on TDCC Performance. *IEEE Transactions on Wireless Communications*, *13*(5), 2422–2431. doi:10.1109/TWC.2014.0204014.131704

Kim, B., & Kong, S.-H. (2015). Ultra-fast L2-CL code acquisition for a dual band GPS receiver. *Journal of Positioning, Navigation, and Timing*, *4*(4), 151–160. doi:10.11003/JPNT.2015.4.4.151

Kong, S.-H. (2014). Fast Multi-Satellite ML Acquisition for A-GPS. *IEEE Transactions on Wireless Communications*, *13*(9), 4935–4946. doi:10.1109/TWC.2014.2327101

Kong, S.-H. (2015, November). SDHT for Fast Detection of Weak GNSS Signals. *IEEE Journal on Selected Areas in Communications*, *33*(11), 2366–2378. doi:10.1109/JSAC.2015.2430291

Li, H., Cui, X., Lu, M., & Feng, Z. (2008, December). Dual-folding based rapid search method for long PN-code acquisition. *IEEE Transactions on Wireless Communications*, *7*(12), 5286–5296. doi:10.1109/TWC.2008.071130

Margaria, D., Dovis, F., & Mulassano, P. (2008). Galileo AltBOC signal multiresolution acquisition strategy. *IEEE Aerospace and Electronic Systems Magazine*, *23*(11), 4–10. doi:10.1109/MAES.2008.4693984

Monnerat, M., Couty, R., Vincent, N., Huez, O., & Chatre, E. (2004). The Assisted GNSS, Technology and Applications. *Proceedings of ION ITM*, 2479-2488.

Pany, T., Winkel, J., Riedl, B., Niedermeier, H., Eissfeller, B., Wrz, T., . . . Jimnez-Baos, D. (2010, December). Experimental Results from an Ultra-Tightly Coupled GPS/Galileo/WiFi/ZigBee/MEMS-IMU Indoor Navigation Test System Featuring Coherent Integration Times of Several Seconds. *Satellite Navigation Technologies and European Workshop on GNSS Signals and Signal Processing (NAVITEC), 2010 5th ESA Workshop on.* doi:10.1109/NAVITEC.2010.5708020

Parkinson, B. W., & Spilker, J. J. (1996). Global Positioning System: Theory and Applications. Aiaa.

Pini, M., Viviani, E., & Lo Presti, L. (2010). GPS L5 Signal Acquisition Exploiting Neumann-Hoffman Code Transitions. *Proceedings of the 2010 International Technical Meeting of The Institute of Navigation*, 765-772.

Progri, I. F., & Bromberg, M. C., & R., M. W. (2004). An enhanced acquisition process of a maximum likelihood GPS receiver. *Proceedings of ION NTM*, 390-398.

Psiaki, M. L. (2004). FFT-based acquisition of GPS L2 civilian CM and CL signals. *Proceedings of ION GNSS*, 646-653.

TR45. (1999, October). *Position Determination Service Standard for Dual-Mode Spread Spectrum Systems.* TIA/EIA/IS-801.

Ta, T. H., Dovis, F., Margaria, D., & Presti, L. L. (2010). Comparative study on joint data/pilot strategies for high sensitivity Galileo E1 open service signal acquisition. *IET Radar, Sonar & Navigation*, *4*(6), 764–779. doi:10.1049/iet-rsn.2009.0256

Ta, T. H., Qaisar, S. U., Dempster, A. G., & Dovis, F. (2012). Partial differential postcorrelation processing for GPS L2C signal acquisition. *IEEE Transactions on Aerospace and Electronic Systems*, *48*(2), 1287–1305. doi:10.1109/TAES.2012.6178062

Van Diggelen, F. (2009). *A-GPS: Assisted GPS, GNSS, and SBAS*. Artech House.

Vecchione, G. A., Palenzuela, D., Toledo, M., de Cceres, J. J. R., Tiley, P., Kubrak, D., . . . Lpez-Risueo, G. (2010, May). DINGPOS, a GNSS-based Multi-sensor Demonstrator for Indoor Navigation: Preliminary Results. *Position Location and Navigation Symposium (PLANS), 2010 IEEE/ION*, 113-121. doi:10.1109/PLANS.2010.5507243

Viterbi, A. J. (1995). *CDMA: Principles of Spread Spectrum Communication*. Addison-Wesley.

Ward, P. W. (1996). GPS Receiver Search Techniques. *Proceedings of Position Location and Navigation Symposium*, 604-611. doi:10.1109/PLANS.1996.509134

Yang, C. (2005). Joint acquisition of CM and CL codes for GPS L2 civil (L2C) signals. *Proceedings of the 61st Annual Meeting of The Institute of Navigation*, 553-562.

Chapter 13
GNSS Vulnerability:
A Taxonomy

Andrew G. Dempster
University of New South Wales, Australia

ABSTRACT

This chapter examines sources of global navigation satellite system (GNSS) vulnerability, identifying the broad range of topics that comes under this heading, and cites some key references in each category area. GNSS vulnerability has been a very productive area for GNSS researchers in recent years and this chapter sets out to be a comprehensive review of the different ways that the operation of GNSS can be degraded by outside influences, from the high (system) to the low (receiver component) level.

INTRODUCTION

Satellite navigation systems are vulnerable to interference of various kinds. This interference has a range of effects and there is a range of remedies available to receiver designers and users. This chapter aims to summarize and provide a taxonomy of the state of knowledge of all sources of satellite navigation vulnerability.

Global Navigation Satellite Systems (GNSS) have been implemented or are being implemented by the US (Global Positioning System – GPS), the USSR, and later Russia (Glonass), Europe (Galileo) and China (Beidou/ Compass). Regional Navigation Satellite Systems (RNSS) are being implemented by China (Beidou/ Compass has regional elements), Japan (Quasi-Zenith Satellite System – QZSS) and India (Indian Regional Navigation Satellite System – IRNSS). Additionally, there are Space-Based Augmentation System (SBAS) satellites that carry navigation payloads that "look like" satellites from other systems, GPS in particular. One thing all these systems have in common is that their received power on the surface of the earth is very weak. For instance, the civilian GPS signal, L1 C/A has a specified minimum received power on the earth's surface of -158.5 dBW which requires an equivalent isotropic radiated power (EIRP) of 26.8 dBW, or power into the antenna of 21.9W (Kaplan, 2006). In other words, about 20W of power is transmitted over 20,000 – 26,000 kilometers. These extremely weak signals can still be used for navigation, as the common in-car navigator proves to the layman. However, their low

DOI: 10.4018/978-1-5225-3528-7.ch013

levels mean that, despite being controlled by strong and stable entities (in the case of GPS for instance, the US military), they are still vulnerable.

This chapter aims to outline the types of vulnerability experienced by GNSS systems and signals.

SOURCES OF GNSS VULNERABILITY

The categories of vulnerability are shown in Figure 1, with arrows indicating a breakdown of more general categories into lower level categories. Figure 1 does not indicate the overall structure of the chapter, because the highest level of Figure 1 includes vulnerabilities to systems that use and augment GNSS. However, it is important here to acknowledge that these vulnerabilities exist, and are important to users and integrators of GNSS systems. Augmentation systems such as space-based augmentation systems (SBAS), continuously operating reference station (CORS) networks, real-time kinematic (RTK) systems, differential GPS (DGPS), and assisted GPS (AGPS) all have vulnerabilities of their own, but are not considered further here. Similarly, systems that integrate GPS can suffer where reported GPS positions can be intercepted and modified (spoofed) (Hackett, 2015). This type of vulnerability is also not considered further here.

We first consider the three segments of a GNSS system: the space segment, the control segment and the user segment. When speaking about GNSS vulnerability, most people think only of signal interference in the user segment. This will be dealt with in a later section. Here we consider the control and space segments.

Figure 1. High-level Categories of where GNSS vulnerability can occur. In other words, vulnerabilities occur at system and segment levels, as well as the better known examples at lower levels. The box at the bottom right is expanded further in Figure 2.

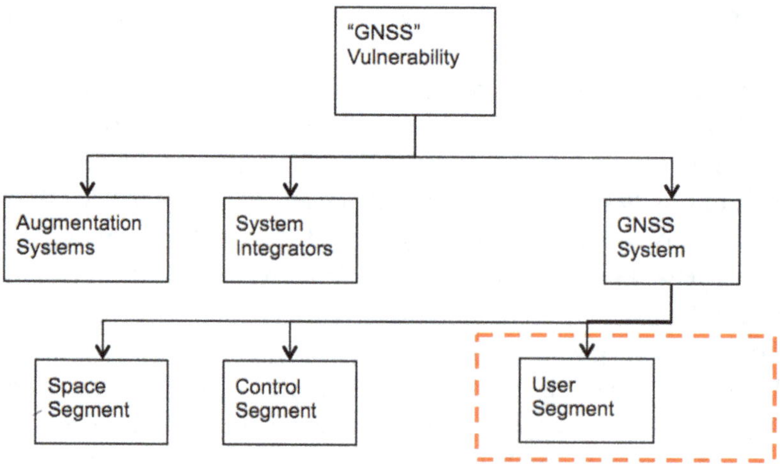

High-Level Vulnerability Categories: Space and Control Segments

Space Segment

Any GNSS consists of a constellation of about 30 satellites. Regional systems have fewer. The vulnerabilities of any given GNSS satellite are similar to those of any satellite: solar radiation, missile attack, nuclear detonation or electromagnetic pulse. The fact that GNSS satellites are in medium earth orbit (MEO) and are usually built by the military (Galileo being the exception) makes most of these risks unlikely, with few or no reports of satellites suffering more than would be expected from this type of loss. All satellites have a limited lifetime, and as a rule, the GPS satellites have outlived their design life, so this cannot really be considered a demonstrated "vulnerability". Glonass satellites, on the other hand, especially early generations, tended not to complete their design lifetime (Sarkar, 2016), which has led to problems, as discussed below.

The risks to constellations, however, have been quite publicly on show. Because of a mismatch between its launch strategy, achieved satellite lifetime, and available funding, Glonass reached a minimum of 6 satellites in 2001, despite it having reached full operating capacity in 1995 (Harvey, 2007). It was another ten years before that constellation became fully operational, i.e. up to that point, Glonass was only really useful when combined with GPS. In 2015, Galileo also had two satellites launched into incorrect orbits that were only partially corrected (Navarro-Reyes, 2015), meaning they were less useful for positioning. Constellation health is also challenged by launch failures such as the spectacular 2013 event where three Glonass satellites were lost (RSW, 2013).

The satellites themselves have had issues with GPS clocks drifting or jumping out of spec in 1997, 2001, 2006 and 2010 (Symmetricon, 2013; Curry, 2010) and transmitting "evil waveforms" in 1993 and 2009 (RAE, 2011), resulting in errors of up to 300km, and/or significant timing errors.

Control Segment

In GPS the main vulnerabilities to the control segment have occurred with uploads of poor quality data, events that happened in 1993, 2000, 2002, 2004 and 2010 (Symmetricon, 2013). The most significant control segment event occurred in April 2014, where the whole Glonass system went offline for several hours (Gibbons, 2014). Also very significant was the January 2016 problem, where the removal of a satellite from the active constellation caused a 13 μsec error in 15 GPS satellites causing problems for up to half a day (Baraniuk, 2016).

Although there have not been reports of hacking the GPS (or other GNSS for that matter) control segment, it is nevertheless protected by GPS Intrusion Reinforcement (IG, 2015).

High-Level Vulnerability Categories: User Segment Signal Interference (System and Natural)

Before being able, in a meaningful sense, to discuss the study of GNSS interference, it is important to categorise the types of interference to, and degradation of, GNSS signals and the ways in which these are studied and dealt with.

Figure 2 shows, in the broadest sense, the different types of interference that can affect satellite navigation receivers. The following sections of this chapter will concentrate on the man-made interference, but before doing that, the other high-level categories are also described here.

Intra-Satellite Interference

Literally, this is a where a satellite signal interferes with itself. In other words, the receiver struggles with a particular signal because of a separate path travelled by that same signal. GNSS users generally refer to this problem as *multipath*, where the same signal finds two or more paths to the receiver. This is a different problem to that experienced by communications receivers, which can exploit multipath. Whereas a communications receiver aims to maximize its signal to noise ratio (SNR) or minimize its bit error rate (BER) by combining signals from all paths (using channel estimation/ equalization), a navigation receiver wishes only to find the line of sight signal to measure its time of arrival, and exclude all other paths.

When despreading the code-division multiple access (CDMA) satellite signal, multipath affects the code phase measured using the code tracking loop in the receiver's *correlator*, causing errors in the region of meters. Similarly, errors also occur in the carrier phase tracking loop, a problem familiar to the communications community as *fading*. Fading is sometimes described using the terms *destructive* and *constructive interference*, a use of the term "interference" which can potentially be confused with

Figure 2. Categories of sources of interference to GNSS

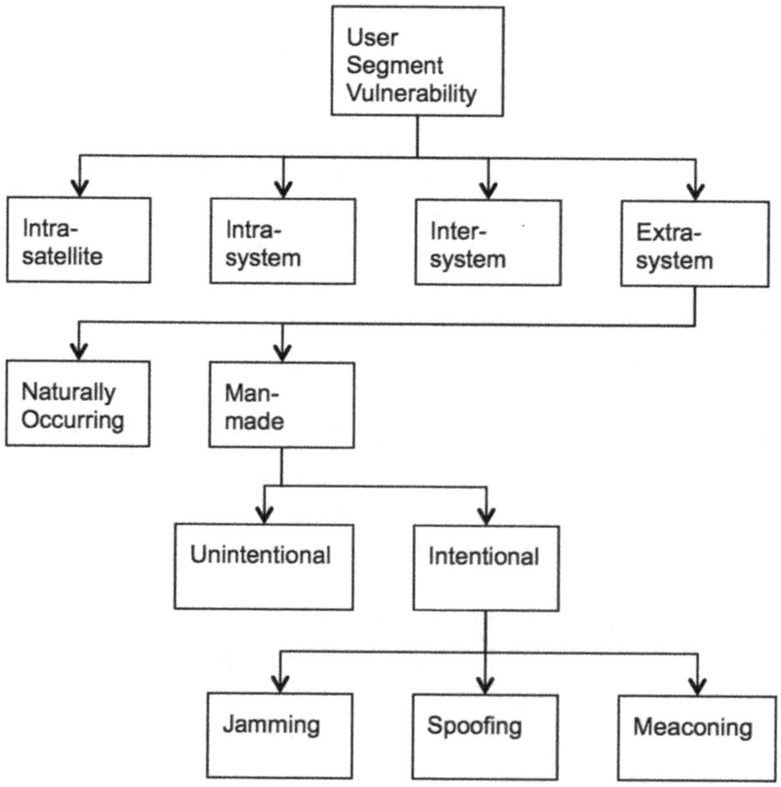

phenomena described later in this chapter. Errors introduced in the carrier phase tracking disturb a survey-grade receiver's ability to perform cm-level positioning.

Remedies for the multipath problem have been investigated for some time, and similar to the area of external interference to be developed later in this chapter, are effectively an entire area of study to themselves. They can be divided into methods internal to the receiver (typically the correlator), and methods outside of the receiver (typically related to antenna design). Of the techniques exploiting correlator redesign are the Narrow Correlator (Van Dierendonck, 1992); the Multipath Elimination Technique (MET) (Novatel, 2000; Townsend, 1994) and Pulse Aperture Correlator (Jones, 2004), both examples of double-delta correlators; strobe or gated correlators (Veitsel, 1998; Garin, 1997) which can also be used with the more complicated Galileo signals (Garin, 2005; Wu 2011), which have non-triangular correlation functions. An innovation that looks at the shape of the code rising edge rather than the correlation function itself is the Vision Correlator (Fenton, 2005). In general, these techniques deal only with single multipath and assume the line-of-sight is the strongest signal. More versatile are techniques that look at each path from the satellite separately, similar to channel estimation in communications systems. Examples are the Multipath Estimating DLL (MEDLL) (van Nee, 1994; van Nee 1992), Modified Rake DLL (MRDLL) (Laxton, 1997), and Fast Iterative Maximum Likelihood Algorithm (FIMLA) (Sahmoudi, 2008).

Multipath can be mitigated by using antennas known as "choke ring" antennas, which have been developed over the years primarily for use with carrier phase receivers, and are now quite sophisticated, for example (Leica, 2009). There are also many schemes for dealing with multipath using antenna arrays, for example (Xu, 2011).

Intra-System Interference

The case of intra-system interference arises when one satellite in a system, say GPS, interferes with another from that same system. The problem is related to the design of the satellite navigation signals and is only an issue where a weak signal is sought in the presence of a strong one. This is similar to the *near-far* problem experienced in communications systems, but not in a literal sense. All navigation satellites are at very similar distances from the receiver, at least in received power terms, so this problem occurs when the signal of interest is attenuated (e.g. by a tree, wall or roof) while the interfering signal is not (it may arrive through a window or free space). It is also known as *Multiple Access Interference* (MAI) and affects all *code-division multiple access* (CDMA) systems to a degree.

To take the example of GPS, its spreading codes are known as *Gold codes*. Each satellite is known by its code number, between 1 and 32. These codes have known auto-correlation and cross-correlation properties. One of these properties is that there is a 24dB cross-correlation margin, meaning that when using the code n in the correlation process, it will produce a peak when correlated with code n (i.e. itself, thus *auto-correlation*) that is 24dB higher than peaks produced by *cross-correlating* with code m ($m \neq n$). Under normal operating conditions, there are other factors that can reduce this margin such as satellite elevation or satellite signal Doppler shift, to as low as 14dB (Kaplan, 2006). What that means is that if a satellite signal is attenuated by more than 14dB, there is a risk that a receiver may acquire and/ or track a different satellite to the one it thinks it is working with.

There are various ways that this problem can be dealt with for GPS L1 (Glennon, 2004; Glennon, 2007a), with the work of Glennon being the most prominent. One example modifies the local code (flipping ones and zeroes) so the output of the integration in the cross-correlation goes to zero (Glennon,

2006a; Glennon, 2007b) (this method has a limit to the number of strong signals that can be mitigated). A second method uses knowledge of the strong signals to subtract them after reconstructing them - critically this needs to happen after the correlation process (Glennon, 2011; Glennon, 2006b); this does not have a limit to the number of satellites that can be mitigated.

Virtually all signals designed after GPS L1 have used codes with much better cross-correlation performance, so this problem does not arise. For GPS L5, a strong signal must be more than 10dB higher in order to cause the same problems as occur for L1 (i.e. its cross-correlation margin is 10dB higher), (van Dierendonck, 2000) and this margin is closer to 20dB for GPS L2CM, the worse of the two L2C codes (Qaisar, 2008).

Inter-System Interference

The idea that one satellite navigation system could interfere with another using the same carrier has been a topic of international negotiations for a long time. The idea that systems must be *interoperable* is one of the central responsibilities of the International Committee on Global Navigation Satellite Systems (ICG) (http://www.unoosa.org/oosa/SAP/gnss/icg.html). However, regardless of any individual satellite being interoperable with the others, a new problem is arising due to the growing numbers of systems sharing carrier frequencies. Under normal circumstances, the GNSS signals sit well below the noise, and when their CDMA signals are despread using the correlator, the *processing gain* from that process results in the signal "popping up" above the noise and an acceptable signal-to-noise ratio (SNR) is achieved. What happens in the presence of a multitude of satellites is that, although each satellite signal is well below the noise, the ensemble adds together to raise the noise floor, degrading the operation of all GNSS. This is a relatively recent problem and discussion and a summary of the situation can be found in (Gibbons, 2011). This is not a problem that has a ready remedy. If the noise floor is raised, there is little that can be done except perhaps use antennas that are more directional than the traditional "isotropic" GNSS antenna – to increase the received power of wanted satellites and attenuate unwanted ones.

Extra-System Interference: Naturally Occurring

Interference from man-made sources other than GNSS signals is really what will be the later focus of this chapter, but before doing so, there is one last type of interference that does not fit into that category: naturally occurring extra-system interference.

The main natural source of radiation that interferes with GNSS is the sun. Solar storms can upset GNSS in several ways. There are three mechanisms: bursts of electromagnetic radiation (travelling at the speed of light, these cannot be "responded to", i.e. the problem arrives at the same time the source is observed), high energy particles (which arrive hours after a solar event) and solar winds (which arrive days after an event). These mechanisms can affect the satellites, the ionosphere, through which the signals pass, or the GNSS receivers. The effects on these system elements by each of the mechanisms are summarized in Table 1. Note the "satellite" effects could be considered to be in the "space segment" category mentioned earlier.

As can be seen, the most serious effects in terms of satellite navigation systems are from particle radiation and solar winds, which can cause expansion of the ionosphere and scintillation. Good reviews can be found of the effects of scintillation on GNSS in (Kintner, 2009; Jensen, 2011) and the more general effects of space weather and how to protect against them in (AMS, 2011).

Table 1. Effects on GNSS system elements of different artifacts of solar storms

	Electromagnetic Radiation	Energetic Particles	Plasma Cloud/Geomagnetic Storms
Satellites	Minor – satellites are hardened against EM pulse	Satellites hardened but electronics could fail for short or long periods, satellite lifetime reduced	-
Ionosphere	Short burst of scintillation	Serious disturbance for periods of hours-days	Serious disturbance for periods of hours-days
Receivers	Similar to jamming for the duration of the EM pulse (10s of seconds)	Minor electronics damage	-

In the lifetime of GPS, there have been a number of serious solar storms, but none that come close to being as powerful as the worst recorded, the "Carrington Event" of 1859, which would cause around US$70billion of damage to the global economy in space assets alone with GPS becoming unusable for days if not permanently (Odenwald, 2006; Odenwald, 2008). A more recent study put the damage figure at US$1-2 trillion with a 4-10 year recovery period (Kerr, 2009). Some examples of solar storms and their effects on GPS are shown in Table 2. A more complete history of solar storms, without specific reference to GPS, can be found in (Pick, 2008). Because solar storms affect the ionosphere, which affects the operation of GNSS receivers, GPS receivers have long been exploited as instruments for observing the ionosphere (see several of the references in Table 2). These ionospheric effects can range from minor right through to long-term loss of signal tracking.

Table 2. Examples of solar storms during the active period of GPS. Example references are shown for each storm, their severity, and the region where the reference observed the storm.

Storm date(s)	Location(s)	GNSS Effects (Minor: Accuracy Affected; Moderate: Tracking Lost; Severe: Extended Periods of Loss of Function)	Reference
31 January 1993	South America	Moderate	(Pi, 1997)
17 September 1995	South America	Minor	(Pi, 1997)
10 January 1997	North America	Minor	(Pi, 1997)
26-28 August 1998	Scandinavia/ South America (global)	Moderate	(Skone, 2001)
14 July 2001(Bastille Day event)	Central America	Minor	(Araujo-Pradure, 2005)
29-31 October 2003(Halloween event)	Northern Latitudes	Severe (WAAS disabled)	(Skone, 2004) (Doherty, 2004)
20 November 2003	Northern Latitudes	Severe (WAAS disabled)	(Skone, 2004) (Doherty, 2004)
10 November 2004	China	Moderate	(Xu, 2007)
8 May 2005	China	Minor	(Li, 2008)
6 December 2006	Russia/ Equatorial latitudes	Severe	(Afraimovich, 2007)
14 February 2011 (Valentine's Day flare)	North America	Minor	(Kneupper, 2011)
12 March 2012	Russia	Minor	(Zacharov, 2016)
17-18 March 2015 (St Patrick's Day storm)	Global	Minor	(Astafyeva, 2015)

High-Level Vulnerability Categories: User Segment Signal Interference (Man-Made)

The previous sections covered vulnerabilities of GNSS that would not usually be termed "interference". That term is almost exclusively reserved for signals produced by man-made electronics.

Man-Made Interference: Unintentional Sources

Many of the famous GPS "jamming" case studies are actually unintentional. In other words, signals were being transmitted with no intention of jamming GPS. The frequencies where such interference can occur are i) the satellite navigation bands themselves, ii) nearby bands but the signals are strong enough to pass through filters in the receiver designed to eliminate such signals, or iii) frequencies that, when passed through a non-linearity, can cause harmonics (typically 3rd) or cross-products that then appear in the satellite navigation bands and appear to the receiver to be the same as the signals in i).

In-Band Unintentional Sources

Probably the best known example of unintentional jamming was when a series of malfunctioning television antennas shut down GPS operation in the area of Moss Landing Harbor near Monterey in California for a period of 37 days. In that case, the antennas were transmitting in the GPS band (Clynch, 2003). This resulted in the failure of the Monterey Aquarium's differential GPS system, and Aquarium staff had to locate the interferer themselves.

A particularly difficult problem relating to in-band jamming arises because GNSS is increasingly being used for applications it was not designed for, such as indoor operation. Thus very weak signals can effectively operate as jammers. Electronic devices that meet their EMC specifications can still produce in-band signals that can prevent GNSS receivers from working when in these difficult environments (Haddrell, 2005; Eurocontrol, 1999).

A separate problem relates to areas in the spectrum that are used by, but not dedicated to, GNSS. An example is the GPS L5/ Galileo E5 band, which is shared with aviation guidance equipment, DME and TACAN (Gao, 2013). These transmit relatively high power signals in a pulsed mode and can prevent acquisition and tracking of L5/ E5 signals. Receivers operating in this band must be designed in such a way that they are less affected by this pulsed interference, as it is unlikely that that equipment will be decommissioned for some time.

There are also other "legal" signals that are transmitted intentionally in the protected L1 band, but not with the intention of interfering (making difficult the choice of taxonomy category for this and the L5 examples above). These include the Japanese IMES system, used for indoor positioning (Dempster, 2009). That system transmits the position of the transmitter in L1, effectively acting like a beacon. "Pseudolites" are another example. These have been used to augment GNSS, especially in the early days when too few satellites were visible. They were accepted to the extent that they have been allocated codes (for "ground transmitters") in the GPS interface specification (IS, 2013; Barnes, 2005). The problems they cause to receivers are similar to the near-far problems identified above. Similarly, for some years, the SNAP Lab at UNSW, where the author is based, retransmitted GPS signals from a rooftop antenna into the laboratory so receivers could be tested indoors with good quality "off-air" signals. If this signal were to leak outside, receivers would give a constant position – that of the rooftop antenna (see meaconing below).

Further to this, despite GNSS frequencies, particularly L1, being protected in most jurisdictions, very low level signals may be allowed to be transmitted. Signals that aren't specified over a single narrow band, i.e. ultra-wideband signals, are potentially problematic for GNSS (Divis, 2015). If the transmitters are numerous enough or close enough to the GNSS receiver, the noise floor would rise, i.e. it would look like a wideband noise jammer, similar to the inter-system problem identified above. This problem is the responsibility of regulators.

Out-of-Band (Nearby) Unintentional Sources

The most "celebrated" source of out-of-band interference was that proposed by *LightSquared*, a 4G LTE communications network. It was initially approved to operate in the band next to GPS L1, but was shown to interfere severely with receivers in tests run by John Deere and others (Deere, 2011; Boulton, 2011). The problem was that the transmissions from a Lightsquared station would be so strong that they would still pass through the protective band-pass filters at the GNSS receiver input and prevent receivers from operating in regions for several kilometers around that station. This scare led to a formalized approach to testing receivers for near-band interference (DOT, 2012). At the time of writing, the successor to Lightsquared, *Ligado*, continues to lobby for use of these adjacent bands.

Out-of-Band (Harmonic) Unintentional Sources

Strong signals transmitted at fractions, typically one third, of the GPS carrier frequencies, have been found to cause interference to GPS receivers. This type of interference has proved remarkably easy to detect. Simply by looking at the transmitter closest to one-third of the L1 frequency in their neighborhood, researchers found, in Sydney, analogue (Tabatabaei Balaei 2007, 2009) and, in Turin, digital (Motella, 2008) television signals that significantly interfered with GPS. The interference was strong enough to cause the automatic gain control to attenuate the wanted signal and pass the interferer, significantly reducing the signal to jamming-plus-noise ratio in different receivers.

Man-Made Interference: Intentional Sources

Figure 2 shows three potential types of intentional interference: Jamming, Spoofing and Meaconing. A jamming attack is intended to disturb or disable a receiver primarily with its signal power. The other two are more sophisticated, intending to fool the receiver to believe it is at a false position or time – either one that is fabricated (spoofing) or from a real location and retransmitted (meaconing).

Intentional Sources: Jamming

GNSS jammers have been around for some time. Their aim is to shut down receivers completely in an area nearby to the jamming transmitter. After the cold war, an Aviaconversia jammer could be bought at airshows for a mere $40,000 (Nordwall, 1997). Now, "personal privacy devices" can be found on various internet sites for tens of dollars. Motivated by a preference not to be tracked, i.e. to protect their privacy, many buy these devices, becoming "casual terrorists" (Dempster, 2011). The most well-known example was that of a van driver who regularly shut down GPS operations (including use for aircraft landings) at Newark airport (Economist, 2011), earning him a $32,000 fine. This is a clear example of unintended consequences of the use of GNSS. Here, tracking a company's assets led to a fear for privacy, and ultimately jamming that threatened airport landings.

Intentional jamming tests in the United Kingdom have shown that a 1.5W noise source can jam over distance of 25km from transmitter source (Grant, 2009).

Jammers have been used for nefarious purposes: by car thieves to prevent tracking of expensive stolen vehicles (Sorrel, 2010), cargo thieves to prevent tracking of shipping containers (FBI, 2014), and by taxi drivers to cheat fares (Collier, 2013). There was a case where a highly-automated port was shut down for seven hours due to jamming (Scott, 2015) which was considered unintentional, but it did draw attention to what could be achieved by "terrorists" or disgruntled employees in an industrial dispute.

Military examples of GNSS jamming go back to the first Gulf war in 1991, when terrestrial Iraqi assets were protected by jammers (Greenemeier, 2016). The strongest and most obvious jamming by the military is North Korea's ongoing jamming of South Korea, especially around Incheon airport, aimed to disrupt air traffic (Reuters, 2016). "Intentional" military jamming also includes cases where the US military has jammed GPS as part of their own testing and taken out local facilities (Coffed, 2014). In fact, the US made a major policy change in its methodology for preventing hostile combatants from using GPS when it ceased applying selective availability (deliberately downgrading L1 C/A accuracy) and replaced it with selective denial (where L1 C/A is jammed in a way that leaves military codes still available for use) (Clinton, 2000).

Intentional Sources: Spoofing

It may at first seem that the main benefits to spoofing GNSS might be to upset positioning. But given that 15 of the 18 critical infrastructure sectors of the economy rely on GNSS timing (Merrill, 2011), it's not surprising that the first real example of making money from spoofing comes from spoofing timing receivers. High-frequency trading now requires trades to be time-stamped with an accuracy of nanoseconds (Flanagan, 2014), and hence GNSS receivers are often used, if not as a primary source of timing, then to discipline atomic clocks. If the stock exchange's clock is derived from GNSS, shifting that time forwards or backwards, depending on whether buying or selling is favorable, can potentially result in big profits (Solon, 2012). (A similar but different fraud can be perpetrated by stripping time stamps and replacing them (Durdon, 2014), which would naturally live in the "System Integrators" box in Figure 1 if the timing was definitely provided by GNSS). Jamming signals are frequently observed at London's stock exchange (Economist, 2013).

Another application of timing spoofing that has been highlighted is when it is applied to the power grid (Shephard, 2012), which uses "synchrophasors" to synchronize AC generation to real time. Shifting a single generator can lead to a trip, in turn leading by cascade to widespread blackouts.

In terms of spoofing GPS position, examples have shown that a superyacht can be fooled to think it is sailing over land (Psiaki, 2016), and separate experiments have been shown to down unmanned aerial vehicles (UAVs) that believe they are still flying above the ground (Fox-Brewster, 2015; IG, 2012). Great publicity was given to the claims that Iran had downed a US stealth drone (Peterson, 2011). Of course, this would only have been possible if the drone was navigating using civilian C/A code GPS signals.

A demonstration of GPS spoofing which on the face of it may seem the most harmless may prove to be quite the opposite. The huge Pokemon Go fad of 2016 led players to develop a spoofer in order to fool their phone to think it was near enough to Pokemons to capture them whereas the phone was at their home (Szczys, 2016). The method of doing it was then made available via Github, so large numbers of low-cost spoofers have now been produced, for this relatively trivial reason. Like the example above where the van driver jammed his GPS to maintain his privacy, this in another example of unintended consequences of the use of GNSS – providing a reward (Pokemon) for being at a given location gives an incentive to fake that location, leading to widespread availability and use of spoofers.

There is a range of levels of sophistication when it comes to spoofers (Dovis, 2015; Jahromi, 2013). A simple architecture simply transmits a GNSS-like signal, possibly increasing its power to take over the receiver. An intermediate architecture uses knowledge of the receiver dynamics to replicate the signals that the receiver sees at the correct delays, before increasing power to take over the receiver. In that case, the receiver could not detect the spoofing by looking at jumps in pseudorange (distance to the satellite) measurements, but it does require the spoofer to have an accurate knowledge of the receiver trajectory. A sophisticated architecture does this from several transmitters in realistic directions with respect to the receiver. That would prevent the receiver from detecting the spoofing by using steerable antennas identifying that all the satellite signals appear from the same direction, as they would in the intermediate case. This progression can be seen as a measures/countermeasures arms race, and it is also clear that the latter two methods are tuned such that only a single receiver can be spoofed.

Intentional Sources: Meaconing

Meaconing, i.e. retransmission of a received/recorded off-air signal, has the disadvantage compared to spoofing that the signal retransmitted corresponds to the position of an antenna that is part of the meaconing equipment, making that equipment (or at least the antenna) easy to find. It is therefore far less likely to be used than spoofing. It does have the advantage, however, of being an "authentic" signal, i.e. if a receiver is observing (possibly without decoding) the presence of the military signal in the quadrature channel in order to authenticate that the received signal is "real", that receiver could detect a spoofer but not a meaconer.

Meaconing, but not necessarily of GNSS signals, has been speculated to be the cause of more than one plane crash. One confirmed case of GNSS meaconing was unintentional, when in 2012 a retransmitter used for aircraft testing in a hangar at Hanover airport affected aircraft up to 1km away.

CONCLUSION

The purpose of this chapter has been to identify sources of vulnerability to GNSS systems. The taxonomy has identified system and segment level vulnerabilities, and the variety of intentional and unintentional sources of interference. It is likely that some of the category decisions made by the author could be contested, and that there are sources and categories omitted. Regardless, hopefully this can also act as a reference list for the various vulnerabilities and offer a starting point for new investigations into this broadening and increasingly relevant area.

REFERENCES

Afraimovich, E. L., Zherebtsov, G. A., & Smol'kov, G. Y. (2007). Total Failure of GPS during a Solar Flare on December 6, 2006. *Doklady Earth Sciences*, *417*(1), 1231–1235. doi:10.1134/S1028334X07080223

AMS. (2011). Satellite Navigation & Space Weather: Understanding the Vulnerability & Building Resilience. American Meteorology Society.

Araujo-Pradere, E. A. (2005). GPS-Derived Total Electron Count Response for the Bastille Day Magnetic Storm of 200 at a Low Mid-latitude Station. *Geofísica Internacional*, *44*(2), 211–218.

Astafyeva, E., Zakharenkova, I. & Forster, M. (2015). Ionospheric response to the 2015 St. Patrick's Day storm: A global multi-instrumental overview. *Journal of Geophysical Research/Space Physics*, *120*(10), 9023-9037.

Baraniuk, C. (2016, February 2). GPS error caused '12 hours of problems' for companies. *BBC World.*

Barnes, J. (2005). The Integration of GPS and Pseudolites for Bridge Monitoring. In F. Sansò (Ed.), *A Window on the Future of Geodesy* (pp. 83–88). Berlin: International Association of Geodesy Symposia, Springer. doi:10.1007/3-540-27432-4_15

Boulton, P. (2011, July). GPS Interference testing: lab, live and lightsquared. *Inside GNSS*, 32-45.

Carrano, C. S., Bridgwood, C. T., & Groves, K. M. (2009). Impacts of the December 2006 solar radio bursts on the performance of GPS. *Radio Science*, *44*(1), 1–12. doi:10.1029/2008RS004071

Clinton, W. (2000, May 1). *Statement By The President Regarding The United States' Decision To Stop Degrading Global Positioning System Accuracy*. The White House.

Clynch, J. R (2003). System Challenge – The Hunt for RFI. *GPS World*, *14*(1), 16–23.

Coffed, J. (2014, February). The threat of GPS jamming: the risk to an information utility. *Exelis.*

Collier, A., & Harris, A. (2013, November 8). Taxi cheats using mobile GPS jammers to steal fares. Herald Sun.

Curry, C. (2010). *Dependency of Communications Systems on PNT Technology*. Chronos Technology White Paper.

Deere. (2011). *LightSquared Interference to GPS and StarFire.* Deere submission to Federal Communications Commission.

Dempster, A. (2009, January). QZSS's Indoor Messaging System: Friend or Foe? *Inside GNSS*, 37-40.

Dempster, A. (2011, March 31). GPS jamming could make you a casual terrorist. *The Conversation.*

Divis, D. A. (2015, August 17). Ultra-wideband Re-emerges as a GNSS Interference Issue. *Inside GNSS.*

Doherty, P., Coster, A. J., & Murtagh, W. (2004). Space weather effects of October–November 2003. *GPS Solutions*, *8*(4), 267–271. doi:10.1007/s10291-004-0109-3

DOT. (2012). GPS Adjacent-Band Compatibility Assessment Plan. US Department of Transportation.

Dovis, F. (2015). *GNSS Interference Threats and Countermeasures*. Artech House.

Durdon, T. (2014, April 19). "Timestamp Fraud": A rigged market explained in one simple animation. *Zero Hedge*.

Economist. (2011, March 10). No jam tomorrow. *The Economist*.

Economist. (2013, July 27). GPS jamming: out of sight. *The Economist*.

Eurocontrol. (1999). *GNSS Frequency Protection Requirements*. Eurocontrol Experimental Centre, EEC Report No. 337.

FBI. (2014, October 2). Cargo Thieves use GPS Jammers to Mask GPS Trackers. FBI Cyber Division.

Fenton, P. C., & Jones, J. (2005). The theory and performance of NovAtel Inc.'s vision correlator. *Proceedings of ION-GNSS*.

Flanagan, T. (2014, December 11). Time Stamps Get Nanosecond Precision. *Markets Media*.

Fox-Brewster, T. (2015, August 8). Watch GPS attacks that can kill DJI drones or bypass White House ban. *Forbes*.

Gao, G., Heng, L., Hornbostel, A., Denks, H., Meurer, M., Walter, T., & Enge, P. (2013). DME/TACAN interference mitigation for GNSS: Algorithms and flight test results. *GPS Solutions, 17*(4), 561–573. doi:10.1007/s10291-012-0301-9

Garin, L., & Rousseau, J. (1997). Enhanced strobe correlator multipath rejection for code and carrier. *Proceedings of ION GPS*, 559-568.

Garin, L. J. (2005). The "shaping correlator," novel multipath mitigation technique applicable to GALILEO BOC(1,1) modulation waveforms in high volume markets. *Proceedings of European Navigation Conference*.

Gibbons, G. (2011, Jauary). GNSS interoperability, not so easy after all. *Inside GNSS*, 28-32.

Gibbons, G. (2013, August 30). FCC fines operator of GPS jammer that affected Newark Airport GBAS. *Inside GNSS*.

Gibbons, G. (2014, April 2). GLONASS suffers temporary systemwide outage; multi-GNSS receiver overcomes problem. *Inside GNSS*.

Glennon, E. P., & Dempster, A. G. (2004). A review of GPS cross correlation mitigation techniques. *Proceedings of IGNSS*.

Glennon, E. P., & Dempster, A. G. (2005). A novel cross correlation mitigation technique. *Proceedings of ION-GNSS*.

Glennon, E. P., & Dempster, A. G. (2006a). Cross correlation mitigation by adaptive orthogonalization using constraints - new results. *Proceedings of ION-GNSS*.

Glennon, E. P., & Dempster, A. G. (2006b). *Apparatus and Method for Mitigation of Cross-Correlation in GPS Systems*. US patent application 11/460,692.

Glennon, E. P., & Dempster, A. G. (2007a). Cross correlation mitigation techniques for software GPS C/A code receivers. *Proceedings of IGNSS.*

Glennon, E. P., & Dempster, A. G. (2011). Delayed PIC for postcorrelation mitigation of continuous wave and multiple access interference in GPS receivers. *IEEE Transactions on Aerospace and Electronic Systems*, *47*(4), 2544–2557. doi:10.1109/TAES.2011.6034650

Grant, A., Williams, P., Ward, N., & Basker, S. (2009). GPS jamming and the impact on maritime navigation. *Journal of Navigation*, *62*(2), 173–187. doi:10.1017/S0373463308005213

Greenemeier, L. (2016, February). GPS and the World's First "Space War". *Scientific American*, 8.

Hackett, R. (2015, August 4). Here's the scary new target hackers are going after. *Fortune.*

Haddrell, A. (2005, March). The enemy inside. *GPS World.*

Harvey, B. (2007). *"Military programs". The Rebirth of the Russian Space Program* (1st ed.). Springer.

IG. (2012, July 16). UAVs vulnerable to civil GPS spoofing. *Inside GNSS.*

IG. (2015, November 30). GPS OCS implements threat protection technology. *Inside GNSS.*

IS. (2013, September 24). *IS-GPS-200H Navstar GPS Space Segment/Navigation User Interfaces.* US DOD.

Jahromi, A. J. (2013). *GNSS Signal Authenticity Verification in the Presence of Structural Interference* (PhD thesis). University of Calgary.

Jensen, A. B. O., & Mitchell, C. (2011, February). GNSS and the ionosphere: What's in store for the next solar maximum? *GPS World*, 40–48.

Jones, J., Fenton, P., & Smith, B. (2004). *Theory and Performance of the Pulse Aperture Correlator.* NovAtel Inc. Retrieved from http://www.novatel.com/Documents/Papers/PAC.pdf

Kaplan, E. D., & Hegarty, C. J. (Eds.). (2006). *Understanding GPS: Principles and Applications* (2nd ed.). Boston: Artech House.

Kerr, R. A. (2009, June). Are we ready for the next solar maximum? No way, say scientists. *Science*, *26*, 1640–1641. doi:10.1126/science.324_1640 PMID:19556483

Kintner, P. M., Humphreys, T., & Hinks, J. (2009, July). GNSS and ionospheric scintillation: how to survive the next solar maximum. *Inside GNSS*, 22-30.

Kneupper, M. R. (2011). *Operating in a GPS Denied Environment in the PACOM Maritime Domain.* DTIC Report ADA546337.

Laxton, M. C., & Devilbiss, S. L. (1997). GPS multipath mitigation during code tracking. *Proceedings of the American Control Conference*, 1429-1433.

Leica. (2009). *Leica AR25 White Paper.* Leica Geosystems. Retrieved from http://www.leica-geosystems.com/common/shared/downloads/inc/downloader.asp?id=10904

Li, G., Ning, B., Zhao, B., Liu, L., Liu, J. Y., & Yumoto, K. (2008). Effects of geomagnetic storm on GPS ionospheric scintillations at Sanya. *Journal of Atmospheric and Solar-Terrestrial Physics*, *70*(7), 1034–1045. doi:10.1016/j.jastp.2008.01.003

Merrill, J. (2011). U. S. Interference Detection & Mitigation Plan (IDM). *51st Meeting of the Civil GPS Service Interface Committee at the Institute of Navigation Conference (GNSS 2011).*

Motella, B., Pinni, M., & Doviz, F. (2008). Investigation on the effect of strong out-of-band signals on global navigation satellite systems receivers. *GPS Solutions*, *12*(2), 77–86. doi:10.1007/s10291-007-0085-5

Navarro-Reyes, D., Castro, R., & Bosch, P. R. (2015). Galileo first foc launch: recovery mission design. *Proceedings of International Symposium on Space Flight Dynamics.*

Nordwall, B. D. (1997, September 29). Aviaconversia is marketing a portable GPS/Glonass jammer. *Aviation Week & Space Technology.*

Novatel. (2000). *Discussions on RF Signal Propagation and Multipath.* Novatel Application Note APN-008 Rev 1.

Odenwald, S. F., & Green, J. L. (2008, July). Bracing the Satellite Infrastructure for a Solar Superstorm. *Scientific American*, 28.

Odenwald, S. F., Green, J. L., & Taylor, W. (2006). Forecasting the impact of an 1859-calibre superstorm on satellite resources. *Advances in Space Research*, *38*(2), 280–297. doi:10.1016/j.asr.2005.10.046

Peterson, S. (2011, December 9). Downed US drone: How iran caught the 'beast'. *The Christian Science Monitor.*

Pi, X., Mannucci, A. J., Lindqwister, U. J., & Ho, C. M. (1997). Monitoring of global Ionospheric irregularities using the worldwide GPS network. *Geophysical Research Letters*, *24*(18), 2283–2286. doi:10.1029/97GL02273

Pick, M., & Vilmer, N. (2008). Sixty-five years of solar radioastronomy: Flares, coronal mass ejections and Sun-Earth connection. *The Astronomy and Astrophysics Review*, *16*(1-2), 1–153. doi:10.1007/s00159-008-0013-x

Psiaki, M., & Humphreys, T. E. (2016, July). Protecting GPS from spoofers is critical to the future of navigation. *IEEE Spectrum*, 29.

Qaisar, S. U., & Dempster, A. G. (2008). Cross-correlation Performance Comparison of L1 & L2C GPS Codes for Weak Signal Acquisition. *Proceedings of IS-GPS/GNSS.*

RAE. (2011). *Global Navigation Satellite Systems: Reliance and Vulnerabilities.* Royal Academy of Engineering.

Reuters. (2016, April 11). South Korea tells U.N. that North Korea GPS jamming threatens boats, planes. *Reuters.*

RSW. (2013). *Russia's Proton crashes with a trio of navigation satellites.* Retrieved 9 Mar 2017, from http://www.russianspaceweb.com/proton_glonass49.html

Sahmoudi, M., & Amin, M. G. (2008). Fast iterative maximum-likelihood algorithm (FIMLA) for multipath mitigation in next generation of GNSS receivers. *IEEE Transactions on Wireless Communications*, *7*(11), 4362–4374. doi:10.1109/T-WC.2008.070700

Sarkar, S., & Bose, A. (2016). Lifetime of revitalized GLONASS satellites: A review. *Proceedings of MDCCT.*

Scott, L. (2015, April). Protecting Position in Critical Operations. *GPS World*, 30.

Shepard, D. P., & Humphreys, T. E. (2012). Evaluation of the vulnerability of phasor measurement units to GPS spoofing attacks. *Proceedings of Sixth Annual IFIP WG 11.10 International Conference on Critical Infrastructure Protection*. doi:10.1016/j.ijcip.2012.09.003

Skone, S. H. (2001). The impact of magnetic storms on GPS receiver performance. *Journal of Geodesy*, *75*(9), 457–468. doi:10.1007/s001900100198

Skone, S. H., Yousuf, R., & Coster, A. (2004). Performance evaluation of the wide Area augmentation system for ionospheric storm events. *Journal of Global Positioning Systems*, *3*(1-2), 251–258. doi:10.5081/jgps.3.1.251

Solon, O. (2012, February 22). GPS 'Spoofers' could be used for high-frequency financial trading fraud. *Wired.*

Sorrel, C. (2010, February 24). Car thieves use GPS jammers to make clean getaway. *Wired.*

Symmetricon. (2013). *Power Utilities: Mitigating GPS Vulnerabilities and Protecting Power Utility Network Timing*. White paper.

Szczys, M. (2016, July 26). *We Declare The Grandmaster Of Pokemon Go GPS Cheats*. Hackaday.

Tabatabaei Balaei, A., & Dempster, A. G. (2009). A statistical inference technique for GPS interference detection. *IEEE Trans Aerospace and Systems*, *45*(4), 1499–1511. doi:10.1109/TAES.2009.5310313

Tabatabaei Balaei, A., Motella, B., & Dempster, A. G. (2007). GPS interference detected in Sydney-Australia. *Proceedings of IGNSS.*

Townsend, B. R., & Fenton, P. (1994). *A practical approach to the reduction of pseudorange multipath errors in a L1 GPS receiver. Proceedings of ION-GPS.*

Van Dierendonck, A. J. (2000). *Mathematical Challenges in GPS: New Opportunities Using New Civil Signals for Multipath Mitigation and Carrier Phase Ambiguity Resolution*. Retrieved from www.ima.umn.edu/talks/workshops/8-16-18.2000/ van-dierendonck/SignalSpecifiationPaper.ppt

Van Dierendonck, A. J., Fenton, P., & Ford, . (1992). Theory and performance of narrow correlator spacing in a GPS receiver. *Navigation*, *39*(3), 265–283. doi:10.1002/j.2161-4296.1992.tb02276.x

van Nee, R. D. J. (1992). The multipath estimating delay lock loop. *Proceedings of IEEE ISSTA*, 39-42.

van Nee, R. D. J. (1994). The multipath estimating delay lock loop: approaching theoretical accuracy limits. *Proceedings of IEEE PLANS*, 246-251. doi:10.1109/PLANS.1994.303320

Veitsel, V. A., Zhdanov, A. V., & Zhodzishsky, M. (1998). The mitigation of multipath errors by strobe correlators in GPS/GLONASS receivers. *GPS Solutions*, *2*(2), 38–45. doi:10.1007/PL00000035

Wu, J., & Dempster, A. G. (2011). "BOC-Gated-PRN" a multipath mitigation technique for BOC(n,n) waveforms. *IEEE Transactions on Aerospace and Electronic Systems*, *47*(2), 1136–1153. doi:10.1109/TAES.2011.5751248

Xu, J. S., Zhu, J., & Li, L. (2007). Effects of a major storm on GPS amplitude scintillations and phase fluctuations at Wuhan in China. *Advances in Space Research*, *39*(8), 1318–1324. doi:10.1016/j.asr.2007.03.004

Xu, Z., Trinkle, M., & Gray, D. A. (2011). A maximum-likelihood based mutual coupling calibration algorithm in the presence of multipath for GPS antenna array. *Proceedings of ION-GNSS*, 1111-1119.

Zakharov, V. I., Yasyukevich, Y. V., & Titova, M. A. (2016). Effect of magnetic storms and substorms on GPS slips at high latitudes. *Cosmic Research*, *54*(1), 20–30. doi:10.1134/S0010952516010147

KEY TERMS AND DEFINITIONS

GNSS: (Global Navigation Satellite System) A constellation of satellites supported by a ground network that provides navigation services across the Earth.

Vulnerability: In addition to the more common usage (openness or susceptibility to attack), a process or situation that can lead to exposure to attack.

Compilation of References

3. GPP TS 25.101 v8.0.0. (2008, January). Radio transmission and reception (FDD)(Release 8).

3. GPP TS 45.010 v8.0.0. (2008, May). Radio subsystem synchronization (Release 8).

Abdi, A., & Kaveh, M. (1999). On the utility of gamma PDF in modeling shadow fading (slow fading). *IEEE 49th Vehicular Technology Conference, 3*, 2308-2312.

Abramowitz, M., & Stegun, I. A. (1970). *Handbook of mathematical functions: with formulas, graphs, and mathematical tables*. Dover Publications.

Abu-Shaban, Z., Zhou, X., & Abhayapala, T. D. (2016). A novel TOA-based mobile localization Technique under mixed LOS/NLOS conditions for cellular networks. *IEEE Transactions on Vehicular Technology, 65*(11), 8841–8853. doi:10.1109/TVT.2016.2517151

Aeberhard, M., Schlichtharle, S., Kaempchen, N., & Bertram, T. (2012). Track-to-track fusion with asynchronous sensors using information matrix fusion for surround environment perception. *IEEE Transactions on Intelligent Transportation Systems, 13*(4), 1717–1726. doi:10.1109/TITS.2012.2202229

Afraimovich, E. L., Zherebtsov, G. A., & Smol'kov, G. Y. (2007). Total Failure of GPS during a Solar Flare on December 6, 2006. *Doklady Earth Sciences, 417*(1), 1231–1235. doi:10.1134/S1028334X07080223

Agrawal, P., & Patwari, N. (2009). Correlated link shadow fading in multi-hop wireless networks. *IEEE Transactions on Wireless Communications, 8*(8), 4024–4036. doi:10.1109/TWC.2009.071293

Ahmed, E.-R. (2002). *Introduction to GPS: The Global Positioning System*. Artech House.

Akos, D. M., Borre, K., & Bertelsen, N. (2007). *A Software-Defined GPS and Galileo Receiver: A Single-Frequency Approach*. Springer.

Akyildiz, I. F., Sun, Z., & Vuran, M. C. (2009). Signal propagation techniques for wireless underground communication networks. *Physical Communication, 2*(3), 167–183. doi:10.1016/j.phycom.2009.03.004

Akyildiz, I. F., Su, W., Sankarasubramaniam, Y., & Cayirci, E. (2002). A survey on sensor networks. *IEEE Communications Magazine, 40*(8), 102–114. doi:10.1109/MCOM.2002.1024422

Alereon Inc. (2017). *Alereon Demonstrates Military Wireless Personal Area Network*. Retrieved March 16, 2017, from: http://www.alereon.com/?page_id=2992

Alippi, C., Bocca, M., Boracchi, G., Patwari, N., & Roveri, M. (2015). RTI Goes Wild: Radio Tomographic Imaging for Outdoor People Detection and Localization. *IEEE Transactions on Mobile Computing*. doi:10.1109/TMC.2015.2504965

Allan, D. W. (1966). Statistics of atomic frequency standards. *IEEE Proceedings*, 221-230.

Amar, A., & Weiss, A. J. (2008), New asymptotic results on two fundamental approaches to mobile terminal location. *Proceedings of International Symposium on Communications, Control and Signal Processing (ISCCSP)*, 1320–1323. doi:10.1109/ISCCSP.2008.4537430

AMS. (2011). Satellite Navigation & Space Weather: Understanding the Vulnerability & Building Resilience. American Meteorology Society.

Amt, J. R. H. (2006) *Methods for aiding height determination in pseudolite-based reference systems using batch Least-Squares Estimation* (Master's Thesis). Air Force Institute of Technology.

Anderson, C. R., Martin, R. K., Walker, T. O., & Thomas, R. W. (2014, February). Radio Tomography for Roadside Surveillance. *IEEE Journal of Selected Topics in Signal Processing*, *8*(1), 66–79. doi:10.1109/JSTSP.2013.2286774

Aoki, H., Schiele, B., & Pentland, A. (1999), Realtime personal positioning system for wearable computers. *Proceedings of 3rd IEEE International Symposium on Wearable Computers ISWC 1999*, 37-44. doi:10.1109/ISWC.1999.806642

Applanix Corporation. (2009). *POS MV V4 User guide*. Author.

Araujo-Pradere, E. A. (2005). GPS-Derived Total Electron Count Response for the Bastille Day Magnetic Storm of 200 at a Low Mid-latitude Station. *Geofísica Internacional*, *44*(2), 211–218.

Arnitz, D., Muehlmann, U., & Witrisal, K. (2012a). Wideband characterization of backscatter channels: Derivations and theoretical background. *IEEE Transactions on Antennas and Propagation*, *60*(1), 257–266. doi:10.1109/TAP.2011.2167923

Arnitz, D., Muehlmann, U., & Witrisal, K. (2012b). Characterization and modeling of UHF RFID channels for ranging and localization. *IEEE Transactions on Antennas and Propagation*, *60*(5), 2491–2501. doi:10.1109/TAP.2012.2189705

Arnold, G. M., & Collins, A. J. (1993). Interpretation of transformed axes in multivariate analysis. *Journal of the Royal Statistical Society. Series C, Applied Statistics*, *42*(2), 381–400.

Arshal, G. (1987). Error equations of inertial navigation. *Journal of Guidance, Control, and Dynamics*, *10*(4), 351–358. doi:10.2514/3.20225

Aspnes, J., Eren, T., Goldenberg, D. K., Morse, A. S., Whiteley, W., Yang, Y. R., & Belhumeur, P. N. et al. (2006). A Theory of Network Localization. *IEEE Transactions on Mobile Computing*, *5*(12), 1663–1678. doi:10.1109/TMC.2006.174

Astafyeva, E., Zakharenkova, I. & Forster, M. (2015). Ionospheric response to the 2015 St. Patrick's Day storm: A global multi-instrumental overview. *Journal of Geophysical Research/Space Physics*, *120*(10), 9023-9037.

Ávila Rodriguez, J. A., Pany, T., & Eissfeller, B. (2004). A theoretical analysis of acquisition algorithms for indoor positioning. *Proceedings of the 2nd ESA Workshop on Satellite Navigation User Equipment Technologies NAVITEC.*

Baccour, N., Koubâa, A., Mottola, L., Zúñiga, M. A., Youssef, H., Boano, C. A., & Alves, M. (2012). Radio link quality estimation in wireless sensor networks: A survey. *ACM Transactions on Sensor Networks*, *8*(4), 34. doi:10.1145/2240116.2240123

Bahl, P., & Padmanabhan, V. (2000). RADAR: An in-building RF-based user location and tracking system. *Proceedings IEEE INFOCOM 2000. Conference on Computer Communications. Nineteenth Annual Joint Conference of the IEEE Computer and Communications Societies*, *2*, 775-784.

Bancroft, J. B., & Lachapelle, G. (2011). Data fusion algorithms for multiple inertial measurement units. *Sensors (Basel)*, *11*(7), 6771–6798. doi:10.3390/s110706771 PMID:22163985

Bao, M.-H. (2004). *Micro mechanical transducers* (2nd ed.). Amsterdam: Elsevier.

Baraniuk, C. (2016, February 2). GPS error caused '12 hours of problems' for companies. *BBC World.*

Barbour, N. M. (2010). *Inertial navigation sensors.* Report, Charles Stark Draper Laboratory (P-4994) Cambridge, MA 02139 USA. Retrieved January 2017, from: http://oai.dtic.mil/oai/oai?verb=getRecord&metadataPrefix=html&identifier=ADA581016

Barbour, N. M., & Schmidt, G. (2001). Inertial sensor technology trends. *IEEE Sensors Journal, 1*(4), 332–339. doi:10.1109/7361.983473

Bar-Itzhack, I. Y. (1977). Navigation computation in terrestrial strapdown inertial navigation systems. *IEEE Transactions on Aerospace and Electronic Systems, 13*(6), 679–689. doi:10.1109/TAES.1977.308509

Barnes, J. (2005). The Integration of GPS and Pseudolites for Bridge Monitoring. In F. Sansò (Ed.), *A Window on the Future of Geodesy* (pp. 83–88). Berlin: International Association of Geodesy Symposia, Springer. doi:10.1007/3-540-27432-4_15

Barnes, J., Rizos, C., & Kanli, M. (2004) *Indoor industrial machine guidance using Locata: A pilot study at BlueScope Steel. 60th Annual Meeting of the U.S. Institute of Navigation,* Dayton, OH.

Barnes, J., Rizos, C., Wang, J., Small, D., Voight, G., & Gambale, N. (2003) LocataNet: The positioning technology of the future? *6th International Symposium on Satellite Navigation Technology Including* Mobile Positioning and Location Services, Melbourne, Australia.

Bar-Shalom, Y. (1981). On the track-to-track correlation problem. *IEEE Transactions on Automatic Control, 26*(2), 571–572. doi:10.1109/TAC.1981.1102635

Bartlett, M., & Kendall, D. G. (1946). The statistical analysis of variance-heterogeneity and the logarithmic transformation. *Supplement to the Journal of the Royal Statistical Society, 8*(1), 128–138. doi:10.2307/2983618

Baruch, B., Doron, M. A., Rosenhouse, J., & Azhari, H. (1999). On direction finding of an emitting source from time delays. *The Journal of the Acoustical Society of America, 106*(6), 3355–3363.

Bastide, F., Julien, O., Macabiau, C., & Roturier, B. (2002). Analysis of L5/E5 Acquisition, Tracking and Data Demodulation Thresholds. *Proceedings of ION GPS,* 2196-2207.

Battiti, R., Nhat, T. L., & Villani, A. (2002). *Location-aware computing: A neural network model for determining location in wireless LANs.* University of Trento, Trento, Italy, Tech. Rep. DIT-02-0083. Retrieved from http://citeseerx.ist.psu.edu/viewdoc/download?doi=10.1.1.20.2366&rep=rep1&type=pdf

Baxter, J., Jones, R., Lin, M., & Olsen, J. (2004). SLLN for weighted independent identically distributed random variables. *Journal of Theoretical Probability, 17*(1), 165–181. doi:10.1023/B:JOTP.0000020480.84425.8d

Baxter, M. J. (1995). Standardization and transformation in principal component analysis, with applications to archaeometry. *Applied Statistics-Journal of the Royal Statistical Society Series C, 44*(4), 513–527.

Beauregard, S. (2006). A helmet-mounted pedestrian dead reckoning system. *Proceedings of International Forum on Applied Wearable Computing IFAWC, 6,* 1–11.

Bebek, O., Suster, M. A., Rajgopal, S., Fu, M. J., Huang, X., Cavusoglu, M. C., & Mastrangelo, C. H. et al. (2010). Personal navigation via high-resolution gait-corrected inertial measurement units. *IEEE Transactions on Instrumentation and Measurement, 59*(11), 3018–3027. doi:10.1109/TIM.2010.2046595

Beck, A., Stoica, P., & Li, J. (2008). Exact and approximate solutions for source localization problems. *IEEE Transactions on Signal Processing, 56*(5), 1770–1778. doi:10.1109/TSP.2007.909342

Bernardos, A., Casar, J., & Tarrio, P. (2010). Real time calibration for RSS indoor positioning systems. *2010 International Conference on Indoor Positioning and Indoor Navigation*, 1-7. doi:10.1109/IPIN.2010.5648231

Bertsch, J., Choudhury, M., Rizos, C., & Kahle, H. G. (2009) *On-the-fly ambiguity resolution for Locata*. International Symposium on GPS/GNSS, Gold Coast, Australia.

Beser, J., & Parkinson, B. W. (1982). The application of NAVSTAR differential GPS in the civilian community. *Navigation Journal of U.S. Institute of Navigation, 29*(2), 107–135. doi:10.1002/j.2161-4296.1982.tb00795.x

Besl, P. J., & Mckay, N. D. (1992). A method for registration of 3-D shapes. *IEEE Transactions on Pattern Analysis and Machine Intelligence, 14*(2), 239–256. doi:10.1109/34.121791

Bialer, O., Raphaeli, D., & Weiss, A. (2013, June). Maximum-likelihood direct position estimation in dense multipath. *IEEE Transactions on Vehicular Technology, 62*(5), 2069–2079. doi:10.1109/TVT.2012.2236895

Bieber, A. M., Brooks, D. W., Harbottle, G., & Sayre, E. V. (1976, February). Application of multivariate techniques to analytical data on aegean ceramics. *Archaeometrys, 18*(1), 59–74. doi:10.1111/j.1475-4754.1976.tb00145.x

Biswas, P., & Ye, Y. (2003). *A distributed method for solving semidefinite programs arising from ad hoc wireless sensor network localization.* Department of Computer Science, Stanford Univ., Stanford, CA, Tech. Rep., Oct.

Biswas, P., Liang, T. C., Toh, K. C., Wang, T. C., & Ye, Y. (2006). Semidefinite programming approaches for sensor network localization with noisy distance measurements. *IEEE Transactions on Automation Science and Engineering, 3*(4), 360–371. doi:10.1109/TASE.2006.877401

Biswas, P., Lian, T. C., Wang, T. C., & Ye, Y. (2006). Semidefinite programming based algorithms for sensor network localization. *ACM Transactions on Sensor Networks, 2*(2), 188–220. doi:10.1145/1149283.1149286

Biswas, P., & Ye, Y. (2003). *Semidefinite programming for ad hoc wireless sensor network localization.* Stanford, CA: Tech. Rep.

Bjorkbom, M., Timonen, J., Yiğitler, H., Kaltiokallio, O., Garcia, J. M., Myrsky, M., & Koivo, H. N. et al. (2013). Localization Services for Online Common Operational Picture and Situation Awareness. *IEEE Access: Practical Innovations, Open Solutions, 1*, 742–757. doi:10.1109/ACCESS.2013.2287302

Bocca, M., Kaltiokallio, O., & Patwari, N. (2013). Radio tomographic imaging for ambient assisted living. In *Evaluating AAL Systems Through Competitive Benchmarking* (pp. 108–130). Springer. doi:10.1007/978-3-642-37419-7_9

Borestein, J., Ojeda, L., & Kwanmuang, S. (2009). *Heuristic reduction of gyro drift in IMU-based personnel tracking systems*. SPIE Defense, Security and Sensing Conference, Orlando, FL.

Borg, I., & Groenen, P. (1997). *Modern multidimensional scaling, Theory and Applications New York*. Springer-Verlag. doi:10.1007/978-1-4757-2711-1

Borio, D., Camoriano, L., & Lo Presti, L. (2008, July). Impact of GPS acquisition strategy on decision probabilities. *IEEE Transactions on Aerospace and Electronic Systems, 44*(3), 996–1011. doi:10.1109/TAES.2008.4655359

Borio, D., O'Driscoll, C., & Lachapelle, G. (2010, January). Composite GNSS signal acquisition over multiple code periods. *IEEE Transactions on Aerospace and Electronic Systems, 46*(1), 193–206. doi:10.1109/TAES.2010.5417156

Born, M., & Wolf, E. (1975). *Principles of optics: electromagnetic theory of propagation, interference and diffraction of light* (5th ed.). Pergamon Press.

Borrás, J., Hatrack, P., & Mandayam, N. B. (1998). Decision theoretic framework for NLOS identification. In *Proceedings of Vehicular Technology Conference* (vol. 2, pp.1583-1587). Piscataway, NJ: IEEE.

Boulton, P. (2011, July). GPS Interference testing: lab, live and lightsquared. *Inside GNSS*, 32-45.

Boyd, S., & Vandenberghe, L. (2004). *Convex optimization.* New York: Cambridge University Press. doi:10.1017/CBO9780511804441

Brajdic, A., & Harle, R. (2013). Walk detection and step counting on unconstrained smartphones. *Proceedings of ACM International Joint Conference on Pervasive and Ubiquitous Computing UbiComp'13*, 225-234. doi:10.1145/2493432.2493449

Brian, G. F., Lionel, G. C., & Kam, W. L. (2002). Locating far-field impulsive sound sources in air by triangulation. *The Journal of the Acoustical Society of America*, *111*(1), 104–116. doi:10.1121/1.1402618 PMID:11831786

Bshara, M., Orguner, U., Gustafsson, F., & Van Biesen, L. (2010, January). Fingerprinting localization in wireless networks based on received-signal-strength measurements: A case study on WiMAX networks. *IEEE Transactions on Vehicular Technology*, *59*(1), 283–294. doi:10.1109/TVT.2009.2030504

Bylemans, I., Weyn, M., & Klepal, M. (2009). Mobile phone-based displacement estimation for opportunistic localisation systems. *Proceedings of 3rd International Conference on Mobile Ubiquitous Computing, Systems, Services and Technologies UBICOMM'09,* 113-118. doi:10.1109/UBICOMM.2009.23

Caceres, M. A., Sottile, F., & Garello, R. (2010). Hybrid GNSS-TOA localization and tracking via cooperative unscented Kalman filter. *IEEE International Symposium on Personal, Indoor and Mobile Radio Communications Workshops*, 272-276. doi:10.1109/PIMRCW.2010.5670378

Caffery, J. J. (1999). *Wireless Location in CDMA Cellular Radio Systems.* Norwell, MA: Kluwer.

Calculation of the Dielectric Properties of Body Tissues in the frequency range 10 Hz - 100 GHz. (2013). Academic Press. http://niremf.ifac.cnr.it/tissprop/htmlclie/htmlclie.php

Candes, E. J., Romberg, J., & Tao, T. (2006). Robust Uncertainty Principles: Exact Signal Reconstruction From Highly Incomplete Frequency Information. *IEEE Transactions on Information Theory*, *52*(2), 489–509. doi:10.1109/TIT.2005.862083

Candes, E. J., & Tao, T. (2006). Near Optimal Signal Recovery from Random Projections: Universal Encoding Strategies. *IEEE Transactions on Information Theory*, *52*(12), 5406–5425. doi:10.1109/TIT.2006.885507

Candes, E. J., & Wakin, M. B. (2008). An Introduction to Compressed Sampling. *IEEE Signal Processing Magazine*, *25*(2), 21–30. doi:10.1109/MSP.2007.914731

Capkun, S., Hamdi, M., & Hubaux, J. P. (2001). GPS-free positioning in mobile ad hoc networks. *Proc. 34th IEEE Hawaii Int. Conf. System Sciences (HICSS-34)*, *34*, 1-10.

Carlson, N. A. (1996) *Federated filter for computer-efficient, near-optimal GPS integration. IEEE Position Location and Navigation Symposium*, Atlanta, GA. doi:10.1109/PLANS.1996.509093

Carlson, N. A., & Berarducci, M. P. (1994). Federated Kalman filter simulation results. *Navigation, Journal of the U.S. Institute of Navigation*, *41*(3), 297–321. doi:10.1002/j.2161-4296.1994.tb01882.x

Carrano, C. S., Bridgwood, C. T., & Groves, K. M. (2009). Impacts of the December 2006 solar radio bursts on the performance of GPS. *Radio Science*, *44*(1), 1–12. doi:10.1029/2008RS004071

Casas, R., Marco, A., Guerrero, J. J., & Falcó, J. (2006). Robust estimator for non-line-of-sight error mitigation in indoor localization. *EURASIP Journal on Applied Signal Processing*, *2006*(1), 1–8. doi:10.1155/ASP/2006/43429

Cassioli, D., Win, M. Z., & Molisch, A. F. (2002). The ultra-wide bandwidth indoor channel: From statistical model to simulations. *IEEE Journal on Selected Areas in Communications*, *20*(6), 1247–1257. doi:10.1109/JSAC.2002.801228

Censi, A. (2008). An ICP variant using a point-to-line metric. *IEEE International Conference on Robotics and Automation.* doi:10.1109/ROBOT.2008.4543181

Cevher, V., Duarte, M., & Baraniuk, R. (2008). Distributed Target Localization Via Spatial Sparsity. *Proceedings of European Signal Processing Conference (EUSIPCO).*

Chan, E., Baciu, G., & Mak, S. (2010). Wi-Fi positioning based on Fourier descriptors. *International Conference on Communications and Mobile Computing, 3*, 545–551.

Chan, F. K. W., & So, H. C. (2009). Efficient weighted multidimensional scaling for wireless sensor network localization. *IEEE Transactions on Signal Processing, 57*(11), 4548–4553. doi:10.1109/TSP.2009.2024869

Chan, F. K. W., So, H. C., & Ma, W. K. (2009). A novel subspace approach for cooperative localization in wireless sensor networks using range measurements. *IEEE Transactions on Signal Processing, 57*(1), 260–269. doi:10.1109/TSP.2008.2005870

Chang, C.-C., & Lin, C.-J. (2011, Apr). LIBSVM: A library for support vector machines. *ACM Transactions on Intelligent Systems and Technology, 2*(3), 27:1–27:27. Retrieved from http://www.csie.ntu.edu.tw/~cjlin/libsvm

Chang, C., & Sahai, A. (2004). Object tracking in a 2D UWB sensor network. *Signals, Systems and Computers, 2004. Conference Record of the Thirty-Eighth Asilomar Conference on*, 1, 1252-1256.

Chan, Y. T., & Ho, K. C. (1994). A simple and efficient estimator for hyperbolic location. *IEEE Transactions on Signal Processing, 42*(8), 1905–1915. doi:10.1109/78.301830

Chazan, D., Zakai, M., & Ziv, J. (1975). Improved lower bounds on signal parameter estimation. *IEEE Transactions on Information Theory, 21*(1), 90–93. doi:10.1109/TIT.1975.1055325

Che, L., Xiong, B., Li, Y., & Wang, Y. (2010). A novel electrostatic-driven tuning fork micromachined gyroscope with a war structure operating at atmospheric pressure. *Journal of Micromechanics and Microengineering, 20*(1), 15–25. doi:10.1088/0960-1317/20/1/015025

Chen, Y., Kleisouris, K., Li, X., Trappe, W., & Martin, R. P. (2009, Feb). A security and robustness performance analysis of localization algorithms to signal strength attacks. *ACM Transactions on Sensor Networks, 5*, 2:1-2:37.

Cheng, J. T., Yang, L., Li, Y., & Zhang, W. (2014). Seamless Outdoor/Indoor Navigation with WIFI/GPS Aided Low Cost Inertial Navigation System. *Elsevier Editorial System(tm) for Physical Communication. SI: Indoor Navigation & Tracking., 13*, 31–43.

Cheng, Z., Wang, Y., & Shen, Y. (2017). Direct position determination of multiple targets via reduced-dimension beamspace. *IEEE International Conference on Communications Workshops*, 1030-1035. doi:10.1109/ICCW.2017.7962794

Chen, P. C. (1999). A non-line-of-sight error mitigation algorithm in location estimation. In *Proc. Wireless Communications and Networking Conference* (vol. 1, pp. 316-320). Piscataway, NJ: IEEE Press.

Chen, R., Pei, L., Liu, J., & Leppäkoski, H. (2012). WLAN and Bluetooth positioning in smart phones. In R. Chen (Ed.), *Ubiquitous positioning and mobile location-based services in smart phones* (pp. 44–68). Hershey, PA: IGI Global. doi:10.4018/978-1-4666-1827-5.ch003

Chen, S. S., Donoho, D. L., & Saunders, M. A. (1999). Atomic Decomposition by Basis Pursuit. *SIAM Journal on Scientific Computing, 20*(1), 33–61. doi:10.1137/S1064827596304010

Chen, Y., & Kobayashi, H. (2002) Signal strength based indoor geolocation. *1st IEEE Int. Conference on Communications*, 1, 436–439.

Chen, Y., Yang, J., Trappe, W., & Martin, R. (2010). Detecting and localizing identity-based attacks in wireless and sensor networks. *IEEE Transactions on Vehicular Technology*, *59*(5), 1–1. doi:10.1109/TVT.2010.2044904

Cheung, K. W., & So, H. C. (2005). A multidimensional scaling framework for mobile location using time-of-arrival measurements. *IEEE Transactions on Signal Processing*, *53*(2), 460–470. doi:10.1109/TSP.2004.840721

Cho, S. Y., & Park, C. G. (2006). MEMS based pedestrian navigation system. *Journal of Navigation*, *59*(1), 135–153. doi:10.1017/S0373463305003486

Choudhury, M., Harvey, B. R., & Rizos, C. (2009b) *Tropospheric correction for Locata when known point ambiguity resolution technique is used in static survey – is it required?* International Symposium on GPS/GNSS, Gold Coast, Australia.

Choudhury, M., Rizos, C., & Harvey, B. R. (2009a) *A survey of techniques and algorithms in deformation monitoring applications and the use of the Locata technology for such applications.* 22nd International Technical Meeting of the Satellite Division of the U.S. Institute of Navigation, Savannah, GA.

CHRobotics. (n.d.). Retrieved 4 18, 2014, from https://www.chrobotics.com/products/inertial-and-orientation-sensors

Clausen, J. (1999). Branch and bound algorithms-principles and examples. Department of Computer Science, University of Copenhagen.

Clinton, W. (2000, May 1). *Statement By The President Regarding The United States' Decision To Stop Degrading Global Positioning System Accuracy*. The White House.

Closas, P., Fernandez-Prades, C., & Fernandez-Rubio, J. A. (2007). Maximum Likelihood Estimation of Position in GNSS. *IEEE Signal Processing Letters*, *14*(5), 359–362. doi:10.1109/LSP.2006.888360

Closas, P., Fernandez-Prades, C., & Fernandez-Rubio, J. A. (2009). Cramer-Rao Bound Analysis of Positioning Approaches in GNSS Receivers. *IEEE Transactions on Signal Processing*, *57*(10), 3775–3786. doi:10.1109/TSP.2009.2025083

Closas, P., Fernandez-Prades, C., Fernandez-Rubio, J. A., Wis, M., Veccione, G., Zanier, F., & Crisci, M. et al. (2015). Evaluation of GNSS Direct Position Estimation in Realistic Multipath Channels. *Proceedings of the ION GNSS International Technical Meeting.*

Clynch, J. R (2003). System Challenge – The Hunt for RFI. *GPS World*, *14*(1), 16–23.

Cobb, H. S. (1997). *GPS pseudolites: Theory, design and applications* (Ph.D thesis). Stanford University, Stanford, CA.

Coffed, J. (2014, February). The threat of GPS jamming: the risk to an information utility. *Exelis.*

Coleri Ergen, S., Tetikol, H., Kontik, M., Sevlian, R., Rajagopal, R., & Varaiya, P. (2014, January). RSSI-fingerprinting-based mobile phone localization with route constraints. *IEEE Transactions on Vehicular Technology*, *63*(1), 423–428. doi:10.1109/TVT.2013.2274646

Collier, A., & Harris, A. (2013, November 8). Taxi cheats using mobile GPS jammers to steal fares. Herald Sun.

Corazza, G. E. (1996, September). On the MAX/TC criterion for code acquisition and its application to DS-SSMA systems. *IEEE Transactions on Communications*, *44*(9), 1173–1182. doi:10.1109/26.536923

Corazza, G. E., & Pedone, R. (2007, November). Generalized and average likelihood ratio testing for post detection integration. *IEEE Transactions on Communications*, *55*(11), 2159–2171. doi:10.1109/TCOMM.2007.908531

Cui, X., Yu, K., & Lu, S. (2015). Evolutionary TDOA-based direction finding methods with 3-D acoustic array. *IEEE Transactions on Instrumentation and Measurement*, *64*(9), 2347–2359. doi:10.1109/TIM.2015.2415051

Cui, X., Yu, K., & Lu, S. (2016). Direction finding for transient acoustic source based on biased TDOA measurement. *IEEE Transactions on Instrumentation and Measurement, 65*(11), 2442–2453. doi:10.1109/TIM.2016.2583224

Cummins, M., & Newman, P. (2008). Accelerated appearance-only SLAM. In *Proceedings of IEEE International Conference on Robotics and Automation* (pp. 1828-1833). IEEE Xplore.

Curry, C. (2010). *Dependency of Communications Systems on PNT Technology.* Chronos Technology White Paper.

Dai, W., Shen, Y., & Win, M. Z. (2017). A computational geometry framework for efficient network localization. *IEEE Transactions Information Theory.*

Dampf, J., Pany, T., Bär, W., Winkel, J., Stöber, C., Fürlinger, K., Closas, P. & Garcia-Molina, J.A. (2015). More than we ever dreamed possible: Processor technology for GNSS software receivers in the year 2015. *Inside GNSS Magazine, 10*(4), 62–72.

Dardari, D., Chong, C. C., & Win, M. (2008). Threshold-based time-of-arrival estimators in UWB dense multipath channels. *IEEE Transactions on Communications, 56*(8), 1366–1378. doi:10.1109/TCOMM.2008.050551

Dardari, D., Closas, P., & Djuric, P. (2015). Indoor tracking: Theory, methods, and technologies. *IEEE Transactions on Vehicular Technology, 64*(4), 1263–1278. doi:10.1109/TVT.2015.2403868

Dardari, D., Conti, A., Ferner, U., Giorgetti, A., & Win, M. Z. (2009). Ranging with ultra wide bandwidth signals in multipath environments. *Proceedings of the IEEE, 97*(2), 404–426. doi:10.1109/JPROC.2008.2008846

De la Torre, A., Peinado, A. M., Segura, J. C., Perez-Cordoba, J. L., Benitez, M. C., & Rubio, A. J. (2005, May). Histogram equalization of speech representation for robust speech recognition. *IEEE Transactions on Speech and Audio Processing, 13*(3), 355–366. doi:10.1109/TSA.2005.845805

Deak, G., Curran, K., Condell, J., Deak, D., & Kiedrowski, P. (2013). Support vector machine classification in a device-free passive localisation (DFPL) scenario. In *Image Processing and Communications Challenges 4* (Vol. 184, pp. 253–260). Berlin: Springer. doi:10.1007/978-3-642-32384-3_31

Decawave. (2017). *Indoor Positioning Systems.* Retrieved March 16, 2017, from http://www.decawave.com/

Deere. (2011). *LightSquared Interference to GPS and StarFire.* Deere submission to Federal Communications Commission.

Della Rosa, F., Leppa, H., Biancullo, S., & Nurmi, J. (2010). Ad-hoc networks aiding indoor calibrations of heterogeneous devices for fingerprinting applications. *International Conference on Indoor Positioning and Indoor Navigation,* 1-6. doi:10.1109/IPIN.2010.5647827

Dellaert, F., Fox, D., Burgard, W., & Thrun, S. (1999). Monte Carlo localization for mobile robots. In *Proceedings of IEEE International Conference Robotics and Automation 1999* (vol. 2, pp. 1322-1328). IEEE.

Dempster, A. (2009, January). QZSS's Indoor Messaging System: Friend or Foe? *Inside GNSS,* 37-40.

Dempster, A. (2011, March 31). GPS jamming could make you a casual terrorist. *The Conversation.*

Dempster, A. P., Laird, N. M., & Rubin, D. B. (1977). Maximum likelihood from incomplete data via the EM algorithm. *Journal of the Royal Statistical Society. Series B. Methodological, 39*(1), 1–38.

Dharanipragada, S., & Padmanabhan, M. (2000). A nonlinear unsupervised adaptation technique for speech recognition. *Proceedings of INTERSPEECH,* 556–559.

DiGiampaolo, E., & Martinelli, F. (2012). A passive UHF-RFID system for the localization of an indoor autonomous vehicle. *IEEE Transactions on Industrial Electronics, 59*(10), 3961–3970. doi:10.1109/TIE.2011.2173091

DiGiampaolo, E., & Martinelli, F. (2014). Mobile robot localization using the phase of passive UHF RFID signals. *IEEE Transactions on Industrial Electronics*, *61*(1), 365–376. doi:10.1109/TIE.2013.2248333

Divis, D. A. (2015, August 17). Ultra-wideband Re-emerges as a GNSS Interference Issue. *Inside GNSS*.

Doherty, L., Pister, K. S. J., & Ghaoui, L. E. (2001). Convex position estimation in wireless sensor networks. *Proceedings - IEEE INFOCOM*, *3*, 1655–1663.

Doherty, P., Coster, A. J., & Murtagh, W. (2004). Space weather effects of October–November 2003. *GPS Solutions*, *8*(4), 267–271. doi:10.1007/s10291-004-0109-3

Dong, F., Chen, Y., Liu, J., Ning, Q., & Piao, S. (2009). A calibration-free localization solution for handling signal strength variance. In *MELT* (pp. 79–90). Berlin: Springer-Verlag. doi:10.1007/978-3-642-04385-7_6

Donoho, D. (2006). Compressed Sensing. *IEEE Transactions on Information Theory*, *52*(4), 1289–1306. doi:10.1109/TIT.2006.871582

DOT. (2012). GPS Adjacent-Band Compatibility Assessment Plan. US Department of Transportation.

Dovis, F., & Ta, T. H. (2012). High sensitivity techniques for GNSS signal acquisition. In Global Navigation Satellite Systems: Signal, Theory and Applications. InTech. doi:10.5772/29453

Dovis, F. (2015). *GNSS Interference Threats and Countermeasures*. Artech House.

Dovis, F., Chiasserini, C. F., Musumeci, L., & Borgiattino, C. (2014). Context-aware peer-to-peer and cooperative positioning. *Proceedings of International Conference on Localization and GNSS 2014 ICL-GNSS 2014*, 1-6.

Durdon, T. (2014, April 19). "Timestamp Fraud": A rigged market explained in one simple animation. *Zero Hedge*.

Durgin, G. D. (2009, June). The Practical Behavior of Various Edge-Diffraction Formulas. *IEEE Antennas & Propagation Magazine*, *51*(3), 24–35. doi:10.1109/MAP.2009.5251189

Economist. (2011, March 10). No jam tomorrow. *The Economist*.

Economist. (2013, July 27). GPS jamming: out of sight. *The Economist*.

El-Sheimy, N., Hou, H., & Niu, X. (2007). Analysis and modeling of inertial sensors using Allan variance. *IEEE Transactions on Instrumentation and Measurement*, *57*(1), 140–149. doi:10.1109/TIM.2007.908635

Eric, M., Dukic, M., & Vucic, D. (2008), Direct Position Estimation of UWB Transmitters in Multipath Conditions. *Proceedings of IEEE International Conference on Ultra-Wide Band (ICUWB)*, 241-244.

Eurocontrol. (1999). *GNSS Frequency Protection Requirements*. Eurocontrol Experimental Centre, EEC Report No. 337.

Fang, S.-H., & Lin, T.-N. (2008). Robust wireless LAN location fingerprinting by SVD-based noise reduction. *2008 3rd International Symposium on Communications, Control and Signal Processing*, 295-298.

Fang, S.-H., Chuang, C.-C., & Wang, C. (2012). Attack-resistant wireless localization using an inclusive disjunction model. *IEEE Transactions on Communications*, *60*(5), 1209–1214. doi:10.1109/TCOMM.2012.040212.100291

Fang, S.-H., & Lin, T.-N. (2008, November). Indoor location system based on discriminant-adaptive neural network in IEEE 802.11 environments. *IEEE Transactions on Neural Networks*, *19*(11), 1973–1978. doi:10.1109/TNN.2008.2005494 PMID:19000967

Fang, S.-H., & Lin, T.-N. (2012, January). Principal component localization in indoor WLAN environments. *IEEE Transactions on Mobile Computing*, *11*(1), 100–110. doi:10.1109/TMC.2011.30

Fang, S.-H., Lin, T.-N., & Lin, P.-C. (2008, May). Location fingerprinting in a decorrelated space. *IEEE Transactions on Knowledge and Data Engineering*, *20*(5), 685–691. doi:10.1109/TKDE.2007.190731

Fang, S.-H., & Wang, C.-H. (2011, March). A dynamic hybrid projection approach for improved Wi-Fi location fingerprinting. *IEEE Transactions on Vehicular Technology*, *60*(3), 1037–1044. doi:10.1109/TVT.2011.2107757

Fang, S.-H., Wang, C.-H., Chiou, S.-M., & Lin, P. (2012). Calibration-free approaches for robust Wi-Fi positioning against device diversity: A performance comparison. *Proc. IEEE Transactions on Vehicular Technology Conference Spring*, 1–5. doi:10.1109/VETECS.2012.6240088

Fang, S.-H., Wang, C.-H., Huang, T.-Y., Yang, C.-H., & Chen, Y.-S. (2012, April). An enhanced ZigBee indoor positioning system with an ensemble approach. *IEEE Communications Letters*, *16*(4), 564–567. doi:10.1109/LCOMM.2012.022112.120131

Faragher, R., & Harle, R. (2014). An analysis of the accuracy of Bluetooth Low Energy for indoor positioning applications. In *Proceedings of 27th International Technical Meeting of The Satellite Division of the Institute of Navigation ION GNSS+ 2014* (pp. 201-210). Institute of Navigation.

Farrell, J. A. (2008). *Aided Navigation, GPS with High Rate Sensors*. New York: McGraw Hill.

Faseth, T., Winkler, M., & Arthaber, H. (2011, Sept.). *The influence of multipath propagation on phase-based narrowband positioning principles in UHF RFID*. Paper presented at Antennas and Propagation in Wireless Communications (APWC), IEEE-APS Topical Conference on, Torino, Italy. doi:10.1109/APWC.2011.6046829

FBI. (2014, October 2). Cargo Thieves use GPS Jammers to Mask GPS Trackers. FBI Cyber Division.

Federal Communications Commission. (2002). *Revision of part 15 of the commission's rules regarding ultra-wideband transmission systems, first report and order* (ET Docket 98-153). Adopted Feb.14, 2002, Released Apr. 22, 2002. Author.

Feng, C., Au, W., Valaee, S., & Tan, Z. (2009). Orientation-aware indoor localization using affinity propagation and compressive sensing. *IEEE International Workshop on Computational Advances in Multi-Sensor Adaptive Processing (CAMSAP)*, 261–264. doi:10.1109/CAMSAP.2009.5413285

Feng, C., Au, W., Valaee, S., & Tan, Z. (2012, December). Received-signal-strengthbased indoor positioning using compressive sensing. *IEEE Transactions on Mobile Computing*, *11*(12), 1983–1993. doi:10.1109/TMC.2011.216

Fenton, P. C., & Jones, J. (2005). The theory and performance of NovAtel Inc.'s vision correlator. *Proceedings of ION-GNSS*.

Fernandez-Prades, C., Closas, P., Vila'Valls, J., Naharro, M. F., & Sanjose, A. T. (2012). Assisted GNSS in LTE-Advanced Networks and its Application to Vector Tracking Loops. *Proceedings of ION ITM*, 1462-1476.

Figuera, C., Rojo-Álvarez, J. L., Mora-Jiménez, I., Guerrero-Curieses, A., Wilby, M., & Ramos-López, J. (2011, July). Time-space sampling and mobile device calibration for WiFi indoor location systems. *IEEE Transactions on Mobile Computing*, *10*(7), 913–926. doi:10.1109/TMC.2011.84

Flanagan, T. (2014, December 11). Time Stamps Get Nanosecond Precision. *Markets Media*.

Flury, B. (1997). *A First Course in Multivariate Statistics*. New York: Springer-Verlag. doi:10.1007/978-1-4757-2765-4

Foucras, M., Ngayap, U., Bacard, F., & Ekambi, B. (2016). Acquisition performance comparison of new generation of GNSS BOC-modulated signals. *Proceedings of ION GNSS*, 107-119.

Fox-Brewster, T. (2015, August 8). Watch GPS attacks that can kill DJI drones or bypass White House ban. *Forbes*.

Foxlin, E. (2005). Pedestrian tracking with shoe-mounted inertial sensors. *IEEE Computer Graphics and Applications, 25*(6), 38–46. doi:10.1109/MCG.2005.140 PMID:16315476

Frey, B. J., & Dueck, D. (2007, February). Clustering by passing messages between data points. *Science, 315*(5814), 2007. doi:10.1126/science.1136800 PMID:17218491

Frieden, B. R. (2004). Science from Fisher information – a unification. Cambridge University Press.

Friis, H. T. (1946). A note on a simple transmission formula. *Proceedings of the IRE, 34*(5), 254–256. doi:10.1109/JRPROC.1946.234568

Gad-el-Hak, M. (2001). Introduction. In M. Gad-el-Hak (Ed.), *The MEMS handbook.* CRC Press. doi:10.1201/9781420050905.sec1

Gao, G., Heng, L., Hornbostel, A., Denks, H., Meurer, M., Walter, T., & Enge, P. (2013). DME/TACAN interference mitigation for GNSS: Algorithms and flight test results. *GPS Solutions, 17*(4), 561–573. doi:10.1007/s10291-012-0301-9

Gao, S., Zhang, S., Wang, G., & Li, Y. (2016). Robust second-order cone relaxation for TW-TOA-based localization with clock imperfection. *IEEE Signal Processing Letters, 23*(8), 1047–1051. doi:10.1109/LSP.2016.2580743

Garcia-Valverde, T., Garcia-Sola, A., Botia, J., & Gomez-Skarmeta, A. (2012). Automatic design of an indoor user location infrastructure using a memetic multi-objective approach. *IEEE Transactions on Systems, Man and Cybernetics. Part C, Applications and Reviews, 42*(5), 704–709. doi:10.1109/TSMCC.2012.2187443

Garin, L. J. (2005). The "shaping correlator," novel multipath mitigation technique applicable to GALILEO BOC(1,1) modulation waveforms in high volume markets. *Proceedings of European Navigation Conference.*

Garin, L., & Rousseau, J. (1997). Enhanced strobe correlator multipath rejection for code and carrier. *Proceedings of ION GPS*, 559-568.

Gauthier, J. P., Glennon, E. P., Rizos, C., & Dempster, A. G. (2013). *Time transfer performance of Locata – Initial results. 45th Annual Precise Time and Time Interval System and Application Meeting*, Bellevue, WA.

Geiger, A., Ziegler, J., & Stiller, C. (2011). Stereoscan: Dense 3D reconstruction in real-time. *IEEE Intelligent Vehicles Symposium (IV).*

Geiger, B., Vogel, C., & Soudan, M. (2012, April). Comparison between ratio detection and threshold comparison for GNSS acquisition. *IEEE Transactions on Aerospace and Electronic Systems, 48*(2), 1772–1779. doi:10.1109/TAES.2012.6178098

Gelb, A. (1974). *Applied . Cambridge, Massachusetts, USA, and London.* UK: The M.I.T. Press.

Geng, Y., He, J., & Deng, H., & Kaveh. (2013, June). *Modeling the effect of human body on TOA ranging for indoor human tracking with wrist mounted sensor.* Paper presented at the International Symposium on Wireless Personal Multimedia Communications (WPMC), Atlantic City, NJ.

Gezici, S. (2008). A survey on wireless position estimation. *Wireless Personal Communications, 44*(3), 263–282. doi:10.1007/s11277-007-9375-z

Gezici, S., Tian, Z., Biannakis, B. G., Kobayashi, H., Molisch, A. F., Poor, H. V., & Sahinoglu, Z. (2005). Localization via ultra-wideband radios: A look at positioning aspects for future sensor networks. *IEEE Transactions on Signal Processing, 22*(4), 70–84. doi:10.1109/MSP.2005.1458289

Ghaboosi, N., & Jamalipour, A. (2012). *Location estimation using geometry of overhearing.* Academic Press.

Ghaddar, M., Talbi, L., Denidni, T. A., & Sebak, A. (2007). A conducting cylinder for modeling human body presence in indoor propagation channel. *IEEE Transactions on Antennas and Propagation*, *55*(11), 3099–3103. doi:10.1109/TAP.2007.908563

Gholami, M. R., Reza, M., Rydstrom, M., & Strom, E. G. (2010). Positioning of Node using Plane Projection onto Convex Sets. *Proceedings of Wireless Communications and Networking Conference (WCNC)*. doi:10.1109/WCNC.2010.5506717

Gholami, M. R., Ström, E. G., Sottile, F., Dardari, D., Conti, A., Gezici, S., & Spirito, M. A. (2010). Static positioning using UWB range measurements. *Proceedings of Future Network and Mobile Summit.*

Gholami, M. R., Tetruashvili, L., Strom, E. G., & Censor, Y. (2013). Cooperative wireless sensor network positioning via implicit convex feasibility. *IEEE Transactions on Signal Processing*, *61*(23), 5830–5840. doi:10.1109/TSP.2013.2279770

Gibbons, G. (2011, Jauary). GNSS interoperability, not so easy after all. *Inside GNSS*, 28-32.

Gibbons, G. (2013, August 30). FCC fines operator of GPS jammer that affected Newark Airport GBAS. *Inside GNSS.*

Gibbons, G. (2014, April 2). GLONASS suffers temporary systemwide outage; multi-GNSS receiver overcomes problem. *Inside GNSS.*

Glennon, E. P., & Dempster, A. G. (2006b). *Apparatus and Method for Mitigation of Cross-Correlation in GPS Systems.* US patent application 11/460,692.

Glennon, E. P., & Dempster, A. G. (2004). A review of GPS cross correlation mitigation techniques. *Proceedings of IGNSS.*

Glennon, E. P., & Dempster, A. G. (2005). A novel cross correlation mitigation technique. *Proceedings of ION-GNSS.*

Glennon, E. P., & Dempster, A. G. (2006a). Cross correlation mitigation by adaptive orthogonalization using constraints - new results. *Proceedings of ION-GNSS.*

Glennon, E. P., & Dempster, A. G. (2007a). Cross correlation mitigation techniques for software GPS C/A code receivers. *Proceedings of IGNSS.*

Glennon, E. P., & Dempster, A. G. (2011). Delayed PIC for postcorrelation mitigation of continuous wave and multiple access interference in GPS receivers. *IEEE Transactions on Aerospace and Electronic Systems*, *47*(4), 2544–2557. doi:10.1109/TAES.2011.6034650

Goshen-Meskin, D., & Bar-Itzhack, I. Y. (1992). Unified approach to inertial navigation system error modelling. *Journal of Guidance, Control, and Dynamics*, *15*(3), 648–653. doi:10.2514/3.20887

Goyal, P., Ribeiro, V., Saran, H., & Kumar, A. (2011). Strap-down pedestrian dead-reckoning system. In *Proceedings of International Conference on Indoor Positioning and Indoor Navigation IPIN'11* (pp. 1-7). IEEE Xplore.

Grant, A., Williams, P., Ward, N., & Basker, S. (2009). GPS jamming and the impact on maritime navigation. *Journal of Navigation*, *62*(2), 173–187. doi:10.1017/S0373463308005213

Greenemeier, L. (2016, February). GPS and the World's First "Space War". *Scientific American*, 8.

Greenspan, R. L. (1995). Inertial navigation technology from 1970–1995. *Navigation*, *42*(1), 165–185. doi:10.1002/j.2161-4296.1995.tb02334.x

Grejner-Brzezinska, D. A., Toth, C. K., Li, L., Park, J., Wang, X., Sun, H., & Zheng, Y. F. et al. (2009). Positioning in GPS-challenged environments: Dynamic sensor network with distributed GPS aperture and internodal ranging signals. In *Proceedings of ION GNSS Conference* (pp. 111-123). Institute of Navigation.

Grejner-Brzezinska, D. A., Toth, C. K., Sun, H., Wang, X., & Rizos, C. (2011). A robust solution to high-accuracy geolocation: Quadruple integration of GPS, IMU, pseudolite, and terrestrial laser scanning. *IEEE Transactions on Instrumentation and Measurement*, *60*(11), 3694–3708. doi:10.1109/TIM.2010.2050981

Grewal, M. S., & Andrews, A. P. (Eds.). (2008). *Kalman filtering: theory and practice using MATLAB* (3rd ed.). Piscataway, NJ: Wiley Blackwell. doi:10.1002/9780470377819

Grewal, M., Weill, S., & Andrews, A. (2007). *Global Positioning Systems, Inertial Navigation, and Integration*. Wiley & Sons Inc. doi:10.1002/0470099720

Griffin, J., & Durgin, G. (2009). Complete link budgets for backscatter-radio and RFID systems. *IEEE Antennas & Propagation Magazine*, *15*(2), 11–25. doi:10.1109/MAP.2009.5162013

Groves, P. D. (2008). *Principles of GNSS, Inertial and Multisensory Integrated Navigation Systems*. Artech House.

Groves, P. D. (2013). *Principles of GNSS, Inertial, and Multisensor Integrated Navigation Systems*. Artech House.

Guo, Q., Bebek, O., Cavusoglu, M. C., Mastrangelo, C. H., & Young, D. J. (2015). A personal navigation system using MEMS-based high-density ground reaction sensor array and inertial measurement unit. In *Proceedings of Transducers - 2015, International Conference on Solid-State Sensors, Actuators and Microsystems* (pp.1077-1080). Piscataway, NJ: IEEE. doi:10.1109/TRANSDUCERS.2015.7181113

Guo, Y., Huang, K., Jiang, N., Guo, X., Li, Y., & Wang, G. (2015, March). An Exponential-Rayleigh Model for RSS-Based Device-Free Localization and Tracking. *IEEE Transactions on Mobile Computing*, *14*(3), 484–494. doi:10.1109/TMC.2014.2329007

Gu, Q. (2005). *RF system design of transceivers for wireless communications*. Springer Science & Business Media.

Gusenbauer, D., Isert, C., & Krösche, J. (2010) Self-contained indoor positioning on off-the-shelf mobile devices. *Proc. ICIPIN*, 1–9. doi:10.1109/IPIN.2010.5646681

Gusi-Amigo, A., Closas, P., Mallat, A., & Vandendorpe, L. (2014), Cramer-Rao Bound Analysis of UWB based Localization Approaches. *Proceedings of IEEE International Conference on Ultra-Wideband (ICUWB)*. doi:10.1109/ICUWB.2014.6958943

Gustafson, D. E., Elwell, J. M., & Soltz, J. A. (2006). Innovative indoor geolocation using RF multipath diversity. In *Proceedings of Position, Location, and Navigation Symposium, 2006 IEEE/ION* (vol.2006, pp.904-912). Piscataway, NJ: IEEE. doi:10.1109/PLANS.2006.1650690

Guvenc, I., & Chong, C. C. (2009). A survey on TOA based wireless localization and NLOS mitigation techniques. *IEEE Communications Surveys and Tutorials*, *11*(3), 107–124. doi:10.1109/SURV.2009.090308

Güvenç, I., Chong, C. C., Watanabe, F., & Inamura, H. (2007). NLOS identification and weighted least-squares localization for UWB systems using multipath channel statistics. *EURASIP Journal on Advances in Signal Processing*, *36*, 1–14.

Guvenc, I., Chong, C. C., Watanabe, F., & Inamura, H. (2008). NLOS identification and weighted least-squares localization for UWB systems using multipath channel statistics. *EURASIP Journal on Advances in Signal Processing*, (1): 36.

Gu, Y., Lo, A., & Niemegeers, I. (2009, January). A survey of indoor positioning systems for wireless personal networks. *EEE Communications Surveys & Tutorials*, *11*(1), 13–32. doi:10.1109/SURV.2009.090103

Hackett, R. (2015, August 4). Here's the scary new target hackers are going after. *Fortune*.

Haddrell, A. (2005, March). The enemy inside. *GPS World*.

Haeberlen, A., Flannery, E., Ladd, A. M., Rudys, A., Wallach, D. S., & Kavraki, L. E. (2004). Practical robust localization over large-scale 802.11 wireless networks. In MobiCom (pp. 70-84). ACM.

Hall, P. (1983). Chi squared approximations to the distribution of a sum of independent random variables. *Annals of Probability*, *11*(4), 1028–1036. doi:10.1214/aop/1176993451

Halperin, D., Hu, W., Sheth, A., & Wetherall, D. (2010, August). Predictable 802.11 packet delivery from wireless channel measurements. *Computer Communication Review*, 41.

Hamilton, B. R., Ma, X., Baxley, R. J., & Matechik, S. M. (2014). Propagation modeling for radio frequency tomography in wireless networks. *IEEE Journal of Selected Topics in Signal Processing*, *8*(1), 55–65. doi:10.1109/JSTSP.2013.2287471

Han, K., & Cho, S. (2010, December). *Advanced LANDMARC with adaptive k-nearest algorithm for RFID location system.* Paper presented at the 2010 2nd IEEE International Conference on Network Infrastructure and Digital Content, Beijing, China. doi:10.1109/ICNIDC.2010.5657852

Han, J., Qian, C., Wang, X., Ma, D., Zhao, J., Xi, W., & Wang, Z. et al. (2016). Twins: Device-free object tracking using passive tags. *IEEE/ACM Transactions on Networking*, *24*(3), 1605–1617. doi:10.1109/TNET.2015.2429657

Han, S., & Wang, J. (2011). Quantization and colored noises error modeling for inertial sensors for GPS/INS integration. *IEEE Sensors Journal*, *11*(6), 1493–1503. doi:10.1109/JSEN.2010.2093878

Hara, S., Anzai, D., Yabu, T., Lee, K., Derham, T., & Zemek, R. (2013). A perturbation analysis on the performance of TOA and TDOA localization in mixed LOS/NLOS environments. *IEEE Transactions on Communications*, *61*(2), 679–689. doi:10.1109/TCOMM.2013.012313.110509

Harmat, A., Trentini, M., & Sharf, I. (2015). Multi-Camera Tracking and Mapping for Unmanned Aerial Vehicles in Unstructured Environments. *Journal of Intelligent & Robotic Systems*, *78*(2), 291–317. doi:10.1007/s10846-014-0085-y

Harper, N. (2010). *Server-Side GPS and Assisted-GPS in Java.* Artech House.

Harrington, R. L., & Dolloff, J. T. (1976) *The inverted range: GPS user test facility. IEEE Position Location and Navigation Symposium*, San Diego, CA.

Hartley, R., & Zisserman, A. (2003). *Multiple View Geometry in Computer Vision.* Cambridge University Press.

Harvey, B. (2007). *"Military programs". The Rebirth of the Russian Space Program* (1st ed.). Springer.

Hashemi, H. (1993). The indoor radio propagation channel. *Proceedings of the IEEE*, *81*(7), 943–968. doi:10.1109/5.231342

Hashemipour, H. R., Roy, S., & Laub, A. J. (1988). Decentralized structures for parallel Kalman filtering. *IEEE Transactions on Automatic Control*, *AC-33*(1), 88–94. doi:10.1109/9.364

Hasnur Rabiain, A., Kealy, A., Retscher, G., Alam, N., Dempster, A., Grejner-Brzezinska, D. A., & Gikas, V. et al. (2013). Cooperative positioning algorithms and techniques for land mobile applications. *Proceedings of 8th International Symposium on Mobile Mapping Technology.*

Hatfield, D. N. (2002). *A report on technical and operational issues impacting the provision of wireless enhanced 911 services.* Federal Communications Commission.

Hegarty, C. J. (2006). Optimal and Near-optimal Detectors for Acquisition of the GPS L5 Signal. *Proceedings of the 2006 National Technical Meeting of The Institute of Navigation*, 717-725.

Hero-III, A. O., Fessler, J. A., & Usman, M. (1996). Exploring estimator bias-variance tradeoffs using the uniform CR bound. *IEEE Transactions on Signal Processing*, *44*(8), 2026–2041. doi:10.1109/78.533723

Herrera-May, A. L., Aguilera-Cortés, L. A., García-Ramírez, P. J., & Manjarrez, E. (2009). Resonant magnetic field sensors based on MEMS technology. *Sensors (Basel)*, *9*(10), 7785–7813. doi:10.3390/s91007785 PMID:22408480

Herrera-May, A. L., Aguilera-Cortés, L. A., García-Ramírez, P. J., Mota-Carrillo, N. B., Padrón-Hernández, W. Y., & Figueras, E. (2011). Development of resonant magnetic field microsensors: Challenges and future applications. In O. Minin (Ed.), *Microsensors* (pp. 65–84). Rijeka, Croatia: InTech.

Hide, C., Botterill, T., & Andreotti, M. (2009). An integrated IMU, GNSS and image recognition sensor for pedestrian navigation. In *Proceedings of ION GNSS Conference*. Institute of Navigation.

Hightower, J., & Boriello, G. (2001). *A survey and taxonomy of location systems for ubiquitous computing. Technical Report*. University of Washington, Department of Computer Science and Engineering.

Holton, J. R. (2004). *An Introduction to Dynamic Meteorology* (4th ed.). Burlington, MD: Elsevier Inc.

Hong, J., & Ohtsuki, T. (2015). Signal eigenvector-based device-free passive localization using array sensor. *IEEE Transactions on Vehicular Technology*, *64*(4), 1354–1363. doi:10.1109/TVT.2015.2397436

Horgan, J., Flannery, F., & Toal, D. (2009). Towards real time vision based UUV navigation using GPU technology. *Oceans*, 2009.

Hossain, A. K. M. M., Van, H. N., & Soh, W.-S. (2010). Utilization of user feedback in indoor positioning system. *Pervasive and Mobile Computing*, *6*(4), 467–481. doi:10.1016/j.pmcj.2010.04.003

Hsieh, S.-J. (2013). Challenges and opportunities in MEMS development, assembly and applications. *Assembly Automation*, *33*(2), 2. doi:10.1108/aa.2013.03333baa.001

Huang, S., & Dissanayake, G. (2007). Convergence and Consistency Analysis for Extended Kalman Filter Based SLAM. *IEEE Transactions on Robotics*, *23*(5), 1036–1049. doi:10.1109/TRO.2007.903811

IEEE Standard for Information Technology. (2007, June). *Telecommunications and information exchange between systems - local and metropolitan area networks - specific requirements - part 11: Wireless LAN medium access control (MAC) and physical layer (PHY) specifications," IEEE Std 802.11-2007 (Revision of IEEE Std 802.11-1999)*. IEEE.

IEEE. (1997). *IEEE Standard Specification Format Guide and Test Procedure for Single-Axis Interferometric Fiber Optic Gyros*. IEEE Standard Specification Format.

IG. (2012, July 16). UAVs vulnerable to civil GPS spoofing. *Inside GNSS*.

IG. (2015, November 30). GPS OCS implements threat protection technology. *Inside GNSS*.

Ihler, A. T., Fisher, J. W., Moses, R. L., & Willsky, A. S. (2005). Nonparametric belief propagation for self-localization of sensor networks. *IEEE Journal on Selected Areas in Communications*, *23*(4), 809–819. doi:10.1109/JSAC.2005.843548

IS. (2013, September 24). *IS-GPS-200H Navstar GPS Space Segment/Navigation User Interfaces*. US DOD.

Jahn, J., Batzer, U., Seitz, J., Patino-Studencka, L., & Boronat, J. (2010). Comparison and evaluation of acceleration based step length estimators for handheld devices. *Proc. ICIPIN*, 1–6. doi:10.1109/IPIN.2010.5646888

Jahromi, A. J. (2013). *GNSS Signal Authenticity Verification in the Presence of Structural Interference* (PhD thesis). University of Calgary.

Jakeman, E. (1982). Fresnel scattering by a corrugated random surface with fractal slope. *JOSA*, *72*(8), 1034–1041. doi:10.1364/JOSA.72.001034

James, M. D. (2008). *Multiarray passive acoustic localization and tracking* (Ph.D. dissertation). Dept. Mech. Eng., Virginia Polytechnic Institute and State Univ., Blacksburg, VA.

Jaradat, M. A. K., & Abdel-Hafez, M. F. (2014). Enhanced, delay dependent, intelligent fusion for INS/GPS navigation system. *IEEE Sensors Journal, 14*(5), 1545–1554. doi:10.1109/JSEN.2014.2298896

Jensen, A. B. O., & Mitchell, C. (2011, February). GNSS and the ionosphere: What's in store for the next solar maximum? *GPS World*, 40–48.

Jiang, W., Li, Y., & Rizos, C. (2013). On-the-fly Locata/inertial navigation system integration for precise maritime application. *Measurement Science & Technology, 24*(10), 105104. doi:10.1088/0957-0233/24/10/105104

Jiménez, A. R., Seco, F., Prieto, J. C., & Guevara, J. (2010) *Indoor Pedestrian Navigation using an INS/EKF framework for Yaw Drift Reduction and a Foot-mounted IMU. 7th IEEE Workshop on Positioning Navigation and Communication*, Dresden, Germany. doi:10.1109/WPNC.2010.5649300

Jin, G., Lu, X., & Park, M. (2006, June). *An indoor localization mechanism using active RFID tag*. Paper presented at the IEEE International Conference on Sensor Networks, Ubiquitous, and Trustworthy Computing (SUTC'06), Taichung.

Jirawimut, R., Ptasinski, P., Garaj, V., Cecelja, F., & Balachandran, W. (2003). A method for dead reckoning parameter correction in pedestrian navigation system. *IEEE Transactions on Instrumentation and Measurement, 52*(1), 209–215. doi:10.1109/TIM.2002.807986

Jo, Y. H., Lee, J. Y., Ha, D. H., & Kang, S. H. (2006). Accuracy enhancement for UWB indoor positioning using ray tracing. In *Proceedings of Position, Location, and Navigation Symposium, 2006 IEEE/ION* (vol.2006, pp.565-568). Piscataway, NJ: IEEE.

Jolliffe, I. (1986). *Principal Component Analysis*. New York: Springer-Verlag. doi:10.1007/978-1-4757-1904-8

Jones, J., Fenton, P., & Smith, B. (2004). *Theory and Performance of the Pulse Aperture Correlator*. NovAtel Inc. Retrieved from http://www.novatel.com/Documents/Papers/PAC.pdf

Juan, J. M., Hernandez-Pajares, M., Sanz, J., Ramos-Bosch, P., Aragon-Angel, A., Orus, R., & Tossaint, M. et al. (2012). Enhanced precise point positioning for GNSS users. *IEEE Transactions on Geoscience and Remote Sensing, 50*(10), 4213–4222. doi:10.1109/TGRS.2012.2189888

Kaemarungsi, K. (2006). Distribution of WLAN received signal strength indication for indoor location determination. *Proc. ISWPC*, 1–6. doi:10.1109/ISWPC.2006.1613601

Kaemarungsi, K., & Krishnamurthy, P. (2004). Modeling of indoor positioning systems based on location fingerprinting. *Proc. IEEE 23rd Annual Joint Conference on Computing and Communication. Soc., 2*, 1012–1022. doi:10.1109/INFCOM.2004.1356988

Kaemarungsi, K., & Krishnamurthy, P. (2004). Properties of indoor received signal strength for WLAN location fingerprinting. *Proc. 1st Annual International Conference on Mobile Ubiquitous Systems Network Servers*, 14–23. doi:10.1109/MOBIQ.2004.1331706

Kak, A. C., & Slaney, M. (1988). *Principles of computerized tomographic imaging*. New York: IEEE.

Kaltiokallio, O., Bocca, M., & Patwari, N. (2012). Enhancing the Accuracy of Radio Tomographic Imaging Using Channel Diversity. *The 9th IEEE International Conference on Mobile Ad hoc and Sensor Systems (MASS), 2012 (pp. 254-262). IEEE.*

Kaltiokallio, O., Bocca, M., & Patwari, N. (2012, Oct). Follow @grandma: Long-term device-free localization for residential monitoring. *IEEE 37th Conference on Local Computer Networks Workshops (LCN Workshops), 2012*, 991-998. doi:10.1109/LCNW.2012.6424092

Kaltiokallio, O., Yiğitler, H., & Jäntti, R. (2017). A Three-State Received Signal Strength Model for Device-Free Localization. *IEEE Transaction on Vehicular Technology.*

Kaltiokallio, O., Bocca, M., & Patwari, N. (2014). A Fade Level-based Spatial Model for Radio Tomographic Imaging. *IEEE Transactions on Mobile Computing, 13*(6), 1159–1172. doi:10.1109/TMC.2013.158

Kaltiokallio, O., Jäntti, R., & Patwari, N. (2017, August). ARTI: An Adaptive Radio Tomographic Imaging System. *IEEE Transactions on Vehicular Technology, 66*(8), 7302–7316. doi:10.1109/TVT.2017.2664938

Kaltiokallio, O., Yiğitler, H., Jäntti, R., & Patwari, N. (2014, April). Non-invasive respiration rate monitoring using a single COTS TX-RX pair. *Proceedings of the 13th International Symposium on Information Processing in Sensor Networks, IPSN-14*, 59-69. doi:10.1109/IPSN.2014.6846741

Kanatas, A., Kountouris, I., Kostaras, G., & Constantinou, P. (1997). A UTD propagation model in urban microcellular environments. *IEEE Transactions on Vehicular Technology, 46*(1), 185–193. doi:10.1109/25.554751

Kanso, M. A., & Rabbat, M. G. (2009). Compressed RF Tomography for Wireless Sensor Networks: Centralized and Decentralized Approaches. *5th IEEE Intl. Conf. on Distributed Computing in Sensor Systems (DCOSS-09).* doi:10.1007/978-3-642-02085-8_13

Kaplan, E. D. (2005). *Understanding GPS: Principles and Applications* (2nd ed.). Artech House.

Kay, S. M. (1993). *Fundamentals of Statistical Signal Processing: Estimation Theory*. Upper Saddle River, NJ: Prentice Hall.

Kay, S. M. (2013). *Fundamentals of statistical signal processing: Practical algorithm development*. Pearson Education.

Kealy, A., & Retscher, G. (2017). MEMS and wireless options in cellular phones for user localization. *Proceedings of ION Pacific PNT.*

Kealy, A., Retscher, G., Toth, C. K., Hasnur-Rabiain, A., Gikas, V., Grejner-Brzezinska, D. A., & Moore, T. et al. (2015). Collaborative navigation as a solution for PNT applications in GNSS challenged environments – Report on field trials of a joint FIG/IAG working group. *Journal of Applied Geodesy, 9*(4), 244–263. doi:10.1515/jag-2015-0014

Keller, J. B. (1962, February). Geometrical Theory of Diffraction. *Journal of the Optical Society of America, 52*(2), 116–130. doi:10.1364/JOSA.52.000116 PMID:14454948

Kerr, R. A. (2009, June). Are we ready for the next solar maximum? No way, say scientists. *Science, 26*, 1640–1641. doi:10.1126/science.324_1640 PMID:19556483

Ke, W., & Wu, L. (2012). Sparsity-Based Multi-Target Direct Positioning Algorithm Based on Joint-Sparse Recovery. *Progress in Electromagnetics Research, 27*, 99–114. doi:10.2528/PIERC11110704

Khodjaev, J., Park, Y., & Malik, A. S. (2010). Survey of NLOS identification and error mitigation problems in UWB-based positioning algorithms for dense environments. *Annals of Telecommunications -. Annales des Télécommunications, 65*(5-6), 301–311. doi:10.1007/s12243-009-0124-z

Kim, B., & Kong, S.-H. (2014, May). Design of FFT-based TDCC for GNSS Acquisition. *IEEE Transactions on Wireless Communications, 13*(5), 2798–2808. doi:10.1109/TWC.2014.040714.131884

Kim, B., & Kong, S.-H. (2014, May). Determination of Detection Parameters on TDCC Performance. *IEEE Transactions on Wireless Communications*, *13*(5), 2422–2431. doi:10.1109/TWC.2014.0204014.131704

Kim, B., & Kong, S.-H. (2015). Ultra-fast L2-CL code acquisition for a dual band GPS receiver. *Journal of Positioning, Navigation, and Timing*, *4*(4), 151–160. doi:10.11003/JPNT.2015.4.4.151

Kim, J. W., Jang, H. J., Hwang, D.-H., & Park, C. (2004). A step, stride and heading determination for the pedestrian navigation system. *Journal of Global Positioning Systems*, *3*(1-2), 273–279. doi:10.5081/jgps.3.1.273

King, T., Kopf, S., Haenselmann, T., Lubberger, C., & Effelsberg, W. (2006). COMPASS: A probabilistic indoor positioning system based on 802.11 and digital compasses. *Proc. WiNTECH*, 34–40. doi:10.1145/1160987.1160995

Kintner, P. M., Humphreys, T., & Hinks, J. (2009, July). GNSS and ionospheric scintillation: how to survive the next solar maximum. *Inside GNSS*, 22-30.

Kjærgaard, M. B. (2007). A taxonomy for radio location fingerprinting. in Proceedings of the 3rd international conference on Location-and context-awareness. *LoCA'07,* 139-156.

Kjærgaard, M. B. (2011, February). Indoor location fingerprinting with heterogeneous clients. *Pervasive and Mobile Computing*, *7*(1), 31–43. doi:10.1016/j.pmcj.2010.04.005

Kjærgaard, M., & Munk, C. (2008). Hyperbolic location fingerprinting: A calibration-free solution for handling differences in signal strength. *Proc. IEEE PerCom*, 110–116.

Klein, D., & Parkinson, B. W. (1984). The use of pseudo-satellites for improving GPS performance. *Navigation, Journal of the U.S. Institute of Navigation*, *31*(4), 303–315. doi:10.1002/j.2161-4296.1984.tb00881.x

Kneupper, M. R. (2011). *Operating in a GPS Denied Environment in the PACOM Maritime Domain.* DTIC Report ADA546337.

Kollár, I. (1994). Bias of mean value and mean square value measurements based on quantized data. *IEEE Transactions on Instrumentation and Measurement*, *43*(5), 733–739. doi:10.1109/19.328894

Kong, S.-H. (2014). Fast Multi-Satellite ML Acquisition for A-GPS. *IEEE Transactions on Wireless Communications*, *13*(9), 4935–4946. doi:10.1109/TWC.2014.2327101

Kong, S.-H. (2015, November). SDHT for Fast Detection of Weak GNSS Signals. *IEEE Journal on Selected Areas in Communications*, *33*(11), 2366–2378. doi:10.1109/JSAC.2015.2430291

Kosba, A. E., Saeed, A., & Youssef, M. (2012). Rasid: A robust WLAN device-free passive motion detection system. *2012 IEEE International Conference on Pervasive Computing and Communications (PerCom)*, Lugano 180-189.

Kotanen, A., Hännikäinen, M., Leppäkoski, H., & Hämälainen, T. D. (2003). Positioning with IEEE 802.11b wireless LAN. *Proceedings of 14th International Symposium on Personal, Indoor and Mobile Radio Communications PIMRC*, 2218-2222. doi:10.1109/PIMRC.2003.1259110

Kschischang, F. R., Frey, B. J., & Loeliger, H. A. (2001). Factor graphs and the sum-product algorithm. *IEEE Transactions on Information Theory*, *47*(2), 498–519. doi:10.1109/18.910572

Kuo, S.-P., & Tseng, Y.-C. (2008, May). A scrambling method for fingerprint positioning based on temporal diversity and spatial dependency. *IEEE Transactions on Knowledge and Data Engineering*, *20*(5), 678–684. doi:10.1109/TKDE.2007.190730

Kuo, S.-P., & Tseng, Y.-C. (2011, February). Discriminant minimization search for largescale RF-based localization systems. *IEEE Transactions on Mobile Computing*, *10*(2), 291–304. doi:10.1109/TMC.2010.67

Kushki, A., Plataniotis, K., & Venetsanopoulos, A. (2007, June). Kernel-based positioning in wireless local area networks. *IEEE Transactions on Mobile Computing*, *6*(6), 689–705. doi:10.1109/TMC.2007.1017

Ladd, A. M. (2002). *Robotics-based location sensing using wireless ethernet.* Proc. Mobile Computing Network.

Ladetto, Q., & Merminod, B. (2002) *Digital magnetic compass and gyroscope integration for pedestrian navigation. 9th Saint Petersburg International Conference on Integrated Navigation Systems*, Saint Petersburg, Russia.

Lagunas, E., Navarro, M., Closas, P., & Najar, M. (2014), Spatial Sparsity Based Direct Positioning for IR-UWB in IEEE 802.15.4a Channels. *Proceedings of IEEE International Conference on Ultra-Wide Band (ICUWB)*. doi:10.1109/ICUWB.2014.6958999

Lakhzouri, A., Lohan, E. S., Hamila, R., & Renfors, M. (2003). Extended Kalman filter channel estimation for line-of-sight detection in WCDMA mobile positioning. *EURASIP Journal on Advances in Signal Processing*, *2003*(13), 1268–1278. doi:10.1155/S1110865703306018

Larsson, E. G., & Danev, D. (2010). Accuracy comparison of LS and squared range LS for source localization. *IEEE Transactions on Signal Processing*, *58*(2), 916–923. doi:10.1109/TSP.2009.2032036

Laxton, M. C., & Devilbiss, S. L. (1997). GPS multipath mitigation during code tracking. *Proceedings of the American Control Conference*, 1429-1433.

Lázaro, A., Girbau, D., & Salinas, D. (2009). Radio link budgets for UHF RFID on multipath environments. *IEEE Transactions on Antennas and Propagation*, *57*(4), 1241–1251. doi:10.1109/TAP.2009.2015818

Lee, H. K., Lee, J. G., Rho, Y. K., & Park, C. G. (1998). Modelling quaternion errors in SDINS: Computer frame approach. *IEEE Transactions on Aerospace and Electronic Systems*, *34*(1), 289–300. doi:10.1109/7.640286

Lee, J. Y., & Scholtz, R. A. (2002). Ranging in a dense multipath environment using an UWB radio link. *IEEE Journal on Selected Areas in Communications*, *20*(9), 1677–1683. doi:10.1109/JSAC.2002.805060

Lee, S., & Mase, K. (2002). Activity and location recognition using wearable sensors. *IEEE Pervasive Computing*, *1*(3), 24–32. doi:10.1109/MPRV.2002.1037719

Leeuw, J. D. (1988). Convergence of the majorization method for multidimensional scaling. *Journal of Classification*, *5*(2), 163–180. doi:10.1007/BF01897162

Leica Geosystems AG. (2010). *Leica Geo Office Datasheet.* Author.

Leica. (2009). *Leica AR25 White Paper*. Leica Geosystems. Retrieved from http://www.leica-geosystems.com/common/shared/downloads/inc/downloader.asp?id=10904

Leinenbach, C., Kattelus, H., & Knechtel, R. (2010). Surface micromachining. In V. Lindroos, M. Tilli, A. Lehto, & T. Motooka (Eds.), *Handbook of silicon based MEMS materials and technologies* (pp. 473–487). William Andrew Applied Science Publisher. doi:10.1016/B978-0-8155-1594-4.00028-0

Leppäkoski, H., Collin, J., & Takala, J. (2013). Pedestrian navigation based on inertial sensors, indoor map, and WLAN signals. *Journal of Signal Processing Systems for Signal, Image, and Video Technology*, *71*(3), 287–296. doi:10.1007/s11265-012-0711-5

Levi, R., & Judd, T. (1999). *Dead reckoning navigational system using accelerometer to measure foot impacts.* US Patent 5583776.

Li, M., Jiang, X., and Guibas, L. (2011). Fingerprinting mobile user positions in sensor networks: Attacks and countermeasures. *IEEE Transactions on Parallel and Distributed Systems.*

Li, X., Zhang, Y., & Amin, M. (2009 May). *Multifrequency-based range estimation of RFID Tags.* Paper presented at the 2009 IEEE International Conference on RFID, Orlando, FL.

Li, Y. (1997). *Decentralized filtering theory and its application to the spacecraft navigation using GPS* (PhD dissertation). Northwestern Polytechnical University, China.

Li, Y., & Rizos, C. (2010), *Seamless navigation through a Locata-enhanced GPS and INS integrated system.* International Symposium on GPS/GNSS, Taipei, Taiwan.

Li, Y., Li, Y., Rizos, C., & Xu, X. S. (2012) Observability Analysis of SINS/GPS during In-Motion Alignment Using Singular Value Decomposition. Advanced Materials Research, 433 – 440, 5918-5923.

Liang, T. C., Wang, T. C., & Ye, Y. (2004). *A gradient search method to round the semidefinite programming relaxation solution for ad hoc wireless sensor network localization.* Sanford University.

Liao, I.-E., & Kao, K.-F. (2008). Enhancing the accuracy of WLAN-based location determination systems using predicted orientation information. *Information Sciences, 178*(4), 1049–1068. doi:10.1016/j.ins.2007.09.025

Liao, L., Chen, W., Zhang, C., Zhang, L., Xuan, D., & Jia, W. (2011, June). Two birds with one stone: Wireless access point deployment for both coverage and localization. *IEEE Transactions on Vehicular Technology, 60*(5), 2239–2252. doi:10.1109/TVT.2011.2109405

Liberti, J. C., & Rappaport, T. S. (1996, Apr). A geometrically based model for line-of-sight multipath radio channels. *Proceedings of Vehicular Technology Conference - VTC, 2*, 844-848. doi:10.1109/VETEC.1996.501430

Lieckfeldt, D., You, J., & Timmermann, D. (2009, Oct.). *Characterizing the influence of human presence on bistatic passive RFID-system.* Paper presented at 2009 IEEE International Conference on Wireless and Mobile Computing, Networking and Communications, Marrakech, Morocco. doi:10.1109/WiMob.2009.64

Lieckfeldt, D., You, J., & Timmermann, D. (2009, October). *Exploiting RF-scatter: human localization with bistatic passive UHF RFID-systems.* Paper presented at the meeting of 2009 IEEE International Conference on Wireless and Mobile Computing, Networking and Communications, Beijing, China. doi:10.1109/WiMob.2009.39

Lien, J., Ferner, U. J., Srichavengsup, W., Wymeersch, H., & Win, M. Z. (2012). A Comparison of Parametric and Sample-Based Message Representation in Cooperative Localization. *International Journal of Navigation and Observation, 2012*, 1–10. doi:10.1155/2012/281592

Li, G., Ning, B., Zhao, B., Liu, L., Liu, J. Y., & Yumoto, K. (2008). Effects of geomagnetic storm on GPS ionospheric scintillations at Sanya. *Journal of Atmospheric and Solar-Terrestrial Physics, 70*(7), 1034–1045. doi:10.1016/j.jastp.2008.01.003

Li, H., Cui, X., Lu, M., & Feng, Z. (2008, December). Dual-folding based rapid search method for long PN-code acquisition. *IEEE Transactions on Wireless Communications, 7*(12), 5286–5296. doi:10.1109/T-WC.2008.071130

Li, L., Stankovic, V., Stankovic, L., Li, L., Cheng, S., & Uttamchandani, D. (2011a). Single pixel optical imaging using a scanning MEMS mirror. *Journal of Micromechanics and Microengineering, 21*(2), 025022. doi:10.1088/0960-1317/21/2/025022

Lingwen, Z., & Zhenhui, T. (2010). A Novel UWB Signal Sampling Method for Localization based on Compressive Sensing. *China Communications, 7*(1), 65–72.

Lin, S., Zhang, J., Zhou, G., Gu, L., Stankovic, J. A., & He, T. (2006). ATPC: Adaptive Transmission Power Control for Wireless Sensor Networks. In *Proceedings of the 4th International Conference on Embedded Networked Sensor Systems* (pp. 223-236). New York: ACM. doi:10.1145/1182807.1182830

Lin, T., Fang, W. S., Tseng, W.-H., Lee, C.-W., & Hsieh, J.-W. (2014, October). A group-discrimination-based access point selection for WLAN fingerprinting localization. *IEEE Transactions on Vehicular Technology*, *63*(8), 3967–3976. doi:10.1109/TVT.2014.2303141

Liu, T., Yang, L., Lin, Q., Guo, Y., & Liu, Y. (2014, April). *Anchor-free backscatter positioning for RFID tags with high accuracy*. Paper presented at IEEE Conference on Computer Communications (INFOCOM), Toronto, Canada. doi:10.1109/INFOCOM.2014.6847960

Liu, Z., Dai, W., & Win, M. Z. (2017). Mercury: An infrastructure-free system for network localization and Navigation. *IEEE Transactions Mobile Computing*.

Liu, H., Darabi, H., Banerjee, P., & Liu, J. (2007, November). Survey of wireless indoor positioning techniques and systems. *IEEE Transactions on System, Management, Cybernetics, Part C, Applications and Reviews*, *37*(6), 1067–1080. doi:10.1109/TSMCC.2007.905750

Liu, H., Gan, Y., Yang, J., Sidhom, S., Wang, Y., Chen, Y., & Ye, F. (2012). *Push the limit of WiFi based localization for smartphones*. Mobicom. doi:10.1145/2348543.2348581

Liu, J. (2012). Hybrid positioning with smart phones. In R. Chen (Ed.), *Ubiquitous positioning and mobile location-based services in smart phones* (pp. 159–194). Hershey, PA: IGI Global. doi:10.4018/978-1-4666-1827-5.ch007

Li, W., Jia, Y., Du, J., & Zhang, J. (2013, July). Distributed multiple-model estimation for simultaneous localization and tracking with NLOS mitigation. *IEEE Transactions on Vehicular Technology*, *62*(6), 2824–2830. doi:10.1109/TVT.2013.2247073

Li, X., He, J., Xu, L., & Wang, Q. (2016). The effect of multipath and NLOS on TOA ranging error and energy based on UWB. In *Proceedings of IEEE International Conference on Consumer Electronics-Taiwan* (pp.1-2). Piscataway, NJ: IEEE. doi:10.1109/ICCE-TW.2016.7520939

Li, X., & Pahlavan, K. (2004). Super-resolution TOA estimation with diversity for indoor geolocation. *IEEE Transactions on Wireless Communications*, *3*(1), 224–234. doi:10.1109/TWC.2003.819035

Li, X., Wang, J., & Li, T. (2013). Seamless positioning and navigation by using geo-referenced images and multi-sensor data. *Sensors (Basel)*, *13*(7), 9047–9069. doi:10.3390/s130709047 PMID:23857267

Li, Y. (2014). Optimal multisensor integrated navigation through information space approach. *Physical Communication. SI: Indoor Navigation and Tracking*, *13*, 44–53.

Li, Y., Chen, X., Coates, M., & Yang, B. (2011). Sequential Monte Carlo Radio-Frequency tomographic tracking. *IEEE International Conference on Acoustics, Speech and Signal Processing (ICASSP)*, 3976-3979. doi:10.1109/ICASSP.2011.5947223

Li, Y., Zheng, Q., Hu, Y., & Xu, Y. (2011b). Micromachined piezoresistive accelerometer based on a asymmetrically gapped cantilever. *Journal of Microelectromechanical Systems*, *20*(1), 83–94. doi:10.1109/JMEMS.2010.2100024

Locata Corporation. (2014). *LocataNet positioning signal interface control document*. Available at http://www.locata.com/wp-content/uploads/2014/07/Locata-ICD-100E.pdf

Loeliger, H. A. (2004). An introduction to factor graphs. *IEEE Signal Processing Magazine*, *21*(1), 28–41. doi:10.1109/MSP.2004.1267047

Lui, G., Gallagher, T., Li, B., Dempster, A. G., & Rizos, C. (2011). Differences in RSSI readings made by different Wi-Fi chipsets: A limitation of WLAN localization. *Proceedings of International Conference on Localization and GNSS*, 53-57. doi:10.1109/ICL-GNSS.2011.5955283

Luo, R., & Lai, C. C. (2014). Multisensor Fusion-Based Concurrent Environment Mapping and Moving Object Detection for Intelligent Service Robotics. *IEEE Transactions on Industrial Electronics*, *61*(8), 4043–4051. doi:10.1109/TIE.2013.2288199

Luo, Z.-Q., Ma, W. K., So, A. M.-C., Ye, Y., & Zhang, S. (2010). Semidefinite relaxation of quadratic optimization problems. *IEEE Signal Processing Magazine*, *27*(3), 20–34. doi:10.1109/MSP.2010.936019

Lv, T., Gao, H., Li, X., Yang, S., & Hanzo, L. (2016). Space-time hierarchical-graph based cooperative localization in wireless sensor networks. *IEEE Transactions on Signal Processing*, *64*(2), 322–334. doi:10.1109/TSP.2015.2480038

Lynen, S., Sattler, T., Bosse, M., Hesch, J., Pollefeys, M., & Siegwart, R. (2015). Get Out of My Lab: Large-scale, Real-Time Visual-Inertial Localization. In L. E. Kavraki, D. Hsu, & J. Buchli (Eds.), *Robotics: Science and Systems XI* (p. 37). Rome, Italy: RSS. doi:10.15607/RSS.2015.XI.037

Mager, B., Patwari, N., & Bocca, M. (2013). Fall detection using RF sensor networks. *IEEE 24th International Symposium on Personal Indoor and Mobile Radio Communications (PIMRC),* London, 2013, pp. 3472-3476. doi: 10.1109/PIMRC.2013.6666749.

Mager, B., Lundrigan, P., & Patwari, N. (2015). Fingerprint-based device-free localization performance in changing environments. *IEEE Journal on Selected Areas in Communications*, *33*(11), 2429–2438. doi:10.1109/JSAC.2015.2430515

Mahtab Hossain, A., Jin, Y., Soh, W.-S., & Van, H. N. (2010). SSD: A robust RF location fingerprint addressing mobile devices' heterogeneity. *IEEE Transactions on Mobile Computing*, *12*(1), 65–77. doi:10.1109/TMC.2011.243

Mahtab Hossain, A., Van, H. N., Jin, Y., & Soh, W.-S. (2007). Indoor localization using multiple wireless technologies. *IEEE International Conference on Mobile Adhoc and Sensor Systems*, 1-8.

Mallat, S. G., & Zhang, Z. (1993). Matching Pursuits with Time-Frequency Dictionaries. *IEEE Transactions on Signal Processing*, *41*(12), 3397–3415. doi:10.1109/78.258082

Mao, G., Fidan, B., & Anderson, B. D. (2007). Wireless sensor network localization techniques. *Computer Networks*, *51*(10), 2529–2553. doi:10.1016/j.comnet.2006.11.018

Marano, S., Gifford, W. M., Wymeersch, H., & Win, M. Z. (2010). NLOS identification and mitigation for localization based on UWB experimental data. *IEEE Journal on Selected Areas in Communications*, *28*(7), 1026–1035. doi:10.1109/JSAC.2010.100907

Margaria, D., Dovis, F., & Mulassano, P. (2008). Galileo AltBOC signal multiresolution acquisition strategy. *IEEE Aerospace and Electronic Systems Magazine*, *23*(11), 4–10. doi:10.1109/MAES.2008.4693984

Marschollek, M., Goevercin, M., Wolf, K.-H., Song, B., Gietzelt, M., Haux, R., & Steinhagen-Thiessen, E. (2008). A performance comparison of accelerometry-based step detection algorithms on a large, non-laboratory sample of healthy and mobility-impaired persons. *Proceedings of 30th Annual International Conference of the IEEE Engineering in Medicine and Biology Society EMBS '08*, 1319-1322. doi:10.1109/IEMBS.2008.4649407

Martin, H., Groves, P., Newman, M., & Faragher, R. (2013). A new approach to better low-cost MEMS IMU performance using sensor arrays. *Proceedings of ION GNSS*, 2125-2142.

Mautz, R. (2012). Indoor positioning technologies. Swiss Geodetic Commission, Geodetic-Geophysical Reports of Switzerland, 86, 134 pgs.

Ma, Y., Zhou, L., Liu, K., & Wang, J. (2014). Iterative phase reconstruction and weighted localization algorithm for indoor RFID-based localization in NLOS environment. *IEEE Sensors Journal*, *14*(2), 597–611. doi:10.1109/JSEN.2013.2286220

Mazuelas, S., Bahillo, A., Lorenzo, R., Fernandez, P., Lago, F., Garcia, E., & Abril, E. et al. (2009). Robust indoor positioning provided by real-time RSSI values in unmodified WLAN networks. *IEEE Journal of Selected Topics in Signal Processing*, *3*(5), 821–831. doi:10.1109/JSTSP.2009.2029191

Medawar, S., Handel, P., & Zetterberg, P. (2013). Approximate maximum likelihood estimation of rician K-Factor and investigation of urban wireless measurements. *IEEE Transactions on Wireless Communications*, *12*(6), 2545–2555. doi:10.1109/TWC.2013.042413.111734

Merrill, J. (2011). U. S. Interference Detection & Mitigation Plan (IDM). *51st Meeting of the Civil GPS Service Interface Committee at the Institute of Navigation Conference (GNSS 2011).*

Mervart, L., Lukes, Z., Rocken, C., & Iwabuchi, T. (2008) *Precise point positioning with ambiguity resolution in real-time.* 21st International Technical Meeting of the Satellite Division of the U.S. Institute of Navigation, Savannah, GA.

Mezentsev, O. A. (2005). *Sensor aiding of HSGPS pedestrian navigation* (PhD Dissertation). Geomatics Engineering, University of Calgary, Canada.

Mian, A., & Law, J. (2010). Geometric optimization of a van der pauw structure based MEMS pressure sensor. *Microsystem Technologies*, *16*(11), 1921–1929. doi:10.1007/s00542-010-1124-7

Mihaylova, L., Angelova, D., Bull, D., & Canagarajah, N. (2011). Localization of mobile nodes in wireless networks with correlated in time measurement noise. *IEEE Transactions on Mobile Computing*, *10*(1), 44–53. doi:10.1109/TMC.2010.132

Misra, P., & Enge, P. (2011). *Global Positioning System.* Signals, Measurements, and Performance.

Moafipoor, S. (2009). *Intelligent personal navigator supported by knowledge-based systems for estimating dead reckoning navigation parameters* (PhD Dissertation). The Ohio State University, USA.

Mok, E., & Retscher, G. (2007). Location determination using WiFi – fingerprinting versus WiFi – trilateration. *Journal of Location Based Services*, *1*(2), 145–159. doi:10.1080/17489720701781905

Mok, E., Retscher, G., Wang, D., & Xia, L. (2011). Use of smartphones for tracking and trip recording. *Proceedings of 8th International Symposium on Location-Based Services*, 137-152.

Mokhtari, H., & Lazaridis, P. (1999, January). Comparative study of lateral profile knife-edge diffraction and ray tracing technique using GTD in urban environment. *IEEE Transactions on Vehicular Technology*, *48*(1), 255–261. doi:10.1109/25.740101

Molisch, A., Balakrishnan, K., Chong, C., Emami, S., Fort, A., Karedal, J., … Siwiak, K. (2004). *P802.15-04-0662-00-004a, IEEE 802.15.4a channel model - Final Report.* IEEE.

Molisch, A. F. (2011). *Wireless communications* (2nd ed.; Vol. 15). John Wiley & Sons.

Monnerat, M., Couty, R., Vincent, N., Huez, O., & Chatre, E. (2004). The Assisted GNSS, Technology and Applications. *Proceedings of ION ITM*, 2479-2488.

Montemerlo, M., Thrun, S., Koller, D., & Wegbreit, B. (2002). FastSLAM: A factored solution to the simultaneous localization and mapping problem. *Proceedings of the National Conference on Artificial Intelligence*, 593-598.

Montillet, J., Roberts, G. W., Hancock, C., Meng, X., Ogundipe, O., & Barnes, J. (2009). Deploying a Locata network to enable precise positioning in urban canyons. *Journal of Geodesy*, *83*(2), 91–103. doi:10.1007/s00190-008-0236-7

Moré, J. J. (1993). Generalizations of the trust region problem, Optimum. *Methods Software*, *2*(3-4), 189–209. doi:10.1080/10556789308805542

Motella, B., Pinni, M., & Doviz, F. (2008). Investigation on the effect of strong out-of-band signals on global navigation satellite systems receivers. *GPS Solutions*, *12*(2), 77–86. doi:10.1007/s10291-007-0085-5

Mrazovac, B., Todorović, B. M., Bjelica, M. Z., & Kukolj, D. (2013). Device-free indoor human presence detection method based on the information entropy of RSSI variations. *Electronics Letters*, *49*(22), 1386–1388. doi:10.1049/el.2013.1041

Müller, G., Friedberger, A., & Knese, K. (2010). Porous silicon based MEMS. In V. Lindroos, M. Tilli, A. Lehto, & T. Motooka (Eds.), *Handbook of silicon based MEMS materials and technologies* (pp. 409–431). William Andrew Applied Science Publisher. doi:10.1016/B978-0-8155-1594-4.00025-5

Mutambara A.G.O. (1998). *Decentralized estimation and control for multisensor systems*. CRC.

Navarro, M., Closas, P., & Najar, M. (2013), Assessment of Direct Positioning for IR-UWB in IEEE 802.15.4a Channels. *Proceedings of IEEE International Conference Ultra-WideBand (ICUWB)*, 55-60. doi:10.1109/ICUWB.2013.6663822

Navarro, M., & Najar, M. (2011). Frequency Domain joint TOA and DOA estimation in IR-UWB. *IEEE Transactions on Wireless Communications*, *10*(10), 3174–3184. doi:10.1109/TWC.2011.072511.090933

Navarro-Reyes, D., Castro, R., & Bosch, P. R. (2015). Galileo first foc launch: recovery mission design. *Proceedings of International Symposium on Space Flight Dynamics*.

Nguyen, T. V., Jeong, Y., Shin, H., & Win, M. Z. (2015). Least square cooperative localization. *IEEE Transactions on Vehicular Technology*, *64*(4), 1318–1330. doi:10.1109/TVT.2015.2398874

Nguyen, T. V., Jeong, Y., Shin, H., & Win, M. Z. (2015). Machine learning for wideband localization. *IEEE Journal on Selected Areas in Communications*, *33*(7), 1357–1380. doi:10.1109/JSAC.2015.2430191

Ni, L., Liu, Y., Lau, Y., & Patil, A. (2003, April). *LANDMARC: indoor location sensing using active RFID*. Paper presented at the First IEEE International Conference on Pervasive Computing and Communications (PerCom 2003), Fort Worth, TX. doi:10.1109/PERCOM.2003.1192765

Nielsen, R. O. (1994). Estimation of azimuth and elevation angles for a plane wave sine wave with a 3-D array. *IEEE Transactions on Signal Processing*, *42*(11), 3274–3276. doi:10.1109/78.330396

Nikitin, P., Martinez, R., Ramamurthy, S., ... (2010, April). *Phase based spatial identification of UHF RFID tags*. Paper presented at the 2010 IEEE International Conference on RFID, Orlando, FL. doi:10.1109/RFID.2010.5467253

Ni, L. M., Liu, Y., Lau, Y. C., & Patil, A. P. (2004). LANDMARC: Indoor location sensing using active RFID. *Wireless Networks*, *10*(6), 701–710. doi:10.1023/B:WINE.0000044029.06344.dd

Nilsson, J. O., Zachariah, D., Skog, I., & Händel, P. (2013). Cooperative localization by dual foot-mounted inertial sensors and inter-agent ranging. *EURASIP Journal on Advances in Signal Processing*, *2013*(1), 164. doi:10.1186/1687-6180-2013-164

Nordwall, B. D. (1997, September 29). Aviaconversia is marketing a portable GPS/Glonass jammer. *Aviation Week & Space Technology*.

Norklit, O., & Andersen, J. B. (1998, June). Diffuse channel model and experimental results for array antennas in mobile environments. *IEEE Transactions on Antennas and Propagation*, *46*(6), 834–840. doi:10.1109/8.686770

Novatel. (2000). *Discussions on RF Signal Propagation and Multipath*. Novatel Application Note APN-008 Rev 1.

Ochieng, W. Y., Quddus, M. A., & Noland, R. B. (2004). Map-matching in complex urban road networks. *Brazilian Journal of Cartography*, *55*(2), 1–14.

Odenwald, S. F., & Green, J. L. (2008, July). Bracing the Satellite Infrastructure for a Solar Superstorm. *Scientific American*, 28.

Odenwald, S. F., Green, J. L., & Taylor, W. (2006). Forecasting the impact of an 1859-calibre superstorm on satellite resources. *Advances in Space Research*, *38*(2), 280–297. doi:10.1016/j.asr.2005.10.046

Osipov, A. V., & Tretyakov, S. A. (2017). *Modern Electromagnetic Scattering Theory with Applications*. John Wiley & Sons, Ltd. doi:10.1002/9781119004639

Pallas, M. A., & Jourdain, G. (1991). Active high resolution time delay estimation for large BT signals. *IEEE Transactions on Signal Processing*, *39*(4), 781–788. doi:10.1109/78.80899

Pany, T., Winkel, J., Riedl, B., Niedermeier, H., Eissfeller, B., Wrz, T., . . . Jimnez-Baos, D. (2010, December). Experimental Results from an Ultra-Tightly Coupled GPS/Galileo/WiFi/ZigBee/MEMS-IMU Indoor Navigation Test System Featuring Coherent Integration Times of Several Seconds. *Satellite Navigation Technologies and European Workshop on GNSS Signals and Signal Processing (NAVITEC), 2010 5th ESA Workshop on*. doi:10.1109/NAVITEC.2010.5708020

Papadakis, S., & Traganitis, A. (2010). *Wireless positioning using the signal strength difference on arrival*. MASS. doi:10.1109/MASS.2010.5663794

Papapostolou, A., & Chaouchi, H. (2009). Orientation-based radio map extensions for improving positioning system accuracy. *Proc. International Conference on Wireless Communication and. Mobile Computing, Connecting the World Wirelessly*, 947–951. doi:10.1145/1582379.1582586

Parkinson, B. W., & Spilker, J. J. (1996). Global Positioning System: Theory and Applications. Aiaa.

Park, J., Kim, Y., & Lee, J. (2012). Waist mounted pedestrian dead-reckoning system. *Proc. ICURAI*, 335–336.

Patnaik, P. B. (1949). The non-central and F-distribution and their applications. *Biometrika*, *36*, 202–232. PMID:18132092

Patwari, N., & Agrawal, P. (2008). Effects of Correlated Shadowing: Connectivity, Localization, and RF Tomography. *IEEE/ACM Int'l Conf. on Information Processing in Sensor Networks (IPSN'08)*, 82-93. doi:10.1109/IPSN.2008.7

Patwari, N., Ash, J. N., Kyperountas, S., Hero, I. I., Moses, R. L., & Correal, N. S. (2005). Locating the nodes: Cooperative localization in wireless sensor networks. *IEEE Signal Processing Magazine*, *22*(4), 54–69. doi:10.1109/MSP.2005.1458287

Patwari, N., Hero, A. O., Perkins, M., Correal, N. S., & O'dea, R. J. (2003). Relative location estimation in wireless sensor networks. *IEEE Transactions on Signal Processing*, *51*(8), 2137–2148. doi:10.1109/TSP.2003.814469

Patwari, N., & Wilson, J. (2010). RF sensor networks for device-free localization: Measurements, models, and algorithms. *Proceedings of the IEEE*, *98*(11), 1961–1973. doi:10.1109/JPROC.2010.2052010

Patwari, N., & Wilson, J. (2011). Spatial models for human motion-induced signal strength variance on static links. *IEEE Transactions on Information Forensics and Security*, *6*(3), 791–802. doi:10.1109/TIFS.2011.2146774

Pearson, E. S. (1959). Note on an Approximation to the Distribution of Non-Central. *Biometrika*, *46*(3/4), 364–364. doi:10.2307/2333533

Perera, L. D., Wijesoma, W. S., & Adams, M. D. (2010). SLAM With Joint Sensor Bias Estimation: Closed Form Solutions on Observability, Error Bounds and Convergence Rates. *IEEE Transactions on Control Systems Technology*, *18*(3), 732–740. doi:10.1109/TCST.2009.2026165

Perez-Ramirez, J., Borah, D., & Voelz, D. (2013, September). Optimal 3-D landmark placement for vehicle localization using heterogeneous sensors. *IEEE Transactions on Vehicular Technology*, *62*(7), 2987–2999. doi:10.1109/TVT.2013.2255072

Perttula, A., Leppakoski, H., Kirkko-Jaakkola, M., Davidson, P., Collin, J., & Takala, J. (2014). Distributed Indoor Positioning System with Inertial Measurements and Map Matching. *IEEE Transactions on Instrumentation and Measurement*, *63*(11), 2682–2695. doi:10.1109/TIM.2014.2313951

Peterson, S. (2011, December 9). Downed US drone: How iran caught the 'beast'. *The Christian Science Monitor*.

Pick, M., & Vilmer, N. (2008). Sixty-five years of solar radioastronomy: Flares, coronal mass ejections and Sun-Earth connection. *The Astronomy and Astrophysics Review*, *16*(1-2), 1–153. doi:10.1007/s00159-008-0013-x

Pinel, N., Bourlier, C., & Saillard, J. (2010). Degree of roughness of rough layers: Extensions of the Rayleigh roughness criterion and some applications. *Progress In Electromagnetics Research B*, *19*, 41–63. doi:10.2528/PIERB09110907

Ping, W., Huailin, R., & Fuhua, F. A Method of Indoor Multi-Path IR-UWB Localization Based on Bayesian Compressed Sensing, *Proceedings of IEEE International Conference on Signal Processing (ICSP)*, 56–59.

Pini, M., Viviani, E., & Lo Presti, L. (2010). GPS L5 Signal Acquisition Exploiting Neumann-Hoffman Code Transitions. *Proceedings of the 2010 International Technical Meeting of The Institute of Navigation*, 765-772.

Pi, X., Mannucci, A. J., Lindqwister, U. J., & Ho, C. M. (1997). Monitoring of global Ionospheric irregularities using the worldwide GPS network. *Geophysical Research Letters*, *24*(18), 2283–2286. doi:10.1029/97GL02273

Poor, H. V. (2013). *An introduction to signal detection and estimation.* Springer Science & Business Media.

Pourhomayoun, M., Fowler, M., & Wu, N. (2012). Spatial Sparsity Based Emitter Localization. *Proceedings of Conference on Information Sciences and Systems (CISS).*

Prasithsangaree, P., Krishnamurthy, P., & Chrysanthis, P. K. (2002). On indoor position location with wireless LANS. *Proceedings of IEEE International Symposium on Personal, Indoor and Mobile Radio Communications*, 720-724. doi:10.1109/PIMRC.2002.1047316

Proakis, J. G., & Salehi, M. (2008). *Digital Communications* (5th ed.). New York, NY: McGraw-Hill.

Progri, I. F., & Bromberg, M. C., & R., M. W. (2004). An enhanced acquisition process of a maximum likelihood GPS receiver. *Proceedings of ION NTM*, 390-398.

Psiaki, M. L. (2004). FFT-based acquisition of GPS L2 civilian CM and CL signals. *Proceedings of ION GNSS*, 646-653.

Psiaki, M., & Humphreys, T. E. (2016, July). Protecting GPS from spoofers is critical to the future of navigation. *IEEE Spectrum*, 29.

Qaisar, S. U., & Dempster, A. G. (2008). Cross-correlation Performance Comparison of L1 & L2C GPS Codes for Weak Signal Acquisition. *Proceedings of IS-GPS/GNSS.*

Quddus, M. A., Ochieng, W. Y., & Noland, R. B. (2007). Current map-matching algorithms for transport applications: State-of-the-art and future research directions. *Transportation Research Part C, Emerging Technologies*, *15*(5), 312–328. doi:10.1016/j.trc.2007.05.002

Quddus, M. A., Ochieng, W. Y., Zhao, L., & Noland, R. B. (2003). A general map matching algorithm for transport telematics applications. *GPS Solutions*, *7*(3), 157–167. doi:10.1007/s10291-003-0069-z

RAE. (2011). *Global Navigation Satellite Systems: Reliance and Vulnerabilities.* Royal Academy of Engineering.

Rampa, V., Savazzi, S., Nicoli, M., & D'Amico, M. (2015, November). Physical Modeling and Performance Bounds for Device-free Localization Systems. *IEEE Signal Processing Letters*, *22*(11), 1864–1868. doi:10.1109/LSP.2015.2438176

Ramsay, J. O. (1982). Some statistical approaches to multidimensional scaling data. *Journal of the Royal Statistical Society. Series A (General), 145*(3), 285–312. doi:10.2307/2981865

Randell, C., Djiallis, C., & Muller, H. (2003). Personal position measurement using dead reckoning. In: *Proceedings of International Symposium on Wearable Computers ISWC '03* (pp. 166-173). IEEE Computer Society.

Randell, C., & Muller, H. (2000). Context awareness by analysing accelerometer data. *Proceedings of International Symposium on Wearable Computers*, 175-176.

Rappaport, T. S. (2002). Wireless Communications: Principles and Practice (2nd ed.). Prentice Hall.

Rappaport, T. (2001). *Wireless Communications: Principles and Practice* (2nd ed.). Upper Saddle River, NJ: Prentice-Hall.

Retscher, G. (2016). Indoor navigation. In E.W. Grafarend (Ed.), Encyclopedia of geodesy. Springer International Publishing Switzerland. doi:10.1007/978-3-319-02370-0_9-1

Retscher, G., & Tatschl, T. (2016). Indoor positioning using Wi-Fi lateration – Comparison of two common range conversion models with two novel differential approaches. In *Proceedings of 2016 4th International Conference on Ubiquitous Positioning, Indoor Navigation and Location Based Services UPINLBS*. IEEE Xplore. doi:10.1109/UPINLBS.2016.7809967

Retscher, G. (2007). Augmentation of indoor positioning systems with a barometric pressure sensor for direct altitude determination in a multi-storey building. *Journal of Cartography and Geographic Information Science CaGIS, 34*(4), 305–310. doi:10.1559/152304007782382909

Retscher, G., & Hofer, H. (2017). Wi-Fi location fingerprinting using an intelligent checkpoint sequence. *Journal of Applied Geodesy, 11*(3). doi:10.1515/jag-2016-0030

Retscher, G., Hofer, H., Kealy, A., Gikas, V., & Obex, F. (2017). Cooperative localization in indoor environments using constrained differential Wi-Fi and UWB measurements. In *Proceedings of ION GNSS++ Conference*. Institute of Navigation.

Retscher, G., & Mok, E. (2004). Sensor fusion and integration using an adapted Kalman filter approach for modern navigation systems. *Survey Review, 37*(292), 439–447. doi:10.1179/sre.2004.37.292.439

Retscher, G., & Thienelt, M. (2004). NAVIO – A navigation and guidance service for pedestrians. *Journal of Global Positioning Systems CPGPS, 3*(1-2), 208–217. doi:10.5081/jgps.3.1.208

Retscher, G., Zhu, M., & Zhang, K. (2012). RFID positioning. In R. Chen (Ed.), *Ubiquitous positioning and mobile location-based services in smart phones* (pp. 69–95). Hershey, PA: IGI Global. doi:10.4018/978-1-4666-1827-5.ch004

Reuters. (2016, April 11). South Korea tells U.N. that North Korea GPS jamming threatens boats, planes. *Reuters*.

Riad, S. M. (1986). The deconvolution problem: An overview. *Proceedings of the IEEE, 74*(1), 82–85. doi:10.1109/PROC.1986.13407

Rizos, C., Roberts, G. W., Barnes, J., & Gambale, N. (2010) Locata: A new high accuracy indoor positioning system. *1st International Conference on Indoor Positioning and Indoor Navigation*, 441 – 447.

Roberts, B., & Pahlavan, K. (2009). Site-specific RSS signature modeling for WiFi localization. *Proc. GLOBECOM*, 1–6. doi:10.1109/GLOCOM.2009.5425740

Robertson, D., & Cipolla, R. (2004). An image-based system for urban navigation. *Proceedings of the British Machine Vision Conference*, 260-272.

Roos, T., Myllymaki, P., & Tirri, H. (2002). A statistical modeling approach to location estimation. *IEEE Transactions on Mobile Computing*, *1*(1), 59–69. doi:10.1109/TMC.2002.1011059

Roy, S., Hashemi, R. H., & Laub, A. J. (1991). Square root parallel Kalman filtering using reduced-order local filters. *IEEE Transactions on Aerospace and Electronic Systems*, *AES-27*(2), 276–289. doi:10.1109/7.78303

RSW. (2013). *Russia's Proton crashes with a trio of navigation satellites.* Retrieved 9 Mar 2017, from http://www.russianspaceweb.com/proton_glonass49.html

Ruan, Y., Willett, P., Marrs, A., Palmieri, F., & Marano, S. (2008). Parctical fusion of quantized measurements via particle filtering. *IEEE Transactions on Aerospace and Electronic Systems*, *AES-44*(1), 15–28. doi:10.1109/TAES.2008.4516986

Rui, L., & Ho, K. C. (2014). Algebraic solution for joint localization and synchronization of multiple sensor nodes in the presence of beacon uncertainties. *IEEE Transactions on Wireless Communications*, *13*(9), 5196–5210. doi:10.1109/TWC.2014.2322053

Ruotsalainen, L. (2013). *Vision-aided pedestrian navigation for challenging GNSS environments* (PhD Dissertation). Publications of the Finnish Geodetic Institute.

Ruotsalainen, L., & Kuusniemi, H. (2012). Visual positioning in a smartphone. In R. Chen (Ed.), *Ubiquitous positioning and mobile location-based services in smart phones* (pp. 130–158). Hershey, PA: IGI Global. doi:10.4018/978-1-4666-1827-5.ch006

Ruotsalainen, L., Kuusniemi, H., & Chen, R. (2011). Visual-aided two-dimensional pedestrian indoor navigation with a smartphone. *Journal of Global Positioning Systems*, *10*(1), 11–18. doi:10.5081/jgps.10.1.11

Rusinkiewicz, S., & Levoy, M. (2001). Efficient Variants of the ICP Algorithm. *Third International Conference on 3D Digital Imaging and Modeling (3DIM).* doi:10.1109/IM.2001.924423

Russ, J. C. (1995). *Image Processing Handbook.* Boca Raton, FL: CRC.

Rydstrom, M., Strom, E. G., & Svensson, A. (2006), Robust Sensor Network Positioning Based on Projections onto Circular and Hyperbolic Convex Sets (POCS). *Proceedings of IEEE Workshop on Signal Processing Advances in Wireless Communications*, 1-5. doi:10.1109/SPAWC.2006.346399

Saeedi, S., & El-Sheimy, N. (2015). Activity recognition using fusion of low-cost sensors on a smartphone for mobile navigation application. *Micromachines*, *6*(8), 1100–1134. doi:10.3390/mi6081100

Sagias, N. C., & Karagiannidis, G. K. (2005). Gaussian class multivariate Weibull distributions: Theory and applications in fading channels. *IEEE Transactions on Information Theory*, *51*, 3608–3619. doi:10.1109/TIT.2005.855598

Sahmoudi, M., & Amin, M. G. (2008). Fast iterative maximum-likelihood algorithm (FIMLA) for multipath mitigation in next generation of GNSS receivers. *IEEE Transactions on Wireless Communications*, *7*(11), 4362–4374. doi:10.1109/TWC.2008.070700

Sallai, J., Hedgecock, W., Volgyesi, P., Nadas, A., Balogh, G., & Ledeczi, A. (2011). Weapon classification and shooter localization using distributed multichannel acoustic sensors. *Journal of Systems Architecture*, *57*(10), 869–885. doi:10.1016/j.sysarc.2011.04.003

Salo, J., Vuokko, L., El-Sallabi, H. M., & Vainikainen, P. (2007). An additive model as a physical basis for shadow fading. *IEEE Transactions on Vehicular Technology*, *56*(1), 13–26. doi:10.1109/TVT.2006.883797

Sanchez, D., Quinteiro, J., Hernandez-Morera, P., & Martel-Jordan, E. (2012). Using data mining and fingerprinting extension with device orientation information for WLAN efficient indoor location estimation. *Proc. Conference on Wireless Mobile Computing, Networking, and Communications*, 77–83. doi:10.1109/WiMOB.2012.6379164

Sankaralingam, S., & Gupta, B. (2010). Determination of Dielectric Constant of Fabric Materials and Their Use as Substrates for Design and Development of Antennas for Wearable Applications. *IEEE Transactions on Instrumentation and Measurement*, *59*(12), 3122–3130. doi:10.1109/TIM.2010.2063090

Sarkar, S., & Bose, A. (2016). Lifetime of revitalized GLONASS satellites: A review. *Proceedings of MDCCT*.

Särkkä, S., Viikari, V., Huusko, M., & Jaakkola, K. (2012). Phase-based UHF RFID tracking with nonlinear kalman filtering and smoothing. *IEEE Sensors Journal*, *12*(5), 904–910. doi:10.1109/JSEN.2011.2164062

Sathyan, T., Humphrey, D., & Hedley, M. (2011). WASP: A system and algorithms for accurate radio localization using low-cost hardware. *IEEE Transactions on Systems, Man, and Cybernetics. Part C*, *41*(2), 211–222.

Saunders, S. R., & Simon, S. R. (1999). *Antennas and Propagation for Wireless Communication Systems* (1st ed.). New York: John Wiley & Sons, Inc.

Savage, P. G. (2007). *Strapdown Analytics.* Strapdown Associates Inc.

Savarese, C., Rabaey, J. M., & Beutel, J. (2001). Location in distributed ad-hoc wireless sensor networks. *IEEE International Conference on Acoustics, Speech, and Signal Processing*, *4*, 2037-2040. doi:10.1109/ICASSP.2001.940391

Savazzi, S., Nicoli, M., Carminati, F., & Riva, M. (2014, February). A Bayesian Approach to Device-Free Localization: Modeling and Experimental Assessment. *IEEE Journal of Selected Topics in Signal Processing*, *8*(1), 16–29. doi:10.1109/JSTSP.2013.2286772

Savic, V., Larsson, E. G., Ferrer-Coll, J., & Stenumgaard, P. (2015). Kernel methods for accurate UWB-based ranging with reduced complexity. *IEEE Transactions on Wireless Communications*, *15*(3), 1783–1793. doi:10.1109/TWC.2015.2496584

Savic, V., & Zazo, S. (2012). Reducing communication overhead for cooperative localization using nonparametric belief propagation. *IEEE Wireless Communications Letters*, *1*(4), 308–311. doi:10.1109/WCL.2012.042512.120172

Saxe, J. B. (1980). *Embeddability of weighted graphs in k-space is strongly NP-hard. Carnegie-Mellon University.* Department of Computer Science.

Scherhäufl, M., Pichler, M., Schimbäck, E., Muller, D. J., Ziroff, A., & Stelzer, A. (2013). Indoor localization of passive UHF RFID tags based on phase-of-arrival evaluation. *IEEE Transactions on Microwave and Techniques*, *61*(12), 4724–4729. doi:10.1109/TMTT.2013.2287183

Schmidt, R. (1986). Multiple emitter location and signal parameter estimation. *IEEE Transactions on Antennas and Propagation*, *34*(3), 276–280. doi:10.1109/TAP.1986.1143830

Scott, L. (2015, April). Protecting Position in Critical Operations. *GPS World*, 30.

Seshadri, V., Zaruba, G., & Huber, M. (2005). A bayesian sampling approach to indoor localization of wireless devices using received signal strength indication. Pervasive Computing and Communications, 75–84.

Shang, Y., & Ruml, W. (2004). Improved MDS-based localization. *INFOCOM 2004. Twenty-third Annual Joint Conference of the IEEE Computer and Communications Societies*, *4*, 2640-2651.

Shang, Y., Ruml, W., Zhang, Y., & Fromherz, M. P. (2003, Jun.). *Localization from mere connectivity.* Paper presented in the 4th international symposium on Mobile ad hoc networking & computing, Annapolis, MD. doi:10.1145/778415.778439

Shang, Y., Rumi, W., Zhang, Y., & Fromherz, M. (2004). Localization from connectivity in sensor networks. *IEEE Transactions on Parallel and Distributed Systems*, *15*(11), 961–974. doi:10.1109/TPDS.2004.67

Sharma, P., & Inoue, A. (2010). Metallic glass. In V. Lindroos, M. Tilli, A. Lehto, & T. Motooka (Eds.), *Handbook of silicon based MEMS materials and technologies* (pp. 447–472). William Andrew Applied Science Publisher. doi:10.1016/B978-0-8155-1594-4.00027-9

Sharp, I., & Yu, K. (2013, August). Enhanced least-squares positioning algorithm for indoor positioning. *IEEE Transactions on Mobile Computing*, *12*(8), 1640–1650. doi:10.1109/TMC.2012.124

Sharp, I., Yu, K., & Guo, Y. J. (2009). GDOP analysis for positioning system design. *IEEE Transactions on Vehicular Technology*, *58*(7), 3371–3382. doi:10.1109/TVT.2009.2017270

Sharp, I., Yu, K., & Hedley, M. (2012). On the GDOP and accuracy for indoor positioning. *IEEE Transactions on Aerospace and Electronic Systems*, *48*(3), 2032–2051. doi:10.1109/TAES.2012.6237577

Sharp, I., Yu, K., & Sathyan, T. (2012). Positional accuracy measurement and error modeling for mobile tracking. *IEEE Transactions on Mobile Computing*, *11*(6), 1021–1032. doi:10.1109/TMC.2011.119

Shen, Y. (2014). *Network localization and navigation: Theoretical framework, efficient operation, and security assurance*. Cambridge, MA: Massachusetts Institute of Technology.

Shen, Y., Wymeersch, H., & Win, M. Z. (2010). Fundamental limits of wideband localization - Part II: Cooperative networks. *IEEE Transactions on Information Theory*, *56*(10), 4981–5000. doi:10.1109/TIT.2010.2059720

Shepard, D. P., & Humphreys, T. E. (2012). Evaluation of the vulnerability of phasor measurement units to GPS spoofing attacks. *Proceedings of Sixth Annual IFIP WG 11.10 International Conference on Critical Infrastructure Protection*. doi:10.1016/j.ijcip.2012.09.003

Shi, W., & Wong, V. (2011 July). *MDS-based localization algorithm for RFID systems*. Paper presented at the 2011 IEEE International Conference on Communications (ICC), Kyoto, Japan. doi:10.1109/icc.2011.5962800

Shimizu, H., Masui, H., Ishii, M., & Sakawa, K. (2000). LOS and NLOS path-loss and delay characteristics at 3.35 GHz in a residential environment. In *Proceedings of Antennas and Propagation Society International Symposium* (vol. 2, pp. 1142-1145). Piscataway, NJ: IEEE. doi:10.1109/APS.2000.875424

Shin, H., Chon, Y., & Cha, H. (2012, November). Unsupervised construction of an indoor floor plan using a smartphone. *IEEE Transactions on Systems, Man and Cybernetics. Part C, Applications and Reviews*, *42*(6), 889–898. doi:10.1109/TSMCC.2011.2169403

Shkel, A. (2001). *Micromachined gyroscopes: Challenges, design solutions, and opportunities. Smart Structures and Materials* (Vol. 4334). SPIE.

SICK. (2008). *Operating instructions: LMS100/111/120 laser measurement systems*. SICK AG.

Silva, B., & Hancke, G. P. (2016). IR-UWB-based non-line-of-sight identification in harsh environments: Principles and challenges. *IEEE Transactions on Industrial Informatics*, *12*(3), 1188–1195. doi:10.1109/TII.2016.2554522

Simon, M. K., & Alouini, M.-S. (2005). *Digital communication over fading channels* (Vol. 95). Wiley.

Skog, I., Nilsson, J. O., & Handel, P. (2014). An open-source multi inertial measurement unit (MIMU) platform. In *Proceedings of International Symposium on Inertial Sensors and Systems* (Vol.20, pp.1-4). Piscataway, NJ: IEEE. doi:10.1109/ISISS.2014.6782523

Skone, S. H. (2001). The impact of magnetic storms on GPS receiver performance. *Journal of Geodesy*, *75*(9), 457–468. doi:10.1007/s001900100198

Skone, S. H., Yousuf, R., & Coster, A. (2004). Performance evaluation of the wide Area augmentation system for ionospheric storm events. *Journal of Global Positioning Systems*, *3*(1-2), 251–258. doi:10.5081/jgps.3.1.251

So, A. M. C., & Ye, Y. (2007). Theory of semidefinite programming for sensor network localization. *Mathematical Programming*, *109*(2-3), 367–384. doi:10.1007/s10107-006-0040-1

Solon, O. (2012, February 22). GPS 'Spoofers' could be used for high-frequency financial trading fraud. *Wired.*

Sorenson, H. W. (1970). Least-squares estimation: From Gauss to Kalman. *Spectrum IEEE*, *7*(7), 63–68. doi:10.1109/MSPEC.1970.5213471

Sorrel, C. (2010, February 24). Car thieves use GPS jammers to make clean getaway. *Wired.*

Sottile, F., Wymeersch, H., Caceres, M. A., & Spirito, M. A. (2011). Hybrid GNSS-terrestrial cooperative positioning based on particle filter. *IEEE Global Telecommunications Conference*, 1-5. doi:10.1109/GLOCOM.2011.6134002

Steinhoff, U., Omerčević, D., Perko, R., Schiele, B., & Leonardis, A. (2007). How computer vision can help in outdoor positioning. In *Proceedings of the European conference on Ambient intelligence 2007* (pp. 124-141). Springer-Verlag Berlin Heidelberg. doi:10.1007/978-3-540-76652-0_8

Stella, M., Russo, M., & Begusic, D. (2007). *Location determination in indoor environment based on RSS fingerprinting and artificial neural network.* Proc. ConTel. doi:10.1109/CONTEL.2007.381886

Stojanović, D., & Stojanović, N. (2014). Indoor localization and tracking: Methods, technologies and research challenges. *Facta Universitatis, Series. Automatic Control and Robotics*, *13*(1), 57–72.

Sun, J., Zhu, X., Zhang, C., & Fang, Y. (2011, March). Rescueme: Location-based secure and dependable VANETs for disaster rescue. *IEEE Journal on Selected Areas in Communications*, *29*(3), 659–669. doi:10.1109/JSAC.2011.110314

Sun, Z., Mao, X., Tian, W., & Zhang, X. (2009). Activity classification and Dead Reckoning for Pedestrian Navigation with Wearable Sensors. *Measurement Science & Technology*, *20*(1), 015203. doi:10.1088/0957-0233/20/1/015203

Symmetricon. (2013). *Power Utilities: Mitigating GPS Vulnerabilities and Protecting Power Utility Network Timing.* White paper.

Szczys, M. (2016, July 26). *We Declare The Grandmaster Of Pokemon Go GPS Cheats.* Hackaday.

Tabatabaei Balaei, A., & Dempster, A. G. (2009). A statistical inference technique for GPS interference detection. *IEEE Trans Aerospace and Systems*, *45*(4), 1499–1511. doi:10.1109/TAES.2009.5310313

Tabatabaei Balaei, A., Motella, B., & Dempster, A. G. (2007). GPS interference detected in Sydney-Australia. *Proceedings of IGNSS.*

Ta, T. H., Dovis, F., Margaria, D., & Presti, L. L. (2010). Comparative study on joint data/pilot strategies for high sensitivity Galileo E1 open service signal acquisition. *IET Radar, Sonar & Navigation*, *4*(6), 764–779. doi:10.1049/iet-rsn.2009.0256

Ta, T. H., Qaisar, S. U., Dempster, A. G., & Dovis, F. (2012). Partial differential postcorrelation processing for GPS L2C signal acquisition. *IEEE Transactions on Aerospace and Electronic Systems*, *48*(2), 1287–1305. doi:10.1109/TAES.2012.6178062

Tenenbaum, J. B., De Silva, V., & Langford, J. C. (2000). A global geometric framework for nonlinear dimensionality reduction. *Science*, *290*(5500), 2319–2323. doi:10.1126/science.290.5500.2319 PMID:11125149

Teng, J., Snoussi, H., Richard, C., & Zhou, R. (2012, June). Distributed variational filtering for simultaneous sensor localization and target tracking in wireless sensor networks. *IEEE Transactions on Vehicular Technology*, *61*(5), 2305–2318. doi:10.1109/TVT.2012.2190631

Teunissen, P. J. G. (2002). *Adjustment Theory – an introduction, Series on Mathematical Geodesy and Positioning*. VSSD.

Tinsley, H. E., & Brown, S. D. (2000). *Handbook of applied multivariate statistics and mathematical modeling*. Academic Press.

Titterton, D. H., & Weston, J. L. (2004). Strapdown Inertial Navigation Technology (2nd ed.). The Institution of Electrical Engineers.

Titterton, D., & Weston, J. (2005). Strapdown inertial navigation technology. *Aerospace & Electronic Systems Magazine IEEE*, *20*(7), 33–34. doi:10.1109/MAES.2005.1499250

Torrieri, D. J. (1984). Statistical theory of passive location systems. *IEEE Transactions on Aerospace and Electronic Systems*, *AES-20*(2), 183–198. doi:10.1109/TAES.1984.310439

Townsend, B. R., & Fenton, P. (1994). *A practical approach to the reduction of pseudorange multipath errors in a L1 GPS receiver. Proceedings of ION-GPS.*

TR45. (1999, October). *Position Determination Service Standard for Dual-Mode Spread Spectrum Systems*. TIA/EIA/IS-801.

Trevisiani, E., & Vitaleti, A. (2004). Cell-ID location technique, limits and benefits: an experimental study. *Proceedings of 6th IEEE Workshop on Mobile Computing Systems and Applications WMCSA 2004*, 51-60. doi:10.1109/MCSA.2004.9

Tropp, J. A. (2004). Greed is Good: Algorithmic Results for Sparse Approximation. *IEEE Transactions on Information Theory*, *50*(10), 2231–2242. doi:10.1109/TIT.2004.834793

Tropp, J. A., & Gilbert, A. C. (2007). Signal Recovery From Random Measurements Via Orthogonal Matching Pursuit. *IEEE Transactions on Information Theory*, *53*(12), 4655–4666. doi:10.1109/TIT.2007.909108

Tsui, A. W., Chuang, Y.-H., & Chu, H.-H. (2009, October). Unsupervised learning for solving RSS hardware variance problem in WiFi localization. *Mobile Networks and Applications*, *14*(5), 677–691. doi:10.1007/s11036-008-0139-0

Tsui, A., Lin, W.-C., Chen, W.-J., Huang, P., & Chu, H.-H. (2010, November). Accuracy performance analysis between war driving and war walking in metropolitan Wi-Fi localization. *IEEE Transactions on Mobile Computing*, *9*(11), 1551–1562. doi:10.1109/TMC.2010.121

Turin, G. L. (1960). The characteristic function of Hermitian quadratic forms in complex normal variables. *Biometrika*, *47*(1-2), 199–201. doi:10.1093/biomet/47.1-2.199

Ubisense. (2017). *Location Intelligence solutions*. Retrieved March 16, 2017, from http://www.ubisense.net/en/

Ulaby, F. T., Long, D. G., Blackwell, W. J., Elachi, C., Fung, A. K., Ruf, C., & Van Zyl, J. (2014). *Microwave radar and radiometric remote sensing* (Vol. 4). University of Michigan Press Ann Arbor.

Urkowitz, H. (1967). Energy detection of unknown deterministic signals. *Proceedings of the IEEE*, *55*(4), 523–531. doi:10.1109/PROC.1967.5573

Vaghefi, R. M., & Buehrer, R. M. (2015). Cooperative localization in NLOS environments using semidefinite programming. *IEEE Communications Letters*, *19*(8), 1382–1385. doi:10.1109/LCOMM.2015.2442580

Vakulya, G., & Simon, G. (2011). Fast adaptive acoustic localization for sensor networks. *IEEE Transactions on Instrumentation and Measurement*, *60*(5), 1820–1829. doi:10.1109/TIM.2011.2108074

Van Dierendonck, A. J. (2000). *Mathematical Challenges in GPS: New Opportunities Using New Civil Signals for Multipath Mitigation and Carrier Phase Ambiguity Resolution*. Retrieved from www.ima.umn.edu/talks/workshops/8-16-18.2000/van-dierendonck/SignalSpecifiationPaper.ppt

Van Dierendonck, A. J., Fenton, P., & Ford, . (1992). Theory and performance of narrow correlator spacing in a GPS receiver. *Navigation*, *39*(3), 265–283. doi:10.1002/j.2161-4296.1992.tb02276.x

Van Diggelen, F. (2009). *A-GPS: Assisted GPS, GNSS, and SBAS*. Artech House.

van Nee, R. D. J. (1992). The multipath estimating delay lock loop. *Proceedings of IEEE ISSTA*, 39-42.

van Nee, R. D. J. (1994). The multipath estimating delay lock loop: approaching theoretical accuracy limits. *Proceedings of IEEE PLANS*, 246-251. doi:10.1109/PLANS.1994.303320

Vaseghi, S. V. (2006). *Advanced Digital Signal Processing and Noise Reduction*. John Wiley & Sons.

Vaughan, R. (2003). *Channels, propagation and antennas for mobile communications*. Iet.

Vaughan, R. G., & Scott, N. L. (1999). Super-resolution of pulsed multipath channels for delay spread characterization. *IEEE Transactions on Communications*, *47*(3), 343–347. doi:10.1109/26.752811

Vecchione, G. A., Palenzuela, D., Toledo, M., de Cceres, J. J. R., Tiley, P., Kubrak, D., . . . Lpez-Risueo, G. (2010, May). DINGPOS, a GNSS-based Multi-sensor Demonstrator for Indoor Navigation: Preliminary Results. *Position Location and Navigation Symposium (PLANS), 2010 IEEE/ION*, 113-121. doi:10.1109/PLANS.2010.5507243

Veitsel, V. A., Zhdanov, A. V., & Zhodzishsky, M. (1998). The mitigation of multipath errors by strobe correlators in GPS/GLONASS receivers. *GPS Solutions*, *2*(2), 38–45. doi:10.1007/PL00000035

Venkatraman, S., & Caffery, J. (2002). Statistical approach to non-line-of-sight BS identification. In *Proceedings of International Symposium on Wireless Personal Multimedia Communications* (vol.1, pp.296-300). Piscataway, NJ: IEEE. doi:10.1109/WPMC.2002.1088180

Viterbi, A. J. (1995). *CDMA: Principles of Spread Spectrum Communication*. Addison-Wesley.

Vogel, L. (2016). *Getting started with Android development – Tutorial*. Retrieved March 2017, from: http://www.vogella.com/tutorials/Android/article.html

Wagner, B., Patwari, N., & Timmermann, D. (2012, Mar.). *Passive RFID tomographic imaging for device-free user localization*, Paper presented at 2012 9th Workshop on Positioning Navigation and Communication (WPNC), Dresden, Germany. doi:10.1109/WPNC.2012.6268750

Wang, C., Wu, H., & Tzeng, N. F. (2007, May). *RFID-based 3-D positioning schemes*. Paper presented at the 26th IEEE International Conference on Computer Communications, Anchorage, AK. doi:10.1109/INFCOM.2007.147

Wang, J., & Shen, Y. (2017). On the Discretization Schemes in Map-aided Indoor Localization. *IEEE Communications Letters*.

Wang, J., Zhang, X., Gao, Q., Yue, H., & Wang, H. (2016). Device-free wireless localization and activity recognition: A deep learning approach. *IEEE Transactions on Vehicular Technology*. DOI: 10.1109/TVT.2016.2635161

Wang, T., Shen, Y., Conti, A., & Win, M. Z. (2017). Network navigation with scheduling: error evolution, *IEEE Transactions Information Theory*.

Wang, G., Chen, H., Li, Y., & Ansari, N. (2014). NLOS error mitigation for TOA-based localization via convex relaxation. *IEEE Transactions on Wireless Communications*, *13*(8), 4119–4131. doi:10.1109/TWC.2014.2314640

Wang, G., So, M. C., & Li, Y. (2016). Robust convex approximation methods for TDOA-based localization under NLOS conditions. *IEEE Transactions on Signal Processing*, *64*(13), 3281–3296. doi:10.1109/TSP.2016.2539139

Wang, J., Chen, X., Fang, D., Wu, C., Yang, Z., & Xing, T. (2015). Transferring- compressive-sensing-based device-free localization across target diversity. *IEEE Transactions on Industrial Electronics*, *62*(4), 2397–2409. doi:10.1109/TIE.2014.2360140

Wang, J., Gao, Q., Wang, H., Cheng, P., & Xin, K. (2015). Device-free localization with multidimensional wireless link information. *IEEE Transactions on Vehicular Technology*, *64*(1), 356–366. doi:10.1109/TVT.2014.2318084

Wang, J., Gao, Q., Yu, Y., Cheng, P., Wu, L., & Wang, H. (2013). Robust device-free wireless localization based on differential RSS measurements. *IEEE Transactions on Industrial Electronics*, *60*(12), 5943–5952. doi:10.1109/TIE.2012.2228145

Wang, J., Ma, Y., Zhao, Y., & Liu, K. (2015). A multipath mitigation localization algorithm based on MDS for passive UHF RFID. *IEEE Communications Letters*, *19*(9), 1652–1655. doi:10.1109/LCOMM.2015.2450217

Wang, N., Zhang, N., & Wang, M. (2006). Wireless sensors in agriculture and food industry—Recent development and future perspective. *Computers and Electronics in Agriculture*, *50*(1), 1–14. doi:10.1016/j.compag.2005.09.003

Wang, Q., Balasingham, I., Zhang, M., & Huang, X. (2011). Improving RSS-based ranging in LOS-NLOS scenario using GMMs. *IEEE Communications Letters*, *15*(10), 1065–1067. doi:10.1109/LCOMM.2011.080811.111087

Wang, Y., & Wu, Y. (2017). An efficient semidefinite relaxation algorithm for moving source localization using TDOA and FDOA measurements. *IEEE Communications Letters*, *21*(1), 80–83. doi:10.1109/LCOMM.2016.2614936

Wang, Z., Liu, H., Xu, S., Bu, X., & An, J. (2015, November). A Diffraction Measurement Model and Particle Filter Tracking Method for RSS-Based DFL. *IEEE Journal on Selected Areas in Communications*, *33*(11), 2391–2403. doi:10.1109/JSAC.2015.2430517

Wang, Z., & Zekavat, S. A. (2012). Omni-directional mobile NLOS identification and localization via multiple cooperative nodes. *IEEE Transactions on Mobile Computing*, *11*(12), 2047–2059. doi:10.1109/TMC.2011.232

Ward, P. W. (1996). GPS Receiver Search Techniques. *Proceedings of Position Location and Navigation Symposium*, 604-611. doi:10.1109/PLANS.1996.509134

Weinberg, M., & Kourepenis, A. (2006). Error sources in in-plane silicon tuning fork gyroscopes. *Journal of Micromechanical Systems*, *15*(3), 479–491. doi:10.1109/JMEMS.2006.876779

Weiss, A. J. (2004). Direct position determination of narrowband radio frequency transmitters. *IEEE Signal Processing Letters*, *11*(5), 513–516. doi:10.1109/LSP.2004.826501

Weiss, A. J. (2011). Direct Geolocation of Wideband Emitters Based on Delay and Doppler. *IEEE Transactions on Signal Processing*, *59*(6), 2513–2521. doi:10.1109/TSP.2011.2128311

Weiss, A. J., & Amar, A. (2005). Direct Position Determination of Multiple Radio Signals. *EURASIP Journal on Advances in Signal Processing*, (1): 653549.

Welch, G., & Bishop, G. (2006). *An introduction to the Kalman filter.* Report, UNC-Chapel Hill TR 95-041.

Wendel, J., Metzger, J., Moenikes, R., Maier, A., & Trommer, G. F. (2006). A performance comparison of tightly coupled GPS/INS navigation systems based on extended and sigma point Kalman filters, *NAVIGATION. Journal of the Institute of Navigation*, *53*(1), 21–31. doi:10.1002/j.2161-4296.2006.tb00368.x

Widrow, B., Kollar, I., & Liu, M.-C. (1996). Statistical theory of quantization. *IEEE Transactions on Instrumentation and Measurement*, *45*(2), 353–361. doi:10.1109/19.492748

William J. Hughes Technical Center. (2014). *Global Positioning System (GPS), Standard Positioning Service (SPS), Performance Analysis Report #86.* William J. Hughes Technical Center, NSTB/WAAS T&E Team.

Wilson, J., & Patwari, N. (2010). Radio Tomographic Imaging with Wireless Networks. *IEEE Transactions on Mobile Computing*, *9*(5), 621–632. doi:10.1109/TMC.2009.174

Wilson, J., & Patwari, N. (2011). See through walls: Motion tracking using variance-based radio tomography networks. *IEEE Transactions on Mobile Computing*, *10*(5), 612–621. doi:10.1109/TMC.2010.175

Wilson, J., & Patwari, N. (2012). A fade-level skew-laplace signal strength model for device-free localization with wireless networks. *IEEE Transactions on Mobile Computing*, *11*(6), 947–958. doi:10.1109/TMC.2011.102

Win, M. Z., Conti, A., Mazuelas, S., Shen, Y., Gifford, W. M., Dardari, D., & Chiani, M. (2011). Network localization and navigation via cooperation. *IEEE Communications Magazine*, *49*(5), 56–62. doi:10.1109/MCOM.2011.5762798

Woodman, O. J. (2007). *An introduction to inertial navigation.* Technical report, University of Cambridge Computer Laboratory, Cambridge CB3 OFD, United Kingdom, UCAM-CL-TR-696, ISSN 1476-2986.

Wright, R. V. S. (1989). *Doing Multivariate Analysis and Prehistory.* Sydney, Australia: University of Sydney Press.

Wu, J., & Dempster, A. G. (2011). "BOC-Gated-PRN" a multipath mitigation technique for BOC(n,n) waveforms. *IEEE Transactions on Aerospace and Electronic Systems*, *47*(2), 1136–1153. doi:10.1109/TAES.2011.5751248

Wymeersch, H., Lien, J., & Win, M. Z. (2009). Cooperative localization in wireless networks. *Proceedings of the IEEE*, *97*(2), 427–450. doi:10.1109/JPROC.2008.2008853

Xia, H., Bertoni, H., Maciel, L., Lindsay-Stewart, A., & Rowe, R. (1993). Radio propagation characteristics for line-of-sight microcellular and personal communications. *IEEE Transactions on Antennas and Propagation*, *41*(10), 1439–1447. doi:10.1109/8.247785

Xiao, Z., Wen, H., Markham, A., Trigoni, N., Blunsom, P., & Frolik, J. (2015). Non-line-of-sight identification and mitigation using received signal strength. *IEEE Transactions on Wireless Communications*, *14*(3), 1689–1702. doi:10.1109/TWC.2014.2372341

Xu, J. S., Zhu, J., & Li, L. (2007). Effects of a major storm on GPS amplitude scintillations and phase fluctuations at Wuhan in China. *Advances in Space Research*, *39*(8), 1318–1324. doi:10.1016/j.asr.2007.03.004

Xu, R., & Wunsch, D. II. (2005). Survey of clustering algorithms. *IEEE Transactions on Neural Networks*, *16*(3), 645–678. doi:10.1109/TNN.2005.845141 PMID:15940994

Xu, Z., Trinkle, M., & Gray, D. A. (2011). A maximum-likelihood based mutual coupling calibration algorithm in the presence of multipath for GPS antenna array. *Proceedings of ION-GNSS*, 1111-1119.

Yacoub, M. D. (2002). The α - μ distribution: a general fading distribution. *13th IEEE International Symposium on Personal, Indoor and Mobile Radio Communications, 2*, 629-633. doi:10.1109/PIMRC.2002.1047298

Yang, L., Chen, Y., Li, X., Xiao, C., Li, M., & Liu, Y. (2014, September). *Tagoram: Real-time tracking of mobile RFID tags to high precision using COTS devices.* Paper presented at the 20th annual international conference on Mobile computing and networking, Maui, HI. doi:10.1145/2639108.2639111

Yang, C. (2005). Joint acquisition of CM and CL codes for GPS L2 civil (L2C) signals. *Proceedings of the 61st Annual Meeting of The Institute of Navigation*, 553-562.

Yang, D., Li, H., & Peterson, G. (2013, January). Compressive Sensing Based Sub-mm Accuracy UWB Positioning Systems: A Space-Time Approach. *Digital Signal Processing*, *23*(1), 340–354. doi:10.1016/j.dsp.2012.07.012

Yang, L., & Giannakis, G. B. (2005). Timing ultra-wideband signals with dirty templates. *IEEE Transactions on Communications*, *53*(11), 1952–1963. doi:10.1109/TCOMM.2005.858663

Yazdi, N., Ayazi, F., & Najafi, K. (1998). Micromachined inertial sensors. *Proceedings of the IEEE*, *86*(8), 1640–1659. doi:10.1109/5.704269

Yiğitler, H., Jäntti, R., & Virrankoski, R. (2014). pRoot: An Adaptable Wireless Sensor-Actuator Hardware Platform. *2014 12th IEEE International Conference on Embedded and Ubiquitous Computing, Milano*, 281-286.

Yiğitler, H., Jäntti, R., Kaltiokallio, O., & Patwari, N. (2017). Detector based Radio Tomographic Imaging. *IEEE Transaction on Mobile Computing*.

Yiğitler, H., Kaltiokallio, O., & Jäntti, R. (2013, Aug). A management framework for device-free localization. *International Joint Conference on Neural Networks (IJCNN)*, 1-8. doi:10.1109/IJCNN.2013.6707131

Yiğitler, H., & Jäntti, R. (2016, March). Experimental Accuracy Assessment of Radio Tomographic Imaging Methods. *IEEE International Conference on Pervasive Computing and Communication Workshops (PerCom Workshops)*, 1-6. doi:10.1109/PERCOMW.2016.7457117

Yiğitler, H., Jäntti, R., & Patwari, N. (2017, April). On Log-Normality of RSSI in Narrowband Receivers Under Static Conditions. *IEEE Signal Processing Letters*, *24*(4), 367–371. doi:10.1109/LSP.2017.2657332

Yin, J., Yang, Q., & Ni, L. (2008, July). Learning adaptive temporal radio maps for signal-strength-based location estimation. *IEEE Transactions on Mobile Computing*, *7*(7), 869–883. doi:10.1109/TMC.2007.70764

Yin, Z., Cui, K., Wu, Z., & Yin, L. (2015). Entropy-based TOA estimation and SVM-based ranging error mitigation in UWB ranging systems. *Sensors (Basel)*, *15*(3), 11701–11724. doi:10.3390/s150511701 PMID:26007726

Youssef, M., & Agrawala, A. (2004, March). Handling samples correlation in the horus system. *INFOCOM 2004. Twenty-third Annual Joint Conference of the IEEE Computer and Communications Societies*, *2*, 1023-1031.

Youssef, V., Agrawala, A., & Udaya Shankar, A. (2003). WLAN location determination via clustering and probability distributions. Pervasive Computing and Communications, 143-150.

Youssef, M., & Agrawala, A. (2008, June). The Horus location determination system. *Wireless Networks*, *14*(3), 357–374. doi:10.1007/s11276-006-0725-7

Youssef, M., Mah, M., & Agrawala, A. (2007). Challenges: device-free passive localization for wireless environments. *ACM Int'l Conf. Mobile Computing and Networking, 2007. MobiCom'07.*, 222-229. doi:10.1145/1287853.1287880

Yu, K., & Dutkiewicz, E. (2012, February). Geometry and motion-based positioning algorithms for mobile tracking in NLOS environments. *IEEE Transactions on Mobile Computing*, *11*(2), 254–263. doi:10.1109/TMC.2011.24

Yu, K., & Dutkiewicz, E. (2013, July). NLOS identication and mitigation for mobile tracking. *IEEE Transactions on Aerospace and Electronic Systems*, *49*(3), 1438–1452. doi:10.1109/TAES.2013.6557997

Yu, K., & Guo, Y. (2009, January). Statistical NLOS identication based on AOA, TOA, and signal strength. *IEEE Transactions on Vehicular Technology*, *58*(1), 274–286. doi:10.1109/TVT.2008.924975

Yu, K., & Oppermann, I. (2004). Positioning. In I. Oppermann, M. Hämäläinen, & J. Iinatti (Eds.), *UWB theory and applications* (pp. 175–196). Hoboken, NJ: John Wiley & Sons, Ltd.

Yu, K., Sharp, I., & Guo, Y. J. (2009). *Ground-Based Wireless Positioning*. IEEE-Wiley Press. doi:10.1002/9780470747988

Zakharov, V. I., Yasyukevich, Y. V., & Titova, M. A. (2016). Effect of magnetic storms and substorms on GPS slips at high latitudes. *Cosmic Research*, *54*(1), 20–30. doi:10.1134/S0010952516010147

Zhang, W., & Kosecka, J. (2006). Image based localization in urban environments. In *3rd International Symposium on 3D Data Processing, Visualization and Transmission* (pp. 33-40). IEEE Computer Society.

Zhang, Y., Liu, W., Fang, Y., & Wu, D. (2006). Secure localization and authentication in ultra-wideband sensor networks. *IEEE Journal on Selected Areas in Communications*, *24*(4), 829–835. doi:10.1109/JSAC.2005.863855

Zhang, Z., Lu, Z., Saakian, V., Qin, X., Chen, Q., & Zheng, L.-R. (2014). Item-level indoor localization with passive UHF RFID based on tag interaction analysis. *IEEE Transactions on Industrial Electronics*, *61*(4), 2122–2135. doi:10.1109/TIE.2013.2264785

Zhao, Y., Liu, Y., & Ni, L. (2007, Oct.). VIRE: active RFID-based localization using virtual reference elimination. Paper presented at the 2007 International Conference on Parallel Processing, Xi'an, China. doi:10.1109/ICPP.2007.84

Zhao, Y., Liu, K., Ma, Y., Gao, Z., Zang, Y., & Teng, J. (2017). Similarity analysis based indoor localization algorithm with backscatter information of Passive UHF RFID tags. *IEEE Sensors Journal*, *17*(1), 185–193. doi:10.1109/JSEN.2016.2624314

Zhao, Y., & Patwari, N. (2015). Robust estimators for variance-based device free localization and tracking. *IEEE Transactions on Mobile Computing*, *14*(10), 2116–2129. doi:10.1109/TMC.2014.2385710

Zhou, B., & Chen, Q. (2016). On the particle-assisted stochastic search mechanism in wireless cooperative localization. *IEEE Transactions on Wireless Communications*, *15*(7), 4765–4777.

Zhou, F., & Shen, Y. (2017). On the outage probability of localization in randomly deployed wireless networks. *IEEE Communications Letters*, *21*(4), 901–904. doi:10.1109/LCOMM.2017.2647919

Zhou, Z., Yang, Z., Wu, C., Shangguan, L., & Liu, Y. (2014). Omnidirectional coverage for device-free passive human detection. *IEEE Transactions on Parallel and Distributed Systems*, *25*(7), 1819–1829. doi:10.1109/TPDS.2013.274

Zinnes, J. L., & MacKay, D. B. (1983). Probabilistic multidimensional scaling: Complete and incomplete data. *Psychometrika*, *48*(1), 27–48. doi:10.1007/BF02314675

Zorn, A. (2002). GPS-aided all-accelerometer inertial navigation. In *Proceedings of International Technical Meeting of Satellite Division of the Institute of Navigation*. Institute of Navigation.

About the Contributors

Mohamed M. Atia received B.S. and M.Sc. degrees in computer systems from Ain Shams University in 2000 and 2006 and a Ph.D. in electrical and computer engineering from Queen's University at Kingston in 2013. He held several full-time positions in industry developing algorithms for natural language processing, online speech recognition, and real-time embedded multi-sensors navigation, guidance, and control systems. He is currently assistant professor in the department of Electronics at Carleton university in Ottawa. Dr. Atia is a recipient of the Alberta Innovate Association Award in 2011, the IEEE excellence in PhD research in 2013, Mitacs Elevate Industrial Postdoc Award in 2014, NSERC PDF Award in 2015, and Queen's University Teaching Award in 2016.

Yang Cai received his B.S. degree in electrical engineering in 2015 from the Department of Electronic Engineering, Tsinghua University, China, where he has been pursuing his master degree since 2015. He was with the Intelligent Sensing Lab, Tsinghua University, China, from 2013-2015. His research interests include communication theory, statistical inference, and control theory. His current research focuses on cooperative localization, resource allocation, and formation control.

Pau Closas received the MS and PhD in Electrical Engineering from the Universitat Politècnica de Catalunya (UPC) in 2003 and 2009, respectively. He also holds a MS degree in Advanced Mathematics and Mathematical Engineering from UPC since 2014. In 2003, he was with the Department of Signal Theory and Communications, UPC, as a Research Assistant. In 2008 he was Research Visitor at the Stony Brook University (SBU), NY, USA. In 2009 he joined the CTTC, where he was Senior Researcher and Head of the Statistical Inference for Communications and Positioning Department. Currently he is Assistant Professor in Electrical and Computer Engineering at Northeastern University. His primary areas of interest include statistical and array signal processing, stochastic filtering, robust statistics, and game theory, with applications to positioning systems, wireless communications, and mathematical biology. He is the recipient of the EURASIP Best PhD Thesis Award 2014, the 9th Duran Farell Award for Technology Research, and the 2016 Institute of Navigation Early Achievement Award.

Xunxue Cui received the B.S. degree in artillery command and the M.S. degree in operation research from Hefei Artillery Academy, Hefei, China, in 1991 and 1996, respectively, and the Ph.D. degree in pattern recognition and artificial intelligence from the University of Science and Technology of China, Hefei, in 2001. He served as a Post-Doctoral Fellow with the Department of Computer Science and Technology, Tsinghua University, Beijing, China, from 2002 to 2004. He is currently a Professor with the Anhui University of Science and Technology, Huainan, China. He holds five authorized patents of invention, and has authored two scientific research monographs and two textbooks. His current research interests include sensor network, target localization (in particular, for acoustic source), swarm intelligent algorithms, and multiobjective optimization. Dr. Cui is a member of the China Computer Federation Technical Committee on Sensor Network.

Andrew Dempster is Director of the Australian Centre for Space Engineering Research (ACSER) in the School of Electrical Engineering and Telecommunications at the University of New South Wales (UNSW). He has a BE and MEngSc from UNSW and a PhD from the University of Cambridge in efficient circuits for signal processing arithmetic. He was system engineer and project manager for the first GPS receiver developed in Australia in the late 80s and has been involved in satellite navigation ever since. His current research interests are in satellite navigation receiver design and signal processing, areas where he has six patents, and new location technologies. He is leading the development of space engineering research at ACSER.

Shih-Hau Fang (M'07-SM'13) received a B.S. from National Chiao Tung University in 1999, an M.S. and a Ph.D. from National Taiwan University, Taiwan, in 2001 and 2009, respectively, all in Communication Engineering. From 2001 to 2007, he was a software architect at Internet Services Division at Chung-Hwa Telecom Company Ltd. In 2009, he joined the Department of Electrical Engineering, Yuan Ze University, where he is currently a full professor. He is also technical advisor to HyXen Technology Company Ltd. Prof. Fang received the YZU Young Scholar Research Award in 2012 and the Project for Excellent Junior Research Investigators, Ministry of Science and Technology in 2013. His research interests include indoor positioning, mobile computing, machine learning, and signal processing. He is a senior member of IEEE.

Genís Floriach received the Electrical Engineering degree from the Polytechnic University of Catalonia (UPC), Barcelona, Spain in 2016. Since 2016, he has been working towards the M. Sc. degree in electrical engineering at Polytechnic University of Catalonia. In 2016 he was awarded with a scholarship of the Spanish government to carry out research in the field of source localization in UWB systems within the Department of Signal Theory and Communications, UPC. His research of interest includes source localization, estimation theory, signal processing and digital communication.

Zheng Gao received the B.Eng. degree in electronic information engineering from Tianjin University, Tianjin, China, in 2015, where he is currently pursuing the M.S. degree with the School of Microelectronics. His current research interests include wireless channel model and indoor localization.

Ricardo García-Gutiérrez received the Electrical Engineering degree from the Polytechnic University of Catalonia (UPC), Barcelona, Spain in 2016. Since 2016, he has been working towards the M. Sc. degree in electrical engineering at Polytechnic University of Catalonia. In 2016, he was awarded with a scholarship of the Spanish Government to carry out research in the field of Compressive Sensing based Localization methods within the Department of Signal Theory and Communications, UPC. His research of interest also includes signal processing, digital communications, speech processing and machine learning.

Riku Jäntti (M'02 - SM'07) is an Associate Professor (tenured) in Communications Engineering and the head of the department of Communications and Networking at Aalto University School of Electrical Engineering, Finland. He received his M.Sc (with distinction) in Electrical Engineering in 1997 and D.Sc (with distinction) in Automation and Systems Technology in 2001, both from Helsinki University of Technology (TKK). Prior to joining Aalto (formerly known as TKK) in August 2006, he was professor pro tem at the Department of Computer Science, University of Vaasa. Prof. Jäntti is a senior member of IEEE and associate editor of IEEE Transactions on Vehicular Technology. He is also IEEE VTS Distinguished Lecturer (Class 2016). The research interests of Prof. Jäntti include radio resource control and optimization for machine type communications, Cloud based Radio Access Networks, spectrum and co-existence management and RF Inference.

Wei Jiang is Associate Professor at the School of Electronic and Information Engineering, Beijing Jiaotong University. She received her PhD degree at the School of Civil and Environmental Engineering, University of New South Wales, Australia. Her current research interest is multi-sensor integration, and in particular the implementation of new navigation and data fusion algorithms.

Ossi Kaltiokallio received the B.Sc. and M.Sc. degrees in electrical engineering from Aalto University, Finland, both in 2011. He is currently a Ph.D. student with the department of Communications and Networking at Aalto University where he holds an ELEC doctoral school position as of January 2014. He is the recipient of SenseApp 2012 and IPSN 2014 conference best paper awards. His current research interests include RF propagation, RSS-based localization and tracking, signal processing, and linear and non-linear Bayesian filtering and smoothing.

Allison Kealy is an Associate Professor in The Department of Infrastructure Engineering at The University of Melbourne Australia. She holds an undergraduate degree in Land Surveying from The University of the West Indies, Trinidad, and a PhD in GPS and Geodesy from the University of Newcastle upon Tyne, UK. Allison's research interests include sensor fusion, Kalman filtering, high precision satellite positioning, GNSS quality control, wireless sensor networks and location based services. Allison is currently the vice president of IAG Commission 4 on Positioning and Applications and the co-chair of the joint IAG Working Group 4.1.1 and FIG Working Group 5.5 on Multi-Sensor Systems.

Seung-Hyun Kong received a B.S. degree in Electronics Engineering from Sogang University, Korea, in 1992, an M.S. degree in Electrical Engineering from Polytechnic University, New York, in 1994, and a Ph.D. degree in Aeronautics and Astronautics from Stanford University, CA, in Jan. 2006. From 1997 to 2004, he was with Samsung Electronics Inc. and Nexpilot Inc., both in Korea, where he worked on developing wireless communication system standards and UMTS mobile positioning technologies. In 2006 and from 2007 to 2009, he was a staff engineer at Polaris Wireless Inc., Santa Clara, and at Qualcomm Inc. (Corp. R&D), San Diego, respectively, where his research was on Assisted-GNSS and wireless positioning technologies. Since 2010, he is with Korea Advanced Institute of Science and Technology (KAIST), where he is currently an associate professor at the CCS Graduate School of Green Transportation. He is a senior member of IEEE, and He serves as an Editor of IET Radar, Sonar and Navigation (RSN), and an Associate Editor of IEEE transactions on Intelligent Transportation Systems (ITS) and IEEE Access. His research interests include signal processing for GNSS, neutral networks for sensing, and vehicular communication systems.

Eva Lagunas (S'09-M'13) received the M.Sc. and Ph.D. degrees in telecommunications engineering from the Polytechnic University of Catalonia (UPC), Barcelona, Spain, in 2010 and 2014, respectively. She was Research Assistant within the Department of Signal Theory and Communications, UPC, from 2009 to 2013. During the summer of 2009 she was a guest research assistant within the Department of Information Engineering, Pisa, Italy, developing TOA estimation algorithms in UWB systems. From November 2011 to May 2012 she held a visiting research appointment at the Center for Advanced Communications (CAC), Villanova University, PA, USA, where she worked on through-wall radar imaging technology. In September 2014, she joined the Interdisciplinary Centre for Security, Reliability and Trust (SnT), University of Luxembourg, as a Research Associate. Her research interest are localization and positioning, compressive sensing, resource allocation and hybrid terrestrial-satellite communications. Dr. Lagunas is an associate editor of EURASIP Journal on Advances in Signal Processing (JASP) since 2015.

Yong Li is currently working with Navextech Technology Ltd, Australia, in charge of MEMS/GNSS based mutisensor systems for indoor/outdoor navigation applications. He was the senior research fellow at the Satellite Navigation and Positioning Lab, the University of New South Wales (UNSW), Australia. He was involved in development of spaceborne GPS receivers for attitude determination and formation flying missions. His current interests include integration of GNSS, INS, and pseudolite (Locata), attitude determination, GNSS receiver technique, and its application to navigation, and optimal estimation/ filtering theory and applications.

Songsheng Lu received his Ph.D. degree in Institute of Plasma Physics Chinese Academy of Sciences, Hefei, China in 2006. During that time, He was involved in the design and manufacture of EAST fast control power supply, acquiring and processing data for fault diagnosis. He is currently an associate professor of New Star Institute of Applied Technology. He is engaged in research of the computer control system theory and application, signal processing, and machine vision.

Yongtao Ma was born in China. He received the B.S. degree in electronic information engineering, and the M.S. and Ph.D. degrees in circuits and systems from Tianjin University, Tianjin, China, in 2001, 2005, and 2009, respectively. He is currently an Associate Professor with the School of Microelectronics, Tianjin University. His current research interests include wireless localization, RFID, cognitive radio and microwave technology.

Montse Nájar Martón received the electrical engineering and the Ph.D. degrees from the Polytechnic University of Catalonia (UPC) in 1991 and 1996, respectively. In 1992, she joined the Department of Signal Theory and Communication, UPC. Since 1997, she is an Associate Professor at UPC. From 2003 to 2006, she was member of the Board of Directors of the Telecommunications School of Barcelona, ETSETB. From 2005 to 2013, she was Research Associate in the Centre Tecnològic de Telecomunicacions de Catalunya (CTTC). She has participated in several EC funded projects (SATURN, PHYDYAS, NEWCOM) as well as national public and private funded projects. Her current research interests include signal processing with application to communication systems, array signal processing and location in wireless systems. She has published over 90 papers in books, international conferences and journals.

Monica Navarro is the Head of the Communication Systems Division and Senior Researcher at the Centre Tecnològic de Telecomunicacions de Catalunya within the Communication Systems Division. She received the MSc degree in Telecommunications Engineering from Universitat Politècnica de Catalunya and the PhD degree in Telecommunications from the Institute for Telecommunications Research, University of South Australia. Prior to join ITR she was a Research Assistant at the Department of Signal Theory and Communications at the UPC, where she worked on the development of fractal shape multiband antennas for wireless cellular communications systems. She has also been part-time lecturer at the Universitat Pompeu Fabra, Barcelona. Her primary areas of interest are on information processing with applications to wireless communications and positioning. Particularly on iterative information processing, adaptive transmissions, coding techniques and random access protocols for massive access. She has strong expertise on new radio air interfaces for cellular networks.

Guenther Retscher is Associate Professor in the Department of Geodesy and Geoinformation at the TU Wien – Vienna University of Technology, Austria. He holds an undergraduate degree in Surveying, a PhD and a Habilitation (venia docendi) in Applied Geodesy from TU Wien with the focus on Mobile Multi-sensor Systems for Personal Navigation and Location-based Services. Guenther's research interests include positioning and navigation with GNSS, location based services, indoor and pedestrian navigation, applications of multi-sensor systems and sensor fusion. Guenther is currently the co-chair of IAG Sub-Commission 4.1 on Emerging Positioning Technologies and GNSS Augmentation and the co-chair of the joint IAG Working Group 4.1.1 and FIG Working Group 5.5 on Multi-Sensor Systems.

Chris Rizos is Professor of Geodesy and Navigation, School of Civil & Environmental Engineering, the University of New South Wales (UNSW), Sydney, Australia. Chris is the past president of the International Association of Geodesy (IAG) 2011-2015, a member of the Governing Board of the International GNSS Service (IGS), and former co-chair of the Multi-GNSS Asia Steering Committee 2009-2016. Chris is a Fellow of the IAG, a Fellow and current president of the Australian Institute of Navigation, a Fellow of the U.S. Institute of Navigation, and an honorary professor of Wuhan University, China. Chris has been researching the technology and applications of GPS and other navigation/positioning systems since 1985, and is an author or co-author of over 650 journal and conference papers.

Ian Sharp is a Senior Consultant on radio positioning systems. He has over thirty years of engineering experience in radio systems. His initial involvement in positioning technology was in aviation, and later in the 1980s with the Interscan microwave landing system (MLS). In the later 1980s to the early 1990s, Ian was the R&D manager for the Quiktrak covert vehicle tracking system. This system is now commercially operating world-wide. From mid 1990s to 2007, Ian worked at the CSIRO mainly on developing experimental radio systems as principal scientist. He was the inventor and architect designer of CSIRO's precision location system (PLS) for sports applications. The PLS has been successfully trialed in Australia and the US. Ian holds a number of patents relating to positioning technology. He is a co-author of the book: Ground-Based Radio positioning (Wiley and IEEE Press, 2009).

Yuan Shen received the Ph.D. degree and the S.M. degree in electrical engineering and computer science from the Massachusetts Institute of Technology (MIT), Cambridge, MA, USA, in 2014 and 2008, respectively, and the B.E. degree (with highest honor) in electronic engineering from Tsinghua University, Beijing, China, in 2005. He is an Associate Professor with the Department of Electronic

Engineering at Tsinghua University. Prior to that, he was a Research Assistant and then Postdoctoral Associate with the Wireless Information and Network Sciences Laboratory at MIT in 2005-2014. He was with the Hewlett-Packard Labs in winter 2009 and the Corporate R&D at Qualcomm Inc. in summer 2008. His research interests include statistical inference, network science, communication theory, information theory, and optimization. His current research focuses on network localization and navigation, inference techniques, resource allocation, and intrinsic wireless secrecy. Professor Shen was a recipient of the Qiu Shi Outstanding Young Scholar Award, the China's Youth 1000-Talent Program, the Marconi Society Paul Baran Young Scholar Award, and the MIT Walter A. Rosenblith Presidential Fellowship. His papers received the IEEE ComSoc Fred W. Ellersick Prize and three Best Paper Awards from the IEEE international conferences. He is elected Vice Chair (2017-2018) and Secretary (2015-2016) for the IEEE ComSoc Radio Communications Committee. He has served as TPC symposium Co-Chair for serveral IEEE conferences and workshops and as Editor for the IEEE Communications Letters.

Yunlong Wang received the B.S. degree in electronic information engineering and M.S. degree in information and communication engineering from Information Engineering University, in 2012 and 2015, respectively. Now he is pursuing his Ph.D. degree in National Digital Switching System Engineering & Technological Research Center, Zhengzhou, China, and also with the Department of Electronic Engineering, Tsinghua University, Beijing China. His current interests focus on array signal processing, source localization, and network optimization.

Kai Wen received the B.Eng. degree in surveying and mapping engineering from China University of Mining and Technology, Xuzhou, China, in 2012 and the M.S. degree in geodesy and geomatics engineering from Wuhan University, Wuhan, China, in 2016. He is currently a Ph.D. student with Wuhan University, researching the integrated positioning technology based on UWB, geomagnetism and IMU for applications in complex environments.

Ling Yang is currently an assistant professor at College of Surveying and Geo-informatics of Tongji University, China. She received her PhD in Surveying and Geo-information Engineering at University of New South Wales, Australia. Her research interests are GNSS/INS+ seamless integrated navigation algorithms and the fault detection and isolation theory for system quality control purpose.

Huseyin Yigitler received the B.Sc. and M.Sc. degrees in electrical and electronics engineering from Middle East Technical University, Ankara, Turkey, in 2004 and 2006 respectively. He is currently a Ph.D. candidate of the department of Communications and Networking at Aalto University School of Electrical Engineering. His research interests include RF propagation, localization and tracking, signal processing, and design and implementation of embedded (wireless) systems.

Yang Zhao received the B.S. degree from Hainan University, Haikou, China, in 2010, and the M.S. and the Ph.D. degrees from Tianjin University, Tianjin, China, in 2012 and 2015, respectively. She is currently a Post-Doctoral Researcher in Tianjin University. Her research interests include wireless localization, pattern recognition and image processing.

Index

Encyclopedia of
Information Science and Technology
Fourth Edition
Mehdi Khosrow-Pour

www.ingramcontent.com/pod-product-compliance
Lightning Source LLC
LaVergne TN
LVHW060252011026
849161LV00011B/2698